THE POWER AND BEAUTY OF ELECTROMAGNETIC FIELDS

IEEE Press
445 Hoes Lane
Piscataway, NJ 08854

THE POWER AND BEAUTY OF ELECTROMAGNETIC FIELDS

F. R. Morgenthaler

IEEE Antenna Propagation Society, *Sponsor*

The IEEE Press Series on Electromagnetic Wave Theory
Andreas C. Cangellaris, *Series Editor*

IEEE
IEEE Press

WILEY
A John Wiley & Sons, Inc., Publication

For general information on our other products and services or for technical support, please contact our Customer Care Department within the United States at (800) 762-2974, outside the United States at (317) 572-3993 or fax (317) 572-4002.

Wiley also publishes its books in a variety of electronic formats. Some content that appears in print may not be available in electronic formats. For more information about Wiley products, visit our web site at www.wiley.com.

Library of Congress Cataloging-in-Publication Data is available.

ISBN: 978-1-118-05757-5

Printed in the United States of America

ePDF ISBN: 978-1-118-11841-2

10 9 8 7 6 5 4 3 2 1

To my loving wife, Barbara,
our splendid daughters,
Ann and Janet,
and their wonderful
children, Sarah, Brian,
Douglas, and Tessie

CONTENTS

Preface xxi

Acknowledgments xxvii

List of Figures xxix

PART I BASIC ELECTROMAGNETIC THEORY

1 Maxwell's Equations 5

1.1 Mathematical notation 5
1.2 Free-space fields and forces 6
Integral form of Maxwell's Equations 6
Units and fundamental constants 8
Linearity and superposition 8
Differential form of Maxwell's Equations 9
1.3 Vector and scalar potentials 10
Lorenz gauge 11
Coulomb gauge 11
1.4 Inhomogeneous wave equations for **E** and **H** 12
1.5 Static fields 12
Integration of Poisson's Equation 13
Electrostatics 14
Magnetostatics 14
1.6 Integration of the inhomogeneous wave equation 15
Current element (Hertzian electric dipole) 16

	Current loop (Hertzian magnetic dipole)	17
1.7	Polarizable, magnetizable, and conducting media	18
	Polarization and Amperian electric currents	19
	Chu formulation	22
	Electrically conducting materials	23
	Perfect conductors	23
	Dielectric and magnetic materials	23
1.8	Boundary conditions	24
	Electric surface charges	24
	Electric surface currents	24
	Conservation of charge	25
1.9	The complex Maxwell Equations	26

2 Quasistatic Approximations **29**

2.1	Quasistatic expansions of a standing wave	30
2.2	Electroquasistatic (*EQS*) fields	31
	Zero-order fields	31
	Boundary conditions	32
	First-order fields	32
2.3	Magnetoquasistatic (*MQS*) fields	33
	Zero-order fields	33
	Boundary conditions	34
	First-order fields	34
2.4	Conduction problems	35
	EQS regime	35
	MQS regime	36
2.5	Laplacian approximations	37

3 Electromagnetic Power, Energy, Stress, and Momentum **39**

3.1	Introduction	39
	Power conversion and force densities	39
	Electromagnetic torque density	40
	Uniqueness of $\mathbf{S}$, W, $\overline{\mathbf{T}}$, and $\mathbf{G}$	41
3.2	The Maxwell–Poynting representation	41
	Maxwell stress tensor	41
	Poynting Theorem	42
3.3	Quasistatic power and energy	43
	Standard form of quasistatic power theorems	43
	Modified form of quasistatic power theorems	44
3.4	Alternative representations	45
	Introduction	45
	An alternate Poynting theorem	46
	An alternate stress-momentum theorem	48
	Alternate (circuit-theory) representation	49
	Electromagnetic force on a moving charge	51
	Alternate accounting of power and momentum	51

		Electromagnetic beauty	52
		Alternate power and energy in the quasistatic limit	53
	3.5	Differences between representations	54
		Uniform plane wave	55
		Hertzian electric dipole (steady-state)	57
4	**Electromagnetic Waves in Free-Space**		**61**
	4.1	Homogeneous waves	61
	4.2	One-dimensional waves	62
		Solutions of the two-dimensional Laplace's Equation	63
	4.3	Harmonic uniform plane waves	63
	4.4	Waves of high symmetry	64
		Spherically symmetric waves	64
		Cylindrically symmetric waves	65
	4.5	Inhomogeneous scalar wave equations	66
		Three-dimensional superposition integrals	66
5	**Electromagnetic Waves in Linear Materials**		**67**
	5.1	Introduction	67
	5.2	Electrically conducting media	67
		Charge-density decay (dielectric relaxation time)	68
		Magnetic diffusion length and skin depth	69
	5.3	Linear dielectric and magnetic media	70
		Uniform plane waves (linearly polarized)	71
		Particle representation of harmonic plane waves	71
6	**Electromagnetic Theorems and Principles**		**77**
	6.1	Introduction	77
	6.2	Complex power and energy theorems	78
		Circuit power	78
		Circuit energy	79
		Complex Poynting Theorem	79
		Complex Alternate-power theorem	80
		Maxwell–Poynting energy theorem	81
		Alternate-energy theorem	82
	6.3	Complex stress theorems	84
		Maxwell–Poynting stress theorem	85
		Alternate-stress theorem	86
	6.4	Complex momentum theorems	86
		Maxwell–Poynting momentum theorem	86
		Alternate-momentum theorem	87
	6.5	Duality	88
		Dual electric and magnetic fields	88
		Dual sources	89
		Dual materials	90
		Boundary conditions	91

	Dual vector potentials	91
	Dual Alternate energy-momentum tensors	92
	Superposition of sources	93
6.6	Uniqueness theorems	94
	Electric and magnetic fields	94
	Laplace's Equation	95
	Sufficiency of the curl and divergence	96
6.7	The equivalence principle	96
6.8	The induction theorem	97
6.9	Babinet's Principle	98
	Complementary structures	98
	Dual structures and their complements	100
6.10	The reciprocity theorem	100

PART II FOUR-DIMENSIONAL ELECTROMAGNETISM

7 Four-Dimensional Vectors and Tensors **105**

7.1	Space–time coordinates	105
7.2	Four-vector electric-current density	106
7.3	Four-vector potential (Lorenz gauge)	106
7.4	Four-Laplacian (wave equation)	107
7.5	Maxwell's Equations and field tensors	107
7.6	The four-dimensional curl operator	109
7.7	Four-dimensional "statics"	110
7.8	Four-dimensional force density	112
7.9	Six-vectors and dual field tensors	113
7.10	Four-vector electric and magnetic fields	113
	Lorentz force on an electric charge	114
	Lorentz force on a magnetic charge	114
	Lorentz invariance of four-vectors	115
7.11	The field tensors and Maxwell's Equations revisited	115
7.12	Linear conductors revisited	116
	Modified Lorenz gauge	117
	Boundary conditions	117

8 Energy-Momentum Tensors **119**

8.1	Introduction	119
	Force and power conversion densities	119
	Electromagnetic torque density	120
8.2	Maxwell–Poynting energy-momentum tensor	121
8.3	Alternate energy-momentum tensors	121
	Dual Alternate energy-momentum tensor	122
	Components of the Alternate tensor	124
	Electromagnetic force on a moving charge	124
8.4	Boundary conditions and gauge considerations	125
8.5	Electromagnetic beauty revisited	126

9 **Dielectric and Magnetic Materials** 129

9.1 Introduction 129
9.2 Maxwell's Equations with polarization and magnetization 130
9.3 Amperian energy-momentum tensors 131
Modified energy-momentum tensors 132
Linear dielectric and magnetic materials 134
Complex Alternate-power theorems 137
Quasistatic approximations 138

10 **Amperian, Minkowski, and Chu Formulations** 141

10.1 Introduction 141
10.2 Maxwell's Equations in the Amperian formulation 141
10.3 Maxwell's Equations in the Minkowski formulation 142
10.4 Maxwell's Equations in the Chu formulation 143
10.5 Energy-momentum tensors and four-force densities 145
Amperian energy momentum and four-force 145
Minkowski energy momentum and four-force 146
Chu energy momentum and four-force 147
10.6 Discussion of force densities 148
10.7 The principle of virtual power 150

PART III ELECTROMAGNETIC EXAMPLES

11 **Static and Quasistatic Fields** 157

11.1 Spherical charge distribution 157
11.2 Electric field in a rectangular slot 158
11.3 Current in a cylindrical conductor 160
Static current 160
Sinusoidal steady-state current 162
11.4 Sphere with uniform conductivity 163
Quasistatic electric-field probe 163
Power and energy 166
Quasistatic magnetic-field probe 168
Power and energy 169
11.5 Quasistatic analysis of a physical resistor 170
Introduction 170
Fields and potentials 171
Equivalent circuits 178
11.6 Magnetic diffusion 179

12 **Uniformly Moving Electric Charges** 183

12.1 Point charge 183
Uniform motion in free-space 183
Motion in a dielectric (Čerenkov radiation) 185
12.2 Surface charges separating at constant velocity 185
Introduction 185

Lorenz gauge 187
Coulomb gauge 188
12.3 Expanding cylindrical surface charge 190
12.4 Expanding spherical surface charge 192

13 Accelerating Charges 195

13.1 Hertzian electric dipole 195
Sinusoidal steady-state 195
Sinusoidal pulse 196
13.2 Hertzian magnetic dipole 200
Sinusoidal pulse 200
13.3 Radiation from an accelerated then decelerated charge 202
Maxwell–Poynting representation 204
Alternate power and energy representation 205

14 Uniform Surface Current 207

14.1 Pulse excitations 207
Step function 207
Exponentially decaying step 209
Step of duration T followed by an exponential decay 210
Pulses of duration T with t^n rise 212
14.2 Resistive-sheet detector 214
14.3 Additional pulse waveforms 217
Gaussian current pulse 217
Zero-average waveforms 218

15 Uniform Line Currents 223

15.1 Axial current step (integral laws) 223
The steady-state solution 229
An assumed zone of field generation 231
The velocity of light 232
The impedance of free-space 233
A modified trial solution 233
The modified trial solution revisited 233
A complete series solution 235
15.2 Axial current step (differential laws) 237
Maxwell–Poynting representation 238
Alternate power and energy representation 239
15.3 Superposition of axial line currents 240
Uniform J_z within a finite radius, r_o 240
Uniform K_z at constant radius, R 242
Uniform K_z on a strip of constant width 245
15.4 Axial current with multiple pulses 246
Trapezoidal wave forms 246
Sine-wave approximation 249
15.5 Fields of a sinusoidal axial current 251

Waves and boundary conditions 251
Convolution integral 252

16 Plane Waves **255**

16.1 Uniform *TEM* plane waves 255
Propagation in free-space 255
Propagation in uniform linear materials 256
16.2 Doppler-shifted *TEM* plane waves 257
16.3 Nonuniform plane waves 258
TE waves 259
TM waves 259
Energy velocities 260
16.4 Skin-depth-limited current in a conductor 261

17 Waves Incident at a Material Interface **263**

17.1 Reflected and transmitted plane waves 263
17.2 *TE* polarization 264
Law of reflection and Snell's Law 265
Critical angle 265
Reflection and transmission coefficients 265
Magnetic Brewster Angle 266
Alternate-power flux 267
17.3 *TM* polarization 267
Reflection and transmission coefficients 267
Brewster Angle 268
Dual Alternate-power flux 268
17.4 Elliptically polarized incident waves 269

18 TEM Transmission Lines **271**

18.1 General time-dependent solutions 271
Source equivalence 272
Power energy and stress momentum 273
18.2 Parallel-plate TEM line in the sinusoidal steady state 274
Infinite line 274
Short-circuit termination at $z = 0$ 277
18.3 TEM tapered-plate "horn" transformer 280
18.4 TEM line with parallel plates of high conductivity 282
Perfectly conducting plates 284
High conductivity revisited (the complete potentials) 285
Application of the complex power theorems 287
18.5 Parallel-plate TEM line loaded with linear material 289

19 Rectangular Waveguide Modes **293**

19.1 Introduction 293
19.2 Periodic potentials and fields 294

19.3 Waveguide dispersion 295
19.4 TE_{nm} modes 296
TE power fluxes 297
19.5 TM_{nm} modes 298
TM power fluxes 298
19.6 Null Alternate-power and Alternate-energy distributions 299
19.7 Uniqueness resolved 300

20 Circular Waveguide Modes 305

20.1 Introduction 305
Waveguide dispersion 306
20.2 TM_{nm} modes 307
TM power fluxes 309
20.3 TE_{nm} modes 310
TE power fluxes 314
20.4 Null Alternate power and energy distributions 323
20.5 Alternate energy momentum and photons 323
TE^{o}_{nm} "circularly polarized" modes 324
TM^{o}_{nm} "circularly polarized" modes 329
Modes of a square waveguide 332

21 Dielectric Waveguides 335

21.1 Introduction 335
21.2 Symmetric TE modes 336
21.3 Antisymmetric TE modes 336
21.4 Dispersion relations 337
Symmetric modes 338
Antisymmetric modes 338

22 Antennas and Diffraction 341

22.1 Introduction 341
22.2 Half-wave dipoles 342
Wire antenna 342
Thin slot in a ground plane 344
22.3 Self-complementary planar antennas 345
22.4 Traveling-wave wire antennas 345
Super-gain and end-fire antennas 347
22.5 The theory of simple arrays 349
Uniform linear arrays 350
Directivity as a function of N, kd, and angle 352
22.6 Diffraction by a rectangular slit 356
Maxwell–Poynting analysis 356
Alternate-representation analysis 359
22.7 Diffraction by a large circular aperture 360
On-axis fields and power 362
Fresnel zones 365

	Off-axis far-field radiation pattern	367
22.8	Diffraction by a small circular aperture	369
22.9	Diffraction by the complementary screen	371
22.10	Paraxial wave equation	372
	Introduction	372
	Gaussian-beam solutions	373
	Higher-order solutions	374
23	**Waves and Resonances in Ferrites**	**377**
23.1	Introduction	377
23.2	Ferrites	378
	Angular momentum and magnetic moments	378
	Constitutive relations	379
	Magnetic resonance	379
23.3	Large-signal equations	380
23.4	Linearized (small-signal) equations	381
	Time-dependent equations	381
	Complex Polder susceptibility and permeability tensors	382
23.5	Uniform precession in a small ellipsoid	383
23.6	Plane wave solutions	384
	Electromagnetic waves	386
	Magnetostatic waves	387
23.7	Small-signal power and energy	388
	Maxwell–Poynting representation	388
	Dual Alternate representation	390
23.8	Small-signal stress and momentum	391
	Maxwell–Poynting representation	391
23.9	Quasiparticle interpretation (magnons)	393
24	**Equivalent Circuits**	**395**
24.1	Receiving circuit of a dipole	395
24.2	*TEM* transmission lines	398
	Basic equations	398
	Power and energy	400
	Lossless, low-loss, and distortionless lines	400
	Reflection coefficient and line impedance	401
	Smith Chart	403
	Impedance matching	404
24.3	Lossless tapered lines	406
24.4	Transients on transmission lines	408
24.5	Plane waves (oblique incidence)	411
	TE waves	411
	TM waves	412
	Summary of parameters	413
24.6	Waveguides	413
	TM modes	415

TE modes 417
24.7 The scattering matrix 418
Single-port 418
N-port junction 419
Lossless junctions 420
24.8 Directional couplers 421
24.9 Resonators 421
Introduction 421
Quality factors 422
Transmission-line resonator 423
Transmission resonator 426
Waveguide resonators 427
Dielectric resonators 429
YIG sphere filter 429

25 Practice Problems 435

25.1 Statics 435
25.2 Quasistatics 448
25.3 Plane waves 458
25.4 Radiation and diffraction 462
25.5 Transmission lines 472
25.6 Waveguides 481
25.7 Junctions and couplers 485
25.8 Resonators 490
25.9 Ferrites 491
25.10 Four-dimensional electromagnetics 496

PART IV BACKMATTER

Summary 505

Electromagnetic Luminaries 511

About the Author 519

Appendix A 521
A.1 Theory of Special Relativity 521
Expanding spherical electromagnetic wave 522
Galilean transformation 522
Lorentz transformation 523
Speed limit of a moving charge 524
Lorentz contraction 525
Four-vector length 526
Particle density 526
Simultaneity and time dilation 526
Addition of velocities 527
Velocity dependence of mass 529

	Force, power, and energy	529
A.2	Transformations between fixed and moving coordinates	530
	Electromagnetic fields and scalars	530
	Sinusoidal steady-state plane waves	533
	Energy-momentum tensors	533
Appendix B		**537**
B.1	The unit step and $u_k(t)$ functions	537
B.2	Three-dimensional vector identities and theorems	538
	Definitions	538
	Basic operations	539
	Curvilinear orthogonal coordinates	539
	Three-space identities	541
	Vector theorems	542
	The Divergence Theorem	542
	Stokes' Theorem	542
B.3	Four-dimensional vector and tensor identities	543
	Definitions	543
	Basic operations	544
B.4	Four-space identities	544
Appendix C		**547**
C.1	Stationary spatially symmetric sources	547
	Spherical symmetry	547
	Cylindrical symmetry with no axial variation	548
	Plane symmetry without planar variation	549
	Superposition of high-symmetry fields	550
C.2	Multipole expansions of static fields	550
	Electrostatics	550
	Magnetostatics	552
C.3	Averaging property of Laplace's Equation	553
C.4	Solutions of Laplace's Equation	554
	Cartesian coordinates	554
	Polar coordinates	555
	Cylindrical coordinates	556
	Spherical coordinates	557
C.5	Laplace's Equation in N dimensions	558
C.6	Ellipsoids in uniform fields	559
	Prolate spheroid ($c > a$)	560
	Oblate spheroid ($c < a$)	560
	Sphere ($c = a$)	561
Appendix D		**563**
D.1	Alternate power, energy, stress, and momentum	563
	Cartesian coordinates	564
	Cylindrical coordinates	565
	Spherical coordinates	566
D.2	Minkowski representations	568
	Maxwell–Poynting–Minkowski representation	568

Alternate–Minkowski representation 569
Dual Alternate–Minkowski representation 570
D.3 Stress-momentum representations of torque 571
Linear isotropic dielectric/magnetic conducting materials 574
Torque contribution from the Alternate-stress integral 575

Appendix E **577**
E.1 Fields of specified charges and currents 577
E.2 Fields of a moving point charge 578
Retarded potentials when the velocity is constant 578
Contour integration 579
Lorentz transformation 580
Liénard–Wiechert Potentials 581
Fields of an accelerated charge 581
E.3 Method of images 583
Infinite ground plane 583
Infinite-length conducting cylinder 584
Conducting sphere 584
Image configurations for magnetic conductors 585
E.4 Characteristic impedances of TEM transmission lines 586
Coaxial line 586
Lecher line 587
Elliptic-function-based transformations 588
Strip with cylindrical shield 588
Coplanar strip line 588
Symmetric strip line 589
Parallel-plate line 590
Microstrip transmission line 592

Appendix F **593**
F.1 Bessel functions 593
Integral definitions of zero-order functions 595
Series solutions and asymptotic approximations 595
Complex Hankel functions 596
Recurrence relations 596
Orthogonality and normalization integrals 596
Wronskian 598
F.2 Chebyshev polynomials 598
Polynomials of the first kind 598
Polynomials of the second kind 599
F.3 Hermite polynomials 600

Appendix G **601**
G.1 Macsyma and Maxima 601
G.2 Macsyma program descriptions 602
Four-Dimensional Vectors and Operators 602
Four-Dimensional Electrodynamics (Free-Space) 603
Four-Dimensional Electrodynamics Workpad 603
G.3 Macsyma notebooks 605

Setup and execution 605
4d-vector and 4d-em demos 606
4d-em workpad 606
G.4 Text of Macsyma/Maxima batch program 608

Appendix H **619**
H.1 Animated fields of surface currents 619
Planar surface current: step pulse with exponential decay 619
Planar surface current: Gaussian pulse 619
Alternate null power flux: interaction of Gaussian pulses 620
Strip surface current, $K_z(t) = K_o u_{-1}(t)$ 620
H.2 Animated fields of a cylindrical volume current, $J_z(t) = J_o u_{-1}(t)$ 620
H.3 Animated fields of a cylindrical surface current, $K_z(t) = K_o u_{-1}(t)$ 621
H.4 Animated fields of line-current transients 622
H.5 Animated field of a radiating Hertzian dipole 623
H.6 Animated beauty-power fluxes of cylindrical waveguide modes 623
H.7 Macsyma animations and graphics 624

References **627**

Index **631**

PREFACE

"The most beautiful thing we can experience is the mysterious. It is the source of all true art and science."

—Albert Einstein

Anyone, who has taken the trouble to explore the rich legacy that James Clerk Maxwell left to the scientific world, cannot fail to be impressed. Because he made such outstanding and profound contributions in multiple disciplines within a life-span of only 48 years, his achievements are all the more remarkable. They include seminal work in thermodynamics, statistical mechanics, and electromagnetic theory, but it is the latter that we here consider.

Although this text was written to instruct advanced undergraduate and first-year graduate students in the basic concepts of classical macroscopic electromagnetic fields, it was done so with the hope of providing new insights into and appreciation for what is surely one of the supreme achievements in science. Certain topics (described below), that do not appear in traditional texts, are deserving of inclusion both in their own right and because they simplify the development of new material concerning electromagnetic power and energy. The title, **The Power and Beauty of Electromagnetic Fields**, was chosen because of its multiple meanings.

During a tour of the campus, more than one M.I.T. student has taken a non-scientist friend to the lobby of the Eastman Laboratories (Building 6) to stand before the marble wall on which mathematical symbols that represent Maxwell's Equations are inscribed in bronze. Usually, the student attempts to convey to his friend just how powerful and useful these compact equations are and how they at first *predicted* (rather than explained) the existence of electromagnetic waves propagating at the speed of light, c. How that, in turn, led to their widespread application – to radio, microwaves, television, light, and x-rays; to electrical generators, motors, transducers, control systems, and power-grids; to integrated-circuits, computers, CD-ROMs, and the Internet; to electrocardiograms, pacemakers, magnetic resonance imaging (MRI), computer aided tomography (CAT), biosensors, and other emerging marvels of biomedical technology. And the list goes on and on .... No one can doubt the utility – the power – of electromagnetic theory.

Mathematicians find beauty and elegance in equations – especially if they have general applicability and can be expressed with brevity; many other scientists and engineers share these feelings as well. Surely, Maxwell's Equations qualify and may be considered beautiful as well as powerful. After all, Albert Einstein pondered their properties and shattered Newtonian concepts of space and time with his Theory of Special Relativity. That in turn led to the introduction (by Minkowski) of four-dimensional space-time (with ict the imaginary fourth-dimension). Recasting electrodynamics in four-dimensions made evident new symmetries that led to even higher levels of understanding and beauty. Finally, the pinnacle was reached when connections to the world of quantumphysics produced Quantum Electrodynamics (QED) – but the last is not a subject for this text.

But there are other meanings of power and beauty as well. In circuit theory, electrical power is voltage times current and flows into and out of network nodes; in electromagnetic theory, the power flux is commonly defined as the Poynting vector, $\mathbf{S} = \mathbf{E} \times \mathbf{H}$ (the vector cross-product of the electric and magnetic fields), which is largely exterior to all highly conducting pathways. Likewise, rather than being localized to the electric charges and currents, the Maxwell field energy density, W, with terms proportional to the scalar products $\mathbf{E} \cdot \mathbf{E}$ and $\mathbf{H} \cdot \mathbf{H}$, is distributed throughout space. These quantities, related to each other at every point by the Poynting Theorem, invite physical interpretation on a per unit area or per unit volume basis. Even at very low-frequencies, that interpretation is very different from the approximate, highly-localized circuit-theory representation.

Nevertheless, as recounted by Julius Stratton [1, pp. 134], the Poynting interpretation was criticized as early as 1902 by H. M. Macdonald [2], and later by George H. Livens [3], and Max Mason and Warren Weaver [4, pp. 264] among other writers. In their thought provoking analogy, Mason and Weaver, *while excepting highly localized regions of space having little or no influence outside of them*,

> "do not believe that 'Where?' is a fair or sensible question to ask concerning energy. Energy is a function of configuration, just as beauty of a certain black-and-white design [such as the 'Tiled-Photons' example] is a function of configuration. [They] see no more reason or excuse for speaking of a spatial energy density than they would for saying, in the case of a design, that its beauty was distributed over it with a certain density. Such a view would lead one to assign to a perfectly blank square inch in one portion of the design a certain amount of beauty, and to an equally blank square inch in another portion a certain different amount of beauty."

Many who have pondered that criticism applaud its cleverness – but consider it a false analogy that does little to advance one's understanding of the issue. After all, they had offered no alternate interpretation of (or replacement for) $\mathbf{S}$ and W. If not in the fields, where does radiating electromagnetic wave power and energy reside? Because circuit-power is confined to the wires that carry electrical currents, it cannot be used to explain radiation without the addition of a rather mysterious "radiation-resistance" to the circuit path. Even with that artifice, when the source-current is turned-off and a

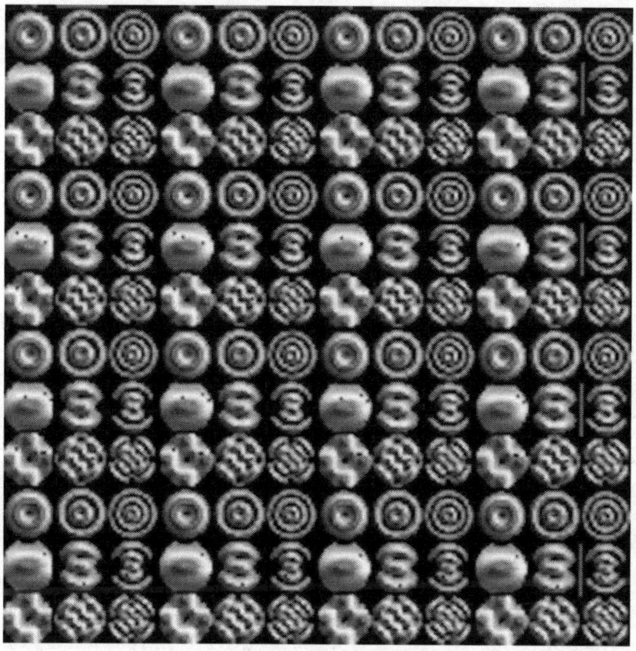

'Tiled-Photons'

transmitting circuit stops radiating, one expects that the energy, that has been radiated, must be located *somewhere* in free-space prior to its eventual detection – possibly very far from the transmitter.

During his junior year as an electrical engineering student at M.I.T., this author recalls an instructor telling the class that "although sophomores believe that electrical power is confined to the wires that carry electrical currents in a circuit, *we* [who have learned about Poynting] *know better*." One skeptical student asked the instructor if his belief in the Poynting vector was strong enough that he would be willing to touch the wires of a high-voltage circuit. Without hesitation, the instructor responded, "of course, provided that you show me how I can do so without putting my hand in the field." We all laughed and became more than a little smug because of our new found superior knowledge – we really thought that we did know better! Nearly forty years later, I learned that we (or at least I) did not; that realization, which came near the end of a long academic career in electromagnetism, was the principal motivation for the writing of this text. The analysis that led to that realization forms an essential component of its content.

Although the "circuit" representation is very useful, conventional wisdom considers it only a low-frequency or quasistatic approximation. In this text, we show that this is not necessarily true and in the process find alternate representations of electromagnetic power and energy that differ from the familiar Poynting theorem values – yet are fully equivalent. The particular choice focussed on features highly-localized power and energy components and emphasizes the circuit rather than the wave nature of these quantities. Moreover, unlike the Poynting vector, this exact representation merges smoothly with well-known quasistatic approximations that have long been used to calculate power flows in both lumped and distributed circuits operating at low-frequencies. As required, the electromagnetic power-conversion density, $\mathbf{E} \cdot \mathbf{J}$ (the dot product of electric-field and current-density), is the same in both the Poynting and all correct alternate representations of electromagnetic power. Maxwell's Equations and the fields they describe are, of course, left unchanged. It is also possible to alter the Maxwell stress-tensor, $\overline{\mathbf{T}}$, and the associated electromagnetic momentum-density, $\mathbf{G}$, in a similar manner without altering the electromagnetic force-density; when all four elements ($\mathbf{S}$, W, $\overline{\mathbf{T}}$, and $\mathbf{G}$) are treated similarly, any of the resulting alternate-representations can replace the Maxwell-Poynting form without approximation. One specific choice is termed the Alternate-representation. For time-harmonic fields, the complex Poynting theorem, energy-theorem, and momentum-theorem all have alternate-counterparts. Electromagnetic angular-momentum also has an alternate counterpart that is shown to connect directly with the spin properties of photons.

For certain electromagnetic problems, the Alternate-representation leads to both conceptual and computational simplicity. It is especially useful when dealing with either antenna radiation or quasistatic fields. For example, the power radiated from a Hertzian electric-dipole and its radiation pattern both can be calculated without first finding the electric and/or magnetic field – not a single curl operation need be performed. For other problems, it is the Maxwell-Poynting representation that is mathematically simpler; having a choice adds both flexibility and insight to the process of problem solving. Some features are surprising; for steady-state single-frequency fields, the free-space Alternate power-flux and energy-density are time-independent. Consequently, Alternate reactive power-flows and energies are banished from free-space and restricted to the locations of the charges and currents.

The text is divided into four-parts: Basic Electromagnetic Theory; Four Dimensional Electromagnetism; Electromagnetic Examples; Backmatter. Part I is devoted to a fairly conventional presentation of the integral and differential forms of Maxwell's Equations in free-space containing electric charges and currents that are subject to Lorentz-forces. Conservation of charge is assumed. The equivalent representation in terms of the magnetic vector-potential and the electric scalar-potential is also given in both Coulomb and Lorenz gauges. Materials with polarization and magnetization that may be electrically-conducting are considered, as are the boundary conditions at material and source interfaces.

When wave-propagation effects are negligible, fields that are mainly electric or magnetic are classified as either electroquasistatic (*EQS*) or magnetoquasistatic (*MQS*). The properties of quasistatic fields and their analysis by approximate methods are developed.

Electromagnetic power, energy, stress, and momentum are presented in both the Maxwell-Poynting and Alternate representations. So too, are complex versions of power and energy theorems that apply to sinusoidal steady-state fields. For linear media, both homogeneous (source free) and inhomogeneous wave equations are studied in one, two, and three dimensions; extensive use is made of both symmetry and the principle of superposition The concepts of electromagnetic duality, equivalence, and Babinet's Principle for complementary structures are other important topics that are included; the uniqueness-theorem, induction-theorem, and reciprocity-theorem are also derived.

All derivations contained in Part I are carried out in conventional three-space coordinates with time as a parameter, but it is actually easier to formulate (and generalize) electrodynamics using four-dimensional representations of both fields and forces. It then follows that power, energy, stress, and momentum are unified in terms of an energy-momentum tensor. In Part II, we introduce these concepts and emphasize the utility of expressing the various tensors in terms of four-vector electric and magnetic fields and the four-vector-potential. However, no prior knowledge of such representations is assumed nor is the reader expected to be familiar with the electrodynamics of moving media which is also discussed. Although these topics must be considered advanced[1], they depend upon straight-forward extensions of the standard vector-calculus – knowledge of which is a prerequisite for almost all texts (including this one) on electromagnetic-theory. Using this approach, an infinity of energy-momentum tensors are found that are fully equivalent to the usual Maxwell-Poynting form; these change the representation of the electromagnetic power, energy, momentum and stress in free-space without altering the four-vector electromagnetic-force density. As noted above, emphasis is given to the particular choice that directly connects with circuit-theory representations of quasistatic-fields. These results are generalized to include the presence of field interactions with matter; special emphasis is given to dielectric and magnetic materials that are linear. Because multiple representations lead to a deeper understanding, four-dimensional electromagnetics is presented in order to complement, illuminate, and (in many cases) simplify, the topics developed in Part I. In addition, the theory certainly possesses great "mathematical-beauty" and so is deserving of inclusion.

In Part III, classic field problems are considered that illustrate how solutions of Maxwell's Equations can be combined in order to satisfy boundary conditions for a wide variety of examples that, in the main, depend upon topics covered in Part I. Solving problems is the best (perhaps the only) way for a student of electromagnetics to master both physical concepts and mathematical techniques. Gradually, one develops physical intuition concerning fields; most find the process challenging, but the rewards great. It is reassuring that, once gained, mastery of the subject will not become obsolete. Certainly, its application to new materials and devices will continue to refresh electromagnetics, but, if history is any guide, the basic field-equations will not be supplanted; Maxwell seems to have got it very right. The specific examples were chosen, not only because the fields themselves are of interest, but because detailed comparisons can be made between the Alternate and Maxwell-Poynting representations. In many cases, the time-averaged distributions agree exactly. However, in others, there are significant differences and not only is Alternate power and energy highly-localized on currents and charges, but, astonishingly, also in regions of free-space. When compared with the usual Poynting energy-density, the propagation of such distributed and localized forms of Alternate-energy at first seems to create grave paradoxes. However, on closer inspection and analysis of the measurement process – by means of which power is detected, all of these are resolved. In other cases, Alternate-power exists in regions free of electric and magnetic fields or vice-versa. These strange results are also reconciled. In addition to the examples, which are worked through in detail, a chapter of over 120 practice problems is included so that the reader can test his/her understanding of the basic concepts and sharpen problem-solving skills. Many of the problems were created by the author for use in both M.I.T. undergraduate and graduate courses; however, new ones have been added as well.

Part IV contains the Backmatter including Summary, Appendices, Bibliography, and Index. A photo gallery of many (unfortunately not all) electromagnetic luminaries is included so that the reader can humanize the science of electromagnetics. As might be expected, many of the pioneers lived interesting and multi-faceted lives; fascinating

[1]If considered too demanding (or when there is insufficient time), Part II may be delayed (or bypassed altogether) until selected chapters of Part III have been studied. These contain topics and illustrative examples that depend mainly on knowledge of Part I. Used in this way, the text can serve both undergraduate and graduate students.

biographical information is readily available both in libraries and on the Internet. Eight appendices complete the text. Appendix-A includes elements of Special-Relativity including space-time coordinates and the Lorentz-transformation. Appendix-B is devoted to three-space vector identities and their four-space vector and tensor equivalents. Appendix-C contains important properties of Laplacian-fields and tabulates solutions of Laplace's Equation. Appendix-D provides expressions for Alternate-power, energy, stress, and momentum in Cartesian, cylindrical, and spherical coordinate systems. Lorentz-torques and electromagnetic angular momentum formulations in both Maxwell-Poynting and Alternate formulations are also included. In Appendix-E, free-space fields associated with pre-specified electric charge and current distributions are considered. The method of images is included, as are formulas for the characteristic impedances of a variety of uniform *TEM* transmission-lines. Appendix-F reviews properties of Bessel functions and both Chebyshev and Hermite polynomials and provides useful recursion formulas and normalization integrals. Appendix-G discusses a very useful computer program: **4d-em.mac** that has built-in knowledge of four-space electrodynamics; it runs under Macsyma and is included on the DVD that accompanies the text. For readers without access to Macsyma, the largely equivalent and freely available Maxima can be substituted and is included. Finally, Appendix-H contains a list of the avi movie files for several of the electromagnetic transients analyzed in the text. These animations are also included and can be viewed with standard media players.

The DVD also includes three electronic versions of the book. [Advanced Level] is the complete text (described above) that is the suitable for graduate courses, reference, and self study. [Introductory Level] is intended for undergraduates; [Intermediate Level] for graduate and well prepared undergraduate students. Both omit Part II and abridge Parts III and IV by including appropriate selections of the [Advanced Level] examples, practice problems, and appendicies. All three Levels are in PDF form suitable for on-line viewing using Adobe Reader; bookmarks and hyperlinks from the table of contents allow convenient browsing. The on-line versions of Appendix-H permit both the Macsyma Notebooks and the animations to be launched directly from icon hyperlinks.

The difference between the energy-momentum tensors in the Alternate and Maxwell-Poynting representation is itself a four-tensor, II^{b}, that produces neither electromagnetic-force nor $\mathbf{E}\cdot\mathbf{J}$ power-density. It is therefore an ephemeral quantity – yet one with components that can be calculated and presented graphically. In a bit of whimsy, that honors the insights of Mason and Weaver, this author has dubbed II^{b} the "electromagnetic-beauty;" this tensor joins electromagnetic-power to form still another set of meanings for the book title. The author believes that graphical rendering of the "beauty-power-flux" of an electromagnetic-field makes that final meaning plausible. Because selected examples are included in the Summary, the reader can either confirm or deny that judgement.

F. R. MORGENTHALER

Wellesley Hills, Massachusetts
November 2010

ACKNOWLEDGMENTS

Over the years, many colleagues and students at M.I.T. have provided stimulating interactions that greatly enriched academic life and helped the author gain some measure of understanding concerning electromagnetic fields. There are too many to name individually, but I am deeply grateful to them all. There are, however, two colleagues who require special mention.

It is a pleasure to acknowledge the influence of Julius A. Stratton whose seminal text: **Electromagnetic Theory** has inspired generations of students by revealing the true beauty of electromagnetism; that underlying the deceptively simple and very elegant Maxwell Equations.

It is an honor to acknowledge the debt owed to Lan Jen Chu who served for many years as mentor, advisor, and friend. He made significant contributions to electromagnetic theory in general, antenna theory in particular, and developed the elegant theory of electrodynamics that has come to be known as the Chu formulation. More than anyone, he encouraged his students to penetrate to the depths of a problem and taught, by example, that the solution of any electromagnetic problem was incomplete without a thorough understanding of the energies and power-flows.

JULIUS ADAMS STRATTON (1901–1994)
Reprinted with permission of the
M.I.T. Museum

LAN JEN CHU (1913–1973)
Reprinted with permission of the
M.I.T. Museum

LIST OF FIGURES

1.1	Surface spanning closed contour	7
1.2	Origin, source point, q, and field point, p	13
1.3	Separated bipolar charge densities	19
1.4	Amperian current loops	20
1.5	Closed surface (normal fields)	24
1.6	Closed contour (tangential fields)	25
6.1	Induction Theorem geometry	97
11.1	Potential contours and **E**-field vectors	160
11.2	Resistor with shield	171
11.3	Equivalent circuits of a physical resistor	179
11.4	Magnetic diffusion through a conducting sheet	181
12.1	Separating planar surface charges	186
12.2	Expanding cylindrical surface charge	190
13.1	Field contours of a radiating Hertzian dipole	198
13.2	Field vectors of a radiating Hertzian dipole	198
13.3	$\mathbf{E} \times \mathbf{H}$ vectors of a radiating Hertzian dipole	199
13.4	$\mathbf{S}^o$ vectors of a radiating Hertzian dipole	199

13.5 The acceleration, velocity, and position of a positive charge 203

14.1 Vector potential and fields of surface current step 208

14.2 Vector potential and fields of surface current pulse 211

14.3 Poynting and Alternate-power fluxes for $t = 5T$ 212

14.4 Current waveforms ($u = t/T$) 212

14.5 Gaussian pulse ($\alpha = 1,\ cT = 5$) 217

14.6 Normalized plots of $A(z/cT),\ S(z/cT),\ S^{\text{o}}(z/cT)$ 218

14.7 $K(t/T)$: saw-tooth "sine" pulse 219

14.8 Normalized plots of $A(u),\ S(u),\ S^{\text{o}}(u)$ 219

14.9 $K(t/T)$: sine pulse 220

14.10 Normalized plots of $A(u),\ S(u),\ S^{\text{o}}(u)$ 220

14.11 $K(t/T)$: saw-tooth "cosine" pulse 221

14.12 Normalized plots of $A(u),\ S(u),\ S^{\text{o}}(u)$ 221

14.13 $K(t/T)$: cosine pulse 222

14.14 Normalized plots of $A(u),\ S(u),\ S^{\text{o}}(u)$ 222

15.1 Axial line current 224

15.2 Linear chain of current sources 224

15.3 Bipolar-charge distribution 226

15.4 Gaussian surface 228

15.5 Circular integration contour 228

15.6 Rectangular integration contour 229

15.7 Magnetic-field pattern 230

15.8 $E_z(r, 12, \infty)$ and $E_z(r, 9, 2)$, $E_z(r, 12, 0)$ for $\delta = .01$ 237

15.9 Current regions 241

15.10 $-E_z(r/R,\ ct/R)$ and $H_\phi(r/R,\ ct/R)$ for $ct/R = .5, 1, 2, 4$ 243

15.11 Wave-front normalization, $F(n)$ 244

15.12 $-E_z(r/R,\ ct/R)$ and $H_\phi(r/R,\ ct/R)$ for $ct/R = .5, 1, 2, 4$ 244

15.13 $-E_z(x, y, t)$ for $ct/w = 6, 12, 18, 24$ 246

15.14 $|\mathbf{H}(x, y, t)|$ for $ct/w = 6, 12, 18, 24$ 247

15.15 Trapezoidal-current transient 247

15.16 Train of current reversals: (a) Fields; (b) Poynting-power flux 249

15.17 Trapezoidal approximation of sin(x) 250

15.18 Fields and Poynting-power flux of trapezoidal "sine" pulses 250

17.1	Plane of oblique incidence	264
18.1	Parallel-plate transmission line	275
18.2	Vector field plot of the exact $\mathbf{A}(x/d, z/d, 0)$ near the short circuit	280
18.3	Tapered-plate transmission line	281
18.4	Parallel-plate transmission line ($\varepsilon = 9\varepsilon_o$, $\mu = 2\mu_o$)	291
19.1	Rectangular waveguide geometry	294
19.2	$TE_{10}^{\square}$ mode: electric and magnetic fields	296
19.3	$TE_{10}^{\square}$ mode: electric surface currents	297
19.4	$TE_{20}^{\square}$ and $TE_{01}^{\square}$ modes: electric and magnetic fields	297
19.5	$TE_{11}^{\square}$ and $TM_{11}^{\square}$ modes: electric and magnetic fields	299
19.6	$\mathrm{Null}_{11}^{\square}$ mode ($a = b$): $S_z^o(x, y)$ or $W^o(x, y)$	300
19.7	Horizontally and vertically stacked waveguides	301
19.8	$TE_{11}^{\square}$ mode Alternate flux, S_z^o, in stacked waveguides	302
19.9	$TE_{11}^{\square}$ mode Poynting flux, $<S_z>$ (independent of stacking)	302
19.10	$TM_{11}^{\square}$ mode Alternate flux, $<S_z^o>$ (single waveguide)	303
19.11	$TM_{11}^{\square}$ mode Poynting flux, $<S_z>$ (single waveguide)	304
20.1	Circular waveguide geometry	306
20.2	TM_{01}^o mode ($k_cR = 2.405$): electric and magnetic fields	308
20.3	Circular-mode divisions ($n = m = 2$)	309
20.4	TM_{01}^o mode ($k_cR = 2.405$): Alternate-power flux, $S_z^o(x, y)$	310
20.5	TM_{01}^o mode ($k_cR = 2.405$): Poynting flux, $<S_z(x, y)>$	310
20.6	TM_{11}^o mode ($k_cR = 3.832$): Alternate-power flux, $S_z^o(x, y)$	311
20.7	TM_{11}^o mode ($k_cR = 3.832$): Poynting flux, $S_z(x, y)$	311
20.8	TM_{02}^o mode ($k_cR = 5.520$): Alternate-power flux, $S_z^o(x, y)$	312
20.9	TM_{02}^o mode ($k_cR = 5.520$): Poynting flux, $S_z(x, y)$	312
20.10	TE_{11}^o mode ($k_cR = 1.841$): electric and magnetic fields	313
20.11	TE_{01}^o mode ($k_cR = 3.832$): electric and magnetic fields	314
20.12	TE_{02}^o mode ($k_cR = 7.016$): electric and magnetic fields	314
20.13	TE_{11}^o mode ($k_cR = 1.841$): Alternate-power fluxes	319
20.14	TE_{11}^o mode ($k_cR = 1.841$): Poynting flux, $<S_z(x, y)>$	320
20.15	TE Alternate-power fluxes: $S_{xy}^o(x, y)$	320
20.16	TE_{21}^o mode ($k_cR = 3.054$): Alternate-power fluxes	321
20.17	TE_{21}^o mode ($k_cR = 3.054$): Poynting flux, $<S_z(x, y)>$	322

20.18 TE_{01}^{o} mode ($k_cR = 3.832$): $S_z^o(x, y) = <S_z(x, y)>$ 322

20.19 TE_{11}^{o} null mode ($k_cR = 1.8412$): $S_z^o(x, y)$ or $W^o(x, y)$ 324

20.20 TM_{01}^{o} null mode ($k_cR = 2.405$): $S_z^o(x, y)$ or $W^o(x, y)$ 324

20.21 $<W^{em}(k_z/k_c)>$ of the TE_{11}^{o} mode 327

20.22 "Circularly polarized" TE_{11}^{o} mode: surface plots of spin densities and angular momenta 328

20.23 Spin densities, $n_+(r) = n_-(r)$, of TE_{nm}^{o} standing ϕ-wave modes 329

20.24 Vector-field plots of $< \mathbf{L}^{spin} >$ for TM_{nm}^{o} modes 331

20.25 Spin and angular-momentum densities of a "circularly polarized" $TE_{01}^{\square}$–$TE_{10}^{\square}$ mode 333

21.1 Dielectric waveguide geometry 336

21.2 $\alpha_x w$ versus $k_x w$ 338

22.1 Directivity (gain) of a half-wave dipole 343

22.2 Self-complementary antennas 346

22.3 Pattern function, $P(\theta, k_z/k)$ 347

22.4 Directivity (without element factor) 348

22.5 Peak directivity, $D_o(k_z/k)$ 348

22.6 Normalized $F(\vartheta, kd)$ for $N = 2$ 352

22.7 Normalized $F(\vartheta, kd)$ for $N = 4$ 353

22.8 Normalized $F(\theta, kd)\sin^2\theta$ for $N = 2$ 353

22.9 Normalized $F(\theta, kd)\sin^2\theta$ for $N = 4$ 354

22.10 Normalized $F(\vartheta, kd)$ (polar plots) for $N = 4$ 354

22.11 Normalized $F(\theta, kd)\sin^2\theta$ (polar plots) for $N = 4$ 355

22.12 Fresnel cosine and sine integrals 358

22.13 Normalized on-axis Fresnel ripples 359

22.14 $S_z^{om}(x/d)/\left(\frac{1}{2}\sqrt{\frac{\varepsilon_o}{\mu_o}}E_o^2\right) kd = 10$ 360

22.15 $S_z^{om}(z/d)/(\frac{1}{2}\sqrt{\frac{\varepsilon_o}{\mu_o}}E_o^2)kd = 10$ 361

22.16 Fresnel ripples (on-axis) 365

22.17 $S_z^o(R/\lambda, z = 5\lambda)$ (normalized) 366

22.18 Fresnel zones (R_n/λ, $z/\lambda = 5$) 366

22.19 Normalized directivity 369

23.1 Magnetic resonance precession 380

23.2 Normalized dispersion: ω/ω_M vs. $k_z/(\omega_M\sqrt{\mu_o\varepsilon})$. 386

23.3	Normalized dispersion: ω/ω_M vs. $k_x/(\omega_M\sqrt{\mu_0\varepsilon})$	386
24.1	Dipole equivalent circuit with Norton source	398
24.2	Dipole equivalent circuit with Thevenin source	398
24.3	Balanced incremental circuit	399
24.4	Unbalanced incremental circuit	399
24.5	Γ plane	402
24.6	Simplified Smith Chart	404
24.7	Smith Chart	405
24.8	Thevenin equivalent circuit with positive-wave voltage source	409
24.9	Thevenin equivalent circuit with negative-wave voltage source	409
24.10	Equivalent circuit (*TE* and *TM* waves)	413
24.11	Equivalent circuit 1	414
24.12	Equivalent circuit 2	414
24.13	N-port junction	419
24.14	Transmission-line resonator	423
24.15	Equivalent coupling: (a) stub and (b) lumped inductor	424
24.16	Transmission resonator	426
24.17	YIG sphere filter	430
B.1	Integral and derivatives of the unit-step function	538
C.1	Equipotentials and electric-field lines of a cylindrical dipole	551
C.2	Contours of $A_\phi(r,\theta)$ created by a tiny current loop	553
C.3	Spheroid geometries	560
E.1	Point-charge geometry	582
E.2	Radial locations of source and image(s)	584
E.3	Image charges/currents ($\sigma_e = \infty$)	585
E.4	Image charges/currents ($\sigma_m = \infty$)	586
E.5	Coaxial transmission line	587
E.6	Lecher transmission line	587
E.7	Conformal mapping function $L(\rho)$	588
E.8	Strip with cylindrical shield	589
E.9	Coplanar strips	589
E.10	Complement of coplanar strips	589
E.11	Symmetric strip line	589
E.12	Parallel-plate line	591
E.13	Microstrip transmission line	592
F.1	Bessel functions of the first kind	594
F.2	Bessel functions of the second kind	594

PART I

BASIC ELECTROMAGNETIC THEORY

INTRODUCTION TO PART I

Michael Faraday, James Clerk Maxwell, and others[1] postulated that the electromagnetic forces generated by electric charge and current densities (ρ and $\mathbf{J}$) are transmitted by electric and magnetic vector fields ($\mathbf{E}$ and $\mathbf{H}$) that exist in the surrounding free-space. This insight led to rapid progress in understanding the interconnections between time-varying sources and these fields and culminated in Maxwell's Equations as set forth in 1864. Their properties led to the prediction of electromagnetic waves traveling at the speed of light. A more complete understanding of the power and energy associated with these fields followed twenty years later.

[1]Originally, electricity and magnetism were separate fields of inquiry, each with its own nomenclature and units. That changed in 1819 when the Danish physicist Hans Christian Oersted (1777–1851) discovered the interaction of electric currents on a compass needle and coined the word "electromagnetism." Following up on this revelation, the French physicist Andre Ampere (1775–1836) studied the forces between electric currents.

The interaction of fields and macroscopic matter is complicated by the fact that (fortunately) materials have very diverse forms that include insulators, metals, and semiconductors. Almost all of these have dielectric properties; a few are strongly magnetic—sometimes even when no external fields are applied. The principal responses of electromagnetic character that are generated in materials are: conduction currents due to the motion of mobile unpaired charges,[2] electric dipoles created by bound-charge pairs of opposite polarity, and magnetic dipoles generated by either tiny current loops or the coupling of the intrinsic magnetic moments of individual electrons (sometimes protons). Three vector fields are found to characterize a wide range of materials: They are the conduction-current density, $\mathbf{J}_u$, the electric-dipole density, $\mathbf{P}$, and the magnetic-dipole density, $\mathbf{M}$. The constitutive laws that relate the material responses to $\mathbf{E}$ and $\mathbf{H}$ are usually nonlinear (especially in the case of ferromagnets), but are often approximated as linear—at least for weak fields.

Part I is devoted to a fairly conventional, but compact, presentation of both the integral and differential forms of Maxwell's Equations in free-space containing electric charges and currents. Rather than follow the historical development, we simply postulate the validity of conservation of charge, the Lorentz-force density, Faraday's Law, and Ampere's Law—as amended by Maxwell. All sources are assumed to be averaged over microscopic distances and times; therefore, the fields that they generate must also be considered macroscopic in nature. An equivalent representation of these fields, in terms of the magnetic vector potential and the electric scalar potential, is also given for both Coulomb and Lorenz gauges.

Next, inhomogeneous wave equations governing the electric and magnetic fields are derived—as are those governing the vector and scalar potentials in both the time and frequency domains. Solutions are found by direct integration of the currents and charges (assumed known). As an important example, the potentials associated with a line current of incremental length are derived; they provide the basis for understanding electromagnetic radiation. Materials with polarization and magnetization that may be electrically conducting are then considered as are the boundary conditions at material and source interfaces. When wave-propagation effects are negligible, fields that are mainly electric or magnetic are classified as either electroquasistatic (*EQS*) or magnetoquasistatic (*MQS*). The equations governing quasistatic fields are derived, and analysis by approximate methods is outlined; the approach taken follows that of earlier M.I.T. textbooks by Fano, Chu, and Adler [5] and Haus and Melcher [6].

As is customary, electromagnetic power, energy, stress, momentum, and angular momentum are defined within the Maxwell–Poynting representation that is based upon the Poynting vector and Maxwell stress tensor. However, the Alternate representation by Morgenthaler [8] is also presented that embraces localized circuit-theory concepts and merges with quasistatic representations, yet is *exact*; the unification of power, energy, stress, and momentum in these, and other representations, is delayed until Part II (where the appropriate mathematical tools are developed). Power-energy theorems for both representations are developed in the time domain. These are used to calculate Poynting power and energy of a uniform plane wave and a radiating electric dipole and compare them with the Alternate counterparts.

[2] In most metals, diffusion currents are negligible because conduction and charge neutrality combine to remove concentration gradients that drive diffusion, In semiconductors, diffusion can be made to dominate; the bipolar transistor depends upon it.

Next, homogeneous wave equations and free-space waves propagating in one, two, and three dimensions are studied in both the time and frequency domains. Waves of high symmetry are given special attention because their superposition can be used to represent inhomogeneous waves generated by arbitrary, but known, sources. Homogeneous waves in linear dielectric and magnetic materials (that may also be conductive) are considered; sinusoidal steady-state (harmonic) waves are given particular attention. In the limit of high conductivity, simple wave propagation is radically altered and replaced by the interrelated concepts of charge relaxation, magnetic diffusion, and skin depth. For loss-free dielectrics, power, energy, stress, and momentum associated with packets of uniform harmonic plane waves are considered from both wave and particle perspectives; the latter is shown to follow a Hamiltonian formulation.

The final chapter of Part I is devoted to important theorems and principles. Frequency-domain theorems involving electromagnetic power, stress, energy, and momentum are derived. These are followed by duality of fields and sources, the uniqueness theorem, the equivalence principle, the induction theorem, Babinet's Principle for complementary structures, and the reciprocity theorem.

When the Part III examples and explanatory texts are included, the overall coverage of the standard wave-propagation issues is similar to that of other texts but is built upon both the Maxwell–Poynting and Alternate representations. The most recent effort [9] (designed for the M.I.T. EECS undergraduate curriculum) is authored by D. H. Staelin, A. W. Morgenthaler, and J. A. Kong; older standards include Adler et al. [10] and Ramo et al. [11].

The text is designed to serve a variety of curriculums. Because much of Part III depends only upon Part I, an undergraduate course could omit the four-dimensional electrodynamics of Part II and include only basic examples; the focus of a particular graduate course might be restricted to advanced electrodynamics, antennas and diffraction, or transmission lines and microwave circuits.

CHAPTER 1

MAXWELL'S EQUATIONS

1.1 MATHEMATICAL NOTATION

- Scalar quantities, such as Φ, are printed in normal type.
- Vectors, such as $\mathbf{E}$ and $\mathbf{H}$, are printed in **bold** type; their components are printed in normal type with subscripts that indicate coordinate directions.
- Unit vectors are printed in **bold** type, but with $\widehat{}$ over the symbol. Thus, $\mathbf{E} = \widehat{\mathbf{x}}E_x + \widehat{\mathbf{y}}E_y + \widehat{\mathbf{z}}E_z$ represents the electric field expressed in Cartesian coordinates as the sum of three orthogonal vectors. Numerical subscripts $1, 2, 3$ often substitute for x, y, z; for example, E_i can be any one of the components.
- The scalar or dot product of two vectors $\mathbf{A}$ and $\mathbf{B}$ is indicated generally by $\mathbf{A} \cdot \mathbf{B}$ or explicitly by either $A_xB_x + A_yB_y + A_zB_z$ or $\sum_{i=1}^{3} A_iB_i$. When (as here) indices are repeated, the summation symbol is often understood to exist and omitted in order to simplify the expression.
- The vector cross product of two vectors $\mathbf{A}$ and $\mathbf{B}$ is indicated by $\mathbf{A} \times \mathbf{B}$.

The Power and Beauty of Electromagnetic Fields, First Edition. F. R. Morgenthaler.

- The del operator is defined by $\nabla = \widehat{\mathbf{x}}\frac{\partial}{\partial x} + \widehat{\mathbf{y}}\frac{\partial}{\partial y} + \widehat{\mathbf{z}}\frac{\partial}{\partial z}$ or in index notation by $\frac{\partial}{\partial x_i}$. It is used to express the gradient of a scalar and the divergence and curl of a vector. These operations are defined in Appendix B, Section B.2.
- Tensor quantities, such as $\overline{\mathbf{T}}$ are printed in **bold** type, but with a bar over the symbol; the components may be expressed in dyadic notation as two adjacent vectors $\mathbf{A}\,\mathbf{B}$ or in component form using normal type. Examples are: $\overline{\mathbf{T}} = \mathbf{A}\,\mathbf{B} = (\widehat{\mathbf{x}}A_x + \widehat{\mathbf{y}}A_y + \widehat{\mathbf{z}}A_z)(\widehat{\mathbf{x}}B_x + \widehat{\mathbf{y}}B_y + \widehat{\mathbf{z}}B_z)$ or $T_{ij} = A_i B_j$. Notice that, in general, $T_{ij} \neq T_{ji}$.
- Four-dimensional vectors, such as $\mathcal{E}$ and $\mathcal{H}$, and dyads, like $\mathcal{A}\,\mathcal{B}$, are printed in calligraphic type.

1.2 FREE-SPACE FIELDS AND FORCES

In free-space regions that contain electric and magnetic field vectors, **E** and **H**, the Lorentz force on a charge, q_i, moving with velocity, $\mathbf{v}_i$ (the subscript enumerates the charge and is not the index of a Cartesian component) is

$$\mathbf{F}_i = q_i(\mathbf{E} + \mathbf{v}_i \times \mu_o \mathbf{H}) \tag{1.1}$$

We employ SI units[1] here and throughout the text.

If there are many charges within a small finite volume, ΔV, the total force within it is

$$\textstyle\sum_i \mathbf{F}_i = (\sum_i q_i)\mathbf{E} + (\sum_i q_i \mathbf{v}_i) \times \mu_o \mathbf{H}$$

When divided by ΔV, Eq. (1.1) reduces to the Lorentz force density,

$$\mathbf{f} = \rho\mathbf{E} + \mathbf{J} \times \mu_o \mathbf{H} \tag{1.2}$$

where $\rho = \lim_{\Delta V \to 0} \frac{\sum_i q_i}{\Delta V}$ is the macroscopic electric charge density and $\mathbf{J} = \lim_{\Delta V \to 0} \frac{\sum_i q_i \mathbf{v}_i}{\Delta V}$ is the macroscopic electric-current density. In these averages, the limiting ΔV must remain large compared to microscopic dimensions. In the event that all of the velocities are equal to **v**, the current, $\mathbf{J} = \rho\mathbf{v}$, is said to be convective. The **E** and **H** fields arise from all of the charges and currents and are themselves macroscopic averages. The equations that follow describe the interactions between these macroscopic quantities.

Integral form of Maxwell's Equations

For a region of free-space containing ρ and **J**, Maxwell's Equations, when expressed in integral form, are

$$\oint_{C_o} \mathbf{H} \cdot d\mathbf{s} - \frac{d}{dt}\int_S \varepsilon_o \mathbf{E} \cdot d\mathbf{a} = \int_S \mathbf{J} \cdot d\mathbf{a} \tag{1.3a}$$

$$\oint_{C_o} \mathbf{E} \cdot d\mathbf{s} + \frac{d}{dt}\int_S \mu_o \mathbf{H} \cdot d\mathbf{a} = 0 \tag{1.3b}$$

[1] In this system of units, the abbreviations used include:
m (meter), s (second), kg (kilogram), C (coulomb), V (volt), A (ampere), F (farad), H (henry), S (siemens), $S^{-1} = \Omega$ (ohms).

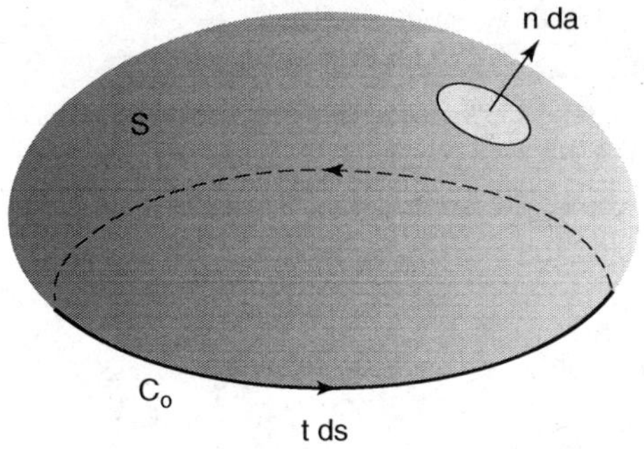

Figure 1.1 Surface spanning closed contour.

where S is an arbitrary surface that spans the closed contour C_o that is depicted in Figure 1.1.

When S is a minimal surface (that of a soap bubble film anchored by the contour), its positive side is defined as that seen from above as we circle the contour in a counter-clockwise direction. The unit vector, $\widehat{\mathbf{n}}$, normal to that surface (or any other that can be formed by simple deformation), points outward from its positive side. The vector differential area of S is defined by $d\mathbf{a} = \widehat{\mathbf{n}}da$; the vector differential distance along the contour, C_o, by $d\mathbf{s} = \widehat{\mathbf{t}}ds$, with $\widehat{\mathbf{t}}$ the unit vector tangential to the contour (and of positive polarity as we circle it). The first of these equations is Ampere's Law as amended by Maxwell to include the time-varying term (renowned as the Maxwell displacement current density); the second is Faraday's Law.

Because the electric charge is conserved, the total current flowing out of the closed-surface, S_o, that encloses an arbitrary volume, V, must equal the negative time-rate of change of the charge contained within it. Therefore,

$$\oint_{S_o} \mathbf{J} \cdot d\mathbf{a} + \frac{d}{dt}\int_V \rho \, dV = 0 \tag{1.4}$$

is the integral form of the Law of Conservation of Charge. If the contour C_o is allowed to shrink to zero as $S \to S_o$, Eq. (1.3a) becomes

$$-\frac{d}{dt}\oint_{S_o} \varepsilon_0 \mathbf{E} \cdot d\mathbf{a} = \oint_{S_o} \mathbf{J} \cdot d\mathbf{a}$$

provided that the magnetic-field tangential to C_o is finite and so cannot contribute to the limiting contour integral. When combined with Eq. (1.4) the result,

$$\frac{d}{dt}\left(\oint_{S_o} \varepsilon_0 \mathbf{E} \cdot d\mathbf{a} - \int_V \rho \, dV\right) = 0$$

can be integrated with respect to time to yield

$$\oint_{S_o} \varepsilon_0 \mathbf{E} \cdot d\mathbf{a} - \int_V \rho \, dV = \text{constant}$$

Consider a very small volume, V, centered upon some chosen point. Because ρ and $\mathbf{E}$ can be forced to vanish in that region for at least one instant of time, the constant must be zero for that and (by extension) *every* region.

It follows that

$$\oint_{S_0} \varepsilon_0 \mathbf{E} \cdot d\mathbf{a} = \int_V \rho \, dV \tag{1.5a}$$

applies to an arbitrary volume. Equation (1.5a) is known as Gauss' Law. Similar arguments applied to Eq. (1.3b) lead to

$$\oint_{S_0} \mu_0 \mathbf{H} \cdot d\mathbf{a} = 0 \tag{1.5b}$$

These equations are often included in the set of Maxwell's Equations, but are not independent. From Eqs. (1.3a) and (1.3b) and *either* Conservation of Charge or Gauss' Law the other equation follows.

Units and fundamental constants

The units of the field vectors are $[E] = Vm^{-1}$and $[H] = Am^{-1}$, those of the electric charge and current densities are $[\rho] = Cm^{-3}$, and $[J] = Am^{-2}$, and that of the Lorentz force density is $[f] = N\ m^{-3}$. The constants μ_0 and ε_0 are, respectively, the permeability and permittivity of free-space. The former is legislated to have the value

$$\mu_0 = 4\pi \times 10^{-7}\, Hm^{-1}$$

whereas the latter is a measured quantity,

$$\varepsilon_0 = 8.854187817 \times 10^{-12}\, Fm^{-1}$$

Because

$$c = \frac{1}{\sqrt{\mu_0 \varepsilon_0}} = 2.99792458 \times 10^8\, ms^{-1} \tag{1.6}$$

(the velocity of light in vacuum) is so close to 3×10^8 m s^{-1}, it is often convenient to approximate the permittivity by the exact value,

$$\varepsilon_0 = \frac{1}{36\pi} \times 10^{-9}\, Fm^{-1}$$

We shall adhere to that practice. Another combination of these constants that is extremely useful is $c\mu_0$ defined as the characteristic impedance of free-space. Given the symbol, η_0, it has the value

$$\eta_0 = c\mu_0 = \sqrt{\frac{\mu_0}{\varepsilon_0}} = 120\pi \simeq 377\, \Omega\ (S^{-1})$$

Linearity and superposition

Because Maxwell's Equations are linear with respect to the fields and the charges and currents, superposition may be applied after dividing the sources into components with respect to either space or time (or some combination). In particular, if

$$\rho = \sum_i \rho_{(i)} \tag{1.7a}$$

$$\mathbf{J} = \sum_i \mathbf{J}_{(i)} \tag{1.7b}$$

and each ith set of sources (not to be confused with a Cartesian index) is chosen so as to satisfy conservation of charge, then

$$\mathbf{E} = \sum_i \mathbf{E}_{(i)} \tag{1.7c}$$

$$\mathbf{H} = \sum_i \mathbf{H}_{(i)} \tag{1.7d}$$

where $\mathbf{E}_{(i)}$ and $\mathbf{H}_{(i)}$ are the Maxwellian fields generated when $\rho_{(i)}$ and $\mathbf{J}_{(i)}$ act alone. The components may be either discrete or differential; in the latter case the summations are replaced by integrals. When it is possible to decompose the sources into elements each providing a high degree of spatial symmetry, finding the field solutions is greatly facilitated. Fields generated by stationary spatially symmetric charge or current sources with either spherical, cylindrical, or planar symmetry are considered in Appendix C, Section C.1. Time-dependent sources are often considered to be the superposition of sinusoidal functions. This gives rise to representations employing Fourier series and Fourier integrals in the respective cases of periodic or aperiodic functions. As we shall learn in later sections of this chapter, representation in terms of complex exponential functions is also a standard technique—applicable to the fields as well as the sources.

Symmetric field-transient example

When a highly symmetric source is suddenly switched on at $t = 0$ and remains constant thereafter, it is sometimes possible to deduce the field solutions using only the integral form of the Maxwell Equations together with symmetry considerations. This method depends solely upon the material that has been presented up to this point and is employed in Part III, Chapter 15, Section 15.1. A more traditional solution of the same problem is presented in Section 15.2; it is based upon the symmetries that are made evident by the differential form of Maxwell's Equations. These are developed and built upon in the remainder of this chapter. Before continuing, the reader is encouraged to jump ahead to the first example in order to gain a preliminary understanding of both propagating electromagnetic transients and the quasistatic fields that follow in their wake. Both the speed of light and the ratio of electric to magnetic radiation field strengths (the characteristic impedance of free-space) emerge from that example.

Differential form of Maxwell's Equations

If the contour C_o is chosen to lie in the $y - z$ plane and it and the surface S spanning the contour are allowed to shrink to vanishingly small dimensions, the contour integral $\oint_{C_o} \mathbf{A} \cdot d\mathbf{s}$ (on a per unit area basis) equals the x component of the curl $\mathbf{A}$. Suitable orientation of the contour (in the $z - x$ and $x - y$ planes) will produce the y and z components. Likewise, when the closed surface shrinks toward a point, the integral of the normal flux of a vector over that surface is (on a per unit volume basis) the divergence of that vector.

The set of integral equations is therefore equivalent to the differential forms

$$\nabla \times \mathbf{H} = \mathbf{J} + \varepsilon_o \frac{\partial \mathbf{E}}{\partial t} \tag{1.8a}$$

$$\nabla \times \mathbf{E} = -\mu_o \frac{\partial \mathbf{H}}{\partial t} \tag{1.8b}$$

$$\nabla \cdot \varepsilon_o \mathbf{E} = \rho \tag{1.8c}$$

$$\nabla \cdot \mu_o \mathbf{H} = 0 \tag{1.8d}$$

where the curl ($\nabla\times$) and divergence ($\nabla\cdot$) operators are written in terms of the gradient operator, ∇ (del). Alternatively, the Divergence Theorem and Stoke's Theorem (Appendix B, Eqs. (B.7) and (B.8)) can be applied to produce the same results. We prove in Chapter 6 that both the curl and divergence of a field vector are determining factors in specifying that vector, and so it is not surprising that Maxwell's Equations constrain these quantities for both **E** and **H**.

Because $\nabla \cdot (\nabla \times \mathbf{H}) = \nabla \cdot \mathbf{J} + \frac{\partial}{\partial t}(\nabla \cdot \varepsilon_o \mathbf{E}) = 0$, the first and third equations imply conservation of electric charge,

$$\nabla \cdot \mathbf{J} + \frac{\partial \rho}{\partial t} = 0 \tag{1.8e}$$

This result also follows directly from Eq. (1.4).

1.3 VECTOR AND SCALAR POTENTIALS

From renewed use of the vector identity, $\nabla \cdot (\nabla \times \mathbf{A}) = 0$, it follows that Eq. (1.8d) is automatically satisfied by

$$\mu_o \mathbf{H} = \nabla \times \mathbf{A} \tag{1.9}$$

When substituted into Eq. (1.8b), the result is

$$\nabla \times \mathbf{E} + \frac{\partial}{\partial t}(\nabla \times \mathbf{A}) = \nabla \times \left(\mathbf{E} + \frac{\partial \mathbf{A}}{\partial t}\right) = 0$$

A second vector identity, $\nabla \times (\nabla \Phi) = 0$, then permits one to express the electric field as

$$\mathbf{E} = -\frac{\partial \mathbf{A}}{\partial t} - \nabla \Phi \tag{1.10}$$

The vector (**A**) and scalar (Φ) potentials (defined with the conventional polarities) are not unique because the set

$$\mathbf{A}' = \mathbf{A} + \nabla \Psi \tag{1.11a}$$

$$\Phi' = \Phi - \frac{\partial \Psi}{\partial t} \tag{1.11b}$$

produces the same values of both **E** and $\mu_o \mathbf{H}$. The primed and unprimed potentials are said to be related by a gauge transformation (set by the scalar function, Ψ). Evidently, the value of $\nabla \cdot \mathbf{A}$ is not unique and may be specified as desired; two possibilities are of special interest.

Lorenz gauge

The Lorenz gauge[2] is defined by

$$\nabla \cdot \mathbf{A} + \frac{1}{c^2}\frac{\partial \Phi}{\partial t} = 0 \tag{1.12}$$

with c given by Eq. (1.6).

If that value is selected and Eqs. (1.9) and (1.10) are substituted into Eqs. (1.8e), the result is

$$\nabla^2 \mathbf{A} - \frac{1}{c^2}\frac{\partial^2 \mathbf{A}}{\partial t^2} = -\mu_0 \mathbf{J} \tag{1.13a}$$

$$\nabla^2 \Phi - \frac{1}{c^2}\frac{\partial^2 \Phi}{\partial t^2} = -\frac{\rho}{\varepsilon_0} \tag{1.13b}$$

where **A** and **J** are assumed to be expressed in Cartesian components. All four components (A_x, A_y, A_z, Φ) are solutions of inhomogeneous wave equations; they are not independent because conservation of charge must be satisfied. The current and charge densities are those existing in free-space and must satisfy Eq. (1.8e); additional sources that model dielectric and magnetic materials are described in Section 1.7 and, more thoroughly, in Chapter 9.

Coulomb gauge

The Coulomb gauge (sometimes called the radiation gauge) is defined by

$$\nabla \cdot \mathbf{A} = 0 \tag{1.14}$$

If that value is selected and Eqs. (1.9) and (1.10) are substituted into Eqs. (1.8e), the result is

$$\nabla^2 \Phi = -\rho/\varepsilon_0 \tag{1.15a}$$

$$\nabla^2 \mathbf{A} = -\mu_0 \left(\mathbf{J} + \varepsilon_0 \frac{\partial \mathbf{E}}{\partial t} \right) \tag{1.15b}$$

$$\nabla^2 \mathbf{A} - \frac{1}{c^2}\frac{\partial^2 \mathbf{A}}{\partial t^2} = -\mu_0 \left(\mathbf{J} - \varepsilon_0 \nabla \frac{\partial \Phi}{\partial t} \right) \tag{1.15c}$$

Equation (1.15a) is Poisson's Equation, which does not contain time derivatives; consequently, the scalar potential at any position must respond *instantaneously and simultaneously* to all changes in the electric charge density, $\rho(\mathbf{r}, t)$. Consideration of Eq. (1.15b) reveals that the same is true of the vector potential, with respect to $\mathbf{J}(\mathbf{r},t) + \varepsilon_0 \frac{\partial \mathbf{E}(\mathbf{r},t)}{\partial t}$ – the current density that includes the Maxwell displacement current. However, when **E** is expressed in terms of both the scalar and vector potentials, the wave equation,

[2]The Lorenz gauge formulated by the Danish mathematician and physicist, Ludwig V. Lorenz (1829–1891), is often incorrectly attributed to Hendrick Lorentz; the latter is noted for the Lorentz transformation, the Lorentz contraction, and his theory of the electron.

Eq. (1.15c), emerges. Of course, the values of **E** and **H** calculated from the Coulomb gauge must (and do) agree with those calculated from the Lorenz (or any other) gauge.

Finally, when $\Phi = 0$ (because, either there is no electric charge anywhere or else Eqs. (1.11) have been used), the Lorenz and Coulomb gauges can be identical, but only in charge-free regions of space. The reader should verify that the requisite Ψ satisfies the homogeneous wave equation, except where $\partial(\nabla \cdot \mathbf{A})/\partial t = -\rho/\varepsilon_o$ causes the Coulomb gauge to fail.

1.4 INHOMOGENEOUS WAVE EQUATIONS FOR E AND H

The electric and magnetic fields, interwoven by Maxwell's Equations, can be decoupled by use of the vector identities,

$$\nabla \times (\nabla \times \mathbf{H}) = \nabla(\nabla \cdot \mathbf{H}) - \nabla^2 \mathbf{H}$$

$$\nabla \times (\nabla \times \mathbf{E}) = \nabla(\nabla \cdot \mathbf{E}) - \nabla^2 \mathbf{E}$$

and substitution of the curls and divergences of **H** and **E**. The result (after reordering partial derivatives) is a pair of inhomogeneous wave equations (expressed in terms of Cartesian components),

$$\nabla^2 \mathbf{H} - \frac{1}{c^2}\frac{\partial^2 \mathbf{H}}{\partial t^2} = -\nabla \times \mathbf{J} \tag{1.16a}$$

$$\nabla^2 \mathbf{E} - \frac{1}{c^2}\frac{\partial^2 \mathbf{E}}{\partial t^2} = \frac{1}{\varepsilon_o}\nabla \rho + \mu_o \frac{\partial \mathbf{J}}{\partial t} \tag{1.16b}$$

The solutions are not independent because, as noted previously, the current and charge densities must satisfy conservation of charge. The velocity of light, c, is given by Eq. (1.6).

An alternate, though less direct, derivation of Eqs. (1.16a) and (1.16b) follows (in the Lorenz gauge) from

$$\nabla \times \left(\nabla^2 \mathbf{A} - \frac{1}{c^2}\frac{\partial^2 \mathbf{A}}{\partial t^2} + \mu_o \mathbf{J}\right) = 0$$

$$\frac{\partial}{\partial t}\left(\nabla^2 \mathbf{A} - \frac{1}{c^2}\frac{\partial^2 \mathbf{A}}{\partial t^2} + \mu_o \mathbf{J}\right) + \nabla\left(\nabla^2 \Phi - \frac{1}{c^2}\frac{\partial^2 \Phi}{\partial t^2} + \frac{\rho}{\varepsilon_o}\right) = 0$$

and (after again interchanging the order of the derivatives) substitution of Eqs. (1.9) and (1.10). Naturally, the final result is independent of the choice of gauge.

1.5 STATIC FIELDS

When there is no time variation in the charges and/or currents, the steady-state fields are also static and related by

$$\nabla^2 \mathbf{E} = \frac{1}{\varepsilon_o}\nabla \rho \tag{1.17a}$$

$$\nabla^2 \Phi = -\frac{\rho}{\varepsilon_o} \tag{1.17b}$$

$$\nabla^2 \mathbf{A} = -\mu_o \mathbf{J} \tag{1.17c}$$

Static electric charge generates electrostatic fields Φ and $\mathbf{E}$; static electric current generates magnetostatic fields, $\mathbf{A}$ and $\mathbf{H}$. When the vectors are expressed in Cartesian coordinates, each of the scalar equations takes the form of Poisson's Equation:

$$\nabla^2 \Psi = -s \tag{1.18}$$

When $s = 0$, the homogeneous equation is termed Laplace's Equation. Solutions in various coordinate systems are discussed in Appendix C, Section C.4; they are of great importance in solving both static and quasistatic boundary-value problems. For now, we assume that s is a known function of position and attempt to integrate Poisson's Equation directly.

Integration of Poisson's Equation

Because Eq. (1.18) is a linear partial differential equation, both the source density, s, and the response function, Ψ, can be expressed as the superposition of differential components that individually satisfy this equation.

When s is a three-dimensional unit impulse (delta function) located at the origin of spherical coordinates, the spherically symmetric solution satisfies

$$\nabla^2 \Psi(r) = \frac{1}{r^2}\frac{\partial}{\partial r}\left(r^2 \frac{\partial \Psi}{\partial r}\right) = -\delta(r) \tag{1.19}$$

with $\delta(r) = \frac{1}{4\pi r^2} u_0(r)$. Here $u_0(r)$ is the derivative of the unit step function, $u_{-1}(r)$ (both are defined in Appendix B, Section B.1, but with origins shifted to $r = 0^+$ so that $\int_0^\infty u_0(r)dr = 1$). It follows that $\Psi = 1/(4\pi r)$ and (with a shift of origin) $d\Psi_p = s_q dV_q/(4\pi r_{qp})$ where, as indicated in Figure 1.2,

$\mathbf{r}_{qp} = \mathbf{r}_p - \mathbf{r}_q$ is the vector between the differential source located at $\mathbf{r}_q$ and the response evaluated at $\mathbf{r}_p$ and $r_{qp} = |\mathbf{r}_p - \mathbf{r}_q|$. After integration, the general solution of Eq. (1.18) is found to be

$$\Psi_p(\mathbf{r}_p) = \int_V \frac{s_q(\mathbf{r}_q) dV_q}{4\pi r_{qp}} \tag{1.20}$$

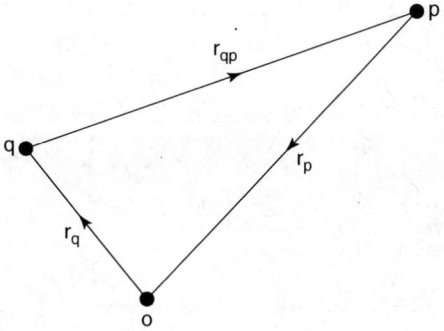

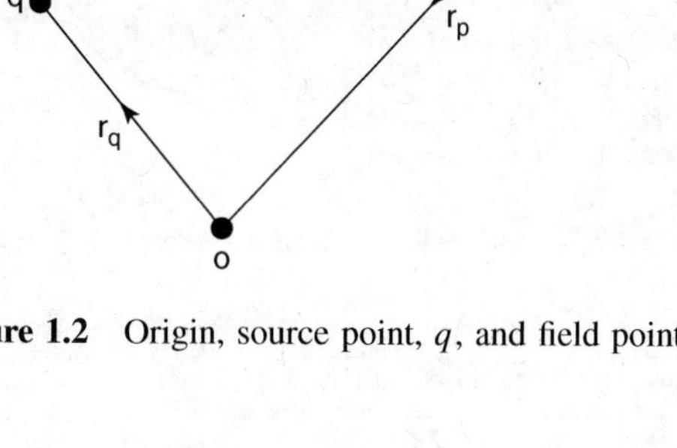

Figure 1.2 Origin, source point, q, and field point, p.

When the field point is distant from all of the sources, $1/r_{qp}$ can be expanded in a series of powers of $1/r_p$. Together these constitute the multipole expansion that is developed in Appendix C, Section C.2; the $1/r_p$ term is the monopole, $1/r_p^2$ the dipole, $1/r_p^3$ the quadrupole, and so on.

Electrostatics

The equations of electrostatics in integral form are

$$\Phi_p(\mathbf{r}_p) = \int_V \frac{\rho_q(\mathbf{r}_q)\, dV_q}{4\pi\varepsilon_o r_{qp}} \tag{1.21a}$$

$$\mathbf{E}_p(\mathbf{r}_p) = \int_V \widehat{\mathbf{r}}_{qp} \frac{\rho_q(\mathbf{r}_q)\, dV_q}{4\pi\varepsilon_o r_{qp}^2} \tag{1.21b}$$

and together form Coulomb's Law.

Sphere of uniform charge

As a simple example, consider a sphere of radius R that contains a uniform volume charge density, ρ_o. Without loss of generality we may define spherical coordinates so that the point p lies on the negative z axis. Then Eq. (1.21a) becomes

$$\Phi(r_p) = \int_0^R \int_0^\pi \frac{\rho_o 2\pi r_q^2\, dr_q \sin\theta_q\, d\theta_q}{4\pi\varepsilon_o \sqrt{r_p^2 + r_q^2 + 2r_p r_q \cos\theta_q}}$$

where r_p may be smaller or larger than R. The result of the two integrations (after dropping the subscript p) is

$$\Phi(r) = \frac{\rho_o}{\varepsilon_o} \begin{cases} \frac{1}{2}R^2 - \frac{1}{6}r^2, & r \le R \\ \dfrac{R^3}{3r}, & r \ge R \end{cases}$$

From the gradient operator, or from Eq. (1.21b), the electrostatic electric field is

$$\mathbf{E}(r) = \widehat{\mathbf{r}} \frac{\rho_o}{3\varepsilon_o} \begin{cases} r, & r \le R \\ \dfrac{R^3}{r^2}, & r \ge R \end{cases}$$

This solution was carried out only to illustrate the procedure; the preferred method of solution is to use symmetry and Gauss' Law to immediately solve for the electric field.

Magnetostatics

The equations of magnetostatics in integral form are (with $\nabla \cdot \mathbf{J} = 0$):

$$\mathbf{A}_p(\mathbf{r}_p) = \int_V \frac{\mu_o \mathbf{J}_q(\mathbf{r}_q)\, dV_q}{4\pi r_{qp}} \tag{1.22a}$$

$$\begin{aligned} \mu_o \mathbf{H}_p(\mathbf{r}_p) = \nabla_p \times \mathbf{A}_p &= \frac{\mu_o}{4\pi} \int_V \nabla_p \times \left[\frac{\mathbf{J}_q(\mathbf{r}_q)}{r_{qp}}\right] dV_q \\ &= \frac{\mu_o}{4\pi} \int_V \nabla_p \left(\frac{1}{r_{qp}}\right) \times \mathbf{J}_q(\mathbf{r}_q) dV_q \end{aligned} \tag{1.22b}$$

Because $\nabla_p \left(\frac{1}{r_{qp}}\right) = -\nabla_q \left(\frac{1}{r_{qp}}\right) = -\frac{\widehat{\mathbf{r}}_{qp}}{r_{qp}^2}$,

$$\mathbf{H}_p(\mathbf{r}_p) = \frac{1}{4\pi} \int_V \frac{\mathbf{J}_q \times \mathbf{r}_{qp}}{r_{qp}^3}\, dV_q \tag{1.22c}$$

Equation (1.22c) is known as the Biot–Savart Law.

Circular current loop

As an important example, we calculate the magnetic field on the axis of a planar circular loop of wire carrying current I_o; the wire radius is assumed to be very small compared to the loop radius, R. Integrating over the area of the wire results in

$$\mathbf{H}_p(\mathbf{r}_p) = \frac{I_o}{4\pi} \oint \frac{\mathbf{ds}_q \times \mathbf{r}_{qp}}{r_{qp}^3} \tag{1.23a}$$

where $\mathbf{ds}_q$ is the differential vector length of the current element. If the current loop lies in the plane $z = 0$ with its center at the origin, the on-axis field (which by symmetry is z-directed) is easily evaluated and found to be

$$\mathbf{H}(r = 0, z) = \widehat{\mathbf{z}} \frac{I_o R^2}{2(R^2 + z^2)^{3/2}} \tag{1.23b}$$

The current direction and **H** obey the right-hand rule (with fingers curled in the direction of the current, the thumb points in the direction of the field).

1.6 INTEGRATION OF THE INHOMOGENEOUS WAVE EQUATION

The inhomogeneous scalar wave equation is defined by

$$\nabla^2 \Psi - \frac{1}{c^2} \frac{\partial^2 \Psi}{\partial t^2} = -s(\mathbf{r}, t) \tag{1.24}$$

If the spatial derivatives dominate over those with respect to time, the inhomogeneous wave equation reduces to Poisson's Equation, and an approximation to the solution is

$$\Psi_p(\mathbf{r}_p, t) \simeq \int_V \frac{s_q(\mathbf{r}_q, t)\, dV_q}{4\pi r_{qp}} \tag{1.25}$$

Such quasistatic solutions are often useful approximations to the exact solution.

Because there is a time delay of r_{qp}/c before Ψ_p can respond to changes in the source located at $\mathbf{r}_q$, the simplest correction to Eq. (1.25) that might be expected to improve the approximation is to replace t with the retarded time, $t - r_{qp}/c$, where

$$r_{qp} = \sqrt{r_q^2 + r_p^2 - 2\mathbf{r}_q \cdot \mathbf{r}_p} \tag{1.26}$$

Remarkably,

$$\Psi_p(\mathbf{r}_p, t) = \int \frac{s_q(\mathbf{r}_q, t - r_{qp}/c)\, dV_q}{4\pi r_{qp}} \tag{1.27}$$

is an *exact* solution of the wave equation. Although this result can be verified by direct substitution, we postpone proofs until Chapter 4, Section 4.5 and Chapter 7, Section 7.7.

In cases where the source term, $s_q(\mathbf{r}_q, t)$ has a sinusoidal time dependence at frequency ω, it is advantageous to introduce complex functions $\underline{s}_q(\mathbf{r}_q, j\omega)$ and $\underline{\Psi}_p(\mathbf{r}_p, j\omega)$ such that

$$s_q(\mathbf{r}_q, t) = Re\{\underline{s}_q(\mathbf{r}_q, j\omega)\exp(j\omega t)\}$$

$$\Psi_p(\mathbf{r}_p, t) = Re\{\underline{\Psi}_p(\mathbf{r}_p, j\omega)\exp(j\omega t)\}$$

where Re { } is the operator that extracts the real part of the expression contained within the curly brackets.

Substitution into Eq. (1.24) produces the complex form of the inhomogeneous wave equation,

$$\nabla^2\underline{\Psi} + k^2\underline{\Psi} = -\underline{s}_q(\mathbf{r}_q) \tag{1.28}$$

where the wavenumber k is defined by

$$k = \omega\sqrt{\mu_0\varepsilon_0} \tag{1.29}$$

Because it is customary to define $\omega = 2\pi f$ and $k = 2\pi/\lambda_0$, the frequency f and free-space wavelength λ_0 are related by

$$f\lambda_0 = c \tag{1.30}$$

From Eq. (1.27), the integral form of the complex solution of Eq. (1.28) is

$$\underline{\Psi}_p(\mathbf{r}_p) = \int \frac{\underline{s}_q(\mathbf{r}_q)\exp(-jkr_{qp})\, dV_q}{4\pi r_{qp}} \tag{1.31}$$

The complex exponential accounts for time retardation through the phase factor kr_{qp}. Integrals of this form are commonly encountered in diffraction and antenna theory; they can often be approximated with simpler forms when the observation point, $\mathbf{r}_p$, is very distant from all source locations, $\mathbf{r}_q$. In such cases, Eq. (1.26) is well-approximated by

$$r_{qp} \simeq r_p - \frac{\mathbf{r}_q \cdot \mathbf{r}_p}{r_p}$$

and the so-called "far-field" solution is

$$\underline{\Psi}_p(\mathbf{r}_p) \simeq \frac{\exp(-jkr_p)}{4\pi r_p}\int_V \underline{s}_q(\mathbf{r}_q)\exp\left(jk\frac{\mathbf{r}_q \cdot \mathbf{r}_p}{r_p}\right) dV_q \tag{1.32}$$

where the amplitude has been safely approximated using, $r_{qp} \simeq r_p$.

We note that in either the time or frequency domain, Ψ can be replaced by the scalar Φ or any Cartesian component of **A**, **E**, or **H** with the corresponding s evaluated from ρ and **J**. Equation (1.32) will be made use of repeatedly in the examples analyzed in Part III, Chapter 22.

Current element (Hertzian electric dipole)

As an important example, we calculate the vector potential of a line current, of length d and magnitude $I_0 \sin\omega t$, that is parallel to the z axis and is "electrically short," that is $d \ll \lambda_0 (kd \ll 1)$. Because we are assuming a steady-state current, it is convenient to employ complex vectors and scalars. With $\underline{s}_q(\mathbf{r}_q) = \underline{J}_z(x_q, y_q, z_q)$ and $\underline{\Psi}_p(\mathbf{r}_p) = \underline{A}_z(\mathbf{r}_p)$, it follows that for all $r_p \gg d$, Eq. (1.32),

$$\underline{A}_z(\mathbf{r}_p) = \mu_0\frac{\exp(-jkr_p)}{4\pi r_p}\iiint \underline{J}_z(x_q, y_q, z_q)\exp(jk\frac{\mathbf{r}_q \cdot \mathbf{r}_p}{r_p})\, dx_q dy_q dz_q$$

can be used. After integrating over the cross section of the current and realizing that $\exp(jk\frac{\mathbf{r}_q \cdot \mathbf{r}_p}{r_p}) \simeq 1$ (because the argument of the exponential never exceeds kd), the result (after dropping the subscript p) is

$$\underline{A}_z(\mathbf{r}) = \mu_o \frac{\exp(-jkr)}{4\pi r} \int_0^{\mathrm{d}} \underline{\mathrm{I}}(z_q)\, dz_q$$

If the current is uniform, the integral is simply $\underline{\mathrm{I}}_o d$; if not, it can be written as $\underline{\mathrm{I}}_o d_{\mathrm{eff}}$ and

$$\underline{A}_z(\mathbf{r}) = \mu_o \frac{\exp(-jkr)}{4\pi r} \underline{\mathrm{I}}_o d_{\mathrm{eff}} \tag{1.33}$$

Because of charge conservation, the current cannot go to zero at the ends of the line element without electric charges being created at the ends and/or along the length according to

$$\frac{\partial \underline{\mathrm{I}}(z)}{\partial z} + j\omega Q'(z) = 0$$

If $\frac{\partial \underline{\mathrm{I}}(z)}{\partial z} = 0$ (except at the ends), point charges of equal magnitude,

$\underline{\mathrm{Q}}_o = \underline{\mathrm{I}}_o/\omega$, but opposite polarity, will exist at the ends and together create an electric dipole of moment, $\underline{\mathrm{Q}}_o d$; otherwise a line-charge density, $Q'(z)$, will exist along the current and form a distributed dipole. In all cases, bipolar charges are the source of a complex scalar potential that satisfies

$$\underline{\Phi}(\mathbf{r}_p) = \frac{\exp(-jkr_p)}{-j\omega 4\pi \varepsilon_o r_p} \int_0^{\mathrm{d}} \frac{\partial \underline{\mathrm{I}}(z_q)}{\partial z_q} \exp\left(jk\frac{\mathbf{r}_q \cdot \mathbf{r}_p}{r_p}\right) dz_q \tag{1.34}$$

This integral is a little tricky to evaluate correctly, so we choose the alternate approach of simply invoking the Lorenz gauge, $\nabla \cdot \underline{\mathbf{A}} + j\omega\mu_o\varepsilon_o\underline{\Phi} = 0$ (which by itself imposes conservation of charge). The result is

$$\begin{aligned}
\underline{\Phi} &= \frac{-1}{j\omega\mu_o\varepsilon_o}\frac{\partial \underline{A}_z(\mathbf{r})}{\partial z} = \frac{-\underline{\mathrm{I}}_o d_{\mathrm{eff}}}{j\omega 4\pi\varepsilon_o}\frac{\partial}{\partial z}\left[\frac{\exp(-jkr)}{r}\right] \\
&= \frac{\underline{\mathrm{I}}_o d_{\mathrm{eff}}}{j\omega 4\pi\varepsilon_o}\left(jk + \frac{1}{r}\right)\frac{\exp(-jkr)}{r}\cos\theta
\end{aligned}$$

The reader should verify that the same result is obtained from Eq. (1.34).

For the $\sin\omega t$ dependence specified at the outset, $\underline{\mathrm{I}}_o = -j\mathrm{I}_o$; therefore the time-dependent versions of the potentials (evaluated by multiplying by $\exp(j\omega t)$ and taking the real part) are

$$A_z(\mathbf{r}, t) = \mu_o I_o d_{\mathrm{eff}} \frac{\sin(\omega t - kr)}{4\pi r} \tag{1.35a}$$

$$\Phi(\mathbf{r}, t) = \sqrt{\frac{\mu_o}{\varepsilon_o}} I_o d_{\mathrm{eff}} \frac{\sin(\omega t - kr) - \dfrac{1}{kr}\cos(\omega t - kr)}{4\pi r}\cos\theta \tag{1.35b}$$

The Hertzian electric dipole is revisited in Chapter 3, Section 3.5 and Part III, Chapter 13, Section 13.1.

Current loop (Hertzian magnetic dipole)

As a second example, consider a small circular loop of radius R that lies in the $x - y$ plane, centered at $z = 0$. A uniform current, $\underline{\mathrm{I}}(t) = \underline{\mathrm{I}}_o \cos\omega t$, circulates around the loop.

We continue to define $k = \omega/c$ and employ spherical coordinates with $\widehat{\phi}$ the unit vector in the azimuthal direction. For $kR \ll 1$, the complex vector and scalar retarded potentials are

$$\underline{\mathbf{A}}(\mathbf{r}_p) = \mu_o \frac{\exp(-jkr_p)}{4\pi r_p} \oint_{\text{loop}} \underline{\mathrm{I}}_o \widehat{\phi}_q \exp\left(jk\frac{\mathbf{r}_q \cdot \mathbf{r}_p}{r_p}\right) ds_q$$

$$\underline{\Phi} = 0$$

The reader is cautioned to remember that although the current is $\mathrm{I}_o\widehat{\phi}$, the unit vector *cannot* simply be moved from inside to outside the integral. Instead, it must be converted to its Cartesian components before the integrations that determine **A** are attempted. Afterwards, the components can be converted to spherical coordinates; of course, only the ϕ component will survive. The electric potential is zero because the time-varying current is uniform and does not produce electric charges. The complex exponential when expanded in a Taylor series becomes

$$\exp\left(jk\frac{\mathbf{r}_q \cdot \mathbf{r}_p}{r_p}\right) \simeq 1 + jk\frac{\mathbf{r}_q \cdot \mathbf{r}_p}{r_p} + \cdots$$

Because the leading term vanishes when integrated around the closed loop, the first-order term must be retained (unlike the case of the electric dipole). The far-field complex vector potential evaluates to

$$\underline{\mathbf{A}}(\mathbf{r}) = jk\left(1 + \frac{1}{jkr}\right) R \frac{\mu_o \underline{\mathrm{I}}_o R}{4} \frac{\exp(-jkr)}{r} \sin\theta \,\widehat{\phi}$$

The time-dependent form is

$$\mathbf{A}(\mathbf{r}) = \widehat{\phi}\frac{\mu_o}{4\pi} \frac{m_o[-kr\sin(\omega t - kr) + \cos(\omega t - kr)]}{r^2} \sin\theta \tag{1.36}$$

where $m_o = \underline{\mathrm{I}}_o \pi R^2$ is defined as the magnetic dipole moment.

The Hertzian magnetic dipole is revisited in Part III, Chapter 13, Section 13.2. We postpone further discussion of complex fields or the introduction of the complex Maxwell's Equations until after the time-domain version has been generalized to include dielectric and magnetic materials.

1.7 POLARIZABLE, MAGNETIZABLE, AND CONDUCTING MEDIA

In the free-space formulation considered so far, only free charges and their associated electric currents are present. When dielectric and magnetic materials are considered, the response of electric and/or magnetic dipoles must be considered. Commonly, these are considered to be paired electric charges (bound to each other), in the case of dielectrics, and small electric-current loops (or their equivalent), in the case of magnetics. Other models (permitted by the equivalence principle) are possible; these include using fictitious magnetic charges of opposing polarities to model magnetic dipoles and/or small current loops carrying fictitious magnetic current to model electric dipoles. Moreover, various superpositions of all four types can be employed. In all models, the individual dipoles are of microscopic or mesoscopic dimensions; this permits one to characterize the material in terms of its polarization (**P**) and magnetization (**M**) vectors. Each is defined as the appropriate macroscopically averaged dipole-moment density. There may be spontaneous values of **M** or **P**; more generally, they arise in response to applied electric or magnetic fields. The response of dielectric materials can often be approximated as linear with respect to the excitation. That of magnetic materials is generally nonlinear and hysteretic, yet linear operation with respect to some operating point is sometimes possible. When the material is deforming and/or accelerating, additional complexity and subtlety is involved.

Because, at present, we wish to continue using a vector-potential formulation, it is convenient to employ a model for materials that is based solely on electric charge and current. The previous formulation can then be generalized by adding additional electric currents to those generated by the free (unpaired) charges. The form of Maxwell's Equations remains unchanged. Other ways of incorporating magnetization into electrodynamic theories are reviewed in Part II; the Chu formulation is of special interest because electric and magnetic dipoles are modelled symmetrically.

Polarization and Amperian electric currents

The dielectric polarization can be modeled as identical charge distributions of opposite polarity, $\pm\rho_o$, displaced from one another by a small vector distance **d**. This is depicted in Figure 1.3 with the separation greatly exaggerated.

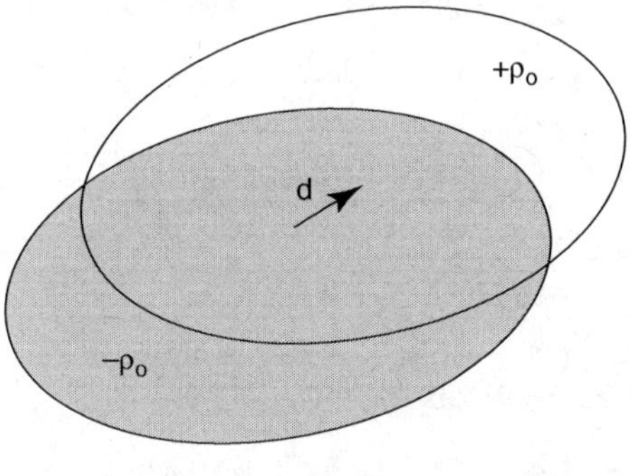

Figure 1.3 Separated bipolar charge densities.

The electric-dipole moment of an incremental volume of the dielectric is the increment $\Delta\mathbf{p} = \rho_o\mathbf{d}\,\Delta V$ [refer to Appendix C, Eq. (C.9)]; therefore

$$\mathbf{p} = \int \mathbf{P}^o dV \tag{1.37}$$

where the polarization vector is defined as $\mathbf{P}^o = \rho_o\mathbf{d}$.[3]

The time rate of change of the polarization produces a current density,

$$\mathbf{J}_{\text{polarization}} = \rho_o\frac{\partial\mathbf{d}}{\partial t} = \frac{\partial\mathbf{P}^o}{\partial t} \tag{1.38}$$

and an associated charge density, $\rho_{\text{polarization}}$. Because the latter is conserved independently of the free charge, we obtain

$$\nabla\cdot\frac{\partial\mathbf{P}^o}{\partial t} = \frac{\partial}{\partial t}(\nabla\cdot\mathbf{P}^o) = -\frac{\partial\rho_{\text{polarization}}}{\partial t}$$

[3]The superscript o is added because we reserve unscripted variables for field quantities in the Chu formulation of electrodynamics. When the material is stationary and nondeformable, there is no difference between corresponding quantities with and without a superscript.

and

$$\rho_{\text{polarization}} = -\nabla \cdot \mathbf{P}^{\text{o}} \tag{1.39}$$

Notice that this charge density is *not* ρ_{o}. Indeed, if $\mathbf{P}^{\text{o}}$ is spatially uniform, $\rho_{\text{polarization}} = 0$, except on the boundaries of the dielectric. There, surface polarization charges reside wherever $\mathbf{P}^{\text{o}} \cdot \widehat{\mathbf{n}}$ is nonzero ($\widehat{\mathbf{n}}$ is the normal to the surface).

For magnetic materials, the magnetization is assumed to arise from tiny Amperian-current loops (shown schematically in Figure 1.4) that produce magnetic-dipole moments.

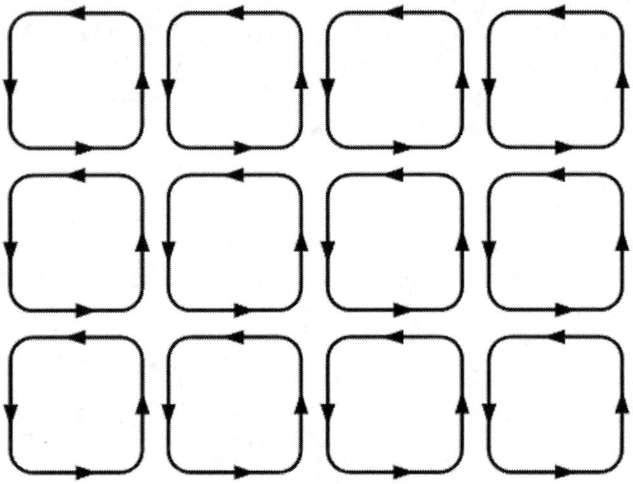

Figure 1.4 Amperian current loops.

If the current circulating around each loop is uniform, no electric charge will be present. Because conservation of Amperian electric charge requires that

$$\nabla \cdot \mathbf{J}_{\text{amperian}} = -\frac{\partial \rho_{\text{amperian}}}{\partial t} = 0$$

it follows that

$$\mathbf{J}_{\text{amperian}} = \nabla \times \mathbf{M}^{\text{o}} \tag{1.40a}$$

$$\rho_{\text{amperian}} = 0 \tag{1.40b}$$

where $\mathbf{M}^{\text{o}}$ is, as yet, an undefined vector.

However, the net magnetic moment, $\mathbf{m}$, of an electric-current distribution is defined as $\int \frac{1}{2}\mathbf{r}' \times \mathbf{J}\, dV$ [refer to Appendix C, Eq. (C.14)], which in this case is

$$\mathbf{m} = \int \frac{1}{2}\mathbf{r}' \times (\nabla \times \mathbf{M}^{\text{o}})\; dV$$

and where the choice of the origin of $\mathbf{r}'$ is arbitrary. The material volume is the sum of increments ΔV; each contributes an incremental magnetic moment $\Delta\mathbf{m}$. As we consider each increment in turn, the origin of $\mathbf{r}' = \mathbf{r}$ can be redefined so as to be centered upon it. Each incremental moment is therefore of the form $\Delta\mathbf{m} = [\lim_{\mathbf{r}\to 0} \frac{1}{2}\mathbf{r} \times (\nabla \times \mathbf{M}^{\text{o}})]\Delta V$.

But

$$\mathbf{r} \times (\nabla \times \mathbf{M}^{\text{o}}) = (\nabla \cdot \mathbf{r} - 1)\mathbf{M}^{\text{o}} + \nabla(\mathbf{M}^{\text{o}} \cdot \mathbf{r}) - \nabla \cdot (\mathbf{M}^{\text{o}}\,\mathbf{r})$$

Therefore, because $\nabla \cdot \mathbf{r} = 3$ and the other terms vanish in the limit, we obtain $\Delta \mathbf{m} = \mathbf{M}^{o} \Delta V$. Consequently, the total magnetic moment of the material volume can also be written as

$$\mathbf{m} = \int \mathbf{M}^{o}\, dV \tag{1.41}$$

with $\mathbf{M}^{o}$ now identified as the magnetization vector (the magnetic-dipole moment per unit volume). If $\mathbf{M}^{o}$ is spatially uniform, $\mathbf{J}_{\text{amperian}} = 0$, except on the boundaries of the magnetic material. There, surface magnetization currents reside wherever $\mathbf{M}^{o} \times \widehat{\mathbf{n}}$ is nonzero ($\widehat{\mathbf{n}}$ is the normal to the surface).

When polarization currents and Amperian currents are added to $\mathbf{J}_{\text{u}}$, the current associated with the free (unpaired) charge, it follows that

$$\mathbf{J}_{\text{total}} = \mathbf{J}_{\text{u}} + \mathbf{J}_{\text{polarization}} + \mathbf{J}_{\text{amperian}} = \mathbf{J}_{\text{u}} + \frac{\partial \mathbf{P}^{o}}{\partial t} + \nabla \times \mathbf{M}^{o} \tag{1.42a}$$

$$\rho_{\text{total}} = \rho_{\text{u}} + \rho_{\text{polarization}} + \rho_{\text{amperian}} = \rho_{\text{u}} - \nabla \cdot \mathbf{P}^{o} + 0 \tag{1.42b}$$

With these field sources, it is customary to express Maxwell's Equations in terms of the field vector $\mathbf{B}/\mu_{o}$ rather than $\mathbf{H}$. With $\mathbf{E}^{o}$ substituted for $\mathbf{E}$ to distinguish the electric field from that used in the Chu formulation, the result is

$$\frac{1}{\mu_{o}} \nabla \times \mathbf{B} - \varepsilon_{o} \frac{\partial \mathbf{E}^{o}}{\partial t} = \mathbf{J}_{\text{total}} = \mathbf{J}_{\text{u}} + \frac{\partial \mathbf{P}^{o}}{\partial t} + \nabla \times \mathbf{M}^{o} \tag{1.43a}$$

$$\varepsilon_{o} \nabla \cdot \mathbf{E}^{o} = \rho_{\text{total}} = \rho_{\text{u}} - \nabla \cdot \mathbf{P}^{o} \tag{1.43b}$$

$$\nabla \times \mathbf{E}^{o} + \frac{\partial \mathbf{B}}{\partial t} = 0 \tag{1.43c}$$

$$\nabla \cdot \mathbf{B} = 0 \tag{1.43d}$$

where

$$\mathbf{B} = \nabla \times \mathbf{A} \tag{1.44a}$$

$$\mathbf{E}^{o} = -\frac{\partial \mathbf{A}}{\partial t} - \nabla \Phi \tag{1.44b}$$

In this Amperian formulation, the Lorentz force density is

$$\mathbf{f} = \rho_{\text{total}} \mathbf{E}^{o} + \mathbf{J}_{\text{total}} \times \mathbf{B} \tag{1.45}$$

It is customary to define

$$\mathbf{D} = \varepsilon_{o} \mathbf{E}^{o} + \mathbf{P}^{o} \tag{1.46a}$$

$$\mathbf{B} = \mu_{o} (\mathbf{H}^{o} + \mathbf{M}^{o}) \tag{1.46b}$$

because, with their use, Maxwell's Equations simplify to the Minkowski form,

$$\nabla \times \mathbf{H}^{o} = \mathbf{J}_{\text{u}} + \frac{\partial \mathbf{D}}{\partial t} \tag{1.47a}$$

$$\nabla \cdot \mathbf{D} = \rho_{\text{u}} \tag{1.47b}$$

$$\nabla \times \mathbf{E}^{o} = -\frac{\partial \mathbf{B}}{\partial t} \tag{1.47c}$$

$$\nabla \cdot \mathbf{B} = 0 \tag{1.47d}$$

Chu formulation

An alternate formulation, due to Lan Jen Chu, uses electric charges to model the polarization and magnetic charges to model the magnetization. In this case, the sources are duals of one another (as discussed in Chapter 6, Section 6.5) and Maxwell's Equations become

$$\nabla \times \mathbf{H} - \varepsilon_o \frac{\partial \mathbf{E}}{\partial t} = \mathbf{J}_e = \mathbf{J}_u + \frac{\partial \mathbf{P}}{\partial t} + \nabla \times (\mathbf{P} \times \mathbf{v}) \tag{1.48a}$$

$$\varepsilon_o \nabla \cdot \mathbf{E} = \rho_e = \rho_u - \nabla \cdot \mathbf{P} \tag{1.48b}$$

$$\nabla \times \mathbf{E} + \mu_o \frac{\partial \mathbf{H}}{\partial t} = -\mathbf{J}_m = -\mu_o \left[\frac{\partial \mathbf{M}}{\partial t} + \nabla \times (\mathbf{M} \times \mathbf{v}) \right] \tag{1.48c}$$

$$\mu_o \nabla \cdot \mathbf{H} = \rho_m = -\mu_o \nabla \cdot \mathbf{M} \tag{1.48d}$$

These are consistent with Eqs. (1.47) provided

$$\mathbf{E}^o = \mathbf{E} + \mu_o \mathbf{M} \times \mathbf{v} \tag{1.49a}$$

$$\mathbf{H}^o = \mathbf{H} - \mathbf{P} \times \mathbf{v} \tag{1.49b}$$

$$\mathbf{P}^o = \mathbf{P} - \mu_o \varepsilon_o \mathbf{M} \times \mathbf{v} \tag{1.49c}$$

$$\mathbf{M}^o = \mathbf{M} + \mathbf{P} \times \mathbf{v} \tag{1.49d}$$

Notice that neither **D** nor **B** requires superscript labeling because

$$\mathbf{D} = \varepsilon_o \mathbf{E} + \mathbf{P} = \varepsilon_o \mathbf{E}^o + \mathbf{P}^o \tag{1.50a}$$

$$\mathbf{B} = \mu_o (\mathbf{H} + \mathbf{M}) = \mu_o (\mathbf{H}^o + \mathbf{M}^o) \tag{1.50b}$$

In the Chu formulation, there are separate Lorentz-force densities acting upon both the electric and magnetic charges:

$$\mathbf{f}_e = \rho_e \mathbf{E} + \mathbf{J}_e \times \mu_o \mathbf{H} \tag{1.51a}$$

$$\mathbf{f}_m = \rho_m \mathbf{E} - \mathbf{J}_m \times \varepsilon_o \mathbf{E} \tag{1.51b}$$

The total force density is their sum which is different from Eq. (1.45) because magnetic charges have been employed; the Minkowski formulation leads to yet another force density. These are all forces of "electromagnetic origin" that are dependent on how the sources are modelled; they must be augmented by mechanical and/or other forces. When done so properly, the *total* force density is independent of the electromagnetic formulation chosen. Despite the explicit velocity dependence, the Chu formulation is often (but not always) simpler to use. However, because we wish to employ the vector-potential, we continue at present with the Amperian formulation (expressed in terms of the Minkowski electric and magnetic fields).

For regions of free-space it follows that the constitutive laws that relate the four field vectors are (with $\mathbf{E}^o = \mathbf{E}$, $\mathbf{H}^o = \mathbf{H}$):

$$\mathbf{D} = \varepsilon_o \mathbf{E} \tag{1.52a}$$

$$\mathbf{B} = \mu_o \mathbf{H} \tag{1.52b}$$

Electrically conducting materials

For linear isotropic (stationary) conductors that obey Ohm's Law,

$$\mathbf{J}_{\mathrm{u}} = \sigma \mathbf{E} \tag{1.53a}$$

where σ is the electrical-conductivity with units of $S\ m^{-1} = A\ V^{-1} m^{-1}$.

For such materials, conservation of charge becomes

$$\nabla \cdot (\sigma \mathbf{E}) + \frac{\partial \rho_{\mathrm{u}}}{\partial t} = 0 \tag{1.53b}$$

These materials may also have dielectric/magnetic properties.

Perfect conductors

A perfect conductor is defined by $\sigma \to \infty$ and therefore unless $\mathbf{J}_{\mathrm{u}}$ is infinite, the electric field inside it must vanish. It follows that current must flow only in surface layers of thickness $\delta \to 0$ and $\mathbf{J}_{\mathrm{u}} \to \infty$ such that there is a finite surface current density, $\mathbf{K}_{\mathrm{u}} = \lim_{\delta \to 0} \mathbf{J}_{\mathrm{u}} \delta$. Because the time rate of change of any magnetic field inside the conductor would produce electric fields, only static magnetic fields are tenable. Nevertheless, we *define* a perfect conductor as one that is *completely field-free*.

Dielectric and magnetic materials

For linear isotropic (stationary) materials, the constitutive laws that relate the four field vectors are

$$\mathbf{D} = \varepsilon \mathbf{E} \tag{1.54a}$$

$$\mathbf{B} = \mu \mathbf{H} \tag{1.54b}$$

or equivalently

$$\mathbf{P} = (\varepsilon - \varepsilon_{\mathrm{o}})\mathbf{E} = \chi_{\mathrm{e}} \varepsilon_{\mathrm{o}} \mathbf{E} \tag{1.55a}$$

$$\mathbf{M} = (\mu/\mu_{\mathrm{o}} - 1)\mathbf{H} = \chi_{\mathrm{m}} \mathbf{H} \tag{1.55b}$$

where ε and μ are, respectively, the dielectric permittivity and the magnetic permeability. The ratios, $\varepsilon/\varepsilon_{\mathrm{o}}$and μ/μ_{o} are defined as the relative permittivity (dielectric constant) and relative permeability. The electric and magnetic susceptibilities are the dimensionless quantities defined by

$$\chi_{\mathrm{e}} = \varepsilon/\varepsilon_{\mathrm{o}} - 1 \tag{1.56a}$$

$$\chi_{\mathrm{m}} = \mu/\mu_{\mathrm{o}} - 1 \tag{1.56b}$$

The reader should again be cautioned that when the material is moving and/or deforming, the values of **E** and **H** in this formulation are not identical to those in the free-space equations. These complications are taken up in Part II of the text.

1.8 BOUNDARY CONDITIONS

The form of Maxwell's Equations reveals that in regions where the charges and currents are continuous functions of position and time, the divergence, curl, and time-derivative operators must all produce continuous functions. This fact requires the electric and magnetic fields to be continuous functions of space and time. However, there may be locations (such as the surface of a conductor or the interface between different materials) where there is an accumulation of charge and/or current that can be modeled as a finite surface density that within zero thickness has infinite volume density. At such locations $\nabla \cdot \mathbf{E}$ and/or $\nabla \times \mathbf{H}$ are infinite; these singularities lead to discontinuities in the respective field. We wish to determine exactly how large these will be.

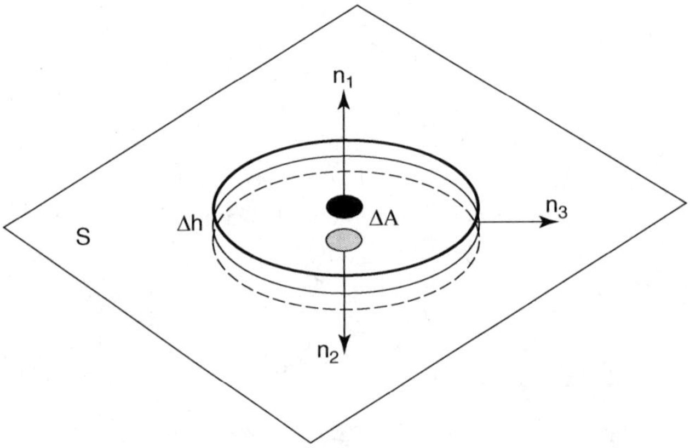

Figure 1.5 Closed surface (normal fields).

Electric surface charges

Assume that an unpaired electric surface charge of density σ_{u}^{s} resides on the surface shown in Figure 1.5. A circular pillbox is constructed with equal top and bottom areas, ΔA, that are located on opposite sides of the surface; the height of the box is Δh. We apply Eqs. (1.5a) and (1.47b) to the closed surface defined by the Gaussian pillbox

$$\oint_{\text{top+bottom+sides}} \mathbf{D} \cdot \widehat{\mathbf{n}}\, da = \int_{\text{area}} \sigma_{\mathrm{u}}^{s}\, da$$

$$[\mathbf{D}^{(1)} \cdot \widehat{\mathbf{n}}_1 + \mathbf{D}^{(2)} \cdot \widehat{\mathbf{n}}_2]\Delta A + \mathbf{D}^{(3)} \cdot \widehat{\mathbf{n}}_3 2\sqrt{\pi\,\Delta A}\Delta h = \sigma_{\mathrm{u}}^{s}\Delta A$$

If $\Delta h \ll \sqrt{\Delta A} \to 0$ and $\mathbf{E}$ is finite, it follows that because $\widehat{\mathbf{n}}_1 = -\widehat{\mathbf{n}}_2 = \widehat{\mathbf{n}}$, we obtain

$$\widehat{\mathbf{n}} \cdot [\mathbf{D}^{(1)} - \mathbf{D}^{(2)}] = \sigma_{\mathrm{u}}^{s} \tag{1.57}$$

Electric surface currents

Assume that an unpaired electric surface current of density $\mathbf{K}_{\mathrm{u}}^{s}$ resides on the horizontal surface shown in Figure 1.6.

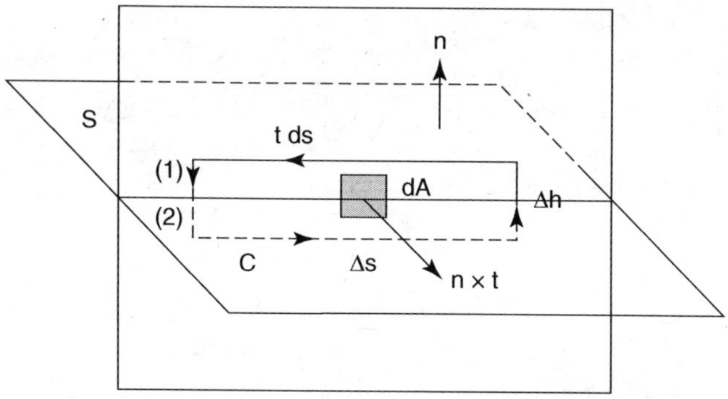

Figure 1.6 Closed contour (tangential fields).

A rectangular area is constructed with equal top and bottom lengths, Δs, that are located on opposite sides of the surface; the height of the rectangle is Δh. We apply Eqs. (1.3a) and (1.47a) to the surface that is the plane of the rectangle:

$$\oint_{\text{perimeter}} \mathbf{H}\cdot\widehat{\mathbf{t}}\, ds = \int_{\text{area}} \mathbf{J}_{\text{u}}\cdot(\widehat{\mathbf{n}}\times\widehat{\mathbf{t}})\, da = \int_{\text{length}} \mathbf{K}_{\text{u}}^{s}\cdot(\widehat{\mathbf{n}}\times\widehat{\mathbf{t}})\, ds$$

$$[\mathbf{H}^{(1)}\cdot\widehat{\mathbf{t}}_1 + \mathbf{H}^{(2)}\cdot\widehat{\mathbf{t}}_2]\Delta s + [\mathbf{H}^{(3)}\cdot\widehat{\mathbf{t}}_3 + \mathbf{H}^{(4)}\cdot\widehat{\mathbf{t}}_4]\Delta h = \mathbf{K}_{\text{u}}^{s}\cdot(\widehat{\mathbf{n}}\times\widehat{\mathbf{t}})\Delta s$$

If $\Delta h \ll \Delta s \to 0$ and $\mathbf{H}$ is finite, it follows that because $\widehat{\mathbf{t}}_1 = -\widehat{\mathbf{t}}_2 = \widehat{\mathbf{t}}$ and $\widehat{\mathbf{n}}\cdot\widehat{\mathbf{t}} = 0$, we have

$$[\mathbf{H}^{(1)} - \mathbf{H}^{(2)}]\cdot\widehat{\mathbf{t}} = \mathbf{K}_{\text{u}}^{s}\cdot(\widehat{\mathbf{n}}\times\widehat{\mathbf{t}}) = (\mathbf{K}_{\text{u}}^{s}\times\widehat{\mathbf{n}})\cdot\widehat{\mathbf{t}} \tag{1.58a}$$

$$\mathbf{H}^{(1)} - \mathbf{H}^{(2)} = \mathbf{K}_{\text{u}}^{s}\times\widehat{\mathbf{n}} \tag{1.58b}$$

$$\widehat{\mathbf{n}}\times[\mathbf{H}^{(1)} - \mathbf{H}^{(2)}] = \mathbf{K}_{\text{u}}^{s} \tag{1.58c}$$

Conservation of charge

If a surface of discontinuity supports σ_{u}^{s} and/or $\mathbf{K}_{\text{u}}^{s}$ and the material on one or both sides supports volume currents, $\mathbf{J}_{\text{u}}$, the boundary conditions must maintain conservation of unpaired electric charge. A Gaussian pillbox is again erected to provide a closed surface over which

$$\nabla\cdot\mathbf{J}_{\text{u}} + \frac{\partial\rho_{\text{u}}}{\partial t} = 0$$

can be integrated. Comparison with Eq. (1.47b) suggests that

$$\mathbf{n}\cdot[\mathbf{J}_{\text{u}}^{(1)} - \mathbf{J}_{\text{u}}^{(2)}] = -\frac{\partial\sigma_{\text{u}}^{s}}{\partial t}$$

However, in this case, $\mathbf{J}_{\text{u}}\to\infty$, and so contributions from the sides cannot be ignored in the limit $\Delta h \to 0$. The correct boundary condition is

$$\mathbf{n}\cdot[\mathbf{J}_{\text{u}}^{(1)} - \mathbf{J}_{\text{u}}^{(2)}] + \nabla_{\Sigma}\cdot\mathbf{K}_{\text{u}}^{s} + \frac{\partial\sigma_{\text{u}}^{s}}{\partial t} = 0 \tag{1.59}$$

where $\nabla_{\Sigma}\cdot\mathbf{K}_{\text{u}}^{s}$ is the two-dimensional surface divergence that accounts for the net outward flow along the surface. It is defined by

$$\nabla_\Sigma \cdot \mathbf{K}_\mathrm{u}^s = \lim_{\Delta A \to 0} \frac{\oint_{C_\mathrm{o}} \mathbf{K}_\mathrm{u}^s \cdot (\mathbf{t} \times \mathbf{n})\, ds}{\Delta A} \tag{1.60}$$

Because polarization charges are separately conserved, this form of boundary condition applies to them as well as to the total current and charge densities. There is no electric charge density associated with the Amperian current density, $\nabla \times \mathbf{M}$; but a surface density,

$$\mathbf{K}_\mathrm{M}^s = \widehat{\mathbf{n}} \times [\mathbf{M}^{(1)} - \mathbf{M}^{(2)}]$$

exists wherever there are discontinuities in the tangential magnetization vector.

Because there are no magnetic charges or magnetic currents, the boundary conditions for **E** and **B** are

$$\mathbf{n} \times [\mathbf{E}^{(1)} - \mathbf{E}^{(2)}] = 0 \tag{1.61a}$$

$$\mathbf{n} \cdot [\mathbf{B}^{(1)} - \mathbf{B}^{(2)}] = 0 \tag{1.61b}$$

It also follows that

$$\mathbf{n} \cdot \varepsilon_\mathrm{o}[\mathbf{E}^{(1)} - \mathbf{E}^{(2)}] = \sigma_\mathrm{total}^s = \sigma_\mathrm{u}^s + \sigma_\mathrm{polarization}^s \tag{1.62a}$$

$$\mathbf{n} \cdot [\mathbf{P}^{(1)} - \mathbf{P}^{(2)}] = -\sigma_\mathrm{polarization}^s \tag{1.62b}$$

$$\mathbf{n} \cdot [\mathbf{H}^{(1)} - \mathbf{H}^{(2)}] = -\mathbf{n} \cdot [\mathbf{M}^{(1)} - \mathbf{M}^{(2)}] \tag{1.62c}$$

1.9 THE COMPLEX MAXWELL EQUATIONS

In the sinusoidal steady state, time-harmonic vectors and scalars can be conveniently expressed as

$$\mathbf{F}(\mathbf{r},t) = Re\left\{\underline{\mathbf{F}}(\mathbf{r}) \exp(j\omega t)\right\}$$

$$\Psi = Re\left\{\underline{\Psi}(\mathbf{r}) \exp(j\omega t)\right\}$$

where the symbol *Re* stands for the real part of the expression in the curly brackets {}, and the vector or scalar that is underscored is a complex quantity that may be a constant or a function of position. Because spatial and temporal derivatives are operations that commute with taking the real part, one may initially suppress the *Re* until the very last stage of analysis and simply insert the bracketed terms into Maxwell's Equations. When this is done, each time derivative will cause multiplication by the factor $j\omega$, but $\exp(j\omega t)$ will remain a common factor in every term; these are cumbersome and can be ignored if we remember to reinsert it before taking the real part.

The complex forms of Eqs. (1.47a)–(1.47d) (with the time dependence suppressed) are therefore

$$\nabla \times \underline{\mathbf{H}}^\mathrm{o} - j\omega\underline{\mathbf{D}} = \underline{\mathbf{J}_\mathrm{u}} \tag{1.63a}$$

$$\nabla \cdot \underline{\mathbf{D}} = \underline{\rho_\mathrm{u}} \tag{1.63b}$$

$$\nabla \times \underline{\mathbf{E}}^\mathrm{o} + j\omega\underline{\mathbf{B}} = 0 \tag{1.63c}$$

$$\nabla \cdot \underline{\mathbf{B}} = 0 \tag{1.63d}$$

The complex forms of Eqs. (1.44a) and (1.44b) are

$$\underline{\mathbf{B}} = \nabla \times \underline{\mathbf{A}} \tag{1.64a}$$

$$\underline{\mathbf{E}}^{\mathrm{o}} = -j\omega\underline{\mathbf{A}} - \nabla\underline{\Phi} \tag{1.64b}$$

In most cases, we assume that the material is stationary and nondeforming; then, the superscripts can be removed from the electric and magnetic fields.

CHAPTER 2

QUASISTATIC APPROXIMATIONS

There are many electromagnetic-field configurations in which propagation effects are negligible. For example, if the time required for the field to travel through a structure is very small compared to the time variation of the sources that generate the fields, then the wave velocity can be considered to be infinite. In such cases, the electromagnetic wave equation is well approximated by Laplace's Equation, and the superposition of Laplacian fields[1] that match the appropriate boundary conditions provides a satisfactory solution. In the limit of slow time variation (low frequency for sinusoidal steady-state fields), it often occurs that the electric field predominates over the magnetic field or vice versa. This leads to the classification of quasistatic fields that are either electroquasistatic (*EQS*) or magnetoquasistatic (*MQS*); references 5 [Chapter 6, p. 213] and 6 both contain treatments of quasistatic analysis, especially for sinusoidal steady-state fields. In some cases involving electrical conduction, the fields may be comparable yet both quasistatic in character. These three types of fields will be considered separately. In all cases the

[1] In some cases, such as those involving piezoelectric or magnetoelastic materials, hybrid waves can propagate with predominatly mechanical energies. Although the weak electric or magnetic fields (associated with what are basically elastic waves) are subject to quasistatic analysis, they are non-Laplacian in character and statements made in this text about Laplacian *EQS* and *MQS* fields do not apply.

The Power and Beauty of Electromagnetic Fields, First Edition. F. R. Morgenthaler.

electric and magnetic fields are expressed as

$$\mathbf{E} = \mathbf{E}_{(0)} + \mathbf{E}_{(1)} + \cdots$$
$$\mathbf{H} = \mathbf{H}_{(0)} + \mathbf{H}_{(1)} + \cdots$$

The main advantage of quasistatic analysis is that for many configurations that model practical devices, in which propagation effects are secondary, very complicated field problems can often be broken into comparatively simple pieces that nevertheless provide a basic understanding of the device operation. In numerous cases, the theoretical analysis required can be carried out by "back of the envelope calculations." When this is possible, the enhanced physical intuition that is gained is of incalculable value. Even when computer simulations are required (or desired), quasistatic analysis is often a valuable tool and guide.

2.1 QUASISTATIC EXPANSIONS OF A STANDING WAVE

To make clear the nature of quasistatic fields, we consider the exact solution of a standing electromagnetic wave of frequency ω and its quasistatic expansion. Linearly polarized fields formed by a pair of equal and opposite harmonic plane waves directed along the z axis are

$$\mathbf{E} = \widehat{\mathbf{x}}E_o \cos(kz)\cos(\omega t)$$
$$\mathbf{H} = \widehat{\mathbf{y}}\frac{1}{c\mu_o}E_o \sin(kz)\sin(\omega t)$$
$$k = \omega\sqrt{\mu_o\varepsilon_o}$$

Near $z = 0$, the field is principally electric in character and so there the expansion is *EQS*. Taylor series expansions of the z variations yield

$$\mathbf{E} = \mathbf{E}_{(0)} + \mathbf{E}_{(2)} + \cdots = \widehat{\mathbf{x}}E_o\cos(\omega t)\left[1 - \frac{1}{2!}(kz)^2 + \frac{1}{4!}(kz)^4 - \cdots\right]$$

$$\mathbf{H} = \mathbf{H}_{(1)} + \mathbf{H}_{(3)} + \cdots = \widehat{\mathbf{y}}\frac{1}{c\mu_o}E_o\sin(\omega t)\left[kz - \frac{1}{3!}(kz)^3 + \cdots\right]$$

where

$$\mathbf{E}_{(n)} = \widehat{\mathbf{x}}E_o(-1)^{n/2}\frac{(kz)^n}{n!}\cos(\omega t) \qquad (n \text{ even})$$

$$\mathbf{H}_{(n)} = \widehat{\mathbf{y}}\frac{1}{c\mu_o}E_o(-1)^{(n-1)/2}\frac{(kz)^n}{n!}\sin(\omega t) \qquad (n \text{ odd})$$

Near $kz = \pi/2$, the dominant field is magnetic and so there the expansion is *MQS*. Although a shift of the z-coordinate origin is convenient, a simpler method is to consider the dual field:

$$\mathbf{H} = \widehat{\mathbf{x}}H_o\cos(kz)\cos(\omega t)$$
$$\mathbf{E} = -\widehat{\mathbf{y}}c\mu_o H_o\sin(kz)\sin(\omega t)$$
$$k = \omega\sqrt{\mu_o\varepsilon_o}$$

for which

$$\mathbf{H} = \mathbf{H}_{(0)} + \mathbf{H}_{(2)} + \cdots = \widehat{\mathbf{x}}H_o\cos(\omega t)\left[1 - \frac{1}{2!}(kz)^2 + \frac{1}{4!}(kz)^4 - \cdots\right]$$

$$\mathbf{E} = \mathbf{E}_{(1)} + \mathbf{E}_{(3)} + \cdots = -\widehat{\mathbf{y}}c\mu_o H_o\sin(\omega t)\left[kz - \frac{1}{3!}(kz)^3 + \cdots\right]$$

and

$$\mathbf{H}_{(n)} = \widehat{\mathbf{x}}H_o(-1)^{n/2}\frac{(kz)^n}{n!}\cos(\omega t) \qquad (n \text{ even})$$

$$\mathbf{E}_{(n)} = -\widehat{\mathbf{y}}c\mu_o H_o(-1)^{(n-1)/2}\frac{(kz)^n}{n!}\sin(\omega t) \qquad (n \text{ odd})$$

Notice that Maxwell's Equations for each order (of either expansion) reduce to

$$\nabla \times \mathbf{E}_{(n)} = -\mu_o\frac{\partial \mathbf{H}_{(n-1)}}{\partial t} \tag{2.1a}$$

$$\nabla \cdot \mathbf{E}_{(n)} = 0 \tag{2.1b}$$

$$\nabla \times \mathbf{H}_{(n)} = \varepsilon_o\frac{\partial \mathbf{E}_{(n-1)}}{\partial t} \tag{2.1c}$$

$$\nabla \cdot \mathbf{H}_{(n)} = 0 \tag{2.1d}$$

and that the time derivatives of the $(n-1)$-order fields contribute to the nth-order fields for $n \geq 1$ (no time derivatives contribute to the zero-order fields). This means that the $(n-1)$ terms act as sources for the nth-order fields.

The usual *EQS* or *MQS* expansion includes only the $n = 0$ and $n = 1$ terms; this implies that higher-order terms are negligible. This is true whenever $|kz| \ll 1$ or $|z| \ll \lambda_o/2\pi$. Note that, with this limitation, both *EQS* and *MQS* fields have spatial dependences that are Laplacian. This is consistent with the fact that the Helmholtz Equations,

$$\nabla^2\mathbf{E} + \omega^2\mu_o\varepsilon_o\mathbf{E} = 0$$

$$\nabla^2\mathbf{H} + \omega^2\mu_o\varepsilon_o\mathbf{H} = 0$$

reduce to Laplace's Equation when $\omega^2\mu_o\varepsilon_o z^2 \ll 1$.

Of course, when the exact solution is available, there is no need to carry out the expansion; the real utility is in dealing with cases where no exact solution is available or it is mathematically complex.

With these simple cases as a preview, we now present the categories and methods of general quasistatic analysis. The examples considered in Part III, Chapter 11, are intended to clarify the details.

2.2 ELECTROQUASISTATIC (*EQS*) FIELDS

Zero-order fields

We assume that the configuration to be analyzed contains only linear isotropic materials and metallic objects. When there is negligible conductivity associated with any dielectric or magnetic material and all metal surfaces are very highly conducting, but do *not* provide closed paths (that would allow circulating current flows), the quasistatic fields are predominantly electric. We use a subscript (0) to indicate the principal fields; therefore

$$\mathbf{D}_{(0)} = \varepsilon \mathbf{E}_{(0)} \tag{2.2a}$$

$$\mathbf{B}_{(0)} = \mu \mathbf{H}_{(0)} = 0 \tag{2.2b}$$

$$\mathbf{J}_{\mathrm{u}(0)} = 0 \tag{2.2c}$$

Maxwell's Equations are then approximated by

$$\nabla \times \mathbf{E}_{(0)} = 0 \tag{2.3a}$$

$$\nabla \cdot \varepsilon \mathbf{E}_{(0)} = \rho_{\mathrm{u}(0)} \tag{2.3b}$$

or equivalently by

$$\mathbf{E}_{(0)} = -\nabla \Phi_{(0)} \tag{2.3c}$$

$$\nabla^2 \Phi_{(0)} + \frac{\nabla \varepsilon}{\varepsilon} \cdot \nabla \Phi_{(0)} = -\frac{1}{\varepsilon} \rho_{\mathrm{u}(0)} \tag{2.3d}$$

In every region with uniform permittivity that does not contain unpaired charge, the zero-order potential satisfies Laplace's Equation,

$$\nabla^2 \Phi_{(0)} = 0$$

Boundary conditions

The total potential must be continuous at all boundaries and match the appropriate equipotential at all perfect conductors. That will automatically insure that the tangential electric field is continuous at all material interfaces and vanishes on the perfect conductors.

Because (by assumption) there is no unpaired electric surface charge at interfaces between materials (except those between material and a perfect conductor), the normal component of $\mathbf{D}_{(0)} = \varepsilon \mathbf{E}_{(0)}$ must also be continuous at every interface—except those that involve a field-free conductor. For the latter type, the discontinuity in $\widehat{\mathbf{n}} \cdot \mathbf{D}_{(0)}$ provides the value of the surface charge, σ_{u}^{s}, on the conducting surface.

First-order fields

After the zero-order fields are found (subject to the boundary conditions at any perfect conductors and/or the interfaces between different dielectric materials), they provide the sources needed to solve for the first-order fields. The latter satisfy

$$\nabla \cdot \mathbf{J}_{\mathrm{u}(1)} = -\frac{\partial \rho_{\mathrm{u}(0)}}{\partial t} \tag{2.4a}$$

$$\nabla \times \mathbf{H}_{(1)} = \varepsilon \frac{\partial \mathbf{E}_{(0)}}{\partial t} + \mathbf{J}_{\mathrm{u}(1)} \tag{2.4b}$$

$$\nabla \cdot \mu \mathbf{H}_{(1)} = 0 \tag{2.4c}$$

Although it is possible to start with either Eq. (2.4a) or (2.4b), it is usually simpler to choose the second. This is because (except on the conducting surfaces) $\mathbf{J}_{\mathrm{u}(1)} = 0$, and only the Maxwell displacement current acts as a source. Any first-order field, $\mathbf{H}_{(1)}$, that satisfies Eq. (2.4b) can serve as the particular solution, but homogeneous fields may be required to satisfy the boundary conditions. In regions with uniform μ, these additional fields satisfy

$$\nabla \times \mathbf{H}'_{(1)} = 0$$

$$\nabla \cdot \mathbf{H}'_{(1)} = 0$$

and so are Laplacian in character because $\mu\mathbf{H}'_{(1)} = \nabla \times \mathbf{A}_{(1)}$ and (for $\nabla \cdot \mathbf{A}_{(1)} = 0$) $\nabla^2 \mathbf{A}_{(1)} = 0$.[2] When the complete magnetic field has been found, any discontinuities in the tangential component imply the existence of the first-order electric surface-current density, $\mathbf{K}^s_{u(1)}$. The latter may also be found from the boundary condition,[3]

$$\nabla_\Sigma \cdot \mathbf{K}^s_{u(1)} + \hat{\mathbf{n}} \cdot \left[\mathbf{J}^{(1)}_{u(1)} - \mathbf{J}^{(2)}_{u(1)}\right] = -\frac{\partial \sigma^s_{u(0)}}{\partial t}$$

where ∇_Σ is the two-dimensional del operator and $\nabla_\Sigma \cdot \mathbf{K}^s_{u(1)}$ is the surface divergence; in this case $\mathbf{J}_{u(1)} = 0$.

The process of approximation can continue with $\frac{\partial}{\partial t}\mathbf{H}_{(n)}$ serving as a source for $\mathbf{E}_{(n+1)}$ and $\frac{\partial}{\partial t}\mathbf{E}_{(n)}$ serving as a source for $\mathbf{H}_{(n+1)}$. The fields are then

$$\mathbf{E} = \mathbf{E}_{(0)} + \mathbf{E}_{(2)} + \cdots$$

$$\mathbf{H} = \mathbf{H}_{(1)} + \mathbf{H}_{(3)} + \cdots$$

where the series is presumed to converge. The *EQS* approximation contains only the zero- and first-order terms; nevertheless, it is often advisable to calculate (or at least estimate) the range of parameters over which $|\mathbf{E}_{(2)}| \ll |\mathbf{E}_{(0)}|$.

2.3 MAGNETOQUASISTATIC (*MQS*) FIELDS

Zero-order fields

We assume that the configuration to be analyzed contains only linear isotropic materials and metallic objects. When there is negligible conductivity associated with any dielectric or magnetic material and all metal surfaces are very highly conducting, but *do* provide closed paths (that allow circulating current flows), the quasistatic fields are predominantly magnetic. We use a subscript (0) to indicate the principal fields; therefore

$$\mathbf{B}_{(0)} = \mu\mathbf{H}_{(0)} = \nabla \times \mathbf{A}_{(0)} \tag{2.5a}$$

$$\mathbf{D}_{(0)} = \varepsilon\mathbf{E}_{(0)} = 0 \tag{2.5b}$$

$$\rho_{u(0)} = 0 \tag{2.5c}$$

$$\Phi_{(0)} = 0 \tag{2.5d}$$

[2]Alternatively, $\mathbf{H}'_{(1)} = \nabla\Psi$ and $\nabla^2\Psi = 0$.

[3]The use of (1) and (2) to label both the quasistatic order and the sides of the surface of discontinuity is confusing, and so in the future the latter will be labeled as *inner* and *outer*.

Maxwell's Equations are then approximated by

$$\nabla \times \mathbf{H}_{(0)} = \mathbf{J}_{u(0)} \tag{2.6a}$$

$$\nabla \cdot \mu\mathbf{H}_{(0)} = 0 \tag{2.6b}$$

$$\nabla \cdot \mathbf{J}_{(0)} = 0 \tag{2.6c}$$

These equations are equivalent to the set

$$\mathbf{B}_{(0)} = \nabla \times \mathbf{A}_{(0)} \tag{2.7a}$$

$$\nabla \cdot \mathbf{A}_{(0)} = 0 \tag{2.7b}$$

$$\nabla^2\mathbf{A}_{(0)} + \left(\frac{\nabla\mu}{\mu}\right) \times [\nabla \times \mathbf{A}_{(0)}] = -\mu\mathbf{J}_{u(0)} \tag{2.7c}$$

Boundary conditions

Although $\mathbf{A}_{(0)}$ is continuous and divergence-free everywhere, *any* particular solution that ignores contributions from currents on the surface of or outside the region of interest can be used and is frequently much easier to calculate.

Because it is divergence-free, the normal component of $\mathbf{B}_{(0)} = \mu\mathbf{H}_{(0)}$ must be continuous at every interface—and therefore vanish at field-free conductors. The discontinuity in $\hat{\mathbf{n}} \times \mathbf{H}_{(0)}$ provides the value of the surface current, $\mathbf{K}^s_{u(0)}$, on any perfectly conducting surface; otherwise, the tangential components of $\mathbf{H}_{(0)}$ are continuous.

First-order fields

After the zero-order fields are found (subject to the boundary conditions at any perfect conductors and/or the interfaces between different magnetic materials), they provide the sources needed to solve for the first-order fields. The latter satisfy

$$\mathbf{E}_{(1)} = -\frac{\partial\mathbf{A}_{(0)}}{\partial t} - \nabla\Phi_{(1)} \tag{2.8a}$$

$$\nabla \cdot \varepsilon\mathbf{E}_{(1)} = \rho_{u(1)} \tag{2.8b}$$

Any first-order field, $\mathbf{E}_{(1)}$, that satisfies Eq. (2.8a) with $\Phi_{(1)} = 0$ can serve as the particular solution, but homogeneous fields may be required to satisfy the boundary conditions. In regions with uniform ε, these additional fields are generated by $\Phi_{(1)}$ that satisfies

$$\nabla^2\Phi_{(1)} = -\frac{\rho_{u(1)}}{\varepsilon}$$

and so, in the absence of unpaired-charge, are Laplacian in character.

The fields are then

$$\mathbf{E} = \mathbf{E}_{(1)} + \mathbf{E}_{(3)} + \cdots$$

$$\mathbf{H} = \mathbf{H}_{(0)} + \mathbf{H}_{(2)} + \cdots$$

where again the series is presumed to converge. The *MQS* approximation contains only the zero- and first-order terms; nevertheless, it is often advisable to calculate (or at least estimate) the range of parameters over which $|\mathbf{H}_{(2)}| \ll |\mathbf{H}_{(0)}|$.

2.4 CONDUCTION PROBLEMS

When there is material with finite conductivity, but it is not part of a path through which zero-order current can flow, $\mathbf{J}_{u(0)} = 0$, $\mathbf{H}_{(0)} = 0$, and the quasistatic fields remain *EQS*. However, $\mathbf{J}_{u(1)} = \sigma\mathbf{E}_{(1)}$ will cause dissipation due to $\frac{\partial \mathbf{E}_{(0)}}{\partial t}$ (which creates a first-order Maxwell displacement current).

When the material is part of a closed path, the configuration is termed a conduction problem, $\mathbf{J}_{(0)} = \sigma\mathbf{E}_{(0)}$, and both $\mathbf{E}_{(0)}$ and $\mathbf{H}_{(0)}$ exist. Depending upon the level of resistance in the path, the conduction may be classified as *EQS* or *MQS*. We consider each regime separately.

EQS regime

Zero-order fields

If the level of resistance is very high, the zero-order current will be small and $\mathbf{E}_{(0)}$ will dominate. The conduction part of the problem is attacked first, subject to continuity of $\Phi_{(0)}$ and the boundary condition $\widehat{\mathbf{n}} \cdot \mathbf{J}_{(0)} = 0$ at the interfaces between conducting and nonconducting regions.

$$\mathbf{J}_{u(0)} = \sigma\mathbf{E}_{(0)} \tag{2.9a}$$

$$\nabla \cdot \mathbf{J}_{u(0)} = 0 \tag{2.9b}$$

$$\mathbf{E}_{(0)} = -\nabla\Phi_{(0)} \tag{2.9c}$$

It follows that within a conductor, we have

$$\rho_{u(0)} = (\varepsilon\nabla\sigma - \sigma\nabla\varepsilon) \cdot \nabla\Phi_{(0)}$$

After $\mathbf{E}_{(0)}$ is found within the conducting region(s), either the entire zero-order magnetic field or those portions of $\mathbf{E}_{(0)}$ that are outside of the conducting region(s) can be found; the order of solution is immaterial.

The zero-order magnetic field satisfies

$$\nabla \times \mathbf{H}_{(0)} = \mathbf{J}_{u(0)} \tag{2.10a}$$

$$\nabla \cdot \mu\mathbf{H}_{(0)} = 0 \tag{2.10b}$$

The zero-order electric field in nonconducting regions satisfies

$$\mathbf{E}_{(0)} = -\nabla\Phi_{(0)}$$

$$\nabla^2\Phi_{(0)} + \frac{\nabla\varepsilon}{\varepsilon} \cdot \nabla\Phi_{(0)} = -\frac{1}{\varepsilon}\rho_{u(0)}$$

First-order fields

After the zero-order fields have been found, they serve as sources that generate the first-order fields. The latter satisfy

$$\nabla \cdot \mathbf{J}_{u(1)} = -\frac{\partial \rho_{\mathrm{u}(0)}}{\partial t} \tag{2.11a}$$

$$\nabla \times \mathbf{H}_{(1)} = \varepsilon \frac{\partial \mathbf{E}_{(0)}}{\partial t} + \mathbf{J}_{u(1)} \tag{2.11b}$$

$$\nabla \cdot \mu \mathbf{H}_{(1)} = 0 \tag{2.11c}$$

and

$$\nabla \times \mathbf{E}_{(1)} = -\mu \frac{\partial \mathbf{H}_{(0)}}{\partial t} \tag{2.12a}$$

$$\nabla \cdot \varepsilon \mathbf{E}_{(1)} = \rho_{u(1)} \tag{2.12b}$$

The total *EQS* fields are

$$\mathbf{E} = \mathbf{E}_{(0)} + \mathbf{E}_{(1)}$$

$$\mathbf{H} = \mathbf{H}_{(0)} + \mathbf{H}_{(1)}$$

The contribution from $\mathbf{E}_{(1)}$ is small but, except at very low frequencies, $\mathbf{H}_{(1)}$ may well dominate over $\mathbf{H}_{(0)}$. This is in agreement with the fact that $\mathbf{H}_{(0)}$, $\mathbf{E}_{(1)} = 0$ in the limit $\sigma = 0$.

MQS regime

If the level of resistance is very low, the current will be large and $\mathbf{H}_{(0)}$ will dominate. The conduction part of the problem is still attacked first, but only after first calculating the skin depth, $\delta = \sqrt{2/(\omega\mu\sigma)}$ (for harmonic fields) or the dominant diffusion length for transient fields. Comparison of these parameters (derivations of which are postponed until Chapter 5, Section 5.2) with the physical dimensions then leads to an appropriate model for the zero-order current.

Conduction with large skin depth

Assume harmonic excitation, $\sigma \gg \omega\varepsilon$, and δ large compared to the pertinent physical dimensions of the conductors. In this case, the equations of *EQS* conduction are still appropriate and the conduction portion(s) of the problem solved first. The principal difference is that for the *MQS* fields, the contribution from $\mathbf{H}_{(1)}$ is small but, except at very low frequencies, $\mathbf{E}_{(1)}$ may well dominate over $\mathbf{E}_{(0)}$. This is in agreement with the limit $\sigma = \infty$ where $\mathbf{E}_{(0)}$, $\mathbf{H}_{(1)} = 0$. However, that limit implies $\delta \to 0$, which violates the assumptions of our analysis. We turn now to that possibility.

Conduction with small skin depth

Assume harmonic excitation, $\sigma \gg \omega\varepsilon$, and δ small compared to the pertinent physical dimensions of the conductors. In this case, it is reasonable to approximate such conductors as being field-free perfect conductors carrying surface currents and surface charges. This leads to the standard *MQS* problem and the solution method outlined earlier. As a final step the surface currents, $\mathbf{K}^s_{u(0)}$, are replaced by volume currents $\mathbf{J}_{u(0)} = \mathbf{K}^s_{u(0)}/\delta$ that are uniform within a distance δ of the surface. Small zero-order electric fields, $\mathbf{J}_{u(0)}/\sigma$, are generated at the surface. These must match the tangential components of the electric fields that are external to the conductor, which is no longer an equipotential surface.

2.5 LAPLACIAN APPROXIMATIONS

In all of these cases, the fields are Laplacian in character because neglecting the time derivatives of the zero-order fields is equivalent to replacing the wave-equation operator with the Laplacian operator.

For quasistatic fields, the time variation is slow enough that the fields have spatial distributions that appear to be static, but with amplitudes that vary with time.[4] If one could make a movie of the field, each frame would approximate a combination of electrostatic and magnetostatic field patterns generated by $-\nabla\phi(\mathbf{r})$ and $\nabla \times \mathbf{a}(\mathbf{r})$. With $\phi(\mathbf{r})$ and $\mathbf{a}(\mathbf{r})$ assumed to be normalized static potentials, the quasistatic potentials are approximated by

$$\Phi = C_\phi(t)\phi(\mathbf{r})$$
$$\mathbf{A} = C_a(t)\mathbf{a}(\mathbf{r})$$

where in uniform regions of material (without conductivity) we obtain

$$C_\phi(t)\nabla^2\phi(\mathbf{r}) = -\frac{1}{\varepsilon}\rho_u$$
$$C_a(t)\nabla^2\mathbf{a}(\mathbf{r}) = -\mu\mathbf{J}_u$$

Conservation of charge is consistent with the gauge condition, $\nabla\cdot\mathbf{A}+\mu\varepsilon\frac{\partial\Phi}{\partial t} = 0$; therefore,

$$C_a(t)\nabla\cdot\mathbf{a}(\mathbf{r}) + \mu\varepsilon\phi(\mathbf{r})\frac{dC_\phi(t)}{dt} = 0$$

and (unless $\nabla\cdot\mathbf{a}(\mathbf{r}) = 0$, $\phi(\mathbf{r}) = 0$), $C_a(t) \sim \frac{dC_\phi(t)}{dt}$.

Accordingly, one can assume that in the nondissipative *EQS* limit we have

$$\Phi = \Phi_{(0)} = C_\phi(t)\phi(\mathbf{r})$$
$$\mathbf{A} = \mathbf{A}_{(1)} = C_a(t)\mathbf{a}(\mathbf{r})$$
$$C_a(t) \sim \frac{dC_\phi(t)}{dt}$$

and in the nondissipative *MQS* limit we obtain

$$\mathbf{A} = \mathbf{A}_{(0)} = C_a'(t)\mathbf{a}(\mathbf{r})$$
$$\Phi = \Phi_{(0)} = 0$$

[4] As a previous footnote warned, this is not true for quasistatic fields that are governed by diffusion or non-electromagnetic wave equations (most often mechanical) that dominate the character of the solution because a movie would reveal that type of motion. Examples include the high-conductivity *MQS* skin-depth regime (where magnetic diffusion dominates) and a piezoelectric wave that is mainly elastic but with an associated weak electric field that, although quasistatic, is forced to propagate at the speed of sound. Plasma waves and magnetostatic waves provide other examples.

CHAPTER 3

ELECTROMAGNETIC POWER, ENERGY, STRESS, AND MOMENTUM

3.1 INTRODUCTION

Power conversion and force densities

Within a region of free-space that may also contain electric charges and/or currents, an important theorem (derived by John Henry Poynting in 1884) relates the power-conversion density, $-\mathbf{E}\cdot\mathbf{J}$ (which if positive represents power flowing *out* of the current), to the Poynting vector, $\mathbf{S}=\mathbf{E}\times\mathbf{H}$, and the scalar, $W=\frac{1}{2}(\varepsilon_0\mathbf{E}\cdot\mathbf{E}+\mu_0\mathbf{H}\cdot\mathbf{H})$. In integral form we have

$$\oint_{S_o}\mathbf{S}\cdot d\mathbf{a}+\frac{d}{dt}\int_{V_o}W dV=-\int_{V_o}\mathbf{E}\cdot\mathbf{J}\,dV \tag{3.1a}$$

where the closed surface S_o encloses the volume V_o. The interpretation given to this integral form of the Poynting Theorem is that the (outward) normal component of $\mathbf{S}$ integrated over any *closed* surface is the *net* electromagnetic power passing out of it (equal to the negative time rate of change of the electromagnetic energy stored within plus the power converted *into* the volume). If $-\mathbf{E}\cdot\mathbf{J}=0$, the theorem is a statement of energy conservation. The differential form of the theorem follows from letting both the

The Power and Beauty of Electromagnetic Fields, First Edition. F. R. Morgenthaler.

volume and the closed surface shrink toward zero; the result

$$\nabla \cdot \mathbf{S} + \frac{\partial W}{\partial t} = -\mathbf{E} \cdot \mathbf{J} \tag{3.1b}$$

invites the interpretation that $\mathbf{S}$ is the local electromagnetic power-flux and W the local electromagnetic energy-density.

Another important theorem, involving the Lorentz force-density, has similar forms for each of its Cartesian components

$$\oint_{S_o} -\mathbf{T}_x \cdot d\mathbf{a} + \frac{d}{dt}\int_{V_o} G_x \, dV = -\int_{V_o} f_x \, dV$$

$$\oint_{S_o} -\mathbf{T}_y \cdot d\mathbf{a} + \frac{d}{dt}\int_{V_o} G_y \, dV = -\int_{V_o} f_y \, dV$$

$$\oint_{S_o} -\mathbf{T}_z \cdot d\mathbf{a} + \frac{d}{dt}\int_{V_o} G_z \, dV = -\int_{V_o} f_z \, dV$$

where $\mathbf{T}_x$,$\mathbf{T}_y$,$\mathbf{T}_z$ are stress-vectors and G_x,G_y,G_z are the Cartesian components of the electromagnetic momentum-density vector , $\mathbf{G}$. It is customary to define the nine components that comprise the three stress-vectors as the electromagnetic stress-tensor, $\overline{\mathbf{T}}$, written in dyadic notation as, $\overline{\mathbf{T}} = \widehat{\mathbf{x}}\mathbf{T}_x + \widehat{\mathbf{y}}\mathbf{T}_y + \widehat{\mathbf{z}}\mathbf{T}_z$. Then

$$\oint_{S_o} -\overline{\mathbf{T}} \cdot d\mathbf{a} + \frac{d}{dt}\int_{V_o} \mathbf{G} \, dV = -\int_{V_o} \mathbf{f} \, dV \tag{3.2a}$$

and, in differential form, can be written either as

$$\nabla \cdot \overline{\mathbf{T}} - \frac{\partial \mathbf{G}}{\partial t} = \mathbf{f} \tag{3.2b}$$

or, using index notation[1] (with summation over the repeated index j assumed), as

$$\frac{\partial T_{ij}}{\partial x_j} - \frac{\partial G_i}{\partial t} = f_i \tag{3.2c}$$

Electromagnetic torque density

The torque density, $\tau = \mathbf{r} \times \mathbf{f}$, that arises from the Lorentz force-density can therefore be expressed as

$$\epsilon_{kij} x_i \left(\frac{\partial T_{j\ell}}{\partial x_\ell} - \frac{\partial G_j}{\partial t} \right) = \epsilon_{kij} \left(\frac{\partial x_i T_{j\ell}}{\partial x_\ell} - \frac{\partial x_i}{\partial x_\ell} T_{j\ell} - \frac{\partial x_i G_j}{\partial t} \right) = \tau_k \tag{3.3a}$$

where ϵ_{kij} is the permutation symbol and $\partial x_i / \partial x_\ell = \delta_{i\ell}$ is the Kronecker delta (both symbols defined in Appendix B, Section B.2). Consequently, the torque density simplifies to

$$\frac{\partial (\epsilon_{kij} x_i T_{j\ell})}{\partial x_\ell} - \epsilon_{kij} T_{ji} - \frac{\partial (\mathbf{r} \times \mathbf{G})_k}{\partial t} = \tau_k \tag{3.3b}$$

[1]The reader should be cautioned that some texts define the divergence of a tensor as $\frac{\partial T_{ji}}{\partial x_j}$. With that convention, the stress tensor would be defined as the transpose, $\overline{\mathbf{T}} = \mathbf{T}_x\widehat{\mathbf{x}} + \mathbf{T}_y\widehat{\mathbf{y}} + \mathbf{T}_z\widehat{\mathbf{z}}$.

which, because $\epsilon_{kij} = -\epsilon_{kji}$, is equivalent to

$$[\nabla \cdot (\mathbf{r} \times \overline{\mathbf{T}}) - \frac{\partial(\mathbf{r} \times \mathbf{G})}{\partial t} - \mathbf{r} \times \mathbf{f}]_k + \epsilon_{kij} T_{ij} = 0 \tag{3.3c}$$

Notice that if the stress tensor is symmetric, $\epsilon_{kij} T_{ij} = 0$. The electromagnetic angular-momentum density is defined as $\mathbf{r} \times \mathbf{G}$.

Uniqueness of S, *W*, $\overline{\mathbf{T}}$, and G

The connection between electromagnetic force and power is revealed because, from Eq. (1.2), it follows that $\mathbf{E} \cdot \mathbf{J} = (\mathbf{E} + \mu_o \mathbf{v} \times \mathbf{H}) \cdot (\mathbf{J} - \rho \mathbf{v}) + \mathbf{f} \cdot \mathbf{v}$, where $\mathbf{v}$ is the velocity of the charge density. This is especially apparent whenever the current density is convective so that $\mathbf{J} = \rho \mathbf{v}$.

We shall learn shortly that (in dyadic notation) $\overline{\mathbf{T}} = \varepsilon_o \mathbf{E}\,\mathbf{E} + \mu_o \mathbf{H}\,\mathbf{H} - W\,(\widehat{\mathbf{x}\,\mathbf{x}} + \widehat{\mathbf{y}\,\mathbf{y}} + \widehat{\mathbf{z}\,\mathbf{z}})$ (the Maxwell stress tensor) and $\mathbf{G} = \mu_o \varepsilon_o \mathbf{S}$ satisfy Eq. (3.2b). Because of their connections to power and energy, $\overline{\mathbf{T}}$ and $\mathbf{G}$ invite interpretation as the local electromagnetic stress tensor and electromagnetic momentum-density vector. Nevertheless, $\mathbf{S}$, W, $\mathbf{G}$, and $\overline{\mathbf{T}}$ are not uniquely specified because, for example, the addition of the curl of an arbitrary vector to $\mathbf{S}$ leaves the divergence of the sum unchanged. Likewise, the addition of any time-independent function of position to W cannot alter the validity of the time-dependent Poynting Theorem (but spatial gradients might generate unwanted static forces). In order to prevent any alteration to either $-\mathbf{E} \cdot \mathbf{J}$ or $\mathbf{f}$, any modifications to $\mathbf{S}$, W, $\mathbf{G}$, and $\overline{\mathbf{T}}$ should be made in a self-consistent manner that maintains the proper interrelationships among them[2]. (Lorentz invariance is discussed in Appendix A.)

These constraints would appear to make the task of finding *nontrivial* modifications of a dynamic nature very difficult—perhaps impossible. On the other hand, certain quasistatic approximations of the Poynting Theorem localize both the power flow and energy to the positions of the electric charges and/or currents. These representations require nontrivial alteration of *both* $\mathbf{S}$ and W. Consequently, it is of interest to understand why this is so and inquire whether such localization can exist within the framework of the *exact* equations. Our previous reexamination of electrodynamics [8] showed that this goal is indeed achievable. Power fluxes and energy densities were found that are distinct alternatives to the Poynting quantities, yet generate identical power-conversion densities. Alternative electromagnetic stress tensors and momentum densities were also found that generate identical force densities. These findings are central to the rationale behind the writing of this text and are derived in Part II.

3.2 THE MAXWELL–POYNTING REPRESENTATION

Maxwell stress tensor

Use of Maxwell's Equations allows $\mathbf{f}$ and $\mathbf{E} \cdot \mathbf{J}$ to be expressed solely in terms of the $\mathbf{E}$ and $\mathbf{H}$ fields. The result is

[2]Attempts by Jeffries [7] to produce alternative power and energy using vector and scalar potentials lead to results inconsistent with the Lorentz force law. Jeffries concludes that Lorentz is incorrect and predicts that measurements of the trajectories of relativistic charged particles will vindicate his new theory. The present author firmly believes that Lorentz will prevail.

$$\mathbf{f} = (\varepsilon_0 \nabla \cdot \mathbf{E})\mathbf{E} + \left(\nabla \times \mathbf{H} - \varepsilon_0 \frac{\partial \mathbf{E}}{\partial t}\right) \times \mu_0 \mathbf{H}$$

$$0 = (\mu_0 \nabla \cdot \mathbf{H})\mathbf{H} + \left(\nabla \times \mathbf{E} + \mu_0 \frac{\partial \mathbf{H}}{\partial t}\right) \times \varepsilon_0 \mathbf{E}$$

where the second equation is a helpful artifice; adding the two together leaves the force density unchanged. The result is

$$\begin{aligned}\mathbf{f} = {} & \varepsilon_0[(\nabla \cdot \mathbf{E})\mathbf{E} + (\nabla \times \mathbf{E}) \times \mathbf{E}] \\ & + \mu_0[(\nabla \cdot \mathbf{H})\mathbf{H} + (\nabla \times \mathbf{H}) \times \mathbf{H}] - \mu_0 \varepsilon_0 \frac{\partial}{\partial t}(\mathbf{E} \times \mathbf{H})\end{aligned}$$

In index notation (with δ_{ij} the Kronecker delta, and summation over repeated indices implied), we have

$$\begin{aligned}[(\nabla \cdot \mathbf{E})\mathbf{E} + (\nabla \times \mathbf{E}) \times \mathbf{E}]_i &= \frac{\partial E_j}{\partial x_j} E_i + \left(\frac{\partial E_i}{\partial x_j} - \frac{\partial E_j}{\partial x_i}\right) E_j \\ &= \frac{\partial}{\partial x_j}\left(E_i E_j - \frac{1}{2} E \cdot E \delta_{ij}\right)\end{aligned}$$

After substituting that equation (and the similar one governing **H**) and switching to dyadic notation (with $\bar{\mathbf{I}} = \widehat{\mathbf{x}\mathbf{x}} + \widehat{\mathbf{y}\mathbf{y}} + \widehat{\mathbf{z}\mathbf{z}}$, the identity tensor), the result is

$$\mathbf{f} = \nabla \cdot \left[\varepsilon_0 \mathbf{E}\mathbf{E} + \mu_0 \mathbf{H}\mathbf{H} - \frac{1}{2}(\varepsilon_0 \mathbf{E} \cdot \mathbf{E} + \mu_0 \mathbf{H} \cdot \mathbf{H})\bar{\mathbf{I}}\right] - \frac{\partial}{\partial t}(\mu_0 \varepsilon_0 \mathbf{E} \times \mathbf{H}) \tag{3.4}$$

which validates Eq. (3.2b)

Because $\bar{\mathbf{T}} = \varepsilon_0 \mathbf{E}\mathbf{E} + \mu_0 \mathbf{H}\mathbf{H} - W\bar{\mathbf{I}}$ is symmetric, the torque density, Eq. (3.3c), reduces to

$$\mathbf{r} \times \mathbf{f} = \nabla \cdot (\mathbf{r} \times \bar{\mathbf{T}}) - \frac{\partial}{\partial t}(\mu_0 \varepsilon_0 \mathbf{r} \times (\mathbf{E} \times \mathbf{H})) \tag{3.5}$$

Poynting Theorem

In a similar manner, the pair of equations

$$-\mathbf{E} \cdot \mathbf{J} = -\mathbf{E} \cdot \left(\nabla \times \mathbf{H} - \varepsilon_0 \frac{\partial \mathbf{E}}{\partial t}\right)$$

$$0 = \mathbf{H} \cdot \left(\nabla \times \mathbf{E} + \mu_0 \frac{\partial \mathbf{H}}{\partial t}\right)$$

can be added to give

$$\begin{aligned}-\mathbf{E} \cdot \mathbf{J} = {} & \mathbf{H} \cdot (\nabla \times \mathbf{E}) - \mathbf{E} \cdot (\nabla \times \mathbf{H}) \\ & + \frac{\partial}{\partial t}\left(\frac{1}{2}\varepsilon_0 \mathbf{E} \cdot \mathbf{E} + \frac{1}{2}\mu_0 \mathbf{H} \cdot \mathbf{H}\right)\end{aligned}$$

and, because $\mathbf{H} \cdot (\nabla \times \mathbf{E}) - \mathbf{E} \cdot (\nabla \times \mathbf{H}) = \nabla \cdot (\mathbf{E} \times \mathbf{H})$, the power conversion density is

$$-\mathbf{E} \cdot \mathbf{J} = \nabla \cdot (\mathbf{E} \times \mathbf{H}) + \frac{\partial}{\partial t}\frac{1}{2}(\varepsilon_0 \mathbf{E} \cdot \mathbf{E} + \mu_0 \mathbf{H} \cdot \mathbf{H}) \tag{3.6}$$

This well-known Poynting Theorem validates Eq. (3.1b).

Because $\overline{\mathbf{T}}$ is the Maxwell stress tensor and $\mathbf{S}$ is the Poynting vector, we shall refer to this set of $\overline{\mathbf{T}}$, $\mathbf{G}$, $\mathbf{S}$, and W as the Maxwell–Poynting representation. Adding superscripts em to denote them as the standard electromagnetic quantities, we obtain

$$\overline{\mathbf{T}}^{\mathrm{em}} = \varepsilon_{\mathrm{o}}\mathbf{E}\,\mathbf{E} + \mu_{\mathrm{o}}\mathbf{H}\,\mathbf{H} - W^{\mathrm{em}}\,\overline{\mathbf{I}} \tag{3.7a}$$

$$\mathbf{G}^{\mathrm{em}} = \mu_{\mathrm{o}}\varepsilon_{\mathrm{o}}\mathbf{S}^{\mathrm{em}} \tag{3.7b}$$

$$\mathbf{S}^{\mathrm{em}} = \mathbf{E} \times \mathbf{H} \tag{3.7c}$$

$$W^{\mathrm{em}} = \frac{1}{2}(\varepsilon_{\mathrm{o}}\mathbf{E}\cdot\mathbf{E} + \mu_{\mathrm{o}}\mathbf{H}\cdot\mathbf{H}) \tag{3.7d}$$

The electromagnetic energy density is the sum of the individual electric and magnetic energy densities,

$$W_{\mathrm{e}} = \frac{1}{2}\varepsilon_{\mathrm{o}}\mathbf{E}\cdot\mathbf{E} \tag{3.8a}$$

$$W_{\mathrm{m}} = \frac{1}{2}\mu_{\mathrm{o}}\mathbf{H}\cdot\mathbf{H} \tag{3.8b}$$

These compact and very elegant results lend themselves to the interpretation that electromagnetic stress, momentum, power, and energy all reside in the fields themselves and are therefore "outside" of the charges and currents that generated those fields. In this sense, the Maxwell–Poynting representation is nonlocal—although, of course, the fields themselves may be quite concentrated near the sources.

When there is dielectric and magnetic material present, the Amperian model, developed in Chapter 1, Section 1.7, includes dielectric polarization charge and currents as well as Amperian electric currents. These combine with the unpaired (free) charge and current to produce ρ_{total} and $\mathbf{J}_{\mathrm{total}}$. In this case, we make the substitutions $\mathbf{E} \to \mathbf{E}^{\mathrm{o}}$ and $\mu_{\mathrm{o}}\mathbf{H} \to \mathbf{B}$ in all of the preceding equations. Alternatively, the Chu representation, described in Section 1.7.2, replaces Amperian current loops with magnetic-charge dipoles. In this case, $\mathbf{E}$ and $\mathbf{H}$ remain unchanged, but separate Lorentz force densities (duals of one another) are required to act upon the electric and magnetic charges. We replace the superscript em by either amp or chu to designate which Maxwell–Poynting set of $\mathbf{S}$, W, $\mathbf{G}$, and $\overline{\mathbf{T}}$ is under discussion.

3.3 QUASISTATIC POWER AND ENERGY

Standard form of quasistatic power theorems

Consideration of $\nabla\cdot\,\mathbf{S}^{\mathrm{em}} = \mathbf{H}\cdot\,\nabla\times\mathbf{E} - \mathbf{E}\cdot\,\nabla\times\mathbf{H}$ within the contexts of electroquasistatics (*EQS*) and magnetoquasistatics (*MQS*) has always made clear the lack of uniqueness of the power flux.

Electroquasistatic approximation

In the case of *EQS*: $\nabla \times \mathbf{E} = 0$, $\mathbf{E} = -\nabla\Phi$, and

$$\nabla \cdot \mathbf{S}^{\text{em}} = \nabla\Phi \cdot (\mathbf{J} + \varepsilon_0 \partial\mathbf{E}/\partial t) = \nabla \cdot \left[\Phi\left(\mathbf{J} + \varepsilon_0 \frac{\partial \mathbf{E}}{\partial t}\right)\right] - \Phi\left(\nabla \cdot \mathbf{J} + \frac{\partial \rho}{\partial t}\right) \tag{3.9}$$

Therefore, because of conservation of charge, $\Phi(\mathbf{J} + \varepsilon_0 \frac{\partial \mathbf{E}}{\partial t})$ is an acceptable substitute for $\mathbf{S}^{\text{em}}$ since it has nearly the same divergence. The differential form of the *EQS* Poynting Theorem can then be written as

$$\nabla \cdot \left[\Phi\left(\mathbf{J} - \varepsilon_0 \frac{\partial}{\partial t}\nabla\Phi\right)\right] + \frac{\partial}{\partial t}\left(\frac{1}{2}\varepsilon_0 \nabla\Phi \cdot \nabla\Phi\right) = \nabla\Phi \cdot \mathbf{J} \tag{3.10}$$

It is seen that when the Maxwell displacement current is either absent or perpendicular to the electric current density, the substitute power flux in the direction of the current is coincident with **J** and zero elsewhere. As expected, the magnetic contribution to W^{em} is missing and the electric energy density is distributed over the quasistatic **E** field.

Magnetoquasistatic approximation

In the case of *MQS*: $\nabla \times \mathbf{H} = \mathbf{J}$, $\mathbf{E} = -\frac{\partial}{\partial t}\mathbf{A} - \nabla\Phi$ and *where there is no current*, $\mathbf{H} = -\nabla\Psi_m$. In such regions, $\Psi_m(\mu_0 \frac{\partial}{\partial t}\mathbf{H})$, the dual of the corresponding *EQS* term, can be used as a substitute flux and the dual of Eq. (3.10) becomes

$$\nabla \cdot \left(\Psi_m \mu_0 \frac{\partial}{\partial t}\nabla\Psi_m\right) + \frac{\partial}{\partial t}\left(\frac{1}{2}\mu_0 \nabla\Psi_m \cdot \nabla\Psi_m\right) = 0 \tag{3.11}$$

More generally, Eqs. (1.9) and (1.13a) can be used and with $\nabla \cdot \mathbf{A} = 0$, $\nabla^2 \mathbf{A} = -\mu_0 \mathbf{J}$. It then follows that the differential form of the *MQS* Poynting Theorem can be written as

$$\nabla \cdot \left(-\frac{\partial \mathbf{A}}{\partial t} \times \mathbf{H}\right) + \frac{\partial}{\partial t}\left(\frac{1}{2}\mu_0 \mathbf{H} \cdot \mathbf{H}\right) = \frac{\partial \mathbf{A}}{\partial t} \cdot \mathbf{J} \tag{3.12}$$

Now it is the electric contribution to W^{em} that is missing and the magnetic energy density is distributed over the quasistatic **H** field. These are the standard quasistatic power theorems as developed in Fano et al. [5, p. 294]. They have proved very useful to generations of students and researchers, but are not the whole story.

Modified form of quasistatic power theorems

Although the *EQS* and *MQS* energy densities are located, respectively, in the **E** and **H** fields, they are not unique and can be moved to the location of the electric charges and currents by simple manipulation.

Equation (3.10) is equivalent to

$$\nabla \cdot \left[\Phi\,\mathbf{J} + \frac{1}{2}\varepsilon_0\left(\frac{\partial \Phi}{\partial t}\nabla\Phi - \Phi\frac{\partial}{\partial t}\nabla\Phi\right)\right] + \frac{\partial}{\partial t}\left(\frac{1}{2}\rho\,\Phi\right) = \nabla\Phi \cdot \mathbf{J} \tag{3.13}$$

whereas, using $\mu_o \mathbf{H} \cdot \mathbf{H} = \mathbf{A} \cdot \mathbf{J} + \nabla \cdot (\mathbf{A} \times \mathbf{H})$ and $\nabla \cdot \mathbf{J} = 0$, Eq. (3.12) can be transformed to

$$\nabla \cdot \left[\Phi \mathbf{J} + \frac{1}{2\mu_o} \left(A_j \nabla \frac{\partial A_j}{\partial t} - \frac{\partial A_j}{\partial t} \nabla A_j \right) \right] + \frac{\partial}{\partial t} \left(\frac{1}{2} \mathbf{A} \cdot \mathbf{J} \right) = \left(\frac{\partial \mathbf{A}}{\partial t} + \nabla \Phi \right) \cdot \mathbf{J} \quad (3.14)$$

(here summation over the repeated index j is assumed for $A_1 = A_x, A_2 = A_y$, and $A_3 = A_z$).

In Eq. (3.13), the substitute electric-energy density is localized to coincide with the electric-charge density, ρ; in Eq. (3.14), the substitute magnetic-energy density is localized to the electric-current density, $\mathbf{J}$. In both cases, these are very different from W^{em}. These modified quasistatic representations are highly compatible with, and useful in making, circuit theory representations of certain electromagnetic field configurations.

Because the power-flux and energy distributions do not gradually shift from the quasistatic versions to those governing the exact fields, the differences cannot simply depend (in the *EQS* case) upon the relative importance of the curl of the $\mathbf{E}$ field. Consequently, we seek another explanation for the behavior and find that *it is possible to generate such localization without having to introduce any approximations to the electromagnetic field equations*. If desired, one can modify the stress and momentum distributions in a similar and unified manner.

3.4 ALTERNATIVE REPRESENTATIONS

Introduction

If $\mu_o \mathbf{H} = \nabla \times \mathbf{A}$ and $\mathbf{E} = -\frac{\partial}{\partial t}\mathbf{A} - \nabla \Phi$ are used to express the Maxwell–Poynting representation in terms of the potentials $\mathbf{A}$ and Φ, nothing is changed except the choice of variables. However, time derivatives now exist in both $\mathbf{S}^{\text{em}}$ and $\overline{\mathbf{T}}^{\text{em}}$ whereas space derivatives exist in both W^{em} and $\mathbf{G}^{\text{em}}$. The question then arises as to whether time derivatives in $\nabla \cdot \mathbf{S}^{\text{em}}$ and $\nabla \cdot \overline{\mathbf{T}}^{\text{em}}$ and space derivatives in $\frac{\partial}{\partial t} W^{\text{em}}$ and $\frac{\partial}{\partial t} \mathbf{G}^{\text{em}}$ can be shifted so as to form a new representation: $\mathbf{S}$, W, $\overline{\mathbf{T}}$, and $\mathbf{G}$ where

$$\nabla \cdot \mathbf{S} + \frac{\partial}{\partial t} W = \nabla \cdot \mathbf{S}^{\text{em}} + \frac{\partial}{\partial t} W^{\text{em}}$$

$$\nabla \cdot \overline{\mathbf{T}} - \frac{\partial}{\partial t} \mathbf{G} = \nabla \cdot \overline{\mathbf{T}}^{\text{em}} - \frac{\partial}{\partial t} \mathbf{G}^{\text{em}}$$

In fact, this can be accomplished (nontrivially) in an infinite number of ways by means of a remarkable identity (derived in Part II, Chapter 8, Section 8.5)

$$\frac{\partial}{\partial t} \nabla \cdot \frac{1}{2} \left[\mathbf{A} \times \mathbf{H} - \varepsilon_o \left(\Phi \mathbf{E} + \Phi \frac{\partial \mathbf{A}}{\partial t} - \frac{\partial \Phi}{\partial t} \mathbf{A} \right) - \frac{1}{\mu_o} g \mathbf{A} \right] = \frac{\partial^2 \beta_{ij}}{\partial x_i \partial x_j} \quad (3.15a)$$

where repeated indices are summed,

$$\beta_{ij} = \frac{1}{\mu_o} \left[E_i A_j - \frac{1}{2} (\mathbf{E} \cdot \mathbf{A} - g\, \Phi)\, \delta_{ij} - \frac{1}{2} \left(\Phi \frac{\partial A_j}{\partial x_i} - \frac{\partial \Phi}{\partial x_i} A_j \right) \right] \quad (3.15b)$$

and the gauge is defined by

$$g = \nabla \cdot \mathbf{A} + \mu_o \varepsilon_o \frac{\partial \Phi}{\partial t} \quad (3.15c)$$

Equation (3.15a) multiplied by an arbitrary scalar can be added to Eq. (3.1b) and/or any of the Cartesian components of Eq. (3.2b). Note, too, that Eq. (3.15a) itself can be expressed as the vanishing divergence of a vector.

One set of possibilities is of special significance. In the following sections, these alternate theorems are developed using a more direct method.

An alternate Poynting theorem

The Poynting Theorem, derived in Section 3.2, is an expansion of the power conversion density, $-\mathbf{E}\cdot\mathbf{J}$, expressed in terms of the electric and magnetic fields. This was accomplished by substituting $\mathbf{J} = \nabla\times\mathbf{H} - \varepsilon_0 \frac{\partial \mathbf{E}}{\partial t}$. An alternate approach, and the underlying basis for Eq. (3.15a), is to substitute $-\mathbf{E} = \frac{\partial \mathbf{A}}{\partial t} + \nabla\Phi$ and carry out the expansion in terms of the vector and scalar potentials. It then follows that

$$-\mathbf{E}\cdot\mathbf{J} = \frac{\partial \mathbf{A}}{\partial t}\cdot\mathbf{J} + \Phi\frac{\partial \rho}{\partial t} + \nabla\cdot(\Phi\mathbf{J}) - \Phi\left(\nabla\cdot\mathbf{J} + \frac{\partial \rho}{\partial t}\right) \tag{3.16}$$

which, because of conservation of charge, can be rewritten as

$$\begin{aligned}-\mathbf{E}\cdot\mathbf{J} &= \nabla\cdot(\Phi\,\mathbf{J}) + \frac{\partial}{\partial t}\left[\frac{1}{2}(\rho\,\Phi + \mathbf{A}\cdot\mathbf{J})\right]\\ &\quad + \frac{1}{2}\left(\Phi\frac{\partial \rho}{\partial t} - \frac{\partial \Phi}{\partial t}\rho\right) + \frac{1}{2}\left(\frac{\partial \mathbf{A}}{\partial t}\cdot\mathbf{J} - \mathbf{A}\cdot\frac{\partial \mathbf{J}}{\partial t}\right)\end{aligned} \tag{3.17}$$

Four-vectors

Although not necessary, it saves some algebraic steps and leads to more compact results if we combine $\mathbf{A}$ and Φ into a vector with four orthogonal components. We define this four-vector as $\mathcal{A} = [A_1, A_2, A_3, A_4]$ with $A_{1,2,3} = A_{x,y,z}$, $A_4 = \alpha\Phi$. We also combine $\mathbf{J}$ and ρ and define a second four-vector $\mathcal{J} = [J_1, J_2, J_3, J_4]$ with $J_{1,2,3} = J_{x,y,z}$, $J_4 = \beta\rho$. The constants α and β are yet to be determined. An equivalent form for these four-vectors is

$$\mathcal{A} = [\mathbf{A},\ \alpha\Phi], \qquad \mathcal{J} = [\mathbf{J},\ \beta\rho]$$

Extending the definition of the dot product to include the fourth component, $\mathcal{A}\cdot\mathcal{J} = \mathbf{A}\cdot\mathbf{J} + \alpha\beta\ \Phi\rho$. The utility of this notation follows if we set

$$\alpha\beta = -1 \tag{3.18}$$

Then

$$\frac{1}{2}\left(\frac{\partial \mathbf{A}}{\partial t}\cdot\mathbf{J} - \mathbf{A}\cdot\frac{\partial \mathbf{J}}{\partial t}\right) + \frac{1}{2}\left(\Phi\frac{\partial \rho}{\partial t} - \frac{\partial \Phi}{\partial t}\rho\right) = \frac{1}{2}\left(\frac{\partial \mathcal{A}}{\partial t}\cdot\mathcal{J} - \mathcal{A}\cdot\frac{\partial \mathcal{J}}{\partial t}\right) \tag{3.19}$$

and

$$\nabla\cdot(\Phi\mathbf{J}) + \frac{\partial}{\partial t}\left[\frac{1}{2}(\rho\,\Phi + \mathbf{A}\cdot\mathbf{J})\right] + \frac{1}{2}\left(\frac{\partial \mathcal{A}}{\partial t}\cdot\mathcal{J} - \mathcal{A}\cdot\frac{\partial \mathcal{J}}{\partial t}\right) = -\mathbf{E}\cdot\mathbf{J} \tag{3.20}$$

In Chapter 1, Section 1.3, the Lorenz-gauge vector and scalar potentials were shown to satisfy

$$\nabla^2 \mathbf{A} - \frac{1}{c^2}\frac{\partial^2 \mathbf{A}}{\partial t^2} = -\mu_o \mathbf{J} \tag{3.21a}$$

$$\nabla^2 \Phi - \frac{1}{c^2}\frac{\partial^2 \Phi}{\partial t^2} = -\frac{\rho}{\varepsilon_o} \tag{3.21b}$$

After multiplying Eq. (3.21b) by α and combining it with Eq. (3.21a), the result is

$$\nabla^2 \mathcal{A} - \frac{1}{c^2}\frac{\partial^2 \mathcal{A}}{\partial t^2} = -\mu_o \mathcal{J} \tag{3.22}$$

provided that

$$\alpha c^2 = \beta \tag{3.23}$$

It follows from Eqs. (3.18) and (3.23) that

$$\alpha = \pm i/c, \qquad \beta = \pm ic$$

We choose the positive sign and substitute Eq. (3.22) into Eq. (3.19); the result is

$$\frac{1}{2}\left(\frac{\partial \mathcal{A}}{\partial t}\cdot \mathcal{J} - \mathcal{A}\cdot\frac{\partial \mathcal{J}}{\partial t}\right) = \frac{1}{2\mu_o}\left[-\frac{\partial \mathcal{A}}{\partial t}\cdot\left(\nabla^2\mathcal{A} - \frac{1}{c^2}\frac{\partial^2\mathcal{A}}{\partial t^2}\right) + \mathcal{A}\cdot\frac{\partial}{\partial t}\left(\nabla^2\mathcal{A} - \frac{1}{c^2}\frac{\partial^2\mathcal{A}}{\partial t^2}\right)\right]$$

or switching to Cartesian index notation, summing over the repeated index, $k = 1, 2, 3, 4$, remembering that $\nabla^2 = \nabla \cdot \nabla$, and making use of Eq. (B.6e), we obtain

$$\begin{aligned}\frac{1}{2}\left(\frac{\partial \mathcal{A}}{\partial t}\cdot \mathcal{J} - \mathcal{A}\cdot\frac{\partial \mathcal{J}}{\partial t}\right) &= \frac{1}{2\mu_o}\left[-\frac{\partial \mathcal{A}_k}{\partial t}\left(\nabla\cdot\nabla\mathcal{A}_k - \frac{1}{c^2}\frac{\partial^2\mathcal{A}_k}{\partial t^2}\right)\right.\\ &\quad \left. + \mathcal{A}_k\frac{\partial}{\partial t}\left(\nabla\cdot\nabla\mathcal{A}_k - \frac{1}{c^2}\frac{\partial^2\mathcal{A}_k}{\partial t^2}\right)\right]\\ &= \nabla\cdot\left[\frac{1}{2\mu_o}\left(\mathcal{A}_k\nabla\frac{\partial\mathcal{A}_k}{\partial t} - \frac{\partial\mathcal{A}_k}{\partial t}\nabla\mathcal{A}_k\right)\right]\\ &\quad + \frac{\partial}{\partial t}\left[\frac{1}{2}\varepsilon_o\left(\frac{\partial\mathcal{A}_k}{\partial t}\frac{\partial\mathcal{A}_k}{\partial t} - \mathcal{A}_k\frac{\partial^2\mathcal{A}_k}{\partial t^2}\right)\right]\end{aligned} \tag{3.24}$$

Finally, substitution of Eq. (3.24) into Eq. (3.20) results in the alternate Poynting theorem,

$$\begin{aligned}&\nabla\cdot\left[\Phi\,\mathbf{J} + \frac{1}{2\mu_o}\left(\mathcal{A}_k\nabla\frac{\partial\mathcal{A}_k}{\partial t} - \frac{\partial\mathcal{A}_k}{\partial t}\nabla\mathcal{A}_k\right)\right]\\ &\quad + \frac{\partial}{\partial t}\left[\frac{1}{2}(\rho\,\Phi + \mathbf{A}\cdot\mathbf{J}) + \frac{1}{2}\varepsilon_o\left(\frac{\partial\mathcal{A}_k}{\partial t}\frac{\partial\mathcal{A}_k}{\partial t} - \mathcal{A}_k\frac{\partial^2\mathcal{A}_k}{\partial t^2}\right)\right] = -\mathbf{E}\cdot\mathbf{J}\end{aligned} \tag{3.25}$$

where $\mathcal{A} = [\mathbf{A}, \frac{i}{c}\Phi]$ and, if desired, Eqs. (3.22) can be substituted into the left-hand side of Eq. (3.25) to replace $\mathbf{J}$ and ρ.

We emphasize that the power flux,

$$\mathbf{S}^{\text{alt}} = \Phi\,\mathbf{J} + \frac{1}{2\mu_o}\left(\mathcal{A}_k\nabla\frac{\partial\mathcal{A}_k}{\partial t} - \frac{\partial\mathcal{A}_k}{\partial t}\nabla\mathcal{A}_k\right) \tag{3.26a}$$

and energy density,

$$W^{\text{alt}} = \frac{1}{2}(\rho\,\Phi + \mathbf{A}\cdot\mathbf{J}) + \frac{1}{2}\varepsilon_o\left(\frac{\partial\mathcal{A}}{\partial t}\cdot\frac{\partial\mathcal{A}}{\partial t} - \mathcal{A}\cdot\frac{\partial^2\mathcal{A}}{\partial t^2}\right) \tag{3.26b}$$

are *not* identical to their Poynting counterparts, $\mathbf{S}^{\text{em}}$ and W^{em}, yet the alternate theorem is *exact*.

An alternate stress-momentum theorem

The stress-momentum theorem, derived in Section 3.2, is an expansion of the Lorentz force density, expressed in terms of the electric and magnetic fields. This was accomplished by substituting $\rho = \varepsilon_o\nabla\cdot\mathbf{E}$ and $\mathbf{J} = \nabla\times\mathbf{H} - \varepsilon_o\frac{\partial\mathbf{E}}{\partial t}$. An alternate approach, similar to that used in the previous section, is to substitute $-\mathbf{E} = \frac{\partial\mathbf{A}}{\partial t} + \nabla\Phi,\ \mu_o\mathbf{H} = \nabla\times\mathbf{A}$ and carry out the expansion in terms of the vector and scalar potentials. It then follows that

$$\mathbf{f} = -\rho\left(\frac{\partial\mathbf{A}}{\partial t} + \nabla\Phi\right) + \mathbf{J}\times(\nabla\times\mathbf{A}) \tag{3.27}$$

Switching to index notation, making use of Appendix B, Eq. (B.61), and again employing the four-vectors $\mathcal{A}$ and $\mathcal{J}$, Eq. (3.27) can be expressed as

$$f_i = -\frac{\partial}{\partial t}(\rho\,A_i) - \frac{\partial}{\partial x_j}\left(A_i\,J_j - \frac{1}{2}\mathcal{A}\cdot\mathcal{J}\delta_{ij}\right) + \frac{1}{2}\left(\frac{\partial\mathcal{A}}{\partial x_i}\cdot\mathcal{J} - \mathcal{A}\cdot\frac{\partial\mathcal{J}}{\partial x_i}\right) \tag{3.28}$$

where summation over $j = 1,2,3$ is implied. After substituting Eq. (3.22) and using an appropriately modified Eq. (3.24), the force density, Eq. (3.28), can be expressed as

$$f_i = \frac{\partial T_{ij}^{\text{alt}}}{\partial x_j} - \frac{\partial G_i^{\text{alt}}}{\partial t} \tag{3.29}$$

where the alternate stress tensor is

$$T_{ij}^{\text{alt}} = -A_i\,J_j + \frac{1}{2}\mathcal{A}\cdot\mathcal{J}\delta_{ij} + \frac{1}{2\mu_o}\left(\mathcal{A}_k\frac{\partial^2\mathcal{A}_k}{\partial x_i\partial x_j} - \frac{\partial\mathcal{A}_k}{\partial x_i}\frac{\partial\mathcal{A}_k}{\partial x_j}\right) \tag{3.30a}$$

and the alternate momentum density is

$$G_i^{\text{alt}} = \rho\,A_i + \frac{1}{2}\varepsilon_o\left(\mathcal{A}_k\frac{\partial^2\mathcal{A}_k}{\partial x_i\partial t} - \frac{\partial\mathcal{A}_k}{\partial x_i}\frac{\partial\mathcal{A}_k}{\partial t}\right) \tag{3.30b}$$

As before, summation over $k = 1,2,3,4$ is implied.

Here, too, we emphasize that $\overline{\mathbf{T}}^{\text{alt}}$ and $\mathbf{G}^{\text{alt}}$ are *not* identical to their Poynting counterparts, $\overline{\mathbf{T}}^{\text{em}}$ and $\mathbf{G}^{\text{em}}$, yet the alternate theorem is *exact*.

Fourth coordinate

Because $\mathcal{A}$ and $\mathcal{J}$ each have a fourth component, it is natural to expect that a fourth coordinate, x_4, might join the other three Cartesian coordinates, $x_1 = x$, $x_2 = y$, and $x_3 = z$. If so, the gradient operator, ∇ (del), must be generalized to

$$\Box = \left[\frac{\partial}{\partial x_1}, \frac{\partial}{\partial x_2}, \frac{\partial}{\partial x_3}, \frac{\partial}{\partial x_4}\right] = \left[\nabla, \frac{\partial}{\partial x_4}\right]$$

It then follows that the four-dimensional divergences of $\mathcal{J}$ and $\mathcal{A}$ can be expressed as

$$\Box \cdot \mathcal{J} = \nabla \cdot \mathbf{J} + \frac{\partial}{\partial x_4}(ic\rho)$$

and

$$\Box \cdot \mathcal{A} = \nabla \cdot \mathbf{A} + \frac{\partial}{\partial x_4}\left(\frac{i}{c}\Phi\right)$$

If x_4 equals ict, the imaginary coordinate introduced by Minkowski to Einstein's Theory of Special Relativity (Appendix A), great simplification occurs. Then, conservation of charge and the Lorenz-gauge condition are, respectively,

$$\Box \cdot \mathcal{J} = 0$$

and

$$\Box \cdot \mathcal{A} = 0$$

Notice also that

$$\Box \cdot \Box = \Box^2 = \nabla^2 - \frac{1}{c^2}\frac{\partial^2}{\partial t^2}$$

is the wave equation operator; therefore Eq. (3.22) can be expressed as

$$\Box^2 \mathcal{A} = -\mu_o \mathcal{J}$$

These, along with additional elements of four-dimensional electrodynamics, are developed fully in Part II.

Alternate (circuit-theory) representation

It is simpler to deal simultaneously with all four of the quantities that make up a representation since together they can be expressed, in space–time coordinates, as a unified energy-momentum tensor. Consequently, we postpone the further derivation of alternative representations and detailed discussions of their properties until Part II; there four-dimensional electrodynamics is developed and the theory generalized to include dielectric and magnetic materials. Nevertheless, *one* free-space representation of particular interest is given below; it is consistent with the results of Section 3.4, valid in an arbitrary gauge, and can be verified by direct substitution.

$$\begin{aligned} S_i^o = {} & \Phi J_i + \frac{1}{2\mu_o}\left(\mathbf{A}\cdot\frac{\partial^2\mathbf{A}}{\partial x_i \partial t} - \frac{\partial \mathbf{A}}{\partial x_i}\cdot\frac{\partial \mathbf{A}}{\partial t}\right) - \frac{1}{2}\varepsilon_o\left(\Phi\frac{\partial^2\Phi}{\partial x_i\partial t} - \frac{\partial\Phi}{\partial x_i}\frac{\partial\Phi}{\partial t}\right) \\ & + \frac{1}{2\mu_o}\left(g\frac{\partial A_i}{\partial t} - \frac{\partial g}{\partial t}A_i\right) \end{aligned} \tag{3.31a}$$

$$\begin{aligned} W^o = {} & \frac{1}{2}(\mathbf{A}\cdot\mathbf{J} + \rho\,\Phi) - \frac{1}{2}\varepsilon_o\left(\mathbf{A}\cdot\frac{\partial^2\mathbf{A}}{\partial t^2} - \frac{\partial\mathbf{A}}{\partial t}\cdot\frac{\partial\mathbf{A}}{\partial t}\right) + \frac{1}{2}\varepsilon_o\frac{1}{c^2}\left[\Phi\frac{\partial^2\Phi}{\partial t^2} - \left(\frac{\partial\Phi}{\partial t}\right)^2\right] \\ & + \frac{1}{2}\varepsilon_o\left(g\frac{\partial\Phi}{\partial t} - \Phi\frac{\partial g}{\partial t}\right) \end{aligned} \tag{3.31b}$$

$$T_{ij}^{\rm o} = -A_i J_j + \frac{1}{2\mu_{\rm o}}\left(\mathbf{A}\cdot\frac{\partial^2\mathbf{A}}{\partial x_i\partial x_j} - \frac{\partial\mathbf{A}}{\partial x_i}\cdot\frac{\partial\mathbf{A}}{\partial x_j}\right) - \frac{1}{2}\varepsilon_{\rm o}\left(\Phi\frac{\partial^2\Phi}{\partial x_i\partial x_j} - \frac{\partial\Phi}{\partial x_i}\frac{\partial\Phi}{\partial x_j}\right)$$
$$+\frac{1}{2}(\mathbf{A}\cdot\mathbf{J} - \rho\Phi)\delta_{ij} + \frac{1}{2\mu_{\rm o}}\left(g\frac{\partial A_j}{\partial x_i} - \frac{\partial g}{\partial x_i}A_j\right) \tag{3.31c}$$

$$G_i^{\rm o} = \rho A_i + \frac{1}{2}\varepsilon_{\rm o}\left(\mathbf{A}\cdot\frac{\partial^2\mathbf{A}}{\partial x_i\partial t} - \frac{\partial\mathbf{A}}{\partial x_i}\cdot\frac{\partial\mathbf{A}}{\partial t}\right) - \frac{1}{2}\varepsilon_{\rm o}\frac{1}{c^2}\left(\Phi\frac{\partial^2\Phi}{\partial x_i\partial t} - \frac{\partial\Phi}{\partial x_i}\frac{\partial\Phi}{\partial t}\right)$$
$$+\frac{1}{2}\varepsilon_{\rm o}\left(\Phi\frac{\partial g}{\partial x_i} - g\frac{\partial\Phi}{\partial x_i}\right) \tag{3.31d}$$

The superscript o is chosen to designate this as the Alternate representation—sometimes referred to as the localized-circuit representation.[3] Notice that when $g = 0$, $\mathbf{S}^{\rm o} = \mathbf{S}^{\rm alt}$, $W^{\rm o} = W^{\rm alt}$, $\overline{\mathbf{T}}^{\rm o} = \overline{\mathbf{T}}^{\rm alt}$, and $\mathbf{G}^{\rm o} = \mathbf{G}^{\rm alt}$. More general formulas that are valid for Cartesian, cylindrical, and spherical coordinate systems are given in Appendix D, Section D.1.

These equations are also valid for materials modeled with Amperian sources except that $\rho \to \rho_{\rm total}$ and $\mathbf{J} \to \mathbf{J}_{\rm total}$. With $\nabla\times\mathbf{A} = \mathbf{B}$ and $-\frac{\partial}{\partial t}\mathbf{A} - \nabla\Phi = \mathbf{E}^{\rm o}$, they provide an *exact* substitute for the Maxwell–Poynting (Amperian) representation.

When $g\frac{\partial\Phi}{\partial t} = \Phi\frac{\partial g}{\partial t}$, it is convenient to divide the Alternate-energy density into electric and magnetic components,

$$W_{\rm e}^{\rm o} = \frac{1}{2}\rho\,\Phi + \frac{1}{2}\varepsilon_{\rm o}\frac{1}{c^2}\left[\Phi\frac{\partial^2\Phi}{\partial t^2} - \left(\frac{\partial\Phi}{\partial t}\right)^2\right] \tag{3.32a}$$

$$W_{\rm m}^{\rm o} = \frac{1}{2}\mathbf{A}\cdot\mathbf{J} - \frac{1}{2}\varepsilon_{\rm o}\left(\mathbf{A}\cdot\frac{\partial^2\mathbf{A}}{\partial t^2} - \frac{\partial\mathbf{A}}{\partial t}\cdot\frac{\partial\mathbf{A}}{\partial t}\right) \tag{3.32b}$$

although such designations can sometimes be misleading when compared to $W_{\rm e}$ and $W_{\rm m}$. Such a division is possible in the Lorenz gauge, $g = 0$, and also in the Coulomb gauge, $g = \frac{1}{c^2}\frac{\partial\Phi}{\partial t}$—except for the latter, $W_{\rm e}^{\rm o} = \frac{1}{2}\rho\,\Phi$, but $W_{\rm m}^{\rm o}$ is unchanged.

When Φ and the Cartesian components of $\mathbf{A}$ are steady-state sinusoidal functions of a single frequency, it follows that $W^{\rm o}$ and $\mathbf{S}^{\rm o}$ are *independent of time* in all regions of free-space. Consequently, Alternate reactive power and energy can exist only where there is electric charge and current. These properties of the Alternate representation (independent of g) are proved in Chapter 6, Section 6.2; they lead to great simplification in many power/energy calculations.

The following relationship is apparent from Eqs. (3.31) when $g = 0$,

$$\mathbf{G}^{\rm o} - \rho\,\mathbf{A} = \frac{1}{c^2}(\mathbf{S}^{\rm o} - \Phi\,\mathbf{J})$$

It follows that, whenever the Lorenz-gauge *Alternate* Poynting vector vanishes in free-space regions, and is therefore coincident with the electric-current density, the momentum density similarly vanishes except where there is electric-charge density; in these cases a true circuit-theory representation emerges. Such localization occurs not only in certain quasistatic problems, but also in many transmission-line configurations. On the other

[3]Not be confused with the superscript used to distinguish the Minkowski fields ($\mathbf{E}^{\rm o}$, $\mathbf{H}^{\rm o}$, $\mathbf{P}^{\rm o}$, $\mathbf{M}^{\rm o}$) from the Chu fields (**E**, **H**, **P**, **M**).

hand, in charge and current free regions, $\mathbf{G}^o = \mu_o\varepsilon_o\mathbf{S}^o$ as expected. In the next subsection, we clarify matters by considering the force on a single moving point charge.

The torque-density theorem, Eq. (3.3c), also has an alternate form,

$$\left[\nabla \cdot (\mathbf{r} \times \overline{\mathbf{T}}^o) - \frac{\partial(\mathbf{r} \times \mathbf{G}^o)}{\partial t} - \mathbf{r} \times \mathbf{f}\right]_k + \epsilon_{kij} T^o_{ij} = 0 \tag{3.33a}$$

Because the nonsymmetric portion of T^o_{ij} is $-A_i J_j + \dfrac{1}{2\mu_0}\left(g\dfrac{\partial A_j}{\partial x_i} - \dfrac{\partial g}{\partial x_i}A_j\right)$, Eq. (3.33a) reduces to

$$\nabla \cdot (\mathbf{r} \times \overline{\mathbf{T}}^o) - \frac{\partial(\mathbf{r} \times \mathbf{G}^o)}{\partial t} + \mathbf{J} \times \mathbf{A} + \frac{1}{2\mu_0}(g\nabla \times \mathbf{A} - \nabla g \times \mathbf{A}) = \mathbf{r} \times \mathbf{f} \tag{3.33b}$$

This simplifies in the Lorenz gauge, $g = 0$. In that case, when integrated over the volume of the current, the term $\mathbf{J} \times \mathbf{A}$ often accounts for nearly all of the total torque. Both Eqs. (3.5) and (3.33b) are discussed at greater length in Appendix D, Section D.3.

Electromagnetic force on a moving charge

The Lorentz force acting upon a moving charge, q (with rest mass, m'), when combined with Newton's Second Law, gives the relativistically correct force law

$$\frac{d}{dt}(\gamma m'\mathbf{v}) = q(\mathbf{E} + \mathbf{v} \times \mu_0\mathbf{H}) = q\left[\left(-\frac{\partial \mathbf{A}}{\partial t} - \nabla\Phi\right) + \mathbf{v} \times (\nabla \times \mathbf{A})\right]$$

Rearrangement produces the equivalent and well-known formula:

$$\frac{d}{dt}(\gamma m'\mathbf{v} + q\mathbf{A}) = -q\left[\nabla(\Phi - \mathbf{A} \cdot \mathbf{v}) + A_j \nabla v_j\right] \tag{3.34}$$

where $\gamma = 1/\sqrt{1 - (\mathbf{v} \cdot \mathbf{v})/c^2}$ is the relativistic factor, $\frac{d\mathbf{A}}{dt}$ is expressed in terms of the substantial derivative, $(\partial/\partial t + \mathbf{v} \cdot \nabla)\mathbf{A}$, and summation over the index $j = 1, 2, 3$ is assumed. Because the value of the vector potential is evaluated at the location of the charge, it is permissible to assign the momentum, $q\mathbf{A}$, to the charge; of course, that value depends upon the global distribution of *all* contributing electromagnetic sources. In this respect, the situation is completely analogous to the electric force, $q\mathbf{E}$.

On a per unit volume basis, the localized momentum $q\mathbf{A}$ can be expressed as the density,

$$\rho A_i = \mu_0\varepsilon_0(\mathbf{E} \times \mathbf{H})_i + \frac{\partial}{\partial x_j}\left[\varepsilon_0(A_i E_j - \frac{1}{2}\mathbf{A} \cdot \mathbf{E}\ \delta_{ij})\right] + \frac{1}{2}\varepsilon_0\left(A_j\frac{\partial E_j}{\partial x_i} - \frac{\partial A_j}{\partial x_i}E_j\right) \tag{3.35}$$

This equation reveals the connection between $\rho\,\mathbf{A}$ and the electromagnetic-field momentum density, $(\mathbf{E} \times \mathbf{H})/c^2$. Equation (3.35) is, in fact, a key to understanding the results that were derived in the previous section.

Alternate accounting of power and momentum

These results both permit and explain alternative models of power and momentum. Consider a point electric charge, q, and a vector line-current element, $\mathbf{I}$, that act as sources of electric and magnetic field. In the normal accounting of electromagnetic power and momentum, all contributions are attributed to the fields which are "outside" of the sources; in the alternate accounting, the contributions, $\Phi\mathbf{I}$ and $q\mathbf{A}$ are respectively assigned to the location of the current and charge. In many quasistatic and some transmission-line problems, the "local" power and momentum are sufficient and little or no "outside" contributions are required. Somewhat similar remarks pertain to the energy. The free-space electromagnetic-momentum density, $(\mathbf{E}\times\mathbf{H})/c^2$, can always be written as $\rho\,\mathbf{A}$, plus a remainder density. A portion of that remainder is expressible as the divergence of a tensor; consequently, the time derivative of the latter can be included with the Maxwell stress tensor. A similar rearrangement of the power flux is possible, in which $\mathbf{E}\times\mathbf{H}$ is written as $\Phi\,\mathbf{J}$ plus a remainder. A portion of the divergence of that remainder is expressible as the time derivative of a scalar; consequently, it can added to the usual electromagnetic energy density. For appropriate choices, the resultant energy density includes the term $\frac{1}{2}(\mathbf{A}\cdot\mathbf{J}+\rho\,\Phi)$ while the resultant stress tensor includes $\frac{1}{2}(\mathbf{A}\cdot\mathbf{J}-\rho\,\Phi)\overline{\mathbf{I}}-\mathbf{A}\;\mathbf{J}$. Such rearrangements produce an Alternate representation comprised of $\overline{\mathbf{T}}^{\mathrm{o}}$, $\mathbf{G}^{\mathrm{o}}$, $\mathbf{S}^{\mathrm{o}}$, and W^{o} that leaves the electromagnetic force and power densities unchanged.

Because the Alternate set has terms that are localized to the electric charges and currents, it lends itself to "circuit-theory" representations especially in the quasistatic field regime. Notice that the accounting scheme just discussed has been applied to *all* of the charge and current density but, of course, it can be applied to only some desired fraction of either. In that way, the representation can be shifted *continuously* between the two extremes. But even more flexibility is possible in that intentional overstatement of "local" contributions can be compensated for by reducing (possibly making negative) the "outer" ones. The set $\overline{\mathbf{T}}^{\mathrm{o}}$, $\mathbf{G}^{\mathrm{o}}$, $\mathbf{S}^{\mathrm{o}}$, and W^{o}, defined by Eqs. (3.31), is referred to in this text as the alternate representation.

Electromagnetic beauty

The differences between the corresponding Alternate and Maxwell–Poynting terms evidently satisfy

$$\nabla\cdot(\mathbf{S}^{\mathrm{o}}-\mathbf{S}^{\mathrm{em}})+\frac{\partial}{\partial t}(W^{\mathrm{o}}-W^{\mathrm{em}})=0 \tag{3.36a}$$

$$\nabla\cdot(\overline{\mathbf{T}}^{\mathrm{o}}-\overline{\mathbf{T}}^{\mathrm{em}})-\frac{\partial}{\partial t}(\mathbf{G}^{\mathrm{o}}-\mathbf{G}^{\mathrm{em}})=0 \tag{3.36b}$$

This set of equations produces neither electromagnetic force nor $\mathbf{E}\cdot\mathbf{J}$ power density. In a bit of whimsy, this author has defined the set the "electromagnetic beauty." Although the concept is ephemeral, the components, such as the "beauty power flux," $\mathbf{S}^{\mathrm{b}}=\mathbf{S}^{\mathrm{o}}-\mathbf{S}^{\mathrm{em}}$, can be calculated and presented graphically.

In any gauge, the power and energy terms of Eq. (3.36a) can be combined by means of Eq. (3.15a) to produce

$$\nabla\cdot\mathbf{S}^{\mathrm{beauty}}=0 \tag{3.37}$$

where

$$\mathbf{S}^{\mathrm{beauty}}=\frac{1}{2\mu_{\mathrm{o}}}\nabla\times\left(\mathbf{A}\times\frac{\partial\mathbf{A}}{\partial t}\right) \tag{3.38}$$

This more restrictive form, related to the spin angular-momentum density, is *not* identical to $\mathbf{S}^{\mathrm{b}}$.

These topics are revisited in Part II, Chapter 8, Section 8.5, where the derivations are greatly facilitated by the use of four-dimensional analysis.

Alternate power and energy in the quasistatic limit

As described in Chapter 2, Section 2.5, for many quasistatic fields the time variation is slow enough that the fields have spatial distributions that appear to be static, but with amplitudes that vary with time. For regions of free-space that may include electric charge and current, one can assume that in the *EQS* limit we have

$$\mathbf{J}_{(0)} = 0$$

$$\Phi = \Phi_{(0)}$$

$$\mathbf{A} = \mathbf{A}_{(1)}$$

$$-\mathbf{E}\cdot\mathbf{J} = \nabla\Phi_{(0)}\cdot\mathbf{J}_{(1)}$$

Therefore the Alternate power and Alternate energy are

$$\begin{aligned} S_i^{\mathrm{o}} = \Phi_{(0)} J_{(1)i} &+ \frac{1}{2\mu_{\mathrm{o}}}\left(\mathbf{A}_{(1)}\cdot\frac{\partial^2\mathbf{A}_{(1)}}{\partial x_i\partial t} - \frac{\partial\mathbf{A}_{(1)}}{\partial x_i}\cdot\frac{\partial\mathbf{A}_{(1)}}{\partial t}\right) \\ &- \frac{1}{2}\varepsilon_{\mathrm{o}}\left(\Phi_{(0)}\frac{\partial^2\Phi_{(0)}}{\partial x_i\partial t} - \frac{\partial\Phi_{(0)}}{\partial x_i}\frac{\partial\Phi_{(0)}}{\partial t}\right) \end{aligned} \tag{3.39a}$$

$$\begin{aligned} W^{\mathrm{o}} = \frac{1}{2}(\mathbf{A}_{(1)}\cdot\mathbf{J}_{(1)} + \rho_{(0)}\Phi_{(0)}) &- \frac{1}{2}\varepsilon_{\mathrm{o}}\left(\mathbf{A}_{(1)}\cdot\frac{\partial^2\mathbf{A}_{(1)}}{\partial t^2} - \frac{\partial\mathbf{A}_{(1)}}{\partial t}\cdot\frac{\partial\mathbf{A}_{(1)}}{\partial t}\right) \\ &+ \frac{1}{2}\varepsilon_{\mathrm{o}}\frac{1}{c^2}\left[\Phi_{(0)}\frac{\partial^2\Phi_{(0)}}{\partial t^2} - \left(\frac{\partial\Phi_{(0)}}{\partial t}\right)^2\right] \end{aligned} \tag{3.39b}$$

Because the time derivative of a field of order n is a quantity of order $(n+1)$, it follows that the terms that depend upon $\mathbf{A}_{(1)}$ produce third-order terms in $\mathbf{S}^{\mathrm{o}}$ and second and fourth-order terms in W^{o}, whereas those that depend upon $\Phi_{(0)}$ produce first-order terms in $\mathbf{S}^{\mathrm{o}}$ and both zero-order and second-order terms in W^{o}. If only the leading terms in both the localized and nonlocalized components are retained, all the terms involving $\mathbf{A}_{(1)}$ can be neglected and Eq. (3.13) is validated.

In the *MQS* limit, we have

$$\mathbf{A} = \mathbf{A}_{(0)}$$

$$\Phi = \Phi_{(1)}$$

$$-\mathbf{E}\cdot\mathbf{J} = \left[\frac{\partial\mathbf{A}_{(0)}}{\partial t} + \nabla\Phi_{(1)}\right]\cdot\mathbf{J}_{(0)}$$

and

$$S_i^o = \Phi_{(1)} J_{(0)i} + \frac{1}{2\mu_o}\left(\mathbf{A}_{(0)} \cdot \frac{\partial^2 \mathbf{A}_{(0)}}{\partial x_i \partial t} - \frac{\partial \mathbf{A}_{(0)}}{\partial x_i} \cdot \frac{\partial \mathbf{A}_{(0)}}{\partial t}\right)$$
$$- \frac{1}{2}\varepsilon_o \left(\Phi_{(1)} \frac{\partial^2 \Phi_{(1)}}{\partial x_i \partial t} - \frac{\partial \Phi_{(1)}}{\partial x_i}\frac{\partial \Phi_{(1)}}{\partial t}\right) \tag{3.40a}$$

$$W^o = \frac{1}{2}(\mathbf{A}_{(0)} \cdot \mathbf{J}_{(0)} + \rho_{(1)}\Phi_{(1)}) - \frac{1}{2}\varepsilon_o \left(\mathbf{A}_{(0)} \cdot \frac{\partial^2 \mathbf{A}_{(0)}}{\partial t^2} - \frac{\partial \mathbf{A}_{(0)}}{\partial t} \cdot \frac{\partial \mathbf{A}_{(0)}}{\partial t}\right)$$
$$+ \frac{1}{2}\varepsilon_o \frac{1}{c^2}\left[\Phi_{(1)} \frac{\partial^2 \Phi_{(1)}}{\partial t^2} - \left(\frac{\partial \Phi_{(1)}}{\partial t}\right)^2\right] \tag{3.40b}$$

If only the leading terms in the localized and nonlocalized components are retained, all the terms involving $\Phi_{(1)}$ can be neglected (except $\Phi_{(1)} J_{(0)i}$), in which case Eq.(3.14) is validated.

It is therefore reassuring to discover that in *both* limits the quasistatic theorems are recovered from

$$\nabla \cdot \mathbf{S}^o + \frac{\partial}{\partial t} W^o = -\mathbf{E} \cdot \mathbf{J}$$

However, even more startling results follow when we recall that when the quasistatic potentials are separable in space and time, we can substitute

$$\Phi_{(0)} = C_\phi(t)\phi(\mathbf{r}), \qquad \mathbf{A}_{(1)} = C_a(t)\mathbf{a}(\mathbf{r})$$

$$\mathbf{A}_{(0)} = C_a'(t)\mathbf{a}'(\mathbf{r}), \qquad \Phi_{(1)} = C_\phi'(t)\phi'(\mathbf{r})$$

The resulting cancellations simplify (in the case of *EQS*) to

$$\mathbf{S}^o = \Phi_{(0)} \mathbf{J}_{(1)} \tag{3.41a}$$

$$W^o = \frac{1}{2}\rho_{(0)}\Phi_{(0)} \tag{3.41b}$$

and (in the case of *MQS*) to

$$\mathbf{S}^o = \Phi_{(1)} \mathbf{J}_{(0)} \tag{3.42a}$$

$$W^o = \frac{1}{2}\mathbf{A}_{(0)} \cdot \mathbf{J}_{(0)} \tag{3.42b}$$

In the case of separable quasistatic fields, *only* the localized terms survive. Because the expansion of any power theorem to first-order includes both zero- and first–order fluxes but only zero-order energy densities, the real surprise is the vanishing of the first-order non-localized Alternate flux. The fact that the second-order nonlocalized Alternate-energy density also disappears is further validation of the usefulness of *EQS* and *MQS* approximations.

3.5 DIFFERENCES BETWEEN REPRESENTATIONS

Part III of this text is devoted to the application of Maxwell's Equations to many different types of electromagnetic problem. Once the complete solution has been found, it is possible to make detailed comparisons of power, energy, stress, and momentum as calculated using both the Maxwell–Poynting and Alternate representation. In this section we preview the differences for two examples: a uniform plane wave and a Hertzian electric dipole driven in the sinusoidal steady state.

Uniform plane wave

Consider a linearly polarized uniform plane wave of frequency ω propagating in free-space in the $+z$ direction. The electric and magnetic fields are assumed to be

$$\mathbf{E} = \widehat{\mathbf{x}} E_o \cos(\omega t - kz) \tag{3.43a}$$

$$\mathbf{H} = \widehat{\mathbf{y}} \sqrt{\frac{\varepsilon_o}{\mu_o}} E_o \cos(\omega t - kz) \tag{3.43b}$$

where $k = \omega\sqrt{\mu_o \varepsilon_o}$. In the Maxwell–Poynting representation,

$$\begin{aligned} \mathbf{S}^{\mathrm{em}} &= \mathbf{E} \times \mathbf{H} = c^2 \mathbf{G}^{\mathrm{em}} \\ &= \widehat{\mathbf{z}} \frac{1}{2} \sqrt{\frac{\varepsilon_o}{\mu_o}} E_o^2 [1 + \cos 2(\omega t - kz)] \end{aligned} \tag{3.44a}$$

$$\begin{aligned} W^{\mathrm{em}} &= \frac{1}{2} \varepsilon_o \mathbf{E} \cdot \mathbf{E} + \frac{1}{2} \mu_o \mathbf{H} \cdot \mathbf{H} \\ &= \frac{1}{2} \varepsilon_o E_o^2 [1 + \cos 2(\omega t - kz)] \end{aligned} \tag{3.44b}$$

$$\overline{\mathbf{T}}^{\mathrm{em}} = \varepsilon_o \mathbf{E}\,\mathbf{E} + \mu_o \mathbf{H}\,\mathbf{H} - W^{\mathrm{em}} \overline{\mathbf{I}} = -\widehat{\mathbf{z}}\widehat{\mathbf{z}}\, W^{\mathrm{em}} \tag{3.44c}$$

Both the power flux and energy density have average and second-harmonic components.[4]

The vector and scalar potentials that generate these fields are not unique because the source currents and/or charges have not been identified. One very simple set of potentials is

$$\mathbf{A} = \widehat{\mathbf{x}} \frac{-E_o}{\omega} \sin(\omega t - kz) \tag{3.45a}$$

$$\Phi = 0 \tag{3.45b}$$

which, when used to calculate the Alternate-representation power flux and energy density, yields

$$\begin{aligned} \mathbf{S}^{\mathrm{o}} &= \widehat{\mathbf{z}} \frac{1}{2\mu_o} \left(\mathbf{A} \cdot \frac{\partial^2 \mathbf{A}}{\partial z\, \partial t} - \frac{\partial \mathbf{A}}{\partial z} \cdot \frac{\partial \mathbf{A}}{\partial t} \right) \\ &= c^2 \mathbf{G}^{\mathrm{o}} = \widehat{\mathbf{z}} \frac{1}{2} \sqrt{\frac{\varepsilon_o}{\mu_o}} E_o^2 \end{aligned} \tag{3.46a}$$

[4]An important exception is the case of circularly polarized plane waves, for which the second-harmonic energies and power of the two linearly polarized components exactly cancel. Such waves carry angular-momentum as well as linear momentum and power.

$$W^{\text{o}} = \frac{-1}{2}\varepsilon_{\text{o}}\left(\mathbf{A}\cdot\frac{\partial^2\mathbf{A}}{\partial t^2} - \frac{\partial\mathbf{A}}{\partial t}\cdot\frac{\partial\mathbf{A}}{\partial t}\right)$$

$$= \frac{1}{2}\varepsilon_{\text{o}}E_{\text{o}}^2 \tag{3.46b}$$

$$\overline{\overline{\mathbf{T}}}^{\text{o}} = \widehat{\mathbf{z}}\widehat{\mathbf{z}}\frac{1}{2\mu_{\text{o}}}\left(\mathbf{A}\cdot\frac{\partial^2\mathbf{A}}{\partial z^2} - \frac{\partial\mathbf{A}}{\partial z}\cdot\frac{\partial\mathbf{A}}{\partial z}\right)$$

$$= -\widehat{\mathbf{z}}\widehat{\mathbf{z}}\frac{1}{2}\varepsilon_{\text{o}}E_{\text{o}}^2 \tag{3.46c}$$

The average terms are identical in both representations, and there are no second-harmonic terms in the Alternate representation (regardless of the state of polarization).

According to quantum theory, the wave is related to photons, each with energy $\hbar\omega$, linear momentum $\hbar k$, and spin angular-momentum $\pm\hbar$ ($2\pi\hbar = h = 6.63\times 10^{-34} Js$, Planck's constant). In the case of linear polarization, equal densities $n_\pm$ of spin ± 1 photons cancel one another and produce no *net* angular momentum. This is not the case for waves that are circularly polarized.[5] In terms of the number density $n = n_+ + n_-$, we have

$$\langle W^{\text{em}}\rangle = W^{\text{o}} = n\hbar\omega \tag{3.47a}$$

$$\langle \mathbf{G}^{\text{em}}\rangle = \mathbf{G}^{\text{o}} = n\hbar k\widehat{\mathbf{z}} \tag{3.47b}$$

$$\langle \mathbf{S}^{\text{em}}\rangle = \mathbf{S}^{\text{o}} = n\hbar\omega c\widehat{\mathbf{z}} \tag{3.47c}$$

$$\langle \overline{\overline{\mathbf{T}}}^{\text{em}}\rangle = \overline{\overline{\mathbf{T}}}^{\text{o}} = -n\hbar\omega\widehat{\mathbf{z}}\widehat{\mathbf{z}} \tag{3.47d}$$

The second-harmonic components must be discarded in the Maxwell–Poynting representation—something that is unnecessary in the Alternate representation.

But this comparison is not the whole story. Another set of potentials that generates the same **E** and **H** (and satisfies the Lorenz gauge) is

$$\mathbf{A} = \widehat{\mathbf{z}}\frac{-E_{\text{o}}}{c}x\cos(\omega t - kz) \tag{3.48a}$$

$$\Phi = -E_{\text{o}}x\cos(\omega t - kz) \tag{3.48b}$$

which, when used to calculate the Alternate representation,

$$\mathbf{S}^{\text{o}} = c^2\mathbf{G}^{\text{o}} = \frac{1}{2\mu_{\text{o}}}\left(A_z\nabla\frac{\partial A_z}{\partial t} - \frac{\partial A_z}{\partial t}\nabla A_z\right) - \frac{1}{2}\varepsilon_{\text{o}}\left(\Phi\nabla\frac{\partial\Phi}{\partial t} - \frac{\partial\Phi}{\partial t}\nabla\Phi\right) \tag{3.49a}$$

[5]The concepts of spin, orbital, and charge angular momentum are contained in the classical macroscopic Maxwell Equations (although, of course, not their quantization). A general treatment of the topic is contained in Appendix D, Section D.3 in both the Maxwell–Poynting and Alternate representations. Derivations of the particle aspects of circular-waveguide modes are carried out in Chapter 20, Section 20.5 using theAlternate representation to calculate $n_\pm$.

$$W^{o} = \frac{-1}{2}\varepsilon_{o}\left[A_{z}\frac{\partial^{2}A_{z}}{\partial t^{2}} - \left(\frac{\partial A_{z}}{\partial t}\right)^{2}\right] + \frac{1}{2}\frac{\varepsilon_{o}}{c^{2}}\left[\Phi\frac{\partial^{2}\Phi}{\partial t^{2}} - \left(\frac{\partial\Phi}{\partial t}\right)^{2}\right] \tag{3.49b}$$

$$T_{ij}^{o} = \frac{1}{2\mu_{o}}\left[A_{z}\frac{\partial^{2}A_{z}}{\partial x_{i}\partial x_{j}} - \frac{\partial A_{z}}{\partial x_{i}}\frac{\partial A_{z}}{\partial x_{j}}\right] - \frac{1}{2}\varepsilon_{o}\left[\Phi\frac{\partial^{2}\Phi}{\partial x_{i}\partial x_{j}} - \frac{\partial\Phi}{\partial x_{i}}\frac{\partial\Phi}{\partial x_{j}}\right] \tag{3.49c}$$

yields $\mathbf{S}^{o} = c^{2}\mathbf{G}^{o} = 0$, $W^{o} = 0$, and $\overline{\overline{\mathbf{T}}}^{o} = 0$ in the *entire region* occupied by **E** and **H**. Because the potentials become unbounded as $|x| \to \infty$, it seems likely that the missing power, momentum, energy, and stress is located on the z-directed currents that are *required* in order to generate A_z. From Eq. (3.31b), with $g = 0$ and using complex fields, the time-averaged total Alternate-energy density[6] is found to be

$$<W_{\text{total}}^{o}> = \tfrac{1}{8}(\underline{\rho}_{u}\underline{\Phi}^{*} + \underline{\rho}_{u}^{*}\underline{\Phi} + \underline{\mathbf{A}}^{*}\cdot\underline{\mathbf{J}}_{u} + \underline{\mathbf{A}}\cdot\underline{\mathbf{J}}_{u}^{*}) + \tfrac{1}{2}\omega^{2}\varepsilon_{o}(\underline{\mathbf{A}}\cdot\underline{\mathbf{A}}^{*} - \mu_{o}\varepsilon_{o}\underline{\Phi\Phi}^{*})$$

and is helpful in clarifying the matter. For the first set of potentials, we have

$$<W_{\text{total}}^{o}> = \tfrac{1}{8}(\underline{A}_{x}^{*}\underline{J}_{ux} + \underline{A}_{x}\underline{J}_{ux}^{*}) + \tfrac{1}{2}\omega^{2}\varepsilon_{o}\underline{A}_{x}\underline{A}_{x}^{*}$$

Although in order to generate $\underline{A}_x$, currents $\underline{J}_{ux}$ must exist *somewhere* (presumably at $z \to -\infty$), the nonlocalized energy density, $\frac{1}{2}\omega^{2}\varepsilon_{o}\underline{\mathbf{A}}\cdot\underline{\mathbf{A}}^{*}$, is in agreement with Eq. (3.47a) and so no additional currents are required.

For the second set of potentials, we have $\underline{\mathbf{A}}\cdot\underline{\mathbf{A}}^{*} - \mu_{o}\varepsilon_{o}\underline{\Phi\Phi}^{*} = 0$ and

$$<W_{\text{total}}^{o}> = \tfrac{1}{8}(\underline{\rho}_{u}\underline{\Phi}^{*} + \underline{\rho}_{u}^{*}\underline{\Phi} + \underline{A}_{z}^{*}\underline{J}_{uz} + \underline{A}_{z}\underline{J}_{uz}^{*})$$

Now, in order to generate $\underline{\Phi}$ and $\underline{A}_z$, both charges and $\underline{J}_{uz}$ are required for *all* z (presumably at $x \to \pm\infty$). But if there is no Alternate power or Alternate energy in free-space, how does power flow to a small conducting probe that surely must dissipate power if it is placed in the path of the propagating wave? The short answer is that currents and charges induced in and on the probe generate their own potentials and so unbalance $\underline{\mathbf{A}}\cdot\underline{\mathbf{A}}^{*} - \mu_{o}\varepsilon_{o}\underline{\Phi\Phi}^{*}$ in and near the probe. *Where required,* Alternate energy (and power) appears. The dissipation density in the probe, $\frac{1}{2}\sigma\underline{\mathbf{E}}\cdot\underline{\mathbf{E}}^{*}$, must be the same regardless of whether the Maxwell–Poynting or Alternate representation is used to calculate it; nor can it matter if the currents used to generate the field are longitudinal or transverse. This matter is revisited in Part III, Chapter 11.

Hertzian electric dipole (steady-state)

The Hertzian electric dipole was considered in Chapter 1, Section 1.6, Here we reconsider the case of a uniform current element, $I(t)d$, that is oriented parallel to the z axis when $I(t) = I_{o}\sin(\omega t)$.[7] With $k = \omega/c$, $\widehat{\mathbf{z}}$ the unit vector along z, r the radius in spherical coordinates, and for $kd \ll 1$, the steady-state vector and scalar retarded potentials are

[6]The Alternate-energy theorem is formally derived in Chapter 6, Section 6.2.

[7]If evaluation of the sinusoidal steady-state electric and magnetic fields were our final goal, it would be advisable to calculate complex scalar and vector quantities and then take their real parts (after multiplication by $\exp(j\omega t)$). Because we wish to compare results in space and time, we choose to work directly in the time domain.

from Eqs. (1.35a) and (1.35b),

$$\mathbf{A} = \hat{\mathbf{z}}\frac{\mu_0 I_0 d \ \sin(\omega t - kr)}{4\pi r} \tag{3.50a}$$

$$\Phi = \frac{I_0 d}{4\pi\varepsilon_0\omega}\left[\frac{kr\sin(\omega t - kr) - \cos(\omega t - kr)}{r^2}\right]\cos\theta \tag{3.50b}$$

The electric and magnetic fields follow from Eqs. (1.9) and (1.10):

$$\mathbf{E} = \frac{c\mu_0 I_0 d}{4\pi kr^3}\left[\begin{array}{c}\hat{\mathbf{r}}2[kr\sin(\omega t - kr) - \cos(\omega t - kr)]\cos\theta \\ +\hat{\theta}[kr\sin(\omega t - kr) - (1 - k^2r^2)\cos(\omega t - kr)]\sin\theta\end{array}\right] \tag{3.51a}$$

$$\mathbf{H} = \hat{\phi}\frac{I_0 d}{4\pi r^2}\left[kr\cos(\omega t - kr) + \sin(\omega t - kr)\right]\sin\theta \tag{3.51b}$$

In the Maxwell–Poynting representation, we have

$$\mathbf{E}\times\mathbf{H} = \frac{c\mu_0 I_0^2 d^2}{32\pi^2 kr^5}\left(\begin{array}{c}\hat{\mathbf{r}}\left[\begin{array}{c}k^3r^3 + (k^3r^3 - 2kr)\cos 2(\omega t - kr) \\ +(2k^2r^2 - 1)\sin 2(\omega t - kr)\end{array}\right]\sin^2\theta \\ +\hat{\theta}\left[\begin{array}{c}(1 - k^2r^2)\sin 2(\omega t - kr) \\ +2kr\ \cos 2(\omega t - kr)\end{array}\right]\sin 2\theta\end{array}\right) \tag{3.52}$$

$$W_e = \frac{\mu_0 I_0^2 d^2}{64\pi^2 k^2 r^6}\left(\begin{array}{c}4\left[\begin{array}{c}k^2r^2(1 - \cos 2(\omega t - kr)) \\ -(1 + 2kr)\sin 2(\omega t - kr)\end{array}\right]\cos^2\theta \\ +\left[\begin{array}{c}(1 - k^2r^2 + k^4r^4) \\ +2kr(k^2r^2 - 1)\sin 2(\omega t - kr) \\ +(k^4r^4 - 3k^2r^2 + 1\ )\cos 2(\omega t - kr)\end{array}\right]\sin^2\theta\end{array}\right) \tag{3.53a}$$

$$W_m = \frac{\mu_0 I_0^2 d^2}{64\pi^2 r^4}\left[\begin{array}{c}(1 + k^2r^2) + 2kr\sin 2(\omega t - kr) \\ +(k^2r^2 - 1)\cos 2(\omega t - kr)\end{array}\right]\sin^2\theta \tag{3.53b}$$

In addition to the time-averaged components, second-harmonic energy and power (directed along $\hat{\theta}$ as well as $\hat{\mathbf{r}}$) are both in evidence. Except in the far field, these are mainly associated with reactive energy and power.

We turn now to the Alternate formulation. In regions where $r \gg d$, there is neither current nor charge and from Eqs. (3.31a) and (3.31b) we have

$$\mathbf{S}^o = \frac{1}{2\mu_0}\left(A_z\frac{\partial\nabla A_z}{\partial t} - \nabla A_z\frac{\partial A_z}{\partial t}\right) - \frac{1}{2}\varepsilon_0\left(\Phi\frac{\partial\nabla\Phi}{\partial t} - \nabla\Phi\frac{\partial\Phi}{\partial t}\right) \tag{3.54a}$$

$$W^o = \frac{1}{2}\varepsilon_0\left[\frac{\partial A_z}{\partial t}\frac{\partial A_z}{\partial t} - A_z\frac{\partial^2 A_z}{\partial t^2} + \frac{1}{c^2}\left(\Phi\frac{\partial^2\Phi}{\partial t^2} - \frac{\partial\Phi}{\partial t}\frac{\partial\Phi}{\partial t}\right)\right] \tag{3.54b}$$

Direct substitution reveals that

$$\mathbf{S}^{\mathrm{o}} = \widehat{\mathbf{r}}\,\frac{1}{2}\mu_{\mathrm{o}} c\; I_{\mathrm{o}}^2\; k^2 d^2 \frac{1-\cos^2\theta}{(4\pi r)^2} \tag{3.55a}$$

$$W^{\mathrm{o}} = \frac{1}{2}\mu_{\mathrm{o}} \frac{I_{\mathrm{o}}^2 d^2}{(4\pi r)^2}\left(k^2 \sin^2\theta - \frac{\cos^2\theta}{r^2}\right) \tag{3.55b}$$

In this representation there is only a radial component which is, as in the plane wave example, independent of time. The constant value of S_r^{o} is equal to the average value of the radial component of $\mathbf{E}\times\mathbf{H}$. The isotropic term in Eq. (3.55a) is due to **A**, that proportional to $\cos^2\theta$ to Φ. The free-space Alternate energy-density is also time-independent. Because second-harmonic terms are absent everywhere in free-space, there can be neither reactive Alternate power nor energy.

As expected, when $kr \gg 1$, the familiar $r^{-2}\sin^2\theta$ radiation term dominates and it is no surprise that it agrees perfectly with the time average of the sum of W_{e} and W_{m} (with each contributing equally). In this limit, $\mathbf{S}^{\mathrm{o}}/W^{\mathrm{o}} \to \widehat{\mathbf{r}}\,\mathrm{c}$.

When $kr \ll 1$, the *negative* contribution[8] must be considered together with $\frac{1}{2}(A_z J_z + \rho\Phi)$. In particular, the capacitive energy, which completely dominates at low frequencies, is localized *at* the point charges, $Q = \pm(\mathrm{I_0}/\omega)\sin\omega t$, that together form the electric dipole. The dipole is where reactive Alternate power and energy is located.

The contrast in simplicity between this and the conventional derivation is striking. This is also true for the case of the Hertzian magnetic dipole considered in Chapter 13, Section 13.2. Fortunately, the advantages are applicable to general antenna theory as we shall learn in Part III, Chapter 22.

[8]Negative energies are permissible because the zero references for both the scalar and vector potentials are not arbitrary but implicitly located at $r \to \infty$.

CHAPTER 4

ELECTROMAGNETIC WAVES IN FREE-SPACE

4.1 HOMOGENEOUS WAVES

In regions where there are no electric currents nor charges, Eqs. (1.13a), (1.13b), (1.16a), and (1.16b) all reduce to homogeneous wave equations of the form:

$$\nabla^2\psi - \frac{1}{c^2}\frac{\partial^2\psi}{\partial t^2} = 0 \tag{4.1}$$

where c is the velocity of light and the scalar ψ is Φ or any of the Cartesian components of **E, H**, and **A**. The set of scalars are not independent because **E** and **H** are related by Maxwell's Equations, and **A** and Φ by both the gauge condition and the requirement that $\mu_0\mathbf{H} = \nabla \times \mathbf{A}$. Although in free-space regions, all solutions of Maxwell's Equations satisfy the wave equation, the reverse is *not* true.

The Power and Beauty of Electromagnetic Fields, First Edition. F. R. Morgenthaler.

Expressed, respectively, in terms of Cartesian coordinates (x, y, z), cylindrical coordinates (r, ϕ, z), and spherical coordinates (r, θ, ϕ)[1], the wave equation is

$$\frac{\partial^2 \psi}{\partial x^2} + \frac{\partial^2 \psi}{\partial y^2} + \frac{\partial^2 \psi}{\partial z^2} - \frac{1}{c^2}\frac{\partial^2 \psi}{\partial t^2} = 0 \tag{4.2a}$$

$$\frac{1}{r}\frac{\partial}{\partial r}\left(r\frac{\partial \psi}{\partial r}\right) + \frac{1}{r^2}\frac{\partial^2 \psi}{\partial \phi^2} - \frac{1}{c^2}\frac{\partial^2 \psi}{\partial t^2} = 0 \tag{4.2b}$$

$$\frac{1}{r^2}\frac{\partial}{\partial r}\left(r^2\frac{\partial \psi}{\partial r}\right) + \frac{1}{r^2 \sin\theta}\frac{\partial}{\partial \theta}\frac{\partial(\sin\theta\,\psi)}{\partial \theta} + \frac{1}{r^2 \sin^2\theta}\frac{\partial^2 \psi}{\partial \phi^2} - \frac{1}{c^2}\frac{\partial^2 \psi}{\partial t^2} = 0 \tag{4.2c}$$

Solutions that exhibit wave-like behavior require that ψ be a function of time and at least one spatial coordinate. The minimum condition leads to one-dimensional waves, which we consider first.

4.2 ONE-DIMENSIONAL WAVES

Solutions with only an x dependence, $\psi(x, t)$, satisfy

$$\frac{\partial^2 \psi}{\partial x^2} - \frac{1}{c^2}\frac{\partial^2 \psi}{\partial t^2} = 0 \tag{4.3}$$

which (because $\frac{\partial^2 \psi}{\partial x \partial t} = \frac{\partial^2 \psi}{\partial t \partial x}$) can be written in operator form as either

$$\left(\frac{\partial}{\partial x} - \frac{1}{c}\frac{\partial}{\partial t}\right)\left(\frac{\partial}{\partial x} + \frac{1}{c}\frac{\partial}{\partial t}\right)\psi = 0$$

or

$$\left(\frac{\partial}{\partial x} + \frac{1}{c}\frac{\partial}{\partial t}\right)\left(\frac{\partial}{\partial x} - \frac{1}{c}\frac{\partial}{\partial t}\right)\psi = 0$$

Evidently, the two first-order differential equations

$$\left(\frac{\partial}{\partial x} \pm \frac{1}{c}\frac{\partial}{\partial t}\right)\psi = 0$$

both satisfy Eq. (4.3). The two independent solutions can be obtained by inspection and superimposed. The result is

$$\psi(x, t) = f_+(x - ct) + f_-(x + ct) \tag{4.4}$$

where the wave $f_+(x - ct)$ is traveling in the $+x$ direction and $f_-(x + ct)$ in the $-x$ direction. Both waves travel at the same speed, c; the shape of each is arbitrary, but unchanged as it moves. Because the x axis can be oriented at will, these solutions apply when the propagation is along an arbitrary fixed direction. The superposition of such one-space-dimensional solutions satisfies the general wave equation.

[1]There is a problem with the common practice (adhered to in this text) of using the symbol r for radial distance in both cylindrical and spherical coordinates. Some authors use r in cylindrical and ρ in spherical coordinates, but that invites confusion with the symbol for electric-charge density. Usually the geometry and/or context under discussion will make clear the meaning; in rare cases when it does not, the symbol r_c will be used for the cylindrical radius.

Solutions of the two-dimensional Laplace's Equation

It is instructive to make the substitution $ct = -iy$ into Eq. (4.4). The result

$$\psi(x,y) = f_+(x+iy) + f_-(x-iy)$$

satisfies

$$\frac{\partial^2\psi}{\partial x^2} + \frac{\partial^2\psi}{\partial y^2} = 0 \tag{4.5}$$

and is therefore the *general solution* of Laplace's Equation in two dimensions. *Any* analytic function of the complex variable, $Z = x + iy$ (or its complex conjugate), is a proper solution; consequently, the real and imaginary parts of such functions are themselves solutions.[2] This result provides the reason why the branch of mathematics dealing with functions of a complex variable (including conformal mapping) is of great theoretical importance in electromagnetic theory.[3] The quantity, ict, introduced here only as a useful artifice, is the fourth space–time coordinate that is central to Special Relativity (Appendix A, Section A.1) and the four-dimensional formulation of electrodynamics developed in Part II.

4.3 HARMONIC UNIFORM PLANE WAVES

If $f_\pm(u)$ are sinusoidal functions of the form $\cos(ku)$, where k is a constant, the result is traveling waves, $\cos[k(x \mp ct)] = \cos(\omega t \mp kx)$ that are periodic in both t and x. Here $\omega = 2\pi f$ is the radian frequency and $k = 2\pi/\lambda_o$ (λ_o is the free-space wavelength.) The dispersion relation can be expressed as either

$$\omega = ck \quad \text{or} \quad f\lambda_o = c$$

The more general solution is of the form

$$\psi(x,t) = C_1\cos(\omega t - kx + \theta_1) + C_2\cos(\omega t + kx - \theta_2) \tag{4.6}$$

where $C_{1,2}$ and $\theta_{1,2}$ are constants. This sinusoidal steady-state solution can also be expressed as

$$\psi(x,t) = Re\{[\underline{C}_1\exp(-jkx) + \underline{C}_2\exp(jkx)]\exp(j\omega t)\} \tag{4.7}$$

Here $j = \sqrt{-1}$, $\underline{C}_{1,2} = C_{1,2}\exp(-j\theta_{1,2})$ are complex constants,[4] and $Re\{\ \}$ implies the real part of the complex quantity that is enclosed within the curly brackets.

[2]Analytic functions $f(Z) = u(x,y) + iv(x,y)$ must satisfy the Cauchy–Riemann relations, $\frac{\partial u}{\partial x} = \frac{\partial v}{\partial y}$ and $\frac{\partial u}{\partial y} = -\frac{\partial v}{\partial x}$ and so both $u(x,y)$ and its conjugate function, $v(x,y)$ automatically satisfy the two-dimensional Laplace's Equation.

[3]For example, Z^n, $\ln(Z)$, and $\exp(KZ)$ (n, a positive or negative integer; K, a complex constant) all provide useful solutions expressed in either Cartesian or polar coordinates when Z is expressed as $x + iy$ or $r\exp(i\theta)$. Because of linearity, these solutions can be superimposed.

[4]We adhere to the normal engineering practice of using j to represent $\sqrt{-1}$ in all expressions arising from complex frequencies and reserve i for the imaginary coordinate, ict.

The generalization of Eq. (4.6) to a sum of traveling waves of the same frequency, each propagating at an arbitrary angle to the x axis, is

$$\psi(\mathbf{r},t) = Re\{\underline{\Psi}(\mathbf{r})\exp(j\omega t)\} \tag{4.8a}$$

$$\underline{\Psi}(\mathbf{r}) = \sum_k \underline{C}_k \exp(-j\mathbf{k}\cdot\mathbf{r}) \tag{4.8b}$$

$$\mathbf{k} = \widehat{\mathbf{x}}k_x + \widehat{\mathbf{y}}k_y + \widehat{\mathbf{z}}k_z \tag{4.8c}$$

where $\omega^2 = c^2\mathbf{k}\cdot\mathbf{k}$. Waves traveling in opposite directions are included by simply reversing the sign of the wave vector, $\mathbf{k}$. Substitution of Eq. (4.8a) into the three-dimensional wave equation results in the time-independent Helmholtz Equation,

$$\nabla^2\underline{\Psi}(\mathbf{r}) + \frac{\omega^2}{c^2}\underline{\Psi}(\mathbf{r}) = 0 \tag{4.9}$$

The complex function, $\underline{\Psi}(\mathbf{r})$, associated with the particular frequency has been expanded in terms of traveling-waves; because of Euler's formula, $\exp(-j\mathbf{k}\cdot\mathbf{r}) = \cos(\mathbf{k}\cdot\mathbf{r}) - j\sin(\mathbf{k}\cdot\mathbf{r})$, an expansion in terms of standing waves, is equally valid.

4.4 WAVES OF HIGH SYMMETRY

Spherically symmetric waves

Waves without any angular dependence can be expressed in spherical coordinates as $\psi(r,t)$; they satisfy

$$\frac{1}{r^2}\frac{\partial}{\partial r}\left(r^2\frac{\partial\psi}{\partial r}\right) - \frac{1}{c^2}\frac{\partial^2\psi}{\partial t^2} = 0 \tag{4.10}$$

If the wave amplitude is written as $\psi = \frac{f(r,t)}{r}$ and substituted into Eq. (4.10), the result is

$$\frac{\partial^2 f}{\partial r^2} - \frac{1}{c^2}\frac{\partial^2 f}{\partial t^2} = 0 \tag{4.11}$$

which has the same form as Eq. (4.3). It therefore follows that

$$\psi(r,t) = \frac{f_+(r-ct) + f_-(r+ct)}{r} \tag{4.12}$$

is the general solution. Harmonic waves can be expressed as

$$\psi(r,t) = \frac{Re\{[\underline{C}_+\exp(-jkr) + \underline{C}_-\exp(jkr)]\exp(j\omega t)\}}{r} \tag{4.13}$$

where Re{ } implies taking the real part of the quantity between the curly brackets. The factor $1/r$ is present because the square of the wave amplitude is inversely proportional to the area of the sphere ($4\pi r^2$).

Impulse-response function

If $2f_+(r-ct) = (r-ct)^{-1}$ and $2f_-(r+ct) = (r+ct)^{-1}$, Eq. (4.12) reduces to

$$\psi(r,t) = \frac{1}{r^2 - c^2t^2} = \frac{1}{r^2 + (ict)^2} \tag{4.14}$$

If an imaginary fourth coordinate, ict, is introduced, this symmetric solution satisfies the four-dimensional Laplace's Equation (Appendix C, Section C.5) and satisfies the wave equation everywhere except where it is singular. If the position of that singularity is shifted to (x', y', z', t'), the solution

$$\psi(\mathbf{r}, t; \mathbf{r}', t') = \frac{1}{(x - x')^2 + (y - y')^2 + (z - z')^2 + [ic(t - t')]^2} \tag{4.15}$$

is proportional to the impulse response and allows integration of the general three-dimensional inhomogeneous wave equation when the source function is specified as a function of the primed coordinates.

Cylindrically symmetric waves

Waves without angular or axial dependencies can be expressed in cylindrical coordinates as $\psi(r, t)$; they satisfy

$$\frac{\partial^2 \psi}{\partial r^2} + \frac{1}{r}\frac{\partial \psi}{\partial r} - \frac{1}{c^2}\frac{\partial^2 \psi}{\partial t^2} = 0 \tag{4.16}$$

Following our success with expressing the spherical wave as $\psi = \frac{f(r,t)}{r}$, we attempt a similar trick and only partially succeed. Because now the square of the wave amplitude is expected to be inversely proportional to the circumference of the circle $(2\pi r_c)$, we try $\psi = \frac{f(r_c,t)}{\sqrt{r_c}}$. Upon substitution, we find that (setting $r_c = r$)

$$\frac{\partial^2 f}{\partial r^2} + \frac{f}{4r^2} - \frac{1}{c^2}\frac{\partial^2 f}{\partial t^2} = 0 \tag{4.17}$$

which is not quite the hoped-for result. Nevertheless, whenever $|\frac{f}{4r^2}| \ll |\frac{\partial^2 f}{\partial r^2}|$, the one-dimensional wave-equation approximation is reasonable; then

$$\psi(r, t) \simeq \frac{f_+(r - ct) + f_-(r + ct)}{\sqrt{r}} \tag{4.18}$$

and in the case of harmonic waves (whenever $2kr \gg 1$) we obtain

$$\psi(r, t) \simeq \frac{Re\{[\underline{C}_+ \exp(-jkr) + \underline{C}_- \exp(jkr)] \exp(j\omega t)\}}{\sqrt{r}} \tag{4.19}$$

These are the asymptotic expansions for the exact cylindrical waves discussed in Appendix F.

Impulse-response function

The static potential associated with a point charge in three dimensions is $\psi \sim \frac{1}{\sqrt{x^2+y^2+z^2}}$. Reversing the trick that turned the wave equation into Laplace's Equation, we convert ψ into a solution of the two-dimensional wave equation by replacing z with ict; the result,

$$\psi \sim \frac{1}{\sqrt{|r_c^2 - c^2 t^2|}} \tag{4.20}$$

is an exact solution of Eq. (4.2b), which satisfies the wave equation everywhere except where it is singular. If the position of that singularity is shifted to (x', y', t'), the solution

$$\psi(\mathbf{r}, t; \mathbf{r}', t') = \frac{1}{\sqrt{(x - x')^2 + (y - y')^2 + [ic(t - t')]^2}} \tag{4.21}$$

is proportional to the generalized impulse response.

4.5 INHOMOGENEOUS SCALAR WAVE EQUATIONS

Three-dimensional superposition integrals

The solution to the inhomogeneous wave equation,

$$\nabla^2\psi - \frac{1}{c^2}\frac{\partial^2\psi}{\partial t^2} = -\xi \tag{4.22}$$

can be found by superimposing the impulse responses to the source differentials $\xi(\mathbf{r}', t')$ $dV'd(ict')$.

For the three-dimensional wave equation, Eq. (4.15) (normalized to a hypersphere) is appropriate and

$$\psi(x, y, z, t) = \int_{-\infty}^{\infty}\int_{-\infty}^{\infty}\int_{-\infty}^{\infty}\int_{-\infty}^{\infty} \frac{\xi(x', y', z', t')dx'dy'dz'd(ict')}{4\pi^2[(x'-x)^2 + (y'-y)^2 + (z'-z)^2 + [ic(t'-t)]^2]} \tag{4.23}$$

All four integrations are on an equal footing only when ict' is considered to be *real* and the entire range is available for the integration. This requires that we consider t' to be a complex variable, and so familiarity with the theory of complex variables and contour integration is essential if one is to become fluent with this approach. Integrals of this form and the implications of introducing complex time will be considered in Part II; for now we simply observe that if integration over the coordinate ict' is carried out first, a Poisson-like integral

$$\psi(x, y, z, t) = \int_{-\infty}^{\infty}\int_{-\infty}^{\infty}\int_{-\infty}^{\infty} \frac{\xi(x', y', z', t - \frac{\sqrt{(x'-x)^2+(y'-y)^2+(z'-z)^2}}{c})dx'dy'dz'}{4\pi\sqrt{(x'-x)^2 + (y'-y)^2 + (z'-z)^2}} \tag{4.24}$$

results except that the time t (a real variable) is replaced by the retarded time, $t - \frac{\sqrt{(x'-x)^2+(y'-y)^2+(z'-z)^2}}{c}$. The calculation (detailed in Part II, Chapter 7, Section 7.7) validates the speculation that led to Eq. (1.27).

CHAPTER 5

ELECTROMAGNETIC WAVES IN LINEAR MATERIALS

5.1 INTRODUCTION

When regions of material replace free-space, additional sources of current and/or charge may be created within those regions and on boundaries between them. Since these are generally unknown in the initial stages of problem solving, direct integration of the inhomogeneous wave equation is not possible. Instead, one usually attempts to find solutions of the homogeneous wave equation—for each of the regions—and then superimpose them in order to satisfy the boundary conditions. In this chapter, we review the equations governing materials that are either electrically conducting or have dielectric and/or magnetic properties. For simplicity, we limit the discussion to linear isotropic materials that are stationary.

5.2 ELECTRICALLY CONDUCTING MEDIA

As discussed in Chapter 1, the constitutive law describing a linear conductor is $\mathbf{J} = \sigma \mathbf{E}$, where σ is the value of the electrical conductivity. If the material is stationary in the

The Power and Beauty of Electromagnetic Fields, First Edition. F. R. Morgenthaler.

laboratory frame and uniform, the conductivity is constant and

$$\nabla^2 \mathbf{H} - \mu_0 \varepsilon_0 \frac{\partial^2 \mathbf{H}}{\partial t^2} = -\nabla \times (\sigma \mathbf{E}) \tag{5.1a}$$

$$\nabla^2 \mathbf{E} - \mu_0 \varepsilon_0 \frac{\partial^2 \mathbf{E}}{\partial t^2} = \frac{1}{\varepsilon_0} \nabla \rho + \sigma \mu_0 \frac{\partial \mathbf{E}}{\partial t} \tag{5.1b}$$

Charge-density decay (dielectric relaxation time)

If the divergence is taken of both sides of these equations, the identity $0 = 0$ results from the first and, after interchanging the order of differentiation and rearranging terms, we have

$$\varepsilon_0 \frac{\partial^2 (\nabla \cdot \mathbf{E})}{\partial t^2} + \sigma \frac{\partial (\nabla \cdot \mathbf{E})}{\partial t} + \nabla \sigma \cdot \frac{\partial \mathbf{E}}{\partial t} = \frac{1}{\mu_0} \nabla^2 \left(\nabla \cdot \mathbf{E} - \frac{1}{\varepsilon_0} \rho \right)$$

from the second. After substituting Eq. (1.8c), and integrating once with respect to time, we find that

$$\frac{\partial \rho}{\partial t} + \frac{\sigma}{\varepsilon_0} \rho + \nabla \sigma \cdot \mathbf{E} = 0$$

where the integration constant has been set equal to zero for reasons identical to those that led to Eq. (1.5a). In regions where σ is spatially uniform, the solution of this charge-relaxation equation is

$$\rho(\mathbf{r}, t) = \rho(\mathbf{r}, 0) \exp(-t/\tau_d)$$

$$\tau_d = \varepsilon_0 / \sigma$$

where τ_d is defined as the dielectric-relaxation time. For normal metals, $\tau_d < 10^{-18} s$ and so, for good conductors, both ρ (and therefore $\nabla \cdot \mathbf{E}$) can be safely approximated as zero. Under these circumstances and using Eq. (1.8c), we obtain

$$\nabla^2 \mathbf{H} - \frac{1}{c^2} \frac{\partial^2 \mathbf{H}}{\partial t^2} = \sigma \mu_0 \frac{\partial \mathbf{H}}{\partial t} \tag{5.2a}$$

$$\nabla^2 \mathbf{E} - \frac{1}{c^2} \frac{\partial^2 \mathbf{E}}{\partial t^2} = \sigma \mu_0 \frac{\partial \mathbf{E}}{\partial t} \tag{5.2b}$$

and both the electric and magnetic fields satisfy the same equation *within* a spatially uniform conductor (the surface of the conductor is a special case because there the normal gradient of σ is infinite). However, their relative strengths can be very different because

$$\nabla \times \mathbf{H} = \sigma \mathbf{E} + \varepsilon_0 \frac{\partial \mathbf{E}}{\partial t}$$

$$\nabla \times \mathbf{E} = -\mu_0 \frac{\partial \mathbf{H}}{\partial t}$$

If $\mathbf{E} = \mathbf{Re}\,\{\underline{\mathbf{E}}_o(\mathbf{r}) \exp(j\omega t)\}$ and $\mathbf{H} = \mathbf{Re}\,\{\underline{\mathbf{H}}_o(\mathbf{r}) \exp(j\omega t)\}$, the complex forms are seen to be

$$\nabla^2 \underline{\mathbf{E}}_o + \omega^2 \mu_0 \varepsilon_0 \underline{\mathbf{E}}_o = j\omega\sigma \mu_0 \underline{\mathbf{E}}_o \tag{5.3a}$$

$$\nabla^2 \underline{\mathbf{H}}_o + \omega^2 \mu_0 \varepsilon_0 \underline{\mathbf{H}}_o = j\omega\sigma \mu_0 \underline{\mathbf{H}}_o \tag{5.3b}$$

and related by

$$\nabla \times \underline{\mathbf{H}}_o = (\sigma + j\omega\varepsilon_o)\underline{\mathbf{E}}_o$$

$$\nabla \times \underline{\mathbf{E}}_o = -j\omega\mu_o\underline{\mathbf{H}}_o$$

Magnetic diffusion length and skin depth

In a uniform highly conducting material, the conduction term dominates and the fields are governed by identical diffusion equations:

$$\nabla^2\mathbf{H} = \sigma\mu_o\frac{\partial \mathbf{H}}{\partial t} \tag{5.4a}$$

$$\nabla^2\mathbf{E} = \sigma\mu_o\frac{\partial \mathbf{E}}{\partial t} \tag{5.4b}$$

and (when $\sigma \gg \omega\varepsilon_o$) by their complex equivalents:

$$\nabla^2\underline{\mathbf{E}}_o = j\omega\sigma\mu_o\underline{\mathbf{E}}_o \tag{5.5a}$$

$$\nabla^2\underline{\mathbf{H}}_o = j\omega\sigma\mu_o\underline{\mathbf{H}}_o \tag{5.5b}$$

In the time domain, transient solutions of the form

$$\mathbf{E}, \mathbf{H} \sim \exp(-t/\tau_k)\cos(\mathbf{k}\cdot\mathbf{r})$$

require that the diffusion time of a particular value of $\mathbf{k}$ satisfies

$$\tau_k = \frac{\sigma\mu_o}{\mathbf{k}\cdot\mathbf{k}}$$

In general a superposition of such terms is required to match boundary conditions; that with the largest value of τ_k dominates the solution and is often written as

$$\tau_{\text{diff}} = \sigma\mu_o\ell_{\text{diff}}^2 \tag{5.6}$$

where τ_{diff} and ℓ_{diff} are, respectively, the diffusion time and diffusion length. Because of charge relaxation, $\mathbf{E}$ (but not $\sigma\mathbf{E}$) is negligible, and $\mathbf{H}$ dominates as it diffuses toward its static equilibrium distribution; for that reason the process is termed magnetic diffusion.

In the frequency domain, steady-state solutions of the form

$$\underline{\mathbf{E}}_o, \underline{\mathbf{H}}_o \sim \exp[-(\boldsymbol{\alpha} + j\boldsymbol{\beta})\cdot\mathbf{r}]$$

are possible, provided that

$$(\boldsymbol{\alpha} + j\boldsymbol{\beta})\cdot(\boldsymbol{\alpha} + j\boldsymbol{\beta}) = j\omega\sigma\mu_o$$

If $\boldsymbol{\alpha}\times\boldsymbol{\beta} = 0$, we have

$$\alpha = |\boldsymbol{\beta}| = \sqrt{\frac{\omega\sigma\mu_o}{2}} = \frac{1}{\delta} \tag{5.7}$$

where δ is defined as the skin depth. When δ is small compared to the appropriate dimension of a conductor, it follows that the current-density, electric, and magnetic field components are all confined near surfaces and are negligible in the interior.

For a perfect conductor ($\sigma = \infty$), the magnetic diffusion time is infinite and the skin depth is zero. These results are consistent with the *definition* that a perfect conductor is always completely free of electric and magnetic fields.[1]

5.3 LINEAR DIELECTRIC AND MAGNETIC MEDIA

Also as discussed in Chapter 1, the constitutive laws for a linear isotropic dielectric and magnetic material that is stationary are

$$\mathbf{D} = \epsilon \mathbf{E}$$

$$\mathbf{B} = \mu \mathbf{H} = \nabla \times \mathbf{A}$$

Therefore, except for replacement of ε and μ for ε_o and μ_o, Maxwell's Equations and Eqs. (4.8a)–(4.8c) are unchanged. If electrical conductivity is also present, the results of the previous section apply after that amendment.

In a region where μ, ε, and σ are uniform, it is often convenient to employ the modified Lorenz gauge,

$$\nabla \cdot \mathbf{A} + \mu\varepsilon \frac{\partial \Phi}{\partial t} + \mu\sigma\Phi = 0 \tag{5.8a}$$

so that $\mathbf{A}$ and Φ satisfy wave equations (including diffusion),

$$\nabla^2 \mathbf{A} - \mu\varepsilon \frac{\partial^2 \mathbf{A}}{\partial t^2} - \mu\sigma \frac{\partial \mathbf{A}}{\partial t} = 0 \tag{5.8b}$$

$$\nabla^2 \Phi - \mu\varepsilon \frac{\partial^2 \Phi}{\partial t^2} - \mu\sigma \frac{\partial \Phi}{\partial t} = 0 \tag{5.8c}$$

that are identical to those governing $\mathbf{E}$ and $\mathbf{H}$. The complex counterparts are

$$\nabla \cdot \underline{\mathbf{A}} + \mu(\sigma + j\omega\varepsilon)\underline{\Phi} = 0 \tag{5.9a}$$

$$\nabla^2 \underline{\mathbf{A}} - j\omega\mu(\sigma + j\omega\varepsilon)\underline{\mathbf{A}} = 0 \tag{5.9b}$$

$$\nabla^2 \underline{\Phi} - j\omega\mu(\sigma + j\omega\varepsilon)\underline{\Phi} = 0 \tag{5.9c}$$

For lossless waves propagating in the sinusoidal steady state, the dispersion relation is altered to $\mathbf{k} \cdot \mathbf{k} = \omega^2 \mu\varepsilon$. That assumption is made in what follows.

[1] *Some* superconductors behave like a perfect conductor in that all magnetic flux is expelled (Meissner Effect) when the superconducting state, $\sigma = \infty$, is reached; for others, the magnetic flux is trapped inside. Even when expelled, the skin depth, δ, remains finite—although, according to the London Equations, very small. [Refer to Chapter 25, Problem 25.2-12].

Uniform plane waves (linearly polarized)

A linearly polarized uniform plane wave of the form

$$\mathbf{E} = \mathbf{u}_1 E_o \cos(\omega t - \mathbf{k}\cdot\mathbf{r} + \theta) \tag{5.10a}$$

$$\mathbf{B} = \mathbf{u}_2 B_o \cos(\omega t - \mathbf{k}\cdot\mathbf{r} + \theta) \tag{5.10b}$$

satisfies Maxwell's Equations provided $\mathbf{u}_1$ and $\mathbf{u}_2$ are orthogonal unit vectors and

$$\begin{aligned}
\mathbf{B}\cdot\mathbf{k} &= 0 \\
\mathbf{E}\times\mathbf{B} &= E_o B_o \,\widehat{\mathbf{k}} \\
B_o &= \mu H_o \\
E_o/H_o &= \eta = \sqrt{\frac{\mu}{\varepsilon}} \\
|\mathbf{k}| &= \omega\sqrt{\mu\varepsilon}
\end{aligned}$$

The phase and group velocities of the wave are defined by

$$\mathbf{v}_{\text{phase}} = \frac{\omega}{\mathbf{k}\cdot\mathbf{k}}\mathbf{k} \tag{5.11}$$

and

$$\mathbf{v}_{\text{group}} = \frac{\partial\omega}{\partial\mathbf{k}} \tag{5.12}$$

Because μ and ε are assumed to be independent of frequency, the two velocities of uniform plane waves are equal (with magnitude $1/\sqrt{\mu\varepsilon}$).

The superposition of such waves (with appropriate polarizations, amplitudes, and phases) allows general elliptically polarized waves to be described. Wave pairs with oppositely directed values of **k** can produce standing waves.

Particle representation of harmonic plane waves

Direct evaluation reveals that the time-averaged values (denoted by $<\ >$) of power flux, energy density, stress tensor, and momentum density are related by

$$<\mathbf{S}> = <W>\mathbf{v}_{\text{group}} \tag{5.13a}$$

$$<\mathbf{G}> = <W>\frac{\mathbf{v}_{\text{phase}}}{\mathbf{v}_{\text{phase}}\cdot\mathbf{v}_{\text{phase}}} \tag{5.13b}$$

$$<\overline{\mathbf{T}}> = -<\mathbf{G}>\mathbf{v}_{\text{group}} \tag{5.13c}$$

even in cases where the material is dispersive and the phase and group velocities of the wave are not equal. These relations also follow from the energy, stress, and momentum theorems derived in Chapter 6, Sections 6.2–6.4.

If one assumes that the energy arises from identical particles distributed with a number density n, each with energy E and linear momentum $\mathbf{p}$, then

$$\mathbf{p} = \frac{E}{\omega}\mathbf{k} \tag{5.14}$$

where, because the Maxwell Equations pertain to macroscopic fields, the number density should be large. Connection with quantum physics is made by choosing $E = \hbar\omega = h\nu$

because then $\mathbf{p} = \hbar\mathbf{k}$ as required for photon propagation. With this choice, we have

$$<W> = n\hbar\omega \tag{5.15a}$$

$$<\mathbf{S}> = \ <W> \mathbf{v}_{\text{group}} = n\hbar\omega\mathbf{v}_{\text{group}} \tag{5.15b}$$

$$<\mathbf{G}> = n\hbar\mathbf{k} \tag{5.15c}$$

$$<\overline{\mathbf{T}}> = -<\mathbf{G}> \mathbf{v}_{\text{group}} = -n\hbar\mathbf{k}\,\mathbf{v}_{\text{group}} \tag{5.15d}$$

Similar but advanced analysis of circular waveguide modes is developed in Chapter 20, Section 20.5, where the spin and orbital angular momentum properties of photons are made evident within the context of the classical macroscopic Maxwell Equations. For linearly polarized waves, n is comprised of equal densities of spin $+1$ and spin -1 photons; consequently, there is no net angular momentum.

Wavepackets

In order to describe spatially localized particles, it is necessary to superimpose uniform plane waves of different frequencies.[2] The result is a wavepacket centered on some frequency, ω_0. If the localization is severe, a very broad band of frequencies is required. If, instead, the spread extends over many wavelengths (of the principal component), the velocity of the packet can be approximated by the group velocity calculated for that central component. If one averages over many periods of ω_0, the power-energy and stress-momentum theorems become

$$\nabla \cdot (n\hbar\omega\mathbf{v}_{\text{group}}) + \frac{\partial}{\partial t}(n\hbar\omega) = -p_{\text{d}} \tag{5.16a}$$

$$\nabla \cdot (-n\hbar\mathbf{k}\,\mathbf{v}_{\text{group}}) - \frac{\partial}{\partial t}(n\hbar\mathbf{k}) = \mathbf{f} \tag{5.16b}$$

where, in general, $\hbar\omega(\mathbf{k}, \mathbf{r}, t)$ serves as the Hamiltonian that describes all possible trajectories of the wavepacket. In the absence of dissipation ($p_{\text{d}} = 0$), these equations may be written as

$$\hbar\omega\left[\nabla \cdot (n\mathbf{v}_{\text{group}}) + \frac{\partial n}{\partial t}\right] + n\left(\frac{\partial}{\partial t} + \mathbf{v}_{\text{group}} \cdot \nabla\right)\hbar\omega = n\left[\frac{\partial \hbar\omega(\mathbf{k}, \mathbf{r}, t)}{\partial t}\right]_{\mathbf{k},\mathbf{r}}$$

$$\hbar\mathbf{k}\left[\nabla \cdot (n\mathbf{v}_{\text{group}}) + \frac{\partial n}{\partial t}\right] + n\left(\frac{\partial}{\partial t} + \mathbf{v}_{\text{group}} \cdot \nabla\right)\hbar\mathbf{k} = -n\left[\frac{\partial \hbar\omega(\mathbf{k}, \mathbf{r}, t)}{\partial \mathbf{r}}\right]_{\mathbf{k},t}$$

or equivalently as either

$$\hbar\omega\left[\nabla \cdot (n\mathbf{v}_{\text{group}}) + \frac{\partial n}{\partial t}\right] + n\left(\frac{d\hbar\omega}{dt} - \left[\frac{\partial \hbar\omega(\mathbf{k}, \mathbf{r}, t)}{\partial t}\right]_{\mathbf{k},\mathbf{r}}\right) = 0 \tag{5.17a}$$

$$\hbar\mathbf{k}\left[\nabla \cdot (n\mathbf{v}_{\text{group}}) + \frac{\partial n}{\partial t}\right] + n\left(\frac{d\hbar\mathbf{k}}{dt} + \left[\frac{\partial \hbar\omega(\mathbf{k}, \mathbf{r}, t)}{\partial \mathbf{r}}\right]_{\mathbf{k},t}\right) = 0 \tag{5.17b}$$

[2] If the constitutive laws of the medium are nonlinear, this type of analysis fails. Nevertheless, linearization about an operating point permits the superposition of small-signal harmonic waves to form particle-like wavepackets often called quasiparticles.

or

$$\frac{d\hbar\omega}{dt} - \left[\frac{\partial\hbar\omega(\mathbf{k},\mathbf{r},t)}{\partial t}\right]_{\mathbf{k},\mathbf{r}} = \left[\frac{\partial\omega(\mathbf{k},\mathbf{r},t)}{\partial\,\mathbf{k}}\right]_{\mathbf{r},t} \cdot \frac{d\hbar\mathbf{k}}{dt} + \left[\frac{\partial\hbar\omega(\mathbf{k},\mathbf{r},t)}{\partial\mathbf{r}}\right]_{\mathbf{k},t} \cdot \frac{d\mathbf{r}}{dt} \tag{5.17c}$$

$$= \mathbf{v}_{\text{group}} \cdot \left(\frac{d\hbar\mathbf{k}}{dt} + [\frac{\partial\hbar\omega(\mathbf{k},\mathbf{r},t)}{\partial\mathbf{r}}]_{\mathbf{k},t}\right)$$

$$\left(\omega - \mathbf{v}_{\text{group}} \cdot \mathbf{k}\right)\left[\nabla \cdot \left(n\mathbf{v}_{\text{group}}\right) + \frac{\partial n}{\partial t}\right] = 0 \tag{5.17d}$$

where

$$\mathbf{v}_{\text{group}} = \left[\frac{\partial\hbar\omega(\mathbf{k},\mathbf{r},t)}{\partial\,\hbar\mathbf{k}}\right]_{\mathbf{r},t} \tag{5.17e}$$

Although we have considered nondispersive wave propagation where

$$(\omega - \mathbf{v}_{\text{group}} \cdot \mathbf{k}) = (\mathbf{v}_{\text{phase}} - \mathbf{v}_{\text{group}}) \cdot \mathbf{k} = 0,$$

assume that the phase and group velocities differ ever so slightly. Then from Eq. (5.17d),

$$\nabla \cdot (n\mathbf{v}_{\text{group}}) + \frac{\partial n}{\partial t} = 0 \tag{5.18}$$

which is a statement of particle conservation.

With $\mathbb{H}(\mathbf{p},\mathbf{r},t) = \hbar\omega(\mathbf{k},\mathbf{r},t)$ serving as the Hamiltonian of the wavepacket, Eqs. (5.17) can be written as

$$\frac{dE}{dt} = \left[\frac{\partial\mathbb{H}(\mathbf{p},\mathbf{r},t)}{\partial t}\right]_{\mathbf{p},\mathbf{r}} \tag{5.19a}$$

$$\frac{d\mathbf{p}}{dt} = -\left[\frac{\partial\mathbb{H}(\mathbf{p},\mathbf{r},t)}{\partial\mathbf{r}}\right]_{\mathbf{p},t} = -[\nabla\mathbb{H}(\mathbf{p},\mathbf{r},t)]_{\mathbf{p},t} \tag{5.19b}$$

$$\mathbf{v}_{\text{group}} = \left[\frac{\partial\mathbb{H}(\mathbf{p},\mathbf{r},t)}{\partial\,\mathbf{p}}\right]_{\mathbf{r},t} \tag{5.19c}$$

These are precisely Hamilton's Equations that describe the trajectory of a particle of energy E, momentum $\mathbf{p}$, and velocity $\mathbf{v}_{\text{group}}$. They apply to wavepackets of moderate localization that propagate in comparatively weak gradients. For a stationary nondeforming dielectric characterized by μ and ε that are both uniform constants, $dE/dt = 0$, $d\mathbf{p}/dt = 0$ and both ω and $\mathbf{k}$ are constant, as expected.

Expressions for E and $\mathbf{p}$ depend upon the representation of power, energy, stress, and momentum. For now, we follow convention and assume the energy density of a modified Maxwell–Poynting representation to be

$$W = \frac{1}{2}\varepsilon\mathbf{E} \cdot \mathbf{E} + \frac{1}{2}\mu\mathbf{H} \cdot \mathbf{H}$$

For the linearly polarized uniform plane wave of Eq. (5.10), one finds that

$$<W> = n\hbar\omega = \frac{1}{2}\sqrt{\mu\varepsilon}E_o H_o \tag{5.20a}$$

$$<G> = n\hbar k = \frac{1}{2}\mu\varepsilon E_o H_o \tag{5.20b}$$

$$<S> = n\hbar\omega \mathrm{v}_g = \frac{1}{2}E_o H_o \tag{5.20c}$$

These results are consistent with

$$W = \frac{1}{2}\mathbf{D}\cdot\mathbf{E} + \frac{1}{2}\mathbf{B}\cdot\mathbf{H} \tag{5.21a}$$

$$\mathbf{G} = \mathbf{D}\times\mathbf{B} = \mu\varepsilon\mathbf{S} \tag{5.21b}$$

$$\mathbf{S} = \mathbf{E}\times\mathbf{H} \tag{5.21c}$$

Equations (5.21) flow from the Minkowski formulation of electrodynamics; it and other formulations are developed fully in Part II, Chapter 10. Despite the simplicity of these results, their obvious correctness in the free-space limit, and their endorsement by many researchers, they have detractors who raise serious questions. The first criticism is that (apart from free-space) frequency-independent values of ε and μ are not physically realizable. Any reasonable model of polarization must contain dipolar charges (with associated masses) connected by forces that simulate some sort of spring. Such mechanical subsystem(s) will certainly lead to dispersive behavior that must be included in any rigorous analysis. This observation is undoubtedly correct, but rather like saying that circuit engineers should never consider a circuit element as ideally capacitive or inductive because real components contain parasitic energy-storage elements of the other type. No doubt this is true, but it has not prevented the use of idealized components in meaningful and practical assessments of complex circuit behavior. Many dielectric materials are nearly dispersionless over wide frequency ranges and we can select wavepackets band-limited within those ranges. A second, more troubling criticism is made by those who insist that even though dispersionless materials are causal—in that they satisfy the Kramers–Krönig relations [12]—they will violate energy conservation unless $\mathbf{G} = \mathbf{S}/c^2$ in the rest frame of the material. Let us see why they think so by applying Hamilton's Equations to the particle representation.

We allow our material (still assumed nondeformable) to accelerate from rest, but consider it at times when its velocity, $\mathbf{v}$, is very small. Since relativistic motion is not an issue, the factor $\gamma = 1/\sqrt{1-(\mathbf{v}\cdot\mathbf{v})/c^2}$ may be set equal to unity. The analysis contained in Appendix A, Section A.1, allows transformation of frequency and wavenumber between the (primed) rest frame and laboratory frame. Correct to first order in the velocity,

$$\omega = \omega' + \mathbf{v}\cdot\mathbf{k}'$$

$$\mathbf{k} = \mathbf{k}' + \frac{\omega'}{c^2}\mathbf{v}$$

$$\omega' = \omega - \mathbf{v}\cdot\mathbf{k}$$

$$\mathbf{k}' = \mathbf{k} - \frac{\omega}{c^2}\mathbf{v}$$

where

$$\omega' = \omega'(\mathbf{k}', t)$$

The first of Hamilton's Equations, Eq. (5.19a), becomes

$$\frac{dE}{dt} = \hbar\left(\frac{\partial\omega}{\partial t}\right)_{\mathbf{k},\mathbf{r}} = \hbar\left[\frac{\partial\omega'}{\partial t} + \mathbf{k}'\cdot\frac{\partial\mathbf{v}}{\partial t} - \frac{\partial\omega'}{\partial\mathbf{k}'}\cdot\left(\frac{\omega}{c^2}\frac{\partial\mathbf{v}}{\partial t}\right)\right]$$

If the rest-frame dispersion relation is independent of time (μ and ε both constants[3]), then (dropping the primes because $\mathbf{v} \to 0$)

$$n\frac{dE}{dt} = \left(n\hbar\mathbf{k} - \frac{n\hbar\omega\mathbf{v}_{\text{group}}}{c^2}\right)\cdot\frac{\partial\mathbf{v}}{\partial t}$$
$$= \left(\mathbf{G} - \frac{\mathbf{S}}{c^2}\right)\cdot\frac{\partial\mathbf{v}}{\partial t} \tag{5.22}$$

Because the acceleration can be applied in an arbitrary direction, insistence that dE/dt vanish, forces $\mathbf{G} = \mathbf{S}/c^2$. Now consider a nonrelativistic particle of mass m that moves at velocity, $\mathbf{v}_{\text{group}}$, that acquires an increment of velocity, $\mathbf{v}$. The kinetic energy becomes

$$E = \frac{1}{2}m(\mathbf{v}_{\text{group}}\cdot\mathbf{v}_{\text{group}} + 2\mathbf{v}_{\text{group}}\cdot\mathbf{v} + \mathbf{v}\cdot\mathbf{v})$$

Although E is unchanged in the limit $\mathbf{v} = 0$, the power

$$\left(\frac{dE}{dt}\right)_{\mathbf{v}=0} = m\mathbf{v}_{\text{group}}\cdot\frac{d\mathbf{v}}{dt} = \mathbf{p}\cdot\frac{d\mathbf{v}}{dt}$$

can be zero or of either sign, depending upon the direction of the acceleration. On the other hand, in free-space there is no possible agent—save the "ether"—to supply or absorb any power. *Only* in that case is the argument that dE/dt must vanish a sound one.

Finally, notice that whenever $\mu\varepsilon \gg \mu_o\varepsilon_o$, the term $\mathbf{S}/c^2$ is a negligible correction to Eq. (5.22) and so the nonrelativistic particle approximation is valid.

[3]If the phase velocity, $1/\sqrt{\mu\varepsilon}$, is a function of time, energy will flow into or out of the wave from the agency causing changes in the permittivity and/or permeability. Such behavior in a voltage-controlled ferroelectric was analyzed many years ago by the present author [13].

CHAPTER 6

ELECTROMAGNETIC THEOREMS AND PRINCIPLES

6.1 INTRODUCTION

This chapter develops important theorems and principles that increase one's general understanding of fields and facilitate the solution of electromagnetic boundary-value problems. First, based upon the complex Maxwell's Equations that describe linear isotropic stationary materials, complex-power fluxes and complex-stress tensors are defined. These lead to the complex-power theorem, energy theorem, complex-stress theorem, and momentum theorem in both the Maxwell–Poynting and Alternate representations. The important concept of duality is then introduced and applied to both the interchange of electric and magnetic fields and the sources that generate them; this leads to the introduction of magnetic charge and magnetic current as an artifice to aid in problem solving. Next, boundary conditions that assure uniqueness of the electric and magnetic fields within a closed surface are considered; the result is the uniqueness theorem. Together, source duality and uniqueness lead to the general equivalence principle that permits substitution of sources outside of (or on) an enclosed region without altering the electric and magnetic fields inside it. The related concept of complementary structures is also introduced and for planar structures made of perfect conductors, Babinet's Principle is derived. The induction theorem and the reciprocity theorem complete the set of topics that are considered.

The Power and Beauty of Electromagnetic Fields, First Edition. F. R. Morgenthaler.

6.2 COMPLEX POWER AND ENERGY THEOREMS

Circuit equations are a consequence of approximations introduced by quasistatic field analysis; therefore, there is a deep connection between electromagnetic and circuit representations of power and energy. Consequently, before developing important theorems that govern the complex power and energy associated with electromagnetic fields, it is helpful to review counterparts that arise from the theory of lumped or distributed circuits.

Circuit power

In linear circuit theory, the network response to excitations of the form e^{st}, where $s = \sigma_\omega + j\omega$ is the complex frequency,[1] is of fundamental importance because with it both transient and sinusoidal steady-state solutions can be determined in the manner described by Guillemin [14, 15]. The ratio of the complex voltage $\underline{V}(s)$ (defined across a terminal pair) to the complex current $\underline{I}(s)$ (flowing into the positive reference terminal) is the complex impedance, $\underline{Z}(s) = 1/\underline{Y}(s)$, where $\underline{Y}(s)$ is the admittance. It follows that (denoting complex conjugation with an asterisk)

$$\frac{1}{2}\underline{V}(s)\,\underline{I}^*(s) = \frac{1}{2}\underline{Z}(s)\underline{I}(s)\underline{I}^*(s) = \frac{1}{2}\underline{Y}^*(s)\underline{V}(s)\underline{V}^*(s) \tag{6.1}$$

For a series *RLC* circuit for which $\underline{Z}(s) = R + Ls + \dfrac{1}{Cs}$, Eq. (6.1) becomes

$$\frac{1}{2}\underline{V}(s)\,\underline{I}^*(s) = \frac{1}{2}R\;\underline{I}(s)\underline{I}^*(s) + 2s\frac{1}{4}L\;\underline{I}(s)\underline{I}^*(s) + 2s^*\frac{1}{4}C\;\underline{V}_c(s)\underline{V}_c^*(s)$$

$$\underline{V}_c(s) = \frac{\underline{I}(s)}{Cs}$$

For a parallel *RLC* circuit for which $\underline{Y}(s) = G + Cs + \dfrac{1}{Ls}$, Eq. (6.1) becomes

$$\frac{1}{2}\underline{V}(s)\,\underline{I}^*(s) = \frac{1}{2}G\;\underline{V}(s)\underline{V}^*(s) + 2s\frac{1}{4}C\;\underline{V}(s)\underline{V}^*(s) + 2s^*\frac{1}{4}L\;\underline{I}_L(s)\underline{I}_L^*(s)$$

$$\underline{I}_L(s) = \frac{\underline{V}(s)}{Ls}$$

Either equation can be expressed as

$$\begin{aligned}\frac{1}{2}\underline{V}(s)\,\underline{I}^*(s) &= P_d(s) + 2sE_m(s) + 2s^*E_e(s) \\ &= P_d(s) + 2\sigma_\omega\,[E_m(s) + E_e(s)] + j2\omega[E_m(s) - E_e(s)]\end{aligned} \tag{6.2}$$

where, when $s = j\omega$, $P_d(s)$ is the time-averaged dissipated power and $E_m(s)$ and $E_e(s)$ are, respectively, the time-averaged magnetic and electric stored energy. It can be shown [14, pp. 510–520] that Eq. (6.2) is valid for *any* linear passive bilateral network. Consequently, when $\sigma_\omega = 0$, the complex power defined by $\frac{1}{2}\underline{V}(j\omega)\,\underline{I}^*(j\omega)$ can be expressed

[1] To avoid confusion with the symbol for conductivity, we use σ_ω rather than the usual definition $Re\{s\} = \sigma$.

in terms of the time averages (denoted by < >) of the dissipated power and the electric and magnetic stored energies as

$$\frac{1}{2}\underline{V}\,\underline{I}^* = \langle P_{\text{dissipated}}\rangle + j2\omega(\langle E_{\text{magnetic}}\rangle - \langle E_{\text{electric}}\rangle) \tag{6.3}$$

The imaginary part is *defined* as the reactive power. This formula is central to understanding the concepts of impedance and admittance because

$$\begin{aligned}\underline{Z}(j\omega) &= R(\omega) + jX(\omega)\\ &= \frac{2}{|\underline{I}(j\omega)|^2}[\langle P_{\text{dissipated}}\rangle + j2\omega(\langle E_{\text{magnetic}}\rangle - \langle E_{\text{electric}}\rangle)]\end{aligned} \tag{6.4}$$

$$\begin{aligned}\underline{Y}(j\omega) &= G(\omega) + jB(\omega)\\ &= \frac{2}{|\underline{V}(j\omega)|^2}[\langle P_{\text{dissipated}}\rangle - j2\omega(\langle E_{\text{magnetic}}\rangle - \langle E_{\text{electric}}\rangle)]\end{aligned} \tag{6.5}$$

Circuit energy

The complex quantity, $\frac{\partial}{\partial\omega}(\frac{1}{2}\underline{V}\,\underline{I}^*)$ has the dimensions of energy, but it is

$$\underline{Q}(j\omega) = \frac{1}{4}\left(\frac{\partial \underline{V}}{\partial\omega}\underline{I}^* + \underline{V}^*\frac{\partial \underline{I}}{\partial\omega}\right)$$

that forms the basis of the energy theorem:

$$\underline{Q} = \frac{1}{4}\frac{\partial \underline{Z}}{\partial\omega}\underline{I}\,\underline{I}^* + \frac{1}{2}R(\omega)\frac{\partial \underline{I}}{\partial\omega}\underline{I}^* = \frac{1}{4}\frac{\partial \underline{Y}}{\partial\omega}\underline{V}\,\underline{V}^* + \frac{1}{2}G(\omega)\frac{\partial \underline{V}}{\partial\omega}\underline{V}^*$$

The imaginary part of $\underline{Q}$ is the total average energy, $\langle E_{\text{total}}\rangle$, which can be expressed (in terms of the current) as

$$\langle E_{\text{total}}\rangle = j\frac{1}{4}R(\omega)\left(\frac{\partial \underline{I}}{\partial\omega}\underline{I}^* - \underline{I}\frac{\partial \underline{I}^*}{\partial\omega}\right) + \frac{1}{4}\frac{\partial \underline{X}}{\partial\omega}\underline{I}\,\underline{I}^*$$

or (in terms of the voltage) as

$$\langle E_{\text{total}}\rangle = j\frac{1}{4}G(\omega)\left(\frac{\partial \underline{V}}{\partial\omega}\underline{V}^* - \underline{V}\frac{\partial \underline{V}^*}{\partial\omega}\right) + \frac{1}{4}\frac{\partial \underline{B}}{\partial\omega}\underline{V}\,\underline{V}^*$$

Complex Poynting Theorem

The complex form of Maxwell Equations (with the $\exp(j\omega t)$ time-dependence suppressed) was derived in Chapter 1, Section 1.9. Similar results apply when the complex frequency, $s = \sigma_\omega + j\omega$, is used; they are

$$\begin{aligned}\nabla\times\underline{\mathbf{H}} - s\underline{\mathbf{D}} &= \underline{\mathbf{J}}_{\text{u}}\\ \nabla\times\underline{\mathbf{E}} + s\underline{\mathbf{B}} &= 0\\ \nabla\cdot\underline{\mathbf{D}} &= \underline{\rho}_{\text{u}}\\ \nabla\cdot\underline{\mathbf{B}} &= 0\end{aligned}$$

In terms of the complex vector and scalar potentials, we have

$$\underline{\mathbf{B}} = \nabla \times \underline{\mathbf{A}}$$
$$\underline{\mathbf{E}} = -s\underline{\mathbf{A}} - \nabla\underline{\Phi}$$

The well-known theorem based upon the complex Poynting vector (usually defined as $\underline{\mathbf{S}}(s) = \frac{1}{2}\underline{\mathbf{E}}\,(s) \times\, \underline{\mathbf{H}}^*(s)$ when $\sigma_\omega = 0$) is

$$\nabla \cdot\, \underline{\mathbf{S}}\,(s) + 2\left(s\frac{1}{4}\underline{\mathbf{B}}(s) \cdot \underline{\mathbf{H}}^*(s) + s^*\frac{1}{4}\underline{\mathbf{E}}(s) \cdot \underline{\mathbf{D}}^*(s)\right) = -\frac{1}{2}\underline{\mathbf{E}}(s) \cdot \underline{\mathbf{J}}_{\text{u}}^*(s) \tag{6.6}$$

The similarity of form between Eqs. (6.2) and (6.6) is evident.

For stationary, linear isotropic (nondispersive) media, $\underline{\mathbf{D}} = \varepsilon\underline{\mathbf{E}}$, $\underline{\mathbf{B}} = \mu\underline{\mathbf{H}}$ and $\underline{\mathbf{J}} = \sigma\underline{\mathbf{E}}$. Separating the real and imaginary parts yields (when $\sigma_\omega = 0$)

$$-\nabla \cdot Re\ \underline{\mathbf{S}} = -Re\nabla \cdot \underline{\mathbf{S}} = \frac{1}{2}\sigma\underline{\mathbf{E}} \cdot \underline{\mathbf{E}}^*$$
$$-\nabla \cdot Im\ \underline{\mathbf{S}} = -Im\nabla \cdot \underline{\mathbf{S}} = 2\omega\left(\frac{1}{4}\mu\underline{\mathbf{H}} \cdot \underline{\mathbf{H}}^* - \frac{1}{4}\varepsilon\underline{\mathbf{E}} \cdot \underline{\mathbf{E}}^*\right)$$

The real part of $-\nabla \cdot \underline{\mathbf{S}}$ equals the time-averaged dissipated-power density; the imaginary part, equals 2ω times the difference between the time-averaged magnetic and electric energy densities. These results are an important and extremely useful part of the Maxwell–Poynting representation; the connection with complex circuit power is apparent.

Complex Alternate-power theorem

There is a corresponding theorem for the complex-power flux expressed in the alternative (localized) representation considered in Chapter 3, Section 3.4. Based solely upon the complex form of conservation of unpaired charge, $\nabla \cdot \underline{\mathbf{J}}_{\text{u}} + j\omega\underline{\rho}_{\text{u}} = 0$, and $\underline{\mathbf{E}} = -j\omega\underline{\mathbf{A}} - \nabla\underline{\Phi}$, one can easily verify the validity of

$$\nabla \cdot \frac{1}{2}\ \underline{\Phi}\underline{\mathbf{J}}_{\text{u}}^* + j2\omega\left(\frac{1}{4}\underline{\mathbf{A}} \cdot \underline{\mathbf{J}}_{\text{u}}^* - \frac{1}{4}\underline{\Phi}\ \underline{\rho}_{\text{u}}^*\right) = -\frac{1}{2}\underline{\mathbf{E}} \cdot \underline{\mathbf{J}}_{\text{u}}^* \tag{6.7}$$

Although this is an interesting equation, it becomes more so when we split $\underline{\mathbf{A}} \cdot \underline{\mathbf{J}}_{\text{u}}^*$ and $\underline{\Phi}\ \underline{\rho}_{\text{u}}^*$ into real and imaginary parts.

The complex vector and scalar potentials satisfy

$$\nabla^2\underline{\mathbf{A}} + \omega^2\mu\varepsilon\underline{\mathbf{A}} - \nabla\underline{g} + j\omega\underline{\Phi}\nabla(\mu\varepsilon) + \frac{\nabla\mu}{\mu} \times (\nabla \times \underline{\mathbf{A}}) = -\mu\underline{\mathbf{J}}_{\text{u}} \tag{6.8a}$$

$$\nabla^2\underline{\Phi} + \omega^2\mu\varepsilon\underline{\Phi} + j\omega\underline{g} + \frac{\nabla\varepsilon}{\varepsilon} \cdot (j\omega\underline{\mathbf{A}} + \nabla\underline{\Phi}) = -\frac{1}{\varepsilon}\underline{\rho}_{\text{u}} \tag{6.8b}$$

These linear equations are the complex forms of Eqs. (1.13a) and (1.13b) generalized to include nonuniform permittivity/permeability and the nonzero complex gauge, $\underline{g} = \nabla \cdot \underline{\mathbf{A}} + j\omega\mu\varepsilon\,\underline{\Phi}$.

If μ and ε are independent of position and $\underline{g} = 0$, the requisite imaginary parts can be expressed as

$$\frac{1}{2}(\underline{\mathbf{A}} \cdot \underline{\mathbf{J}}_{\mathrm{u}}^{*} - \underline{\mathbf{A}}^{*} \cdot \underline{\mathbf{J}}_{\mathrm{u}}) = \frac{1}{2\mu}(\underline{\mathbf{A}}^{*} \cdot \nabla^{2}\underline{\mathbf{A}} - \underline{\mathbf{A}} \cdot \nabla^{2}\underline{\mathbf{A}}^{*})$$
$$= \frac{\partial}{\partial x_k}\left[\frac{1}{2\mu}\left(\underline{\mathbf{A}}^{*} \cdot \frac{\partial \underline{\mathbf{A}}}{\partial x_k} - \frac{\partial \underline{\mathbf{A}}^{*}}{\partial x_k} \cdot \underline{\mathbf{A}}\right)\right]$$

$$\frac{1}{2}(\underline{\Phi}\, \underline{\rho}_{\mathrm{u}}^{*} - \underline{\Phi}^{*} \underline{\rho}_{\mathrm{u}}) = \frac{1}{2}\varepsilon(\underline{\Phi}^{*}\nabla^{2}\underline{\Phi} - \underline{\Phi}\, \nabla^{2}\underline{\Phi}^{*})$$
$$= \frac{\partial}{\partial x_k}\left[\frac{1}{2}\varepsilon\left(\underline{\Phi}^{*} \cdot \frac{\partial \underline{\Phi}}{\partial x_k} - \frac{\partial \underline{\Phi}^{*}}{\partial x_k} \cdot \underline{\Phi}\right)\right]$$

(where summation over $k = 1, 2, 3$ is implied). These terms, and those that arise when $\underline{g}$ is nonzero and μ and ε are nonuniform, can be combined with Eq. (6.7) to produce the sought after theorem,

$$\nabla \cdot \underline{\mathbf{S}}^{\mathrm{o}} + j2\omega\left[\frac{1}{8}(\underline{\mathbf{A}} \cdot \underline{\mathbf{J}}_{\mathrm{u}}^{*} + \underline{\mathbf{A}}^{*} \cdot \underline{\mathbf{J}}_{\mathrm{u}}) - \frac{1}{8}(\underline{\Phi}\, \underline{\rho}_{\mathrm{u}}^{*} + \underline{\Phi}^{*}\, \underline{\rho}_{\mathrm{u}})\right] = -\frac{1}{2}\underline{\mathbf{E}} \cdot \underline{\mathbf{J}}_{\mathrm{u}}^{*} \tag{6.9}$$

where $\underline{\mathbf{S}}^{\mathrm{o}}$ is the complex Alternate-power vector,

$$\underline{\mathbf{S}}^{\mathrm{o}} = \frac{1}{2}\,\underline{\Phi}\underline{\mathbf{J}}_{\mathrm{u}}^{*} - j\omega\frac{1}{4\mu}(\underline{A}_k \nabla \underline{A}_k^{*} - \underline{A}_k^{*}\nabla \underline{A}_k) + j\omega\frac{1}{4}\varepsilon\,(\underline{\Phi}\nabla\underline{\Phi}^{*} - \underline{\Phi}^{*}\nabla\underline{\Phi})$$
$$+ j\omega\frac{1}{4}\left[\frac{1}{\mu}(\underline{g}^{*}\underline{\mathbf{A}} - \underline{g}\underline{\mathbf{A}}^{*}) + (\underline{\mathbf{A}}^{*}\underline{\mathbf{A}} - \underline{\mathbf{A}}\,\underline{\mathbf{A}}^{*}) \cdot \nabla\left(\frac{1}{\mu}\right)\right] \tag{6.10}$$

However, the reader is cautioned that Eq. (6.10) is valid only where the material is at rest and the constitutive constants, μ and ε, are isotropic. That restriction is removed in Part II, Chapter 9, where both time-dependent and complex Alternate power theorems are generalized.

In current-free regions, and independent of $\underline{g}$, $\underline{\mathbf{S}}^{\mathrm{o}}$ is a *real* vector; therefore, *except for the circuit term*, there is *no reactive Alternate power*. Because the *difference* in the time-averaged magnetic and electric energy density contributions includes only the *localized* components, it is obvious that, *in the Alternate representation,* the reactive energy density also vanishes in source-free regions. Both Alternate reactive power and energy are, of course, permitted (indeed, usually required) in the regions where there is charge and current.

Maxwell–Poynting energy theorem

Based upon the circuit-energy formula and the complex Poynting vector, the energy theorem (in the Maxwell–Poynting representation) is derived from the complex vector,

$$-4\underline{\mathbf{Q}} = \left(\underline{\mathbf{E}}^{*} \times \frac{\partial \underline{\mathbf{H}}}{\partial \omega} + \frac{\partial \underline{\mathbf{E}}}{\partial \omega} \times \underline{\mathbf{H}}^{*}\right)$$

After taking the divergence and substituting the complex Maxwell's Equations, the result is

$$-4\nabla\cdot\underline{\mathbf{Q}} = \frac{\partial\underline{\mathbf{H}}}{\partial\omega}\cdot(j\omega\underline{\mathbf{B}}^*) - \underline{\mathbf{E}}^*\cdot\frac{\partial}{\partial\omega}(\underline{\mathbf{J}}_{\mathrm{u}} + j\omega\underline{\mathbf{D}})$$
$$+\underline{\mathbf{H}}^*\cdot\frac{\partial}{\partial\omega}\left(-j\omega\underline{\mathbf{B}}\right) - \frac{\partial\underline{\mathbf{E}}}{\partial\omega}\cdot(\underline{\mathbf{J}}_{\mathrm{u}}^* - j\omega\underline{\mathbf{D}}^*)$$

The medium is linear and isotropic (and stationary and nondeforming), but we allow temporal dispersion, so that the constitutive laws may be frequency-dependent,

$$\underline{\mathbf{D}} = \varepsilon(\omega)\underline{\mathbf{E}}$$
$$\underline{\mathbf{B}} = \mu(\omega)\underline{\mathbf{H}}$$
$$\underline{\mathbf{J}}_{\mathrm{u}} = \sigma(\omega)\underline{\mathbf{E}}$$

After substitution and division by the factor of (-4), we obtain

$$-\frac{1}{4}\nabla\cdot\left(\underline{\mathbf{E}}^*\times\frac{\partial\underline{\mathbf{H}}}{\partial\omega} + \frac{\partial\underline{\mathbf{E}}}{\partial\omega}\times\underline{\mathbf{H}}^*\right) = \frac{1}{2}\sigma\frac{\partial\underline{\mathbf{E}}}{\partial\omega}\cdot\underline{\mathbf{E}}^* + \frac{1}{4}\frac{\partial\sigma}{\partial\omega}\underline{\mathbf{E}}\cdot\underline{\mathbf{E}}^*$$
$$+j\frac{1}{4}\left[\frac{\partial(\omega\varepsilon)}{\partial\omega}\underline{\mathbf{E}}\cdot\underline{\mathbf{E}}^* + \frac{\partial(\omega\mu)}{\partial\omega}\underline{\mathbf{H}}\cdot\underline{\mathbf{H}}^*\right] \quad (6.11)$$

the final form of the theorem emerges. The imaginary part of the right-hand side is the total average energy density,

$$\langle W_{\mathrm{total}}\rangle = \frac{1}{4}\left[\frac{\partial(\omega\varepsilon)}{\partial\omega}\underline{\mathbf{E}}\cdot\underline{\mathbf{E}}^* + \frac{\partial(\omega\mu)}{\partial\omega}\underline{\mathbf{H}}\cdot\underline{\mathbf{H}}^*\right] + j\frac{1}{4}\sigma\left(\frac{\partial\underline{\mathbf{E}}^*}{\partial\omega}\cdot\underline{\mathbf{E}} - \frac{\partial\underline{\mathbf{E}}}{\partial\omega}\cdot\underline{\mathbf{E}}^*\right) \quad (6.12)$$

When $\sigma = 0$, the total becomes

$$\langle W_{\mathrm{total}}\rangle = \frac{1}{4}\left[\varepsilon(\omega)\underline{\mathbf{E}}\cdot\underline{\mathbf{E}}^* + \mu(\omega)\underline{\mathbf{H}}\cdot\underline{\mathbf{H}}^*\right] + \frac{1}{4}\,\omega\left[\frac{\partial\varepsilon(\omega)}{\partial\omega}\underline{\mathbf{E}}\cdot\underline{\mathbf{E}}^* + \frac{\partial\mu(\omega)}{\partial\omega}\underline{\mathbf{H}}\cdot\underline{\mathbf{H}}^*\right] \quad (6.13)$$

If μ and ε are independent of frequency, the familiar result emerges; if they are not, there is at least one other type of energy (often mechanical) masquerading as an electromagnetic component. The details are hidden in the constitutive law(s). It is a common error to interpret $\frac{1}{4}\,\varepsilon(\omega)\underline{\mathbf{E}}\cdot\underline{\mathbf{E}}^*$ and $\frac{1}{4}\mu(\omega)\underline{\mathbf{H}}\cdot\underline{\mathbf{H}}^*$ as, respectively, the electric and magnetic portions of the total[2]. In the complex Poynting Theorem, the term $\frac{1}{4}\left[\mu(\omega)\underline{\mathbf{H}}\cdot\underline{\mathbf{H}}^* - \varepsilon(\omega)\underline{\mathbf{E}}\cdot\underline{\mathbf{E}}^*\right]$ is therefore *not* simply the difference between the average magnetic and electric energy densities.

Alternate-energy theorem

Just as the complex Poynting Theorem has both Maxwell–Poynting and Alternate-representation forms, so too, does the energy theorem. Study of the power theorems and

[2]For example, a cold uniform (neutral) plasma can be modeled with $\mu = \mu_0$ and $\varepsilon = \varepsilon_0(1 - \omega_p^2/\omega^2)$ where ω_p is the plasma frequency. In this case, $\frac{1}{4}\,\varepsilon\underline{\mathbf{E}}\cdot\underline{\mathbf{E}}^*$ can be zero or *negative,* whereas $\frac{1}{4}\frac{\partial(\omega\varepsilon)}{\partial\omega}\underline{\mathbf{E}}\cdot\underline{\mathbf{E}}^* = \frac{1}{4}\varepsilon_0(1 + \omega_p^2/\omega^2)\underline{\mathbf{E}}\cdot\underline{\mathbf{E}}^*$ is the sum of two positive-definite terms: the true average electric-energy density and the average kinetic-energy density of the charged particles that make up the plasma.

Eq. (6.11) suggests that for materials (with uniform constitutive constants), the complex vector

$$4\underline{\mathbf{Q}}^{\mathrm{o}} = \underline{\Phi}^* \frac{\partial \underline{\mathbf{J}}_{\mathrm{u}}}{\partial \omega} + \frac{\partial \underline{\Phi}}{\partial \omega}\underline{\mathbf{J}}_{\mathrm{u}}^* + \frac{j\omega}{\mu}\left(\underline{A}_i^* \nabla \frac{\partial \underline{A}_i}{\partial \omega} - \frac{\partial \underline{A}_i}{\partial \omega}\nabla \underline{A}_i^*\right)$$
$$- j\omega\varepsilon\left(\underline{\Phi}^* \nabla \frac{\partial \underline{\Phi}}{\partial \omega} - \frac{\partial \underline{\Phi}}{\partial \omega}\nabla \underline{\Phi}^*\right)$$

(summation on i is implied) may be appropriate (one should be careful not to confuse the vector or tensor indices with the j that represents $\sqrt{-1}$). We begin evaluation of the divergence and find that

$$\nabla \cdot 4\underline{\mathbf{Q}}^{\mathrm{o}} = \nabla \cdot \left(\underline{\Phi}^* \frac{\partial \underline{\mathbf{J}}_{\mathrm{u}}}{\partial \omega} + \frac{\partial \underline{\Phi}}{\partial \omega}\underline{\mathbf{J}}_{\mathrm{u}}^*\right) + \frac{j\omega}{\mu}\left(\underline{A}_i^* \nabla^2 \frac{\partial \underline{A}_i}{\partial \omega} - \frac{\partial \underline{A}_i}{\partial \omega}\nabla^2 \underline{A}_i^*\right)$$
$$- j\omega\varepsilon\left(\underline{\Phi}^* \nabla^2 \frac{\partial \underline{\Phi}}{\partial \omega} - \frac{\partial \underline{\Phi}}{\partial \omega}\nabla^2 \underline{\Phi}^*\right)$$

Based upon the gauge $\underline{g} = 0$, the complex equations needed for further evaluation are, with $k^2 = \omega^2 \mu\varepsilon$ (and $\nabla k = 0$),

$$\nabla^2 \frac{\partial \underline{\mathbf{A}}}{\partial \omega} + k^2 \frac{\partial \underline{\mathbf{A}}}{\partial \omega} + 2k\frac{\partial k}{\partial \omega}\underline{\mathbf{A}} = -\frac{\partial(\mu \underline{\mathbf{J}}_{\mathrm{u}})}{\partial \omega} \tag{6.14a}$$

$$\nabla^2 \frac{\partial \underline{\Phi}}{\partial \omega} + k^2 \frac{\partial \underline{\Phi}}{\partial \omega} + 2k\frac{\partial k}{\partial \omega}\underline{\Phi} = -\frac{\partial(\underline{\rho}_{\mathrm{u}}/\varepsilon)}{\partial \omega} \tag{6.14b}$$

$$\frac{\partial \underline{\mathbf{E}}}{\partial \omega} = -j\left(\underline{\mathbf{A}} + \omega \frac{\partial \underline{\mathbf{A}}}{\partial \omega}\right) - \nabla \frac{\partial \underline{\Phi}}{\partial \omega} \tag{6.14c}$$

$$\nabla \cdot \frac{\partial \underline{\mathbf{J}}_{\mathrm{u}}}{\partial \omega} = -j\left(\underline{\rho}_{\mathrm{u}} + \omega \frac{\partial \underline{\rho}_{\mathrm{u}}}{\partial \omega}\right) \tag{6.14d}$$

With these substitutions, the result is

$$-\nabla \cdot \underline{\mathbf{Q}}^{\mathrm{o}} = \frac{1}{4}\left(\frac{\partial \underline{\mathbf{E}}}{\partial \omega} \cdot \underline{\mathbf{J}}_{\mathrm{u}}^* + \underline{\mathbf{E}}^* \cdot \frac{\partial \underline{\mathbf{J}}_{\mathrm{u}}}{\partial \omega}\right) + j\frac{1}{4}\left(\underline{\mathbf{A}} \cdot \underline{\mathbf{J}}_{\mathrm{u}}^* + \underline{\rho}\underline{\Phi}^*\right)$$
$$+ j\omega\frac{1}{8}\left(\frac{1}{\varepsilon}\frac{\partial \varepsilon}{\partial \omega}\underline{\rho}_{\mathrm{u}}\underline{\Phi}^* + \frac{1}{\mu}\frac{\partial \mu}{\partial \omega}\underline{\mathbf{A}}^* \cdot \underline{\mathbf{J}}_{\mathrm{u}}\right) + j\frac{1}{4}\omega k\frac{\partial k}{\partial \omega}\left(\frac{\underline{\mathbf{A}} \cdot \underline{\mathbf{A}}^*}{\mu} - \varepsilon \underline{\Phi}\underline{\Phi}^*\right)$$

As in the Maxwell–Poynting form of the theorem, it is the *imaginary part* of the right-hand side that is the time average of the total Alternate energy density. Therefore,

$$<W_{\mathrm{total}}^{\mathrm{o}}> = j\frac{1}{4}\sigma\left(\frac{\partial \underline{\mathbf{E}}^*}{\partial \omega} \cdot \underline{\mathbf{E}} - \frac{\partial \underline{\mathbf{E}}}{\partial \omega} \cdot \underline{\mathbf{E}}^*\right)$$
$$+ \frac{1}{8}\left[\frac{1}{\varepsilon}\frac{\partial(\omega\varepsilon)}{\partial \omega}\left(\underline{\rho}_{\mathrm{u}}\underline{\Phi}^* + \underline{\rho}_{\mathrm{u}}^*\underline{\Phi}\right) + \frac{1}{\mu}\frac{\partial(\omega\mu)}{\partial \omega}(\underline{\mathbf{A}}^* \cdot \underline{\mathbf{J}}_{\mathrm{u}} + \underline{\mathbf{A}} \cdot \underline{\mathbf{J}}_{\mathrm{u}}^*)\right]$$
$$+ \frac{1}{2}\omega k\frac{\partial k}{\partial \omega}\left(\frac{\underline{\mathbf{A}} \cdot \underline{\mathbf{A}}^*}{\mu} - \varepsilon \underline{\Phi}\underline{\Phi}^*\right) \tag{6.15}$$

When the material is nondispersive and lossless, both μ and ε are independent of frequency, $\sigma = 0$, and

$$<W^{\mathrm{o}}_{\mathrm{total}}> = \frac{1}{8}(\underline{\rho}_{\mathrm{u}}\underline{\Phi}^* + \underline{\rho}^*_{\mathrm{u}}\underline{\Phi} + \underline{\mathbf{A}}^* \cdot \underline{\mathbf{J}}_{\mathrm{u}} + \underline{\mathbf{A}} \cdot \underline{\mathbf{J}}^*_{\mathrm{u}}) + \frac{1}{2}\omega^2\varepsilon(\underline{\mathbf{A}} \cdot \underline{\mathbf{A}}^* - \mu\varepsilon\underline{\Phi}\underline{\Phi}^*) \tag{6.16}$$

This last result agrees with the time average of W^{o} as evaluated from Eq. (3.31b), when $g = 0$, and μ and ε are substituted for μ_{o} and ε_{o}.

For quasistatic fields, the terms proportional to ω^2 are second-order and higher and therefore negligible. Then, as expected, the average Alternate-energy density is completely localized to the unpaired currents and charges. Even when the frequency is high enough to invalidate quasistatics, the Alternate energy vanishes in all current- and charge-free regions wherever $\underline{\mathbf{A}} \cdot \underline{\mathbf{A}}^* = \mu\varepsilon\underline{\Phi}\underline{\Phi}^*$.

When the material is isotropic and lossless and the potentials represent a uniform plane wave with propagation vector $\mathbf{k}$, the quantity $k\frac{\partial k}{\partial\omega}$ can be replaced by $\mathbf{k}\cdot\frac{\partial\mathbf{k}}{\partial\omega}$. Then in charge- and current-free regions and also making use of Eq. (6.10) with $\underline{g} = 0$, we obtain

$$<W^{\mathrm{o}}_{\mathrm{total}}> = \frac{\partial\mathbf{k}}{\partial\omega} \cdot <\mathbf{S}^{\mathrm{o}}> \tag{6.17a}$$

$$<\mathbf{S}^{\mathrm{o}}> = \frac{1}{2}\omega\mathbf{k}\left(\frac{\underline{\mathbf{A}} \cdot \underline{\mathbf{A}}^*}{\mu} - \varepsilon\underline{\Phi}\underline{\Phi}^*\right) = <W^{\mathrm{o}}_{\mathrm{total}}>\mathbf{v}_{\mathrm{group}} \tag{6.17b}$$

$$\mathbf{v}_{\mathrm{group}} = \frac{\partial\omega}{\partial\mathbf{k}} \tag{6.17c}$$

The corresponding result is true in the Maxwell–Poynting representation, but not quite as evident.

6.3 COMPLEX STRESS THEOREMS

In the preceding section, the complex field equations were used to derive power theorems. Because we have already learned that each Cartesian component of the electromagnetic force and momentum density, together with the corresponding Maxwell stress vector, satisfies a "Poynting theorem," it follows that there should be three additional theorems involving the complex-force. These can be combined through the introduction of a complex stress tensor. Here, we seek complex-stress theorems in both the Maxwell–Poynting and Alternate representations. Before beginning, it is helpful to review the complex force of a simple mechanical system.

Newton's second law of motion applied to a single particle gives the change in momentum, $\mathbf{p}$, to an applied force as $\mathbf{F} = \frac{d}{dt}\mathbf{p}$. If the force is sinusoidal for all time, it may be expressed as $\mathbf{F} = Re\{\underline{\mathbf{F}}\exp(j\omega t)\}$. The complex force, $\underline{\mathbf{F}} = j\omega\underline{\mathbf{p}}$, has no time average, nor does the momentum. In the case of electromagnetic fields, forces are quadratic functions and so both average forces and average momentum can result in the sinusoidal steady state. Therefore, we expect to develop complex-stress theorems that relate the densities of these types of average (with components of the momentum density multiplied by $j\omega$).

Maxwell–Poynting stress theorem

The complex stress is defined as

$$\underline{T}_{ij} = \frac{1}{2}\left[\underline{E}_i\underline{D}_j^* + \underline{H}_i\underline{B}_j^* - \frac{1}{2}(\underline{\mathbf{E}}\cdot\underline{\mathbf{D}}^* + \underline{\mathbf{H}}\cdot\underline{\mathbf{B}}^*)\delta_{ij}\right] \tag{6.18}$$

The divergence evaluates to

$$\begin{aligned}\frac{\partial \underline{T}_{ij}}{\partial x_j} &= \frac{1}{2}(\underline{\rho}_{\rm u}^*\underline{\mathbf{E}} + \underline{\mathbf{J}}_{\rm u}\times\underline{\mathbf{B}}^*)_i \\ &\quad + \frac{1}{4}\left(\frac{\partial\underline{\mathbf{E}}}{\partial x_i}\cdot\underline{\mathbf{D}}^* - \underline{\mathbf{E}}\cdot\frac{\partial\underline{\mathbf{D}}^*}{\partial x_i} + \frac{\partial\underline{\mathbf{H}}}{\partial x_i}\cdot\underline{\mathbf{B}}^* - \underline{\mathbf{H}}\cdot\frac{\partial\underline{\mathbf{B}}^*}{\partial x_i}\right) \\ &\quad + j2\omega\frac{1}{4}(\underline{\mathbf{D}}\times\underline{\mathbf{B}}^* + \underline{\mathbf{D}}^*\times\underline{\mathbf{B}})_i \end{aligned} \tag{6.19}$$

and is comprised of the complex Lorentz-force density, components due to gradients of the complex fields, and $j2\omega$ times the time-averaged Minkowski momentum density. As expected, the real part of the divergence equals the time-averaged total force density.[3] In keeping with the form of the complex Poynting Theorem, the imaginary part can be cast as $j2\omega$ times the *difference* between the average Minkowski momentum density and that due to the charges, currents, and gradients.

Uniform plane waves

It is instructive to apply this result to the case of a uniform harmonic plane wave propagating in a region of uniform (isotropic) dissipationless material (with $k = |\mathbf{k}| = \omega\sqrt{\mu\varepsilon}$). Then $\underline{\mathbf{J}}_{\rm u}$, $\underline{\rho}_{\rm u}$, $\nabla\mu$, and $\nabla\varepsilon$ are all zero, and

$$\underline{\mathbf{D}} = \varepsilon\underline{\mathbf{E}}, \qquad \underline{\mathbf{B}} = \mu\underline{\mathbf{H}}$$

$$\underline{\mathbf{E}} = \underline{\mathbf{E}}_{\rm o}\exp(-j\mathbf{k}\cdot\mathbf{r})$$

$$\underline{\mathbf{H}} = \underline{\mathbf{H}}_{\rm o}\exp(-j\mathbf{k}\cdot\mathbf{r})$$

$$\underline{\mathbf{E}}_{\rm o}\cdot\mathbf{k} = 0, \qquad \underline{\mathbf{H}}_{\rm o}\cdot\mathbf{k} = 0$$

$$\underline{\mathbf{E}}_{\rm o}\times\underline{\mathbf{H}}_{\rm o}^* = \frac{\mathbf{k}}{\omega}\mu\underline{\mathbf{H}}_{\rm o}\cdot\underline{\mathbf{H}}_{\rm o}^* = \frac{\mathbf{k}}{\omega}\varepsilon\underline{\mathbf{E}}_{\rm o}\cdot\underline{\mathbf{E}}_{\rm o}^*$$

Because $\dfrac{\partial}{\partial x_i} = \dfrac{x_i}{r}\dfrac{\partial}{\partial r}$ and $\dfrac{x_i}{r}k = k_i$, the complex tensor, $\underline{T}_{ij}$ is independent of position. It follows that Eq. (6.19) evaluates to

$$-j\frac{1}{2}\mathbf{k}(\varepsilon\underline{\mathbf{E}}_{\rm o}\cdot\underline{\mathbf{E}}_{\rm o}^* + \mu\underline{\mathbf{H}}_{\rm o}\cdot\underline{\mathbf{H}}_{\rm o}^*) + j\omega\frac{1}{2}\mu\varepsilon(\underline{\mathbf{E}}_{\rm o}\times\underline{\mathbf{H}}_{\rm o}^* + \underline{\mathbf{E}}_{\rm o}^*\times\underline{\mathbf{H}}_{\rm o}) = 0$$

We therefore verify the important relationship:

$$<\mathbf{G}> = \frac{\mathbf{k}}{\omega}<W> \tag{6.20a}$$

[3]This is the total force of electromagnetic origin as *defined* by the Minkowski formulation. It is not the total force density acting within the material. Refer to Chapter 9 for a detailed discussion of electromagnetic forces and references to the previous literature.

which, together with

$$<\mathbf{S}> = \frac{\partial \omega}{\partial \mathbf{k}} <W> \tag{6.20b}$$

permits a particle representation of plane waves; in free-space, these are photons.

Alternate-stress theorem

For uniform isotropic materials and maintaining $\underline{g} = 0$, the complex Alternate-stress tensor can be defined as

$$\underline{T}^{\text{o}}_{ij} = -\frac{1}{2}\underline{A}_i\underline{J}^*_j + \frac{1}{4}(\underline{\mathbf{A}} \cdot \underline{\mathbf{J}}^*_{\text{u}} - \underline{\Phi}\underline{\rho}^*_{\text{u}})\delta_{ij} + \frac{1}{4\mu}\left(\underline{\mathbf{A}} \cdot \frac{\partial^2 \underline{\mathbf{A}}^*}{\partial x_i \partial x_j} - \frac{\partial \underline{\mathbf{A}}}{\partial x_i} \cdot \frac{\partial \underline{\mathbf{A}}^*}{\partial x_j}\right) - \frac{1}{4}\varepsilon\left(\underline{\Phi}\frac{\partial^2 \underline{\Phi}^*}{\partial x_i \partial x_j} - \frac{\partial \underline{\Phi}}{\partial x_i} \cdot \frac{\partial \underline{\Phi}^*}{\partial x_j}\right)$$

The divergence evaluates to

$$\frac{\partial \underline{T}^{\text{o}}_{ij}}{\partial x_j} = \left[\begin{array}{c} \frac{1}{2}(\underline{\rho}^*_{\text{u}}\underline{\mathbf{E}} + \underline{\mathbf{J}}^*_{\text{u}} \times \underline{\mathbf{B}})_i \\ +\frac{1}{4}\left[\frac{1}{\mu}\left(\frac{\partial \underline{\mathbf{A}}}{\partial x_j} \cdot \frac{\partial^2 \underline{\mathbf{A}}^*}{\partial x_i \partial x_j} - \frac{\partial \underline{\mathbf{A}}^*}{\partial x_j} \cdot \frac{\partial^2 \underline{\mathbf{A}}}{\partial x_i \partial x_j}\right) - \varepsilon\left(\frac{\partial \underline{\Phi}}{\partial x_j}\frac{\partial^2 \underline{\Phi}^*}{\partial x_i \partial x_j} - \frac{\partial \underline{\Phi}^*}{\partial x_j}\frac{\partial^2 \underline{\Phi}}{\partial x_i \partial x_j}\right)\right] \\ +\frac{k^2}{4}\left[\frac{1}{\mu}\left(\frac{\partial \underline{\mathbf{A}}}{\partial x_i} \cdot \underline{\mathbf{A}}^* - \underline{\mathbf{A}} \cdot \frac{\partial \underline{\mathbf{A}}^*}{\partial x_i}\right) - \varepsilon\left(\frac{\partial \underline{\Phi}}{\partial x_i}\underline{\Phi}^* - \underline{\Phi}\frac{\partial \underline{\Phi}^*}{\partial x_i}\right)\right] \end{array}\right] \tag{6.21}$$

The real part is solely due to the Lorentz-force density; this is a consequence of our assuming that the gradients of the permeability and permittivity are zero. When such gradients exist, it is easier to use the Maxwell–Poynting form than seek the generalization of Eq. (6.21). Finally, notice that $\frac{i}{\omega}\frac{1}{2}\underline{\rho}^*_{\text{u}}\underline{\mathbf{E}}$ contains the complex localized momentum density, $\frac{1}{2}\underline{\mathbf{A}}\underline{\rho}^*_{\text{u}}$.

6.4 COMPLEX MOMENTUM THEOREMS

The complex-field equations and their derivatives with respect to frequency were used to derive energy theorems that were extensions of the complex-power theorems. A similar approach is now employed to derive momentum theorems (in both the Maxwell–Poynting and Alternate representations) from extensions of the complex stress of the previous section. Complex vectors $\underline{\mathbf{Q}}$ (related to $\underline{\mathbf{S}}$) are the basis of the energy theorems; here we employ complex tensors $\overline{\underline{\mathbf{Q}}}$ (related to $\overline{\underline{\mathbf{T}}}$).

Maxwell–Poynting momentum theorem

With $\underline{Q}_{ij}$ defined by

$$\underline{Q}_{ij} = \frac{1}{4}\left(\frac{\partial \underline{E}_i}{\partial \omega}\underline{D}^*_j + \underline{E}^*_i\frac{\partial \underline{D}_j}{\partial \omega} + \frac{\partial \underline{H}_i}{\partial \omega}\underline{B}^*_j + \underline{H}^*_i\frac{\partial \underline{B}_j}{\partial \omega}\right) - \frac{1}{8}\left(\frac{\partial \underline{\mathbf{E}}}{\partial \omega} \cdot \underline{\mathbf{D}}^* + \underline{\mathbf{E}}^* \cdot \frac{\partial \underline{\mathbf{D}}}{\partial \omega} + \frac{\partial \underline{\mathbf{H}}}{\partial \omega} \cdot \underline{\mathbf{B}}^* + \underline{\mathbf{H}}^* \cdot \frac{\partial \underline{\mathbf{B}}}{\partial \omega}\right)\delta_{ij}$$

the divergence is given by

$$\frac{\partial \underline{Q}_{ij}}{\partial x_j} = \begin{bmatrix} \frac{1}{4}\left(\underline{\rho}_u^* \frac{\partial \underline{E}_i}{\partial \omega} + \frac{\partial \underline{\rho}_u}{\partial \omega}\underline{E}_i^*\right) + \frac{1}{4}\left(\frac{\partial \underline{\mathbf{J}}_u}{\partial \omega} \times \underline{\mathbf{B}}^* + \underline{\mathbf{J}}_u^* \times \frac{\partial \underline{\mathbf{B}}}{\partial \omega}\right)_i \\ +\frac{1}{8}\left(\frac{\partial^2 \underline{\mathbf{E}}}{\partial x_i \partial \omega} \cdot \underline{\mathbf{D}}^* - \frac{\partial \underline{\mathbf{E}}}{\partial \omega} \cdot \frac{\partial \underline{\mathbf{D}}^*}{\partial x_i} + \frac{\partial \underline{\mathbf{E}}^*}{\partial x_i} \cdot \frac{\partial \underline{\mathbf{D}}}{\partial \omega} - \underline{\mathbf{E}}^* \cdot \frac{\partial^2 \underline{\mathbf{D}}}{\partial x_i \partial \omega}\right) \\ +\frac{1}{8}\left(\frac{\partial^2 \underline{\mathbf{H}}}{\partial x_i \partial \omega} \cdot \underline{\mathbf{B}}^* - \frac{\partial \underline{\mathbf{H}}}{\partial \omega} \cdot \frac{\partial \underline{\mathbf{B}}^*}{\partial x_i} + \frac{\partial \underline{\mathbf{H}}^*}{\partial x_i} \cdot \frac{\partial \underline{\mathbf{B}}}{\partial \omega} - \underline{\mathbf{H}}^* \cdot \frac{\partial^2 \underline{\mathbf{B}}}{\partial x_i \partial \omega}\right) \\ +j\frac{1}{4}(\underline{\mathbf{D}} \times \underline{\mathbf{B}}^* + \underline{\mathbf{D}}^* \times \underline{\mathbf{B}})_i \end{bmatrix}$$

The imaginary part of this divergence is the time-averaged *total* momentum density; that may include contributions from the free charges and currents. If μ and ε are frequency dependent, the total may also include additional terms associated with the underlying physics that renders the material dispersive. In regions where the free current and free charge is zero, we have $\underline{\mathbf{D}} = \varepsilon(\omega)\ \underline{\mathbf{E}}$ and $\underline{\mathbf{B}} = \mu(\omega)\ \underline{\mathbf{H}}$; and with summation over the repeated index, i, assumed, we obtain

$$<\mathbf{G}_{\text{total}}> = \begin{bmatrix} \frac{1}{4}\mu(\omega)\varepsilon(\omega)(\underline{\mathbf{E}} \times \underline{\mathbf{H}}^* + \underline{\mathbf{E}}^* \times \underline{\mathbf{H}}) \\ +j\frac{1}{8}\left[\frac{\partial \varepsilon}{\partial \omega}(\underline{E}_i^* \nabla \underline{E}_i - \underline{E}_i \nabla \underline{E}_i^*) + \frac{\partial \mu}{\partial \omega}(\underline{H}_i^* \nabla \underline{H}_i - \underline{H}_i \nabla \underline{H}_i^*)\right] \\ +j\frac{1}{4}\left[\nabla\varepsilon\left(\frac{\partial \underline{\mathbf{E}}}{\partial \omega} \cdot \underline{\mathbf{E}}^* - \underline{\mathbf{E}} \cdot \frac{\partial \underline{\mathbf{E}}^*}{\partial \omega}\right) + \nabla\mu\left(\frac{\partial \underline{\mathbf{H}}}{\partial \omega} \cdot \underline{\mathbf{H}}^* - \underline{\mathbf{H}} \cdot \frac{\partial \underline{\mathbf{H}}^*}{\partial \omega}\right)\right] \end{bmatrix} \tag{6.22}$$

For uniform plane waves (propagating in a uniform material), all complex fields are proportional to $\exp(-j\mathbf{k} \cdot \mathbf{r})$ with $\mathbf{k} \cdot \mathbf{k} = \omega^2 \mu\varepsilon$. It therefore follows that

$$<\mathbf{G}_{\text{total}}> = \frac{1}{4}\frac{\partial(\omega\mu\varepsilon)}{\partial \omega}(\underline{\mathbf{E}} \times \underline{\mathbf{H}}^* + \underline{\mathbf{E}}^* \times \underline{\mathbf{H}}) \tag{6.23}$$

As with the average total energy density, dispersion indicates that there are physical mechanisms[4] hidden in the material (generally of a mechanical nature) that are coupled to the electric and/or magnetic dipoles. In addition to the coupling provided to the electromagnetic fields, there may be other important aspects that require further study.

Alternate-momentum theorem

In a like-manner, and still assuming that the material is uniform, and $\underline{g} = 0$, the complex Alternate-stress tensor, $\overline{\underline{\mathbf{Q}}}^{\text{o}}$, is defined by

[4]In addition to the plasma discussed in a previous footnote, examples include: ferrite, piezoelectric, piezomagnetic, electroelastic, and magnetoelastic materials. If dissipation is present, the frequency-dependent permittivity and/or permeability will be complex and often nonisotropic parameters.

$$\underline{Q}^{o}_{ij} = \left[\begin{array}{c} -\frac{1}{4}\left(\frac{\partial \underline{A}_i}{\partial \omega}\underline{J}_j^* + \underline{A}_i^* \frac{\partial \underline{J}_j}{\partial \omega}\right) \\ +\frac{1}{8}\left(\frac{\partial \underline{\mathbf{A}}}{\partial \omega}\cdot \underline{\mathbf{J}}_{\mathrm{u}}^* + \underline{\mathbf{A}}^* \frac{\partial \underline{\mathbf{J}}_{\mathrm{u}}}{\partial \omega} - \frac{\partial \underline{\Phi}}{\partial \omega}\underline{\rho}_{\mathrm{u}}^* - \underline{\Phi}^* \frac{\partial \underline{\rho}_{\mathrm{u}}}{\partial \omega}\right)\delta_{ij} \\ +\frac{1}{8\mu}\left(\frac{\partial \underline{\mathbf{A}}}{\partial \omega}\cdot \frac{\partial^2 \underline{\mathbf{A}}^*}{\partial x_i \partial x_j} + \underline{\mathbf{A}}^* \cdot \frac{\partial^3 \underline{\mathbf{A}}}{\partial \omega \partial x_i \partial x_j} - \frac{\partial^2 \underline{\mathbf{A}}}{\partial \omega \partial x_i}\cdot \frac{\partial \underline{\mathbf{A}}^*}{\partial x_j} - \frac{\partial \underline{\mathbf{A}}^*}{\partial x_i}\cdot \frac{\partial^2 \underline{\mathbf{A}}}{\partial \omega \partial x_j}\right) \\ -\frac{1}{8}\varepsilon\left(\frac{\partial \underline{\Phi}}{\partial \omega}\frac{\partial^2 \underline{\Phi}^*}{\partial x_i \partial x_j} + \underline{\Phi}^* \frac{\partial^3 \underline{\Phi}}{\partial \omega \partial x_i \partial x_j} - \frac{\partial^2 \underline{\Phi}}{\partial \omega \partial x_i}\frac{\partial \underline{\Phi}^*}{\partial x_j} - \frac{\partial \underline{\Phi}^*}{\partial x_i}\frac{\partial^2 \underline{\Phi}}{\partial \omega \partial x_j}\right) \end{array} \right] \tag{6.24}$$

As was the case with the Alternate-energy theorem, Eqs. (6.14) are required to evaluate $\nabla \cdot \underline{\overline{\mathbf{Q}}}^{o}$, which in summed index notation is

$$\frac{\partial \underline{Q}^{o}_{ij}}{\partial x_j} = \left[\begin{array}{c} j\frac{1}{8}\underline{A}_i^*\underline{\rho}_{\mathrm{u}} + \frac{1}{4}\left(\underline{\rho}_{\mathrm{u}}^* \frac{\partial \underline{E}_i}{\partial \omega} + \frac{\partial \underline{\rho}_{\mathrm{u}}}{\partial \omega}\underline{E}_i^*) + \frac{1}{4}(\frac{\partial \underline{\mathbf{J}}_{\mathrm{u}}}{\partial \omega}\times \underline{\mathbf{B}}^* + \underline{\mathbf{J}}_{\mathrm{u}}^* \times \frac{\partial \underline{\mathbf{B}}}{\partial \omega}\right)_i \\ -\frac{1}{8\mu}\frac{\partial \mu}{\partial \omega}\left(\underline{\mathbf{A}}^*\cdot \frac{\partial \underline{\mathbf{J}}_{\mathrm{u}}}{\partial x_i} - \frac{\partial \underline{\mathbf{A}}^*}{\partial x_i}\cdot \underline{\mathbf{J}}_{\mathrm{u}}\right) - \frac{1}{8\varepsilon}\frac{\partial \varepsilon}{\partial \omega}\left(\underline{\Phi}^* \frac{\partial \underline{\rho}_{\mathrm{u}}}{\partial x_i} - \frac{\partial \underline{\Phi}^*}{\partial x_i}\underline{\rho}_{\mathrm{u}}\right) \\ +\frac{1}{8}\frac{\partial(\omega^2\mu\varepsilon)}{\partial \omega}\left[\frac{1}{\mu}\left(\underline{\mathbf{A}}\cdot \frac{\partial \underline{\mathbf{A}}^*}{\partial x_i} - \frac{\partial \underline{\mathbf{A}}}{\partial x_i}\cdot \underline{\mathbf{A}}^*) - \varepsilon(\underline{\Phi}\frac{\partial \underline{\Phi}^*}{\partial x_i} - \frac{\partial \underline{\Phi}}{\partial x_i}\underline{\Phi}^*\right)\right] \end{array} \right] \tag{6.25}$$

The *imaginary part* of this divergence is the time-averaged *total* momentum density; this may include contributions from the free charges and currents. If μ and ε are frequency dependent, the total may also include additional terms associated with the underlying physics that renders the material dispersive.

In regions where the free current and free charge is zero, we have

$$\begin{aligned} <\mathbf{G}^{o}_{\mathrm{total}}>_i &= -j\frac{1}{8}\frac{\partial(\omega^2\mu\varepsilon)}{\partial \omega}\left[\frac{1}{\mu}\left(\underline{\mathbf{A}}\cdot \frac{\partial \underline{\mathbf{A}}^*}{\partial x_i} - \frac{\partial \underline{\mathbf{A}}}{\partial x_i}\cdot \underline{\mathbf{A}}^*\right) - \varepsilon\left(\underline{\Phi}\frac{\partial \underline{\Phi}^*}{\partial x_i} - \frac{\partial \underline{\Phi}}{\partial x_i}\underline{\Phi}^*\right)\right] \\ &= \frac{k}{\omega}\frac{\partial k}{\partial \omega}(\mathbf{S}^{o}_{\mathrm{total}})_i \end{aligned} \tag{6.26}$$

As with the Alternate power flux and energy density, the *nonlocalized* average total momentum density vanishes whenever $\underline{\mathbf{A}} = \sqrt{\mu\varepsilon}\underline{\Phi}\widehat{\mathbf{u}}$ and $\widehat{\mathbf{u}}$ is a unit vector of fixed direction. In the general case, but when the material is nondispersive, $<\mathbf{G}^{o}_{\mathrm{total}}> = \mu\varepsilon\mathbf{S}^{o}_{\mathrm{total}}$.

6.5 DUALITY

Dual electric and magnetic fields

In regions of free-space where there is neither charge nor current, Maxwell's Equations are symmetric in the sense that, if **E** and **H** satisfy the equations, either of the substitutions (columns)

$$\left\| \begin{array}{cc} \mathbf{E} \rightarrow \pm \mathbf{H}' & \mathbf{H} \rightarrow \mp \mathbf{E}' \\ \mu_o \rightarrow \varepsilon_o & \varepsilon_o \rightarrow \mu_o \\ \varepsilon_o \rightarrow \mu_o & \mu_o \rightarrow \varepsilon_o \end{array} \right\| \tag{6.27}$$

leave the source-free equations unchanged and reveal that $\mathbf{E}'$ and $\mathbf{H}'$ are also proper solutions of Maxwell's Equations (Eqs. (1.8a) and (1.8b) are merely interchanged). The primed and unprimed fields are said to be duals of one another.[5] The switching of the permittivity and permeability places the dual fields in a different universe; this can be avoided by altering the substitutions to either of

$$\left\| \mathbf{E} \to \pm\sqrt{\frac{\mu_o}{\varepsilon_o}}\mathbf{H}' \qquad \mathbf{H} \to \mp\sqrt{\frac{\varepsilon_o}{\mu_o}}\mathbf{E}' \right\| \tag{6.28}$$

Less general forms of duality also exist that may prove useful in certain situations. For example, if $\widehat{\mathbf{u}}$ is a unit vector of fixed direction and both $\mathbf{E}$ and $\mathbf{H}$ are two-dimensional fields transverse to $\widehat{\mathbf{u}}$, then μ_o and ε_o also remain unchanged and either of

$$\left\| \mathbf{E} \to \pm\sqrt{\frac{\mu_o}{\varepsilon_o}}\widehat{\mathbf{u}} \times \mathbf{H}' \qquad \mathbf{H} \to \mp\sqrt{\frac{\varepsilon_o}{\mu_o}}\widehat{\mathbf{u}} \times \mathbf{E}' \right\| \tag{6.29}$$

produces dual fields that are rotated with respect to each other by 90 degrees. We shall make use of this result in connection with Babinet's Principle.

Dual sources

If electric-charge and electric-current densities are present in the free-space region, duality fails unless dual currents and dual charges exist. Without prejudging their physical existence, let us add (mathematically) magnetic-charge and magnetic-current densities (ρ_m and $\mathbf{J}_m$) and, for clarity, add the subscript e to the electric-charge and electric-current densities. The modified Maxwell Equations become

$$\nabla \times \mathbf{H} - \varepsilon_o \frac{\partial \mathbf{E}}{\partial t} = \mathbf{J}_e \tag{6.30a}$$

$$\nabla \times \mathbf{E} + \mu_o \frac{\partial \mathbf{H}}{\partial t} = -\mathbf{J}_m \tag{6.30b}$$

$$\nabla \cdot \varepsilon_o \mathbf{E} = \rho_e \tag{6.30c}$$

$$\nabla \cdot \mu_o \mathbf{H} = \rho_m \tag{6.30d}$$

As a consequence of these equations, electric and magnetic charges are conserved separately:

$$\nabla \cdot \mathbf{J}_e + \frac{\partial \rho_e}{\partial t} = 0 \tag{6.31a}$$

$$\nabla \cdot \mathbf{J}_m + \frac{\partial \rho_m}{\partial t} = 0 \tag{6.31b}$$

Equation (6.31b) validates our sign conventions.

To be consistent, there must also be a Lorentz force on magnetic charge and magnetic current. In particular,

$$\mathbf{f}_e = \rho_e \mathbf{E} + \mathbf{J}_e \times \mu_o \mathbf{H} \tag{6.32a}$$

$$\mathbf{f}_m = \rho_m \mathbf{H} - \mathbf{J}_m \times \varepsilon_o \mathbf{E} \tag{6.32b}$$

[5]How this fact is taken depends upon one's view of the world. An optimist will be delighted because every solution learned means twice as much knowledge has been gained; a pessimist will be discouraged because there are twice as many problems that he/she cannot solve!

The generalization of Eqs. (6.27) and (6.28) are

$$\left\| \begin{array}{ll} \mathbf{E} \to \pm\mathbf{H}' & \mathbf{H} \to \mp\mathbf{E}' \\ \mathbf{J}_\mathrm{e} \to \pm\mathbf{J}'_\mathrm{m} & \rho_\mathrm{e} \to \pm\rho'_\mathrm{m} \\ \mathbf{J}_\mathrm{m} \to \mp\mathbf{J}'_\mathrm{e} & \rho_\mathrm{m} \to \mp\rho'_\mathrm{e} \\ \mu_\mathrm{o} \to \varepsilon_\mathrm{o} & \varepsilon_\mathrm{o} \to \mu_\mathrm{o} \\ \varepsilon_\mathrm{o} \to \mu_\mathrm{o} & \mu_\mathrm{o} \to \varepsilon_\mathrm{o} \end{array} \right\| \tag{6.33}$$

and

$$\left\| \begin{array}{ll} \mathbf{E} \to \pm\sqrt{\frac{\mu_\mathrm{o}}{\varepsilon_\mathrm{o}}}\mathbf{H}' & \mathbf{H} \to \mp\sqrt{\frac{\varepsilon_\mathrm{o}}{\mu_\mathrm{o}}}\mathbf{E}' \\ \mathbf{J}_\mathrm{e} \to \pm\sqrt{\frac{\varepsilon_\mathrm{o}}{\mu_\mathrm{o}}}\mathbf{J}'_\mathrm{m} & \mathbf{J}_\mathrm{m} \to \mp\sqrt{\frac{\mu_\mathrm{o}}{\varepsilon_\mathrm{o}}}\mathbf{J}'_\mathrm{e} \\ \rho_\mathrm{e} \to \pm\sqrt{\frac{\varepsilon_\mathrm{o}}{\mu_\mathrm{o}}}\rho'_\mathrm{m} & \rho_\mathrm{m} \to \mp\sqrt{\frac{\mu_\mathrm{o}}{\varepsilon_\mathrm{o}}}\rho'_\mathrm{e} \\ \mu_\mathrm{o} \to \mu_\mathrm{o} & \varepsilon_\mathrm{o} \to \varepsilon_\mathrm{o} \\ \varepsilon_\mathrm{o} \to \varepsilon_\mathrm{o} & \mu_\mathrm{o} \to \mu_\mathrm{o} \end{array} \right\| \tag{6.34}$$

Dual materials

When dielectric and magnetic materials are present, the flux vectors, **B** and **D**, are duals expressed in terms of either the Chu or Minkowski fields:

$$\mathbf{B} = \mu_\mathrm{o}(\mathbf{H} + \mathbf{M}) = \mu_\mathrm{o}(\mathbf{H}^\mathrm{o} + \mathbf{M}^\mathrm{o})$$

$$\mathbf{D} = \varepsilon_\mathrm{o}\mathbf{E} + \mathbf{P} = \varepsilon_\mathrm{o}\mathbf{E}^\mathrm{o} + \mathbf{P}^\mathrm{o}$$

For stationary materials the superscript o can be omitted. For linear isotropic materials, the permeability and permittivity, μ and ε, are also duals. If there is dissipation that obeys Ohm's Law, $\mathbf{J}_\mathrm{eu} = \sigma_\mathrm{e}\mathbf{E}$, its dual, $\mathbf{J}_\mathrm{mu} = \sigma_\mathrm{m}\mathbf{H}$, must then be introduced and

$$\left\| \begin{array}{ll} \mathbf{D} \to \pm\mathbf{B}' & \mathbf{B} \to \mp\mathbf{D}' \\ \mathbf{P} \to \pm\mu_\mathrm{o}\mathbf{M}' & \mu_\mathrm{o}\mathbf{M} \to \mp\mathbf{P}' \\ \sigma_\mathrm{e} \to \sigma_\mathrm{m} & \sigma_\mathrm{m} \to \sigma_\mathrm{e} \\ \mu \to \varepsilon & \varepsilon \to \mu \\ \varepsilon \to \mu & \mu \to \varepsilon \end{array} \right\| \tag{6.35}$$

Notice that, because of the way electric and magnetic dipole moments are defined, it is **P** and $\mu_\mathrm{o}\mathbf{M}$ that are duals.

If, instead, μ_0 and ε_0 are left unchanged, we obtain

$$\left\| \begin{array}{cc} \mathbf{D} \to \pm\sqrt{\frac{\varepsilon_0}{\mu_0}}\mathbf{B}' & \mathbf{B} \to \mp\sqrt{\frac{\mu_0}{\varepsilon_0}}\mathbf{D}' \\ \mathbf{P} \to \pm\sqrt{\mu_0\varepsilon_0}\mathbf{M}' & \mathbf{M} \to \mp\mathbf{P}'/\sqrt{\mu_0\varepsilon_0} \\ \sigma_e \to \frac{\varepsilon_0}{\mu_0}\sigma_m & \sigma_m \to \frac{\mu_0}{\varepsilon_0}\sigma_e \\ \mu \to \frac{\mu_o}{\varepsilon_o}\varepsilon & \varepsilon \to \frac{\varepsilon_o}{\mu_o}\mu \\ \varepsilon \to \frac{\varepsilon_o}{\mu_o}\mu & \mu \to \frac{\mu_o}{\varepsilon_o}\varepsilon \end{array} \right\| \tag{6.36}$$

where now it is **P** and $\mathbf{M}/c$ that are are duals. In any event, perfect magnetic conductors ($\sigma_m = \infty$) are required in order to complete duality. As in the case of perfect electric conductors, the skin depth ($\delta_m = \sqrt{2/\omega\varepsilon\sigma_m}$) is indeterminate when $\omega = 0$; we again *define* the limit as zero.

Boundary conditions

Because both electric-surface-charge and electric-surface-current densities (σ_e^s, K_e^s) are permitted, their duals must be included. The magnetic-surface-charge and magnetic-surface-current densities are designated by σ_m^s and K_m^s. There are unpaired (free) components and bound (dipolar) components

The set of boundary conditions that apply to the unpaired sources and generalize those of Chapter 1, Section 1.8 are

$$\mathbf{n} \times [\mathbf{H}^{(1)} - \mathbf{H}^{(2)}] = \mathbf{K}_{eu}^s \tag{6.37a}$$

$$\mathbf{n} \times [\mathbf{E}^{(1)} - \mathbf{E}^{(2)}] = -K_{mu}^s \tag{6.37b}$$

$$\mathbf{n} \cdot [\mathbf{D}^{(1)} - \mathbf{D}^{(2)}] = \sigma_{eu}^s \tag{6.37c}$$

$$\mathbf{n} \cdot [\mathbf{B}^{(1)} - \mathbf{B}^{(2)}] = \sigma_{mu}^s \tag{6.37d}$$

$$\mathbf{n} \cdot [\mathbf{J}_{eu}^{(1)} - \mathbf{J}_{eu}^{(2)}] + \nabla_\Sigma \cdot \mathbf{K}_{eu}^s + \frac{\partial \sigma_{eu}^s}{\partial t} = 0 \tag{6.37e}$$

$$\mathbf{n} \cdot [\mathbf{J}_{mu}^{(1)} - \mathbf{J}_{mu}^{(2)}] + \nabla_\Sigma \cdot \mathbf{K}_{mu}^s + \frac{\partial \sigma_{mu}^s}{\partial t} = 0 \tag{6.37f}$$

Dual vector potentials

If $\mathbf{J}_e = 0$ and $\rho_e = 0$, we obtain

$$\nabla \times \mathbf{E} + \mu_0 \frac{\partial \mathbf{H}}{\partial t} = -\mathbf{J}_m$$

$$\mu_0 \nabla \cdot \mathbf{H} = \rho_m$$

$$\nabla \times \mathbf{H} - \varepsilon_o \frac{\partial \mathbf{E}}{\partial t} = 0$$

$$\varepsilon_o \nabla \cdot \mathbf{E} = 0$$

Although the conventional vector and scalar potentials are no longer valid, their duals, $\mathbf{A}_m$ and Φ_m, can be introduced. Then,

$$\varepsilon_o \mathbf{E} = -\nabla \times \mathbf{A}_m$$

$$\mathbf{H} = -\frac{\partial \mathbf{A}_m}{\partial t} - \nabla \Phi_m$$

and (in the dual Lorenz gauge)

$$\nabla \cdot \mathbf{A}_m + \mu_o \varepsilon_o \frac{\partial \Phi_m}{\partial t} = 0$$

$$\nabla^2 \mathbf{A}_m - \mu_o \varepsilon_o \frac{\partial^2 \mathbf{A}_m}{\partial t^2} = -\varepsilon_o \mathbf{J}_m$$

$$\nabla^2 \Phi_m - \mu_o \varepsilon_o \frac{\partial^2 \Phi_m}{\partial t^2} = -\frac{1}{\mu_o} \rho_m$$

Dual Alternate energy-momentum tensors

With the introduction of the magnetic charge and magnetic current, it follows that there will be an Alternate formulation of power, energy, stress, and momentum that is the dual of that which exists for electric charge and electric current. Based upon the choices,

$$\left\| \begin{array}{c} \mathbf{J} \rightarrow \mathbf{J}_m \\ \rho \rightarrow \rho_m \\ \mathbf{A} \rightarrow \mathbf{A}_m \\ \Phi \rightarrow \Phi_m \\ \mu_o \rightarrow \varepsilon_o \\ \varepsilon_o \rightarrow \mu_o \end{array} \right\| \tag{6.38}$$

or, if μ_o and ε_o are left unchanged,

$$\left\| \begin{array}{c} \mathbf{J} \rightarrow \sqrt{\frac{\varepsilon_o}{\mu_o}} \mathbf{J}_m \\ \rho \rightarrow \sqrt{\frac{\varepsilon_o}{\mu_o}} \rho_m \\ \mathbf{A} \rightarrow \sqrt{\frac{\mu_o}{\varepsilon_o}} \mathbf{A}_m \\ \Phi \rightarrow \sqrt{\frac{\mu_o}{\varepsilon_o}} \Phi_m \end{array} \right\| \tag{6.39}$$

the new Alternate quantities (indicated by the superscript om) can be written down by inspection using

$$\left\| \begin{matrix} \mathbf{S}^{\mathrm{o}} \to \mathbf{S}^{\mathrm{om}} \\ W^{\mathrm{o}} \to W^{\mathrm{om}} \\ \mathbf{G}^{\mathrm{o}} \to \mathbf{G}^{\mathrm{om}} \\ T^{\mathrm{o}}_{ij} \to T^{\mathrm{om}}_{ij} \end{matrix} \right\| \tag{6.40}$$

There are, of course, dual versions of the complex power, energy, and momentum theorems given in Sections 6.2–6.4.

Superposition of sources

When the fields are generated by *either* electric charge/current *or* magnetic charge/current, but *not* both, then only one of the fields (**H** or **E**) is everywhere divergence-free. This fact permits one kind of vector and scalar potential to be used globally; these potentials, in turn, generate the corresponding type of Alternate power, energy, stress, and momentum. But, what if both electric and magnetic charge/current is employed to simplify the analysis of the fields? Then neither $\mathbf{A}_{\mathrm{e}}$ nor $\mathbf{A}_{\mathrm{m}}$ can be defined everywhere[6] and it would appear that the Alternate representation is fatally flawed. Of course, there is no similar difficulty with the Maxwell–Poynting representation, which depends upon the local field values and not on what type of source generated them. The way out of the dilemma is simple: Because Maxwell's Equations are linear, we use the principle of superposition to divide the fields into two sets; one generated by $\mathbf{J}_{\mathrm{e}}$ and ρ_{e} produces $\mathbf{A}_{\mathrm{e}}, \Phi_{\mathrm{e}}$, $\mathbf{E}_{\mathrm{e}}$ and $\mathbf{H}_{\mathrm{e}}$; the other generated by $\mathbf{J}_{\mathrm{m}}$ and ρ_{m} produces $\mathbf{A}_{\mathrm{m}}$, Φ_{m}, $\mathbf{E}_{\mathrm{m}}$, and $\mathbf{H}_{\mathrm{m}}$. The total fields are the sums: $\mathbf{E} = \mathbf{E}_{\mathrm{e}} + \mathbf{E}_{\mathrm{m}}$ and $\mathbf{H} = \mathbf{H}_{\mathrm{e}} + \mathbf{H}_{\mathrm{m}}$. Notice that because power, energy, stress, and momentum all are quadratic functions, there are cross-terms to be considered. With the Poynting vector, this is straightforward, but the Alternate representation is more problematic. Although the fluxes $\mathbf{S}^{\mathrm{o}}$ and $\mathbf{S}^{\mathrm{om}}$ can substitute, respectively, for $\mathbf{E}_{\mathrm{e}} \times \mathbf{H}_{\mathrm{e}}$ and $\mathbf{E}_{\mathrm{m}} \times \mathbf{H}_{\mathrm{m}}$, what substitution, if any, should be made for the cross-terms $\mathbf{E}_{\mathrm{e}} \times \mathbf{H}_{\mathrm{m}}$ and $\mathbf{E}_{\mathrm{m}} \times \mathbf{H}_{\mathrm{e}}$ requires careful consideration. The flux, $\mathbf{S}^{\mathrm{ox}}$, and energy, W^{ox}, must satisfy the crossed-field power theorem,

$$\nabla \cdot \mathbf{S}^{\mathrm{ox}} + \frac{\partial W^{\mathrm{ox}}}{\partial t} = -\mathbf{E}_{\mathrm{m}} \cdot \mathbf{J}_{\mathrm{e}} - \mathbf{H}_{\mathrm{e}} \cdot \mathbf{J}_{\mathrm{m}} \tag{6.41}$$

In a like manner, the stress tensor, $\overline{\overline{\mathbf{T}}}^{\mathrm{ox}}$, and momentum density, $\mathbf{G}^{\mathrm{ox}}$, must satisfy

$$\nabla \cdot \overline{\overline{\mathbf{T}}}^{\mathrm{ox}} - \frac{\partial \mathbf{G}^{\mathrm{ox}}}{\partial t} = \rho_{\mathrm{e}} \mathbf{E}_{\mathrm{m}} + \mathbf{J}_{\mathrm{e}} \times \mu_{\mathrm{o}} \mathbf{H}_{\mathrm{m}} + \rho_{\mathrm{m}} \mathbf{H}_{\mathrm{e}} - \mathbf{J}_{\mathrm{m}} \times \varepsilon_{\mathrm{o}} \mathbf{E}_{\mathrm{e}} \tag{6.42}$$

[6]Naturally, either type of vector potential can be defined within divergence-free regions, but one must be aware that problems can occur at the boundaries of regions that contain divergences.

One set of choices, valid in free-space when the Lorenz-gauge is assumed for both sets of potentials, leads to the following Alternate representation:

$$\mathbf{S}^{\mathrm{ox}} = \frac{\partial \mathbf{A}_{\mathrm{m}}}{\partial t} \times \frac{\partial \mathbf{A}_{\mathrm{e}}}{\partial t} + \mathbf{A}_{\mathrm{m}} \times \frac{\partial^2 \mathbf{A}_{\mathrm{e}}}{\partial t^2} - \mathbf{A}_{\mathrm{e}} \times \frac{\partial^2 \mathbf{A}_{\mathrm{m}}}{\partial t^2}$$
$$+ \mathrm{c}^2[\mathbf{A}_{\mathrm{e}} \times \nabla(\nabla \cdot \mathbf{A}_{\mathrm{m}}) - \mathbf{A}_{\mathrm{m}} \times \nabla(\nabla \cdot \mathbf{A}_{\mathrm{e}}) + (\nabla \times \mathbf{A}_{\mathrm{e}}) \times (\nabla \times \mathbf{A}_{\mathrm{m}})] \quad (6.43a)$$

$$W^{\mathrm{ox}} = \mathbf{A}_{\mathrm{m}} \cdot \left(\nabla \times \frac{\partial \mathbf{A}_{\mathrm{e}}}{\partial t}\right) - \mathbf{A}_{\mathrm{e}} \cdot \left(\nabla \times \frac{\partial \mathbf{A}_{\mathrm{m}}}{\partial t}\right) \quad (6.43b)$$

and

$$T_{ij}^{\mathrm{ox}} = \begin{bmatrix} \varepsilon_{\mathrm{o}} E_{\mathrm{m}i} E_{\mathrm{e}j} + \mu_{\mathrm{o}} H_{\mathrm{e}i} H_{\mathrm{m}j} \\ + A_{\mathrm{m}i} \left(\nabla \times \frac{\partial \mathbf{A}_{\mathrm{e}}}{\partial t}\right)_j - A_{\mathrm{e}i} \left(\nabla \times \frac{\partial \mathbf{A}_{\mathrm{m}}}{\partial t}\right)_j \\ + \frac{\partial \Phi_{\mathrm{e}}}{\partial x_i} (\nabla \times \mathbf{A}_{\mathrm{m}})_j - \frac{\partial \Phi_{\mathrm{m}}}{\partial x_i} (\nabla \times \mathbf{A}_{\mathrm{e}})_j \\ [\nabla \cdot (\mathbf{A}_{\mathrm{e}} \times \nabla \Phi_{\mathrm{m}} - \mathbf{A}_{\mathrm{m}} \times \nabla \Phi_{\mathrm{e}}) - W^{\mathrm{ox}}] \delta_{ij} \end{bmatrix} \quad (6.44a)$$

$$G_i^{\mathrm{ox}} = \begin{bmatrix} \mu_{\mathrm{o}} \varepsilon_{\mathrm{o}} (\mathbf{E}_{\mathrm{e}} \times \mathbf{H}_{\mathrm{m}} - \mathbf{E}_{\mathrm{m}} \times \mathbf{H}_{\mathrm{e}})_i \\ + \mathbf{A}_{\mathrm{e}} \cdot \frac{\partial}{\partial x_i} (\nabla \times \mathbf{A}_{\mathrm{m}}) - \mathbf{A}_{\mathrm{m}} \cdot \frac{\partial}{\partial x_i} (\nabla \times \mathbf{A}_{\mathrm{e}}) \end{bmatrix} \quad (6.44b)$$

In most cases, the Alternate representation is not particularly helpful unless the cross-terms vanish.

Notice that if $\mathbf{J}_{\mathrm{u}} = 0$, it is possible to *simultaneously* define both vector potentials, $\mathbf{B} = \nabla \times \mathbf{A}_{\mathrm{e}}$ and $\mathbf{D} = -\nabla \times \mathbf{A}_{\mathrm{m}}$, and use either one to define the Alternate representation.

6.6 UNIQUENESS THEOREMS

Electric and magnetic fields

Assume that Maxwell's Eqs. (1.8a)–(1.8d) apply in a region of free-space (where there may be *independent* sources) that is bounded by a closed surface, S_{o}. The question we pose is: what boundary conditions on S_{o} will guarantee that the electric and magnetic fields are unique everywhere *inside* the surface? If we assume that they are not unique, there is at least one pair of fields, $\mathbf{E}_1$,$\mathbf{H}_1$ and $\mathbf{E}_2$,$\mathbf{H}_2$, each of which satisfies these equations for the same set of sources. It follows that the difference fields defined by

$$\mathbf{E}_d = \mathbf{E}_1 - \mathbf{E}_2$$

$$\mathbf{H}_d = \mathbf{H}_1 - \mathbf{H}_2$$

satisfy the source-free Maxwell Equations and thus also satisfy the Poynting Theorem, which in integral form is

$$\oint_{S_o} \mathbf{E}_d \times \mathbf{H}_d \cdot \widehat{\mathbf{n}}\, da + \frac{d}{dt}\int_V \frac{1}{2}(\varepsilon_o \mathbf{E}_d \cdot \mathbf{E}_d + \mu_o \mathbf{H}_d \cdot \mathbf{H}_d)\, dV = 0$$

Therefore, if $\widehat{\mathbf{n}} \times \mathbf{E}$ *or* $\widehat{\mathbf{n}} \times \mathbf{H}$ is specified at every point on the surface, either $\widehat{\mathbf{n}} \times \mathbf{E}_d = 0$ or $\widehat{\mathbf{n}} \times \mathbf{H}_d = 0$. It makes no difference which, because

$$\mathbf{E}_d \times \mathbf{H}_d \cdot \widehat{\mathbf{n}} = (\widehat{\mathbf{n}} \times \mathbf{E}_d) \cdot \mathbf{H}_d = -(\widehat{\mathbf{n}} \times \mathbf{H}_d) \cdot \mathbf{E}_d = 0$$

and it follows that because no Poynting power (associated with the difference fields) can bring energy into (or out of) the volume, the total energy

$$\int_V \frac{1}{2}(\varepsilon_o \mathbf{E}_d \cdot \mathbf{E}_d + \mu_o \mathbf{H}_d \cdot \mathbf{H}_d)\, dV$$

must therefore be constant. Consider for a moment a perfectly conducting closed surface that forms an empty cavity. Since $\widehat{\mathbf{n}} \times \mathbf{E} = 0$ is the operative boundary condition, can electric and magnetic fields exist inside the cavity? Yes indeed, provided that they are associated with natural resonance modes; also, more than one mode might be excited at the same time. No matter, the total energy will be constant—but for how long? If there is the slightest bit of dissipative matter anywhere within the cavity that can come in contact with the fields, the mode(s) will decay exponentially in time; the same argument is true for the difference fields. In that case, $\mathbf{E}_d$ and $\mathbf{H}_d$ must vanish at every point within the volume, and *uniqueness is assured in the steady-state whenever the tangential component of* **either E or H** *is specified at every point on the entire closed surface*. This theorem can easily be generalized to include passive linear dielectric and magnetic materials that may include dissipation.

Laplace's Equation

A scalar potential, Φ, satisfies Laplace's Equation in a volume that is bounded by a closed surface, S_o. We wish to find boundary conditions that will ensure a unique potential within the volume. We first assume that the potential is *not* unique; then there are at least two distinct Laplacian solutions, $\Phi_1$and Φ_2, that satisfy the *same* boundary conditions. Because of linearity, the difference between the two, $\Phi_d = \Phi_1 - \Phi_2$, satisfies

$$\nabla^2 \Phi_d = \nabla^2 \Phi_1 - \nabla^2 \Phi_2 = 0$$

Applying the divergence theorem,

$$\int_V \nabla \cdot \mathbf{A}\, dV = \oint_{S_o} \mathbf{A} \cdot \mathbf{n}\, da$$

to the vector $\mathbf{A} = \Phi_d \nabla \Phi_d$, where

$$\nabla \cdot (\Phi_d \nabla \Phi_d) = \Phi_d \nabla^2 \Phi_d + \nabla \Phi_d \cdot \nabla \Phi_d = |\nabla \Phi_d|^2$$

and $\nabla \Phi_d \cdot \mathbf{n} \doteq \dfrac{\partial \Phi_d}{\partial n}$, leads to

$$\oint_{S_o} \Phi_d \frac{\partial \Phi_d}{\partial n}\, da = \int_V |\nabla \Phi_d|^2\, dV$$

It follows that if *either* Φ *or* $\dfrac{\partial \Phi}{\partial n}$ is specified at every point on the surface, then $\Phi_d \dfrac{\partial \Phi_d}{\partial n} = 0$ everywhere on S_o and

$$\int_V |\nabla \Phi_d|^2 \, dV = 0$$

Because the integrand is positive definite, $\nabla \Phi_d = 0$ everywhere within the volume and at most Φ_1 and Φ_2 can differ by a constant (which must be zero if Φ was specified on even one point of the surface). In any event, the gradient field associated with Φ is unique, provided that *either the potential or its normal derivative is specified everywhere on the closed surface.*

Sufficiency of the curl and divergence

We are now in a position to prove that if the curl *and* divergence of a vector within a volume are *both* specified and if *either* the normal *or* tangential component is specified on the *entire* enclosing surface, the vector within that volume is unique. Assume the contrary and that $\mathbf{A}_1$and $\mathbf{A}_2$ are distinct; then $\mathbf{A}_d = \mathbf{A}_1 - \mathbf{A}_2$ is itself curl- and divergence-free and so $\mathbf{A}_d = -\nabla \Phi_d$ with $\nabla^2 \Phi_d = 0$. Because, everywhere on the surface, either $\widehat{\mathbf{n}} \cdot \mathbf{A}_d = \frac{\partial \Phi_d}{\partial n} = 0$ or $\widehat{\mathbf{n}} \times \mathbf{A}_d = 0$, the solution $\Phi_d = \mathit{constant}$ throughout the volume is a Laplacian solution that fits the boundary conditions and so, by the theorem just proved, $\mathbf{A}_d = 0$. Consider now that $\widehat{\mathbf{n}} \times \mathbf{A}$ is specified on S_a and $\widehat{\mathbf{n}} \cdot \mathbf{A}$ on the remaining surface, S_b. Although Φ_d is an equipotential on each section of S_a, these are not necessarily the *same* equipotential unless all are connected; if disconnected, uniqueness fails because $\nabla \Phi_d$ may be nonzero.

It is sometimes said that if the curl and divergence of a vector are specified *everywhere,* the vector is unique. That is tantamount to saying that if Laplace's Equation applies *everywhere,* there are no sources and so the solution is at most a constant. In these cases, care must be taken that there are no sources hiding at infinity. It is safer to apply the theorem to a finite volume and then let the enclosing surface expand toward infinity.

6.7 THE EQUIVALENCE PRINCIPLE

Consider a set of sources located outside of a source-free volume that may include passive linear materials; the fields everywhere are assumed known. Assume now that a second set of sources replaces the first, but generates fields on the surface enclosing the volume, such that either the electric or magnetic tangential component is identical to that initially present. By the uniqueness theorem, the steady-state **E** and **H** fields everywhere inside are identical[7] and so it follows that, with respect to the fields *inside* the enclosed volume, both sets of sources are equivalent. The fields outside can be very different in the two cases.

One set of equivalent sources is very easy to calculate; we simply set the outside fields to zero and use the boundary conditions given in Section 6.5 to calculate the surface currents (electric and magnetic) required to terminate $\widehat{\mathbf{n}} \times \mathbf{H}$ and $\widehat{\mathbf{n}} \times \mathbf{E}$ and the surface charges (electric and magnetic) required to terminate $\widehat{\mathbf{n}} \cdot \mathbf{D}$ and $\widehat{\mathbf{n}} \cdot \mathbf{B}$. It must be stressed that this is a convenient mathematical artifice that can be used even though

[7]Notice that, in general, the second set of sources produce vector and scalar potentials that are different —even though they generate the same **E** and **H** fields. This is why sources outside of a perfectly conducting region can affect the distribution of Alternate power and energy without affecting the fields inside.

magnetic charges and currents are physically unobtainable. This approach is advantageous when the tangential fields are known exactly—or to a good approximation, as is often the case in the analysis of large aperture antennas. Another important application of the equivalence principle is in the modeling of the magnetization of a magnetic material. If the latter is considered to be composed of very tiny grains that are separated by free-space, one is principally concerned with obtaining the correct macroscopic magnetic field outside of the grains. On the basis of the underlying physics, it is certainly reasonable to assume that the magnetic moment associated with a grain is due to an electric current circulating in tiny loops. However, by the equivalence principle we can replace part or all of the current loops with bound magnetic-charge pairs (of opposite polarity) that form magnetic dipoles; these are analogous to the electric dipoles of a dielectric. When all of the loops are replaced, the Chu formulation of electrodynamics emerges. Criticism that the magnetic charges are nonphysical misses the essential nature of the equivalence principle.

6.8 THE INDUCTION THEOREM

The Equivalence Principle can be used to develop a method useful in solving scattering problems—particularly those involving large perfectly conducting objects with smooth surfaces. We assume that the fields incident upon such an object, $\mathbf{E}^i$ and $\mathbf{H}^i$ (often a uniform plane wave), are known; the problem is to solve for the scattered fields, $\mathbf{E}^s$ and $\mathbf{H}^s$, that are created by the electric currents and charges that are induced on the surface of the conductor(s). Consider any closed surface, S_o, such as the one shown in Figure 6.1, that surrounds the scattering region, S.

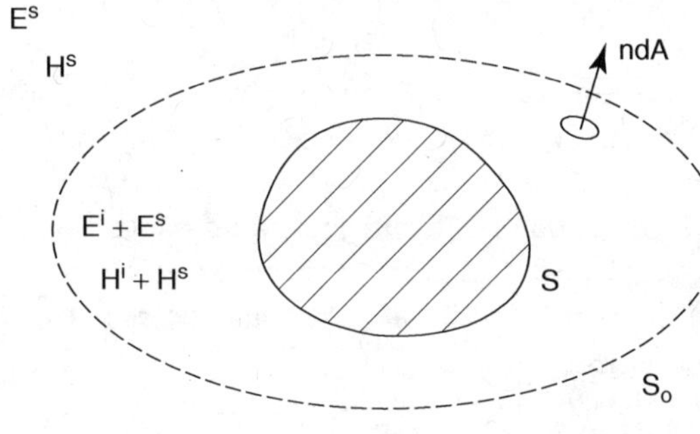

Figure 6.1 Induction Theorem geometry.

Since *outside* of the surface, we are interested only in calculating the scattered field, let us *there* extinguish $\mathbf{E}^i$ and $\mathbf{H}^i$ (without altering *any* of the fields inside of S_o). This is possible if we use the Equivalence Principle to mount fictitious sources on the surface that are consistent with the fields on both sides; the proper electric and magnetic surface currents and surface charges will insure that the proper tangential components of the actual fields ($\mathbf{E}^i + \mathbf{E}^s$ and $\mathbf{H}^i + \mathbf{H}^s$) will be present on the *inside* of the surface. The required surface currents are

$$\mathbf{K}_e^s = \widehat{\mathbf{n}} \times \mathbf{H}^i$$

$$\mathbf{K}_m^s = -\widehat{\mathbf{n}} \times \mathbf{E}^i$$

These sources together with the conductors form a separate field problem, that if solved, will provide both the actual electric currents and charges on the surface of the scatterer(s) and the scattered fields (outside of S_o). This replacement of the incident field with the set of currents constitutes the induction theorem.

No simplification has yet occurred, but if we allow S_o to approach the surface of the scattering object, one may be possible. Assume for a moment that the surface of the object is the perfectly conducting plane $z = 0$, the surface S_o is the plane $z = -\delta_o$, and the incident field is a linearly polarized plane wave propagating in the $+z$ direction. In that event, $\mathbf{K}_e^s$ and $\mathbf{K}_m^s$ are orthogonal vectors. The solution of the scattering problem can be affected by the method of images. With the conducting plane removed, the image pairs, $\mathbf{K}_e^s(-\delta_o) - \mathbf{K}_e^s(+\delta_o)$ and $\mathbf{K}_m^s(-\delta_o) + \mathbf{K}_m^s(+\delta_o)$, create fields for $z < 0$ that satisfy the boundary conditions of a perfect conductor at $z = 0$; naturally, there are no fields for $z > 0$. In addition, the correct total field exists in the region $-\delta_o < z < 0$, and only the scattered field exists in the region $z < -\delta_o$. This solution persists in the limit $\delta_o \to 0$ where $\mathbf{K}_e^s \to 0$ and $\mathbf{K}_m^s \to -\widehat{\mathbf{n}} \times 2\mathbf{E}^i$. These results hold generally, provided that the surfaces are smooth and have large curvatures (compared to any relevant wavelength). This form of the induction theorem is of practical use. When the scattering object is small compared to the relevant wavelength, the method is not applicable; but then, quasistatic analysis can be used to find the induced dipole moments (electric and/or magnetic) (when the object is a small sphere, refer to Part III, Chapter 11, Section 11.4), which in turn can be used to evaluate the radiation fields of Hertzian dipoles.

6.9 BABINET'S PRINCIPLE

Complementary structures

Consider a very thin sheet of electrical conductivity, $\sigma_e(x, y)$, that coincides with the plane $z = 0$. Free-space exists both above and below the sheet. Assume further that at every point, the value of σ_e is either zero or infinity. Although the pattern may be general, it is often a large conducting plane containing one or more holes or slots that serve as apertures through which an electromagnetic plane wave can be diffracted. The complementary structure is formed by interchanging the regions of zero and infinite conductivity; that function is defined as $\sigma_e^c(x, y)$.

Assume that a uniform plane wave (originating at $z = -\infty$) is normally incident upon the sheet; as it interacts, electric currents and charges are induced on both sides. These are the actual sources of the scattered fields, $\mathbf{E}^s$ and $\mathbf{H}^s$, for both positive and negative z. Applying the induction theorem to the volume between $-\delta_o < z < \delta_o$ extinguishes the incident wave for $|z| > \delta_o$ and replaces it with induction sources on the planes $z = \pm\delta_o$:

$$\mathbf{K}_e^s = \widehat{\mathbf{z}} \times [\mathbf{H}^i(x, y, \delta_o) - \mathbf{H}^i(x, y, -\delta_o)]$$

$$\mathbf{K}_m^s = \widehat{\mathbf{z}} \times [\mathbf{E}^i(x, y, -\delta_o) - \mathbf{E}^i(x, y, \delta_o)]$$

where

$$\mathbf{E}^i = -\sqrt{\frac{\mu_o}{\varepsilon_o}}\widehat{\mathbf{z}} \times \mathbf{H}^i$$

Both sets of currents are odd functions of the z coordinate and, *together with the conducting screen between them*, constitute the scattering problem; these are *not* image currents.

In regions where the screen is transparent ($\sigma_e = 0$), the two components of each type of current cancel as $\delta_o \to 0$. Only the scattered fields $\mathbf{E}^s$ and $\mathbf{H}^s$ exist for $|z| > \delta_o$. Each set of currents, when considered separately, creates a set of scattered fields consistent with the symmetry of those sources; the results can then be superimposed.

To proceed further, consider two sets of fields that, for $|z| > \delta_o$, are solutions of the free-space Maxwell's Equations.

Type I

$$\begin{bmatrix} E_x(x,y,z,t) = +E_x(x,y,-z,t) \\ E_y(x,y,z,t) = +E_y(x,y,-z,t) \\ E_z(x,y,z,t) = -E_z(x,y,-z,t) \\ \\ H_x(x,y,z,t) = -H_x(x,y,-z,t) \\ H_y(x,y,z,t) = -H_y(x,y,-z,t) \\ H_z(x,y,z,t) = +H_z(x,y,-z,t) \end{bmatrix}$$

and

Type II

$$\begin{bmatrix} E_x(x,y,z,t) = -E_x(x,y,-z,t) \\ E_y(x,y,z,t) = -E_y(x,y,-z,t) \\ E_z(x,y,z,t) = +E_z(x,y,-z,t) \\ \\ H_x(x,y,z,t) = +H_x(x,y,-z,t) \\ H_y(x,y,z,t) = +H_y(x,y,-z,t) \\ H_z(x,y,z,t) = -H_z(x,y,-z,t) \end{bmatrix}$$

Each preserves the plane symmetry of the screen. The first set has transverse electric and magnetic fields that are, respectively, even and odd functions of z; in the second set, the symmetries are reversed. For each side of the screen, $\mathbf{K}_e^s$ and $\mathbf{K}_m^s$ together with their images (after removal of the screen and the non-image currents on the other side) generate the scattered fields. Therefore the magnetic currents and their images *add* and generate a field of Type I; the electric currents and their images *cancel* to generate a field of Type II – but magnitude zero. On the perfectly conducting areas, $\widehat{\mathbf{z}} \times (\mathbf{E}^i + \mathbf{E}^s) = 0$; on the open areas, $\widehat{\mathbf{z}} \times \sqrt{\frac{\mu_0}{\varepsilon_0}}\mathbf{H}^s = 0$.

Now consider a second diffraction problem that involves scattering from the complementary screen, $\sigma_e^c(x,y)$, that is excited by $\mathbf{E}^{i\prime}$ and $\mathbf{H}^{i\prime}$, duals of the original incident field, where

$$\mathbf{E}^{i\prime} = \sqrt{\frac{\mu_0}{\varepsilon_0}}\mathbf{H}^i \tag{6.45a}$$

$$\mathbf{H}^{i\prime} = -\sqrt{\frac{\varepsilon_0}{\mu_0}}\mathbf{E}^i \tag{6.45b}$$

These, too, generate Type I fields, $\mathbf{E}^{s\prime}$ and $\mathbf{H}^{s\prime}$. On the perfectly conducting areas of the complementary structure, $\widehat{\mathbf{z}} \times (\mathbf{E}^{i\prime} + \mathbf{E}^{s\prime}) = 0$; on the open areas, $\widehat{\mathbf{z}} \times \sqrt{\frac{\mu_0}{\varepsilon_0}}\mathbf{H}^{s\prime} = 0$. These two diffraction problems can be made to correspond exactly, provided that, for $z > 0$,

$$\mathbf{E}^i + \mathbf{E}^s \leftrightarrow \sqrt{\frac{\mu_0}{\varepsilon_0}}\mathbf{H}^{s\prime} \tag{6.46a}$$

$$\mathbf{E}^{i\prime} + \mathbf{E}^{s\prime} \leftrightarrow -\sqrt{\frac{\mu_0}{\varepsilon_0}}\mathbf{H}^s \tag{6.46b}$$

This result is known as Babinet's Principle. Simply stated, the solution of plane wave diffraction by a thin perfectly conducting screen also provides the solution of diffraction by the complementary screen. As a check, consider the case where the entire screen is perfectly conducting. Then $\mathbf{E}^s = -\mathbf{E}^i$ and $\mathbf{H}^{s\prime} = 0$; also $\mathbf{H}^s = -\mathbf{H}^i$ and $\mathbf{E}^{s\prime} = 0$. This agrees with the analysis of the complementary screen which has no conducting regions and therefore cannot scatter the incident wave.

Dual structures and their complements

The dual structure and the dual complement formed by replacing $\sigma_e(x, y)$ with $\sigma_m(x, y)$ and $\sigma_e^c(x, y)$ with $\sigma_m^c(x, y)$ are useful concepts. When the induction theorem is applied to the dual excitation, the set of electric currents and their images now *add* to generate a field of Type II; it is the magnetic currents and their images that *cancel* to generate a field of Type I—but magnitude zero. On the areas where $\sigma_m(x, y) = \infty$, the boundary conditions are $\widehat{\mathbf{z}} \times (\mathbf{H}^{i\prime} + \mathbf{H}^{s\prime}) = 0$; on the open areas, $\widehat{\mathbf{z}} \times \sqrt{\frac{\varepsilon_0}{\mu_0}}\mathbf{E}^{s\prime} = 0$. If we restrict comparisons to only one side of the screen, the dual structure, $\sigma_m(x, y)$, can be used to model the complementary structure, $\sigma_e^c(x, y)$; likewise, the dual complement, $\sigma_m^c(x, y)$, can be used to model the original structure, $\sigma_e(x, y)$.

6.10 THE RECIPROCITY THEOREM

An electromagnetic environment is assumed to be characterized by linear dielectric and magnetic materials that may be anisotropic and include resistive dissipation. The complex constitutive laws are, therefore,

$$\underline{D}_i = \sum_{j=1}^{3} \varepsilon_{ij}\underline{E}_j$$

$$\underline{B}_i = \sum_{j=1}^{3} \mu_{ij}\underline{H}_j$$

$$\underline{J}_i = \sum_{j=1}^{3} \sigma_{ij}\underline{E}_j + \underline{J}_{Si}$$

where $\underline{\mathbf{J}}_S(j\omega)$ is the current source density that generates the fields. In what follows the environment is kept fixed.

Assume that $\underline{\mathbf{J}}_S^a(j\omega)$ is a complex electric-current density source that generates the set of complex fields labeled by superscript a; they satisfy the complex Maxwell Equations:

$$\nabla \times \underline{\mathbf{E}}^a = -j\omega\underline{\mathbf{B}}^a$$

$$\nabla \times \underline{\mathbf{H}}^a = \underline{\mathbf{J}}^a + j\omega\underline{\mathbf{D}}^a$$

Assume further that $\underline{\mathbf{J}}_S^b(j\omega)$ is a second source that generates the set of fields labeled by b; they satisfy

$$\nabla \times \underline{\mathbf{E}}^b = -j\omega\underline{\mathbf{B}}^b$$

$$\nabla \times \underline{\mathbf{H}}^b = \underline{\mathbf{J}}^b + j\omega\underline{\mathbf{D}}^b$$

Then it follows that

$$\begin{aligned}\nabla \cdot (\underline{\mathbf{E}}^a \times \underline{\mathbf{H}}^b - \underline{\mathbf{E}}^b \times \underline{\mathbf{H}}^a) \\ = -j\omega(\underline{\mathbf{H}}^b \cdot \underline{\mathbf{B}}^a - \underline{\mathbf{H}}^a \cdot \underline{\mathbf{B}}^b + \underline{\mathbf{E}}^a \cdot \underline{\mathbf{D}}^b - \underline{\mathbf{E}}^b \cdot \underline{\mathbf{D}}^a) - \underline{\mathbf{E}}^a \cdot \underline{\mathbf{J}}^b + \underline{\mathbf{E}}^b \cdot \underline{\mathbf{J}}^a\end{aligned} \tag{6.47}$$

This looks like the difference of two complex Poynting Theorems, but it is not because the vector cross-products involve both the a and b sets of fields and no complex conjugation operations are involved. When the constitutive laws (assumed identical for both sets of fields) are inserted, the result is

$$\begin{aligned}\nabla \cdot (\underline{\mathbf{E}}^a \times \underline{\mathbf{H}}^b - \underline{\mathbf{E}}^b \times \underline{\mathbf{H}}^a) = \underline{\mathbf{J}}_S^a \cdot \underline{\mathbf{E}}^b - \underline{\mathbf{J}}_S^b \cdot \underline{\mathbf{E}}^a \\ + \sum_{i=1}^{3}\sum_{j=1}^{3}\left(j\omega[(\mu_{ij} - \mu_{ji})\underline{H}_i^a\underline{H}_j^b - (\varepsilon_{ij} - \varepsilon_{ji})\underline{E}_i^a\underline{E}_j^b] - (\sigma_{ij} - \sigma_{ji})\underline{E}_i^a\underline{E}_j^b\right)\end{aligned} \tag{6.48}$$

In almost all cases,[8] the linear constitutive laws are reciprocal. Assuming this to be true, Eq. (6.48) reduces to

$$\nabla \cdot (\underline{\mathbf{E}}^a \times \underline{\mathbf{H}}^b - \underline{\mathbf{E}}^b \times \underline{\mathbf{H}}^a) = \underline{\mathbf{J}}_S^a \cdot \underline{\mathbf{E}}^b - \underline{\mathbf{J}}_S^b \cdot \underline{\mathbf{E}}^a$$

The integral form of this equation is

$$\oint_{S_0} (\underline{\mathbf{E}}^a \times \underline{\mathbf{H}}^b - \underline{\mathbf{E}}^b \times \underline{\mathbf{H}}^a) \cdot \widehat{\mathbf{n}}\, da = \int_V (\underline{\mathbf{J}}_S^a \cdot \underline{\mathbf{E}}^b - \underline{\mathbf{J}}_S^b \cdot \underline{\mathbf{E}}^a)\, dV$$

Now let the closed surface expand to infinity. Because the current sources are each assumed to be of finite extent and separated by a finite distance, it follows that both the a and b fields must diminish as $1/r$ (or even more rapidly). Since the surface area of the left-hand-side integral increases as r^2, it is not obvious that the integral vanishes, but it

[8]A very important exception occurs with magnetized ferrite materials (considered in Chapter 23)—for which $\mu_{ij} \neq \mu_{ji}$. In this case, reciprocity fails. This has led to the development of very useful nonreciprocal devices (especially in the microwave frequency range) called isolators, circulators, and gyrators.

does so for the following reason. In the far field, both the a and b fields are transverse with amplitudes and polarizations related by

$$\underline{\mathbf{H}}^a \rightarrow \sqrt{\frac{\varepsilon_0}{\mu_0}}\widehat{\mathbf{r}} \times \underline{\mathbf{E}}^a$$

$$\underline{\mathbf{H}}^b \rightarrow \sqrt{\frac{\varepsilon_0}{\mu_0}}\widehat{\mathbf{r}} \times \underline{\mathbf{E}}^b$$

Therefore

$$\underline{\mathbf{E}}^a \times \underline{\mathbf{H}}^b = \underline{\mathbf{E}}^b \times \underline{\mathbf{H}}^a = \sqrt{\frac{\varepsilon_0}{\mu_0}}\underline{\mathbf{E}}^a \cdot \underline{\mathbf{E}}^b\widehat{\mathbf{r}}$$

and

$$\int_V \underline{\mathbf{J}}_S^a \cdot \underline{\mathbf{E}}^b dV = \int_V \underline{\mathbf{J}}_S^b \cdot \underline{\mathbf{E}}^a dV \tag{6.49}$$

The equality of these integrals, called "actions," is a statement of the Lorentz reciprocity theorem. The volumes of the integrals can be restricted to those of the respective currents. If the latter are essentially line currents, of lengths ℓ^a and ℓ^b, it follows that

$$\underline{I}^a\widehat{\mathbf{z}}_a \cdot \int_0^{\ell^a} \underline{\mathbf{E}}^b dz_a = \underline{I}^b\widehat{\mathbf{z}}_b \cdot \int_0^{\ell^b} \underline{\mathbf{E}}^a dz_b \tag{6.50a}$$

$$\underline{I}^a\underline{V}^a = \underline{I}^b\underline{V}^b \tag{6.50b}$$

where we have defined the voltages in terms of the line integrals of the components of $\underline{\mathbf{E}}^b$ and $\underline{\mathbf{E}}^a$ that are parallel to the currents. Because the overall system is linear, it is expected that the complex voltages and currents are related by circuit theory like impedances; in general,

$$\underline{V}^a = Z_{aa}(j\omega)\underline{I}^a + Z_{ab}(j\omega)\underline{I}^b \tag{6.51a}$$

$$\underline{V}^b = Z_{ba}(j\omega)\underline{I}^a + Z_{bb}(j\omega)\underline{I}^b \tag{6.51b}$$

If $\underline{I}^a$ is energized when $\underline{I}^b = 0$, $\underline{V}^b = Z_{ba}(j\omega)\underline{I}^a$; if $\underline{I}^b$ is energized when $\underline{I}^a = 0$, $\underline{V}^a = Z_{ab}(j\omega)\underline{I}^b$. It follows from Eq. (6.50b) that

$$Z_{ab}(j\omega) = Z_{ba}(j\omega) \tag{6.52}$$

as expected from the theory of reciprocal networks that arise from either lumped or distributed circuit elements.

PART II

FOUR-DIMENSIONAL ELECTROMAGNETISM

INTRODUCTION TO PART II

The classical electromagnetic theory, developed in Part I, is based upon the Maxwell Equations expressed in spatial coordinates with time as a parameter. Both space and time are expressed in terms of real variables, but we have already seen that it is sometimes advantageous to express the wave equation in terms of four coordinates: $x_1 = x$, $x_2 = y$, $x_3 = z$, $x_4 = ict$. If these are all considered real, the wave equation is replaced by Laplace's Equation in four dimensions. Then, static-like solutions appear, but at the price of our having to deal with either imaginary time or functions of complex variables.

In Part II, this approach is developed more fully and not just considered a "trick." The space–time coordinates so defined obey the Lorentz transformation, support Special Relativity, and lead to four-dimensional vectors and tensors. The addition of a fourth component to the del operator supports the definitions of the four-dimensional gradient, divergence, and curl. These lead to generalizations of standard vector identities and, as noted, the four-dimensional Laplacian operator generates the electromagnetic wave equation. All electromagnetic field vectors are shown to have four vector counterparts that are Lorentz-invariant. In terms of these, conservation of charge and the Lorenz gauge have particularly simple representations. Maxwell's Equations can be generated from the four divergences of appropriate antisymmetric field tensors. Alternatively, the six independent components of these tensors are defined as six vectors comprised of **E** and **H**. These mathematical tools are developed and form the basis of the theory of four-dimensional electromagnetism. They are used to derive energy-momentum tensors for both the Maxwell–Poynting and Alternate representations that govern free-space and materials that model dipolar media with polarization charges and Amperian currents. The four-dimensional electrodynamics of Minkowski and Chu are also discussed and contrasted to the Amperian formulation; particular attention is paid to differences among the four-vector force densities of "electromagnetic origin." Power theorems in the Amperian, Minkowski, and Chu forms of the Maxwell–Poynting representation as well as in the Alternate representation are developed for linear materials.

CHAPTER 7

FOUR-DIMENSIONAL VECTORS AND TENSORS

7.1 SPACE–TIME COORDINATES

Despite valiant attempts by Michelson and Morley to discover a medium through which free-space electromagnetic waves propagate, no "ether" has ever been found; had the attempts been successful, Maxwell's Equations would have required amendment. As it is, they do not depend upon any particular choice of coordinate system, nor upon whether the choice selected is moving at velocity, **v**, or is at rest with respect to some arbitrary reference frame—commonly referred to as laboratory coordinates. As a consequence, solutions of the wave equation propagate at the velocity, $c = 1/\sqrt{\mu_0 \varepsilon_0}$, *independent* of **v**. This is in marked contrast to other wave equations, notably those governing sound waves, where the medium through which the propagation takes place is physically identifiable and its state of motion alters the speed of sound in the same way a swimmer is affected by water currents. The resolution of this paradox required the revelation (provided by Albert Einstein) that space and time must be interconnected in a special way and, moreover, that the speed of ponderable matter is limited to c. For charged particles, this speed limit is a consequence of Maxwell's Equations and, together with Galilean and Lorentz transformations, is discussed in Appendix A.

The Power and Beauty of Electromagnetic Fields, First Edition. F. R. Morgenthaler.

Consider a spherical wave front (originating at the time, $t = 0$ and Cartesian coordinates $x = y = z = 0$) described in the laboratory frame by the locus,

$$x^2 + y^2 + z^2 - c^2t^2 = 0$$

and in a moving frame (with the origins defined so as to coincide at $t = t' = 0$), by

$$(x')^2 + (y')^2 + (z')^2 - c^2(t')^2 = 0$$

The two sets of coordinates (x, y, z, t) and (x', y', z', t') are related by a Lorentz transformation. In general, motion of the primed coordinate system will cause time in the two systems to differ.

Because $x^2 + y^2 + z^2 + (ict)^2$ is Lorentz-invariant, it is useful to define four-dimensional space–time coordinates ($x_1 = x,\ x_2 = y,\ x_3 = z,\ x_4 = ict$) and is instructive to apply them to various electromagnetic equations. In so doing, we follow the example of Hermann Minkowski.

7.2 FOUR-VECTOR ELECTRIC-CURRENT DENSITY

We begin with conservation of charge, and express $\nabla \cdot \mathbf{J} + \dfrac{\partial \rho}{\partial t} = 0$ as

$$\frac{\partial}{\partial x} J_x + \frac{\partial}{\partial y} J_y + \frac{\partial}{\partial z} J_z + \frac{\partial}{\partial ict}(ic\rho) = 0$$

This equation can be considered the divergence of a four-dimensional current vector defined and denoted by

$$\mathcal{J} = [J_x,\ J_y,\ J_z,\ ic\rho] = [\mathbf{J},\ ic\rho] \tag{7.1}$$

The gradient del operator can also be generalized to four dimensions with

$$\Box = \left[\nabla,\ \frac{\partial}{\partial x_4}\right]$$

Combining the notations produces

$$\Box \cdot \mathcal{J} = 0 \tag{7.2}$$

These and the other four-dimensional representations introduced later on are described in Appendix B, Section B.3. Useful four-vector identities, that complement the standard three-dimensional forms, are also included.

7.3 FOUR-VECTOR POTENTIAL (LORENZ GAUGE)

In a similar way, expressing $\nabla \cdot \mathbf{A} + \dfrac{1}{c^2}\dfrac{\partial \Phi}{\partial t} = 0$ as

$$\frac{\partial}{\partial x} A_x + \frac{\partial}{\partial y} A_y + \frac{\partial}{\partial z} A_z + \frac{\partial}{\partial\, ict}\left(\frac{i}{c}\Phi\right) = 0$$

invites the definition of a four-vector potential,

$$\mathcal{A} = \left[\mathbf{A},\ \frac{i}{c}\Phi\right] \tag{7.3}$$

The Lorenz-gauge condition can then be expressed as

$$\square \cdot \mathcal{A} = 0 \tag{7.4}$$

Notice that these definitions of $\mathcal{J}$ and $\mathcal{A}$ agree with the four-vectors introduced in Part I, Chapter 3, Section 3.4.

7.4 FOUR-LAPLACIAN (WAVE EQUATION)

In three dimensions, the divergence of the gradient of a scalar, Ψ, can be written in any of the forms

$$\nabla \cdot (\nabla\Psi) = (\nabla \cdot \nabla)\,\Psi = \nabla^2\Psi$$

In four dimensions, these become

$$\square \cdot (\square\Psi) = (\square \cdot \square)\,\Psi = \square^2\Psi$$

The generalization of the three-dimensional Laplacian operator, $\nabla \cdot \nabla = \nabla^2$, is therefore

$$\square \cdot \square = \square^2 = \nabla^2 - \frac{1}{c^2}\frac{\partial^2}{\partial t^2}$$

When applied to any scalar function of space and time, this four-Laplacian operator generates the wave equation. It follows that the wave equations governing **A** and Φ, Eqs. (1.13a) and (1.13b), can be combined as

$$\square^2\mathcal{A} = -\mu_0\mathcal{J} \tag{7.5}$$

7.5 MAXWELL'S EQUATIONS AND FIELD TENSORS

We turn now to the Maxwell curl equation

$$\nabla \times \mathbf{H} - \varepsilon_0\frac{\partial}{\partial t}\mathbf{E} = \mathbf{J}$$

expressed in Cartesian coordinates (with $E_{x,y,z} = E_{1,2,3}$, $H_{x,y,z} = H_{1,2,3}$ and $J_{x,y,z} = J_{1,2,3}$). The three components become

$$\frac{\partial}{\partial x_1}(0) + \frac{\partial}{\partial x_2}(H_3) + \frac{\partial}{\partial x_3}(-H_2) + \frac{\partial}{\partial x_4}(-ic\varepsilon_0 E_1) = J_1 \tag{7.6a}$$

$$\frac{\partial}{\partial x_1}(-H_3) + \frac{\partial}{\partial x_2}(0) + \frac{\partial}{\partial x_3}(H_1) + \frac{\partial}{\partial x_4}(-ic\varepsilon_0 E_2) = J_2 \tag{7.6b}$$

$$\frac{\partial}{\partial x_1}(H_2) + \frac{\partial}{\partial x_2}(-H_3) + \frac{\partial}{\partial x_3}(0) + \frac{\partial}{\partial x_4}(-ic\varepsilon_0 E_3) = J_3 \tag{7.6c}$$

where zeros have been added so that the left-hand side of each equation can be recognized as the four-divergence of a four-vector; the right-hand sides are the Cartesian components of **J**. It appears that a fourth equation involving J_4 belongs to this set; Gauss' Law supplies it when written as

$$\frac{\partial}{\partial x_1}(ic\varepsilon_o E_1) + \frac{\partial}{\partial x_2}(ic\varepsilon_o E_2) + \frac{\partial}{\partial x_3}(ic\varepsilon_o E_3) + \frac{\partial}{\partial x_4}(0) = J_4 = ic\rho \tag{7.6d}$$

Using index notation, Eqs. (7.6d) can be combined as

$$\sum_{j=1}^{4} \frac{\partial}{\partial x_j}\mathcal{G}_{ij} = \mathcal{J}_i \qquad (i = 1, 2, 3, 4)$$

The sixteen elements $\mathcal{G}_{ij}$ constitute an antisymmetric field tensor

$$\mathcal{G} = \begin{bmatrix} 0 & H_3 & -H_2 & -ic\varepsilon_o E_1 \\ -H_3 & 0 & H_1 & -ic\varepsilon_o E_2 \\ H_2 & -H_1 & 0 & -ic\varepsilon_o E_3 \\ ic\varepsilon_o E_1 & ic\varepsilon_o E_2 & ic\varepsilon_o E_3 & 0 \end{bmatrix} \tag{7.7}$$

When the remaining Maxwell Equations,

$$\nabla \times \mathbf{E} + \mu_o \frac{\partial}{\partial t}\mathbf{H} = 0$$

$$\mu_o \nabla \cdot \mathbf{H} = 0$$

are treated in a similar fashion, the result is

$$\frac{\partial}{\partial x_1}(0) + \frac{\partial}{\partial x_2}(-E_3) + \frac{\partial}{\partial x_3}(E_2) + \frac{\partial}{\partial x_4}(-ic\mu_o H_1) = 0 \tag{7.8a}$$

$$\frac{\partial}{\partial x_1}(E_3) + \frac{\partial}{\partial x_2}(0) + \frac{\partial}{\partial x_3}(-E_1) + \frac{\partial}{\partial x_4}(-ic\mu_o H_2) = 0 \tag{7.8b}$$

$$\frac{\partial}{\partial x_1}(-E_2) + \frac{\partial}{\partial x_2}(E_3) + \frac{\partial}{\partial x_3}(0) + \frac{\partial}{\partial x_4}(-ic\mu_o H_3) = 0 \tag{7.8c}$$

and

$$\frac{\partial}{\partial x_1}(ic\mu_o H_1) + \frac{\partial}{\partial x_2}(ic\mu_o H_2) + \frac{\partial}{\partial x_3}(ic\mu_o H_3) + \frac{\partial}{\partial x_4}(0) = 0 \tag{7.8d}$$

Using index notation, Eqs. (7.8d) can be combined as

$$\sum_{j=1}^{4} \frac{\partial}{\partial x_j}\mathcal{K}_{ij} = 0 \qquad (i = 1, 2, 3, 4)$$

The sixteen elements $\mathcal{K}_{ij}$ constitute a second antisymmetric field tensor

$$\mathcal{K} = \begin{bmatrix} 0 & -E_3 & E_2 & -ic\mu_0 H_1 \\ E_3 & 0 & -E_1 & -ic\mu_0 H_2 \\ -E_2 & E_1 & 0 & -ic\mu_0 H_3 \\ ic\mu_0 H_1 & ic\mu_0 H_2 & ic\mu_0 H_3 & 0 \end{bmatrix} \tag{7.9}$$

In terms of the field tensors $\mathcal{G}$ and $\mathcal{K}$, Maxwell's Equations can be expressed as

$$\Box \cdot \mathcal{G} = \mathcal{J} \tag{7.10a}$$

$$\Box \cdot \mathcal{K} = 0 \tag{7.10b}$$

Notice that because $\mathcal{G}$ is antisymmetric, we have $\Box \cdot (\Box \cdot \mathcal{G}) = \Box \cdot \mathcal{J} = 0$ and conservation of charge is automatically satisfied.

If electric and magnetic field components are expressed in terms of the vector and scalar potentials, the result is

$$\mu_0 \mathcal{G} = \begin{bmatrix} 0 & \frac{\partial \mathcal{A}_2}{\partial x_1} - \frac{\partial \mathcal{A}_1}{\partial x_2} & \frac{\partial \mathcal{A}_3}{\partial x_1} - \frac{\partial \mathcal{A}_1}{\partial x_3} & \frac{\partial \mathcal{A}_4}{\partial x_1} - \frac{\partial \mathcal{A}_1}{\partial x_4} \\ \frac{\partial \mathcal{A}_1}{\partial x_2} - \frac{\partial \mathcal{A}_2}{\partial x_1} & 0 & \frac{\partial \mathcal{A}_3}{\partial x_2} - \frac{\partial \mathcal{A}_2}{\partial x_3} & \frac{\partial \mathcal{A}_4}{\partial x_2} - \frac{\partial \mathcal{A}_2}{\partial x_4} \\ \frac{\partial \mathcal{A}_1}{\partial x_3} - \frac{\partial \mathcal{A}_3}{\partial x_1} & \frac{\partial \mathcal{A}_2}{\partial x_3} - \frac{\partial \mathcal{A}_3}{\partial x_2} & 0 & \frac{\partial \mathcal{A}_4}{\partial x_3} - \frac{\partial \mathcal{A}_3}{\partial x_4} \\ \frac{\partial \mathcal{A}_1}{\partial x_4} - \frac{\partial \mathcal{A}_4}{\partial x_1} & \frac{\partial \mathcal{A}_2}{\partial x_4} - \frac{\partial \mathcal{A}_4}{\partial x_2} & \frac{\partial \mathcal{A}_3}{\partial x_4} - \frac{\partial \mathcal{A}_4}{\partial x_3} & 0 \end{bmatrix} \tag{7.11}$$

7.6 THE FOUR-DIMENSIONAL CURL OPERATOR

Because of the form of Eq. (7.11), it is convenient (almost imperative) to introduce a four-dimensional curl operator so that it generates such an antisymmetric tensor. In particular, if $\mathcal{A}$ is *any* four-vector, one defines

$$(\Box \times \mathcal{A})_{ij} = \frac{\partial \mathcal{A}_j}{\partial x_i} - \frac{\partial \mathcal{A}_i}{\partial x_j} \qquad (i, j = 1, 2, 3, 4) \tag{7.12}$$

When $\mathcal{A} = [\mathbf{A}, \frac{i}{c}\Phi]$ and Eqs. (1.9) and (1.10) are used, the four-curl (expressed in terms of **E** and **H**) can be identified as

$$\Box \times \mathcal{A} = \mu_0 \mathcal{G} \tag{7.13}$$

Notice that this four-curl is a tensor quantity and *not* some four-vector that might be expected to generalize $\nabla \times \mathbf{A}$.

Upon taking the four-divergences of both sides, the result in index notation (summation over repeated indices implied) is

$$\frac{\partial}{\partial x_j}\left(\frac{\partial \mathcal{A}_j}{\partial x_i} - \frac{\partial \mathcal{A}_i}{\partial x_j}\right) = \frac{\partial}{\partial x_i}\left(\frac{\partial \mathcal{A}_j}{\partial x_j}\right) - \frac{\partial^2 \mathcal{A}_i}{\partial x_j \partial x_j} = \mu_0 J_i \tag{7.14a}$$

which can be written as

$$\Box \cdot (\Box \times \mathcal{A}) = \Box(\Box \cdot \mathcal{A}) - \Box^2 \mathcal{A} = \mu_0 \mathcal{J} \tag{7.14b}$$

This form is valid for any gauge; if the Lorenz gauge is chosen, Eq. (7.5) is recovered.

If the four-curl operation is performed on both sides of Eq. (7.14b), we obtain

$$\Box \times [\Box(\Box \cdot \mathcal{A})] - \Box \times (\Box^2 \mathcal{A}) = \Box \times (\mu_0 \mathcal{J})$$

Employing the four-dimensional identity $\Box \times \Box\Psi = 0$ and interchanging the order of differentiation results in

$$\Box^2(\Box \times \mathcal{A}) = -\mu_0 \Box \times \mathcal{J}$$

Finally, after substitution of Eq. (7.13), it follows that the field tensor, $\mathcal{G}$, satisfies the wave equation,

$$\Box^2 \mathcal{G} = -\,\Box \times \mathcal{J} \tag{7.15}$$

Because of the antisymmetric form of $\mathcal{G}$, there are only six independent components of Eq. (7.15); these are the Cartesian components of Eqs. (1.16a) and (1.16b).

7.7 FOUR-DIMENSIONAL "STATICS"

Because the four-dimensional Laplacian operator,

$$\Box \cdot \Box = \Box^2 = \nabla^2 - \frac{1}{c^2}\frac{\partial^2}{\partial t^2} \tag{7.16}$$

generates the wave equation, the formulation of electrodynamics problems in terms of four-space "statics" is both possible and fruitful.

In three-space,

$$\nabla^2(1/r_{qp}) = -\delta(r_{qp}) \tag{7.17}$$

produces a delta-function singularity, $\delta(r_{qp}) = \dfrac{1}{4\pi r_{qp}^2} u_0(r_{qp})$, at the point $r_{qp} = 0$ where p denotes the point of observation, q the point of differentiation or integration, with $\mathbf{r}_{qp} = \mathbf{r}_p - \mathbf{r}_q$, and $r_{qp}^2 = (x_p - x_q)^2 + (y_p - y_q)^2 + (z_p - z_q)^2$. This formula allows one to integrate Poisson's Equation (Part I, Chapter 1, Section 1.5) and so evaluate (within statics or quasistatics) Φ or any Cartesian component of $\mathbf{A}, \mathbf{E}$, or $\mathbf{H}$.

In four-space,

$$\Box^2(1/\mathcal{R}^{qp} \cdot \mathcal{R}^{qp}) = -\delta(|\mathcal{R}^{qp}|) \tag{7.18}$$

produces a delta-function singularity at the four-space point $\mathcal{R}^{qp} = 0$ where $\mathcal{R}^{qp} \cdot \mathcal{R}^{qp} = r_{qp}^2 + [ic(t_p - t_q)]^2$. In three-space, that point is the surface of a sphere of radius $c|t_p - t_q|$. This formula allows one to integrate the inhomogeneous wave equation for Φ or any Cartesian component of $\mathbf{A}, \mathbf{E}$, or $\mathbf{H}$.

As one example, the four-vector potential, $\mathcal{A}(\mathbf{r}_p, t)$, that satisfies the inhomogeneous wave equation, Eq. (7.5), can be expressed, in terms of the four-current density, $\mathcal{J}(\mathbf{r}_q, t_q)$, as

$$\mathcal{A}(\mathbf{r}_p, t_p) = \int_{-\infty}^{\infty}\int_{-\infty}^{\infty}\int_{-\infty}^{\infty}\int_{-\infty}^{\infty} \frac{\mu_0 \mathcal{J}(\mathbf{r}_q, t_q) dx_1^q dx_2^q dx_3^q dx_4^q}{4\pi^2 \mathcal{R}^{qp}\cdot\mathcal{R}^{qp}} \tag{7.19}$$

where both $\mathcal{J}$ and $\mathcal{A}$ are described in terms of Cartesian components. (The extra factor of π arises because the four-dimensional delta function must be integrated over a hypersphere.) Now (using indexed-coordinate notation) p denotes the four-point of observation $[\mathbf{r}_p,\ ict_p] = [x_1^p, x_2^p, x_3^p, x_4^p]$, q the four-point of integration $[\mathbf{r}_q,\ ict_q] = [x_1^q, x_2^q, x_3^q, x_4^q]$, and $\mathcal{R}^{qp} = [\mathbf{r}_p - \mathbf{r}_q,\ ic(t_p - t_q)]$[1]. This formula is the four-dimensional equivalent of Poisson's Equation. Such an integral was encountered previously in Part I, Chapter 4, Section 4.5, where it was observed that all four integrations are on an equal footing only when x_4^q is considered to be *real* and the entire range is available for the integration. This requires that we consider t_q to be a complex variable, $t_q = t_q' - it_q''$, and integrate t_q'' from $-\infty$ to $+\infty$. Although this may at first seem perplexing, at least we are spared having to integrate t_q' from the infinite past to the infinite future! The value of $t_q' = t_p$ can take on any time up to and including the present; for convenience and without loss of generality, the clock is often reset so that $t_p = 0$.

The integrations can be carried out in any desired order; if that over x_4^q is carried out first, we obtain

$$\mathcal{A}(\mathbf{r}_p, t_p) = \int_{-\infty}^{\infty}\int_{-\infty}^{\infty}\int_{-\infty}^{\infty}\int_{-\infty}^{\infty} \frac{\mu_0 \mathcal{J}(\mathbf{r}_q, t_q) dx_1^q dx_2^q dx_3^q d\left(\dfrac{x_4^q - x_4^p}{r_{qp}}\right)}{4\pi^2 r_{qp}\left[1 + \left(\dfrac{x_4^q - x_4^p}{r_{qp}}\right)^2\right]}$$

Because $\dfrac{1}{(u+i)(u-i)}$ is a delta function with normalization, $\int_{-\infty}^{\infty} \dfrac{du}{1+u^2} = \pi$, it follows that $\int_{-\infty}^{\infty} f(u) \dfrac{du}{1+u^2} = \pi f(\pm i)$ (the sign depending upon which pole is encircled by the closed contour used to evaluate the integral). The case at hand, $u = \dfrac{x_4^q - x_4^p}{r_{qp}} = \pm i$, corresponds to $t_q = t_p \pm \dfrac{r_{qp}}{c}$. We choose the minus sign on the basis of causality. The result (expressed in terms of the retarded time) is

$$\mathcal{A}(\mathbf{r}_p, t_p) = \int_{-\infty}^{\infty}\int_{-\infty}^{\infty}\int_{-\infty}^{\infty} \frac{\mu_0 \mathcal{J}\left(\mathbf{r}_q, t - \dfrac{|\mathbf{r}_p - \mathbf{r}_q|}{c}\right) dx_1^q dx_2^q dx_3^q}{4\pi |\mathbf{r}_p - \mathbf{r}_q|} \tag{7.20}$$

which validates Eq. (1.27).

If the four-curl operator is applied to both sides of Eq. (7.19), it may be brought inside the quadruple integral since the differentiations are with respect to $\mathbf{r}_p$ and t_p and affect only $\mathcal{R}^{qp}$. Written in terms of the four-cross operator defined by Eq. (B.13), the result is

$$\Box \times \mathcal{A}(\mathbf{r}_p, t_p) = \int_{-\infty}^{\infty}\int_{-\infty}^{\infty}\int_{-\infty}^{\infty}\int_{-\infty}^{\infty} \frac{\mu_0 \left[\mathcal{J}(\mathbf{r}_q, t_q) \times \mathcal{R}^{qp}\right] dx_1^q dx_2^q dx_3^q dx_4^q}{2\pi^2 \left(\mathcal{R}^{qp}\cdot\mathcal{R}^{qp}\right)^2} \tag{7.21}$$

[1]The p and q labels are helpful in keeping track of the coordinates but are a little cumbersome. An alternate notation is $x_i^p = x_i$, $x_i^q = x_i'$, $t_p = t$, and $t_q = t'$.

Equation (7.21) is the four-dimensional equivalent of the Biot–Savart Law of magnetostatics (Part I, Chapter 1, Section 1.5); according to Eq. (7.13), it generates all components of the **E** and **B** fields. Substituting $\xi = x_4^q - x_4^p = ic(t_q - t_p)$, the $i,j = 4$ components of Eq. (7.21) reduce to

$$\mathbf{E}_p(\mathbf{r}_p, t_p) = \int_{-\infty}^{\infty}\int_{-\infty}^{\infty}\int_{-\infty}^{\infty}\int_{-\infty}^{\infty} \frac{\left[\rho_q(\mathbf{r}_p - \mathbf{r}_q) + \mathbf{J}_q(t_q - t_p)\right]\, dx_1^q dx_2^q dx_3^q d\xi}{2\pi^2\varepsilon_0\left(r_{qp}^2 + \xi^2\right)^2} \tag{7.22a}$$

and the $i,j \neq 4$ components to

$$\mathbf{H}_p(\mathbf{r}_p, t_p) = \int_{-\infty}^{\infty}\int_{-\infty}^{\infty}\int_{-\infty}^{\infty}\int_{-\infty}^{\infty} \frac{\mathbf{J}_q \times (\mathbf{r}_p - \mathbf{r}_q)\, dx_1^q dx_2^q dx_3^q d\xi}{2\pi^2\left(r_{qp}^2 + \xi^2\right)^2} \tag{7.22b}$$

These results determine the exact fields (in free-space) when charge and current distributions (that satisfy conservation of charge) are specified. For both of these integrals, r_p and t_p are held constant; evaluations must take into account that, in general, the variables ρ_q and $\mathbf{J}_q$ are functions of $\mathbf{r}_q$ and $t_q = t_p - i\xi/c$.

This approach is often more convenient than evaluating $\mathcal{A}$ from Eq. (7.19) and then performing the four-curl operation. It is one of several methods developed in Appendix E and used in Part III to solve for the fields of a charge moving at constant velocity. This point is discussed in Stratton [1, pp. 470–476] in connection with the derivation of the radiation fields of an accelerating charge as carried out by Sommerfeld [16]. The approximate result for nonrelativistic motion is derived and the general Sommerfeld solution given in Section E.2.

Notice that if $\mathbf{J}_q = 0$ (everywhere and for all time), ρ_q is a static distribution, then $\mathbf{H}_p = 0$ and

$$\mathbf{E}_p(\mathbf{r}_p) = \int_{-\infty}^{\infty}\int_{-\infty}^{\infty}\int_{-\infty}^{\infty} \frac{\rho_q(\mathbf{r}_p - \mathbf{r}_q)\, dx_1^q dx_2^q dx_3^q}{4\pi\varepsilon_0|\mathbf{r}_p - \mathbf{r}_q|^3}$$

agrees with Eq. (1.21b), Coulomb's Law.

On the other hand, if $\rho_q = 0$ (everywhere and for all time) and **J** is a static distribution $(\nabla \cdot \mathbf{J} = 0)$, then $\mathbf{E}_p = 0$ and

$$\mathbf{H}_p(\mathbf{r}_p) = \int_{-\infty}^{\infty}\int_{-\infty}^{\infty}\int_{-\infty}^{\infty} \frac{\mathbf{J}_q \times (\mathbf{r}_p - \mathbf{r}_q)\, dx_1^q dx_2^q dx_3^q}{4\pi|\mathbf{r}_p - \mathbf{r}_q|^3}$$

agrees with Eq. (1.22c), the Biot–Savart Law.

In these cases, either ρ_q or $\mathbf{J}_q$ is independent of ξ and use has been made of

$$\int_{-\infty}^{\infty} \frac{d\xi}{\left(r_{qp}^2 + \xi^2\right)^2} = \frac{\pi}{2|r_{qp}|^3}$$

7.8 FOUR-DIMENSIONAL FORCE DENSITY

The Lorentz force density can also be expressed using the four-dimensional notation and concepts. It is easy to verify that, using index notation,

$$\sum_{j=1}^{4} \mu_0 \mathcal{G}_{ij}\mathcal{J}_j = f_i, \qquad i = 1, 2, 3$$

reproduces Eq. (1.2). However, this equation certainly invites one to evaluate the term $i = 4$ and consider it the fourth component of a four-vector. The result, $f_4 = \frac{i}{c}\mathbf{E} \cdot \mathbf{J}$, is proportional to the power-conversion density. Therefore, the four-force density defined as

$$\mathcal{F} = \left[\rho\mathbf{E} + \mathbf{J} \times \mu_o\mathbf{H},\ \frac{i}{c}\mathbf{E} \cdot \mathbf{J}\right] \tag{7.23a}$$

can be expressed as

$$\mathcal{F} = \mu_o\mathcal{G} \cdot \mathcal{J} = (\Box \times \mathcal{A}) \cdot \mathcal{J} \tag{7.23b}$$

All of the basic equations presented in Part I have now been reexpressed in four-dimensional form. The four-dimensional curl of $\mathcal{A}$ is seen to play a pivotal role in electromagnetics. In the next chapter, we consider in detail the relationships among force, power, and energy and develop energy-momentum tensors of both conventional and novel form.

7.9 SIX-VECTORS AND DUAL FIELD TENSORS

Because they are antisymmetric, the field tensors $\mathcal{G}$ and $\mathcal{K}$ are both completely defined by the six Cartesian components of **E** and **H**. Each tensor may therefore be considered a six-vector. An often used notation is to enclose the pair of vectors (separated by a semicolon) between curly braces. Therefore,

$$\mathcal{G} = \{\mathbf{H};\quad -ic\varepsilon_o\mathbf{E}\} \tag{7.24a}$$

$$\mathcal{K} = \{-\mathbf{E};\quad -ic\mu_o\mathbf{H}\} \tag{7.24b}$$

It is useful to define the dual six-vector by interchanging the two vectors from which the original six-vector has been constructed. If a superscript dagger (†) is used to indicate the dual, we obtain

$$\mathcal{G}^\dagger = \{-ic\varepsilon_o\mathbf{E};\ \mathbf{H}\} = ic\varepsilon_o\mathcal{K} \tag{7.25a}$$

$$\mathcal{K}^\dagger = \{-ic\mu_o\mathbf{H};\ -\mathbf{E}\} = -ic\mu_o\mathcal{G} = -ic(\Box \times \mathcal{A}) \tag{7.25b}$$

Because $(\mathcal{K}^\dagger)^\dagger = \mathcal{K}$, we have

$$\mathcal{K} = -ic(\Box \times \mathcal{A})^\dagger \tag{7.26}$$

Equation (7.10b) is automatically satisfied, because as noted in Appendix B, Section B.3,

$$\Box \cdot (\Box \times \mathcal{A})^\dagger = 0$$

is a four-vector identity. Note that complex conjugation is *not* performed when a dual six-vector is created.

7.10 FOUR-VECTOR ELECTRIC AND MAGNETIC FIELDS

Although the field tensors, $\mathcal{G}$ and $\mathcal{K}$, contain the components of the electric and magnetic fields, it is only natural to try and find four-vector generalizations of **E** and **H**. Denoted by $\mathcal{E}$ and $\mathcal{H}$, they must properly describe the Lorentz force on moving charges.

Lorentz force on an electric charge

We calculate the four-vector force on an electric charge, q_e, by first considering a convective electric-current density, $\mathbf{J}_e = \rho_e \, \mathbf{v}$. The four-current density is

$$\mathcal{J}_e = [\rho_e \mathbf{v},\ ic\rho_e] = \rho_e[\mathbf{v},\ ic] \tag{7.27}$$

Because $\mathcal{J}_e$ is Lorentz-invariant,

$$\mathcal{J}_e \cdot \mathcal{J}_e = \rho_e^2(\mathbf{v}\cdot\mathbf{v} - c^2) \tag{7.28}$$

must be independent of the velocity. If ρ_e' is defined as the value when $\mathbf{v} = 0$, it follows that

$$\rho_e = \gamma\rho_e' \tag{7.29}$$

$$\frac{1}{\gamma} = \sqrt{1 - \frac{\mathbf{v}\cdot\mathbf{v}}{c^2}} \tag{7.30}$$

The increase in ρ_e is due to the Lorentz contraction along the direction of motion; this reduces the volume containing a fixed amount of charge by $1/\gamma$.

From Eq. (7.23b). the four-force density is

$$\mathcal{F}_e = \mu_o \mathcal{G} \cdot \rho_e' \gamma[\mathbf{v},\ ic] \tag{7.31}$$

If the differential volume in the rest frame is $\Delta V'$ and $\rho_e' \Delta V' = q_e$, we obtain

$$\mathcal{F}_e \Delta V' = q_e \mathcal{E} = q_e \mu_o \mathcal{G} \cdot \gamma[\mathbf{v},\ ic] \tag{7.32}$$

This total four-force, $q_e\mathcal{E}$, leads to the four-vector electric field,

$$\mathcal{E} = \mu_o \mathcal{G} \cdot \mathcal{V} = (\Box \times \mathcal{A}) \cdot \mathcal{V} \tag{7.33a}$$

$$= \gamma[\mathbf{E} + \mathbf{v} \times \mu_o \mathbf{H},\ \frac{i}{c}\mathbf{E}\cdot\mathbf{v}] \tag{7.33b}$$

The four-vector continuum velocity, $\mathcal{V} = [\gamma\mathbf{v},\ i\gamma c]$, is derived in Appendix B, Section B.3. It, too, is Lorentz-invariant with $\mathcal{V} \cdot \mathcal{V} = -c^2$.

Lorentz force on a magnetic charge

If magnetic charge, q_m, existed, a dual Lorentz force would be required to act upon it. Because the roles of **E** and **H** would be reversed, it follows that the derivation just completed can be repeated with $(q_m, \rho_m, \varepsilon_o \mathcal{K})$ substituted for $(q_e, \rho_e, \mu_o \mathcal{G})$. The result is the four-vector magnetic field,

$$\mathcal{H} = \varepsilon_o \mathcal{K} \cdot \mathcal{V} = \frac{-i}{c\mu_o}(\Box \times \mathcal{A})^\dagger \cdot \mathcal{V} \tag{7.34a}$$

$$= \gamma[\mathbf{H} - \mathbf{v} \times \varepsilon_o \mathbf{E},\ \frac{i}{c}\mathbf{H}\cdot\mathbf{v}] \tag{7.34b}$$

This result does not depend upon the value of q_m (which was introduced solely as an artifice). The value can be set equal to zero so as to agree with physical observation, but it can be revived whenever we need sources to generate dual fields. We will revisit the concept of magnetic charge later when we consider the equivalence principle and materials with dielectric and/or magnetic properties.

Notice that both $\mathcal{E} \cdot \mathcal{V} = 0$ and $\mathcal{H} \cdot \mathcal{V} = 0$.

Lorentz invariance of four-vectors

All of the four-vectors that have been introduced are Lorentz-invariant. This means that their *magnitudes* are independent of the inertial frame used to express them. It is convenient to consider a pair of frames. One is chosen to be the fixed laboratory coordinate set, x, y, z, t; the other is chosen to be, any set x', y', z', t' that is moving, with respect to the first, at a constant velocity, $\mathbf{v}$. A Lorentz transformation, derived in Appendix A, Section A.1, connects the coordinates between the two frames; it is "built-in" to each four-vector. In Section A.2, Lorentz invariance is used to deduce how the underlying three-vectors and scalars transform between the moving and stationary frames. This is carried out for $\mathcal{A}$, $\mathcal{J}$, $\mathcal{V}$, $\mathcal{E}$, $\mathcal{H}$, as well as for four-vectors that will be introduced, in Chapter 9, to account for dipoles within polarizable and magnetizable material.

7.11 THE FIELD TENSORS AND MAXWELL'S EQUATIONS REVISITED

Although the field tensors $\mathcal{G}$ and $\mathcal{K}$ are conveniently expressed in terms of the components of the three-space electric and magnetic fields, they can also be expressed in terms of the four-vector fields $\mathcal{E}$, $\mathcal{H}$, and $\mathcal{V}$. As shown in Morgenthaler [17], equivalent representations are

$$\mathcal{G} = \frac{-i}{c}(\mathcal{H} \times \mathcal{V})^{\dagger} - \varepsilon_0(\mathcal{E} \times \mathcal{V}) \tag{7.35a}$$

$$\mathcal{K} = \frac{i}{c}(\mathcal{E} \times \mathcal{V})^{\dagger} - \mu_0(\mathcal{H} \times \mathcal{V}) \tag{7.35b}$$

where the cross product of $\mathcal{E}$ and $\mathcal{V}$ is the antisymmetric tensor, $\mathcal{E} \times \mathcal{V} = \mathcal{E}\mathcal{V} - \mathcal{V}\mathcal{E}$.

In this form it is easy to check previous results. For example, because of orthogonality and $\mathcal{V} \cdot \mathcal{V} = -c^2$,

$$\mu_0\mathcal{G} \cdot \mathcal{V} = \frac{-i}{c}(\mu_0\mathcal{H} \times \mathcal{V})^{\dagger} \cdot \mathcal{V} - \mu_0\varepsilon_0[\mathcal{E}(\mathcal{V} \cdot \mathcal{V}) - \mathcal{V}(\mathcal{E} \cdot \mathcal{V})] = \mathcal{E}$$

$$\varepsilon_0\mathcal{K} \cdot \mathcal{V} = \frac{i}{c}(\varepsilon_0\mathcal{E} \times \mathcal{V})^{\dagger} \cdot \mathcal{V} - \mu_0\varepsilon_0[\mathcal{H}(\mathcal{V} \cdot \mathcal{V}) - \mathcal{V}(\mathcal{H} \cdot \mathcal{V})] = \mathcal{H}$$

In free-space, Maxwell's Equations can be expressed as

$$\Box \cdot \mathcal{G} = \Box \cdot \left[\frac{-i}{c}(\mathcal{H} \times \mathcal{V})^{\dagger} - \varepsilon_0(\mathcal{E} \times \mathcal{V})\right] = \mathcal{J} \tag{7.36a}$$

$$\Box \cdot \mathcal{K} = \Box \cdot \left[\frac{i}{c}(\mathcal{E} \times \mathcal{V})^{\dagger} - \mu_0(\mathcal{H} \times \mathcal{V})\right] = 0 \tag{7.36b}$$

These can be rewritten as

$$\begin{aligned} -\frac{i}{c}(\Box \times \mathcal{H})^{\dagger} \cdot \mathcal{V} - (\mathcal{V} \cdot \Box)(\varepsilon_0\mathcal{E}) + (\Box \cdot \varepsilon_0\mathcal{E})\mathcal{V} - \mathcal{J} \\ = \frac{i}{c}\mathcal{H} \cdot (\Box \times \mathcal{V})^{\dagger} + \varepsilon_0\mathcal{E}(\Box \cdot \mathcal{V}) - (\varepsilon_0\mathcal{E} \cdot \Box)\mathcal{V} \end{aligned}$$

$$\begin{aligned} -\frac{i}{c}(\Box \times \mathcal{E})^{\dagger} \cdot \mathcal{V} + (\mathcal{V} \cdot \Box)(\mu_0\mathcal{H}) - (\Box \cdot \mu_0\mathcal{H})\mathcal{V} \\ = \frac{i}{c}\mathcal{E} \cdot (\Box \times \mathcal{V})^{\dagger} - \mu_0\mathcal{H}(\Box \cdot \mathcal{V}) + (\mu_0\mathcal{H} \cdot \Box)\mathcal{V} \end{aligned}$$

When $\square \mathcal{V} = 0$, the right-hand sides of both these latter equations vanish; then

$$-\frac{i}{c}(\square \times \mathcal{H})^{\dagger} \cdot \mathcal{V} - \varepsilon_0 \frac{\partial \mathcal{E}}{\partial \tau} = \mathcal{J} - \rho' \mathcal{V} \tag{7.37a}$$

$$-\frac{i}{c}(\square \times \mathcal{E})^{\dagger} \cdot \mathcal{V} + \mu_0 \frac{\partial \mathcal{H}}{\partial \tau} = 0 \tag{7.37b}$$

where $\frac{\partial}{\partial \tau} = \mathcal{V} \cdot \square = \gamma(\frac{\partial}{\partial t} + \mathbf{v} \cdot \nabla)$ is the derivative with respect to proper time (that in the rest frame) and ρ' is the rest frame value of the electric-charge density. It also follows that

$$\square \cdot (\varepsilon_0 \mathcal{E}) = \rho' = \frac{-\mathcal{J} \cdot \mathcal{V}}{c^2} \tag{7.38a}$$

$$\square \cdot (\mu_0 \mathcal{H}) = 0 \tag{7.38b}$$

It is reassuring to see the familiar form of Maxwell's Equations shining through the four-dimensional formulation. Notice, too, that from Eqs. (7.33) and (7.34) we obtain

$$\mathcal{E} = (\square \times \mathcal{A}) \cdot \mathcal{V} = -\frac{\partial \mathcal{A}}{\partial \tau} - \square \Phi' - \mathcal{A} \cdot \square \mathcal{V} \tag{7.39a}$$

$$\mu_0 \mathcal{H} = -\frac{i}{c}(\square \times \mathcal{A})^{\dagger} \cdot \mathcal{V} \tag{7.39b}$$

The four-vector Lorentz-force density, Eq. (7.23a), can also be expressed in terms of $\mathcal{E}$ and $\mathcal{H}$; the result is

$$\mathcal{F} = \left[\frac{-i}{c}(\mu_0 \mathcal{H} \times \mathcal{V})^{\dagger} - \mu_0 \varepsilon_0 (\mathcal{E} \times \mathcal{V})\right] \cdot \mathcal{J} \tag{7.40}$$

Using the identities of Appendix B, Eqs. (B.16), this equation can be rewritten as

$$\mathcal{F} = \rho' \mathcal{E} + \frac{-i}{c}(\mathcal{J} \times \mu_0 \mathcal{H})^{\dagger} \cdot \mathcal{V} + \frac{\mathcal{E} \cdot \mathcal{J}}{c^2} \mathcal{V} \tag{7.41a}$$

$$= \left[\rho \mathbf{E} + \mathbf{J} \times \mu_0 \mathbf{H},\ \frac{i}{c} \mathbf{E} \cdot \mathbf{J}\right] \tag{7.41b}$$

(again with $\rho' = \frac{-\mathcal{J} \cdot \mathcal{V}}{c^2}$).

7.12 LINEAR CONDUCTORS REVISITED

The previous results can be applied to a linear material with electrical-conductivity, σ'(as measured in the rest frame). For simplicity, we assume that there are no electric or magnetic dipoles; the inclusion of intrinsic polarization and/or magnetization is deferred until Chapter 9. In the rest frame of a conductor, the current density, $\mathcal{J} = \sigma' \mathcal{E}$, can be expressed using Eq. (7.39a). The wave equation governing $\mathcal{A}$ is therefore

$$\square(\square \cdot \mathcal{A}) - \square^2 \mathcal{A} = \mu_0 \sigma' \square \times \mathcal{A} \cdot \mathcal{V}$$

which can be rewritten as

$$-\square(\square \cdot \mathcal{A} - \mu_0 \sigma' \mathcal{A} \cdot \mathcal{V}) + \square^2 \mathcal{A} - \mu_0 \sigma' \frac{\partial \mathcal{A}}{\partial \tau} = \mu_0 [\sigma' \mathcal{A} \cdot \square \mathcal{V} + \mathcal{A} \cdot \mathcal{V} \square \sigma'] \tag{7.42}$$

Modified Lorenz gauge

Selecting the gauge,

$$\Box \cdot \mathcal{A} = \mu_0 \sigma' \mathcal{A} \cdot \mathcal{V} = -\mu_0 \sigma' \Phi' \tag{7.43}$$

simplifies Eq. (7.42) to

$$\Box^2 \mathcal{A} - \mu_0 \sigma' \frac{\partial \mathcal{A}}{\partial \tau} = \mu_0 [\sigma' \mathcal{A} \cdot \Box \mathcal{V} + \mathcal{A} \cdot \mathcal{V} \Box \sigma'] \tag{7.44}$$

For a uniform nonaccelerating conductor, the result is a homogeneous equation; the diffusion term replaces the current density and modifies the normal wave equation. Source terms are generated by gradients of the conductivity whenever the rest-frame electric potential is nonzero.

Boundary conditions

At surfaces of the conductor, the spatial gradient of σ' is infinite in the direction of the normal. Because $\partial \sigma'/\partial t = 0$, there is no discontinuity in either the scalar potential or the tangential component of the vector potential. However, because

$$\frac{\partial^2 A_n}{\partial n^2} = -\mu_0 \Phi' \frac{\partial \sigma'}{\partial n} \to \infty$$

is the term that matches the singularity, it follows that

$$\Phi^{(1)} = \Phi^{(2)} \tag{7.45a}$$

$$\mathbf{n} \times [\mathbf{A}^{(1)} - \mathbf{A}^{(2)}] = 0 \tag{7.45b}$$

$$\frac{\partial A_n^{(1)}}{\partial n} - \frac{\partial A_n^{(2)}}{\partial n} = \mu_0 \Phi' [\sigma'^{(1)} - \sigma'^{(2)}] \tag{7.45c}$$

CHAPTER 8

ENERGY-MOMENTUM TENSORS

8.1 INTRODUCTION

Force and power conversion densities

In Chapter 3, we explored the relationships among electromagnetic force, power, energy, stress, and momentum. When expressed in terms of the electric and magnetic fields, these led to the Maxwell stress tensor and the Poynting vector.

When Eqs. (3.4) and (3.6) are rewritten in terms of four-dimensional coordinates,

$$\nabla \cdot \overline{\mathbf{T}} + \frac{\partial(-ic\mathbf{G})}{\partial(ict)} = \Box \cdot [\overline{\mathbf{T}}, \ -ic\mathbf{G}] = \mathbf{f} \tag{8.1a}$$

$$\nabla \cdot \left(\frac{\mathbf{S}}{ic}\right) + \frac{\partial W}{\partial(ict)} = \Box \cdot \left[-\frac{i}{c}\mathbf{S}, \ W\right] = \frac{i}{c}\mathbf{E} \cdot \mathbf{J} \tag{8.1b}$$

it is seen that the Poynting Theorem supplies the fourth equation of a set that generates the Lorentz four-force density,

$$\mathcal{F} = \left[\mathbf{f}, \ \frac{i}{c}\mathbf{E} \cdot \mathbf{J}\right]$$

The Power and Beauty of Electromagnetic Fields, First Edition. F. R. Morgenthaler.

Evidently, Eqs. (8.1a) and (8.1b) can expressed as the four-dimensional divergence of an energy-momentum tensor, II, defined by the 4×4 array:

$$\mathrm{II} = \begin{bmatrix} \overline{\mathbf{T}} & \vdots & -ic\mathbf{G} \\ \cdots & \vdots & \cdots \\ -\frac{i}{c}\mathbf{S} & \vdots & W \end{bmatrix} \tag{8.2}$$

so that

$$\Box \cdot \mathrm{II} = \mathcal{F} \tag{8.3a}$$

or, using index notation (with summation over repeated indices assumed),

$$\frac{\partial\, \mathrm{II}_{ij}}{\partial x_j} = \mathcal{F}_i \qquad (i, j = 1, 2, 3, 4) \tag{8.3b}$$

Electromagnetic torque density

The four-space generalization of the torque density, $\mathbf{r} \times \mathbf{f}$, discussed in Chapter 3, Section 3.1, is based upon the antisymmetric tensor, $\mathcal{R} \times \mathcal{F}$, where

$$\mathcal{R} = [\mathbf{r},\ ict]$$

After substitution of Eq. (8.3a) into the cross product and making use of $\frac{\partial R_i}{\partial x_j} = \delta_{ij}$, the four-dimensional Kronecker delta, the result in index notation (with summation over k implied) is

$$\frac{\partial}{\partial x_k}(\mathcal{R}_i \mathrm{II}_{jk} - \mathrm{II}_{ik}\mathcal{R}_j) + \mathrm{II}_{ij} - \mathrm{II}_{ji} = (\mathcal{R} \times \mathcal{F})_{ij} \tag{8.4}$$

The four-vector torque density is defined as

$$\mathcal{T} = \frac{-i}{c}(\mathcal{R} \times \mathcal{F})^{\dagger} \cdot \mathcal{V} \tag{8.5}$$

and, because $\mathcal{T} \cdot \mathcal{V} = 0$, the rest-frame value is

$$\mathcal{T}' = [\mathbf{r}' \times \mathbf{f}', 0]$$

The three-space result given by Eq. (3.3c), repeated here for convenience,

$$[\nabla \cdot (\mathbf{r} \times \overline{\mathbf{T}}) - \frac{\partial(\mathbf{r} \times \mathbf{G})}{\partial t} - \mathbf{r} \times \mathbf{f}]_k + \epsilon_{kij} T_{ij} = 0$$

is consistent with Eqs. (8.4) and (8.5) (refer to Problem 25.10-7). Notice that if the stress tensor is symmetric, $\epsilon_{kij} T_{ij} = 0$.

The connection between the angular-momentum density, $\mathbf{r} \times \mathbf{G}$, and the spin, orbit, and charge angular momenta associated with photons and charges is explored in Appendix D, Section D.3. In turn, these connections clarify the properties of the circular waveguide modes analyzed in Chapter 20, Section 20.5.

8.2 MAXWELL–POYNTING ENERGY-MOMENTUM TENSOR

Because $\mathcal{F} = \mu_o \mathcal{G} \cdot \mathcal{J} = \mu_o \mathcal{G} \cdot (\Box \cdot \mathcal{G}) + \varepsilon_o \mathcal{K} \cdot (\Box \cdot \mathcal{K})$, one can verify (using Appendix B, Eq. (B.18b)) that

$$\mathrm{II}^{\mathrm{em}} = \begin{bmatrix} \overline{\mathbf{T}}^{\mathrm{em}} & \vdots & -ic\mathbf{G}^{\mathrm{em}} \\ \cdots & \vdots & \cdots \\ -\frac{i}{c}\mathbf{S}^{\mathrm{em}} & \vdots & W^{\mathrm{em}} \end{bmatrix} \tag{8.6}$$

can be written, in terms of the field tensors, as

$$\begin{aligned} \mathrm{II}^{\mathrm{em}} &= \frac{1}{2}(\mu_o \mathcal{G} \cdot \mathcal{G} + \varepsilon_o \mathcal{K} \cdot \mathcal{K}) \\ &= \frac{1}{2\mu_o}\left[(\Box \times \mathcal{A}) \cdot (\Box \times \mathcal{A}) - (\Box \times \mathcal{A})^{\dagger} \cdot (\Box \times \mathcal{A})^{\dagger}\right] \end{aligned} \tag{8.7a}$$

Since $\Box \times (\mathcal{A} + \Box\Psi) = \Box \times \mathcal{A}$, the Maxwell–Poynting representation is clearly independent of the gauge.

Equation (8.7a) can also be expressed as

$$\begin{aligned} \mathrm{II}^{\mathrm{em}} &= \varepsilon_o \mathcal{E}\,\mathcal{E} + \mu_o \mathcal{H}\,\mathcal{H} - \frac{1}{2}(\varepsilon_o \mathcal{E} \cdot \mathcal{E} + \mu_o \mathcal{H} \cdot \mathcal{H})\left(\mathcal{I} + 2\frac{\mathcal{V}\,\mathcal{V}}{c^2}\right) \\ &\quad + \frac{i}{c^3}\left[\mathcal{V}\,(\mathcal{E} \times \mathcal{H})^{\dagger} \cdot \mathcal{V} + (\mathcal{E} \times \mathcal{H})^{\dagger} \cdot \mathcal{V}\,\mathcal{V}\right] \end{aligned} \tag{8.7b}$$

where $\mathcal{I}$ is the identity tensor for four dimensions. Because all of these forms are equivalent, the value of the three-vector velocity (contained in $\mathcal{V}$, $\mathcal{E}$, and $\mathcal{H}$) must (and does) cancel when Eq. (8.7b) is evaluated.

Because $\mathrm{II}^{\mathrm{em}}$ is symmetric, it follows that Eq. (8.4) reduces to

$$\frac{\partial}{\partial x_k}(\mathcal{R}_i \mathrm{II}^{\mathrm{em}}_{jk} - \mathrm{II}^{\mathrm{em}}_{ik}\mathcal{R}_j) = (\mathcal{R} \times \mathcal{F})_{ij}$$

8.3 ALTERNATE ENERGY-MOMENTUM TENSORS

It is helpful to express

$$\mathcal{F} = (\Box \times \mathcal{A}) \cdot \mathcal{J}$$

and

$$\Box \cdot (\Box \times \mathcal{A}) = \Box(\Box \cdot \mathcal{A}) - \Box^2 \mathcal{A} = \mu_o \mathcal{J}$$

in index notation as

$$\mathcal{F}_i = \left(\frac{\partial \mathcal{A}_j}{\partial x_i} - \frac{\partial \mathcal{A}_i}{\partial x_j}\right)\mathcal{J}_j \tag{8.8}$$

and

$$\frac{\partial}{\partial x_j}\left(\frac{\partial \mathcal{A}_j}{\partial x_i} - \frac{\partial \mathcal{A}_i}{\partial x_j}\right) = \frac{\partial}{\partial x_i}\left(\frac{\partial \mathcal{A}_j}{\partial x_j}\right) - \frac{\partial^2 \mathcal{A}_i}{\partial x_j \partial x_j} = \mu_o \mathcal{J}_i \tag{8.9}$$

Here summation over repeated indices is implied; in this case $j = 1, 2, 3, 4$; the free index, i, takes on values $1, 2, 3, 4$.

Equation (8.8) can be rewritten as

$$\mathcal{F}_i = \frac{1}{2}\frac{\partial(\mathcal{A}_j\mathcal{J}_j)}{\partial x_i} - \frac{\partial(\mathcal{A}_i\mathcal{J}_j)}{\partial x_j} + \frac{1}{2\mu_o}\left[\frac{\partial \mathcal{A}_j}{\partial x_i}(\mu_o\mathcal{J}_j) - \mathcal{A}_j\frac{\partial(\mu_o\mathcal{J}_j)}{\partial x_i}\right] + \mathcal{A}_i\frac{\partial \mathcal{J}_j}{\partial x_j}$$

What is most interesting is the fact that, after invoking charge conservation ($\Box \cdot \mathcal{J} = 0$) and substituting Eq. (8.9), the right-hand side of this equation can be manipulated to become the four-divergence of a *new* tensor, II°, defined by

$$\begin{aligned}\text{II}^{\text{o}}_{ij} &= \frac{1}{2}\mathcal{A} \cdot \mathcal{J}\,\delta_{ij} - \mathcal{A}_i\,\mathcal{J}_j \\ &+ \frac{1}{2\mu_o}\left[\mathcal{A} \cdot \frac{\partial^2 \mathcal{A}}{\partial x_i \partial x_j} - \frac{\partial \mathcal{A}}{\partial x_i} \cdot \frac{\partial \mathcal{A}}{\partial x_j} + \frac{\partial \mathcal{A}_j}{\partial x_i}(\Box \cdot \mathcal{A}) - \mathcal{A}_j \frac{\partial}{\partial x_i}(\Box \cdot \mathcal{A})\right]\end{aligned} \tag{8.10}$$

We designate this as *the* Alternate energy-momentum tensor that is *distinct* from II^{em} yet has the *identical* four-divergence and therefore produces exactly the *same* four-force density, $\mathcal{F}$. If desired, Eq. (8.9) can be substituted into Eq. (8.10) in order to express the latter solely in terms of the four-dimensional vector potential.

If suitably normalized, any linear combination of II^{em} and II° can serve as an alternative energy-momentum tensor; the possibilities are limitless. However, we consider these two as worthy of special consideration; II^{em}, because power and energy are associated solely with **E** and **H**; II°, because at low frequencies they are associated mainly with ρ and **J** and a circuit-theory interpretation is not only permitted, but encouraged. For harmonic fields, the Maxwell–Poynting representation is one of waves whereas the Alternate representation is one of particles (charges and photons).

Because II° is, in general, nonsymmetric, it follows that Eq. (8.4) reduces to

$$\frac{\partial}{\partial x_k}(\mathcal{R}_i \text{II}^{\text{o}}_{jk} - \text{II}^{\text{o}}_{ik}\mathcal{R}_j) + (\mathcal{J} \times \mathcal{A})_{ij} + \frac{1}{2\mu_o}(g\,\Box \times \mathcal{A} - \Box g \times \mathcal{A})_{ij} = (\mathcal{R} \times \mathcal{F})_{ij} \tag{8.11}$$

where the gauge is defined by

$$g = \Box \cdot \mathcal{A}$$

Equations (8.10) and (8.11) are both simplified if the Lorenz-gauge is chosen (refer to Problem 25.10-7) so we normally set $g = 0$. When the Coulomb gauge is required, $\nabla \cdot \mathbf{A} = 0$ ($g = \frac{1}{c^2}\frac{\partial \Phi}{\partial t}$). For a conducting medium, $g = \mu_o\sigma'\mathcal{A} \cdot \mathcal{V} = -\mu_o\sigma'\Phi'$.

Dielectric and magnetic materials are considered in the next chapter; in the Amperian formulation both polarization and magnetization electric currents are added to $\mathcal{J}$.

Dual Alternate energy-momentum tensor

According to the principle of duality that was discussed in Part I, Chapter 6, Section 6.5, the substitutions

$\mathbf{J}$	$\rightarrow$	$\mathbf{J}_{\text{m}}$
ρ	$\rightarrow$	ρ_{m}
$\mathbf{A}$	$\rightarrow$	$\mathbf{A}_{\text{m}}$
Φ	$\rightarrow$	Φ_{m}
μ_o	$\rightarrow$	ε_o
ε_o	$\rightarrow$	μ_o

(where $\mathbf{J}_{\text{m}}$ and ρ_{m} are magnetic current and charge densities) create the dual free-space fields (denoted with primes)

$\mathbf{E}$	$\rightarrow$	$\mathbf{H}'$
$\mathbf{H}$	$\rightarrow$	$-\mathbf{E}'$

and the dual Lorentz-force density,

$$\| \ \rho \mathbf{E} + \mathbf{J} \times \mu_0 \mathbf{H} \quad \to \quad \rho_m \mathbf{H}' - \mathbf{J}_m \times \varepsilon_0 \mathbf{E}' \ \|$$

It is advantageous to define the following four-vectors for the dual fields:

$$\mathcal{J}_m = [\mathbf{J}_m, \ ic\rho_m]$$

$$\mathcal{A}_m = \left[\mathbf{A}_m, \ \frac{i}{c}\Phi_m \right]$$

$$\mathcal{F}_{mi} = \left(\frac{\partial \mathcal{A}_{mj}}{\partial x_i} - \frac{\partial \mathcal{A}_{mi}}{\partial x_j} \right) \mathcal{J}_{mj}$$

Conservation of magnetic charge requires

$$\Box \cdot \mathcal{J}_m = 0$$

The dual of Eq. (8.9) is

$$\Box(\Box \cdot \mathcal{A}_m) - \Box^2 \mathcal{A}_m = \varepsilon_0 \mathcal{J}_m$$

When set, the Lorenz gauge is

$$\Box \cdot \mathcal{A}_m = 0$$

The dual Alternate energy-momentum tensor, II_{ij}^{om}, and the associated torque tensor can be written down by inspection:

$$\begin{aligned} \text{II}_{ij}^{om} &= \frac{1}{2}\mathcal{A}_m \cdot \mathcal{J}_m \, \delta_{ij} - \mathcal{A}_{mi}\mathcal{J}_{mj} \\ &+ \frac{1}{2\varepsilon_0}\left[\mathcal{A}_m \cdot \frac{\partial^2 \mathcal{A}_m}{\partial x_i \partial x_j} - \frac{\partial \mathcal{A}_m}{\partial x_i} \cdot \frac{\partial \mathcal{A}_m}{\partial x_j} + \frac{\partial \mathcal{A}_{mj}}{\partial x_i}(\Box \cdot \mathcal{A}_m) - \mathcal{A}_{mj}\frac{\partial}{\partial x_i}(\Box \cdot \mathcal{A}_m) \right] \end{aligned} \tag{8.12}$$

and

$$\begin{aligned} &\frac{\partial}{\partial x_k}(\mathcal{R}_i \text{II}_{jk}^{om} - \text{II}_{ik}^{om}\mathcal{R}_j) + (\mathcal{J}_m \times \mathcal{A}_m)_{ij} + \frac{1}{2\varepsilon_0}(g_m \Box \times \mathcal{A}_m - \Box g_m \times \mathcal{A}_m)_{ij} \\ &\quad = (\mathcal{R} \times \mathcal{F}_m)_{ij} \end{aligned} \tag{8.13}$$

where

$$g_m = \Box \cdot \mathcal{A}_m$$

In the remainder of this chapter, we will focus primarily on the Alternate energy-momentum tensor, but similar formulas apply to the dual tensor. Either set of formulas can be used as the basis of four-dimensional electrodynamics, provided that the sources are either electric or magnetic charges–but not both.[1]

[1]When *both* types of charge are involved, the superpostion discussed in Part I, Chapter 6, Section 6.5 is required.

Components of the Alternate tensor

The four alternate-components (stress, momentum-density, power-flux, and energy-density) of Eq. (8.10) are defined by

$$\text{II}^{\text{o}} = \begin{bmatrix} \overline{\mathbf{T}}^{\text{o}} & \vdots & -ic\mathbf{G}^{\text{o}} \\ \cdots & \vdots & \cdots \\ -\frac{i}{c}\mathbf{S}^{\text{o}} & \vdots & W^{\text{o}} \end{bmatrix} \tag{8.14}$$

where (for i, j; 1, 2, 3),

$$T_{ij}^{\text{o}} = \frac{1}{2}(\mathbf{A}\cdot\mathbf{J} - \rho\,\Phi)\delta_{ij} - A_i J_j + \frac{1}{2\mu_{\text{o}}}\left(\mathcal{A}\cdot\frac{\partial^2 \mathcal{A}}{\partial x_i \partial x_j} - \frac{\partial \mathcal{A}}{\partial x_i}\cdot\frac{\partial \mathcal{A}}{\partial x_j} + g\frac{\partial A_j}{\partial x_i} - \frac{\partial g}{\partial x_i}A_j\right) \tag{8.15a}$$

$$G_i^{\text{o}} = \rho A_i + \frac{1}{2}\varepsilon_{\text{o}}\left(\mathcal{A}\cdot\frac{\partial^2 \mathcal{A}}{\partial x_i \partial t} - \frac{\partial \mathcal{A}}{\partial x_i}\cdot\frac{\partial \mathcal{A}}{\partial t} + \frac{\partial g}{\partial x_i}\Phi - g\frac{\partial \Phi}{\partial x_i}\right) \tag{8.15b}$$

$$S_i^{\text{o}} = \Phi J_i + \frac{1}{2\mu_{\text{o}}}\left(\mathcal{A}\cdot\frac{\partial^2 \mathcal{A}}{\partial x_i \partial t} - \frac{\partial \mathcal{A}}{\partial x_i}\cdot\frac{\partial \mathcal{A}}{\partial t} + g\frac{\partial A_i}{\partial t} - \frac{\partial g}{\partial t}A_i\right) \tag{8.15c}$$

$$W^{\text{o}} = \frac{1}{2}(\mathbf{A}\cdot\mathbf{J} + \rho\Phi) - \frac{1}{2}\varepsilon_{\text{o}}\left(\mathcal{A}\cdot\frac{\partial^2 \mathcal{A}}{\partial t^2} - \frac{\partial \mathcal{A}}{\partial t}\cdot\frac{\partial \mathcal{A}}{\partial t} + \frac{\partial g}{\partial t}\Phi - g\frac{\partial \Phi}{\partial t}\right) \tag{8.15d}$$

Whenever g is independent of $\mathbf{A}$, as in the Lorenz, Coulomb, or stationary-conductor gauge, Eq. (8.15d) is separable into electric and magnetic components, $W^{\text{o}} = W_{\text{e}}^{\text{o}}(\Phi) + W_{\text{m}}^{\text{o}}(\mathbf{A})$.

It is interesting that, although II^{em} is a symmetric tensor, II° is not, except wherever $g = 0$ and $\mathcal{A} \times \mathcal{J} = 0$. The latter condition is met when both $\mathbf{J} \times \mathbf{A} = 0$ and $\Phi\mathbf{J} = \text{c}^2\rho\mathbf{A}$. In such cases, the general relationship,

$$\mathbf{G}^{\text{o}} - \rho\mathbf{A} = \frac{1}{c^2}(\mathbf{S}^{\text{o}} - \Phi\mathbf{J}) \tag{8.16}$$

implies that $\mathbf{S}^{\text{o}} = \text{c}^2\mathbf{G}^{\text{o}}$ in these and all free-space regions. The connections between the asymmetry of II°, the electromagnetic torque density, and the spin angular momentum of photons are profound; as already mentioned, they are discussed in Appendix D, Section D.3 and applied to circular waveguide modes in Chapter 20, Section 20.5.

Equation (8.16), encountered previously in Part I, Chapter 3, Section 3.4, is clarified by consideration of the force on a moving charge. We revisit the derivation, contained in Section 3.4, from the perspective of four-dimensional analysis.

Electromagnetic force on a moving charge

The Lorentz-force acting upon a moving charge, q (with rest mass, m'), when combined with Newton's Second Law, gives in four dimensions

$$\frac{d}{dt'}(m'\mathcal{V}) = \left[\frac{d}{dt'}(\gamma m'\mathbf{v}),\ \frac{i}{c}\frac{d}{dt'}(\gamma m' c^2)\right] = q\mathcal{E} \tag{8.17}$$

The fourth component reveals Einstein's $E = \gamma m' c^2 = mc^2$, but our present concern is with the three-space components. Because $\frac{d}{dt'} = \gamma \frac{d}{dt}$, the relativistically correct force law is

$$\frac{d}{dt}(\gamma m' \mathbf{v}) = q(\mathbf{E} + \mathbf{v} \times \mu_0 \mathbf{H}) = q\left[\left(-\frac{\partial \mathbf{A}}{\partial t} - \nabla\Phi\right) + \mathbf{v} \times (\nabla \times \mathbf{A})\right]$$

Rearrangement produces the equivalent and well-known formula:

$$\frac{d}{dt}(\gamma m' \mathbf{v} \; + \; q\mathbf{A}) = -q\left[\nabla(\Phi - \mathbf{A} \cdot \mathbf{v}) + A_j \nabla \mathrm{v}_j\right] \tag{8.18}$$

where $\frac{d\mathbf{A}}{dt}$ is expressed in terms of the substantial derivative, $(\partial/\partial t + \mathbf{v} \cdot \nabla)\mathbf{A}$, and summation over the index $j = 1, 2, 3$ is assumed. We again make the point that because the value of the vector potential is evaluated at the location of the charge, it is permissible to assign the momentum, $q\mathbf{A}$, to the charge; of course, that value depends upon the global distribution of *all* contributing electromagnetic sources.

On a per-unit volume basis, the localized momentum $q\mathbf{A}$ can be expressed as the density,

$$\rho A_i = \mu_0 \varepsilon_0 (\mathbf{E} \times \mathbf{H})_i + \frac{\partial}{\partial x_j}\left[\varepsilon_0 (A_i E_j - \frac{1}{2}\mathbf{A} \cdot \mathbf{E}\, \delta_{ij})\right] + \frac{1}{2}\varepsilon_0 \left(A_j \frac{\partial E_j}{\partial x_i} - \frac{\partial A_j}{\partial x_i} E_j\right) \tag{8.19}$$

For a magnetic charge, q_m, the duals of Eqs. (8.17) and (8.18) are required and lead to the localized momentum, $q_m \mathbf{A}_m$, and the dual of Eq. (8.19),

$$\rho_m A_{mi} = \mu_0 \varepsilon_0 (\mathbf{E} \times \mathbf{H})_i + \frac{\partial}{\partial x_j}\left[\mu_0 (A_{mi} H_j - \frac{1}{2}\mathbf{A}_m \cdot \mathbf{H} \delta_{ij})\right] + \frac{1}{2}\mu_0 \left(A_{mj} \frac{\partial H_j}{\partial x_i} - \frac{\partial A_{mj}}{\partial x_i} H_j\right) \tag{8.20}$$

These equations reveal the connection between $\rho\mathbf{A}$, $\rho_m \mathbf{A}_m$ and the electromagnetic-field momentum density, $(\mathbf{E} \times \mathbf{H})/c^2$, and make understandable the alternate accounting of power and momentum described in Part I, Chapter 3, Section 3.4.

8.4 BOUNDARY CONDITIONS AND GAUGE CONSIDERATIONS

From Eq. (7.7), the four-dimensional curl of $\mathcal{A}$ is (in six-vector notation)

$$\square \times \mathcal{A} = \left\{\mu_0 \mathbf{H}; \; -\frac{i}{c}\mathbf{E}\right\} \tag{8.21}$$

Therefore, assuming that the gauge (set by $\square \cdot \mathcal{A}$) and the fields $\mathbf{E}$ and $\mathbf{H}$ are everywhere finite, it follows that $\mathcal{A}$ must be piecewise continuous throughout space and time.[2]

[2]There are, of course, exceptional cases in which the electric and/or magnetic field is properly infinite or conveniently modeled as such. These include regions located arbitrarily close to charge or current singularities and, in some cases, wave fronts created by discontinuities of current arising from very rapidly accelerating electric charge distributions.

Consequently, even when **E** and **H** are restricted to a well-defined region (for example, by means of perfectly conducting boundaries), the potentials may have fringing components outside that region. In such cases, II° may be nonzero in the electric and magnetic field-free regions.

It is often relatively easy to find $\mathcal{A}_p$, a particular solution that, while generating the correct electric and magnetic fields in all regions (except *on* one or more of the boundaries between them), violates the continuity imposed by a consistent gauge.[3]

In each region, the missing homogeneous component, $\mathcal{A}_h$, generates no fields and thus satisfies $\Box \times \mathcal{A}_h = 0$; it follows that $\mathcal{A}_h = \Box\Psi_h$ where Ψ_h is a scalar function. If the Lorenz gauge is selected, $\Box \cdot \mathcal{A}_h = \Box^2\Psi_h = 0$ (except on boundaries where $\mathcal{A}_p$ is discontinuous). Connection between regions requires continuity of $(\mathcal{A}_p + \Box\Psi_h)$ through all boundaries. Because in each region,

$$\left[\Box \times (\mathcal{A}_p + \mathcal{A}_h)\right] \cdot \mathcal{J} = \left(\Box \times \mathcal{A}_p\right) \cdot \mathcal{J} = \mathcal{F} \tag{8.22}$$

the correct four-vector force density can be obtained *without* including the homogeneous potential, provided that limiting values of $\mathcal{A}$ can be evaluated correctly where there are surface currents and/or surface charges. It is then permissible to use $\mathcal{A}_p$ in Eqs. (8.10). However, the spatial distributions and the degree of circuit localization may be altered by the omission of $\mathcal{A}_h$. This is demonstrated in the Part III examples of a *TEM* parallel-plate transmission line and of TE_{nm} modes of a circular waveguide.

8.5 ELECTROMAGNETIC BEAUTY REVISITED

Because the difference $\text{II}^{\text{o}} - \text{II}^{\text{em}}$ has no four-dimensional divergence, it or any scaled version can be added to II^{em} *without altering the force*, $\mathcal{F}$, in any way. Within limits, that difference can therefore be used to redistribute (redefine) power flows and energy according to an individuals whim or intuition. It seems fitting to use this result to establish the concept of "electromagnetic beauty" and honor the thought-provoking comments of Mason and Weaver, cited in the preface. Accordingly, we define $\text{II}^{\text{b}} = \text{II}^{\text{o}} - \text{II}^{\text{em}}$ to be the "electromagnetic-beauty" tensor with

$$\Box \cdot \text{II}^{\text{b}} = 0 \tag{8.23}$$

Both stress momentum and power energy are embraced by the concept. Although it produces neither electromagnetic force nor $\mathbf{E} \cdot \mathbf{J}$ power density, and is therefore an ephemeral quantity, II^{b} is not identical to nor should it be confused with the "zilch" introduced by Lipkin [18].

From Eqs. (8.15b) and (8.19), we have

$$\left(\mathbf{G}^{\text{o}} - \frac{\mathbf{E} \times \mathbf{H}}{c^2}\right)_i = \frac{\partial \alpha_{ij}}{\partial x_j} \tag{8.24}$$

[3] As a single example, consider a region completely enclosed by a perfectly conducting surface. If all currents and charges are (and always were) confined to the interior, the exterior is field-free, and $\mathcal{A}_p = 0$ is the simplest possible choice for the particular solution (and obviously satisfies both the Lorenz and Coulomb gauges). However, if $\mathcal{A}_p$ has any nonzero components at the innersurface of the boundary, one or more of **E**, $\mu_o\mathbf{H}$, and $\Box \cdot \mathcal{A}$ will become infinite *on* the surface. This is prevented by adding homogeneous solutions, $\mathcal{A}_h$, that fringe outside (and perhaps inside) the surface so as to restore continuity at the boundary without altering the electric and magnetic fields, except on the surface where any previous infinite values have now been removed.

with the α-tensor defined by

$$\alpha_{ij} = \varepsilon_o \left[A_i E_j - \frac{1}{2}\mathbf{A}\cdot\mathbf{E}\,\delta_{ij} + \frac{1}{2}\left(\Phi\frac{\partial A_j}{\partial x_i} - \frac{\partial \Phi}{\partial x_i} A_j \right) \right] \tag{8.25}$$

Based upon Eqs. (8.15c), (8.15d), and $g = \Box \cdot \mathcal{A}$, we have

$$(\mathbf{S}^o - \mathbf{E}\times\mathbf{H})_i = \left[\nabla\times(\Phi\mathbf{H}) + \frac{\partial}{\partial t}\left(\mathbf{A}\times\mathbf{H} - \varepsilon_o\Phi\mathbf{E} - \frac{1}{2\mu_o} g\mathbf{A} \right) \right]_i - \frac{\partial \beta_{ij}}{\partial x_j} \tag{8.26a}$$

and

$$W^o - W^{em} = \nabla\cdot\frac{1}{2}\left[\varepsilon_o(\Phi\mathbf{E} + \frac{\partial\Phi}{\partial t}\mathbf{A} - \Phi\frac{\partial \mathbf{A}}{\partial t}) - \mathbf{A}\times\mathbf{H} \right] \tag{8.26b}$$

where

$$\beta_{ij} = \frac{1}{\mu_o}\left[E_i A_j - \frac{1}{2}(\mathbf{E}\cdot\mathbf{A} - g\Phi)\delta_{ij} - \frac{1}{2}\left(\Phi\frac{\partial A_j}{\partial x_i} - \frac{\partial \Phi}{\partial x_i} A_j \right) \right] \tag{8.27}$$

with δ_{ij} the Kronecker delta.

The "electromagnetic-beauty" theorem,

$$\nabla\cdot(\mathbf{S}^o - \mathbf{E}\times\mathbf{H}) + \frac{\partial}{\partial t}(W^o - W^{em}) = 0 \tag{8.28}$$

when evaluated from Eqs. (8.26a) and (8.26b), results in

$$\frac{\partial}{\partial t}\nabla\cdot\frac{1}{2}\left[\mathbf{A}\times\mathbf{H} - \varepsilon_o\left(\Phi\mathbf{E} + \Phi\frac{\partial\mathbf{A}}{\partial t} - \frac{\partial\Phi}{\partial t}\mathbf{A} \right) - \frac{1}{\mu_o} g\mathbf{A} \right] = \frac{\partial^2 \beta_{ij}}{\partial x_i \partial x_j} \tag{8.29}$$

where summation over both repeated indices $(i, j; 1, 2, 3)$ is assumed. This result is applicable to any set of fields $(\mathbf{A}, \Phi, \mathbf{E}, \mathbf{H}, \mathbf{J}, \rho)$ that satisfy Maxwell's Equations. Notice that this equation, which validates Eq. (3.15a), can be rewritten in the form of a conservation theorem:

$$\nabla\cdot\mathbf{S}^b + \frac{\partial}{\partial t}W^b = 0 \tag{8.30}$$

Notice, too, that as expected, $\mathbf{S}^b$ and W^b are not unique since the latter can be made to vanish by simple rearrangement. If that choice is made, (and the superscript b replaced by *beauty* to so signify), it follows that (independent of g) we have

$$\mathbf{S}^{beauty} = \frac{1}{2\mu_o}\nabla\times\left(\mathbf{A}\times\frac{\partial\mathbf{A}}{\partial t} \right) \tag{8.31}$$

which, of course, is divergence-free. When $\Phi = 0$, Eq. (8.31) can be expressed as $\mathbf{S}^{beauty} = \frac{1}{2}c^2\nabla\times\mathbf{L}^{spin}$ (where $\mathbf{L}^{spin}$ is the spin angular-momentum density derived in Appendix D, Section D.3).

Evidently, one can add *any* arbitrarily scaled version of Eq. (8.30) to the conventional Poynting Theorem and interpret the resultant divergence and time derivative as the "true" electromagnetic power flux and energy density. It bears repeating that of the multitude of possible choices, those based upon Eqs. (8.7) and (8.10) represent two extremes worthy of special attention: the first because it expresses the field representation, the second because it emphasizes the circuit representation. Naturally, there are similar expressions involving dual Alternate energy-momentum tensors.

CHAPTER 9

DIELECTRIC AND MAGNETIC MATERIALS

9.1 INTRODUCTION

In Chapter 1, Section 1.7, dielectric and magnetic materials are characterized by electric and/or magnetic dipoles that are generated by polarization currents and Amperian currents. Because we are using a vector potential formulation, it is convenient to model both the polarization and magnetization in terms of electric charges and currents. When these are added to $\mathbf{J}_\text{u}$, the current associated with the free (unpaired) charge, the result (repeated for convenience) is

$$\mathbf{J}_\text{total} = \mathbf{J}_\text{u} + \mathbf{J}_\text{polarization} + \mathbf{J}_\text{amperian} = \mathbf{J}_\text{u} + \frac{\partial \mathbf{P}^\text{o}}{\partial t} + \nabla \times \mathbf{M}^\text{o} \tag{9.1a}$$

$$\rho_\text{total} = \rho_\text{u} + \rho_\text{polarization} + \rho_\text{amperian} = \rho_\text{u} - \nabla \cdot \mathbf{P}^\text{o} + 0 \tag{9.1b}$$

The total four-vector electric current is then

$$\mathcal{J}_\text{total} = [\mathbf{J}_\text{total},\ ic\rho_\text{total}] \tag{9.2}$$

As previously mentioned, each component of electric charge is separately conserved, as is their total.

The Power and Beauty of Electromagnetic Fields, First Edition. F. R. Morgenthaler.

9.2 MAXWELL'S EQUATIONS WITH POLARIZATION AND MAGNETIZATION

We continue to employ a four-dimensional formulation and introduce four-vectors: $\mathcal{E}$, $\mathcal{B}$, $\mathcal{P}$, $\mathcal{M}$, and $\mathcal{V}$ used to represent the electric field, magnetic induction, polarization, magnetization, and the continuum velocity of the material. As before,

$$\mathcal{V} = \left[\gamma \mathbf{v},\ i\gamma c\right]$$

$$\gamma = \frac{1}{\sqrt{1 - \dfrac{\mathbf{v}\cdot\mathbf{v}}{c^2}}}$$

Because we have chosen to model the material sources as purely electric currents and charges,[1] we initially choose the three-space vectors $\mathbf{E}^{\mathrm{o}}$, $\mathbf{B}$, $\mathbf{P}^{\mathrm{o}}$, and $\mathbf{M}^{\mathrm{o}}$ with which to express

$$\mathcal{E} = (\Box \times \mathcal{A}) \cdot \mathcal{V} = \left[\gamma(\mathbf{E}^{\mathrm{o}} + \mathbf{v}\times\mathbf{B}),\ i\gamma\frac{\mathbf{B}\cdot\mathbf{v}}{c}\right] \tag{9.3a}$$

$$\mathcal{B} = -\frac{i}{c}(\Box \times \mathcal{A})^{\dagger} \cdot \mathcal{V} = \left[\gamma\left(\mathbf{B} - \frac{\mathbf{v}\times\mathbf{E}^{\mathrm{o}}}{c^2}\right),\ i\gamma\frac{\mathbf{B}\cdot\mathbf{v}}{c}\right] \tag{9.3b}$$

$$\mathcal{P} = \left[\gamma\left(\mathbf{P}^{\mathrm{o}} - \frac{\mathbf{v}\times\mathbf{M}^{\mathrm{o}}}{c^2}\right),\ i\gamma\frac{\mathbf{P}^{\mathrm{o}}\cdot\mathbf{v}}{c}\right] \tag{9.3c}$$

$$\mathcal{M} = \left[\gamma(\mathbf{M}^{\mathrm{o}} + \mathbf{v}\times\mathbf{P}^{\mathrm{o}}),\ i\gamma\frac{\mathbf{M}^{\mathrm{o}}\cdot\mathbf{v}}{c}\right] \tag{9.3d}$$

It is convenient to define two additional four-vectors:

$$\mathcal{D} = \varepsilon_{\mathrm{o}}\mathcal{E} + \mathcal{P}$$

$$\mathcal{H} = \mathcal{B}/\mu_{\mathrm{o}} - \mathcal{M}$$

These four-vectors satisfy $\mathcal{E}\cdot\mathcal{V} = \mathcal{B}\cdot\mathcal{V} = \mathcal{P}\cdot\mathcal{V} = \mathcal{M}\cdot\mathcal{V} = \mathcal{D}\cdot\mathcal{V} = \mathcal{H}\cdot\mathcal{V} = 0$.

Maxwell's Equations can then be written as

$$\Box\cdot(\Box \times \mathcal{A}) = \Box\cdot\left[\frac{-i}{c}(\mathcal{B}\times\mathcal{V})^{\dagger} - \frac{\mathcal{E}\times\mathcal{V}}{c^2}\right] = \mu_{\mathrm{o}}\mathcal{J}_{\mathrm{total}} \tag{9.4a}$$

$$ic\Box\cdot(\Box \times \mathcal{A})^{\dagger} = \Box\cdot\left[\frac{-i}{c}(\mathcal{E}\times\mathcal{V})^{\dagger} + \mathcal{B}\times\mathcal{V}\right] = 0 \tag{9.4b}$$

where

$$\mathcal{J}_{\mathrm{total}} = \mathcal{J}_{\mathrm{u}} + \Box\cdot\left[\mathcal{P}\times\mathcal{V} + \frac{-i}{c}(\mathcal{M}\times\mathcal{V})^{\dagger}\right] \tag{9.4c}$$

[1] This leads to an Amperian current model of the magnetization. Comparisons with other four-dimensional formulations of electrodynamics used to model deformable polarizable and magnetizable media are postponed until Chapter 10.

Again, the subscript u denotes the unpaired (free) charge and current densities. The polarization and Amperian electric-current densities, as well as the polarization charge density, are all in evidence.

Upon substitution of the four-vectors, the three-space equations are obtained; they are

$$\nabla \times \mathbf{B} - \frac{1}{c^2}\frac{\partial \mathbf{E}^{\mathrm{o}}}{\partial t} = \mu_{\mathrm{o}}\mathbf{J}_{\mathrm{total}}$$

$$\nabla \cdot \varepsilon_{\mathrm{o}}\mathbf{E}^{\mathrm{o}} = \rho_{\mathrm{total}}$$

$$\nabla \times \mathbf{E}^{\mathrm{o}} + \frac{\partial \mathbf{B}}{\partial t} = 0$$

$$\nabla \cdot \mathbf{B} = 0$$

which agree exactly with Eqs. (1.43).

9.3 AMPERIAN ENERGY-MOMENTUM TENSORS

Because

$$(\Box \times \mathcal{A}) = -\frac{i}{c}(\mathcal{B} \times \mathcal{V})^{\dagger} - \frac{\mathcal{E} \times \mathcal{V}}{c^2} \tag{9.5}$$

the new four-dimensional electromagnetic force density, $\mathcal{F} = \mathcal{F}^{\mathrm{amp}}$, is represented as both

$$\mathcal{F}^{\mathrm{amp}} = (\Box \times \mathcal{A}) \cdot \mathcal{J}_{\mathrm{total}} \tag{9.6a}$$

$$= -\frac{\mathcal{J}_{\mathrm{total}} \cdot \mathcal{V}}{c^2}\mathcal{E} + \left(-\frac{i}{c}\mathcal{J}_{\mathrm{total}} \times \mathcal{B}\right)^{\dagger} \cdot \mathcal{V} + \frac{\mathcal{E} \cdot \mathcal{J}_{\mathrm{total}}}{c^2}\mathcal{V} \tag{9.6b}$$

$$= \left[\rho_{\mathrm{total}}\mathbf{E}^{\mathrm{o}} + \mathbf{J}_{\mathrm{total}} \times \mathbf{B},\ \frac{i}{c}\mathbf{E}^{\mathrm{o}} \cdot \mathbf{J}_{\mathrm{total}}\right] \tag{9.6c}$$

and the four-divergence of the modified energy-momentum tensor, $\mathrm{II}^{\mathrm{em}} \to \mathrm{II}^{\mathrm{amp}}$. The latter is defined by

$$\mathrm{II}^{\mathrm{amp}} = \varepsilon_{\mathrm{o}}\mathcal{E}\,\mathcal{E} + \frac{1}{\mu_{\mathrm{o}}}\mathcal{B}\,\mathcal{B} - \frac{1}{2}\left(\varepsilon_{\mathrm{o}}\mathcal{E} \cdot \mathcal{E} + \frac{1}{\mu_{\mathrm{o}}}\mathcal{B} \cdot \mathcal{B}\right)\left(\mathcal{I} + 2\frac{\mathcal{V}\,\mathcal{V}}{c^2}\right)$$
$$+ \frac{i}{c}\varepsilon_{\mathrm{o}}\left[\mathcal{V}\,(\mathcal{E} \times \mathcal{B})^{\dagger} \cdot \mathcal{V} + (\mathcal{E} \times \mathcal{B})^{\dagger} \cdot \mathcal{V}\,\mathcal{V}\right] \tag{9.7a}$$

(where $\mathcal{I}$ is the identity tensor for four dimensions) or, equivalently, as

$$\mathrm{II}^{\mathrm{amp}} = \begin{bmatrix} \varepsilon_{\mathrm{o}}\mathbf{E}^{\mathrm{o}}\,\mathbf{E}^{\mathrm{o}} + \frac{1}{\mu_{\mathrm{o}}}\mathbf{B}\,\mathbf{B} - \frac{1}{2}(\varepsilon_{\mathrm{o}}\mathbf{E}^{\mathrm{o}} \cdot \mathbf{E}^{\mathrm{o}} + \frac{1}{\mu_{\mathrm{o}}}\mathbf{B} \cdot \mathbf{B})\bar{\mathbf{I}} & \vdots & -\frac{i}{c\mu_{\mathrm{o}}}\mathbf{E}^{\mathrm{o}} \times \mathbf{B} \\ \cdots\cdots\cdots & \vdots & \cdots\cdots \\ -\frac{i}{c\mu_{\mathrm{o}}}\mathbf{E}^{\mathrm{o}} \times \mathbf{B} & \vdots & \frac{1}{2}(\varepsilon_{\mathrm{o}}\mathbf{E}^{\mathrm{o}} \cdot \mathbf{E}^{\mathrm{o}} + \frac{1}{\mu_{\mathrm{o}}}\mathbf{B} \cdot \mathbf{B}) \end{bmatrix} \tag{9.7b}$$

The form of the Maxwell–Poynting energy-momentum tensor, Eq. (8.6), is unchanged except for the replacement of $\mu_0\mathbf{H}$ by $\mathbf{B}$ and $\mathbf{E}$ by $\mathbf{E}^o$; the Alternate tensor, Eq. (8.10), is unchanged except that $\mathcal{J}$ is replaced by $\mathcal{J}_{\text{total}}$. In either representation, the three-space stress-momentum and power-energy theorems are

$$\nabla\cdot\overline{\mathbf{T}} - \frac{\partial\mathbf{G}}{\partial t} = \rho_{\text{total}}\mathbf{E}^o + \mathbf{J}_{\text{total}}\times\mathbf{B} \tag{9.8}$$

and

$$\nabla\cdot\mathbf{S} + \frac{\partial W}{\partial t} = -\mathbf{E}^o\cdot\mathbf{J}_{\text{total}} \tag{9.9}$$

Modified energy-momentum tensors

The material source terms can be combined with $\mathbf{S}^{\text{amp}} = \mathbf{E}^o\times\mathbf{B}/\mu_0$ and W^{amp} to create the modified Poynting theorem

$$\begin{aligned}\nabla\cdot(\mathbf{E}^o\times\mathbf{H}^o) &+ \frac{\partial W^{\text{mink}}}{\partial t}\\ &= -\mathbf{E}^o\cdot\mathbf{J}_u - \frac{1}{2}\left\{\mathbf{E}^o\cdot\frac{\partial\mathbf{P}^o}{\partial t} - \frac{\partial\mathbf{E}^o}{\partial t}\cdot\mathbf{P}^o + \mathbf{B}\cdot\frac{\partial\mathbf{M}^o}{\partial t} - \frac{\partial\mathbf{B}}{\partial t}\cdot\mathbf{M}^o\right\}\end{aligned} \tag{9.10a}$$

where, with the use of Eqs. (9.7b), the Minkowski energy density is defined by

$$W^{\text{mink}} = \frac{1}{2}\mathbf{E}^o\cdot(\varepsilon_0\mathbf{E}^o + \mathbf{P}^o) + \frac{1}{2}\mathbf{B}\cdot\left(\frac{\mathbf{B}}{\mu_0} - \mathbf{M}^o\right) = \frac{1}{2}\mathbf{E}^o\cdot\mathbf{D} + \frac{1}{2}\mathbf{B}\cdot\mathbf{H}^o \tag{9.10b}$$

or they can be combined with $\mathbf{S}^o$ and W^o to create the modified Alternate Poynting theorem

$$\nabla\cdot\mathbf{S}^{\text{ou}} + \frac{\partial W^{\text{ou}}}{\partial t} = -\mathbf{E}^o\cdot\mathbf{J}_u - \frac{1}{2}\left\{\mathbf{E}^o\cdot\frac{\partial\mathbf{P}^o}{\partial t} - \frac{\partial\mathbf{E}^o}{\partial t}\cdot\mathbf{P}^o + \mathbf{B}\cdot\frac{\partial\mathbf{M}^o}{\partial t} - \frac{\partial\mathbf{B}}{\partial t}\cdot\mathbf{M}^o\right\} \tag{9.11a}$$

where

$$\begin{aligned}S_i^{\text{ou}} &= \Phi J_{ui} + \frac{1}{2\mu_0}\left(\mathcal{A}\cdot\frac{\partial^2\mathcal{A}}{\partial x_i\,\partial t} - \frac{\partial\mathcal{A}}{\partial x_i}\cdot\frac{\partial\mathcal{A}}{\partial t}\right)\\ &+ \frac{1}{2\mu_0}\left[\Phi\frac{\partial(\Box\cdot\mathcal{A})}{\partial x_i} - \frac{\partial\Phi}{\partial x_i}(\Box\cdot\mathcal{A})\right]\\ &+ \left[\frac{1}{2}\left(\frac{\partial\mathbf{A}}{\partial t}\times\mathbf{M}^o - \mathbf{A}\times\frac{\partial\mathbf{M}^o}{\partial t}\right) + \frac{1}{2}\left(\Phi\frac{\partial\mathbf{P}^o}{\partial t} - \frac{\partial\Phi}{\partial t}\mathbf{P}^o\right)\right]_i\end{aligned} \tag{9.11b}$$

and

$$\begin{aligned}W^{\text{ou}} &= \frac{1}{2}(\mathbf{A}\cdot\mathbf{J}_u + \rho_u\Phi) - \frac{1}{2}\varepsilon_0\left(\mathcal{A}\cdot\frac{\partial^2\mathcal{A}}{\partial t^2} - \frac{\partial\mathcal{A}}{\partial t}\cdot\frac{\partial\mathcal{A}}{\partial t}\right)\\ &+ \frac{1}{2}\varepsilon_0\left[(\Box\cdot\mathcal{A})\frac{\partial\Phi}{\partial t} - \Phi\frac{\partial(\Box\cdot\mathcal{A})}{\partial t}\right] + \frac{1}{2}\left(\mathbf{A}\cdot\frac{\partial\mathbf{P}^o}{\partial t} - \frac{\partial\mathbf{A}}{\partial t}\cdot\mathbf{P}^o\right)\end{aligned} \tag{9.11c}$$

As required, the right-hand sides of these two Poynting theorems are identical.

For nonaccelerating material governed by linear-reciprocal[2] constitutive laws (with time-independent parameters), the curly bracketed terms on the right-hand sides of Eqs. (9.10a) or (9.11a) vanish. Transfer of power into or out of the electromagnetic system, which includes the electric and magnetic dipoles, is then solely dependent upon $-\mathbf{E}\cdot\mathbf{J}_u$.

In a similar manner, the material source terms can be combined with $\overline{\mathbf{T}}^{amp}$ and $\mathbf{G}^{amp} = \varepsilon_o\mathbf{E}^o \times \mathbf{B}$ to create the modified stress-momentum theorem

$$\nabla\cdot\left[\mathbf{E}^o\,\mathbf{D} + \mathbf{H}^o\,\mathbf{B} - \frac{1}{2}(\mathbf{E}^o\cdot\mathbf{D} + \mathbf{H}^o\cdot\mathbf{B})\overline{\mathbf{I}}\right] - \frac{\partial}{\partial t}(\mathbf{D}\times\mathbf{B})$$
$$= \rho_u\mathbf{E}^o + \mathbf{J}_u\times\mathbf{B} + \frac{1}{2}(P_k^o\nabla E_k^o - E_k^o\nabla P_k^o) + \frac{1}{2}(M_k^o\nabla B_k - B_k\nabla M_k^o) \tag{9.12}$$

or combined with $\overline{\mathbf{T}}^o$ and $\mathbf{G}^o$ to create

$$\nabla\cdot\left(\overline{\mathbf{T}}^o - \left(\frac{\partial\mathbf{A}}{\partial t} + \nabla\Phi\right)\mathbf{P}^o - \mathbf{M}^o\ \nabla\times\mathbf{A} + \frac{1}{2}\left[\left(\frac{\partial\mathbf{A}}{\partial t} + \nabla\Phi\right)\cdot\mathbf{P}^o + \mathbf{M}^o\cdot\nabla\times\mathbf{A}\right]\overline{\mathbf{I}}\right)$$
$$-\frac{\partial}{\partial t}\left[\mathbf{G}^o + \mathbf{P}^o\times(\nabla\times\mathbf{A})\right]$$
$$= \rho_u\mathbf{E}^o + \mathbf{J}_u\times\mathbf{B} + \frac{1}{2}(P_k^o\nabla E_k^o - E_k^o\nabla P_k^o) + \frac{1}{2}(M_k^o\nabla B_k - B_k\nabla M_k^o) \tag{9.13}$$

or, equivalently,

$$\nabla\cdot\overline{\mathbf{T}}^{ou} - \frac{\partial}{\partial t}\mathbf{G}^{ou} = \rho_u\mathbf{E}^o + \mathbf{J}_u\times\mathbf{B} + \frac{1}{2}(P_k^o\nabla E_k^o - E_k^o\nabla P_k^o) + \frac{1}{2}(M_k^o\nabla B_k - B_k\nabla M_k^o) \tag{9.14a}$$

where

$$T_{ij}^{ou} = \begin{bmatrix} -A_iJ_{uj} + \frac{1}{2}(\mathbf{A}\cdot\mathbf{J}_u - \rho_u\Phi)\delta_{ij} + \frac{1}{2\mu_o}\left(\mathcal{A}\cdot\frac{\partial^2\mathcal{A}}{\partial x_i\partial x_j} - \frac{\partial\mathcal{A}}{\partial x_i}\cdot\frac{\partial\mathcal{A}}{\partial x_j}\right) \\ +\frac{1}{2\mu_o}\left[(\Box\cdot\mathcal{A})\frac{\partial A_j}{\partial x_i} - A_j\frac{\partial(\Box\cdot\mathcal{A})}{\partial x_i}\right] - A_i(\nabla\times\mathbf{M}^o)_j - M_i^o(\nabla\times\mathbf{A})_j \\ -\frac{\partial\Phi}{\partial x_i}P_j^o + \frac{1}{2}\left[\nabla\cdot(\Phi\mathbf{P}^o) + \mathbf{A}\cdot(\nabla\times\mathbf{M}^o) + \mathbf{M}^o\cdot(\nabla\times\mathbf{A})\right]\delta_{ij} \end{bmatrix} \tag{9.14b}$$

and

$$G_i^{ou} = \begin{bmatrix} \rho_uA_i + \frac{1}{2}\varepsilon_o\left(\mathcal{A}\cdot\frac{\partial^2\mathcal{A}}{\partial x_i\partial t} - \frac{\partial\mathcal{A}}{\partial x_i}\cdot\frac{\partial\mathcal{A}}{\partial t}\right) \\ +\frac{1}{2}\varepsilon_o\left[\Phi\frac{\partial(\Box\cdot\mathcal{A})}{\partial x_i} - (\Box\cdot\mathcal{A})\frac{\partial\Phi}{\partial x_i}\right] + \frac{1}{2}\left(\mathbf{P}^o\cdot\frac{\partial\mathbf{A}}{\partial x_i} - \frac{\partial\mathbf{P}^o}{\partial x_i}\cdot\mathbf{A}\right) \end{bmatrix} \tag{9.14c}$$

With $\mathbf{P}^o = \mathbf{D} - \varepsilon_o\mathbf{E}^o$ and $\mathbf{M}^o = \mathbf{B}/\mu_o - \mathbf{H}^o$, the modified energy-momentum tensor constructed from Eqs. (9.10a) and (9.12) is known as the Minkowski tensor and will be

[2]If the constitutive laws are linear but nonreciprocal, the bracketed terms do not necessarily vanish. Magnetized ferrite material operating in a linearized region is one such example.

revisited in Chapter 10; its Alternate form is constructed from Eqs. (9.11a) and either (9.13) or (9.14a). These Minkowski representations and their dual forms are summarized in Appendix D, Section D.2.

Linear dielectric and magnetic materials

Isotropic media

Assume the constitutive laws governing the polarization and magnetization are linear and isotropic[3] (in the rest frame), so that

$$\mathcal{D} = \varepsilon' \mathcal{E}$$
$$\mathcal{B} = \mu' \mathcal{H}$$

and also assume they are uniform in space and time ($\Box\varepsilon' = 0$; $\Box\mu' = 0$). Then the prior free-space analysis can be used provided the replacements $\varepsilon_o \to \varepsilon'$ and $\mu_o \to \mu'$ are made in *all* formulas (including γ) that describe either the Maxwell–Poynting or the Alternate representation. The Lorenz-gauge condition for the modified system is then $\nabla \cdot \mathbf{A} + \mu'\varepsilon' \frac{\partial \Phi}{\partial t} = 0$. This is tantamount to declaring that our measured value of the speed of light, c, was in error and we are simply correcting it. (The new value c' might even be greater than c.) If (as expected) the material has boundaries and is surrounded by free-space, then $c' \le c$. It may, however, still be convenient to employ the modified gauge in the material region.

In general, the polarization and magnetization components of the total current density are from Eq. (9.4c),

$$\Box \cdot \left[\mathcal{P} \times \mathcal{V} + \frac{-i}{c} (\mathcal{M} \times \mathcal{V})^\dagger \right]$$

When subtracted from $\Box \cdot (\Box \times \mathcal{A}) = \mu_o \mathcal{J}_{\text{total}}$, the result is

$$\Box \cdot \left[\Box \times \mathcal{A} - \mu_o \mathcal{P} \times \mathcal{V} - \frac{-i}{c} (\mu_o \mathcal{M} \times \mathcal{V})^\dagger \right] = \mu_o \mathcal{J}_u \tag{9.15}$$

For linear isotropic media,

$$\mathcal{P} = (\varepsilon' - \varepsilon_o)\mathcal{E} = (\varepsilon' - \varepsilon_o)\,(\Box \times \mathcal{A}) \cdot \mathcal{V} \tag{9.16a}$$

$$\mathcal{M} = \left(\frac{1}{\mu_o} - \frac{1}{\mu'} \right) \mathcal{B} = \left(\frac{1}{\mu_o} - \frac{1}{\mu'} \right) \frac{-i}{c} (\Box \times \mathcal{A})^\dagger \cdot \mathcal{V} \tag{9.16b}$$

$$\mathcal{J}_u = \sigma' \mathcal{E} = \sigma' \,(\Box \times \mathcal{A}) \cdot \mathcal{V} \tag{9.16c}$$

[3] When the rest-frame constitutive laws are linear but anisotropic, one or more of the permittivity, permeability, and conductivity is a tensor quantity. In such cases, the coupled pair of four-vectors are not, in general, parallel. Although derivations that generalize the results of this section are straightforward, they are somewhat tedious and are therefore omitted.

Therefore

$$\Box\cdot\left[\Box\times\mathcal{A}+\left(1-\frac{\varepsilon'}{\varepsilon_0}\right)\frac{[(\Box\times\mathcal{A})\cdot\mathcal{V}]\times\mathcal{V}}{c^2}+\left(1-\frac{\mu_0}{\mu'}\right)\frac{\left(\left[(\Box\times\mathcal{A})^{\dagger}\cdot\mathcal{V}\right]\times\mathcal{V}\right)^{\dagger}}{c^2}\right]$$

$$=\mu_0\mathcal{J}_u \tag{9.17}$$

which may be simplified through use of the Appendix B identity, Eq. (B.18a) (with $\mathcal{Q}=\Box\times\mathcal{A}$ and $\mathcal{C}=\mathcal{V}$). The final result is

$$\Box\cdot\frac{\mu_0}{\mu'}\left[\Box\times\mathcal{A}+\left(1-\frac{\mu'\varepsilon'}{\mu_0\varepsilon_0}\right)\frac{[(\Box\times\mathcal{A})\cdot\mathcal{V}]\times\mathcal{V}}{c^2}\right]=\mu_0\sigma'(\Box\times\mathcal{A})\cdot\mathcal{V} \tag{9.18}$$

which applies to accelerating and deforming materials.

Uniform nondeforming and nonaccelerating materials

If $\Box\sigma'=\Box\mu'=\Box\varepsilon'=0$ and $\Box\mathcal{V}=0$, Eq. (9.18) becomes

$$\Box\left[(\Box\cdot\mathcal{A})+(\mu_0\varepsilon_0-\mu'\varepsilon')\frac{\partial(\mathcal{A}\cdot\mathcal{V})}{\partial\tau}-\mu'\sigma'\mathcal{A}\cdot\mathcal{V}\right]-\Box^2\mathcal{A}$$

$$+(\mu_0\varepsilon_0-\mu'\varepsilon')\left[-\frac{\partial^2\mathcal{A}}{\partial\tau^2}+\left[\frac{\partial(\Box\cdot\mathcal{A})}{\partial\tau}-\Box^2(\mathcal{A}\cdot\mathcal{V})\right]\mathcal{V}\right]+\mu'\sigma'\frac{\partial\mathcal{A}}{\partial\tau}=0 \tag{9.19}$$

where we recall that $\dfrac{\partial}{\partial\tau}=\mathcal{V}\cdot\Box$.

The gauge that simplifies Eq. (9.19) is obviously

$$\Box\cdot\mathcal{A}=\left(\mu'\varepsilon'-\mu_0\varepsilon_0\right)\frac{\partial}{\partial\tau}(\mathcal{A}\cdot\mathcal{V})+\mu'\sigma'\mathcal{A}\cdot\mathcal{V} \tag{9.20}$$

and with that choice, Eq. (9.19) reduces to

$$\Box^2\mathcal{A}-\left(\mu'\varepsilon'-\mu_0\varepsilon_0\right)\frac{\partial^2\mathcal{A}}{\partial\tau^2}-\mu'\sigma'\frac{\partial\mathcal{A}}{\partial\tau}=0 \tag{9.21}$$

These equations are valid when the material is in uniform motion at any velocity **v**. For the material at rest, we have

$$\nabla\cdot\mathbf{A}+\mu'\varepsilon'\frac{\partial\Phi}{\partial t}+\mu'\sigma'\Phi=0 \tag{9.22a}$$

$$\Box\cdot\mathcal{A}=-\left(\mu'\varepsilon'-\mu_0\varepsilon_0\right)\frac{\partial\Phi}{\partial t}-\mu'\sigma'\Phi \tag{9.22b}$$

and

$$\left(\nabla^2-\mu'\varepsilon'\frac{\partial^2}{\partial t^2}-\mu'\sigma'\frac{\partial}{\partial t}\right)\mathcal{A}=0 \tag{9.23}$$

Both **A** and Φ satisfy the same wave-diffusion equation. Notice that the constant controlling the diffusion, $\mu'\sigma'$, is proportional to the magnetic permeability but is unaffected by the permittivity. When $\mu'\varepsilon'=\mu_0\varepsilon_0$, Eq. (9.22b) reduces to the gauge derived earlier for a conductor without polarization or magnetization. When there is no dissipation, the diffusion term vanishes and Eq. (9.23) becomes the wave equation, but with the slower propagation velocity $1/\sqrt{\mu'\varepsilon'}$ replacing c. When conduction dominates the Maxwell-Displacement current ($\sigma' \gg \omega\varepsilon'$), the equation reduces to the ordinary diffusion equation.

With $g' = \nabla \cdot \mathbf{A} + \mu'\varepsilon'\frac{\partial\Phi}{\partial t}$, the modified Alternate power flux and Alternate-energy density simplify to

$$S_i^{\mathrm{ou}} = \Phi\sigma' E_i + \frac{1}{2\mu'}\left(\mathbf{A} \cdot \frac{\partial^2 \mathbf{A}}{\partial x_i \partial t} - \frac{\partial \mathbf{A}}{\partial x_i} \cdot \frac{\partial \mathbf{A}}{\partial t}\right)$$
$$- \frac{1}{2}\varepsilon'\left(\Phi\frac{\partial^2\Phi}{\partial x_i \partial t} - \frac{\partial\Phi}{\partial x_i}\frac{\partial\Phi}{\partial t}\right) + \frac{1}{2\mu'}\left(\frac{\partial A_i}{\partial t}g' - A_i\frac{\partial g'}{\partial t}\right) + \Delta S_i \tag{9.24a}$$

and

$$W^{\mathrm{ou}} = \frac{1}{2}(\mathbf{A} \cdot \sigma' \mathbf{E} + \rho_{\mathrm{u}}\Phi) - \frac{1}{2}\varepsilon'\left(\mathbf{A} \cdot \frac{\partial^2 \mathbf{A}}{\partial t^2} - \frac{\partial \mathbf{A}}{\partial t} \cdot \frac{\partial \mathbf{A}}{\partial t}\right)$$
$$+ \frac{1}{2}\mu'\varepsilon'^2\left[\Phi\frac{\partial^2\Phi}{\partial t^2} - \left(\frac{\partial\Phi}{\partial t}\right)^2\right] + \frac{1}{2}\varepsilon'\left(\frac{\partial\Phi}{\partial t}g' - \Phi\frac{\partial g'}{\partial t}\right) + \Delta W \tag{9.24b}$$

where

$$\Delta \mathbf{S} = -\frac{1}{2}(\varepsilon' - \varepsilon_{\mathrm{o}})\left(\Phi\frac{\partial^2 \mathbf{A}}{\partial t^2} - \frac{\partial^2\Phi}{\partial t^2}\mathbf{A}\right) - \frac{1}{2}\left(\frac{1}{\mu_{\mathrm{o}}} - \frac{1}{\mu'}\right)\nabla \times \left(\mathbf{A} \times \frac{\partial \mathbf{A}}{\partial t}\right)$$

$$\Delta W = \frac{1}{2}(\varepsilon' - \varepsilon_{\mathrm{o}})\nabla \cdot \left(\Phi\frac{\partial \mathbf{A}}{\partial t} - \frac{\partial\Phi}{\partial t}\mathbf{A}\right)$$

are components which vanish when $\mu' = \mu_{\mathrm{o}}$ and $\varepsilon' = \varepsilon_{\mathrm{o}}$. However, since $\nabla \cdot \Delta\mathbf{S} + \partial\Delta W/\partial t = 0$, both excess terms can simply be ignored.

A similar approach can be taken in order to evaluate the modified Alternate-stress tensor and Alternate-momentum density; the result (after neglecting the excess terms) is

$$T_{ij}^{\mathrm{ou}} = -A_i J_{\mathrm{u}j} + \frac{1}{2}(\mathbf{A} \cdot \mathbf{J}_{\mathrm{u}} - \rho_{\mathrm{u}}\Phi)\delta_{ij} + \frac{1}{2\mu'}\left(\mathbf{A} \cdot \frac{\partial^2 \mathbf{A}}{\partial x_i \partial x_j} - \frac{\partial \mathbf{A}}{\partial x_i} \cdot \frac{\partial \mathbf{A}}{\partial x_j}\right)$$
$$- \frac{1}{2}\varepsilon'\left(\Phi\frac{\partial^2\Phi}{\partial x_i \partial x_j} - \frac{\partial\Phi}{\partial x_i}\frac{\partial\Phi}{\partial x_j}\right) + \frac{1}{2\mu'}\left(g'\frac{\partial A_j}{\partial x_i} - A_j\frac{\partial g'}{\partial x_i}\right) \tag{9.25a}$$

and

$$G_i^{\mathrm{ou}} = \rho_{\mathrm{u}}A_i + \frac{1}{2}\varepsilon'\left(\mathbf{A} \cdot \frac{\partial^2 \mathbf{A}}{\partial x_i \partial t} - \frac{\partial \mathbf{A}}{\partial x_i} \cdot \frac{\partial \mathbf{A}}{\partial t}\right) - \frac{1}{2}\mu'\varepsilon'^2\left(\Phi\frac{\partial^2\Phi}{\partial x_i \partial t} - \frac{\partial\Phi}{\partial x_i}\frac{\partial\Phi}{\partial t}\right)$$
$$+ \frac{1}{2}\varepsilon'\left(\Phi\frac{\partial g'}{\partial x_i} - g'\frac{\partial\Phi}{\partial x_i}\right) \tag{9.25b}$$

For the gauge of Eq. (9.22b), $g' = -\mu'\sigma'\Phi$, and the redefined-gauge terms vanish from W^{ou} and $\mathbf{G}^{\mathrm{ou}}$ (but not from $\mathbf{S}^{\mathrm{ou}}$ and T_{ij}^{ou}).

When the velocity, **v**, is nonzero, Eq. (9.21) becomes

$$\left\{\nabla^2 + \gamma^2 \left[\left(\nabla_{\shortparallel} + \frac{\mathbf{v}}{c^2}\frac{\partial}{\partial t}\right) \cdot \left(\nabla_{\shortparallel} + \frac{\mathbf{v}}{c^2}\frac{\partial}{\partial t}\right) - \mu'\varepsilon' \left(\frac{\partial}{\partial t} + \mathbf{v}\cdot\nabla\right)^2 \right] \right.$$
$$\left. + \gamma\mu'\sigma' \left(\frac{\partial}{\partial t} + \mathbf{v}\cdot\nabla\right)\right\} \underline{\mathcal{A}} = 0 \tag{9.26}$$

where $\nabla_{\shortparallel}$ and $\nabla_{\perp}$ are, respectively, the components of ∇ that are parallel and perpendicular to the velocity.

One additional observation is in order. For lossless harmonic plane waves, we have $\underline{\mathcal{A}} \sim \exp\left[j(\omega t - \mathbf{k}\cdot\mathbf{r})\right]$; therefore the dispersion relation that follows from Eq. (9.26) is (in terms of $\mathbf{k}_{\shortparallel}$ and $\mathbf{k}_{\perp}$)

$$\mathbf{k}_{\perp}\cdot\mathbf{k}_{\perp} + \gamma^2 \left(\mathbf{k}_{\shortparallel} - \frac{\omega}{c^2}\mathbf{v}\right)\cdot\left(\mathbf{k}_{\shortparallel} - \frac{\omega}{c^2}\mathbf{v}\right) - \mu'\varepsilon'(\omega - \mathbf{k}\cdot\mathbf{v})^2 = 0$$

In rest-frame coordinates, $\underline{\mathcal{A}} \sim \exp\left[j(\omega' t' - \mathbf{k}'\cdot\mathbf{r}')\right]$; therefore

$$\mathbf{k}'\cdot\mathbf{k}' - \mu'\varepsilon'\omega'^2 = 0$$

These results are consistent with the Lorentz transformation derived in Appendix A, Section A.1.

Complex Alternate-power theorems

Assuming that an $\exp(j\omega t)$ time-dependence permits valid solutions $\underline{\mathcal{A}} = [\underline{\mathbf{A}}, \frac{i}{c}\underline{\Phi}]$, the complex Alternate power flux presented in Chapter 6, Section 6.2, Eq. (6.10) generalizes to

$$\underline{\mathbf{S}}^{\text{o}} = \frac{1}{2}\,\underline{\Phi}\,\underline{\mathbf{J}}^*_{\text{total}} - j\omega\frac{1}{4\mu_{\text{o}}}(\underline{\mathcal{A}}_k\,\nabla\underline{\mathcal{A}}^*_k - \underline{\mathcal{A}}^*_k\,\nabla\underline{\mathcal{A}}_k + \underline{g}\,\underline{\mathbf{A}}^* - \underline{g}^*\underline{\mathbf{A}})$$
$$\text{(summation over } k = 1,2,3,4) \tag{9.27}$$

where $\underline{g} = \square\cdot\underline{\mathcal{A}}$ and the complex form of Eq. (9.2) applies. With $\square = [\nabla, \frac{-i}{c}j\omega]$, the complex Alternate power theorem is obtained by taking the divergence of $\underline{\mathbf{S}}^{\text{o}}$ and then substituting $\underline{\mathcal{A}}_k\,\square\cdot\square\underline{\mathcal{A}}^*_k - \underline{\mathcal{A}}^*_k\,\square\cdot\square\underline{\mathcal{A}}_k$ for $\underline{\mathcal{A}}_k\,\nabla^2\underline{\mathcal{A}}^*_k - \underline{\mathcal{A}}^*_k\,\nabla^2\underline{\mathcal{A}}_k$.

Because $\square\cdot\square\,\underline{\mathcal{A}} = \square(\square\cdot\underline{\mathcal{A}}) - \mu_{\text{o}}\underline{\mathcal{J}}_{\text{total}}$, it follows that

$$\nabla\cdot\underline{\mathbf{S}}^{\text{o}} + j\omega\frac{1}{4}(\underline{\mathcal{A}}\cdot\underline{\mathcal{J}}^*_{\text{total}} + \underline{\mathcal{A}}^*\cdot\underline{\mathcal{J}}_{\text{total}}) = -\frac{1}{2}\underline{\mathbf{E}}^{\text{o}}\cdot\underline{\mathbf{J}}^*_{\text{total}} \tag{9.28a}$$

Equation (9.28a) is equivalent to

$$\nabla\cdot\underline{\mathbf{S}}^{\text{o}} + j2\omega\left[\frac{1}{8}(\underline{\mathbf{A}}\cdot\underline{\mathbf{J}}^*_{\text{total}} + \underline{\mathbf{A}}^*\cdot\underline{\mathbf{J}}_{\text{total}}) - \frac{1}{8}(\underline{\Phi}\,\underline{\rho}^*_{\text{total}} + \underline{\Phi}^*\,\underline{\rho}_{\text{total}})\right]$$
$$= -\frac{1}{2}\underline{\mathbf{E}}^{\text{o}}\cdot\underline{\mathbf{J}}^*_{\text{total}} \tag{9.28b}$$

This theorem applies to nondeforming nonaccelerating uniform linear materials. However, for lossless isotropic materials at rest where the constitutive constants, μ' and ε', are spatially uniform, one can make the replacements: $\underline{\mathbf{E}}^{\text{o}} \to \underline{\mathbf{E}}$, $\underline{\mathbf{P}} \to 0$, $\underline{\mathbf{M}} \to 0$, $\underline{\mathbf{J}}_{\text{total}} \to \underline{\mathbf{J}}_{\text{u}}$,

$\underline{\rho}_{\text{total}} = \underline{\rho}_{\text{u}}$, $\mu_{\text{o}} \to \mu'$, $\varepsilon_{\text{o}} \to \varepsilon'$, and $\underline{g} \to \underline{g}' = \nabla \cdot \underline{\mathbf{A}} + j\omega\mu'\varepsilon' \,\underline{\Phi}$. Then, the Alternate reactive power flux and reactive energy density vanish except where $\underline{\mathbf{J}}_{\text{u}}$ and $\underline{\rho}_{\text{u}}$ are nonzero. Notice that the modified Eq. (9.28b) validates Eq. (6.9). When the dielectric is conductive, $\varepsilon_{\text{o}} \to \varepsilon' - j\sigma/\omega$, but, more generally, dissipation in the material requires both μ' and ε' to be frequency-dependent complex constants.

Quasistatic approximations

Each order of the quasistatic field expansion can be expressed as a function of position multiplied by a time-dependent coefficient. In addition, the expansion is usually truncated after the first order. Accordingly, with

$$\mathbf{J}(\mathbf{r}, t) = C_{a0}(t)\,\mathbf{j}_0(\mathbf{r}) + C_{a1}(t)\,\mathbf{j}_1(\mathbf{r})$$

$$\rho(\mathbf{r}, t) = C_{\phi 0}(t)\rho_0(\mathbf{r}) + C_{\phi 1}(t)\rho_1(\mathbf{r})$$

and

$$\nabla^2 \mathbf{a}_k = -\mu\,\mathbf{j}_k$$

$$\nabla^2 \phi_k = -\rho_k/\varepsilon$$

the quasistatic fields are

$$\mathbf{A}(\mathbf{r}, t) = C_{a0}\mathbf{a}_0(\mathbf{r}) + C_{a1}\mathbf{a}_1(\mathbf{r}) \tag{9.29a}$$

$$\Phi(\mathbf{r}, t) = C_{\phi 0}\phi_0(\mathbf{r}) + C_{\phi 1}\phi_1(\mathbf{r}) \tag{9.29b}$$

$$\mathbf{B}(\mathbf{r}, t) = C_{a0}\nabla \times \mathbf{a}_0 + C_{a1}\nabla \times \mathbf{a}_1 \tag{9.29c}$$

$$\mathbf{E}^{\text{o}}(\mathbf{r}, t) = -\left[\frac{dC_{a0}}{dt}\mathbf{a}_0 + C_{\phi 0}\nabla\phi_0 + \frac{dC_{a1}}{dt}\mathbf{a}_1 + C_{\phi 1}\nabla\phi_1\right] \tag{9.29d}$$

The Lorenz-gauge conditions are, for zero and first order,

$$\nabla \cdot \mathbf{a}_0 = 0$$

$$C_{a1}\nabla \cdot \mathbf{a}_1 + \frac{dC_{\phi 0}}{dt}\mu\varepsilon\,\phi_0 = 0$$

For nondissipative *EQS* fields, $C_{a0}(t) = C_{\phi 1}(t) = 0$. For nondissipative *MQS* fields, $C_{\phi 0}(t) = C_{a1}(t) = 0$.

In conduction problems, all four coefficients are nonzero, but in high- and low-resistance limits, the fields are approximately *EQS* or *MQS*. In such cases, a single term suffices, and the spatial distribution of each field is static-like with an amplitude that varies with time.

For stationary materials that are linear and reciprocal (but not necessarily isotropic), the superscript o can be dropped and

$$\mathbf{P} = (\overline{\varepsilon} - \varepsilon_{\text{o}}\overline{\mathbf{I}}) \cdot \mathbf{E}, \qquad P_{ij} = (\varepsilon_{ij} - \varepsilon_{\text{o}}\,\delta_{ij})E_j \tag{9.30a}$$

$$\mathbf{M} = (\mu_{\text{o}}^{-1}\overline{\mathbf{I}} - \overline{\mu}^{\,-1}) \cdot \mathbf{B}, \qquad M_{ij} = (\mu_{\text{o}}^{-1}\delta_{ij} - \mu_{ij}^{-1})B_j \tag{9.30b}$$

where $\overline{\varepsilon}$ ($\varepsilon_{ij} = \varepsilon_{ji}$) is the permittivity tensor, $\overline{\mu}^{\,-1}$ ($\mu_{ij}^{-1} = \mu_{ji}^{-1}$) is the inverse permeability tensor, and $\overline{\mathbf{I}}$ (δ_{ij}) is the identity tensor.

It follows that (suppressing the order subscripts) the modified Alternate power flux, Eq. (9.11b), and the energy density, Eq. (9.11c), reduce to

$$\mathbf{S}^{\text{ou}} = \Phi \mathbf{J}_{\text{u}} + \frac{1}{2}\left(\frac{dC_a}{dt}\frac{dC_\phi}{dt} - \frac{d^2C_a}{dt^2}C_\phi\right)\phi(\overline{\varepsilon} - \varepsilon_{\text{o}}\overline{\mathbf{I}}) \cdot \mathbf{a} \tag{9.31a}$$

and

$$\begin{aligned} W^{ou} &= \frac{1}{2}(\mathbf{A} \cdot \mathbf{J}_{\text{u}} + \rho_{\text{u}}\Phi) + \frac{1}{2}\left[\left(\frac{dC_a}{dt}\right)^2 - \frac{d^2C_a}{dt^2}C_a\right]\mathbf{a} \cdot \overline{\varepsilon} \cdot \mathbf{a} \\ &\quad - \frac{1}{2c^2}\varepsilon_{\text{o}}\left[\left(\frac{dC_\phi}{dt}\right)^2 - \frac{d^2C_\phi}{dt^2}C_\phi\right]\phi^2 \\ &\quad + \frac{1}{2}\left(\frac{dC_a}{dt}C_\phi - C_a\frac{dC_\phi}{dt}\right)\mathbf{a} \cdot (\overline{\varepsilon} - \varepsilon_{\text{o}}\overline{\mathbf{I}}) \cdot \nabla\phi \end{aligned} \tag{9.31b}$$

Conservation of charge and the Lorenz gauge both require that $C_a \sim \frac{dC_\phi}{dt}$. Consequently, all of the nonlocalized terms in Eqs. (9.31a) and (9.31b) have time dependences that are second order or higher. They are negligible within the quasistatic approximation, hence localization of the Alternate power and energy with respect to the *unpaired* currents and charges is complete. It is this fact which permits valid circuit representations to be made from calculations of the potentials, currents, and charges in the localized regions. A description of the terminal characteristics of such circuits requires calculation of the *total* electric and magnetic energies; the latter can be evaluated from either the Alternate or Poynting representation, whichever is more convenient. Because the scalar and vector potentials are continuous functions at material boundaries, there may be "fringing" components—even outside perfectly conducting enclosures. It is therefore often more difficult to find *complete* solutions for $\mathbf{A}$ and Φ than it is for $\mathbf{E}$ and $\mathbf{B}$. *Within quasistatics*, the evaluation of $\int \frac{1}{2}\Phi\rho_{\text{u}}\, dV$ and/or $\int \frac{1}{2}\mathbf{A} \cdot \mathbf{J}_{\text{u}}\, dV$ can be accomplished *without* including the correct homogeneous solutions. This is because $\nabla\Phi_{\text{homogeneous}}$, $\nabla \times \mathbf{A}_{\text{homogenous}}$, and $\nabla \cdot \mathbf{J}_{\text{u}}$ are all zero and the homogeneous components do not contribute to the integrals.[4] These matters are clarified in the quasistatic examples analyzed in Part III, Chapter 11.

[4]The homogeneous potentials may have nonzero gradients or curls on boundaries where surface charge and/or current is present. However, these may be ignored, provided that limiting values of the potentials can be evaluated correctly on such boundaries.

CHAPTER 10

AMPERIAN, MINKOWSKI, AND CHU FORMULATIONS

10.1 INTRODUCTION

In this chapter we compare the Amperian, the Minkowski, and the Chu formulations of four-dimensional electrodynamics—all used to model materials that include polarization and/or magnetization vectors. Such materials may be deforming and accelerating and have linear or nonlinear constitutive laws (that connect these vectors to the electric and magnetic fields).

10.2 MAXWELL'S EQUATIONS IN THE AMPERIAN FORMULATION

The Amperian formulation is repeated here to facilitate comparison with the Minkowski and Chu formulations. The three-space equations (Chapter 1, Section 1.7),

$$\nabla \times \frac{\mathbf{B}}{\mu_o} - \varepsilon_o \frac{\partial \mathbf{E}^o}{\partial t} = \mathbf{J}_{\text{total}} = \mathbf{J}_u + \frac{\partial \mathbf{P}^o}{\partial t} + \nabla \times \mathbf{M}^o$$

$$\nabla \cdot \varepsilon_o \mathbf{E}^o = \rho_{\text{total}} = \rho_u - \nabla \cdot \mathbf{P}^o$$

$$\nabla \times \mathbf{E}^o + \frac{\partial \mathbf{B}}{\partial t} = 0$$

$$\nabla \cdot \mathbf{B} = 0$$

The Power and Beauty of Electromagnetic Fields, First Edition. F. R. Morgenthaler.

have already been cast in four-dimensional form (Chapter 9, Section 9.2) as

$$\Box \cdot (\Box \times \mathcal{A}) = \Box \cdot \left[-\frac{i}{c} (\mathcal{B} \times \mathcal{V})^\dagger - \frac{\mathcal{E} \times \mathcal{V}}{c^2} \right] = \mu_0 \mathcal{J}_{\text{total}} \tag{10.1a}$$

$$\Box \cdot [ic(\Box \times \mathcal{A})^\dagger] = \Box \cdot \left[-\frac{i}{c} (\mathcal{E} \times \mathcal{V})^\dagger + \mathcal{B} \times \mathcal{V} \right] = 0 \tag{10.1b}$$

$$\mathcal{J}_{\text{total}} = \mathcal{J}_{\text{u}} + \Box \cdot \left[\mathcal{P} \times \mathcal{V} + \frac{-i}{c} (\mathcal{M} \times \mathcal{V})^\dagger \right] = [\mathbf{J}_{\text{total}},\ ic\rho_{\text{total}}] \tag{10.1c}$$

where

$$\mathcal{B} = -\frac{i}{c} (\Box \times \mathcal{A})^\dagger \cdot \mathcal{V} \tag{10.2a}$$

$$\mathcal{E} = (\Box \times \mathcal{A}) \cdot \mathcal{V} = -\frac{\partial \mathcal{A}}{\partial \tau} - \Box \Phi' - \mathcal{A} \cdot \Box \mathcal{V} \tag{10.2b}$$

and $\Phi' = -\mathcal{A} \cdot \mathcal{V}$ is the potential evaluated in the rest frame. As expected, $\mathcal{B}$ is generated by the curl and $\mathcal{E}$ by the combined time-derivative and gradient operations. The polarization and magnetization four-vectors are

$$\mathcal{P} = \left[\gamma \left(\mathbf{P}^{\text{o}} - \frac{\mathbf{v} \times \mathbf{M}^{\text{o}}}{c^2} \right),\ i\gamma \frac{\mathbf{P}^{\text{o}} \cdot \mathbf{v}}{c} \right] \tag{10.3a}$$

$$\mathcal{M} = \left[\gamma (\mathbf{M}^{\text{o}} + \mathbf{v} \times \mathbf{P}^{\text{o}}),\ i\gamma \frac{\mathbf{M}^{\text{o}} \cdot \mathbf{v}}{c} \right] \tag{10.3b}$$

which reveals that (in this representation) $\mathbf{v} \times \mathbf{M}^{\text{o}}$ creates polarization and $\mathbf{v} \times \mathbf{P}^{\text{o}}$ creates magnetization.

10.3 MAXWELL'S EQUATIONS IN THE MINKOWSKI FORMULATION

Equation (10.1a) can be rearranged as

$$\Box \cdot \left[-\frac{i}{c} \left[\left(\frac{\mathcal{B}}{\mu_0} - \mathcal{M} \right) \times \mathcal{V} \right]^\dagger - (\varepsilon_0 \mathcal{E} + \mathcal{P}) \times \mathcal{V} \right] = \mathcal{J}_{\text{u}} \tag{10.4}$$

Two additional four-vectors now in evidence are

$$\mathcal{H} = \frac{\mathcal{B}}{\mu_0} - \mathcal{M} = \left[\gamma (\mathbf{H}^{\text{o}} + \mathbf{v} \times \mathbf{D}),\ i\gamma \frac{\mathbf{H}^{\text{o}} \cdot \mathbf{v}}{c} \right] \tag{10.5a}$$

$$\mathcal{D} = \varepsilon_0 \mathcal{E} + \mathcal{P} = \left[\gamma \left(\mathbf{D} - \frac{\mathbf{v} \times \mathbf{H}^{\text{o}}}{c^2} \right),\ i\gamma \frac{\mathbf{D} \cdot \mathbf{v}}{c} \right] \tag{10.5b}$$

With these, Eqs. (10.4) and (10.1b) can be rewritten in Minkowski form as

$$\Box \cdot \left[-\frac{i}{c} (\mathcal{H} \times \mathcal{V})^\dagger - \mathcal{D} \times \mathcal{V} \right] = \mathcal{J}_{\text{u}} \tag{10.6a}$$

$$\Box \cdot \left[-\frac{i}{c} (\mathcal{E} \times \mathcal{V})^\dagger + \mathcal{B} \times \mathcal{V} \right] = 0 \tag{10.6b}$$

or, using vector identities found in Appendix B, Section B.3, as

$$-\frac{i}{c}\left(\Box \times \mathcal{H}\right)^{\dagger} \cdot \mathcal{V} - (\mathcal{V} \cdot \Box)\,\mathcal{D} + (\Box \cdot \mathcal{D})\,\mathcal{V} - \mathcal{J}_{\mathrm{u}}$$

$$= \frac{i}{c}\mathcal{H} \cdot (\Box \times \mathcal{V})^{\dagger} + \mathcal{D}(\Box \cdot \mathcal{V}) - (\mathcal{D} \cdot \Box)\mathcal{V}$$

$$-\frac{i}{c}\left(\Box \times \mathcal{E}\right)^{\dagger} \cdot \mathcal{V} + (\mathcal{V} \cdot \Box)\,\mathcal{B} - (\Box \cdot \mathcal{B})\,\mathcal{V}$$

$$= \frac{i}{c}\mathcal{E} \cdot (\Box \times \mathcal{V})^{\dagger} - \mathcal{B}(\Box \cdot \mathcal{V}) + (\mathcal{B} \cdot \Box)\mathcal{V}$$

When $\Box\mathcal{V} = 0$, the right-hand sides of both these latter equations vanish; then

$$-\frac{i}{c}\left(\Box \times \mathcal{H}\right)^{\dagger} \cdot \mathcal{V} - \frac{\partial \mathcal{D}}{\partial \tau} = \mathcal{J}_{\mathrm{u}} - \rho'_{\mathrm{u}}\,\mathcal{V} \tag{10.7a}$$

$$-\frac{i}{c}\left(\Box \times \mathcal{E}\right)^{\dagger} \cdot \mathcal{V} + \frac{\partial \mathcal{B}}{\partial \tau} = 0 \tag{10.7b}$$

where $\frac{\partial}{\partial \tau} = \mathcal{V} \cdot \Box = \gamma\left(\frac{\partial}{\partial t} + \mathbf{v} \cdot \nabla\right)$ is the derivative with respect to proper time (that in the rest frame) and ρ'_{u} is the rest frame value of the unpaired electric-charge density. It also follows that

$$\Box \cdot \mathcal{D} = \rho'_{\mathrm{u}} = \frac{-\mathcal{J}_{\mathrm{u}} \cdot \mathcal{V}}{c^2} \tag{10.7c}$$

$$\Box \cdot \mathcal{B} = 0 \tag{10.7d}$$

It is reassuring to see that the familiar form of Maxwell's Equations is not obscured by the four-dimensional formulation.

The three-space Maxwell's Equations in the Minkowski formulation (without restriction on $\Box\mathcal{V}$) are

$$\nabla \times \mathbf{H}^{\mathrm{o}} - \frac{\partial \mathbf{D}}{\partial t} = \mathbf{J}_{\mathrm{u}} \tag{10.8a}$$

$$\nabla \cdot \mathbf{D} = \rho_{\mathrm{u}} \tag{10.8b}$$

$$\nabla \times \mathbf{E}^{\mathrm{o}} + \frac{\partial \mathbf{B}}{\partial t} = 0 \tag{10.8c}$$

$$\nabla \cdot \mathbf{B} = 0 \tag{10.8d}$$

Only the unpaired charge and current densities serve as sources for the four vector fields.

10.4 MAXWELL'S EQUATIONS IN THE CHU FORMULATION

All of these equations are completely consistent with the formulation developed by Lan Jen Chu [5, Appendix One, p. 453] that can be expressed in terms of the *same* four-vectors, but different three-vectors. The relationships between the Amperian ($\mathbf{E}^o$, $\mathbf{B}$, $\mathbf{P}^o$,$\mathbf{M}^o$), Minkowski ($\mathbf{E}^o$, $\mathbf{B}$, $\mathbf{D}$, $\mathbf{H}^o$), and Chu ($\mathbf{E}$, $\mathbf{H}$, $\mathbf{P}$, $\mathbf{M}$) fields are given by

$$\mathbf{E}^o = \mathbf{E} - \mu_0 \mathbf{v} \times \mathbf{M} \tag{10.9a}$$

$$\mathbf{H}^o = \mathbf{H} + \mathbf{v} \times \mathbf{P} \tag{10.9b}$$

$$\mathbf{P}^o = \mathbf{P} + \frac{\mathbf{v} \times \mathbf{M}}{c^2} \tag{10.9c}$$

$$\mathbf{M}^o = \mathbf{M} - \mathbf{v} \times \mathbf{P} \tag{10.9d}$$

$$\mathbf{D} = \varepsilon_0 \mathbf{E}^o + \mathbf{P}^o = \varepsilon_o \mathbf{E} + \mathbf{P} \tag{10.9e}$$

$$\mathbf{B} = \mu_0(\mathbf{H}^o + \mathbf{M}^o) = \mu_o(\mathbf{H} + \mathbf{M}) \tag{10.9f}$$

In terms of the Chu fields, the four-vectors,

$$\mathcal{E} = \left[\gamma(\mathbf{E} + \mathbf{v} \times \mu_0 \mathbf{H}),\ i\gamma\frac{\mathbf{E}\cdot\mathbf{v}}{c}\right] \tag{10.10a}$$

$$\mathcal{H} = \left[\gamma(\mathbf{H} - \mathbf{v} \times \varepsilon_0 \mathbf{E}),\ i\gamma\frac{\mathbf{H}\cdot\mathbf{v}}{c}\right] \tag{10.10b}$$

$$\mathcal{P} = \left[\frac{\mathbf{P}}{\gamma} + \gamma\frac{(\mathbf{P}\cdot\mathbf{v})\mathbf{v}}{c^2},\ i\gamma\frac{\mathbf{P}\cdot\mathbf{v}}{c}\right] \tag{10.10c}$$

$$\mathcal{M} = \left[\frac{\mathbf{M}}{\gamma} + \gamma\frac{(\mathbf{M}\cdot\mathbf{v})\mathbf{v}}{c^2},\ i\gamma\frac{\mathbf{M}\cdot\mathbf{v}}{c}\right] \tag{10.10d}$$

are unchanged, although it is of particular interest to observe that the polarization, $\mathbf{P}$, contributes only to $\mathcal{P}$ while the magnetization, $\mathbf{M}$, contributes only to $\mathcal{M}$. These results follow from modeling materials as free-space plus dipoles created from electric and magnetic charge. In the laboratory frame, $\mathcal{P} = n'q'\mathcal{L}$ with $n' = n/\gamma$, $q' = q$, and $\mathcal{L} = [\mathbf{L} + \gamma^2\frac{\mathbf{L}\cdot\mathbf{v}}{c^2}\mathbf{v}, i\gamma^2\frac{\mathbf{L}\cdot\mathbf{v}}{c}]$, the four-vector dipole length derived in Appendix A, Eq. (A.20). The three-vector polarization density is defined as $\mathbf{P} = nq\mathbf{L}$. Because the magnetic dipoles are modeled in a similar manner, $\mu_0\mathcal{M}$ is the exact dual of $\mathcal{P}$. In terms of these four-vectors, we have

$$\Box\cdot\left[-\frac{i}{c}(\mathcal{H}\times\mathcal{V})^\dagger - \varepsilon_0\,\mathcal{E}\times\mathcal{V}\right] = \mathcal{J}_e = \mathcal{J}_u + \Box\cdot[\mathcal{P}\times\mathcal{V}] \tag{10.11a}$$

$$\Box\cdot\left[\frac{i}{c}(\mathcal{E}\times\mathcal{V})^\dagger - \mu_0\,\mathcal{H}\times\mathcal{V}\right] = \mathcal{J}_m = \Box\cdot[\mu_0\mathcal{M}\times\mathcal{V}] \tag{10.11b}$$

In this case, the right-hand side of the second equation can be interpreted as a magnetic-current density, $\mathcal{J}_m$, that is analogous to the electric-current density, $\mathcal{J}_e$. (In general, the latter is *not* identical to the Amperian electric-current density.) In three-space, these equations are given by

$$\nabla \times \mathbf{H} - \varepsilon_o \frac{\partial \mathbf{E}}{\partial t} = \mathbf{J}_e = \mathbf{J}_u + \frac{\partial \mathbf{P}}{\partial t} + \nabla \times (\mathbf{P} \times \mathbf{v}) \quad (10.12a)$$

$$\nabla \cdot \varepsilon_o \mathbf{E} = \rho_e = \rho_u - \nabla \cdot \mathbf{P} \quad (10.12b)$$

$$\nabla \times \mathbf{E} + \mu_o \frac{\partial \mathbf{H}}{\partial t} = -\mathbf{J}_m = -\mu_o \left[\frac{\partial \mathbf{M}}{\partial t} + \nabla \times (\mathbf{M} \times \mathbf{v}) \right] \quad (10.12c)$$

$$\nabla \cdot \mu_o \mathbf{H} = \rho_m = -\nabla \cdot \mu_o \mathbf{M} \quad (10.12d)$$

The four-dimensional equations (10.11a) and (10.11b) represent the Chu formulation with clarity and brevity and form a part of the book cover design.

10.5 ENERGY-MOMENTUM TENSORS AND FOUR-FORCE DENSITIES

The Maxwell Equations expressed in the Amperian, Minkowski, or Chu formulations are completely equivalent because they differ only in the rearrangement of terms and all four-vector fields $\mathcal{E}$, $\mathcal{H}$, $\mathcal{P}$, $\mathcal{M}$, $\mathcal{D}$, $\mathcal{B}$ from which they are constructed are unchanged by the choice of three-vector components. On the other hand, the associated energy-momentum tensors (that generalize the Maxwell–Poynting free-space tensor) are not equivalent; consequently, the force and power-conversion densities differ among the three formulations. This is to be expected because what are considered to be the sources of the fields are different in each case (possible because of the equivalence principle) and *the forces act directly upon these sources* and only indirectly on the masses that make up the bulk of the material. How that force is transmitted from the dipoles and the fixed and mobile charges to the lattice(s) of nuclei in a solid is a separate question that will have different answers, depending upon the specific material, its state (plasma, gas, liquid, or solid), and the model(s) used to approximate the relevant underlying physics.

Amperian energy momentum and four-force

Force and power densities in the Amperian formulation are

$$\mathcal{F}^{\text{amp}} = \Box \cdot \text{II}^{\text{amp}} = \left[-\frac{\mathcal{J}_{\text{total}} \cdot \mathcal{V}}{c^2} \mathcal{E} + \left(-\frac{i}{c} \mathcal{J}_{\text{total}} \times \mathcal{B} \right)^{\dagger} \cdot \mathcal{V} + \frac{\mathcal{E} \cdot \mathcal{J}_{\text{total}}}{c^2} \mathcal{V} \right]$$
$$= \left[\rho_{\text{total}} \mathbf{E} + \mathbf{J}_{\text{total}} \times \mathbf{B},\ \frac{i}{c} \mathbf{E} \cdot \mathbf{J}_{\text{total}} \right] \quad (10.13)$$

where II^{amp}, the energy-momentum tensor, is

$$\text{II}^{\text{amp}} = \left[\begin{array}{l} \varepsilon_o \mathcal{E}\,\mathcal{E} + \frac{1}{\mu_o} \mathcal{B}\,\mathcal{B} - \frac{1}{2} \left(\varepsilon_o \mathcal{E} \cdot \mathcal{E} + \frac{1}{\mu_o} \mathcal{B} \cdot \mathcal{B} \right) (\mathcal{I} + 2 \frac{\mathcal{V}\,\mathcal{V}}{c^2}) \\ + \frac{i}{c} \left[\mathcal{V}\, (\varepsilon_o \mathcal{E} \times \mathcal{B})^{\dagger} \cdot \mathcal{V} + (\varepsilon_o \mathcal{E} \times \mathcal{B})^{\dagger} \cdot \mathcal{V}\,\mathcal{V} \right] \end{array} \right]$$

$$= \begin{bmatrix} \varepsilon_0 \mathbf{E}^o \mathbf{E}^o + \frac{1}{\mu_0}\mathbf{B}\,\mathbf{B} - \frac{1}{2}(\varepsilon_0 \mathbf{E}^o \cdot \mathbf{E}^o + \frac{1}{\mu_0}\mathbf{B}\cdot\mathbf{B})\,\bar{\mathbf{I}} & \vdots & -ic\varepsilon_0 \mathbf{E}^o \times \mathbf{B} \\ \cdots\cdots\cdots\cdots & \vdots & \cdots\cdots\cdots \\ -ic\varepsilon_0 \mathbf{E}^o \times \mathbf{B} & \vdots & \frac{1}{2}(\varepsilon_0 \mathbf{E}^o \cdot \mathbf{E}^o + \frac{1}{\mu_0}\mathbf{B}\cdot\mathbf{B}) \end{bmatrix} \tag{10.14}$$

The current circulating in each Amperian-loop contributes momentum to the density, $\varepsilon_0 \mathbf{E}^o \times \mathbf{B}$; it also effects the power flux, $\frac{1}{\mu_0}\mathbf{E}^o \times \mathbf{B}$.

Minkowski energy momentum and four-force

The Minkowski energy-momentum tensor and its alternate forms were first encountered in Chapter 9, Section 9.3. Here the Minkowski force density, $\mathcal{F}^{\text{mink}}$, is expressed as

$$\mathcal{F}^{\text{mink}} = \Box \cdot \text{II}^{\text{mink}}$$

$$= \begin{bmatrix} -\frac{\mathcal{J}_u \cdot \mathcal{V}}{c^2}\mathcal{E} + \left(-\frac{i}{c}\mathcal{J}_u \times \mathcal{B}\right)^{\dagger} \cdot \mathcal{V} + \frac{\mathcal{E} \cdot \mathcal{J}_u}{c^2}\mathcal{V} \\ +\frac{1}{2}\left[\mathcal{D}_k\,\Box\mathcal{E}_k - \mathcal{E}_k\,\Box\mathcal{D}_k + \mathcal{B}_k\,\Box\mathcal{H}_k - \mathcal{H}_k\,\Box\mathcal{B}_k\right] \\ -\frac{i}{c}\left[\,(\mathcal{D} \times \mathcal{B})^{\dagger} \cdot \mathcal{V} - (\varepsilon_0\mathcal{E} \times \mu_0\mathcal{H})^{\dagger} \cdot \mathcal{V}\right]_k \Box\mathcal{V}_k \end{bmatrix}$$

$$= \begin{bmatrix} \rho_u \mathbf{E}^o + \mathbf{J}_u \times \mathbf{B} + \frac{1}{2}(D_k \nabla E_k^o - E_k^o \nabla D_k + B_k \nabla H_k^o - H_k^o \nabla B_k), \\ \frac{i}{c}[\mathbf{E}^o \cdot \mathbf{J}_u + \frac{1}{2}(\mathbf{E}^o \cdot \frac{\partial \mathbf{D}}{\partial t} - \mathbf{D} \cdot \frac{\partial \mathbf{E}^o}{\partial t} + \mathbf{H}^o \cdot \frac{\partial \mathbf{B}}{\partial t} - \mathbf{B} \cdot \frac{\partial \mathbf{H}^o}{\partial t})] \end{bmatrix} \tag{10.15}$$

where summation over $k = 1, 2, 3, 4$ is implied for four-vectors and $k = 1, 2, 3$ for three-vectors. The energy-momentum tensor, II^{mink}, is similarly expressed as

$$\text{II}^{\text{mink}} = \begin{bmatrix} \mathcal{E}\,\mathcal{D} + \mathcal{H}\,\mathcal{B} - \frac{1}{2}(\mathcal{E}\cdot\mathcal{D} + \mathcal{H}\cdot\mathcal{B})(\mathcal{I} + 2\frac{\mathcal{V}\,\mathcal{V}}{c^2}) \\ +\frac{i}{c}\left[\mathcal{V}\,(\varepsilon_0\mathcal{E} \times \mu_0\mathcal{H})^{\dagger} \cdot \mathcal{V} + (\mathcal{D} \times \mathcal{B})^{\dagger} \cdot \mathcal{V}\,\mathcal{V}\right] \end{bmatrix}$$

$$= \begin{bmatrix} \mathbf{E}^o \mathbf{D} + \mathbf{H}^o \mathbf{B} - \frac{1}{2}(\mathbf{E}^o \cdot \mathbf{D} + \mathbf{H}^o \cdot \mathbf{B})\bar{\mathbf{I}} & \vdots & -ic\mathbf{D} \times \mathbf{B} \\ \cdots\cdots\cdots\cdots & \vdots & \cdots\cdots\cdots \\ \frac{-i}{c}\mathbf{E}^o \times \mathbf{H}^o & \vdots & \frac{1}{2}(\mathbf{E}^o \cdot \mathbf{D} + \mathbf{H}^o \cdot \mathbf{B}) \end{bmatrix} \tag{10.16}$$

In this formulation, the power flux is $\mathbf{E}^o \times \mathbf{H}^o$ and the momentum density $\mathbf{D} \times \mathbf{B}$. Some authors are critical of the fact that the tensor is nonsymmetric. This point has already been discussed in Part I, Chapter 5, Section 5.3. Notice that the component of $\mathcal{F}^{\text{mink}}$,

$$-\frac{i}{c}\left[\,(\mathcal{D} \times \mathcal{B})^{\dagger} \cdot \mathcal{V} - (\varepsilon_0\mathcal{E} \times \mu_0\mathcal{H})^{\dagger} \cdot \mathcal{V}\right]_k \Box\mathcal{V}_k$$

produces

$$\Delta \mathcal{F}^{\text{mink}} = (\mu'\varepsilon' - \mu_0\varepsilon_0)\mathbf{E} \times \mathbf{H} \cdot \frac{\partial \mathbf{v}}{\partial t} \tag{10.17}$$

in an accelerating medium when $\mathcal{D} = \varepsilon'\mathcal{E}$, $\mathcal{B} = \mu'\mathcal{H}$, and $\mathbf{v} \simeq 0$. This term, seemingly missing from the Minkowski three-space power density, is actually hidden in

$$-\frac{1}{2}\left(\mathbf{E}^o \cdot \frac{\partial \mathbf{D}}{\partial t} - \mathbf{D} \cdot \frac{\partial \mathbf{E}^o}{\partial t} + \mathbf{H}^o \cdot \frac{\partial \mathbf{B}}{\partial t} - \mathbf{B} \cdot \frac{\partial \mathbf{H}^o}{\partial t}\right)$$

because, to first order in the velocity, we have

$$\mathbf{D} \simeq \varepsilon'\mathbf{E}^o + \mathbf{v} \times (\mu'\varepsilon' - \mu_0\varepsilon_0)\mathbf{H}^o$$

$$\mathbf{B} \simeq \mu'\mathbf{H}^o - \mathbf{v} \times (\mu'\varepsilon' - \mu_0\varepsilon_0)\mathbf{E}^o$$

Equation (10.17) was previously encountered in Eq. (5.22) and is therefore consistent with the Hamiltonian formulation of particles used to represent uniform plane waves propagating in a linear (nondeforming) dielectric/magnetic material. In this connection, notice that in anisotropic media, $\mathbf{D} \times \mathbf{B}$ remains parallel to the wavevector $\mathbf{k}$ (and hence the particle momentum) while $\mathbf{E}^o \times \mathbf{H}^o$ will (and should) deviate from that direction whenever the phase and group velocities are misaligned.

Chu energy momentum and four-force

The equivalence principle requires that there be a Lorentz force on *both* moving electric and magnetic charges and is satisfied because, in the Chu formulation, the four-force density of electromagnetic origin is

$$\mathcal{F}^{\text{chu}} = \Box \cdot \text{II}^{\text{chu}} = \begin{bmatrix} -\dfrac{\mathcal{J}_e \cdot \mathcal{V}}{c^2}\mathcal{E} + \left(-\dfrac{i}{c}\mathcal{J}_e \times \mu_0\mathcal{H}\right)^{\dagger} \cdot \mathcal{V} + \dfrac{\mathcal{E} \cdot \mathcal{J}_e}{c^2}\mathcal{V} \\ \\ -\dfrac{\mathcal{J}_m \cdot \mathcal{V}}{c^2}\mathcal{H} + \left(\dfrac{i}{c}\mathcal{J}_m \times \varepsilon_0\mathcal{E}\right)^{\dagger} \cdot \mathcal{V} + \dfrac{\mathcal{H} \cdot \mathcal{J}_m}{c^2}\mathcal{V} \end{bmatrix}$$

$$= \left[\rho_e\mathbf{E} + \mathbf{J}_e \times \mu_0\mathbf{H} + \rho_m\mathbf{H} - \mathbf{J}_m \times \varepsilon_0\mathbf{E},\ \frac{i}{c}(\mathbf{E} \cdot \mathbf{J}_e + \mathbf{H} \cdot \mathbf{J}_m)\right] \tag{10.18}$$

where II^{chu}, the free-space energy-momentum tensor, is

$$\text{II}^{\text{chu}} = \begin{bmatrix} \varepsilon_0\mathcal{E}\,\mathcal{E} + \mu_0\mathcal{H}\,\mathcal{H} - \frac{1}{2}(\varepsilon_0\mathcal{E} \cdot \mathcal{E} + \mu_0\mathcal{H} \cdot \mathcal{H})\left(\mathcal{I} + 2\dfrac{\mathcal{V}\,\mathcal{V}}{c^2}\right) \\ \\ +\dfrac{i}{c^3}\left[\mathcal{V}\,(\mathcal{E} \times \mathcal{H})^{\dagger} \cdot \mathcal{V} + (\mathcal{E} \times \mathcal{H})^{\dagger} \cdot \mathcal{V}\,\mathcal{V}\right] \end{bmatrix}$$

$$= \left[\begin{array}{c:c} \varepsilon_o\mathbf{E}\,\mathbf{E} + \mu_0\mathbf{H}\,\mathbf{H} - \frac{1}{2}(\varepsilon_0\mathbf{E} \cdot \mathbf{E} + \mu_0\mathbf{H} \cdot \mathbf{H})\,\bar{\mathbf{I}} & \dfrac{-i}{c}\mathbf{E} \times \mathbf{H} \\ \cdots\cdots\cdots & \cdots\cdots \\ \dfrac{-i}{c}\mathbf{E} \times \mathbf{H} & \frac{1}{2}(\varepsilon_0\mathbf{E} \cdot \mathbf{E} + \mu_0\mathbf{H} \cdot \mathbf{H}) \end{array}\right] \tag{10.19}$$

Because magnetic charges (instead of circulating currents) are used to model magnetic dipoles, there are no material contributions to the momentum density. The tensor is therefore symmetric and identical in form to that of free-space.

10.6 DISCUSSION OF FORCE DENSITIES

The symmetry of the Chu formulation is strikingly simple and often computationally advantageous. Because Eq. (10.18) differs from both its Minkowski and Amperian counterparts, controversies swirled around the Chu force density for magnetic materials soon after it was introduced. These joined other controversies that were long-standing; many are recounted and critiqued by Penfield and Haus (P-H) in their important monograph [19]. In general, all of the formulations are incomplete and must be augmented by considering additional material subsystems. These are found through consideration of the detailed constitutive laws. All three formalisms can then be made to provide consistent *total* force densities[1] – although the portions defined as the electromagnetic force will continue to differ[2]. Indeed, since we have shown that the models are built out of the *same* four-vectors, it is hardly possible for it to be otherwise. In the Chu formulation, $\mathcal{E}$ and $\mathcal{H}$ are considered the fundamental field quantities with $\mathcal{B}$ and $\mathcal{D}$ useful auxiliary vectors; in the Amperian formulation, $\mathcal{E}$ and $\mathcal{B}$ are considered the fundamental field quantities with $\mathcal{H}$ and $\mathcal{D}$ the auxiliary vectors. In both cases, $\mathcal{P}$ and $\mathcal{M}$ model the material response and there are *four* independent four-vectors linearly related by Maxwell's Equations. ($\mathcal{E}$, $\mathcal{B}$, $\mathcal{H}$, and $\mathcal{D}$ in the Minkowski formulation.) Chu modeled macroscopic material as grains of matter surrounded by free-space which contains **E** and **H** fields. That choice leads to $\mathbf{E} \times \mathbf{H}$ as the electromagnetic Poynting flux, which he believed was necessary, and produces a simpler set of three-space equations when the material is moving and deforming. As long as the free-space **E** and **H** fields are "outside" of the matter, the equivalence principle allows substitution of one set of "inside" sources for another. Since magnetic charges can be used to create magnetic dipoles just as well as Amperian electric currents, we are permitted to make that substitution.[3]

Although with respect to Maxwell's Equations, the choice is merely one of rearrangement, it is important to reiterate that what is *defined* as the four-vector force density of *electromagnetic origin* will be different. Inevitably, such a force is incomplete in any formulation of electrodynamics, until the connections among the four vector fields are specified. These connections may lead one to deduce the existence of other forces. Additional information, usually in the form of constitutive laws, is required. Finding the connections is often quite difficult—especially when they are nonlinear, nonlocal, dissipative, and possibly hysteretic—but often easier to accomplish within the Chu formulation. As long as the *total* force density is correct, it does not matter how the whole is apportioned.

Some of the force may lead to both kinetic and potential energy transfers to mechanical and/or electric and magnetic strains; dissipation will certainly affect thermodynamic properties such as temperature and entropy. The form and specification of the constitutive laws, which are *assumed* to provide a reasonable model, will govern the connections between various subsystems within the material. What is considered a separate subsystem in a particular electromagnetic formulation may be included in the energy-momentum

[1] Of course, consistency is no guarantee of correctness.

[2] Tellegen [20] calculated the electromagnetic force on both an Amperian current-loop and a magnetic-charge dipole and found a small (but fundamental) difference. P-H show why Tellegen's analysis is incomplete and his conclusion incorrect.

[3] An even more general formulation is possible, in which electric and magnetic charges create *some* of the electric and magnetic dipoles while the remainder are created by Amperian magnetic and electric currents.

tensor of another. It is only after those connections have been studied (possibly by the principle of "virtual power" discussed below) that the total force density will emerge. Because materials with electromagnetic properties are nearly endless in their variety, it is futile to expect that a study of Maxwell's Equations by themselves can ever produce the "correct" force density.

Orbital and/or spin angular momentum at the atomic level is widely acknowledged to be the physical source of magnetization. This causes dynamic constitutive laws that affect the material subsystems. Can the magnetic-charge model, devoid of these dynamic properties, represent the correct macroscopic electrodynamics? Because the physics of a ferrimagnetic material, with nonlinear constitutive laws that lead to both temporal and spatial dispersion, is very well understood, ferrites provide an interesting test of the Chu model; they were considered in the *MQS* limit by P-H and in a full electrodynamic treatment by the present author [17]. In contradistinction to P-H, nonzero momentum in the rest frame of the material is predicted by the latter analysis [refer to Part III, Chapter 23, Section 23.8 for a derivation of the small-signal version of this momentum density]; because both theories employ the Chu formulation, the origin of the disagreement lies elsewhere [21]. Nevertheless, our analysis has shown that the angular momentum, together with its associated kinetic energy, can readily coexist with the magnetic charge model.

In the context of the present discussion concerning Alternate power and Alternate stress, it does not matter that the four-vector force density, Eq. (10.13), is incomplete, only that the electromagnetic and Alternate energy-momentum tensors both lead to the *same* result. Because we require a formulation in terms of $\mathcal{A}$, the variables $\mathcal{E}$ and $\mathcal{B}$ and the Amperian energy-momentum tensor are the obvious choice, although the Minkowski version is also very useful; which set of three-vector fields and what model should be employed to solve for the electromagnetic fields is a matter of personal preference.

The forces are assumed to act upon the electric/magnetic charges and currents; they in turn are transferred to the particle masses of the material which are themselves coupled due to elastic, electro-elastic, magneto-elastic, or other effects. In a complex solid, there may be multiple sublattices (denoted by i), each described by separate continuum averages of the rest-frame values of particle density and mass density, n'_i and $n'_i m'_i$, and where the continuum velocities of the material are $\mathcal{V}_{(i)}$, One of these sublattices may contain most of the mass, but others may have important reservoirs of kinetic energy that supply momentum even though the primary lattice is at rest. An example is the angular momentum of the electrons orbiting around heavy nuclei. In most cases, the orbital kinetic energy simply contributes to the total internal energy of the solid, but can be ignored because the macroscopic average of the angular momentum is zero. In this respect it is similar to the incoherent vibrational motion of the nuclei that contribute to the temperature of the solid. In materials with spontaneous magnetization, the associated angular-momentum is organized; if it can freely precess, the orbital and or spin kinetic-energy reservoir can provide interesting and technically important magnetic-resonance interactions that can convert a portion of the angular momentum to linear momentum. Such materials are the exception; they are considered in Part III, Chapter 23.

Newton's Law of motion and conservation of mass require, respectively, that for a single lattice system without dissipation we have

$$\Box \cdot (n'm'\mathcal{V}\mathcal{V}) = \mathcal{F}_{\text{total}} = \Box \cdot \Pi_{\text{total}}$$

and

$$\Box \cdot (n'm'\mathcal{V}) = 0$$

We again emphasize that what has been defined as the electromagnetic force density is, in general, only part of $\mathcal{F}_{\text{total}}$.

10.7 THE PRINCIPLE OF VIRTUAL POWER

The four-vectors

$$\mathcal{V} \cdot \text{II} = [\gamma(\mathbf{S} + \mathbf{v} \cdot \overline{\mathbf{T}}),\ i\gamma c(W - \mathbf{G} \cdot \mathbf{v})]$$

and

$$\text{II} \cdot \mathcal{V} = [\gamma(\mathrm{c}^2\mathbf{G} + \overline{\mathbf{T}} \cdot \mathbf{v}),\ i\gamma c(W - \frac{\mathbf{S} \cdot \mathbf{v}}{\mathrm{c}^2})]$$

when evaluated in the rest frame, are respectively $[\mathbf{S}',\ icW']$ and $[c^2\mathbf{G}',\ icW']$. It also follows that

$$\mathcal{V} \cdot (\Box \cdot \text{II}) = \Box \cdot (\mathcal{V} \cdot \text{II}) - \text{II}_{ij} \frac{\partial \mathcal{V}_i}{\partial x_j} = \mathcal{F} \cdot \mathcal{V}$$

With substitution and rearrangement, we obtain

$$\Box' \cdot [\mathbf{S}',\ icW'] = \mathcal{F} \cdot \mathcal{V} + \gamma \left(T_{ij} \frac{\partial \mathrm{v}_i}{\partial x_j} - \mathbf{G} \cdot \frac{\partial \mathbf{v}}{\partial t} \right) + \frac{\partial \gamma}{\partial x_j}(S_j + \mathrm{v}_i T_{ij}) + \frac{\partial \gamma}{\partial t}(W - \mathbf{G} \cdot \mathbf{v})$$

or, equivalently,

$$\left(\frac{\partial W'}{\partial t'} \right)_{\mathrm{o}} = (\mathcal{F} \cdot \mathcal{V})_{\mathrm{o}} + \left(T_{ij} \frac{\partial \mathrm{v}_i}{\partial x_j} - \mathbf{G} \cdot \frac{\partial \mathbf{v}}{\partial t} - \frac{\partial S'_j}{\partial x'_j} \right)_{\mathrm{o}} \tag{10.20}$$

where the subscript o denotes evaluation at $\mathbf{v} = 0$.

It is not necessary that II be the complete energy-momentum tensor; it can be that associated with a particular subsystem. Assume that one has sufficient information about the rest-frame value W' to make an expansion of the left-hand side of Eq. (10.20). The expansion must match the form of the right-hand side; therefore if the subsystem is dissipationless, we have $\mathcal{F} \cdot \mathcal{V} = 0$, and it should be possible to identify a divergence that defines $\mathbf{S}'$, and the coefficients of $\frac{\partial \mathrm{v}_i}{\partial x_j}$ and $\frac{\partial \mathbf{v}}{\partial t}$ that, respectively, define T_{ij}, and $\mathbf{G}$. P-H call this procedure the principle of "virtual power" and use it to verify results that, in most cases, have been obtained by other means. They do caution that components of T_{ij} and $\mathbf{G}$ inherently orthogonal to $\frac{\partial \mathrm{v}_i}{\partial x_j}$ and $\frac{\partial \mathbf{v}}{\partial t}$ will be missed.

Another potential danger is the fact that declaring a particular subsystem to be dissipationless is not always consistent with the correct underlying physics—especially when the subsystem contains accelerating electric or magnetic dipoles. If, nevertheless, one insists that there is no dissipation, the mathematics will try to accommodate, but one must expect certain paradoxes. This is particularly true in the case of magnetic materials that can undergo magnetic resonance because then the spontaneous magnetization is coupled to an intrinsic angular-momentum density over and above that of the lattice of masses (nuclei) that constitute the solid. Without dissipation (including radiation), the magnetization vector cannot align with an applied magnetic field and so would be in a perpetual state of precession. Consequently, one should not be surprised if a trial model of W_0' and $\mathbf{S}_0'$ does not accurately account for power flowing into or out of the subsystem due to acceleration. Any omissions can lead to erroneous conclusions about the form (or even existence) of **G**.

PART III

ELECTROMAGNETIC EXAMPLES

INTRODUCTION TO PART III

Part III of the text is devoted to the application of Maxwell's Equations to many different types of electromagnetic problem. In the process, the basic principles of Part I are reviewed and methods of problem solving that depend upon linearity, superposition, and the use of boundary conditions are taught by example. Once the complete solution of the fields and potentials of an electromagnetic problem has been found, it is instructive to compare the electromagnetic power, energy, stress, and momentum as calculated in both the Maxwell–Poynting and Alternate representation. Such comparisons are made for a variety of classic field problems.

Chapter 11 begins with the static analysis of charges and currents and quasistatic analysis of a small conducting sphere placed in a region of either electric or magnetic field. This is followed with a thorough analysis of a physical resistor inside a perfectly conducting shield. In these examples, the benefits of employing the Alternate representation of power and energy are clearly shown. The final example deals with magnetic diffusion through an electrically conducting plate. In Chapter 12, a point charge (and charge distributions with planar, cylindrical, and spherical symmetries) moving at constant velocity are analyzed. In Chapter 13, accelerated charges are considered starting with the Hertzian electric-dipole driven in the sinusoidal steady state (already studied in Part I). Except for the absence of free-space reactive Alternate power and reactive Alternate energy, there are few surprises – save that the power/energy calculations are so simple. Additional understanding is gained when the same dipole current is modulated with a unit step function and the entire transient studied; also considered is a magnetic dipole and a single radiating charge experiencing constant acceleration, followed by constant deceleration.

The next set of examples, considered in Chapter 14, involve fields produced by uniform current transients. Nondispersive waves generated by a spatially uniform planar surface current are considered. Striking differences between the Maxwell–Poynting and Alternate representations are revealed that are strongly dependent on the pulse waveform; closer study provides both resolution of the apparent paradoxes and beneficial insights. In Chapter 15, enhanced perspectives of localized power and energy are applied to a uniform line source of infinite length, which generates a current step. The superposition of these solutions allows a variety of related problems to be solved as well.

Normal and Doppler-shifted uniform plane waves, as well as nonuniform plane waves, are considered in Chapter 16; reflection and transmission of oblique incidence plane waves are studied in Chapter 17. The *TEM* mode of uniform transmission lines is considered in Chapter 18. The case of a parallel-plate *TEM* transmission line filled with linear isotropic dielectric and magnetic material is included to illustrate under what circumstances the Alternate power and energy remain localized to the free or unpaired charges and currents and when instead they shift to the polarization and/or magnetization currents distributed throughout the volume of the material. The *TE* and *TM* modes of rectangular and circular waveguides are considered, respectively, in Chapters 19 and 20. The Alternate energy momentum of circular waveguide modes, with clockwise or counterclockwise phase progression, is shown to connect directly with the spin properties of photons. Dielectric waveguides are analyzed in Chapter 21.

Antenna and diffraction problems are considered in Chapter 22. The radiation characteristics of a half-wave dipole are derived for both a wire antenna and its complement—a thin slot cut in an infinite ground plane. Diffraction by a rectangular slit of infinite length and width, d, is considered when $d/\lambda \gg 1$. The solution is represented by the superposition of uniform and nonuniform plane waves and introduces the Fresnel-Integral functions. On-axis diffraction by a large circular aperture and its complement are the next topics; they provide examples that employ source duality, the induction theorem, Fresnel zones, and Babinet's Principle. Off-axis diffraction in the far-field region is also solved for holes that are both large and small compared to the incident wavelength. Finally, the paraxial approximation is introduced and Gaussian beams, confined in two or three dimensions, are shown to be solutions of the paraxial wave equation.

In Chapter 23, magnetic resonance modes and wave propagation in a magnetically saturated ferrite material are considered. The nonlinear constitutive law is linearized to model small-signal operation and found to be both nonreciprocal and frequency-dispersive. Small-signal power-energy and stress-momentum theorems are sought and found in both Maxwell–Poynting and Alternate representations. These form the basis of understanding how quasiparticles (magnons) can represent magnetostatic waves.

The final examples are contained in Chapter 24; they involve equivalent circuits that represent dipole antennas, transmission lines, waveguides, junctions, and resonators.

Smith-Chart analysis and scattering-matrix concepts are also presented. Where possible, each of the examples in Part III is broken up into a series of problems that are already understood. The principles of linearity and superposition facilitate this approach. The practice problems and exercises (solutions not provided) in the final chapter, Chapter 25, are grouped by topic and intended to test a student's mastery of the material at both fundamental and advanced levels.

CHAPTER 11

STATIC AND QUASISTATIC FIELDS

When all fields are static, Alternate power, energy, and momentum are localized to the charges and currents; in such cases, the differences between the Maxwell–Poynting and Alternate representations are the most striking. We first consider two very simple examples: a uniform electric-charge density confined to a spherical volume and a uniform electric current flowing parallel to the axis of an infinite cylindrical conductor of circular cross section. Free-space exists outside the regions of charge or current.

11.1 SPHERICAL CHARGE DISTRIBUTION

Positive electric-charge density, ρ_o, is uniformly distributed throughout the volume of a sphere of radius R. From Gauss' Law the electric field (radial) is given by

$$\mathbf{E} = \widehat{\mathbf{r}}\frac{\rho_o}{3\varepsilon_o}\begin{cases} r & (r \leq R) \\ \dfrac{R^3}{r^2} & (r \geq R) \end{cases} \tag{11.1}$$

The associated scalar potential (which must be continuous at $r = R$) is

$$\Phi = \frac{\rho_o}{6\varepsilon_o}\begin{cases} 3R^2 - r^2 & (r \leq R) \\ \dfrac{2R^3}{r} & (r \geq R) \end{cases} \tag{11.2}$$

The Power and Beauty of Electromagnetic Fields, First Edition. F. R. Morgenthaler.

The magnetic field and vector potential are both zero. These results were previously obtained in Part I, Chapter 1, Section 1.5.

There is neither power flux nor momentum density in either the Maxwell–Poynting or Alternate representation, but the respective electric-energy densities and stress tensors are

$$W_e = \frac{1}{2}\varepsilon_0 \mathbf{E}\cdot\mathbf{E}$$

$$W^o = \frac{1}{2}\rho_0\Phi$$

and

$$\overline{\mathbf{T}}^{\text{em}} = \varepsilon_0 \mathbf{E}\,\mathbf{E} - W_e \overline{\mathbf{I}}$$

$$T^o_{ij} = -\frac{1}{2}\varepsilon_0\left(\Phi\frac{\partial^2\Phi}{\partial x_i\partial x_j} - \frac{\partial\Phi}{\partial x_i}\frac{\partial\Phi}{\partial x_j}\right) - W^o\delta_{ij}$$

Because of radial symmetry, the last equation is equivalent to

$$T^o_{ij} = \frac{1}{2}\varepsilon_0\left[\left(\frac{\partial\Phi}{\partial r}\right)^2 - \Phi\frac{\partial^2\Phi}{\partial r^2} + \Phi\frac{1}{r}\frac{\partial\Phi}{\partial r}\right]\frac{x_i x_j}{r^2} - \left(\frac{1}{2}\varepsilon_0\Phi\frac{1}{r}\frac{\partial\Phi}{\partial r} + W^o\right)\delta_{ij}$$

It follows that the total energy,

$$\text{Energy} = \frac{4\pi\rho_0^2 R^5}{15\varepsilon_0} = \frac{\frac{3}{5}Q^2}{4\pi\varepsilon_0 R} \tag{11.3}$$

is the same in either representation, where $Q = \frac{4}{3}\pi R^3\rho_0$ is the total charge. Direct evaluation also reveals that 5/6 of the Poynting energy, but none of the Alternate energy, resides outside of the sphere.

Although the net Lorentz force on the distribution is zero, the radial density, $\rho_0\mathbf{E}$, attempts to blow apart the charge distribution and needs to be balanced by opposing mechanical forces. The net upward force on the northern hemisphere of charge can be found by integrating the $+z$ component of the Lorentz-force density, $\rho_0\mathbf{E}$, over that half-volume or equivalently by evaluating

$$\oint \overline{\mathbf{T}}\cdot d\mathbf{a}$$

using either representation of the stress tensor. The result is

$$\mathbf{F} = \hat{\mathbf{z}}\frac{\pi}{12\varepsilon_0}\rho_0^2 R^4 = \hat{\mathbf{z}}\frac{3Q^2}{64\pi\varepsilon_0 R^2} \tag{11.4}$$

The southern hemisphere has an equal, but opposite, force acting upon it. Notice that for a fixed magnitude Q, both Eq. (11.3) and (11.4) diverge in the limit of a point charge, $R \to 0$.

11.2 ELECTRIC FIELD IN A RECTANGULAR SLOT

Consider a region of free-space bounded by perfectly conducting planes located at $y = \pm d/2$ and $x = \ell$; all three are maintained at ground potential, $\Phi = 0$. The potential of the plane, $x = 0$, is specified to be $V(y) = V_0 \cos(\pi y/d)$ over the region $|y| \leq d/2$. The boundaries constitute a rectangular slot of width d and depth ℓ that is assumed to be independent of z. We wish to calculate the electric potential and **E** field inside of the slot.

Within the slot, the potential is a solution of Laplace's Equation and, as proved in Part I, Chapter 6, Section 6.6, is unique because $\Phi(x, y)$ is specified on all four boundaries. Solutions that are product functions of Cartesian coordinates are discussed in Appendix C, Section C.4 Those appropriate to the problem at hand are of the form

$$\Phi(x, y) = \sum_k [C_1(k)\cos(ky) + C_2(k)\sin(ky)][C_3(k)\exp(kx) + C_4(k)\exp(-kx)]$$

$$(-d/2 \leq y \leq +d/2,\ 0 < x \leq \ell)$$

Matching boundary conditions on the slot surface, $x = \ell$, leads to

$$C_3(k) = -C_4(k)\exp(-2k\ell)$$

Matching boundary conditions on the slot surfaces, we obtain

$$y = \pm d/2 \quad \text{leads to} \quad C_2(k) = 0 \text{ and } k = (2m-1)\pi/d, \quad m = 1, 2, 3, \ldots$$

It follows that, because $2\sinh(u) = \exp(u) - \exp(-u)$ and with $C(m) = 2\exp(k\ell)$ $C_1(k)C_4(k)$, we have

$$\Phi(x, y) = \sum_{m=1}^{\infty} C(m)\cos\left[\frac{(2m-1)\pi y}{d}\right]\sinh\left[\frac{(2m-1)\pi(\ell - x)}{d}\right]$$

In general, matching the final boundary, $x = 0$, requires that any even function $\Phi(0, y)$ be expanded in terms of a Fourier cosine series; in this case, only one term, $m = 1$, is required. Consequently, the final form of the solution can be expressed as

$$\Phi(x, y) = V_0 \cos\left(\frac{\pi y}{d}\right)\frac{\sinh[\frac{\pi(\ell - x)}{d}]}{\sinh(\frac{\pi\ell}{d})}$$

The electric field, found by taking the gradient, is therefore

$$\mathbf{E}(\mathbf{x}, \mathbf{y}) = \frac{\frac{\pi}{d}V_0}{\sinh(\frac{\pi\ell}{d})}\left\{\widehat{\mathbf{x}}\cos\left(\frac{\pi y}{d}\right)\cosh\left[\frac{\pi(\ell - x)}{d}\right] + \widehat{\mathbf{y}}\sin\left(\frac{\pi y}{d}\right)\sinh\left[\frac{\pi(\ell - x)}{d}\right]\right\} \tag{11.5}$$

If $\ell \gg d$, the field decays rapidly along x; this is also true in cases where $\Phi(0, y)$ generates values of $m > 1$. Because the higher harmonics decay ever more rapidly, the *form* of the field near the bottom of a deep slot remains that of Eq. (11.5). A gray-scale contour plot of $\Phi(x/d,\ y/d)$ superimposed on a vector-field plot of $\mathbf{E}(x/d,\ y/d)$ is shown in Figure 11.1 for $\ell = 1.5d$.

If, instead, the potential had been specified along a cylindrical surface of constant radius (independent of z), the analysis would have been based upon product functions of polar coordinates (Appendix C, Section C.4).

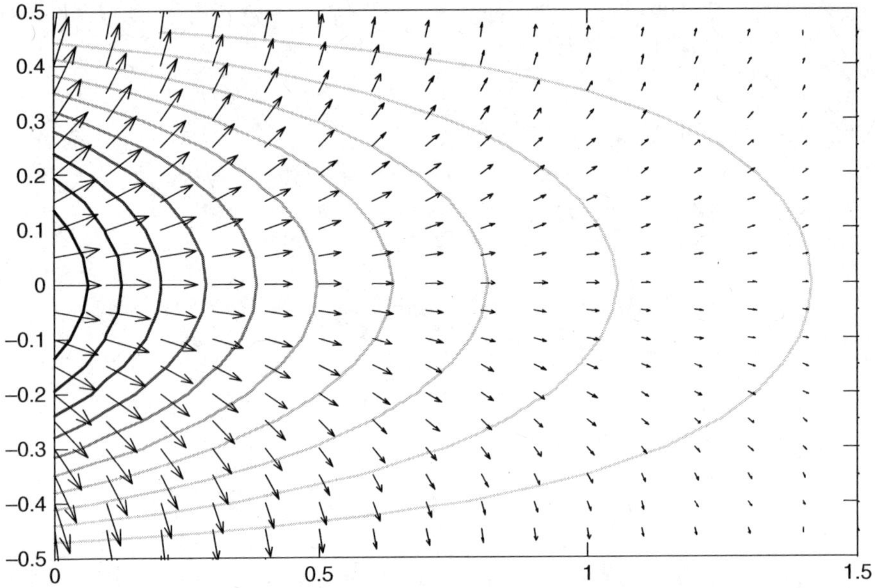

Figure 11.1 Potential contours and **E**-field vectors.

11.3 CURRENT IN A CYLINDRICAL CONDUCTOR

Static current

The static-current density, **J**, flowing parallel to the axis of a conducting cylinder is governed by

$$\nabla^2 \left(\frac{\mathbf{J}}{\sigma} \right) = 0$$

and will therefore be uniformly distributed over the cross section if the conductivity, σ, is uniform. If the cylinder has radius R_o and is parallel to the z axis, we have

$$\mathbf{J} = \hat{\mathbf{z}} \frac{I_o}{\pi R_o^2} = \sigma \mathbf{E}$$

where I_o is the total current.

The magnetic field and the vector and scalar potentials are

$$\mathbf{H} = \hat{\phi} \frac{I_o}{2\pi R_o} \begin{cases} \dfrac{r}{R_o} & (r \le R_o) \\[2ex] \dfrac{R_o}{r} & (r \ge R_o) \end{cases}$$

$$\mathbf{A} = \hat{\mathbf{z}} \left[A_o + \frac{\mu_o I_o}{2\pi R_o} \begin{cases} -\dfrac{r^2}{2R_o} & (r \le R_o) \\[2ex] R_o \left[\ln\left(\dfrac{R_o}{r} \right) - \frac{1}{2} \right] & (r \ge R_o) \end{cases} \right]$$

$$\Phi = \frac{-I_o}{\pi R_o^2 \sigma} z \qquad \text{(all } r\text{)}$$

Within the cylinder, the Lorentz-force density is

$$\mathbf{f} = \mathbf{J} \times \mu_o \mathbf{H} = -\widehat{\mathbf{r}} \frac{\mu_o I_o^2}{2\pi^2 R_o^4} r$$

The respective power fluxes are

$$\mathbf{S} = \mathbf{E} \times \mathbf{H}$$

$$\mathbf{S}^o = \Phi \mathbf{J}$$

which evaluate to

$$\mathbf{S} = -\widehat{\mathbf{r}} \frac{I_o^2}{2\sigma\pi^2 R_o^3} \begin{cases} \dfrac{r}{R_o} & (r \le R_o) \\ \dfrac{R_o}{r} & (r \ge R_o) \end{cases}$$

$$\mathbf{S}^o = \widehat{\mathbf{z}} \frac{-I_o^2}{\pi^2 R_o^4 \sigma} z \, u_{-1}(R_o - r)$$

These fluxes are very different, but as required, both divergences are equal to $-\mathbf{E} \cdot \mathbf{J}$. Outside of the conductor, $\mathbf{S}^o$ and $\nabla \cdot \mathbf{S}$ both vanish; $\mathbf{S}$ does not.

There is no momentum density in either representation, but the respective energy densities and stress tensors are

$$W^{em} = \frac{1}{2}\mu_o \mathbf{H} \cdot \mathbf{H} + \frac{1}{2}\varepsilon_o \mathbf{E} \cdot \mathbf{E}$$

$$W^o = \frac{1}{2}\mathbf{A} \cdot \mathbf{J}$$

and

$$T_{ij}^{em} = \mu_o H_i H_j + \varepsilon_o E_i E_j - W^{em} \delta_{ij}$$

$$T_{ij}^o = -A_i J_j + \frac{1}{2}\mathbf{A} \cdot \mathbf{J}\, \delta_{ij} + \frac{1}{2}\mu_o \left(\mathbf{A} \cdot \frac{\partial^2 \mathbf{A}}{\partial x_i \partial x_j} - \frac{\partial \mathbf{A}}{\partial x_i} \cdot \frac{\partial \mathbf{A}}{\partial x_j} \right) - \frac{1}{2}\varepsilon_o \left(\Phi \frac{\partial^2 \Phi}{\partial x_i \partial x_j} - \frac{\partial \Phi}{\partial x_i} \frac{\partial \Phi}{\partial x_j} \right)$$

The divergence of either tensor produces the correct Lorentz-force density.

If

$$\frac{1}{\sigma} \ll \sqrt{\frac{\mu_o}{\varepsilon_o}} R_o$$

the electric energy and stress components are negligible (except very near $r = 0$). Then

$$W^{em} = \frac{1}{2}\mu_o \mathbf{H} \cdot \mathbf{H}$$

$$W^o = \frac{1}{2}\mathbf{A} \cdot \mathbf{J}$$

and

$$T_{ij}^{\text{em}} = \mu_0 H_i H_j - W^{\text{em}} \delta_{ij}$$

$$T_{ij}^{\text{o}} = -A_i J_j + \frac{1}{2} \mathbf{A} \cdot \mathbf{J}\, \delta_{ij} + \frac{1}{2} \mu_0 \left(\mathbf{A} \cdot \frac{\partial^2 \mathbf{A}}{\partial x_i \partial x_j} - \frac{\partial \mathbf{A}}{\partial x_i} \cdot \frac{\partial \mathbf{A}}{\partial x_j} \right)$$

In the Alternate representation, the constant A_o is not arbitrary; it must be chosen to satisfy the boundary conditions. For example, if a cylindrical shield is placed concentric to the conductor, at radius R, to provide a return path for the current, $A_z(R) = 0$, and

$$A_o = \frac{\mu_0 I_0}{2\pi} \left[\ln \left(\frac{R}{R_o} \right) + \frac{1}{2} \right]$$

In that case, the total energy (per unit length along z) is $\frac{1}{2} L' I_0^2$, where

$$L' = \frac{\mu_0}{2\pi} \left[\ln \left(\frac{R}{R_o} \right) + \frac{1}{4} \right]$$

is the inductance per unit length. In the Maxwell–Poynting representation the ln term is called the external inductance; the remainder, the internal inductance. These terms arise from the magnetic field outside or inside the conductor. Of course, in the Alternate (circuit) representation, *both* are internal inductances.

Sinusoidal steady-state current

If the total z-directed current is modulated in time so that

$$I_z = I_0 \cos \omega t$$

the current distribution, J_z, depends upon the ratio δ / R_o, where $\delta = \sqrt{2/(\omega \mu_0 \sigma)}$ (the skin depth defined in Chapter 5, Section 5.2). The rationale behind the approximations that follow can be found in Chapter 16, Section 16.4.

If ω is a low frequency such that $\delta / R_o \gg 1$, the current density remains uniform throughout the conductor and the previous analyses of **H**, **A**, and Φ remain valid; they constitute the zero-order fields of a quasistatic conduction problem that is *MQS* if $\sigma \sqrt{\frac{\mu_0}{\varepsilon_0}} R_o \gg 1$ and *EQS* if $\sigma \sqrt{\frac{\mu_0}{\varepsilon_0}} R_o \ll 1$.

If ω is a high frequency such that $\delta / R_o \ll 1$, the current density (and therefore **E** and **H**) is nearly zero inside the conductor *except* within a layer that is approximately equal to δ. In that case,

$$\mathbf{J} = \sigma \mathbf{E} = \begin{cases} \hat{\mathbf{z}} \dfrac{I_0}{2\pi R_o \delta} \cos \omega t & (R_o - \delta < r \le R_o) \\ 0 & (r \le R_o - \delta) \end{cases}$$

and a reasonable approximation is to replace **J** by a surface electric-current density, $\mathbf{K} = \mathbf{J}\delta$, at $r = R_o$. The fields are *MQS*, with **A** and **H** outside the conductor having the same forms as before. In this approximation, $\mathbf{H} = 0$ inside the conductor and $\mathbf{E} = 0$ everywhere.

11.4 SPHERE WITH UNIFORM CONDUCTIVITY

We now turn to fields that are time-varying, but well approximated by the quasistatic approximation. A small uniformly conducting sphere is placed in a region of free-space that contains either an electric or magnetic field. The radius, R, conductivity, σ, and time variation of the field are all such that the field may be considered to be zero-order *EQS* or *MQS*.

Quasistatic electric-field probe

Zero-order fields

In this *EQS* case, for $r \gg R$, the outer fields approach

$$\mathbf{E}_{(0)} = \widehat{\mathbf{z}} E_o(t)$$

$$\mathbf{H}_{(0)} = 0$$

The zero-order fields satisfy

$$\nabla \cdot \mathbf{J}_{(0)} = 0$$

$$\mathbf{J}_{(0)} = \sigma \mathbf{E}_{(0)}$$

$$\nabla \times \mathbf{J}_{(0)} = \nabla \sigma \times \mathbf{E}_{(0)}$$

$$\nabla \times \mathbf{E}_{(0)} = 0$$

$$\varepsilon_o \nabla \cdot \mathbf{E}_{(0)} = \rho_{(0)}$$

$$\nabla \times \mathbf{H}_{(0)} = \mathbf{J}_{(0)}$$

$$\nabla \cdot \mathbf{H}_{(0)} = 0$$

Because of charge relaxation, we have $\rho_{(0)} = 0$, except on the surface of the sphere. The boundary conditions at $r = R$ are:

$$\widehat{\mathbf{r}} \cdot \mathbf{J}^{\text{inner}}_{(0)} = 0$$

$$\widehat{\mathbf{r}} \cdot \varepsilon_o [\mathbf{E}^{\text{outer}}_{(0)} - \mathbf{E}^{\text{inner}}_{(0)}] = \sigma^s_{(0)}$$

$$\widehat{\mathbf{r}} \times [\mathbf{E}^{\text{outer}}_{(0)} - \mathbf{E}^{\text{inner}}_{(0)}] = 0$$

where the superscripts inner and outer indicate on which side of the surface the boundary field resides.[1]

Following the prescription for solving conduction problems, the conduction-current density is the first field to be solved for. Because outside the sphere $\mathbf{J}_{(0)} = 0$ and inside it $\nabla \times \mathbf{J}_{(0)} = \nabla \sigma \times \mathbf{E}_{(0)} = 0$ (except possibly at $r = R$) $\mathbf{J}_{(0)}$ is a Laplacian field. By

[1]When dealing with quasistatic fields, we avoid using superscripts (1) and (2), which might be confused with the order of the field. In this case, *outer* = (1) and *inner* = (2).

inspection, $\mathbf{J}_{(0)} = 0$ satisfies all boundary conditions and by the uniqueness theorem is the *only* solution. Then for $r < R$, $\mathbf{E}_{(0)} = 0$ and since $\mathbf{E}_{(0)}^{\text{outer}}$ has only a radial component at the surface, $\nabla\sigma \times \mathbf{E}_{(0)} = 0$ everywhere. Because $\mathbf{E}_{(0)}$ is a conservative field,

$$\mathbf{E}_{(0)} = -\nabla\Phi_{(0)}$$

$$\nabla^2\Phi_{(0)} = -\frac{1}{\varepsilon_o}\rho_{(0)} = 0 \qquad \text{(except on } r = R)$$

The solution consists of a uniform field and dipole field; it satisfies all boundary conditions and so is unique.

$$\Phi_{(0)} = \begin{cases} 0 & (r \le R) \\ -E_o(t)\left(r - \dfrac{R^3}{r^2}\right)\cos\theta & (r \ge R) \end{cases}$$

$$\mathbf{E}_{(0)} = \begin{cases} 0 & (r < R) \\ E_o(t)\left[\widehat{\mathbf{r}}\left[1 + 2\left(\dfrac{R}{r}\right)^3\right]\cos\theta + \widehat{\theta}\left[-1 + \left(\dfrac{R}{r}\right)^3\right]\sin\theta\right] & (r \ge R) \end{cases}$$

$$\sigma^s_{(0)} = 3\varepsilon_o E_o(t)\cos\theta \qquad (r = R)$$

With $\mathbf{J}_{(0)} = 0$, everywhere there is no source for either $\mathbf{H}_{(0)}$ or $\mathbf{A}_{(0)}$; therefore,

$$\mathbf{A}_{(0)} = 0$$

$$\mathbf{H}_{(0)} = 0$$

complete the zero-order fields.

First-order fields

The first-order fields satisfy

$$\nabla \times \mathbf{H}_{(1)} = \sigma\mathbf{E}_{(1)} + \varepsilon_o\frac{\partial\mathbf{E}_{(0)}}{\partial t}$$

$$\nabla \cdot \mathbf{H}_{(1)} = 0$$

$$\nabla \times \mathbf{E}_{(1)} = 0$$

$$\nabla \cdot \sigma\mathbf{E}_{(1)} = \frac{-\partial\rho_{(0)}}{\partial t} = 0 \qquad \text{(except on } r = R)$$

$$\varepsilon_o\nabla \cdot \mathbf{E}_{(1)} = \rho_{(1)} = 0 \qquad \text{(except on } r = R)$$

and are subject to the boundary conditions at $r = R$:

$$\nabla_\Sigma \cdot \mathbf{K}^s_{(1)} - \widehat{\mathbf{r}} \cdot \mathbf{J}^{\text{inner}}_{(1)} + \frac{\partial\sigma^s_{(0)}}{\partial t} = 0$$

$$\widehat{\mathbf{r}} \cdot \varepsilon_o[\mathbf{E}^{\text{outer}}_{(1)} - \mathbf{E}^{\text{inner}}_{(1)}] = \sigma^s_{(1)}$$

$$\hat{\mathbf{r}} \times [\mathbf{E}_{(1)}^{\text{outer}} - \mathbf{E}_{(1)}^{\text{inner}}] = 0$$

$$\hat{\mathbf{r}} \cdot \mu_o[\mathbf{H}_{(1)}^{\text{outer}} - \mathbf{H}_{(1)}^{\text{inner}}] = 0$$

$$\hat{\mathbf{r}} \times [\mathbf{H}_{(1)}^{\text{outer}} - \mathbf{H}_{(1)}^{\text{inner}}] = \mathbf{K}_{(1)}^{s}$$

Unless $\sigma = \infty$, in which case all fields inside the sphere are zero, $\mathbf{K}_{(1)}^{s} = 0$; and in the *EQS* limit, $\mathbf{J}_{(1)}$ is uniform inside the sphere. Assuming this to be true, the scalar and vector potentials, which satisfy the nth-order gauge condition

$$\nabla \cdot \mathbf{A}_{(n)} + \mu_o \left[\varepsilon_o \frac{\partial \Phi_{(n-1)}}{\partial t} + \sigma \Phi_{(n)} \right] = 0,$$

themselves satisfy (except on $r = R$)

$$\nabla^2 \Phi_{(1)} = -\frac{1}{\varepsilon_o} \rho_{(1)} = 0$$

$$\nabla^2 \mathbf{A}_{(1)} = -\mu_o \varepsilon_o \frac{\partial \mathbf{E}_{(0)}}{\partial t}$$

$$\nabla \cdot \mathbf{A}_{(1)} + \mu_o \left[\varepsilon_o \frac{\partial \Phi_{(0)}}{\partial t} + \sigma \Phi_{(1)} \right] = 0$$

with

$$\mathbf{E}_{(1)} = -\nabla \Phi_{(1)}$$

$$\mu_o \mathbf{H}_{(1)} = \nabla \times \mathbf{A}_{(1)}$$

The solutions are

$$\Phi_{(1)} = \frac{-3\varepsilon_o}{\sigma} \frac{dE_o}{dt} \cos\theta \begin{cases} r & (r \le R) \\ \dfrac{R^3}{r^2} & (r \ge R) \end{cases}$$

$$\mathbf{E}_{(1)} = \frac{3\varepsilon_o}{\sigma} \frac{dE_o}{dt} \begin{cases} \hat{\mathbf{r}} \cos\theta - \hat{\theta} \sin\theta & (r \le R) \\ -\left(\dfrac{R}{r}\right)^3 (\hat{\mathbf{r}} 2\cos\theta + \hat{\theta} \sin\theta) & (r \ge R) \end{cases}$$

$$\sigma_{(1)}^{s} = \frac{-9\varepsilon_o^2}{\sigma} \frac{dE_o}{dt} \cos\theta \qquad (r = R)$$

$$\mathbf{H}_{(1)} = \hat{\phi}\, \varepsilon_o \frac{dE_o}{dt} \sin\theta \begin{cases} \dfrac{3r}{2} & (r \le R) \\ \left(\dfrac{r}{2} + \dfrac{R^3}{r^2}\right) & (r \ge R) \end{cases}$$

Although we do not require $\mathbf{A}_{(1)}$ for calculation of the quasistatic Alternate power and Alternate energy, we include it to complete the first-order analysis.

$$\mathbf{A}_{(1)} = \frac{\mu_o \varepsilon_o}{20} \frac{dE_o}{dt} \begin{cases} \begin{bmatrix} \hat{\mathbf{r}}\left[3r^2(5\cos^2\theta + 1) + 20R^2\right]\cos\theta \\ -\hat{\theta}\left[3r^2(5\cos^2\theta - 3) + 20R^2\right]\sin\theta \end{bmatrix} & (r \le R) \\ \begin{bmatrix} \hat{\mathbf{r}}\left[5r^2(3\cos^2\theta - 1) + 8\dfrac{R^5}{r^3} + 20\dfrac{R^3}{r}\right]\cos\theta \\ -\hat{\theta}\left[5r^2(3\cos^2\theta - 1) - 4\dfrac{R^5}{r^3} + 20\dfrac{R^3}{r}\right]\sin\theta \end{bmatrix} & (r \ge R) \end{cases}$$

Homogeneous terms (that do not alter $\mathbf{H}_{(1)}$) are required to ensure continuity at $r = R$. The particular solution of the second-order electric field is given by

$$\mathbf{E}_{(2)} = -\frac{\partial \mathbf{A}_{(1)}}{\partial t} - \nabla \Phi_{(2)}$$

But, because $\frac{\partial \mathbf{A}_{(1)}}{\partial t}$ is continuous, no additional fields are required and both $\sigma^s_{(2)} = 0$ and $\Phi_{(2)} = 0$. We are now in a position to validate the *EQS* assumption that $\mathbf{J}_{(2)} = \sigma \mathbf{E}_{(2)}$ is negligible compared to $\mathbf{J}_{(1)} = \hat{\mathbf{z}} 3\, \varepsilon_o \frac{dE_o}{dt}$. Of the two conditions,

$$\sigma \mu_o \left| \frac{d^2 E_o}{dt^2} \left[3r^2(5\cos^2\theta + 1) + 20R^2\right] \right| \ll 60 \left| \frac{dE_o}{dt} \right|$$

$$\sigma \mu_o \left| \frac{d^2 E_o}{dt^2} \left[3r^2(5\cos^2\theta - 3) + 20R^2\right] \right| \ll 60 \left| \frac{dE_o}{dt} \right|$$

the first is the most restrictive when $r = R$, $\theta = 0, \pi$ and leads to

$$\sigma \mu_o R^2 \left| \frac{d^2 E_o}{dt^2} \right| \ll \frac{30}{19} \left| \frac{dE_o}{dt} \right|$$

For sinusoidal fields of frequency ω, the condition is $\omega \sigma \mu_o R^2 \ll 2$ or, in terms of the skin depth, $R \ll \delta$ as expected.

Power and energy

The zero- through second-order terms of the Maxwell–Poynting representation are

$$\mathbf{S}^{\text{em}}_{(0)} = \mathbf{E}_{(0)} \times \mathbf{H}_{(0)} = 0$$

$$\nabla \cdot \mathbf{S}^{\text{em}}_{(1)} + \frac{\partial W^{\text{em}}_{(0)}}{\partial t} = 0$$

$$W^{\text{em}}_{(0)} = \frac{1}{2} \varepsilon_o \mathbf{E}_{(0)} \cdot \mathbf{E}_{(0)}$$

$$\mathbf{S}^{\text{em}}_{(1)} = \mathbf{E}_{(0)} \times \mathbf{H}_{(1)}$$

$$\nabla \cdot \mathbf{S}^{\text{em}}_{(2)} + \frac{\partial W^{\text{em}}_{(1)}}{\partial t} = -\sigma \mathbf{E}_{(1)} \cdot \mathbf{E}_{(1)}$$

$$W^{\text{em}}_{(1)} = \varepsilon_o \mathbf{E}_{(0)} \cdot \mathbf{E}_{(1)}$$

$$\mathbf{S}^{\text{em}}_{(2)} = \mathbf{E}_{(1)} \times \mathbf{H}_{(1)} + \mathbf{E}_{(0)} \times \mathbf{H}_{(2)}$$

Although evaluation of $\mathbf{H}_{(2)}$ is required in order to fully evaluate the second-order Poynting Theorem, it may be shown that the equivalent theorem,

$$\nabla \cdot [\mathbf{E}_{(1)} \times \mathbf{H}_{(1)}] = -\mathbf{E}_{(1)} \cdot \left[\sigma \mathbf{E}_{(1)} + \varepsilon_o \frac{\partial \mathbf{E}_{(0)}}{\partial t}\right]$$

allows one to avoid that evaluation.

The Alternate representation is very much simpler because the leading terms are the first-order energy density (localized to the surface charge) and the second-order power flux. The results are

$$\mathbf{S}^{o}_{(0)} = \Phi_{(0)} \mathbf{J}_{(0)} = 0$$

$$W^{o}_{(0)} = \frac{1}{2} \rho_{(0)} \Phi_{(0)} = 0$$

$$\mathbf{S}^{o}_{(1)} = \Phi_{(0)} \mathbf{J}_{(1)} - \frac{\varepsilon_o}{2} \left[\Phi_{(0)} \nabla \frac{\partial \Phi_{(0)}}{\partial t} - \frac{\partial \Phi_{(0)}}{\partial t} \nabla \Phi_{(0)}\right] = 0$$

$$\nabla \cdot \mathbf{S}^{o}_{(2)} + \frac{\partial W^{o}_{(1)}}{\partial t} = -\sigma \mathbf{E}_{(1)} \cdot \mathbf{E}_{(1)}$$

$$W^{o}_{(1)} = \frac{1}{2} \rho_{(0)} \Phi_{(1)} = \frac{1}{2} \sigma^{s}_{(0)} \Phi_{(1)} u_0(r - R)$$

$$= \frac{9\varepsilon_o^2}{2\sigma} R \cos^2 \theta E_o \frac{dE_o}{dt} u_0(r - R)$$

$$\mathbf{S}^{o}_{(2)} = \Phi_{(1)} \mathbf{J}_{(1)} - \frac{\varepsilon_o}{2} \left[\Phi_{(0)} \nabla \frac{\partial \Phi_{(1)}}{\partial t} - \frac{\partial \Phi_{(0)}}{\partial t} \nabla \Phi_{(1)}\right] - \frac{\varepsilon_o}{2} \left[\Phi_{(1)} \nabla \frac{\partial \Phi_{(0)}}{\partial t} - \frac{\partial \Phi_{(1)}}{\partial t} \nabla \Phi_{(0)}\right]$$

$$\mathbf{S}^{o}_{(2)} = \begin{cases} -\hat{\mathbf{z}} \dfrac{9\varepsilon_o^2}{\sigma} \left(\dfrac{dE_o}{dt}\right)^2 z & (r \leq R) \\ -\hat{\mathbf{r}} \dfrac{9\varepsilon_o^2}{2\sigma} \left[\left(\dfrac{dE_o}{dt}\right)^2 - E_o \dfrac{d^2 E_o}{dt^2}\right] \dfrac{R^3}{r^2} \cos^2 \theta & (r \geq R) \end{cases}$$

It is instructive to compare the values of $\mathbf{S}^{o}_{(2)}$ and $\mathbf{E}_{(1)} \times \mathbf{H}_{(1)}$.

$$\mathbf{E}_{(1)} \times \mathbf{H}_{(1)} = \begin{cases} \dfrac{-9\varepsilon_o^2}{2\sigma} \left(\dfrac{dE_o}{dt}\right)^2 r \sin \theta (\hat{\mathbf{r}} \sin \theta + \hat{\theta} \cos \theta) & (r \leq R) \\ \dfrac{-3\varepsilon_o^2}{\sigma} \left(\dfrac{dE_o}{dt}\right)^2 \dfrac{R^3}{r^2} \sin \theta \left[\hat{\mathbf{r}} \left(\tfrac{1}{2} + \dfrac{R^3}{r^3}\right) \sin \theta \right. \\ \left. \quad - \hat{\theta} \left(1 + \dfrac{R^3}{2r^3}\right) \cos \theta \right] & (r \geq R) \end{cases}$$

The Alternate flux is z-directed inside and purely radial and divergence-free outside the sphere; the outer component is proportional to $\cos^2 \theta$ and discontinuous at the surface of the sphere because of the Alternate-energy surface density. Notice that for sinusoidal fields, the Alternate flux is independent of time for $r > R$; both average and reactive components exist for $r < R$ because there is first-order current density inside the sphere.

In the Maxwell–Poynting representation, there is no surface-energy density and the radial component of the Poynting vector is continuous and proportional to $\sin^2\theta$. After integrating the flux over the surface, the total Poynting power entering the sphere is

$$P^{\text{em}} = \frac{9\varepsilon_0^2}{\sigma}\left(\frac{dE_o}{dt}\right)^2 \frac{4}{3}\pi R^3$$

The corresponding total Alternate power entering is

$$P^{\text{o}} = \frac{9\varepsilon_0^2}{2\sigma}\left[\left(\frac{dE_o}{dt}\right)^2 - E_o\frac{d^2E_o}{dt^2}\right]\frac{4}{3}\pi R^3$$

and the total Alternate energy on the surface of the sphere is

$$U^{\text{o}}_{\text{energy}} = \frac{9\varepsilon_0^2}{2\sigma}E^{\text{o}}\frac{dE_o}{dt}\frac{4}{3}\pi R^3$$

There is agreement because

$$P^{\text{o}} + \frac{d}{dt}U^{\text{o}}_{\text{energy}} = P^{\text{em}} = \sigma\mathbf{E}_{(1)}\cdot\mathbf{E}_{(1)}$$

Notice that for sinusoidal fields or pulses of duration T, which start at $t = 0$ and where $E_o(0) = E_o(T) = 0$, the time-averaged values of P^{em} and P^{o} agree because no net power flows from the surface-energy. Other time variations reveal some surprises. For example, if $E_o(t) = Ct^n$ (for n a positive integer), the ratio of the two representations of power that enter the sphere from free-space is

$$\frac{P^{\text{o}}}{P^{\text{em}}} = \frac{1}{2n}$$

while

$$\frac{\frac{d}{dt}U^{\text{o}}_{\text{energy}}}{P^{\text{em}}} = 1 - \frac{1}{2n}$$

For large n, we have the free-space Alternate flux $\to 0$ and the dissipated power is almost entirely due to the time rate of change of the surface energy! The power theorems give the correct results, but how is the energy supplied to the surface? This apparent paradox arises again in Chapter 14, where it will be fully explained.

Quasistatic magnetic-field probe

Zero-order fields

In this *MQS* case, for $r \gg R$, the outer fields approach

$$\mathbf{H}_{(0)} = \hat{\mathbf{z}}H_o(t)$$

$$\mathbf{A}_{(0)} = \hat{\phi}\frac{1}{2}\mu_o r H_o(t)\sin\theta$$

With $\rho_{(0)} = 0$ everywhere there is no source for either $\mathbf{E}_{(0)}$ or $\Phi_{(0)}$; therefore,

$$\mathbf{E}_{(0)} = 0$$

$$\Phi_{(0)} = 0$$

complete the zero-order fields.

First-order fields

$$\mathbf{E}_{(1)} = -\frac{\partial \mathbf{A}_{(0)}}{\partial t} - \nabla\Phi_{(1)}$$

$$\mathbf{H}_{(1)} = \frac{\sigma\mu_o}{30}\frac{dH_o}{dt}\begin{cases} \widehat{\mathbf{r}}(3r^2 - 5R^2)\cos\theta + \widehat{\theta}(5R^2 - 6r^2)\sin\theta & (r < R) \\ \dfrac{-R^5}{r^3}(2\widehat{\mathbf{r}}\cos\theta + \widehat{\theta}\sin\theta) & (r > R) \end{cases}$$

$$\mathbf{A}_{(1)} = \widehat{\phi}\frac{-\sigma\mu_o^2}{30}\frac{dH_o}{dt}\sin\theta\begin{cases} \dfrac{1}{2}r(5R^2 - 3r^2) & (r < R) \\ \dfrac{R^5}{r^2} & (r > R) \end{cases}$$

$$\Phi_{(1)} = 0$$

Second-order fields

$$\mathbf{E}_{(2)} = \frac{-\partial \mathbf{A}_{(1)}}{\partial t} - \nabla\Phi_{(2)}$$

$$\Phi_{(2)} = 0$$

Power and energy

$$\nabla\cdot\mathbf{S}^{\text{em}}_{(1)} + \frac{\partial W^{\text{em}}_{(0)}}{\partial t} = 0$$

$$W^{\text{em}}_{(0)} = \frac{1}{2}\mu_o\mathbf{H}_{(0)}\cdot\mathbf{H}_{(0)}$$

$$\mathbf{S}^{\text{em}}_{(1)} = \mathbf{E}_{(1)}\times\mathbf{H}_{(0)}$$

$$\nabla\cdot\mathbf{S}^{\text{em}}_{(2)} + \frac{\partial W^{\text{em}}_{(1)}}{\partial t} = -\sigma\mathbf{E}_{(1)}\cdot\mathbf{E}_{(1)}$$

$$W^{\text{em}}_{(1)} = \mu_o\mathbf{H}_{(0)}\cdot\mathbf{H}_{(1)}$$

$$\mathbf{S}^{\text{em}}_{(2)} = \mathbf{E}_{(1)}\times\mathbf{H}_{(1)} + \mathbf{E}_{(2)}\times\mathbf{H}_{(0)}$$

$$\mathbf{S}^{\text{em}}_{(2)} = \frac{\sigma\mu_o^2}{60}r\sin\theta\begin{cases} \widehat{\mathbf{r}}\left[H_o\dfrac{d^2H_o}{dt^2}(5R^2 - 3r^2) + \left(\dfrac{dH_o}{dt}\right)^2(5R^2 - 6r^2)\right]\sin\theta \\ \qquad (r \le R) \\ +\,\widehat{\theta}\left[H_o\dfrac{d^2H_o}{dt^2} + \left(\dfrac{dH_o}{dt}\right)^2\right](5R^2 - 3r^2)\cos\theta \end{cases}$$

$$\mathbf{S}^{\text{em}}_{(2)} = \frac{\sigma\mu_0^2}{60}\sin\theta\frac{R^5}{r^2}\left\{\begin{array}{l}\hat{\mathbf{r}}\left[2H_o\frac{d^2H_o}{dt^2} - \left(\frac{dH_o}{dt}\right)^2\right]\sin\theta \\ \qquad\qquad\qquad (r \geq R) \\ +\hat{\theta}\left[H_o\frac{d^2H_o}{dt^2} + \left(\frac{dH_o}{dt}\right)^2\right]2\cos\theta\end{array}\right.$$

$$\nabla\cdot\mathbf{S}^{\text{o}}_{(1)} + \frac{\partial W^{\text{o}}_{(0)}}{\partial t} = 0$$

$$W^{\text{o}}_{(0)} = 0$$

$$\mathbf{S}^{\text{o}}_{(1)} = \frac{1}{2\mu_0}\left[A_{\phi(0)}\frac{\partial}{\partial t}(\nabla A_{\phi(0)}) - \frac{\partial}{\partial t}(A_{\phi(0)})\nabla A_{\phi(0)}\right] = 0$$

$$\nabla\cdot\mathbf{S}^{\text{o}}_{(2)} + \frac{\partial W^{\text{o}}_{(1)}}{\partial t} = -\sigma\mathbf{E}_{(1)}\cdot\mathbf{E}_{(1)}$$

$$W^{\text{o}}_{(1)} = \frac{1}{2}\mathbf{A}_{(0)}\cdot\mathbf{J}_{(1)} = \left\{\begin{array}{ll}\frac{1}{2}\sigma A_{\phi(0)}E_{\phi(1)} & (r \leq R) \\ 0 & (r \geq R)\end{array}\right.$$

$$\mathbf{S}^{\text{o}}_{(2)} = \frac{1}{2\mu_0}\left[A_{\phi(0)}\frac{\partial}{\partial t}\left(\nabla A_{\phi(1)}) - \frac{\partial}{\partial t}(A_{\phi(0)}\right)\nabla A_{\phi(1)} + A_{\phi(1)}\frac{\partial}{\partial t}(\nabla A_{\phi(0)}) - \frac{\partial}{\partial t}(A_{\phi(1)})\nabla A_{\phi(0)}\right]$$

$$\mathbf{S}^{\text{o}}_{(2)} = \hat{\mathbf{r}}\frac{\sigma\mu_0^2}{40}\left[H_o\frac{d^2H_o}{dt^2} - \left(\frac{dH_o}{dt}\right)^2\right]\sin^2\theta\left\{\begin{array}{ll}r^3 & (r \leq R) \\ \frac{R^5}{r^2} & (r \geq R)\end{array}\right.$$

In this *MQS* case, because there is no Alternate surface energy, both $\mathbf{S}^{\text{o}}$ and $\mathbf{S}^{\text{em}}$ are continuous at the surface of the sphere. In general, the instantaneous values differ; the time averages do not, provided that $<H_o(t)> = 0$.

11.5 QUASISTATIC ANALYSIS OF A PHYSICAL RESISTOR

Introduction

Quasistatic analysis of a physical resistor allows one to gain insight into its operation and frequency limitations and provides useful design information. We assume that the device is intended to function as a resistor that maintains a specified value of ohms over a wide range of frequencies and be largely unaffected by parasitic inductance and/or capacitance effects. A wire wound resistor is rejected because of its unacceptable inductance. In order to keep the resistance constant, we require the current distribution to be sensibly independent of frequency. This implies that the skin depth should be large compared to the appropriate physical dimension that determines the DC current flow. Rather than try to minimize the electric and magnetic energy storage, we attempt to equalize them. The geometry that we choose to analyze is shown in axial section in Figure 11.2. It consists of a solid cylinder of material characterized by uniform conductivity, σ, permittivity, ε_o, and permeability, μ_o; the length is ℓ, the radius is R. A perfectly conducting shield surrounds the resistor. It is a can-shaped coaxial cylinder of the same length, with radius $R_o > R$ and a flat circular "bottom;" the latter is bonded to one end of the resistor; a perfectly conducting circular electrode of radius R is bonded to the other end. The bonds are assumed to have no contact resistance; DC current can flow between the circular electrode at one end of the resistive cylinder and the coaxial shield.

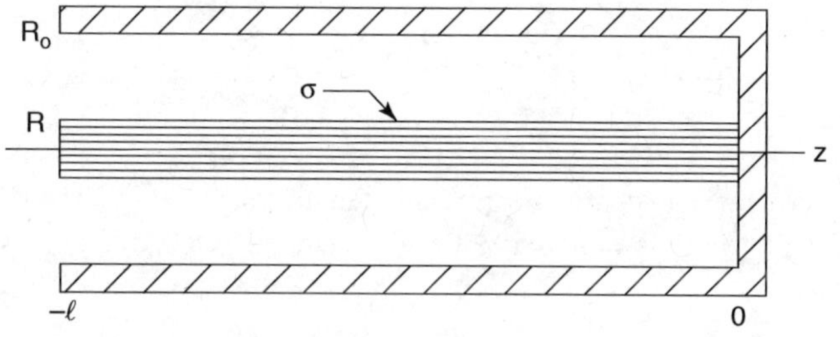

Figure 11.2 Resistor with shield.

Fields and potentials

Zero-order terms

As with any conduction problem, the analysis begins by finding the current density – subject to the boundary conditions. The zero-order current density in the cylindrical region is divergence-free, therefore $J_{r(0)}(r = R, z) = 0$. The end electrodes are perfectly conducting, therefore $E_{r(0)}(r, z = 0) = E_{r(0)}(r, z = -\ell) = 0$. Because $\mathbf{J}_{(0)} = \sigma \mathbf{E}_{(0)}$ and the conductivity is uniform, the zero-order current density is also curl-free and so it is a Laplacian field. By inspection,

$$\mathbf{J}_{(0)} = \hat{\mathbf{z}} \begin{cases} \sigma E_o(t) & (0 \le r \le R) \\ 0 & (R \le r \le R_o) \end{cases}$$

satisfies all these conditions; the uniqueness theorem proves that it is the only solution. (The actual level of current is imposed by a source connected between the electrodes.) At this point, we may either solve for the remainder of $\mathbf{E}_{(0)}$ or first find $\mathbf{H}_{(0)}$. We choose the former and find that the zero-order scalar potential,

$$\Phi_{(0)} = \begin{cases} -zE_o(t) & (0 \le r \le R) \\ -z\dfrac{\ln\left(\dfrac{R_o}{r}\right)}{\ln\left(\dfrac{R_o}{R}\right)} E_o(t) & (R \le r \le R_o) \end{cases} \tag{11.6}$$

is Laplacian, with the shield and end electrode an equipotential (for convenience set equal to zero). As required, it is continuous at $r = R$. Nevertheless, until the source located at $z = -\ell$ has been specified, the potential between resistor and shield is not unique (although any additions cannot alter the inner field). One can assume that either (a) distributed electric-field sources are connected that are consistent with $\Phi_{(0)}(r, z = -\ell)$ and so impose uniqueness or (b) distributed magnetic-field sources (that maintain $\nabla \cdot \mathbf{J}_{(0)}$) are present that do not constrain the zero-order potential. These alternatives assume that either a voltage source or current source is applied at $z = -\ell$. To simplify the analysis, we assume the former and that for $R \leq r \leq R_o$, $E_r(r, z = -\ell) \sim 1/r$; then Eq. (11.6) is unique and

$$\mathbf{E}_{(0)} = \begin{cases} \widehat{\mathbf{z}} E_o(t) & (0 \leq r \leq R) \\ \dfrac{E_o(t)}{\ln(\frac{R_o}{R})} [\, \widehat{\mathbf{z}} \ln(R_o/r) \ - \widehat{\mathbf{r}} \frac{z}{r}] & (R \leq r \leq R_o) \end{cases} \tag{11.7}$$

Zero-order electric surface charges terminate the normal components of $\mathbf{E}_{(0)}$ on the perfect conductors and at the boundary of the resistor; the density, $\sigma^s_{(0)}$, is given by

$$\sigma^s_{(0)} = \begin{cases} -\varepsilon_o E_o(t) \dfrac{z}{R \ln\left(\dfrac{R_o}{R}\right)} & (r = R) \\ \varepsilon_o E_o(t) \dfrac{z}{R_o \ln\left(\dfrac{R_o}{R}\right)} & (r = R_o) \\ -\varepsilon_o E_o(t) & (z = 0,\ 0 \leq r \leq R) \\ -\varepsilon_o E_o(t) \dfrac{\ln\left(\dfrac{R_o}{r}\right)}{\ln\left(\dfrac{R_o}{R}\right)} & (z = 0,\ R \leq r \leq R_o) \\ \varepsilon_o E_o(t) & (z = -\ell,\ 0 \leq r \leq R) \end{cases} \tag{11.8}$$

The zero-order magnetic-field is divergence-free and satisfies $\nabla \times \mathbf{H}_{(0)} = \mathbf{J}_{(0)}$. From symmetry considerations, there is no ϕ-dependence; and because over the range $-\ell < z < 0$, $\mathbf{J}_{(0)}$ is independent of z, we have $H_{r(0)} = H_{z(0)} = 0$,

$$\frac{1}{r} \frac{\partial}{\partial r} (r H_{\phi(0)}) = J_{z(0)}$$

and

$$\mathbf{H}_{(0)} = \widehat{\phi} \frac{\sigma E_o(t)}{2} \begin{cases} r & (0 \leq r \leq R) \\ \dfrac{R^2}{r} & (R \leq \ r \leq R_o) \end{cases} \tag{11.9}$$

At the surfaces of the perfect conductors (and the field source), discontinuities in the tangential components are created by the electric surface-current densities:

$$\mathbf{K}^s_{(0)} = \frac{\sigma E_o(t)}{2} \begin{cases} \hat{\mathbf{r}} r & (z = 0,\ 0 \le r \le R) \\ \hat{\mathbf{r}} \dfrac{R^2}{r} & (z = 0,\ R \le r \le R_o) \\ -\hat{\mathbf{z}} \dfrac{R^2}{R_o} & (r = R_o) \\ -\hat{\mathbf{r}} \dfrac{R^2}{r} & (z = -\ell,\ R \le r \le R_o) \\ -\hat{\mathbf{r}} r & (z = -\ell,\ 0 \le r \le R) \end{cases} \tag{11.10}$$

This completes the analysis of the zero-order terms except for the magnetic vector potential that we will need for the Alternate representation. A valid particular solution, which ignores the contributions from the radial components of $\mathbf{K}^s_{(0)}$, is

$$\mathbf{A}_{(0)} = \hat{\mathbf{z}} \frac{-\mu_0 \sigma E_o(t)}{4} \begin{cases} r^2 - R^2 \left[1 + 2 \ln \left(\dfrac{R_o}{R} \right) \right] & (0 \le r \le R) \\ 2R^2 \ \ln \left(\dfrac{r}{R_o} \right) & (R \le r \le R_o) \end{cases}$$

Because the homogeneous component can be expressed as $\nabla \Psi_{(0)}$ (except on the surface currents) and $\nabla \times \nabla \Psi_{(0)} = 0$, it follows that the zero-order magnetic field found from $\mu_0 \mathbf{H}_{(0)} = \nabla \times \mathbf{A}_{(0)}$ must, and does, agree with Eq. (11.9).

First-order terms

We turn now to the first-order terms and, by adding the homogeneous term to $\mathbf{A}_{(0)}$, find that

$$\mathbf{E}_{(1)} = -\frac{\partial \mathbf{A}_{(0)}}{\partial t} - \nabla \left[\frac{\partial \Psi_{(0)}}{\partial t} + \Phi^{\text{total}}_{(1)} \right]$$

The homogeneous terms are generated, respectively, by $\frac{\partial \mathbf{K}^s_{(0)}}{\partial t}$ and $\sigma^s_{(1)}$; they can be combined as

$$\Phi_{(1)} = \frac{\partial \Psi_{(0)}}{\partial t} + \Phi^{\text{total}}_{(1)} \tag{11.11}$$

so that

$$\mathbf{E}_{(1)} = -\nabla \Phi_{(1)} + \hat{\mathbf{z}} \frac{\mu_0 \sigma}{4} \frac{dE_o(t)}{dt} \begin{cases} r^2 - R^2 + 2R^2 \ln \left(\dfrac{R}{R_o} \right) & (0 \le r \le R) \\ 2R^2 \ln \left(\dfrac{r}{R_o} \right) & (R \le r \le R_o) \end{cases} \tag{11.12}$$

where in the conductor $\mathbf{J}_{(1)} = \sigma \mathbf{E}_{(1)}$ and at $r = R_-$ the radial component $J_{r(1)}$ must match $\frac{\partial \sigma^s_{(0)}}{\partial t}$. Any discontinuity in the normal component of $\varepsilon_0 \mathbf{E}_{(1)}$ at $r = R$ or on the perfectly conducting walls produces the first-order surface-charge density, $\sigma^s_{(1)}$.

The first-order potential is a solution of Laplace's Equation in cylindrical coordinates. Reference to Appendix C, Section C.4, and the form of $\sigma^s_{(0)}$ reveals that

$$\Phi_{(1)} = \frac{\varepsilon_0 \dfrac{dE_o(t)}{dt}}{3\sigma R^2 \ln^2\left(\dfrac{R_o}{R}\right)} \begin{cases} \ln\left(\dfrac{R_o}{R}\right)(z^3 - \frac{3}{2}zr^2) + \frac{3}{2}z(R_o^2 - R^2) & (0 \le r \le R) \\ \ln\left(\dfrac{R_o}{r}\right)(z^3 - \frac{3}{2}zr^2) + \frac{3}{2}z(R_o^2 - r^2) & (R \le r \le R_o) \end{cases} \tag{11.13}$$

will match $\frac{\partial \sigma^s_{(0)}}{\partial t}$, but is not exactly an equipotential at $z = -\ell$, $0 \le r \le R$—as required. Rather than add the additional terms required to make it so, we observe that if $R^2 \ll [\frac{2}{3}\ell^2 \ln(R_o/R) + R_o^2]/[\ln(R_o/R) + 1]$, the r^2 variation of $\Phi_{(1)}$ over the electrode is negligible. Accordingly, we assume that this inequality is satisfied and accept Eq. (11.13) as valid.

The final step is to solve for the first-order magnetic field and the associated first-order surface current densities. There is only $H_{\phi(1)}$, which must satisfy

$$\frac{1}{r}\frac{\partial(rH_{\phi(1)})}{\partial r} = \begin{cases} \sigma E_{z(1)} + \varepsilon_0 \dfrac{\partial E_{z(0)}}{\partial t} & (0 \le r \le R) \\ \varepsilon_0 \dfrac{\partial E_{z(0)}}{\partial t} & (R \le r \le R_o) \end{cases}$$

Separately integrating $\frac{\partial(rH_{\phi(1)})}{\partial r}$ in each region and matching the values of $H_{\phi(1)}$ at $r = R$ leads to

$$\mathbf{H}_{(1)} = \hat{\phi}\,\frac{\varepsilon_0 \dfrac{-dE_o(t)}{dt}}{2\ln(R_o/R)} \begin{cases} \left[\dfrac{z^2}{R^2} - \dfrac{1}{4}\dfrac{r^2}{R^2} + \dfrac{1}{2}\left(\dfrac{R_o^2}{R^2} - 1\right)\Big/\ln\left(\dfrac{R_o}{R}\right) - \ln\left(\dfrac{R_o}{R}\right)\right] r \\ \quad + \dfrac{1}{2}\dfrac{\mu_0}{\varepsilon_0}\sigma^2 \ln\left(\dfrac{R_o}{R}\right)\left[\left(\ln\left(\dfrac{R_o}{R}\right) + \dfrac{1}{2}\right)R^2 - \dfrac{1}{4}r^2\right] r & (0 \le r \le R) \\ r\left[\ln\left(\dfrac{r}{R_o}\right) - \dfrac{1}{2}\right] + (z^2 + \frac{1}{4}R^2)/r + \frac{1}{2}(R_o^2 - R^2)\Big/\left[r \ln\left(\dfrac{R_o}{R}\right)\right] \\ \quad + \dfrac{1}{2}\dfrac{\mu_0}{\varepsilon_0}\sigma^2 \ln\left(\dfrac{R_o}{R}\right)\left[\ln\left(\dfrac{R_o}{R}\right) + \dfrac{1}{4}\right]\dfrac{R^4}{r} & (R \le r \le R_o) \end{cases} \tag{11.14}$$

The discontinuities in the tangential values of $\mathbf{H}_{(1)}$ at the conducting walls allow calculation of $\mathbf{K}^s_{(1)}$.

We have omitted finding $\mathbf{A}_{(1)}$ for reasons that will become clear later; otherwise the quasistatic analysis of the fields is complete. Although the zero-order terms are quite simple, that is not as true of the first-order fields. Nevertheless, what would be a much more difficult problem has been reduced to a superposition of functions that satisfy either Poisson's or Laplace's Equation.

Maxwell--Poynting representation

The Poynting vector and electromagnetic energy density, expanded to first-order terms, are

$$\mathbf{S}^{\text{em}} = \mathbf{E}_{(0)} \times \mathbf{H}_{(0)} + [\mathbf{E}_{(1)} \times \mathbf{H}_{(0)} + \mathbf{E}_{(0)} \times \mathbf{H}_{(1)}]$$

$$W^{\text{em}} = \frac{1}{2}[\varepsilon_o \mathbf{E}_{(0)} \cdot \mathbf{E}_{(0)} + \mu_o \mathbf{H}_{(0)} \cdot \mathbf{H}_{(0)}] + \varepsilon_o \mathbf{E}_{(0)} \cdot \mathbf{E}_{(1)} + \mu_o \mathbf{H}_{(0)} \cdot \mathbf{H}_{(1)}$$

but notice that neither the zero-order nor first-order Poynting theorem,

$$\nabla \cdot \mathbf{S}^{\text{em}}_{(0)} = -\mathbf{E}_{(0)} \cdot \mathbf{J}_{(0)}$$

and

$$\nabla \cdot \mathbf{S}^{\text{em}}_{(1)} + \frac{\partial}{\partial t} W^{\text{em}}_{(0)} = -[\mathbf{E}_{(0)} \cdot \mathbf{J}_{(1)} + \mathbf{E}_{(1)} \cdot \mathbf{J}_{(0)}]$$

requires calculation of the first-order energy density. At the field source, the Poynting flux is z-directed; integration over the annular area $\pi(R_o^2 - R^2)$ gives the input power to the device. It is necessary to connect the field amplitudes to voltages and currents in order to make an equivalent circuit representation. At sufficiently low frequencies, the zero-order fields and flux dominate; this circuit is a simple resistance. However, the first-order power flux may be dominated by either $\mathbf{E}_{(0)} \times \mathbf{H}_{(1)}$ or $\mathbf{E}_{(1)} \times \mathbf{H}_{(0)}$, depending upon whether the zero-order energy is principally electric or magnetic in character. In the first case, $\frac{1}{2}\varepsilon_o \mathbf{E}_{(0)} \cdot \mathbf{E}_{(0)}$ and $\mathbf{E}_{(0)} \times \mathbf{H}_{(1)}$ suffice, the fields are *EQS*, and the equivalent circuit adds a capacitance in parallel with the resistor. In the second, $\frac{1}{2}\mu_o \mathbf{H}_{(0)} \cdot \mathbf{H}_{(0)}$ and $\mathbf{E}_{(1)} \times \mathbf{H}_{(0)}$ suffice, the fields are *MQS*, and the equivalent circuit adds an inductance in series with the resistor. The connections cannot be otherwise, because there must be a path for the DC current to flow through.

In the general case, it is often advantageous to use the integral form of the complex Poynting Theorem,

$$-\oint \frac{1}{2}\underline{\mathbf{E}} \times \underline{\mathbf{H}}^* \cdot \mathbf{n}\, da = \int \frac{1}{2}\sigma \underline{\mathbf{E}} \cdot \underline{\mathbf{E}}^* dV + j2\omega \int \left(\frac{1}{4}\mu_o \underline{\mathbf{H}} \cdot \underline{\mathbf{H}}^* - \frac{1}{4}\varepsilon_o \underline{\mathbf{E}} \cdot \underline{\mathbf{E}}^*\right) dV$$

and approximate the complex power into the device terminals as

$$\frac{1}{2}\underline{V}\underline{I}^* = \frac{1}{2}Z(j\omega)\underline{I}\underline{I}^* = \frac{1}{2}Y^*(j\omega)\underline{V}\underline{V}^*.$$

It then follows that either

$$Z(j\omega) = \frac{2}{|\underline{I}|^2}\left[\int \frac{1}{2}\sigma \underline{\mathbf{E}} \cdot \underline{\mathbf{E}}^* dV + j2\omega \int \left(\frac{1}{4}\mu_o \underline{\mathbf{H}} \cdot \underline{\mathbf{H}}^* - \frac{1}{4}\varepsilon_o \underline{\mathbf{E}} \cdot \underline{\mathbf{E}}^*\right) dV\right]$$

or

$$Y(j\omega) = \frac{2}{|\underline{V}|^2}\left[\int \frac{1}{2}\sigma \underline{\mathbf{E}} \cdot \underline{\mathbf{E}}^* dV - j2\omega \int \left(\frac{1}{4}\mu_o \underline{\mathbf{H}} \cdot \underline{\mathbf{H}}^* - \frac{1}{4}\varepsilon_o \underline{\mathbf{E}} \cdot \underline{\mathbf{E}}^*\right) dV\right]$$

may be used to relate the average dissipated power and difference between average stored magnetic and electric energies to the circuit impedance or admittance. We learn that if the difference in energies is zero, the circuit is purely resistive. This can happen at a specific frequency with a resonant LC circuit, but it remains to be seen if such a balance can be maintained over a wide frequency range. Because the electric and magnetic fields are distributed throughout the structure of the physical resistor, the integrations are necessary in order that connection be made with lumped circuit parameters. Although this is a straightforward step, we turn instead to the Alternate (circuit) representation of power and energy.

Alternate (circuit) representation

The results of Part I, Chapter 3, Section 3.4 have shown that, within the quasistatic approximation, only the localized terms of the Alternate power flux and energy densities survive and

$$\mathbf{S}^{o} = \Phi_{(0)}\mathbf{J}_{(0)} + [\Phi_{(0)}\mathbf{J}_{(1)} + \Phi_{(1)}^{\text{total}}\mathbf{J}_{(0)}]$$

$$W^{o} = \frac{1}{2}[\Phi_{(0)}\rho_{(0)} + \mathbf{A}_{(0)}^{\text{total}} \cdot \mathbf{J}_{(0)}]$$
$$+ \frac{1}{2}[\Phi_{(0)}\rho_{(1)} + \Phi_{(1)}^{\text{total}}\rho_{(0)} + \mathbf{A}_{(0)}^{\text{total}} \cdot \mathbf{J}_{(1)} + \mathbf{A}_{(1)}^{\text{total}} \cdot \mathbf{J}_{(0)}]$$

As with the Maxwell–Poynting representation, the zero-order and first-order power fluxes obey separate theorems:

$$\nabla \cdot \mathbf{S}_{(0)}^{o} = -\mathbf{E}_{(0)} \cdot \mathbf{J}_{(0)}$$

and

$$\nabla \cdot \mathbf{S}_{(1)}^{o} + \frac{\partial}{\partial t}W_{(0)}^{o} = -[\mathbf{E}_{(0)} \cdot \mathbf{J}_{(1)} + \mathbf{E}_{(1)} \cdot \mathbf{J}_{(0)}]$$

Because neither of these involve the first-order energies, there is no need to calculate $\mathbf{A}_{(1)}^{\text{total}}$ or $\rho_{(1)}$. However, before proceeding further, there is a point that requires clarification. The zero-order vector potential, $\mathbf{A}_{(0)}^{\text{total}} = \mathbf{A}_{(0)} + \nabla\Psi_{(0)}$, includes the homogeneous contribution (due to the wall currents $\mathbf{K}_{(0)}^{s}$) and $\Phi_{(1)}^{\text{total}}$ is given by Eq. (11.11). With those substitutions and because $\nabla \cdot \mathbf{J}_{(0)} = 0$, the first-order theorem becomes

$$\nabla \cdot \left(\Phi_{(0)}\mathbf{J}_{(1)} + \left[\Phi_{(1)} - \frac{\partial \Psi_{(0)}}{\partial t}\right]\mathbf{J}_{(0)}\right)$$
$$+ \frac{\partial}{\partial t}\frac{1}{2}[\Phi_{(0)}\rho_{(0)} + \mathbf{A}_{(0)} \cdot \mathbf{J}_{(0)} + \nabla \cdot (\Psi_{(0)}\mathbf{J}_{(0)})] = -[\mathbf{E}_{(0)} \cdot \mathbf{J}_{(1)} + \mathbf{E}_{(1)} \cdot \mathbf{J}_{(0)}]$$

The order of the divergence operation and time derivative can be interchanged; this allows the energy term to become part of the power flux. An equivalent first-order theorem is

$$\nabla \cdot \left(\Phi_{(0)}\mathbf{J}_{(1)} + \Phi_{(1)}\mathbf{J}_{(0)} + \frac{1}{2}\left[\Psi_{(0)}\frac{\partial \mathbf{J}_{(0)}}{\partial t} - \frac{\partial \Psi_{(0)}}{\partial t}\mathbf{J}_{(0)}\right]\right)$$
$$+ \frac{\partial}{\partial t}\frac{1}{2}[\Phi_{(0)}\rho_{(0)} + \mathbf{A}_{(0)} \cdot \mathbf{J}_{(0)}] = -[\mathbf{E}_{(0)} \cdot \mathbf{J}_{(1)} + \mathbf{E}_{(1)} \cdot \mathbf{J}_{(0)}]$$

But, $\Psi_{(0)} = E_{o}(t)f(r,z)$ and $\mathbf{J}_{(0)} = \hat{\mathbf{z}}E_{o}(t)g(r,z)$; therefore, the resulting zero-order energy density and first-order power flux are the *same* as if $\Psi_{(0)}$ were set equal to zero. Since, in general, the calculation of $\Psi_{(0)}$ is formidable, this result is of great importance.

Using the Alternate representation, the input power to the physical resistor is very easy to calculate because

$$\text{Input power} = \int_0^R \mathbf{S}^{\rm o}(r, -\ell, t) \cdot \hat{\mathbf{z}} 2\pi r\, dr$$
$$= \left(\Phi_{(0)} [J_{z(0)} \pi R^2 + \int_0^R J_{z(1)} 2\pi r\, dr] + J_{z(0)} \int_0^R \Phi_{(1)} 2\pi r\, dr \right)_{z=-\ell}$$

requires integration only over the radius of the conducting cylinder. Because $R^2 \ll [\frac{2}{3}\ell^2 \ln(R_{\rm o}/R) + R_{\rm o}^2]/[\ln(R_{\rm o}/R) + 1]$, both $J_{z(1)}$ and $\Phi_{(1)}$ have negligible radial variation at $z = -\ell$ and

$$\Phi_{(0)} J_{z(0)} \pi R^2 = \frac{\sigma \pi R^2}{\ell} [E_{\rm o}(t)\ell]^2$$

$$\Phi_{(0)} \int_0^R J_{z(1)} 2\pi r dr = \frac{\pi \varepsilon_{\rm o} \ell}{\ln\left(\dfrac{R_{\rm o}}{R}\right)} [E_{\rm o}(t)\ell] \frac{d[E_{\rm o}(t)\ell]}{dt}$$
$$+ \frac{\mu_{\rm o}}{2\pi} \left[\ln\left(\frac{R_{\rm o}}{R}\right)\right] \ell\ [\sigma E_{\rm o}(t)\pi R^2] \frac{d[\sigma E_{\rm o}(t)\pi R^2]}{dt}$$

$$J_{z(0)} \int_0^R \Phi_{(1)} 2\pi r dr = \frac{-\pi \varepsilon_{\rm o} \ell}{3 \ln\left(\dfrac{R_{\rm o}}{R}\right)} [E_{\rm o}(t)\ell] \frac{d[E_{\rm o}(t)\ell]}{dt}$$

which can be written as

$$\text{Input power} = V_{(0)} I_{(0)} + V_{(0)} I_{(1)} + V_{(1)} I_{(0)}$$
$$= G V_{(0)}^2 + C V_{(0)} \frac{dV_{(0)}}{dt} + L\, I_{(0)} \frac{dI_{(0)}}{dt}$$

with

$$V_{(0)} = E_{\rm o}(t)\ell$$

$$I_{(0)} = \sigma E_{\rm o}(t) \pi R^2$$

$$V_{(1)} = L \frac{dI_{(0)}}{dt}$$

$$I_{(1)} = C \frac{dV_{(0)}}{dt}$$

and

$$G = \frac{\sigma \pi R^2}{\ell} \tag{11.15}$$

$$L = \frac{\mu_{\rm o}}{2\pi} \left[\ln\left(\frac{R_{\rm o}}{R}\right)\right] \ell \tag{11.16}$$

$$C = \frac{2\pi \varepsilon_{\rm o} \ell}{3 \ln\left(\dfrac{R_{\rm o}}{R}\right)} \tag{11.17}$$

The circuit-element values L and C also follow directly from evaluation of the total zero-order energies:

$$\frac{1}{2}LI_{(0)}^2 = \int \frac{1}{2}\mathbf{A}_{(0)} \cdot \mathbf{J}_{(0)}\, dV$$

$$= \frac{\mu_o[\sigma E_o(t)]^2}{4}\ell \int_0^R \left(-r^2 + R^2\left[1 + 2\ln\left(\frac{R_o}{R}\right)\right]\right) \pi r\, dr$$

$$\frac{1}{2}CV_{(0)}^2 = \int \frac{1}{2}\rho_{(0)}\Phi_{(0)}\, dV = \int \frac{1}{2}\sigma_{(0)}^s \Phi_{(0)} 2\pi R\, dz$$

$$= \frac{\varepsilon_o E_o^2(t)\pi}{\ln\left(\frac{R_o}{R}\right)} \int_0^\ell z^2\, dz$$

Notice that because $\mathbf{A}_{(0)} \cdot \mathbf{K}_{(0)}^s = 0$ and $\sigma_{(0)}^s \Phi_{(0)} = 0$ (except on the surface of the cylinder), the wall currents and charges to not contribute. The value of inductance, $L = \frac{\mu_o}{2\pi}[\ln(R_o/R) + \frac{1}{4}]\ell$ is slightly more accurate than Eq. (11.16), because it includes the contribution from the inner field that was neglected by the approximations we made in calculating $\Phi_{(1)}$. The value of the capacitance is identical to Eq. (11.17). The value of C is $1/3$ the value that would result for a coaxial capacitor (without the conducting plane at $z = 0$). Because of the short circuit, the potential varies as z and the electric-energy density varies as z^2; the average of the latter is therefore $\ell^2/3$.

The complex form of the Alternate-power theorem can also be used if desired. It, too, is simpler to evaluate than the complex Poynting Theorem because the energies are localized to the charges and currents. The final results are equivalent.

Equivalent circuits

There are an infinite number of networks that can be synthesized to yield the input power (correct to first order). However, canonical networks have only three circuit elements; as shown in Figure 11.3, there are two possibilities: (a) the C in parallel with the series combination of G^{-1} and L or (b) the L in series with the parallel combination of G and C.

The input impedances for these cases are

$$Z(j\omega) = \frac{1 + j\omega LG}{G(1 - \omega^2 LC) + j\omega C} \qquad \text{case (a)}$$

$$Z(j\omega) = \frac{(1 - \omega^2 LC) + j\omega LG}{G + j\omega C} \qquad \text{case (b)}$$

They are equivalent provided $\omega^2 LC = \frac{1}{3}\omega^2 \mu_o \varepsilon_o \ell^2 = \frac{1}{3}(\frac{2\pi}{\lambda_o}\ell)^2 \ll 1$. The validity of quasistatic analysis depends upon the second-order circuit power being negligible, which requires this inequality to be met. In network (a), the first-order current flows only through C, but both the zero-order and first-order voltage are across it. In network (b), the first-order voltage is across only L, but both the zero-order and first-order current flow through it. In each case, the product $V_{(1)}I_{(1)}$ is negligible second-order power.

In the event that $LG \ll C/G$, the inductance is negligible, the quasistatic fields are *EQS*, and the two-element network is the resistance in parallel with C. In the opposite

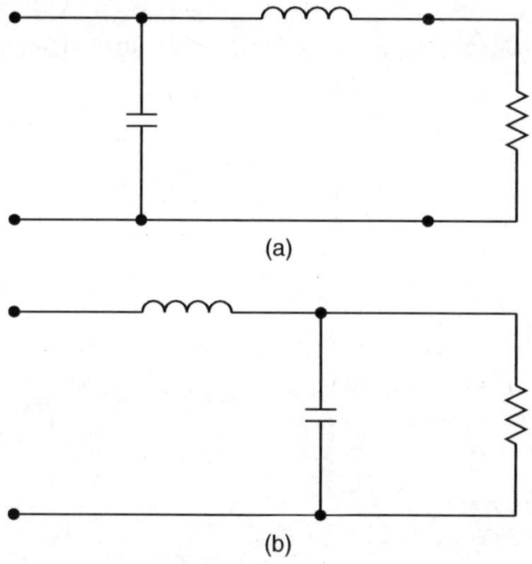

Figure 11.3 Equivalent circuits of a physical resistor.

limit, $LG \gg C/G$, the quasistatic fields are *MQS*, the capacitance is negligible, and the two-element network is the resistance in series with L.

Notice, however, that if the two time constants of the circuit, LG and C/G, are equal, the frequency dependence of the numerator cancels that of the denominator and $Y(j\omega) = 1/Z(j\omega) = G$. This condition is

$$G = \sqrt{\frac{C}{L}} = \sqrt{\frac{\varepsilon_0}{3\mu_0}} \frac{2\pi}{\ln\left(\frac{R_o}{R}\right)} = \frac{\sigma\pi R^2}{\ell}$$

and is exactly what one would expect from transmission-line theory. The LC circuit is a one-term approximation to a transmission line that has a characteristic admittance, $\sqrt{C/L}$. When the line is terminated with a conductance of the same value, there are no reflected waves and the input admittance is independent of frequency. The electric and magnetic energy can therefore be balanced over a wide range of frequencies. As noted earlier, the skin depth inequality, $\delta = \sqrt{\frac{2}{\omega\mu_0\sigma}} \gg R$, creates an additional frequency limitation. This can be modified by hollowing out the resistor so that it is actually a cylindrical thin film of thickness, Δ. In this case, the reader should verify that $\delta \gg \sqrt{R\Delta}$ and $G = \frac{\sigma 2\pi R\Delta}{\ell}$.

11.6 MAGNETIC DIFFUSION

An infinite sheet of conducting material with uniform conductivity, σ, and permeability, μ, extends for $0 < x < d$; it is surrounded by free-space. For $t < 0$, and $x > 0$, there are no fields present. At $t = 0$, an approximately uniform *MQS* magnetic field, $\mathbf{H}_{(0)} = \hat{\mathbf{y}}H_0$, is created in the free-space region, $x < 0$. We wish to find the *MQS* magnetic and electric fields that develop in the regions $x > 0$ for $t > 0$ under the assumption that d/c, the electromagnetic transit time, is negligibly small compared to the magnetic-diffusion time, Eq. (5.6).

The conductivity is large enough that the conduction current dominates the Maxwell-Displacement current. Under these conditions, the results of Chapter 5, Section 5.2 apply and

$$\nabla^2 H_y - \mu\sigma \frac{\partial H_y}{\partial t} = 0 \tag{11.18a}$$

$$\sigma E_z = \frac{\partial H_y}{\partial x} \tag{11.18b}$$

The approximate solutions outside the conductor are Laplacian and uniform; within the conductor they are of the form

$$H_y(x,t) = H_o - \sum_{i=1}^{\infty} C_i \exp(-\alpha_i t)\sin(k_i x + \theta_i) \tag{11.19a}$$

$$\sigma E_z(x,t) = -\sum_{i=1}^{\infty} k_i C_i \exp(-\alpha_i t)\cos(k_i x + \theta_i) \tag{11.19b}$$

$$\mu\sigma\alpha_i = k_i^2 \tag{11.19c}$$

At $x = d$, the zero-order electric field is continuous and zero; the boundary condition $\frac{\partial H_y}{\partial x}|_{x=d} = 0$ is automatically satisfied if for every value of i,

$$\theta_i = 0$$

$$k_i = \frac{(2i-1)\pi}{2d}, \qquad i = 1, 2, 3, \cdots$$

The magnetic field must vanish inside the conductor at $t = 0$; therefore

$$\sum_{i=1}^{\infty} C_i \sin(k_i x) = H_o \tag{11.20a}$$

The set of coefficients that is the solution of this Fourier sine series is

$$C_i = \frac{4H_o}{\pi\,(2i-1)}, \qquad i = 1, 2, 3, \ldots \tag{11.20b}$$

For $x \le 0$, the *MQS* electric field is uniform and continuous with a value

$$E_z(t) = -\frac{2H_o}{\sigma d} \sum_{i=1}^{\infty} \exp(-\alpha_i t)$$

Because $\alpha_i = \left[\frac{(2i-1)\pi}{2d}\right]^2 / (\mu\sigma)$, the terms with large values of i decay very rapidly and the diffusion is dominated by $i = 1$. For $\mu = \mu_o$, $\sigma = 10^8\ S\ m^{-1}$, and $d = 1\ cm$, the numerical value of this time constant is $\tau = 1/\alpha_1 = 5.1 \times 10^{-3} s$, which is very long compared to the electromagnetic transit time of $3.3 \times 10^{-11}\ s$.

In Figure 11.4, the normalized magnetic field is plotted for a 30-term series approximation as a function of x/d for the times $t/\tau = 0, 1, 5, 10$. The ripple in the $t = 0$ plot demonstrates Gibb's phenomenon [26].

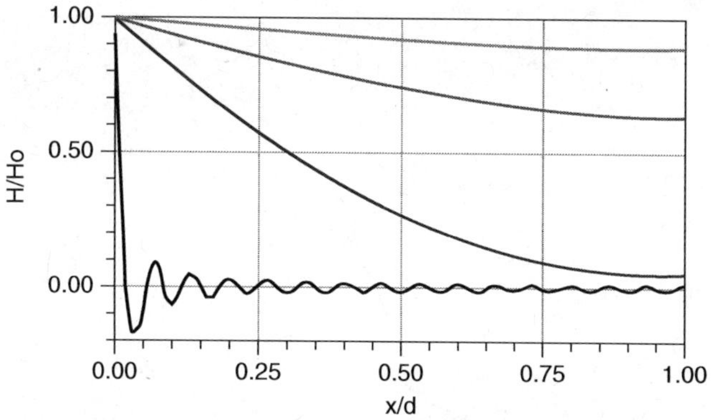

Figure 11.4 Magnetic diffusion through a conducting sheet.

At $t = 0$, the field step at $x = 0$ has very high frequency content and so J_z is initially a surface current confined by zero skin depth. As time progresses, the effective frequency and skin depth, Eq. (5.7), respectively approach zero and infinity; this allows the magnetic field to spread uniformly through the conductor. A crude estimate can be made by setting "ω" $= \frac{\pi}{2}/t$ and "δ" $= \sqrt{\frac{2}{\text{"}\omega\text{"}\mu\sigma}}$; then "$\delta$"$/d = .7\sqrt{t/\tau}$. Spatial uniformity would require the latter ratio to be at least 2 or 3 and so we expect the transient to extend for at least 10τ (50 *ms* in this example); this is borne out by the numerical calculations.

CHAPTER 12

UNIFORMLY MOVING ELECTRIC CHARGES

12.1 POINT CHARGE

Uniform motion in free-space

The Liénard–Wiechert vector and scalar potentials associated with an electric charge, q, moving at constant velocity, $\mathbf{v} = \widehat{\mathbf{z}}v$, are (from Appendix E, Section E.2)

$$\Phi = \frac{q}{4\pi\varepsilon_o\sqrt{(1-\frac{v^2}{c^2})(x^2+y^2)+(z-vt)^2}} \tag{12.1a}$$

$$\mathbf{A} = \widehat{z}\frac{v\Phi}{c^2} \tag{12.1b}$$

The electric and magnetic fields generated by the charge satisfy Eqs. (1.9) and (1.10); they are given by

$$\mathbf{E} = \frac{q}{4\pi\varepsilon_o}\left(1-\frac{v^2}{c^2}\right)\frac{\widehat{\mathbf{x}}x+\widehat{\mathbf{y}}y+\widehat{\mathbf{z}}(z-vt)}{[(1-\frac{v^2}{c^2})(x^2+y^2)+(z-vt)^2]^{3/2}} \tag{12.2a}$$

The Power and Beauty of Electromagnetic Fields, First Edition. F. R. Morgenthaler.

$$\mathbf{H} = \varepsilon_0 \mathbf{v} \times \mathbf{E} = \frac{q\ v}{4\pi}\left(1 - \frac{v^2}{c^2}\right) \frac{\widehat{\mathbf{y}}x - \widehat{\mathbf{x}}y}{[(1 - \frac{v^2}{c^2})(x^2 + y^2) + (z - vt)^2]^{3/2}} \tag{12.2b}$$

The power flux vectors in the conventional and Alternate representations (in the free-space regions) are

$$\mathbf{E} \times \mathbf{H} = \frac{q^2\ v}{(4\pi)^2\varepsilon_0}\left(1 - \frac{v^2}{c^2}\right)^2 \frac{(v\,t - z)(\widehat{\mathbf{x}}\,x + \widehat{\mathbf{y}}\,y) + \widehat{\mathbf{z}}\,(x^2 + y^2)}{\left[\left(1 - \frac{v^2}{c^2}\right)(x^2 + y^2) + (z - vt)^2\right]^3} \tag{12.3a}$$

$$\mathbf{S}^o = -\frac{1}{2}\varepsilon_0\left(1 - \frac{v^2}{c^2}\right)\left(\Phi\frac{\partial \nabla\Phi}{\partial t} - \nabla\Phi\frac{\partial \Phi}{\partial t}\right) \tag{12.3b}$$

The latter evaluates to

$$\mathbf{S}^o = \frac{q^2\ v}{(4\pi)^2\varepsilon_0}\left(1 - \frac{v^2}{c^2}\right) \frac{\begin{bmatrix} (1 - \frac{v^2}{c^2})(z - vt)(\widehat{\mathbf{x}}\,x + \widehat{\mathbf{y}}\,y) \\ -\widehat{\mathbf{z}}\frac{1}{2}[(1 - \frac{v^2}{c^2})(x^2 + y^2) - (z - vt)^2] \end{bmatrix}}{\left[\left(1 - \frac{v^2}{c^2}\right)(x^2 + y^2) + (z - vt)^2\right]^3} \tag{12.4}$$

Unlike Eq. (12.3a), the on-axis z component of Eq. (12.4) does *not* vanish. Also, the two transverse components have opposite signs! Nevertheless, for a sphere of arbitrary radius centered on the charge, the integral over either of these fluxes vanish as expected.

The energy densities in the conventional and new representations in the free-space regions are

$$W^{em} = \frac{1}{2}\frac{q^2}{(4\pi)^2\varepsilon_0}\left(1 - \frac{v^2}{c^2}\right)^2 \frac{(z - vt)^2 + (1 + \frac{v^2}{c^2})(x^2 + y^2)}{[(z - vt)^2 + (1 - \frac{v^2}{c^2})(x^2 + y^2)]^3} \tag{12.5a}$$

$$W^o = \frac{1}{2}\rho\Phi\left(1 - \frac{v^2}{c^2}\right) + \frac{1}{2}\frac{q^2}{(4\pi)^2\varepsilon_0}\left(1 - \frac{v^2}{c^2}\right)\frac{v^2}{c^2} \frac{(z - vt)^2 - (1 - \frac{v^2}{c^2})(x^2 + y^2)}{[(z - vt)^2 + (1 - \frac{v^2}{c^2})(x^2 + y^2)]^3} \tag{12.5b}$$

The first term of W^o is the localized contribution.

The differences between Eqs. (12.5a) and (12.5b) are made especially clear when the velocity, v, is zero. Then, $W^{em} = \frac{1}{2}\frac{q^2}{(4\pi)^2\varepsilon_0 r^4}$ whereas $W^o = \frac{1}{2}\rho\Phi$. If these densities are integrated over the volume between radii r_1 and r_2 (both centered on the charge), the total energies are very different, yet each satisfies the integral form of the appropriate Poynting theorem because the electromagnetic energy is time-independent. The apparent contradiction in the value of the total energy that occurs in the limit $r_1 \to 0$ is resolved when the point-source contribution is included.

When v is nonzero (but nonrelativistic), W^o is still negligible *in free-space* (*EQS* applies with $\nabla \cdot \mathbf{S}^o \simeq 0$). Expressed in terms of spherical coordinates centered on the charge, Eq. (12.4) reduces to

$$\mathbf{S}^{o} \simeq \frac{q^2 v}{(4\pi)^2 \varepsilon_0} \frac{(\widehat{\mathbf{r}}\, \cos\theta\ + \widehat{\theta}\, \sin\theta)}{2r^4} \tag{12.6}$$

and should be contrasted with the nonrelativistic Poynting vector,

$$\mathbf{S} \simeq \frac{-\, q^2 v}{(4\pi)^2 \varepsilon_0} \frac{\widehat{\theta}\, \sin\theta}{r^4} \tag{12.7}$$

The divergence of $\mathbf{S}$ is nonzero because it must account for the time rate of change of W^{em}.

Motion in a dielectric (Čerenkov radiation)

If the charge is moving with respect to a linear isotropic uniform material (such as a dielectric liquid) characterized by μ and ε, all elements of the previous analysis can be used provided that $\mu_0 \to \mu$, $\varepsilon_0 \to \varepsilon$, and $c^2 \to \frac{1}{\mu\varepsilon}$. Because the velocity can exceed the speed of light in the material (but *not* $1/\sqrt{\mu_0 \varepsilon_0}$), the factor $(1 - \mu\varepsilon\, \mathrm{v}^2)$ can vanish or become negative. This creates the possibility of electric and magnetic shock waves (Čerenkov radiation) [22] on the cones,

$$\sqrt{x^2 + y^2} = \frac{|z - \mathrm{v}t|}{\sqrt{(\mu\varepsilon\, \mathrm{v}^2 - 1)}} \qquad \left(\frac{1}{\sqrt{\mu\varepsilon}} \leq \mathrm{v} < \mathrm{c}\right)$$

The power flux and energy densities also diverge on these cones, regardless of which representation is chosen.

12.2 SURFACE CHARGES SEPARATING AT CONSTANT VELOCITY

Introduction

For time $t < 0$, assume that two parallel planes of surface charge (shown in Figure 12.1) are arbitrarily close to one another and that the charges are of opposite polarity. Except between the planes, no fields of any type are present. At $t \geq 0$, the planes (located in Cartesian coordinates at $z = \pm \mathrm{v}t$) move in opposite directions (normal to the planes) at constant velocities $\pm v$ so that their separation is $2vt$. The attraction between the planes is overcome by the application of mechanical forces that will be considered later. We assume that polarities are such that the electric field between the planes is $\mathbf{E} = -\widehat{\mathbf{z}} E_o$, where E_o is a positive constant; the surface charge densities at $z = \pm \mathrm{v}t$ are, respectively, $\sigma_s = \pm \varepsilon_o E_o$. The electric-charge density, ρ, consists of these two separating surface charges,

$$\rho = |\sigma_s|\, [u_0(z - vt) - u_0(z + vt)] \tag{12.8}$$

where $u_0(z)$ is the unit-impulse function located at $z = 0$. The associated convective electric-current density, $\mathbf{J} = \rho \mathbf{v}$, is therefore positive for both sheets and is given by

$$\mathbf{J} = \widehat{\mathbf{z}}\, \mathrm{v}\, |\sigma_s|\, [u_0(z - vt) + u_0(z + vt)] \tag{12.9}$$

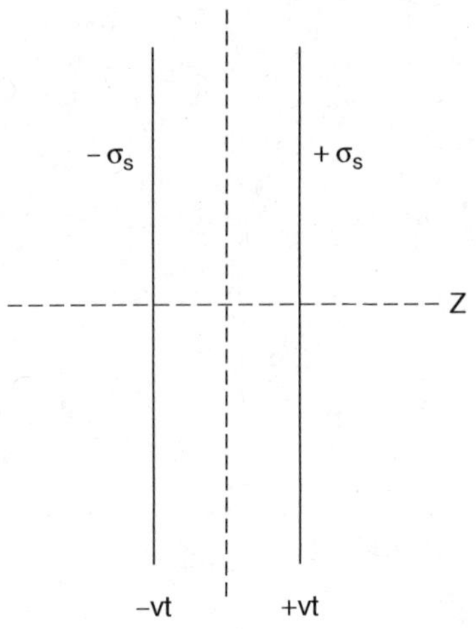

Figure 12.1 Separating planar surface charges.

By inspection, the complete set of **E** and **H** fields is

$$\mathbf{E} = \begin{cases} -\hat{\mathbf{z}}E_o & (|z| \le vt) \\ 0 & (|z| > vt) \end{cases} \tag{12.10a}$$

$$\mathbf{H} = 0 \qquad (\text{all } z) \tag{12.10b}$$

The electric-force density,

$$\mathbf{f} = \rho\mathbf{E} + \mathbf{J} \times \mu_o\mathbf{H} = -\hat{\mathbf{z}}\frac{1}{2}\varepsilon_o E_o^2\,[u_0(z - vt) - u_0(z + vt)] \tag{12.11}$$

consists of impulses on each charge plane ($\frac{1}{2}E_o$ is the average field at each plane); these must be balanced by mechanical forces in order to allow the separation to occur. Notice that the electrical power density is expressed as $\mathbf{E} \cdot \mathbf{J} = \mathbf{f} \cdot \mathbf{v}$.

In the Maxwell–Poynting representation, we have

$$W_e = \frac{1}{2}\varepsilon_o E_o^2 \qquad (|z| \le vt) \tag{12.12a}$$

$$W_m = 0 \tag{12.12b}$$

$$\mathbf{S} = \mathbf{E} \times \mathbf{H} = 0 \tag{12.12c}$$

Since the total electric energy (per unit area) is $\varepsilon_o E_o^2 vt$, the mechanical power (per unit area) required to maintain the separation is $\varepsilon_o E_o^2 v$. As expected, the opposite pressures applied to each plane are therefore $\frac{1}{2}\varepsilon_o E_o^2$. Let us now turn to the Alternate representation which requires that we calculate both the vector and scalar potentials. For completeness, we consider both Lorenz and Coulomb gauges.

Lorenz gauge

For negative time, sources and fields are everywhere zero; for positive time, ρ and $\mathbf{J}$ generate, respectively, Φ and $\mathbf{A}$. Since these *both* obey the wave equation, the regions $|z| > ct$ have not yet experienced the transient, so there $\Phi = 0$, $\mathbf{A} = 0$. Elsewhere, because the current is z-directed, $\mathbf{A} = \widehat{\mathbf{z}}A_z(z,t)$; therefore $\nabla \times \mathbf{A} = 0$. Between the planes of charge, both $\Phi(z,t)$ and $A_z(z,t)$ are of the form $C_1 z + C_2 t$; outside, linear combinations of $z \mp ct$ satisfy the wave equation, can match the inner functions, and automatically vanish at $|z| = ct$. The appropriate combinations that satisfy the gauge condition, $g = \frac{\partial A_z}{\partial z} + \frac{1}{c^2}\frac{\partial \Phi}{\partial t} = 0$, and are piecewise-continuous result in

$$A_z(z, t \geq 0) = \begin{cases} 0 & (z \leq -ct) \\ \mathrm{v}\dfrac{z + ct}{c^2 - v^2}E_\mathrm{o} & (-\mathrm{c}t \leq z \leq -vt) \\ \dfrac{\mathrm{v}}{c + v}E_\mathrm{o}t & (-\mathrm{v}t \leq z \leq vt) \\ -\mathrm{v}\dfrac{z - ct}{c^2 - v^2}E_\mathrm{o} & (\mathrm{v}t \leq z \leq ct) \\ 0 & (ct \leq z) \end{cases} \tag{12.13a}$$

$$\Phi(z, t \geq 0) = \begin{cases} 0 & (z \leq -ct) \\ -\mathrm{vc}\dfrac{z + ct}{c^2 - v^2}E_\mathrm{o} & (-ct \leq z \leq -vt) \\ \dfrac{c}{c + v}E_\mathrm{o}z & (-vt \leq z \leq vt) \\ -vc\dfrac{z - ct}{c^2 - v^2}E_\mathrm{o} & (vt \leq z \leq ct) \\ 0 & (ct \leq z) \end{cases} \tag{12.13b}$$

From Eqs. (3.31a) and (3.31b), the Alternate-power flux and Alternate-energy density are

$$S_z^\mathrm{o}(z,t) = \Phi J_z - \frac{1}{2}\varepsilon_\mathrm{o}\left[\Phi\frac{\partial^2\Phi}{\partial z\,\partial t} - \frac{\partial\Phi}{\partial z}\frac{\partial\Phi}{\partial t}\right] + \frac{1}{2\mu_\mathrm{o}}\left[A_z\frac{\partial^2 A_z}{\partial z\,\partial t} - \frac{\partial A_z}{\partial z}\frac{\partial A_z}{\partial t}\right] \tag{12.14a}$$

$$W^\mathrm{o}(z,t) = W_e^\mathrm{o}(z,t) + W_m^\mathrm{o}(z,t) \tag{12.14b}$$

where

$$W_e^\mathrm{o}(z,t) = \frac{1}{2}\Phi\rho + \frac{1}{2}\frac{\varepsilon_\mathrm{o}}{c^2}\left[\Phi\frac{\partial^2\Phi}{\partial t^2} - \left(\frac{\partial\Phi}{\partial t}\right)^2\right] \tag{12.14c}$$

$$W_m^{\mathrm{o}}(z,t) = \frac{1}{2}A_zJ_z + \frac{1}{2}\varepsilon_{\mathrm{o}}\left[\left(\frac{\partial A_z}{\partial t}\right)^2 - A_z\frac{\partial^2 A_z}{\partial t^2}\right] \tag{12.14d}$$

We observe that both electric and magnetic energies exist in this representation. It is instructive to evaluate these terms separately

$$\frac{1}{2}\Phi\rho = \frac{1}{2}\frac{c}{c+\mathrm{v}}\varepsilon_{\mathrm{o}}E_{\mathrm{o}}^2 vt\ [u_0(z-vt)+u_0(z+vt)] \tag{12.15a}$$

$$\frac{1}{2}\frac{\varepsilon_{\mathrm{o}}}{c^2}\left[\Phi\frac{\partial^2\Phi}{\partial t^2} - \left(\frac{\partial\Phi}{\partial t}\right)^2\right] = \frac{1}{2}\frac{-cv}{c^2-v^2}\frac{v}{c+v}\varepsilon_{\mathrm{o}}E_{\mathrm{o}}^2 vt\ [u_0(z-vt)+u_0(z+vt)] \tag{12.15b}$$

$$\frac{1}{2}A_zJ_z = \frac{1}{2}\frac{v}{c+v}\varepsilon_{\mathrm{o}}E_{\mathrm{o}}^2 vt\ [u_0(z-vt)+u_0(z+vt)] \tag{12.15c}$$

$$\frac{1}{2}\varepsilon_{\mathrm{o}}\left[\left(\frac{\partial A_z}{\partial t}\right)^2 - A_z\frac{\partial^2 A_z}{\partial t^2}\right] = \left[\begin{array}{l}\frac{1}{2}\frac{v^2}{c^2-v^2}\frac{v}{c+v}\varepsilon_{\mathrm{o}}E_{\mathrm{o}}^2 vt\ [u_0(z-vt)+u_0(z+vt)] \\ +\frac{1}{2}\varepsilon_{\mathrm{o}}E_{\mathrm{o}}^2\left(\frac{v}{c+v}\right)^2 u_{-1}(vt-|z|)\end{array}\right] \tag{12.15d}$$

where $u_{-1}(z)$ is the unit step function.

The magnetic energies are smaller than the electric energies by the factor v/c and so can be ignored for nonrelativistic velocities; the *EQS* energy-density, $\frac{1}{2}\Phi\rho$, is then a completely adequate approximation. However, for any value of $v \le c$, the exact localized energies sum to

$$\frac{1}{2}(\Phi\rho + A_zJ_z) = \frac{1}{2}\varepsilon_{\mathrm{o}}E_{\mathrm{o}}^2 vt\ [u_0(z-vt)+u_0(z+vt)] \tag{12.16}$$

whereas the other terms (which also include impulses) sum to

$$\frac{1}{2}\varepsilon_{\mathrm{o}}E_{\mathrm{o}}^2\left(\frac{v}{c+v}\right)^2(u_{-1}(vt-|z|) - vt\ [u_0(z-vt)+u_0(z+vt)]) \tag{12.17}$$

When integrated over all z, the total Alternate energy (per unit area) is $\varepsilon_{\mathrm{o}}E_{\mathrm{o}}^2\mathrm{v}t$, which agrees with the total Maxwell–Poynting energy.

The total Alternate-power flux evaluates to

$$\begin{aligned} S_z^{\mathrm{o}} &= \frac{1}{2}\varepsilon_{\mathrm{o}}E_{\mathrm{o}}^2 v^2 t\ [u_0(z-vt)+u_0(z+vt)] \\ &= v\frac{1}{2}(\Phi\rho + A_zJ_z) \end{aligned} \tag{12.18}$$

Coulomb gauge

Poisson's Equation governing the scalar potential is

$$\frac{\partial^2\Phi}{\partial z^2} = -\frac{\rho}{\varepsilon_{\mathrm{o}}} = -E_{\mathrm{o}}[u_0(z-vt)-u_0(z+vt)] \tag{12.19a}$$

which can be integrated twice to produce the solution

$$\Phi(z,t \geq 0) = \begin{cases} -E_o vt & (z \leq -vt) \\ E_o z & (-vt \leq z \leq vt) \\ E_o vt & (vt \leq z) \end{cases} \tag{12.19b}$$

which is necessarily antisymmetric with respect to z.

The wave equation governing A_z is

$$\frac{\partial^2 A_z}{\partial z^2} - \frac{1}{c^2}\frac{\partial^2 A_z}{\partial t^2} = \frac{1}{c^2}\frac{\partial^2 \Phi}{\partial z \partial t} = 0 \tag{12.20}$$

Deprived of a source, $A_z = 0$ which satisfies the gauge-condition, $\frac{\partial A_z}{\partial z} = 0$. For positive time, this potential extends throughout all space, but yields the proper value of **E**.

From Eqs. (3.31b) and (3.31a) we have

$$W_e^o(z,t) = \frac{1}{2}\Phi\rho \tag{12.21a}$$

$$W_m^o(z,t) = \frac{1}{2}A_z J_z + \frac{1}{2}\varepsilon_o\left[\left(\frac{\partial A_z}{\partial t}\right)^2 - A_z\frac{\partial^2 A_z}{\partial t^2}\right] \tag{12.21b}$$

and

$$S_z^o(z,t) = \begin{cases} \Phi J_z - \frac{1}{2}\varepsilon_o\left[\Phi\frac{\partial^2\Phi}{\partial z\partial t} - \frac{\partial\Phi}{\partial z}\frac{\partial\Phi}{\partial t}\right] \\ +\frac{1}{2\mu_o}\left[A_z\frac{\partial^2 A_z}{\partial z\partial t} - \frac{\partial A_z}{\partial z}\frac{\partial A_z}{\partial t}\right] \\ +\frac{1}{2}\varepsilon_o\left[\frac{\partial A_z}{\partial t}\frac{\partial\Phi}{\partial t} - A_z\frac{\partial^2\Phi}{\partial t^2}\right] \end{cases} \tag{12.22}$$

Consequently, in this gauge there is no magnetic energy and

$$W^o(z,t) = W_e^o = \frac{1}{2}\Phi\rho = \frac{1}{2}\varepsilon_o E_o^2 vt\ [u_0(z - vt) + u_0(z + vt)] \tag{12.23a}$$

$$S_z^o(z,t) = \frac{1}{2}\varepsilon_o E_o^2 v^2 t\,[u_0(z - vt) - u_0(z + vt)] = vW^o(z,t) \tag{12.23b}$$

For this particular example, the Maxwell–Poynting representation is the simplest because $\mathbf{S} = 0$. However, regardless of which representation (or which gauge) is used, we obtain

$$\nabla \cdot \mathbf{S} + \frac{\partial W}{\partial t} = -\mathbf{E}\cdot\mathbf{J} = \frac{1}{2}\varepsilon_o E_o^2 v\,[u_0(z - vt) + u_0(z + vt)] \tag{12.24}$$

as required.

12.3 EXPANDING CYLINDRICAL SURFACE CHARGE

Assume that a total electric charge, Q' (per unit length), is spread uniformly over the surface of a cylinder of infinite length and radius, $r_o \to 0$ for $t < 0$ and $r_o = vt$ for $t > 0$. The geometry is shown in Figure 12.2.

We again ignore the mechanical force that is required to sustain the motion and focus on the electromagnetic fields which are calculated by inspection. There is no magnetic field anywhere; the electric field is zero for $r < vt$ and is radially directed with magnitude $Q'/(2\pi\varepsilon_0 r)$ for $r > vt$. In the normal representation, W^{em} is zero inside the cylinder and has only an electric contribution outside of it; nowhere is there a Poynting vector, **S**. The continual decrease in the total electric stored energy is, of course, balanced by work done by the mechanical force.

Now consider the Alternate representation; we begin by calculating **A** and Φ in the Lorenz gauge. For $t < 0$, there is no current, hence no **A**; the scalar potential is $\Phi = -\frac{Q'}{2\pi\varepsilon_0}\ln r$. Homogeneous solutions that produce neither electric nor magnetic fields, but which are required to match the boundary conditions imposed by continuity of the potentials, satisfy

$$\mathbf{A}_h = \nabla\Psi \tag{12.25a}$$

$$\Phi_h = -\frac{\partial\Psi}{\partial t} \tag{12.25b}$$

where Ψ satisfies the homogeneous wave equation, and for this configuration we have

$$\Psi = -C_0' t - C_1' t \ln r - C_2'\left[t\cosh^{-1}\left(\frac{ct}{r}\right) - \frac{\sqrt{c^2t^2 - r^2}}{c}\right]. \tag{12.26}$$

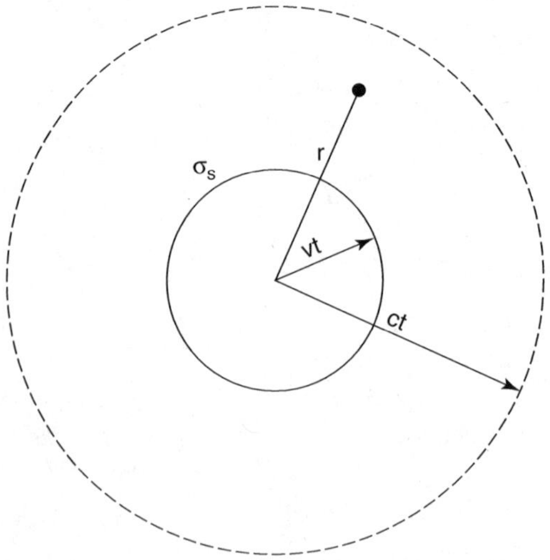

Figure 12.2 Expanding cylindrical surface charge.

For $t > 0$, the potentials are found to be

$$\Phi = \begin{cases} C_0 + C_1 \ln r + C_2 \cosh^{-1}\left(\frac{ct}{r}\right) & (r \leq vt) \\ -\frac{Q'}{2\pi\varepsilon_0} \ln r + C_3 \cosh^{-1}\left(\frac{ct}{r}\right) & (vt \leq r \leq ct) \\ -\frac{Q'}{2\pi\varepsilon_0} \ln r & (ct \leq r) \end{cases} \tag{12.27a}$$

and

$$\mathbf{A} = \widehat{\mathbf{r}} \begin{cases} -C_1 \frac{t}{r} + C_2 \frac{\sqrt{c^2t^2 - r^2}}{cr} & (r \leq vt) \\ C_3 \frac{\sqrt{c^2t^2 - r^2}}{cr} & (vt \leq r \leq ct) \\ 0 & (ct \leq r) \end{cases} \tag{12.27b}$$

Continuity of both $\mathbf{A}$ and Φ at $r = vt$ and $r = ct$ for all t is satisfied if

$$C_0 = \frac{Q'}{2\pi\varepsilon_0} \frac{c}{\sqrt{c^2 - v^2}} \cosh^{-1}\left(\frac{c}{v}\right)$$

$$C_1 = C_2 = -\frac{Q'}{2\pi\varepsilon_0}$$

$$C_3 = \frac{Q'}{2\pi\varepsilon_0}\left(\frac{c}{\sqrt{c^2 - v^2}} - 1\right)$$

and naturally lead to the same electric and magnetic fields that were deduced by inspection. Although the transient (which spreads at the speed of light) generates no fields, it is required to ensure continuity of both the vector and scalar potentials. At $r = 0$, the vector potential vanishes (as required) and the scalar potential evaluates to $\Phi(0, t) = \frac{-Q'}{2\pi\varepsilon_0} \ln(v_{\text{eff}}\, t)$ where

$$v_{\text{eff}} = \frac{2c}{\left[c/v + \sqrt{(c/v)^2 - 1}\right]^{\frac{c/v}{\sqrt{(c/v)^2 - 1}}}}$$

For nonrelativistic velocities, $v/c \ll 1$, $v_{\text{eff}} \rightarrow v$ as expected; for $v \rightarrow c$, $v_{\text{eff}} \rightarrow 2c/e$.

For $0 \leq r \leq vt$, the nonlocalized Alternate power flux and energy density are found to be

$$S_r^{\text{o}} = \left(\frac{cQ'}{2\pi}\right)^2 \frac{\left(\begin{array}{c} r^2\left[\frac{c}{\sqrt{c^2 - v^2}} \cosh^{-1}(c/v) - \cosh^{-1}\left(\frac{ct}{r}\right) - \ln(r) - 1\right] \\ + ct(\sqrt{c^2t^2 - r^2} - c^2t^2 + r^2) \end{array}\right)}{2c\varepsilon_0 r(c^2t^2 - r^2)^{3/2}} \tag{12.28a}$$

$$W^{\text{o}} = \left(\frac{Q'}{2\pi}\right)^2 \frac{\left(\begin{array}{c} r^2 ct\left[\frac{c}{\sqrt{c^2 - v^2}} \cosh^{-1}(c/v) - \cosh^{-1}\left(\frac{ct}{r}\right) - \ln(r)\right] \\ - (2c^2t^2 - r^2)(ct - \sqrt{c^2t^2 - r^2}) \end{array}\right)}{2\varepsilon_0 r^2(c^2t^2 - r^2)^{3/2}} \tag{12.28b}$$

For $vt \leq r \leq ct$,

$$S_r^{\text{o}} = \left(\frac{cQ'}{2\pi}\right)^2 \frac{\left(\begin{array}{c}\left[2\left(\frac{c}{\sqrt{c^2-v^2}}-1\right)-\frac{v^2}{c^2-v^2}\right]\left[r^2\cosh^{-1}\left(\frac{ct}{r}\right)-ct\sqrt{c^2t^2-r^2}\right] \\ +\left(\frac{c}{\sqrt{c^2-v^2}}-1\right)\left(r^2\ln(r)+r^2-c^2t^2\right)\end{array}\right)}{2c\varepsilon_{\text{o}}r(c^2t^2-r^2)^{3/2}} \tag{12.29a}$$

$$W^{\text{o}} = \left(\frac{Q'}{2\pi}\right)^2 ct \frac{\left(\begin{array}{c}\left[\left(2(\frac{c}{\sqrt{c^2-v^2}}-1\right)-\frac{v^2}{c^2-v^2}\right]\left[r^2\cosh^{-1}(ct/r)-ct\sqrt{c^2t^2-r^2}\right] \\ +\left(\frac{c}{\sqrt{c^2-v^2}}-1\right)r^2\ln(r)\end{array}\right)}{2\varepsilon_{\text{o}}r^2(c^2t^2-r^2)^{3/2}} \tag{12.29b}$$

For $ct \leq r$, we have

$$S_r^{\text{o}} = 0 \tag{12.30a}$$

$$W^{\text{o}} = 0 \tag{12.30b}$$

In addition, localized components, $\Phi\mathbf{J}$ and $\frac{1}{2}\rho\Phi + \frac{1}{2}\mathbf{A}\cdot\mathbf{J}$, exist at $r = vt$ the only location where ρ and $\mathbf{J} = \widehat{\mathbf{r}}\,\rho\text{v}$ are nonzero; there $\rho = \frac{Q'}{2\pi r}u_0(r - vt)$.

Although the Alternate representation often leads to relative simplification, this is certainly not so in this case. However, if the Coulomb gauge ($\nabla\cdot\mathbf{A} = 0$) is used, $\mathbf{A} = 0$ everywhere and

$$\Phi = \begin{cases} \frac{-Q'}{2\pi\varepsilon_0}\ln(vt) & (r \leq vt) \\ \frac{-Q'}{2\pi\varepsilon_0}\ln r & (vt \leq r) \end{cases} \tag{12.31}$$

Because of the shift of gauge, $g = \frac{1}{c^2}\frac{\partial\Phi}{\partial t}$, and from Eq. (3.31a), $\mathbf{S}^{\text{o}} = \Phi\mathbf{J} + \frac{1}{2}\varepsilon_0(\frac{\partial\Phi}{\partial t}\nabla\Phi - \Phi\nabla\frac{\partial\Phi}{\partial t})$. This flux is nonzero only at the surface charge, $r = vt$. Localization is complete because $W^{\text{o}} = \frac{1}{2}\rho\Phi$ and, as before, $\mathbf{J} = \widehat{\mathbf{r}}\,\rho\text{v}$.

12.4 EXPANDING SPHERICAL SURFACE CHARGE

Assume that a total electric charge, Q, is spread uniformly over the surface of a sphere of radius, $r_{\text{o}} = 0$ for $t < 0$ and $r_{\text{o}} = vt$ for $t > 0$. With r now representing the spherical radius, Figure 12.2 also serves to describe this geometry.

We again ignore the mechanical force that is required to sustain the motion and focus on the electromagnetic fields which, as in the previous example, can be calculated by inspection. There is no magnetic field anywhere; the electric field is zero for $r < vt$ and radially directed with magnitude $Q/(4\pi\varepsilon_{\text{o}}r^2)$ for $r > vt$. In the normal representation, W^{em} is zero inside the sphere and has only an electric contribution outside of it; nowhere is there a Poynting vector, **S**. The continual decrease in the total electric stored energy is, of course, balanced by work done by the mechanical force.

Now consider the Alternate representation; we begin by calculating **A** and Φ in the Lorenz gauge. For $t < 0$, there is no current, hence no **A**; the scalar potential is $\Phi = Q/(4\pi\varepsilon_0 r)$. Homogeneous solutions that produce neither electric nor magnetic fields, but which are required to match the boundary conditions imposed by continuity of the potentials, satisfy

$$\mathbf{A}_h = \nabla\Psi \tag{12.32a}$$

$$\Phi_h = -\frac{\partial\Psi}{\partial t} \tag{12.32b}$$

where Ψ satisfies the homogeneous wave equation and is of the form

$$\Psi = -C_0'\left[\frac{t}{r}\ln\left(\frac{ct+r}{ct-r}\right) + \frac{\ln\left(c^2t^2 - r^2\right)}{c}\right] - C_1'\frac{t}{r} \tag{12.33}$$

For $t > 0$, the potentials are found to be

$$\mathbf{A} = \hat{\mathbf{r}}\begin{cases} C_0\dfrac{t\ln\left(\dfrac{ct+r}{ct-r}\right) - \frac{2r}{c}}{r^2} & (r \le vt) \\ C_1\dfrac{t}{r^2} & (vt \le r \le ct) \\ 0 & (ct \le r) \end{cases} \tag{12.34a}$$

and

$$\Phi = \begin{cases} C_0\dfrac{\ln\left(\dfrac{ct+r}{ct-r}\right)}{r} & (r \le vt) \\ \left(\dfrac{Q}{4\pi\varepsilon_0} + C_1\right)\dfrac{1}{r} & (vt \le r \le ct) \\ \dfrac{Q}{4\pi\varepsilon_0 r} & (ct \le r) \end{cases} \tag{12.34b}$$

The constants C_0 and C_1 are found by imposing continuity of both **A** and Φ (Notice that $\mathbf{A} \to 0$ at the origin.) Actually, components of the inner-potential functions are also required in the region $vt \le r \le ct$ in order to accomplish the match at $r = ct$; however, because of the logarithmic infinity, the amplitudes of these terms are vanishingly small. The required constants are

$$C_0 = \frac{Q}{8\pi\varepsilon_0}\frac{c}{v}$$

$$C_1 = \frac{Q}{8\pi\varepsilon_0}\left[\frac{c}{v}\ln\left(\frac{c+v}{c-v}\right) - 2\right]$$

and naturally lead to the same electric and magnetic fields that were deduced by inspection. Although the transient (which spreads at the speed of light) generates no fields, it is required to ensure continuity of both the vector and scalar potentials.

Evaluation of the Alternate power and Alternate energy reveals that there is neither $\mathbf{S}^o$ for $r > vt$ nor W^o for $r > ct$. However, there is time-independent energy within the range $vt \leq r \leq ct$ and *both* $\mathbf{S}^o$ and W^o exist for $r \leq vt$. On the surface charge, localized components of the power flux and energy density are dominant. In the inner region, we get

$$S_r^o = 2C_0^2 \frac{2rct - (c^2t^2 - r^2)\ln\left|\dfrac{ct+r}{ct-r}\right|}{c\mu_0 r^2(c^2t^2 - r^2)^2} \tag{12.35a}$$

$$W^o = \frac{1}{2}\varepsilon_0 C_0^2 \frac{\left[(c^2t^2 - r^2)\ln\left|\dfrac{ct+r}{ct-r}\right| - 2rct\right]^2 + 4r^4}{r^4(c^2t^2 - r^2)^2} \tag{12.35b}$$

When evaluated at $r = 0$, we obtain $S_r^o = 0$, $W^o = Q^2/(8\pi^2\varepsilon_0 c^2 v^2 t^4)$, $A_r = 0$, and $\Phi = Q/(4\pi\varepsilon_0 vt)$. In some locations, there is power and energy where there are no electromagnetic fields and vice versa!

As in the previous two examples, the Alternate representation does not lead to simplification. However, in the Coulomb gauge ($\nabla \cdot \mathbf{A} = 0$), $\mathbf{A} = 0$ everywhere and

$$\Phi = \begin{cases} \dfrac{Q}{4\pi\varepsilon_0 vt} & (r \leq vt) \\[2ex] \dfrac{Q}{4\pi\varepsilon_0 r} & (vt \leq r) \end{cases} \tag{12.36}$$

As before, because of the shift of gauge, the Alternate power flux and energy density must be calculated from Eqs. (3.31); then

$$\mathbf{S}^o = \Phi\mathbf{J} + \frac{1}{2}\varepsilon_0\left(\frac{\partial\Phi}{\partial t}\nabla\Phi - \Phi\nabla\frac{\partial\Phi}{\partial t}\right)$$

Localization is complete because $W^o = \frac{1}{2}\rho\Phi$ and, as before, $\mathbf{J} = \hat{\mathbf{r}}\,\rho v$—now with $\rho = \frac{Q}{4\pi r^2}u_0(r - vt)$.

CHAPTER 13

ACCELERATING CHARGES

13.1 HERTZIAN ELECTRIC DIPOLE

Sinusoidal steady-state

The Hertzian electric dipole considered in Chapters 1 and 3 consists of a uniform current element, $I(t)d$, that is oriented parallel to the z axis with $I(t) = I_o \sin(\omega t)$. In addition, $kd \ll 1$, where $k = \omega/c$. The vector and scalar potentials were derived in Section 1.6; the electric and magnetic fields were derived in Section 3.5. All that remains is to calculate the total radiated power and the directivity of the dipole. The time-averaged radiated power is calculated from either $\mathbf{S}^{em}$ or $\mathbf{S}^{o}$. The latter has only an average component,

$$S_r^o = \sqrt{\frac{\mu_o}{\varepsilon_o}} I_o^2 \frac{(kd)^2}{32\pi^2 r^2} \sin^2\theta$$

Therefore,

$$\begin{aligned} < P_{\text{radiated}} > &= \int_0^{\pi} 2\pi r^2 S_r^o \sin\theta \, d\theta = \sqrt{\frac{\mu_o}{\varepsilon_o}} \frac{(kd)^2}{12\pi} I_o^2 \\ &= \frac{1}{2} R_{\text{rad}} I_o^2 \end{aligned}$$

The Power and Beauty of Electromagnetic Fields, First Edition. F. R. Morgenthaler.

where the radiation resistance is defined as

$$R_{\text{rad}} = \sqrt{\frac{\mu_0}{\varepsilon_0}}\frac{(kd)^2}{6\pi} = 20(kd)^2 \text{ ohms}$$

The directivity is defined by dividing the actual radiated power flux by that which would be radiated isotropically. It follows that

$$D(\theta, \phi) = \frac{\sqrt{\frac{\mu_0}{\varepsilon_0}} I_0^2 \frac{(kd)^2}{32\pi^2 r^2}\sin^2\theta}{\frac{< P_{\text{radiated}} >}{4\pi r^2}} = \frac{3}{2}\sin^2\theta$$

Sinusoidal pulse

If the sinusoidal current is turned on at $t = 0$ so that $\mathrm{I}(t) = I_0 \sin(\omega t + \phi_0)u_{-1}(t)$, with $u_{-1}(t)$ the unit step function (defined in Appendix B, Section B.1) and ϕ_0 a constant, it follows that the retarded potentials become

$$\mathbf{A} = \widehat{\mathbf{z}}\frac{\mu_0 I_0 d}{4\pi r}\sin(\omega t - kr + \phi_0)u_{-1}\left(t - \frac{r}{c}\right) \tag{13.1a}$$

$$\Phi = \frac{I_0 d}{4\pi\varepsilon_0 \omega r^2}\begin{bmatrix} kr\sin(\omega t - kr + \phi_0) \\ -\cos(\omega t - kr + \phi_o) + \cos\phi_o \end{bmatrix}\cos\theta \; u_{-1}\left(t - \frac{r}{c}\right) \tag{13.1b}$$

where r is still the radius in spherical coordinates, $\widehat{\mathbf{z}}$ remains the unit vector along z, and $kd \ll 1$.

We have assumed that the current is uniform over the effective length d, and therefore charges $\pm Q$ appear only at the ends. These cannot appear instantaneously and so $\mathrm{Q}(t) = \mathrm{I}_0/\omega\,[\cos\phi_0 - \cos(\omega t + \phi_0)]\,u_{-1}(t)$. The first term is the source of the constant dipole potential for $r < ct$. Of course, when linear dissipation is included, the pair of constant charges experiences exponential decay and both $\mathbf{A}$ and Φ are appropriately modified (Chapter 25, Problem 25.4-18).

Notice that, unless $\sin\phi_0 = 0$, both A_z and Φ are discontinuous at $r = ct$. But, if the current suddenly steps to the value $I_0 \sin\phi_0$, the values of $\nabla \times \mathbf{A}$ and $\nabla\Phi$ are infinite at the discontinuities.[1] In this case, the complete electric and magnetic fields become

$$\mathbf{E} = \frac{c\mu_0 I_0 d}{4\pi k r^3}\begin{pmatrix} \widehat{\mathbf{r}}\,2\begin{bmatrix} kr\sin(\omega t - kr + \phi_0) \\ +\cos\phi_0 - \cos(\omega t - kr + \phi_o) \end{bmatrix}\cos\theta \cdot u_{-1}\left(t - \frac{r}{c}\right) \\ +\widehat{\theta}\begin{bmatrix} \begin{bmatrix} kr\sin(\omega t - kr + \phi_0) + \cos\phi_0 \\ +(k^2r^2 - 1)\cos(\omega t - kr + \phi_o) \end{bmatrix} u_{-1}\left(t - \frac{r}{c}\right) \\ +\frac{kr^2}{c}\sin\phi_0 \cdot u_0\left(t - \frac{r}{c}\right) \end{bmatrix}\sin\theta \end{pmatrix} \tag{13.2a}$$

[1]Mobile charge associated with the current step will experience a step in velocity. It is the infinite acceleration that produces these singularities in the radiation field.

$$\mathbf{H} = \widehat{\phi}\frac{I_o d}{4\pi r^2}\left(\begin{bmatrix} kr\cos(\omega t - kr + \phi_o) \\ +\sin(\omega t - kr + \phi_o) \end{bmatrix} u_{-1}\left(t - \frac{r}{c}\right) \atop +\frac{r}{c}\sin\phi_o \cdot u_0\left(t - \frac{r}{c}\right)\right)\sin\theta \tag{13.2b}$$

As predicted, both E_θ and H_ϕ are infinite on the expanding wave front. But, to create these singularities, the dipole must deliver infinite power and energy[2]; consequently, we eliminate the current discontinuity by choosing $\sin\phi_o = 0$. Notice that, even then, the electric-field transient retains memory of its initiation – due to the static electric-dipole potential generated for $r < ct$ (that persists because dissipation was neglected). It is interesting that the radial component of **E** vanishes precisely on the wave boundary for all r – not just in the far-field region. With $\phi_o = 0$, and multiplication by $u_{-1}(t - \frac{r}{c})$ implied, the power fluxes and energy densities are

$$\mathbf{E}\times\mathbf{H} = \frac{c\mu_o I_o^2 d^2}{32\pi^2 k r^5}\left(\begin{array}{l} \widehat{\mathbf{r}}\begin{bmatrix} k^3r^3 + (k^3r^3 - 2kr)\cos 2(\omega t - kr) \\ +(2k^2r^2 - 1)\sin 2(\omega t - kr) \\ +2[kr\cos(\omega t - kr) + \sin(\omega t - kr)] \end{bmatrix}\sin^2\theta \\ +\widehat{\theta}\begin{bmatrix} 2kr\,[\cos 2(\omega t - kr) - \cos(\omega t - kr)] \\ +(1 - k^2r^2)\sin 2(\omega t - kr) \\ -2\sin(\omega t - kr) \end{bmatrix}\sin 2\theta \end{array}\right) \tag{13.3a}$$

$$W_e = \frac{\mu_o I_o^2 d^2}{32\pi^2 k^2 r^6}\left(\begin{array}{l} \begin{bmatrix} 1 + kr\sin(\omega t - kr) \\ +\left(k^2r^2 - 1\right)\cos(\omega t - kr) \end{bmatrix}^2 \sin^2\theta \\ +4\,[kr\sin(\omega t - kr) - \cos(\omega t - kr) + 1]^2\cos^2\theta \end{array}\right) \tag{13.3b}$$

$$W_m = \frac{\mu_o I_o^2 d^2}{32\pi^2 r^4}[kr\cos(\omega t - kr) + \sin(\omega t - kr)]^2\sin^2\theta \tag{13.3c}$$

and

$$\mathbf{S}^o = \widehat{\mathbf{r}}\,\frac{\mu_o c\, I_o^2 k d^2}{32\pi^2 r^3}\left[kr\,\sin^2(\theta) - \sin(\omega t - kr)\cos^2(\theta)\right] \tag{13.4a}$$

$$W^o = \frac{\mu_o I_o^2 d^2}{32\pi^2 r^4}\left(\begin{array}{c} k^2r^2\sin^2\theta \\ +[\cos(\omega t - kr) - kr\sin(\omega t - kr) - 1]\cos^2\theta \end{array}\right) \tag{13.4b}$$

On the wave front, and regardless of whether $r = ct$ is in the near or far field,

$$\mathbf{E}\times\mathbf{H} = 2\mathbf{S}^o = \widehat{\mathbf{r}}cW^{em} \tag{13.5a}$$

[2]One might conclude that, since the wave-front Poynting flux is infinite for zero duration, integration will lead to finite radiation energy. However, since the product of two impulses is involved, the integral, over time, is itself singular.

$$W^{em} = 2W^{o} = \frac{\mu_o I_o^2 k^2 d^2}{16\pi^2 r^2} \sin^2\theta \tag{13.5b}$$

The electric field is shown plotted as a contour plot in Figure 13.1 and as a vector plot in Figure 13.2. The width and height of both plots are approximately 5λ; the region close to the origin is intentionally excluded. Notice the reversal of direction of the field lines

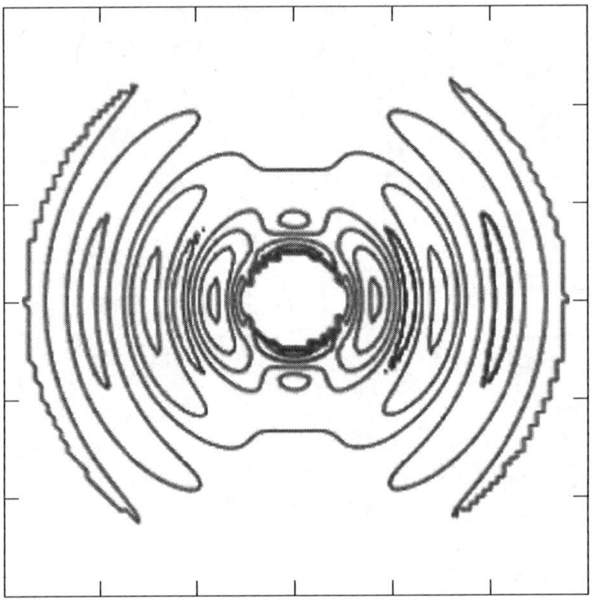

Figure 13.1 Field contours of a radiating Hertzian dipole.

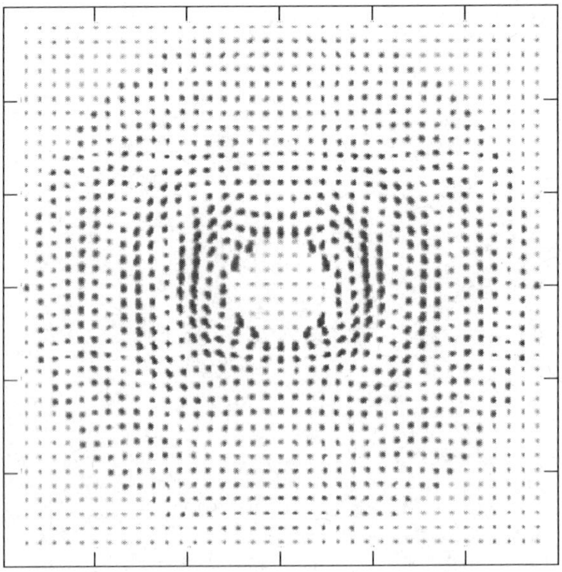

Figure 13.2 Field vectors of a radiating Hertzian dipole.

and how, in the radiation region, they close upon themselves rather than terminating on the dipole.

Vector plots of the Poynting flux and the Alternate power flux are shown in Figures 13.3 and 13.4 (and on the back book cover). The ripples of the former reveal the

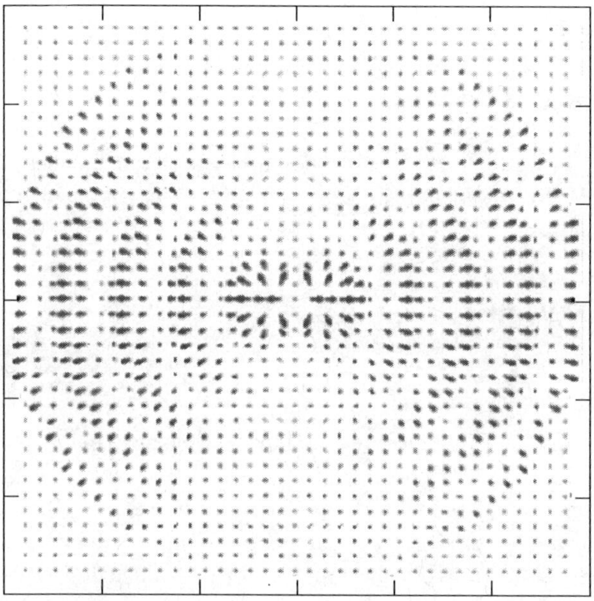

Figure 13.3 $\mathbf{E} \times \mathbf{H}$ vectors of a radiating Hertzian dipole.

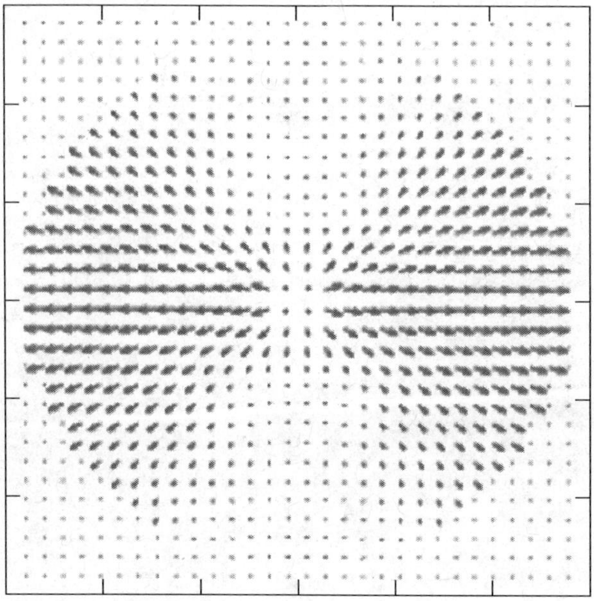

Figure 13.4 $\mathbf{S}^{o}$ vectors of a radiating Hertzian dipole.

wave nature of the power flow; the straight-line trajectories of the latter reveal the particle (photon) nature.

For steady-state harmonic fields of a single frequency, the Alternate power and energy are time-independent while the Maxwell–Poynting power and energy contain second-harmonic components. This observation is in accord with the respective complex power theorems. However, in this transient solution, components at the fundamental are present in *both* the Maxwell–Poynting and Alternate formulations; these are due to the static electric dipole that becomes negligible when $kr \gg 1$. Although, as expected, the time-averaged values of $\mathbf{S}$ and $\mathbf{S}^{o}$(as well as W^{em} and W^{o}) agree, the values on the wave front differ by a factor of two. One final observation concerns the *negative* average term in W^{o} that is proportional to $I_o^2(\cos^2\theta)/r^4$—and which appears in the steady-state analysis contained in Chapter 1. Because the amplitude is independent of frequency, it was unexpected. However, in the limit $\omega \to 0$, this term is canceled in the transient solution and, in free-space, both $\mathbf{S}^{o}$ and W^{o} vanish.[3]

13.2 HERTZIAN MAGNETIC DIPOLE

Sinusoidal pulse

A small circular-loop magnetic dipole, excited in the sinusoidal steady state, was considered in Chapter 1, Section 1.6. Here we consider a sinusoidal pulse excitation. As before, the loop has a radius R that lies in the $x-y$ plane, centered at $z=0$. A uniform current, $I(t) = I_o[\cos(\omega t + \phi_o) - \cos(\phi_o)]u_{-1}(t)$ circulates around the loop creating a magnetic flux proportional to $I(t)$. To avoid the infinities in the radiated power and energy, we again choose $\phi_o = 0$. We continue to define $k = \omega/c$, employ spherical coordinates with $\widehat{\phi}$ the unit vector in the azimuthal direction, and imply multiplication by $u_{-1}(t - \frac{r}{c})$.

For $kR \ll 1$, the vector and scalar retarded potentials are

$$\mathbf{A} = \widehat{\phi}\frac{\mu_o I_o R^2}{4}\left[\frac{-kr\sin(\omega t - kr) + \cos(\omega t - kr) - 1}{r^2}\right]\sin\theta \qquad (13.6a)$$

$$\Phi = 0 \qquad (13.6b)$$

The reader is again cautioned to remember that although the current is $J_\phi\widehat{\phi}$, the unit vector cannot simply be moved from inside to outside the integral! Instead, it must be converted to its Cartesian components before the integrations that determine $\mathbf{A}$ are attempted. Afterwards, the components can be converted to spherical coordinates; of course, only the ϕ-component will survive. When there is dissipation, the current-step term will decay at a rate governed by the appropriate time constant, as expressed in terms of the retarded time.

The electric and magnetic fields follow from Eqs. (1.9) and (1.10),

$$\mathbf{E} = \widehat{\phi}\frac{c\mu_o I_o\, kR^2}{4\widehat{\mathbf{r}}\,2}\left[kr\cos(\omega t - kr) + \sin(\omega t - kr)\right]\sin\theta \qquad (13.7a)$$

[3]Of course, in the dc limit, the dipole cannot radiate at all because $I_o = 0$.

$$\mathbf{H} = \frac{-I_o R^2}{4r^3}\begin{pmatrix} \widehat{\mathbf{r}}2[kr\sin(\omega t - kr) + 1 - \cos(\omega t - kr)]\cos\theta \\ +\widehat{\theta}[kr\sin(\omega t - kr) + 1 - (1 - k^2r^2)\cos(\omega t - kr)]\sin\theta \end{pmatrix} \quad (13.7b)$$

Figures 13.1 and 13.2 are also applicable to this transient, except that now the plots represent the magnetic field.

In the Maxwell–Poynting representation, we have

$$\mathbf{E}\times\mathbf{H} = \frac{c\mu_o I_o^2 kR^4}{32r^5}\begin{pmatrix} \widehat{\mathbf{r}}\begin{bmatrix} k^3r^3 + (k^3r^3 - 2kr)\cos 2(\omega t - kr) \\ +(2k^2r^2 - 1)\sin 2(\omega t - kr) \\ +2[kr\cos(\omega t - kr) + \sin(\omega t - kr)] \end{bmatrix}\sin^2\theta \\ +\widehat{\theta}\begin{bmatrix} 2kr\ [\cos 2(\omega t - kr) - \cos(\omega t - kr)] \\ +(2 - k^2r^2)\sin 2(\omega t - kr) \\ -2\sin(\omega t - kr) \end{bmatrix}\sin 2\theta \end{pmatrix} \quad (13.8a)$$

$$W_e = \frac{\mu_o I_o^2\ R^4}{64r^6}\begin{pmatrix} \left[1 + kr\sin(\omega t - kr) + \left(k^2r^2 - 1\right)\cos(\omega t - kr)\right]^2\sin^2\theta \\ +4\,[kr\sin(\omega t - kr) - \cos(\omega t - kr) + 1]^2\cos^2\theta \end{pmatrix} \quad (13.8b)$$

$$W_m = \frac{\mu_o I_o^2\ k^2R^4}{64r^4}[kr\cos(\omega t - kr) + \sin(\omega t - kr)]^2\sin^2\theta \quad (13.8c)$$

In addition to the time-averaged components, first- and second-harmonic energy and power (directed along $\widehat{\theta}$ as well as $\widehat{\mathbf{r}}$) are both in evidence. Figure 13.3 is applicable to the Poynting fluxes of either type of dipole.

We turn now to the Alternate formulation. In regions where $r \gg R$, there is neither current nor charge and from Eqs. (3.31a) and (3.31b) we obtain

$$\mathbf{S}^o = \frac{1}{2\mu_o}\left(A_\phi\frac{\partial\nabla A_\phi}{\partial t} - \nabla A_\phi\frac{\partial A_\phi}{\partial t}\right) \quad (13.9a)$$

$$W^o = \frac{1}{2}\varepsilon_o\left(\frac{\partial A_\phi}{\partial t}\frac{\partial A_\phi}{\partial t} - A_\phi\frac{\partial^2 A_\phi}{\partial t^2}\right) \quad (13.9b)$$

Direct substitution reveals that

$$\mathbf{S}^o = \widehat{\mathbf{r}}\,\frac{\mu_o c\ I_o^2\ k^3R^4}{32r^3}[kr + \sin(\omega t - kr)]\sin^2\theta \quad (13.10a)$$

$$W^o = \frac{\mu_o I_o^2 k^2R^4}{32r^4}[k^2r^2 + kr\sin(\omega t - kr) + 1 - \cos(\omega t - kr)]\sin^2\theta \quad (13.10b)$$

As with the electric dipole, there is only a radial component of $\mathbf{S}^o$, but notice that Eq. (13.10a) differs from Eq. (13.4a) in the range $kr < 1$. However, except for that range, Figure 13.4 also applies to the magnetic dipole.

The second-harmonic components are still missing, although components at the fundamental are again present in *both* the Maxwell–Poynting and Alternate formulations;

these are due to the static magnetic dipole that becomes negligible when $kr \gg 1$. Most of the comments that apply to the electric dipole also apply here.

When $kr \ll 1$, the r^{-4} energy contribution must be considered together with $\frac{1}{2}A_\phi J_\phi$; the latter term is present in the limit $\omega \to 0$. This inductive energy is localized to the current loop that produces the magnetic dipole of moment, $\hat{\mathbf{z}} I_o \pi R^2$.

Finally, we observe that since magnetic and electric dipoles are duals, so too, are their fields. Here the magnetic dipole arises from the electric-current loop, but we could have used magnetic charge and current to model the fields by generating $\mathbf{A}_m$ and Φ_m (identical in form to Eqs. (13.1a) and (13.1b)). Power and energy would then be expressed in the dual Alternate representation. In that case, Figure 13.4 applies exactly.

13.3 RADIATION FROM AN ACCELERATED THEN DECELERATED CHARGE

Consider a positive charge q, located at a distance $z = -d$ from the origin of spherical coordinates. The charge is motionless for time $t < -T$; it experiences constant acceleration, $+a_o$, in the positive z direction for $-T \le t < 0$ and experiences constant deceleration, $-a_o$, along z for $0 < t \le T$. Further assume that the peak velocity is nonrelativistic ($a_o T \ll c$) and $d = \frac{1}{2}a_o T^2$ so that at $t = T$ we have $z = +d$. Consequently, the position of the charge, $z_o(t)$, is antisymmetric in time and given by

$$z_o(t) = \begin{cases} -\frac{1}{2}a_o T^2 & (t \le -T) \\ \frac{1}{2}a_o t(2T + t) & (-T \le t \le 0) \\ \frac{1}{2}a_o t(2T - t) & (0 \le t \le T) \\ \frac{1}{2}a_o T^2 & (T \le t) \end{cases} \tag{13.11}$$

The acceleration, velocity, and position of the charge are sketched in Figure 13.5.

If we place a pair of stationary charges ($\pm q$) at the origin, the fields are not affected, but we can use linearity and superposition to consider the source to be the sum of a positive charge at the origin and an electric dipole with moment $qz_o(t)$ that points along the z axis. For all locations where the spherical radius satisfies $r \gg \frac{1}{2}|z_o(t)|$, the dipole can be considered a point dipole (located at $r = 0$) without incurring significant error. The moving charge also creates a vector potential that, in turn, modifies the dipole potential. Because, in the Lorenz gauge, we have

$$\frac{\partial A_z}{\partial r}\cos\theta + \frac{1}{c^2}\frac{\partial \Phi}{\partial t} = 0$$

the approximate retarded potentials, derived in Appendix E, Eqs. (E.9a) and (E.9b), are

$$A_z = \frac{\mu_o q v^*}{4\pi r} \tag{13.12a}$$

$$\Phi = \frac{q}{4\pi\varepsilon_o}\left(\frac{1}{r} + \frac{z_o^* + \mathrm{v}^* \frac{r}{c}}{r^2}\cos\theta\right) \tag{13.12b}$$

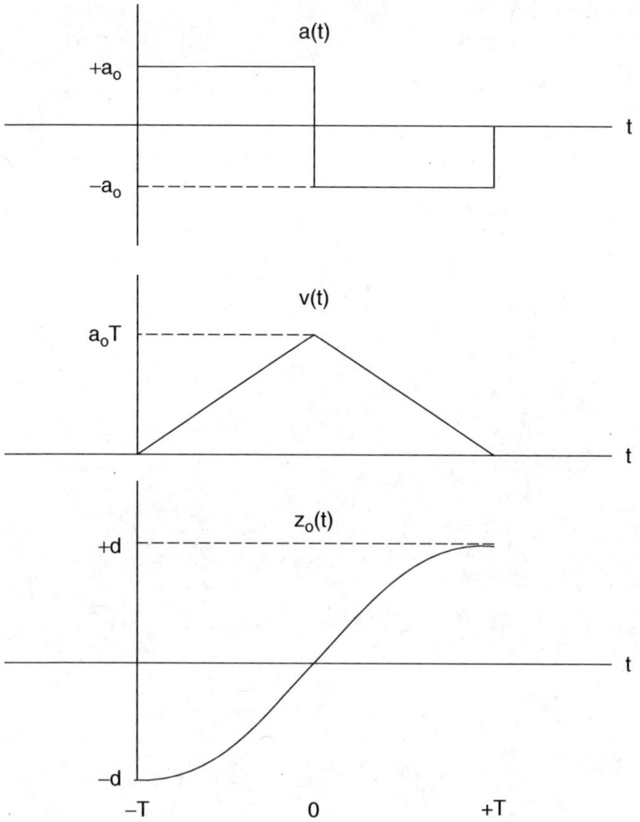

Figure 13.5 The acceleration, velocity, and position of a positive charge.

where $t^* = t - \frac{r}{c}$. During the acceleration $(-T < t < 0)$, we have

$$z_0^* = -d + \frac{1}{2}a_0(T + t^*)^2 = \frac{1}{2}a_0 t^*(2T + t^*)$$

$$v^* = a_0(T + t^*)$$

During the deceleration $(0 < t < T)$, we obtain

$$z_0^* = d - \frac{1}{2}a_0(T - t^*)^2 = \frac{1}{2}a_0 t^*(2T - t^*)$$

$$v^* = a_0(T - t^*)$$

Potentials valid for all values of retarded time can be expressed in terms of unit-step functions; the results are

$$A_z = \frac{\mu_0 q a_0}{4\pi r}\left(\begin{array}{l}\left(T + t - \frac{r}{c}\right)\ [u_{-1}(r - ct) - u_{-1}(r - ct - cT)] \\ + \left(T - t + \frac{r}{c}\right)[u_{-1}(r - ct + cT) - u_{-1}(r - ct)]\end{array}\right) \tag{13.13a}$$

$$\Phi = \frac{q}{4\pi\varepsilon_o}\left(\frac{1}{r} + \frac{a_o\cos\theta}{2\,r^2}\begin{bmatrix} [t(2T+t) - \frac{r^2}{c^2}]\left[u_{-1}(r-ct) - u_{-1}(r-ct-cT)\right] \\ + [t(2T-t) + \frac{r^2}{c^2}]\left[u_{-1}(r-ct+cT) - u_{-1}(r-ct)\right] \\ + T^2\,[u_{-1}(ct-r-cT) - u_{-1}(r-ct-cT)] \end{bmatrix}\right) \tag{13.13b}$$

Consequently, the electric and magnetic fields (neglecting the static electric dipole terms) are as follows:

For $c(t+T) < r < c(t-T)$,

$$\mathbf{E} = \mathbf{E}_0 = \hat{\mathbf{r}}\frac{q}{4\pi\varepsilon_o}\frac{1}{r^2} \tag{13.14a}$$

$$\mathbf{H} = \mathbf{H}_0 = 0 \tag{13.14b}$$

For $-cT < ct < r < c(t+T)$,

$$\mathbf{E} = \mathbf{E}_1 = \begin{pmatrix} \hat{\mathbf{r}}\left[\frac{q}{4\pi\varepsilon_o r^2} + q\mu_o a_o\frac{c^2 t(2T+t) - r^2}{4\pi r^3}\cos\theta\right] \\ +\hat{\theta}\,q\mu_o a_o\frac{c^2 t(2T+t) + r^2}{8\pi r^3}\sin\theta \end{pmatrix} \tag{13.15a}$$

$$\mathbf{H} = \mathbf{H}_1 = \hat{\phi}\frac{q a_o(T+t)}{4\pi r^2}\sin\theta \tag{13.15b}$$

For $0 < c(t-T) < r < ct$,

$$\mathbf{E} = \mathbf{E}_2 = \begin{pmatrix} \hat{\mathbf{r}}\left[\frac{q}{4\pi\varepsilon_o r^2} + q\mu_o a_o\frac{c^2 t(2T-t) + r^2}{4\pi r^3}\cos\theta\right] \\ +\hat{\theta}\,q\mu_o a_o\frac{c^2 t(2T-t) - r^2}{8\pi r^3}\sin\theta \end{pmatrix} \tag{13.16a}$$

$$\mathbf{H} = \mathbf{H}_2 = \hat{\phi}\frac{q a_o(T-t)}{4\pi r^2}\sin\theta \tag{13.16b}$$

As expected, the dominant fields are proportional to $1/r$ in the radiation zone ($r \simeq ct \gg cT$). In that zone, the tangential components E_θ and H_ϕ reverse polarity at the wave front, $r = ct$, which is created when the acceleration of the charge reverses. The significant result is that *acceleration of charge produces radiation.*

Maxwell–Poynting representation

The corresponding Poynting fluxes and energy densities in the Maxwell–Poynting representation are

$$\mathbf{S}_0 = 0 \tag{13.17a}$$

$$\mathbf{S}_1 = \frac{q^2 a_0 (T+t)}{16\pi^2 \varepsilon_0 r^5} \begin{pmatrix} \hat{\mathbf{r}} \frac{a_0}{2} \left[t(2T+t) + \frac{r^2}{c^2} \right] \sin^2\theta \\ -\hat{\theta} \left[r + a_0 t(2T+t) - \frac{r^2}{c^2} \cos\theta \right] \sin\theta \end{pmatrix} \tag{13.17b}$$

$$\mathbf{S}_2 = \frac{q^2 a_0 (T-t)}{16\pi^2 \varepsilon_0 r^5} \begin{pmatrix} \hat{\mathbf{r}} \frac{a_0}{2} \left[t(2T-t) - \frac{r^2}{c^2} \right] \sin^2\theta \\ -\hat{\theta} \left[r + a_0 t(2T-t) + \frac{r^2}{c^2} \cos\theta \right] \sin\theta \end{pmatrix} \tag{13.17c}$$

and

$$W_0^{em} = \frac{q^2}{32\pi^2 \varepsilon_0} \frac{1}{r^4} \tag{13.18a}$$

$$W_1^{em} = \frac{q^2}{32\pi^2 \varepsilon_0 r^6} \begin{pmatrix} [r + a_0 \left(t^2 + 2T\,t - \frac{r^2}{c^2} \right) \cos\theta]^2 \\ + a_0^2 \left[\frac{r^2}{c^2} (T+t)^2 + \frac{1}{4} \left(t^2 + 2Tt + \frac{r^2}{c^2} \right)^2 \right] \sin^2\theta \end{pmatrix} \tag{13.18b}$$

$$W_2^{em} = \frac{q^2}{32\pi^2 \varepsilon_0 r^6} \begin{pmatrix} [r - a_0 (t^2 - 2T\,t - \frac{r^2}{c^2}) \cos\theta]^2 \\ + a_0^2 \left[\frac{r^2}{c^2} (T-t)^2 + \frac{1}{4} \left(t^2 - 2Tt + \frac{r^2}{c^2} \right)^2 \right] \sin^2\theta \end{pmatrix} \tag{13.18c}$$

Alternate power and energy representation

In the Alternate-power representation, the corresponding quantities are

$$\mathbf{S}_0^o = 0 \tag{13.19a}$$

$$\mathbf{S}_1^o = \frac{\mu_0 q^2 a_0}{32\pi^2 r^4} \begin{pmatrix} \hat{\mathbf{r}} \left(a_0 r \left[\frac{r}{c} - (T+t)\cos^2\theta \right] + c^2 (T+t) \cos\theta \right) \\ -\hat{\theta}\, c^2 (T+t) \sin\theta \end{pmatrix} \tag{13.19b}$$

$$\mathbf{S}_2^o = \frac{\mu_0 q^2 a_0}{32\pi^2 r^4} \begin{pmatrix} \hat{\mathbf{r}} \left(a_0 r \left[\frac{r}{c} + (T-t)\cos^2\theta \right] + c^2 (T-t) \cos\theta \right) \\ -\hat{\theta}\, c^2 (T-t) \sin\theta \end{pmatrix} \tag{13.19c}$$

and

$$W_0^o = 0 \tag{13.20a}$$

$$W_1^o = \frac{\mu_0 q^2 a_0}{64\pi^2 r^4} \left(a_0 \left[\frac{2r^2}{c^2} - \left(t^2 + 2Tt + 2T^2 + \frac{r^2}{c^2} \right) \cos^2\theta \right] + 2r\cos\theta \right) \tag{13.20b}$$

$$W_2^{\mathrm{o}} = \frac{\mu_{\mathrm{o}} q^2 a_{\mathrm{o}}}{64\pi^2 r^4} \left(a_{\mathrm{o}} \left[\frac{2r^2}{c^2} - \left(t^2 - 2Tt + 2T^2 + \frac{r^2}{c^2} \right) \cos^2\theta \right] - 2r\cos\theta \right) \tag{13.20c}$$

But, this is not the whole story. The operators $\nabla\left(\frac{\partial}{\partial t}\right)$ and $\frac{\partial^2}{\partial t^2}$, when applied to the components of **A** (and Φ), can sometimes produce approximate infinities at locations where the potentials are nonzero and the electric and magnetic fields remain finite. In such cases both the Alternate-energy density and the Alternate-power flux have infinities and thus localized contributions whereas their Maxwell–Poynting counterparts do not. In the present example, this occurs with both A_z and Φ at $r = ct > 0$, where the effect of the sudden reversal of the acceleration causes tangential field discontinuities that produce the impulses,

$$(S_r^{\mathrm{o}})_{\mathrm{local}} = c(W^{\mathrm{o}})_{\mathrm{local}} = \frac{\mu_{\mathrm{o}} q^2 a_{\mathrm{o}}^2 T}{(4\pi r)^2} \sin^2\theta \; u_0(r - ct) \tag{13.21}$$

Each impulse is seen to contribute 50% of the total radiated power/energy when integrated through the spherical surface of radius ct. These contributions are required because, in the radiation zone, the nonlocalized components satisfy $S_r^{\mathrm{o}}/S_r = W^{\mathrm{o}}/W^{\mathrm{em}} = 1/2$. It follows that the total energy radiated from the charge is $\frac{\mu_{\mathrm{o}} q^2 a_{\mathrm{o}}^2 T}{3\pi c}$ (in agreement with the Maxwell–Poynting calculation).

If the motion of the charge is made periodic in time (by alternating constant positive and negative acceleration for intervals $2T$), the solutions just given can be superimposed with appropriate polarity and shift of the time origin. In this case, the localized Alternate-power and Alternate-energy impulses occur periodically in time created whenever the acceleration reverses.

If the reversal of the acceleration is made continuous, but still fairly abrupt, the impulses will be replaced by finite peaks in the Alternate-power flux and energy density.

CHAPTER 14

UNIFORM SURFACE CURRENT

14.1 PULSE EXCITATIONS

A uniform surface current of infinite extent, $\mathbf{K} = \widehat{\mathbf{y}}K_y(t)$, is located at $z = 0$ and surrounded by free-space. This sheet current produces vector and scalar potentials $\mathbf{A} = \widehat{\mathbf{y}}A_y(z,t)$ and (in the Lorenz gauge), $\Phi = 0$; these in turn generate $\mathbf{E} = \widehat{\mathbf{y}}E_y(z,t)$ and $\mathbf{H} = \widehat{\mathbf{x}}H_x(z,t)$. Because the exact solution is available for an arbitrary excitation, this configuration allows one to study the differences between the Maxwell–Poynting and Alternate-power formulations as a function of pulse shape. There are a number of surprises that, at first, seem rather startling.

Step function

We begin with a simple current step

$$K_y(t) = K_o u_{-1}(t) \tag{14.1}$$

Because this current generates electric and magnetic fields that are everywhere finite, the value of A_y must be continuous and therefore vanish where $|z| = ct$. Consequently,

$$A_y(z,t) = \frac{\mu_o K_o}{2}\left(ct - |z|\right) u_{-1}\left(t - \frac{|z|}{c}\right) \tag{14.2a}$$

The Power and Beauty of Electromagnetic Fields, First Edition. F. R. Morgenthaler.

and from Eqs. (1.9) and (1.10), we obtain

$$E_y = -\frac{\partial A_y}{\partial t} = \frac{-c\mu_o K_o}{2} u_{-1}\left(t - \frac{|z|}{c}\right) \tag{14.2b}$$

$$H_x = \frac{-1}{\mu_o}\frac{\partial A_y}{\partial z} = \frac{z}{|z|}\frac{K_o}{2} u_{-1}\left(t - \frac{|z|}{c}\right) \tag{14.2c}$$

In Figure 14.1, these fields are plotted as functions of z for the time t. Although the steady-state value of the magnetic field is expected from magnetostatics, the persistent electric field is puzzling until one realizes that the transient wave front is never "far" from the infinite plane of current. For a current strip of finite width (w), considered in Chapter 15, Section 15.3, the *MQS* electric field approaches zero as $1/t$ when $ct \gg w$.

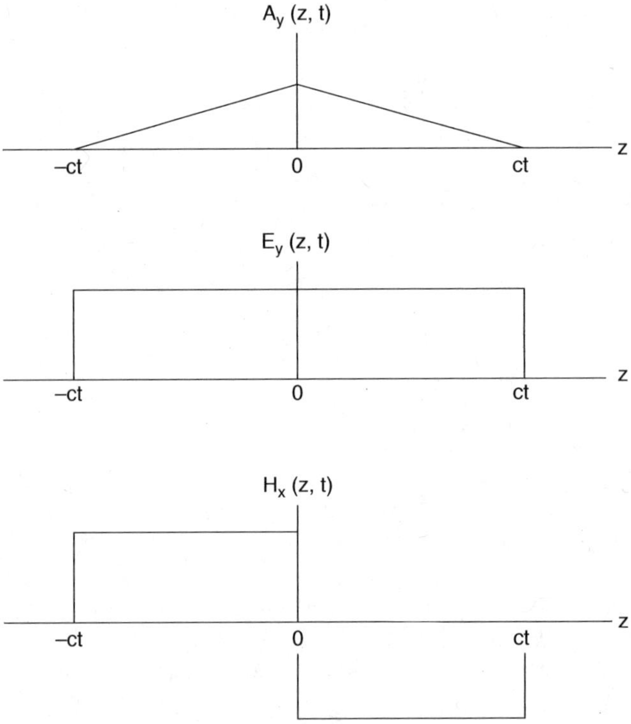

Figure 14.1 Vector potential and fields of surface current step.

The Maxwell–Poynting representation of power and energy is

$$S_z = -E_y H_x = \frac{z}{|z|}\frac{c\mu_o K_o^2}{4} u_{-1}\left(t - \frac{|z|}{c}\right) \tag{14.3a}$$

$$W^{em} = W_e + W_m = \frac{\mu_o K_o^2}{4} u_{-1}\left(t - \frac{|z|}{c}\right) \tag{14.3b}$$

The Alternate formulation, based upon Eqs. (3.31a) and (3.31b), is

$$S_z^o = \frac{z}{|z|}\frac{c\mu_o K_o^2}{8} u_{-1}\left(t - \frac{|z|}{c}\right) \tag{14.4a}$$

$$W^{o} = \frac{1}{2}A_y J_y + \frac{\mu_o K_o^2}{8} u_{-1}\left(t - \frac{|z|}{c}\right) \tag{14.4b}$$

$$\frac{1}{2}A_y J_y = \frac{\mu_o K_o^2}{4} ct\ u_{-1}(t) \tag{14.4c}$$

It is rather a shock to find that $S_z^o = \frac{1}{2}S_z$. Although the time rate of change of the Alternate-energy density that is localized on the plane $z = 0$ supplies the missing power, by what mechanism can it be transferred to some remote detector? Also troubling is the fact that if the current is turned off after some time, T, the localized energy density (with value $\frac{1}{4}\mu_o K_o^2 cT$) will disappear while for all r, such that $cT < r < ct$, the Alternate-power flux and Alternate-energy density cannot possibly have learned of the shut-off! As a first step in understanding this peculiar situation, let us alter $K_y(t)$ so that the step is followed by an (adjustable) exponential decay that allows us to turn off the current either gradually or abruptly.

Exponentially decaying step

We modify the current step to be

$$K_y(t) = K_o \exp(-\alpha t) u_{-1}(t) \qquad (\alpha \geq 0) \tag{14.5}$$

which leads to

$$A_y(z,t) = \frac{\mu_o K_o}{2} \exp\left[-\alpha\left(t - \frac{|z|}{c}\right)\right](ct - |z|)u_{-1}\left(t - \frac{|z|}{c}\right) \tag{14.6a}$$

and

$$E_y = \frac{-c\mu_o K_o}{2} \exp\left[-\alpha\left(t - \frac{|z|}{c}\right)\right] u_{-1}\left(t - \frac{|z|}{c}\right) \tag{14.6b}$$

$$H_x = \frac{z}{|z|}\frac{K_o}{2} \exp\left[-\alpha\left(t - \frac{|z|}{c}\right)\right] u_{-1}\left(t - \frac{|z|}{c}\right) \tag{14.6c}$$

The Maxwell–Poynting representation of power and energy is now

$$S_z = \frac{z}{|z|}\frac{c\mu_o K_o^2}{4} \exp\left[-2\alpha\left(t - \frac{|z|}{c}\right)\right] u_{-1}\left(t - \frac{|z|}{c}\right) \tag{14.7a}$$

$$W^{\text{em}} = \frac{\mu_o K_o^2}{4} \exp\left[-2\alpha\left(t - \frac{|z|}{c}\right)\right] u_{-1}\left(t - \frac{|z|}{c}\right) \tag{14.7b}$$

and the Alternate formulation,

$$S_z^o = \frac{z}{|z|}\frac{c\mu_o K_o^2}{8} \exp\left[-\alpha\left(t - \frac{|z|}{c}\right)\right] u_{-1}\left(t - \frac{|z|}{c}\right) \tag{14.8a}$$

$$W^{o} = \frac{1}{2}A_y J_y + \frac{\mu_o K_o^2}{8} \exp\left[-\alpha\left(t - \frac{|z|}{c}\right)\right] u_{-1}\left(t - \frac{|z|}{c}\right) \tag{14.8b}$$

$$\frac{1}{2}A_y J_y = \frac{\mu_o K_o^2}{4} ct\ \exp(-2\alpha t)\, u_{-1}(t) \tag{14.8c}$$

When $\alpha = 0$, the results of the previous section are recovered.

Although the Maxwell–Poynting and Alternate energies are now distributed differently, direct evaluation verifies that the total energy is independent of the representation. For $\alpha t \gg 1$, the localized Alternate energy is negligible and it is easy to compare the free-space values. Although at $|z| = ct$ the ratio of $W^{\mathrm{o}}/W^{\mathrm{em}} = 1/2$, independent of α, the ratio of the respective decay rates is also one-half. It therefore follows that the *total* free-space energies are identical. We postpone until later the question of how the power dissipated in a remote detector located at $z = z_{\mathrm{o}}$ can be the same function of time when calculated using the two different representations.

Step of duration *T* followed by an exponential decay

We next modify the current to be a step of constant amplitude and duration T followed by an exponential decay:

$$K_y(t) = K_{\mathrm{o}}\left[u_{-1}(t) + (\exp\left[-\alpha(t-T)\right] - 1)\, u_{-1}(t-T)\ \right] \qquad (\alpha \geq 0) \tag{14.9}$$

This current generates

$$\begin{aligned} A_y(z,t) = \frac{c\mu_{\mathrm{o}}K_{\mathrm{o}}}{2\alpha}\Bigg\{&\alpha\left(t - \frac{|z|}{c}\right)u_{-1}\left(t - \frac{|z|}{c}\right) \\ &+\left(1 + \alpha\left(T - t + \frac{|z|}{c}\right) - \exp\left[-\alpha\left(t - T - \frac{|z|}{c}\right)\right]\right)u_{-1}\left(t - T - \frac{|z|}{c}\right)\Bigg\} \end{aligned} \tag{14.10a}$$

and

$$\begin{aligned} E_y &= \frac{-c\mu_{\mathrm{o}}K_{\mathrm{o}}}{2}\Bigg\{u_{-1}\left(t - \frac{|z|}{c}\right) \\ &\quad + \left(\exp\left[-\alpha\left(t - T - \frac{|z|}{c}\right)\right] - 1\right)u_{-1}\left(t - T - \frac{|z|}{c}\right)\Bigg\} \end{aligned} \tag{14.10b}$$

$$\begin{aligned} H_x &= \frac{z}{|z|}\frac{K_{\mathrm{o}}}{2}\Bigg\{u_{-1}\left(t - \frac{|z|}{c}\right) \\ &\quad + \left(\exp\left[-\alpha\left(t - T - \frac{|z|}{c}\right)\right] - 1\right)u_{-1}\left(t - T - \frac{|z|}{c}\right)\Bigg\} \end{aligned} \tag{14.10c}$$

In Figure 14.2, these fields are plotted as functions of z for $\alpha \to \infty$ and times $t < T$, $t = T$, and $t > T$.

The Maxwell–Poynting representation of power and energy is now

$$\begin{aligned} S_z &= \frac{z}{|z|}\frac{c\mu_{\mathrm{o}}K_{\mathrm{o}}^2}{4}\Bigg\{u_{-1}\left(t - \frac{|z|}{c}\right) \\ &\quad + \left(\exp\left[-2\alpha\left(t - T - \frac{|z|}{c}\right)\right] - 1\right)u_{-1}\left(t - T - \frac{|z|}{c}\right)\Bigg\} \end{aligned} \tag{14.11a}$$

$$\begin{aligned} W^{\mathrm{em}} &= \frac{\mu_{\mathrm{o}}K_{\mathrm{o}}^2}{4}\Bigg\{u_{-1}\left(t - \frac{|z|}{c}\right) \\ &\quad + \left(\exp\left[-2\alpha\left(t - T - \frac{|z|}{c}\right)\right] - 1\right)u_{-1}\left(t - T - \frac{|z|}{c}\right)\Bigg\} \end{aligned} \tag{14.11b}$$

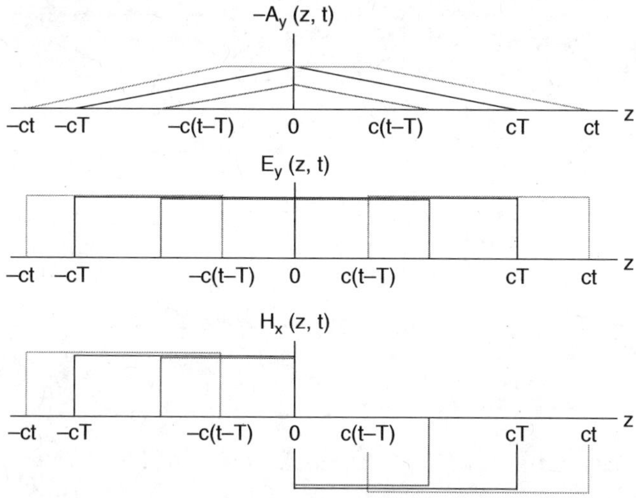

Figure 14.2 Vector potential and fields of surface current pulse.

while the Alternate formulation evaluates to

$$S_z^{\mathrm{o}} = \frac{z}{|z|}\frac{c\mu_{\mathrm{o}}K_{\mathrm{o}}^2}{8}\left\{u_{-1}\left(t-\frac{|z|}{c}\right)\right.$$
$$\left.+\left[(1+\alpha T)\exp\left[-\alpha\left(t-T-\frac{|z|}{c}\right)\right]-1\right]u_{-1}\left(t-T-\frac{|z|}{c}\right)\right\} \qquad (14.12a)$$

$$W^{\mathrm{o}} = \frac{1}{2}A_yJ_y + \frac{\mu_{\mathrm{o}}K_{\mathrm{o}}^2}{8}\left\{u_{-1}\left(t-\frac{|z|}{c}\right)\right.$$
$$\left.+\left[(1+\alpha T)\exp\left[-\alpha\left(t-T-\frac{|z|}{c}\right)\right]-1\right]u_{-1}\left(t-T-\frac{|z|}{c}\right)\right\} \qquad (14.12b)$$

$$\frac{1}{2}A_yJ_y = \frac{c\mu_{\mathrm{o}}K_{\mathrm{o}}^2}{4}\left\{t\,u_{-1}(t)\right.$$
$$\left.+\left(\left[\frac{1}{\alpha}+T-\frac{1}{\alpha}\exp\left(\alpha\,[T-t]\right)\right]\exp\left(\alpha\,[T-t]\right)-t\right)u_{-1}(t-T)\right\} \qquad (14.12c)$$

When $T=0$, this case reduces to the previous one. As the current pulse is turned off more and more abruptly, $\alpha \to \infty$ and $\frac{d}{dt}(\frac{1}{2}A_yJ_y)$ develops a temporal impulse at $t=T$. Even in that limit, conservation of energy is maintained because the peak values of both S_z^{o} and W^{o} are increased by the factor $1+\alpha T$. Therefore, both Alternate power and Alternate energy develop impulses at their trailing edges, $z=\pm c(t-T)$, which propagate symmetrically as localized components that contribute 50% of the total. All temporal impulses coincide at the location of the current sheet when $t=T$. These fluxes are plotted in Figure 14.3 for the cases $\alpha T:1,4$ when $t=5T$ and $cT=1$.

For all three pulse waveforms that have been studied, the ratio $\frac{W^{\mathrm{o}}}{W^{\mathrm{em}}}$ is $\frac{1}{2}$ at $|z|=ct$. We shall learn in the next section (where we alter the initial rise of the current) that this is a consequence of the current starting as a step function. With appropriate pulse shape, the ratio can be made even smaller!

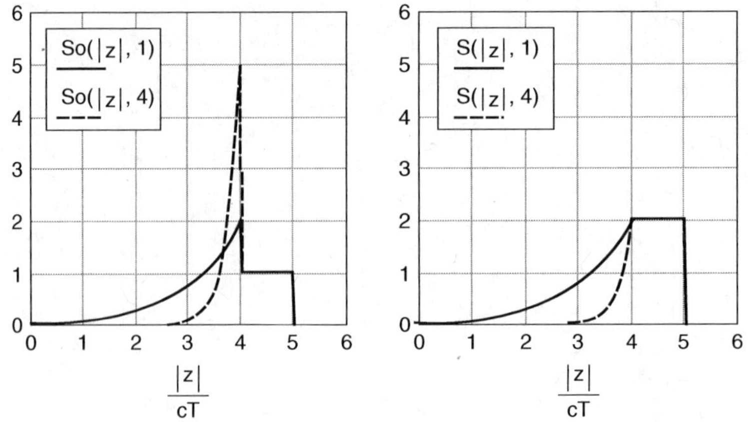

Figure 14.3 Poynting and Alternate-power fluxes for $t = 5T$.

Pulses of duration T with t^n rise

We consider the family of surface-current pulses, all of duration T,

$$K_y(t,n) = K_o \left(\frac{t}{T}\right)^n [u_{-1}(t) - u_{-1}(t-T)] \qquad (n \geq 0) \tag{14.13}$$

The first four members with integer values of n $(0, 1, 2, 3)$ are shown (normalized) in Figure 14.4.

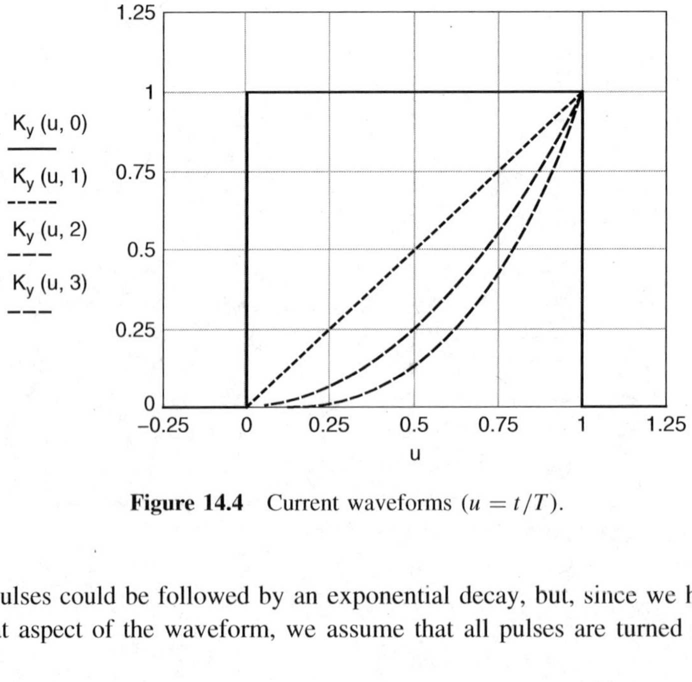

Figure 14.4 Current waveforms ($u = t/T$).

These pulses could be followed by an exponential decay, but, since we have already studied that aspect of the waveform, we assume that all pulses are turned off abruptly

($\alpha \to \infty$). Then,

$$A_y(z,t) = \frac{\mu_o K_o}{2} \frac{cT}{(n+1)} \left\{ \left(\frac{ct - |z|}{cT} \right)^{n+1} \left[u_{-1}\left(t - \frac{|z|}{c}\right) - u_{-1}\left(t - T - \frac{|z|}{c}\right) \right] + u_{-1}\left(t - T - \frac{|z|}{c}\right) \right\} \tag{14.14a}$$

and

$$E_y = \frac{-c\mu_o K_o}{2} \left(\frac{ct - |z|}{cT} \right)^n \left[u_{-1}\left(t - \frac{|z|}{c}\right) - u_{-1}\left(t - T - \frac{|z|}{c}\right) \right] \tag{14.14b}$$

$$H_x = \frac{z}{|z|} \frac{K_o}{2} \left(\frac{ct - |z|}{cT} \right)^n \left[u_{-1}\left(t - \frac{|z|}{c}\right) - u_{-1}\left(t - T - \frac{|z|}{c}\right) \right] \tag{14.14c}$$

The Maxwell–Poynting representation of power and energy produces

$$S_z = \frac{z}{|z|} \frac{c\mu_o K_o^2}{4} \left(\frac{ct - |z|}{cT} \right)^{2n} \left[u_{-1}\left(t - \frac{|z|}{c}\right) - u_{-1}\left(t - T - \frac{|z|}{c}\right) \right] \tag{14.15a}$$

$$W^{em} = \frac{\mu_o K_o^2}{4} \left(\frac{ct - |z|}{cT} \right)^{2n} \left[u_{-1}\left(t - \frac{|z|}{c}\right) - u_{-1}\left(t - T - \frac{|z|}{c}\right) \right] \tag{14.15b}$$

and the Alternate formulation

$$S_z^o = \frac{z}{|z|} \frac{c\mu_o K_o^2}{8(n+1)} \left[\left(\frac{ct - |z|}{cT} \right)^{2n} \left[u_{-1}\left(t - \frac{|z|}{c}\right) - u_{-1}\left(t - T - \frac{|z|}{c}\right) \right] + Tu_0\left(t - T - \frac{|z|}{c}\right) \right] \tag{14.16a}$$

$$W^o = \frac{1}{2} A_y J_y + \frac{\mu_o K_o^2}{8(n+1)} \left[\left(\frac{ct - |z|}{cT} \right)^{2n} \left[u_{-1}\left(t - \frac{|z|}{c}\right) - u_{-1}\left(t - T - \frac{|z|}{c}\right) \right] + Tu_0\left(t - T - \frac{|z|}{c}\right) \right] \tag{14.16b}$$

$$\frac{1}{2} A_y J_y = \frac{\mu_o K_o^2}{4(n+1)} ct \left(\frac{t}{T} \right)^{2n} [u_{-1}(t) - u_{-1}(t - T)] u_0(z) \tag{14.16c}$$

Notice that, in addition to the localized energy on the current sheet, localized Alternate power and Alternate energy in free-space are generated by $A_y \frac{\partial^2 A_y}{\partial z \partial t}$ and $A_y \frac{\partial^2 A_y}{\partial t^2}$[1]. These impulses, which previously invited disbelief, but are now at least partly understood, occur at the moment the current pulse is complete; they are required to ensure that Alternate energy is conserved. As noted previously, at time T, the energy impulse $\frac{1}{2} A_y K_y u_0(z)$ splits with one-half traveling in each of the positive and negative z directions. This remarkable fact would seem to imply grave contradictions with the Maxwell–Poynting

[1]Such localization also occurs in the example of the accelerated then decelerated charge that is discussed in Chapter 13, Section 13.3.

formulation because, for $t < T$, $S_z^{\mathrm{o}} = \frac{1}{2(n+1)} S_z$, and (in free-space) the energy per unit volume is given by $W^{\mathrm{o}} = \frac{1}{2(n+1)} W^{\mathrm{em}}$. For a rectangular pulse ($n = 0$) the reduction factor is $\frac{1}{2}$, but approaches zero for very large n. While it is true that for $t > T$ the Alternate-energy impulse restores the missing energy to free-space, it occurs at the last possible moment and so cannot possibly affect earlier portions of the distribution. Assume that a resistive sheet located at $z = z_{\mathrm{o}} < cT$ is used to detect the $+z$-directed wave. How can the reduced flux, $S_z^{\mathrm{o}} = \frac{1}{2(n+1)} S_z$, result in *identical* power dissipation within the sheet for *all* $t < \frac{z_{\mathrm{o}}}{c} + T$ (*before* the impulse arrives)? Also, what happens when $z_{\mathrm{o}} > cT$ and the energy and power impulses at the end of the pulse finally do arrive? Certainly the actual power dissipation cannot become infinite for even the briefest instant of time. We next calculate the power dissipation in the sheet and show that there are no contradictions. Although the analysis can be made for any of the pulses that have been considered, we continue with the waveform under discussion.

14.2 RESISTIVE-SHEET DETECTOR

Assume that a very thin sheet of electrical conductivity, σ, and thickness, δ, is located at the plane, $z = z_{\mathrm{o}}$. For convenience, we define $z' = z - z_{\mathrm{o}}$ and $t' = t - z_{\mathrm{o}}/c$ so that the new origins are located at the position of the sheet and the time of arrival of the pulse.

The detector causes a reflected vector-potential wave, A_y^-, to be created for $z' < 0$ as well as a transmitted wave, A_y^t, for $z' > 0$. In terms of the reflection coefficient, Γ, these waves are

$$A_y^- = \Gamma \frac{\mu_{\mathrm{o}} K_{\mathrm{o}} c T}{2(n+1)} \left[\left(\frac{ct' + z'}{cT} \right)^{n+1} \left[u_{-1}\left(t' + \frac{z'}{c}\right) - u_{-1}\left(t' - T + \frac{z'}{c}\right) \right] + u_{-1}\left(t' - T + \frac{z'}{c}\right) \right] \tag{14.17a}$$

$$A_y^t = (1+\Gamma) \frac{\mu_{\mathrm{o}} K_{\mathrm{o}} c T}{2(n+1)} \left[\left(\frac{ct' - z'}{cT} \right)^{n+1} \left[u_{-1}\left(t' - \frac{z'}{c}\right) - u_{-1}\left(t' - T - \frac{z'}{c}\right) \right] + u_{-1}\left(t' - T - \frac{z'}{c}\right) \right] \tag{14.17b}$$

When the incident wave, Eq. (14.14a), is included, A_y is continuous through the detector with the value

$$A_y(0,t') = (1+\Gamma) \frac{\mu_{\mathrm{o}} K_{\mathrm{o}} c T}{2(n+1)} \left[\left(\frac{t'}{T} \right)^{n+1} [u_{-1}(t') - u_{-1}(t' - T)] + u_{-1}(t' - T) \right] \tag{14.18}$$

The electric field within the sheet is therefore

$$E_y(0,t') = E_\sigma(t') = -(1+\Gamma) \frac{c\mu_{\mathrm{o}} K_{\mathrm{o}}}{2} \left(\frac{t'}{T} \right)^n [u_{-1}(t') - u_{-1}(t' - T)] \tag{14.19}$$

and generates both an effective surface current, $K_y(t') = \sigma\delta E_\sigma(t')$, and the accompanying localized energy density, $\frac{1}{2} A_y(0,t') K_y(t') u_0(z')$, where

$$\frac{1}{2} A_y(0,t') K_y(t') = -\frac{1}{2} \sigma\delta T \frac{(1+\Gamma)^2}{n+1} \left(\frac{c\mu_{\mathrm{o}} K_{\mathrm{o}}}{2} \right)^2 \left(\frac{t'}{T} \right)^{2n+1} [u_{-1}(t') - u_{-1}(t' - T)] \tag{14.20}$$

Although this localized energy does not appear on the resistive sheet until $t = T$ $(t' = 0)$, and thereafter builds up as $(t'/T)^{2n+1}$, its time derivative immediately starts to supply the power missing from S_z^{o}. Application of the integral form of the normal and Alternate–Poynting theorem to a volume of thickness, δ, that contains unit area of the sheet-detector leads, in turn, to

$$S_z(0+,t) - S_z(0-,t) = -E_\sigma K_y(t') \tag{14.21}$$

and

$$S_z^{\text{o}}(0+,t) - S_z^{\text{o}}(0-,t) + \frac{d}{dt'}\left[\frac{1}{2}A_y(0,t')K_y(t')\right] = -E_\sigma K_y(t') \tag{14.22}$$

These equations become

$$\left[(1+\Gamma)^2 - (1-\Gamma^2)\right](S_z)^+ = 2\Gamma(1+\Gamma)\,(S_z)^+ = -E_\sigma K_y(t') \tag{14.23}$$

and

$$2\Gamma(1+\Gamma)\left(S_z^{\text{o}}\right)^+ + \frac{d}{dt'}\left[\frac{1}{2}A_y(0,t')K_y(t')\right] = -E_\sigma K_y(t') \tag{14.24a}$$

$$2\Gamma(1+\Gamma)\left(S_z^{\text{o}}\right)^+ + \frac{1}{2}A_y(0,t')\frac{d}{dt'}K_y(t') = -\frac{1}{2}E_\sigma K_y(t') \tag{14.24b}$$

Notice that whenever $\frac{d}{dt'}K_y(t') = 0$ (as it does for $n = 0$ – except at the pulse boundaries), the net (free-space) Alternate-power flux term contributes $\frac{1}{2}$ of the total; in general, the factor is $\frac{1}{2(n+1)}$.

From Eqs. (14.15a) and (14.16a), the incident Poynting and Alternate-power fluxes are

$$(S_z)^+ = \frac{c\mu_{\text{o}}K_{\text{o}}^2}{4}\left(\frac{t'}{T}\right)^{2n}\left[u_{-1}(t') - u_{-1}(t'-T)\right] \tag{14.25a}$$

$$\left(S_z^{\text{o}}\right)^+ = \frac{c\mu_{\text{o}}K_{\text{o}}^2}{8(n+1)}\left(\frac{t'}{T}\right)^{2n}\left[u_{-1}(t') - u_{-1}(t'-T) + Tu_0(t'-T)\right] \tag{14.25b}$$

In the Maxwell–Poynting representation, only the Poynting flux contributes to the power dissipation within the resistive sheet. This requires that

$$\Gamma(1+\Gamma)\frac{c\mu_{\text{o}}K_{\text{o}}^2}{2} = -\sigma\delta(1+\Gamma)^2\left(\frac{c\mu_{\text{o}}K_{\text{o}}}{2}\right)^2 \tag{14.26}$$

In the Alternate representation, *both* the Alternate flux and the time rate of change of the localized energy contribute to the power dissipation within the resistive sheet. This requires that (prior to $t' = T$)

$$\Gamma(1+\Gamma)\frac{c\mu_{\text{o}}K_{\text{o}}^2}{4(n+1)} - \frac{(2n+1)}{2(n+1)}\sigma\delta(1+\Gamma)^2\left(\frac{c\mu_{\text{o}}K_{\text{o}}}{2}\right)^2 = -\sigma\delta(1+\Gamma)^2\left(\frac{c\mu_{\text{o}}K_{\text{o}}}{2}\right)^2 \tag{14.27}$$

As required, both Eqs. (14.26) and (14.27) lead to the same result (independent of n), namely

$$\Gamma = -\frac{1}{2}c\mu_{\text{o}}\sigma\delta(1+\Gamma)$$

The reflection coefficient of the vector potential is identical to that of the electric field and agrees with the value calculated from transmission-line theory,

$$\Gamma = \frac{1 - Z_0 Y_L}{1 + Z_0 Y_L} = \frac{-c\mu_0 \sigma \delta}{2 + c\mu_0 \sigma \delta} \tag{14.28}$$

where $Y_L = \frac{1}{c\mu_0} + \sigma\delta$ and $Z_0 = c\mu_0$. It is noteworthy that maximum power is transferred to the sheet not when the sheet conductance (per square), $\sigma\delta$, is $\frac{1}{c\mu_0} = \sqrt{\frac{\varepsilon_0}{\mu_0}}$ $(\Gamma = -\frac{1}{3})$, but rather when $c\mu_0\sigma\delta$ is 2 $(\Gamma = -\frac{1}{2})$.

In the Maxwell–Poynting representation, there is no localized energy and so the Poynting flux must provide all of the power. In the Alternate representation, the time rate of change of the localized energy provides (for $t' < T$) $\frac{2n+1}{2(n+1)}$ of the total power with the Alternate-power flux providing the remaining fraction, $\frac{1}{2(n+1)}$. It is interesting that for a rectangular pulse ($n = 0$) each contribution is one-half of the total, but perhaps even more interesting that as $n \to \infty$, the portion propagated through free-space approaches zero! Finally, notice that at $t' = T$, the Alternate-flux impulse,

$$2\Gamma(1+\Gamma)\frac{c\mu_0 K_0^2}{8(n+1)} T u_0(t' - T)$$

arrives to *exactly cancel* that created by the time rate of change of the localized energy at $t' = T$ [caused by $\frac{dK_y}{dt'} = -K_0 u_0(t' - T)$]. It therefore follows that (irrespective of the value of Γ) the power dissipated in the detector remains finite and is proportional to $(\frac{t'}{T})^{2n}$ over the duration of the pulse in agreement with the Maxwell–Poynting representation; no singularity in the dissipation occurs at $t' = T$. Although there are no inconsistencies between the Maxwell–Poynting and Alternate representation, the free-space fluxes are very different. As n increases, the slower rise of the current pulse becomes increasingly quasistatic in character (low frequencies dominate) and $S_z^o/S_z \to 0$; the abrupt turn-off of the pulse generates high-frequency components and the consequent impulse in S_z^o at the trailing edge of the pulse. The localized energy created in the detector appears as if by magic during the interval $0 < t' < T$; only at time $t' = T$ is the trick explained and the apparent paradox resolved. Neither the pulse shape nor the value of the reflection coefficient affects the agreement because, when $\Phi = 0$, the Poynting and Alternate incident plane-wave power fluxes are related by

$$(S_z)^+ = 2\left(S_z^o\right)^+ - \frac{1}{\mu_0} A_y^+ \frac{\partial^2 A_y^+}{\partial z\, \partial t}$$

Although the fact that (except for the impulse) $S_z^o = \frac{1}{2(n+1)} S_z$ has been reconciled, there is still another apparent paradox because the very first of the examples considered (the Hertzian dipole in the sinusoidal steady state) did not lead to any reduction factor.[2] Why not? The answer is that the sinusoidal steady-state excitation produces localized energy and associated power that are purely reactive and thus without average values. Consequently, the free-space Alternate power and Alternate energy must deliver the correct average value *at all times*. (The time rate of change of the localized energies—whether at source or detector—cannot contribute to the average power.) For arbitrary

[2]However, the same dipole excited with a sinusoidal current step creates a wave front where $A = 0$, $\Phi = 0$, and $\mathbf{S} = 2\mathbf{S}^o$.

nonsinusoidal excitations, the terms $A_j \nabla\left(\frac{\partial A_j}{\partial t}\right)$, $\mathbf{A} \cdot \frac{\partial^2 \mathbf{A}}{\partial t^2}$ and, in case the scalar potential is nonzero, $\Phi \nabla\left(\frac{\partial \Phi}{\partial t}\right)$ and $\Phi \frac{\partial^2 \Phi}{\partial t^2}$ may have singularities. Their existence insures that, once the complete pulse has been generated, the average power and energy will agree in both the Alternate and Maxwell–Poynting representations. For the family of current pulses just studied, the impulses occur at the trailing edge of the pulse, where $\mathbf{A}$ is not zero.

In the event that the surface-current pulse, $K_y(t)$, is piecewise continuous within the interval $0 < t < T$, but with zero average value, the vector potential $\mathbf{A}(ct - |z|)$ satisfies the conditions $\mathbf{A}(0) = \mathbf{A}(cT) = 0$. In this case, even if one or both of $K_y(0)$ and $K_y(T)$ are nonzero, $\mathbf{S}^{\mathrm{o}}$ and W^{o} will not contain impulses.[3] In the next section, we consider pulse waveforms with these properties after examining a single current pulse that is a Gaussian.

14.3 ADDITIONAL PULSE WAVEFORMS

Gaussian current pulse

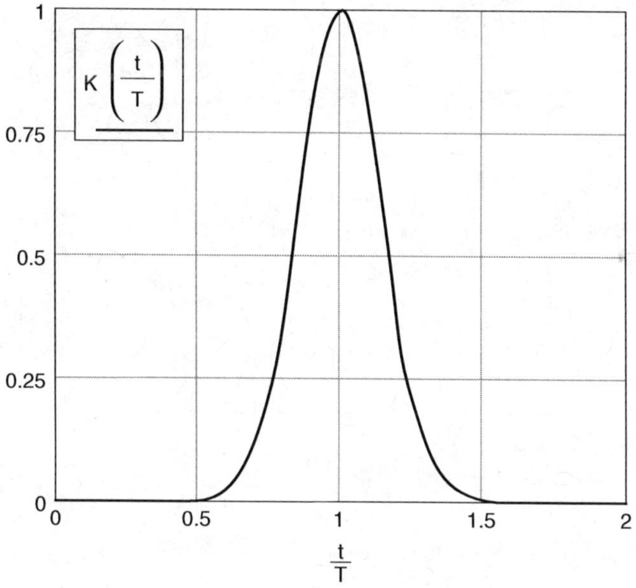

Figure 14.5 Gaussian pulse ($\alpha = 1,\ cT = 5$).

The Gaussian pulse shown in Figure 14.5 is assumed to be of the form

$$K(t) = K_0 \exp[-\alpha c^2 (T - t)^2]\ u_{-1}(t)$$

where $\alpha(cT)^2$ is large enough to make $K(0)$ negligible; consequently, the step function can be omitted. For $z > 0$, the vector potential and electric and magnetic fields are

$$A(z, ct) = \tfrac{1}{4}\mu_0 K_0 \sqrt{\frac{\pi}{\alpha}}[1 - erf(\sqrt{\alpha}(cT + z - ct))]$$

[3] However, if $K_y(t)$ contains step discontinuities within the pulse interval, both S_z^{o} and W^{o} will contain impulses at those locations.

$$E_y(z, ct) = -\tfrac{1}{2}c\mu_0 K_0 \exp[-\alpha(z + cT - ct)^2]$$
$$H_x(z, ct) = \tfrac{1}{2}K_0 \exp[-\alpha(z + cT - ct)^2]$$

where the error function, erf, is defined in Appendix B, Section B.1.

It follows that the Poynting and Alternate power fluxes are given by

$$S_z(z, ct) = \tfrac{1}{4}c\mu_0 K_0^2 \exp[-2\alpha(z + cT - ct)^2]$$
$$S_z^o(z, ct) = \tfrac{1}{8}c\mu_0 K_0^2\{\exp(-2\alpha(z + cT - ct)^2) + \sqrt{\pi\alpha}(z + cT - ct)\exp(-\alpha(z + cT - ct)^2)[erf(\sqrt{\alpha}(z + cT - ct)) - 1]\}$$

The normalized vector potential and these fluxes are plotted in Figure 14.6 as a function of $z/(cT)$ for $t = 2T$ when $\alpha = 1, cT = 5$.

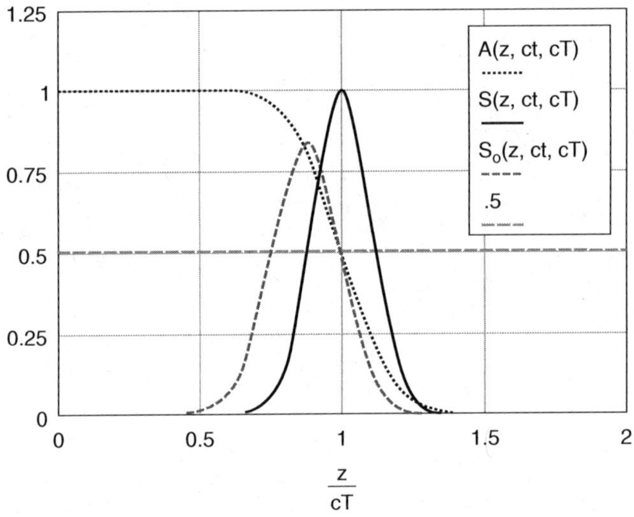

Figure 14.6 Normalized plots of $A(z/cT)$, $S(z/cT)$, $S^o(z/cT)$.

Although the areas of the two flux pulses are equal, notice that the peaks in the power are not coincident. As expected from the analysis of the previous section, the value of the Poynting flux exceeds that of the Alternate flux when the current is increasing; the reverse is true when the current is decreasing. The transition occurs when $z/cT = .928$.

Zero-average waveforms

In the remainder of these examples, the pulses are all of duration $2T$ with zero average. It is convenient to use the normalized variables $\frac{t}{T}$ and $u = \frac{ct-z}{cT}$ to plot the normalized surface current, vector potential, Poynting flux, and Alternate flux. As required, the average values of the power fluxes are equal, but note the sensitivity of S^o to the nonsinusoidal components of **K**.

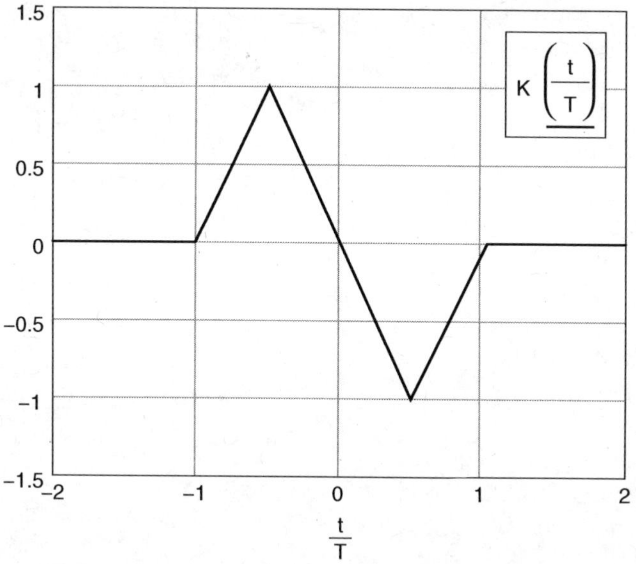

Figure 14.7 $K(t/T)$: saw-tooth "sine" pulse.

Sawtooth sine-wave current

$$A(u) = u_{-1}(1-|u|)\begin{cases} \frac{1}{4} - \frac{1}{2}u^2 & \left(|u| < \frac{1}{2}\right) \\ \frac{1}{2}(|u|-1)^2 & \left(|u| > \frac{1}{2}\right) \end{cases}$$

$$S(u) = u_{-1}(1-|u|)\begin{cases} 4u^2 & \left(|u| < \frac{1}{2}\right) \\ 4(|u|-1)^2 & \left(|u| > \frac{1}{2}\right) \end{cases}$$

$$S^{o}(u) = u_{-1}(1-|u|)\begin{cases} \frac{1}{2} + u^2 & \left(|u| < \frac{1}{2}\right) \\ (|u|-1)^2 & \left(|u| > \frac{1}{2}\right) \end{cases}$$

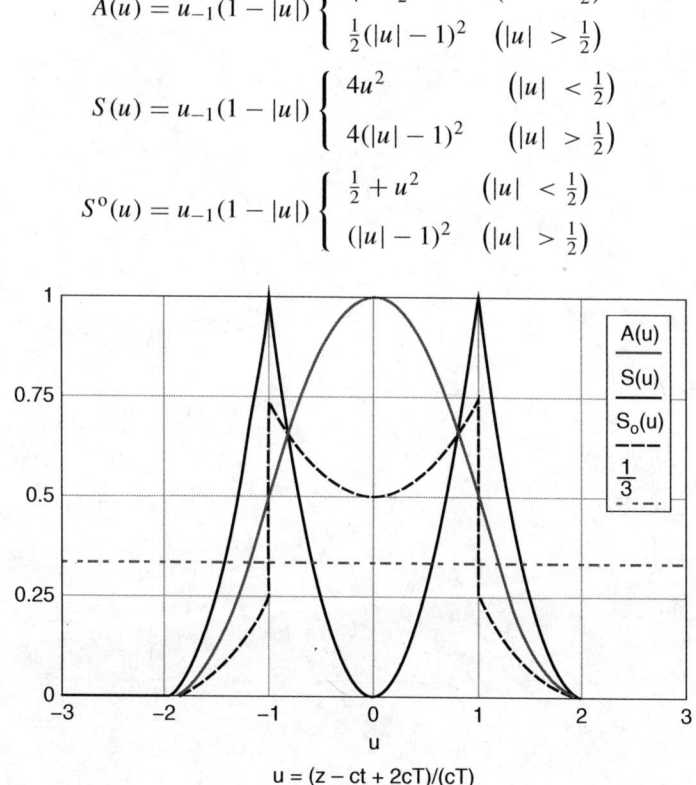

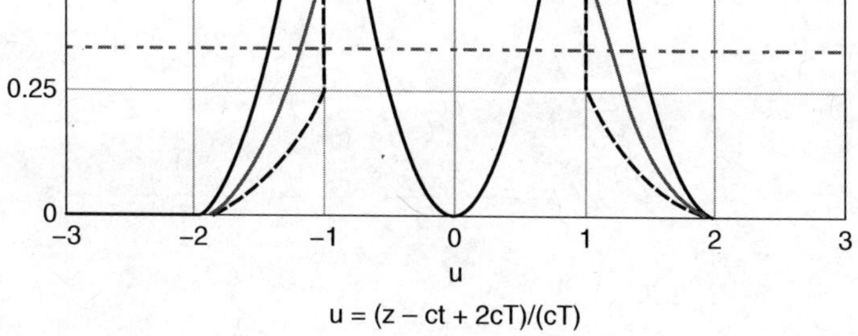

Figure 14.8 Normalized plots of $A(u)$, $S(u)$, $S^{o}(u)$.

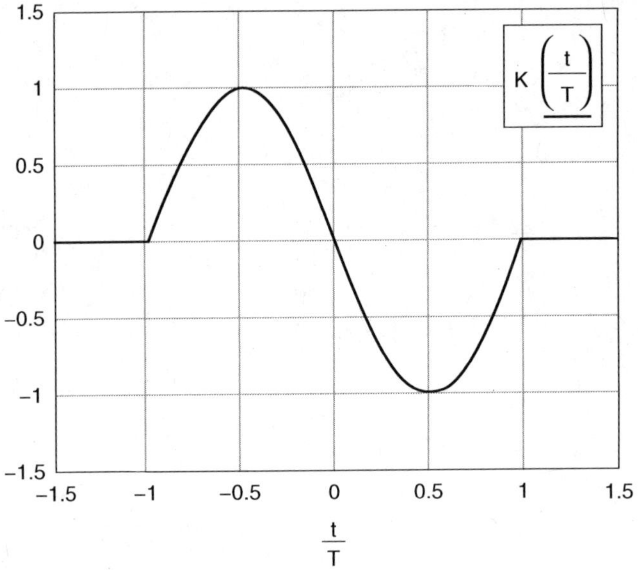

Figure 14.9 $K(t/T)$: sine pulse.

Single-cycle sine-wave current

$$A(u) = \frac{1 - \cos[\pi(1 - |u|)]}{2\pi} u_{-1}(1 - |u|)$$

$$S(u) = \sin^2[\pi(1 - |u|)]\, u_{-1}(1 - |u|)$$

$$S^{\circ}(u) = \frac{1 - \cos[\pi(1 - |u|)]}{2\pi} u_{-1}(1 - |u|)$$

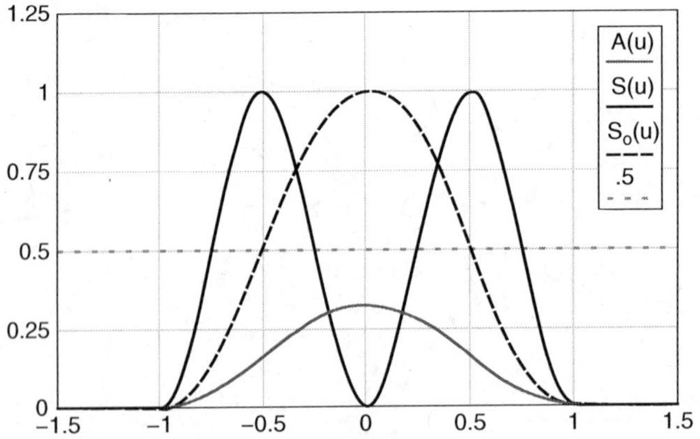

Figure 14.10 Normalized plots of $A(u)$, $S(u)$, $S^{\circ}(u)$.

Sawtooth cosine wave current

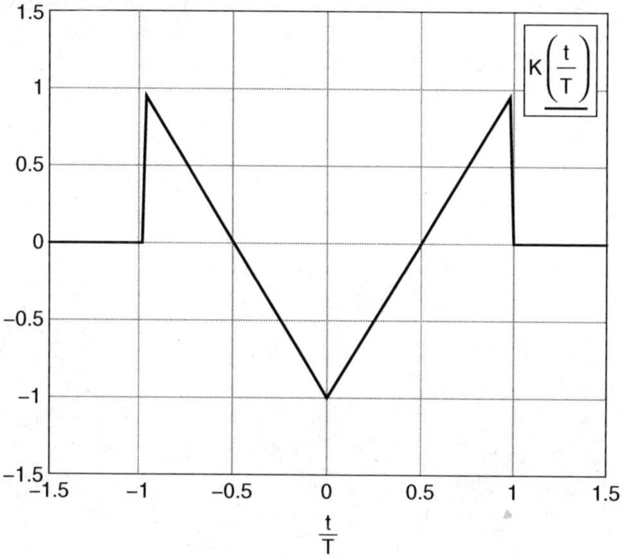

Figure 14.11 $K(t/T)$: saw-tooth "cosine" pulse.

$$A(u) = u(|u| - 1)u_{-1}(1 - |u|)$$

$$S(u) = (1 - 2|u|)^2 u_{-1}(1 - |u|)$$

$$S^{\mathrm{o}}(u) = \left(u^2 - |u| + \frac{1}{2}\right) u_{-1}(1 - |u|)$$

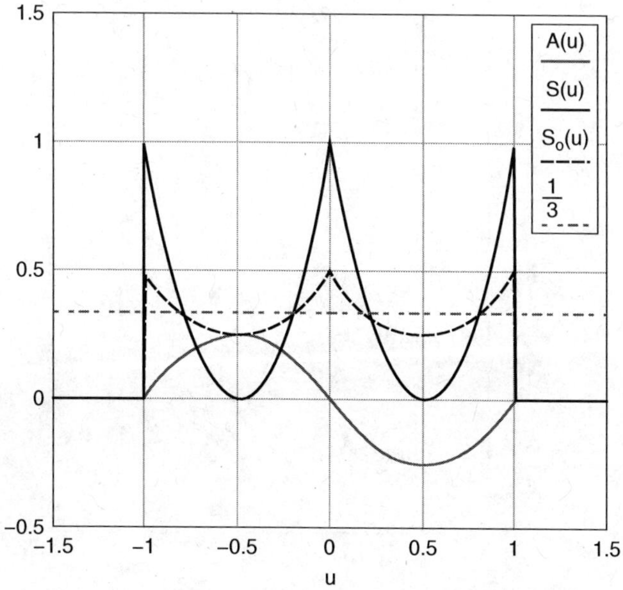

Figure 14.12 Normalized plots of $A(u)$, $S(u)$, $S^{\mathrm{o}}(u)$.

Single-cycle cosine-wave current

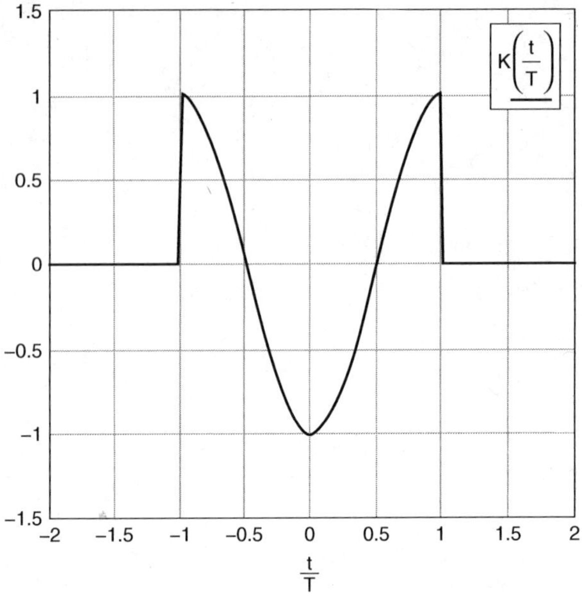

Figure 14.13 $K(t/T)$: cosine pulse.

$$A(u) = -\frac{\sin(\pi u)}{\pi} u_{-1}(1 - |u|)$$

$$S = \cos^2(\pi u)\, u_{-1}(1 - |u|)$$

$$S^{\mathrm{o}} = \tfrac{1}{2} u_{-1}(1 - |u|)$$

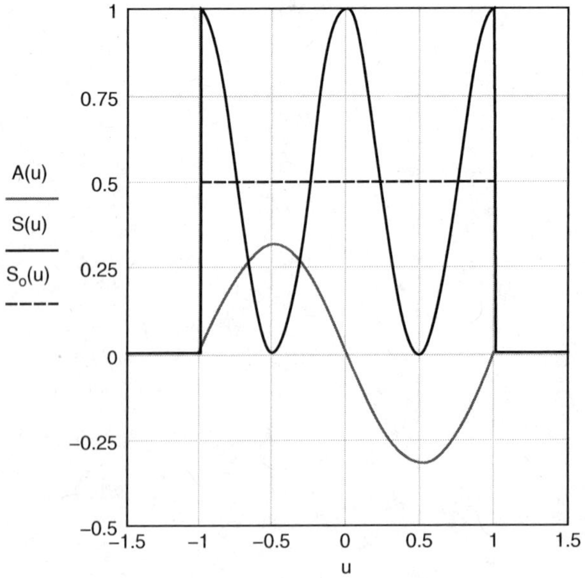

Figure 14.14 Normalized plots of $A(u)$, $S(u)$, $S^{\mathrm{o}}(u)$.

CHAPTER 15

UNIFORM LINE CURRENTS

This chapter considers the problem of a uniform line current of infinite extent when the current waveform is a step function. First, the integral form of Maxwell's Equations and considerations of symmetry without the use of the wave equation are employed. This gives rise to a set of approximate solutions that embraces both radiation and *MQS* fields and in the limit is exact; the process of solution provides significant physical insight. This treatment is followed by a more conventional one based upon the differential forms of Maxwell's Equations; the exact solution of the wave equation is obtained directly. In addition, both Poynting and Alternate power fluxes and energy densities are evaluated and compared. In subsequent sections, these line-current solutions are superimposed in space and time to find the solutions of an axial current step uniformly distributed over either the volume or the surface of a circular cylinder and uniformly distributed over a rectangular strip. A line current excited by multiple pulses or a sinusoidal waveform is also considered.

15.1 AXIAL CURRENT STEP (INTEGRAL LAWS)

We attempt to solve the problem of a uniform line current of infinite extent when excited by a step function by using the integral form of Maxwell's Equations and considerations

The Power and Beauty of Electromagnetic Fields, First Edition. F. R. Morgenthaler.

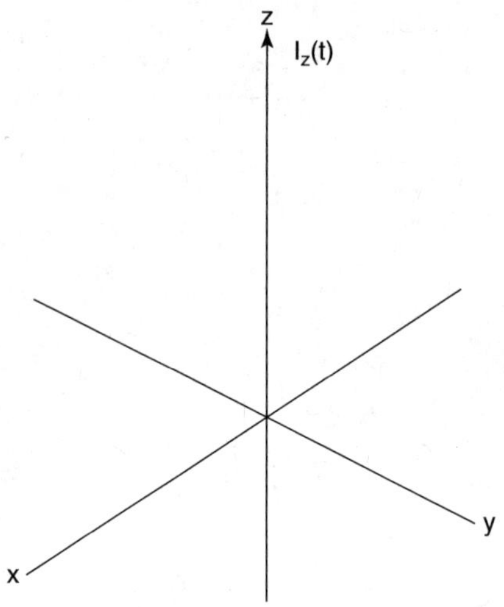

Figure 15.1 Axial line current.

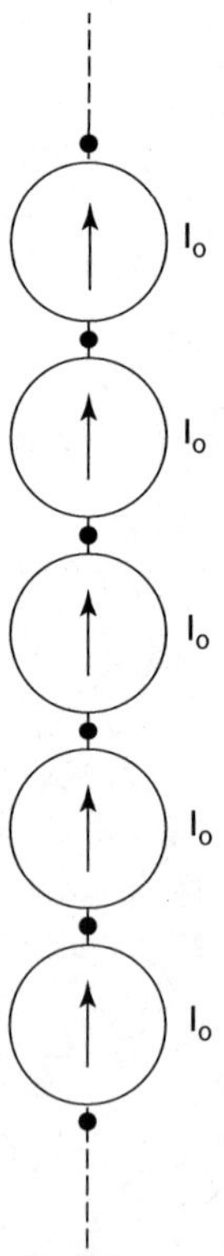

Figure 15.2 Linear chain of current sources.

of symmetry. The geometry of the source is defined in Figure 15.1; an equivalent circuit made from ideal current sources that form the line current is shown in Figure 15.2.

The axial current can also be modeled by a bipolar pair of convective line-charge densities, which is electrically neutral.

$$\mathbf{J} = \rho_{(1)}\mathbf{v}_{(1)} + \rho_{(2)}\mathbf{v}_{(2)}$$

$$\rho = \rho_{(1)} + \rho_{(2)} = 0$$

One of the pair can be considered at rest. To particularize, consider a very long thin cylinder of length 2ℓ and radius r_o that is filled with a uniform charge density ρ_o that is independent of time. If the cylinder is oriented parallel to the z axis and all of the charge is moving at the same velocity, $\widehat{\mathbf{z}}\,\mathrm{v}_z(t)$, then the charge density and the convective current density that is generated by its motion are given by

$$\left.\begin{aligned} \rho_{(1)} &= \rho_o \\ \mathbf{J}_{(1)} &= \widehat{\mathbf{z}}\,\rho_o\,\mathrm{v}_z(t) \end{aligned}\right\} \qquad -\ell + \int \mathrm{v}_z(t)\,dt \le z \le \ell + \int \mathrm{v}_z(t)\,dt$$

Now consider that a second charge distribution of opposite polarity $(-\rho_o)$ is superimposed on the first, but is not moving.

$$\left.\begin{aligned} \rho_{(2)} &= -\rho_o \\ \mathbf{J}_{(2)} &= 0 \end{aligned}\right\} \qquad -\ell + \le z \le \ell$$

The resultant electric charge and current densities are

$$\rho = \begin{cases} -\rho_o & \left(-\ell \le z \le -\ell + \int \mathrm{v}_z(t)\,dt\right) \\ 0 & \left(-\ell + \int v_z(t)\,dt \le z \le \ell\right) \\ +\rho_o & \left(\ell \le z \le \ell + \int v_z(t)\,dt\right) \end{cases} \tag{15.1a}$$

$$\mathbf{J} = \widehat{\mathbf{z}}\rho_o v_z(t) \qquad \left(-\ell + \int \mathrm{v}_z(t)\,dt \le z \le \ell + \int v_z(t)\,dt\right) \tag{15.1b}$$

These distributions are depicted in Figure 15.3, assuming that both ρ_o and $v_z(t) = v_o$ are positive constants. The two charge distributions are shown at the left of the figure (displaced for clarity); their superposition is indicated at the right. Since $J_z(t) = \rho_o \mathrm{v}_o$ is a constant current density, the net charges near the ends, $z = \pm\ell$, are $\pm Q(t)$, where $Q(t) = \mathrm{I}_o t = \rho_o \pi r_o^2 v_o t$ is the charge contained in each nonoverlapping cylindrical volume of length $v_o t$.

If the electric charge distributions were static, they would produce an electric field that, for $|z| \ll \ell$, could be approximated as that due to bipolar point charges, $\pm Q$, located at the ends, $z = \pm\ell$. Then (making use of Appendix C, Eq. (C.1a)),

$$E_z(r,z) = \frac{-Q}{4\pi\varepsilon_o}\left[\frac{\ell - z}{\left[(\ell - z)^2 + r^2\right]^{\frac{3}{2}}} + \frac{\ell + z}{\left[(\ell + z)^2 + r^2\right]^{\frac{3}{2}}}\right]$$

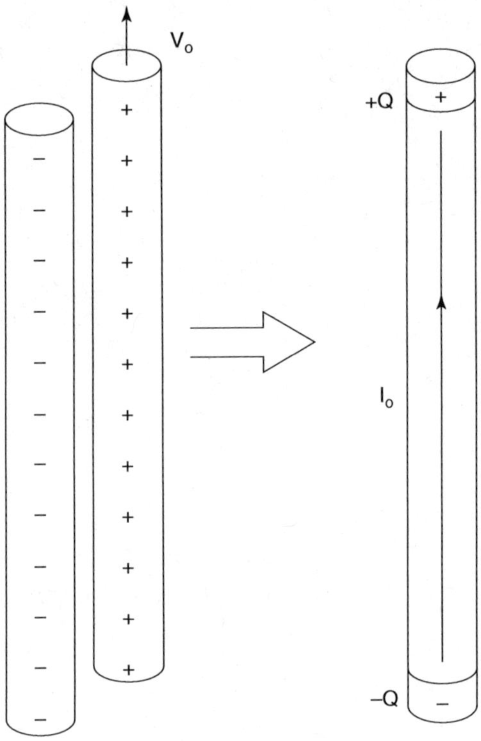

Figure 15.3 Bipolar-charge distribution.

$$E_r(r,z) = \frac{Q}{4\pi\varepsilon_0}\left[\frac{r}{\left[(\ell - z)^2 + r^2\right]^{\frac{3}{2}}} - \frac{r}{\left[(\ell + z)^2 + r^2\right]^{\frac{3}{2}}}\right]$$

$$\mathbf{E}(r,z,t) \simeq \widehat{\mathbf{z}} \lim_{\ell\to\infty} \frac{-\int \mathrm{I}_z(t)\, dt}{2\pi\varepsilon_0 \ell^2} \to 0 \qquad (r, |z| \ll \ell)$$

Even though $Q(t)$ is not static, it is nonzero only for $t > 0$. Therefore, because the fields propagate at velocity c, the final result remains valid for all finite times. Consequently, for finite values of r and z, one can safely set $\rho = 0$ and assume that only the electric-current density generates the fields during and after the transient.

The full set of the Maxwell Equations in integral form is repeated for convenience:

$$\oint_{C_o} \mathbf{H}\cdot\mathbf{ds} = \int_S \mathbf{J}\cdot\mathbf{da} + \frac{d}{dt}\int_S \varepsilon_o\mathbf{E}\cdot\mathbf{da} \tag{15.2a}$$

$$\oint_{C_o} \mathbf{E}\cdot\mathbf{ds} = -\frac{d}{dt}\int_S \mu_o\mathbf{H}\cdot\mathbf{da} \tag{15.2b}$$

$$\oint_{S_o} \mu_o\mathbf{H}\cdot\mathbf{da} = 0 \tag{15.3a}$$

$$\oint_{S_o} \varepsilon_o\mathbf{E}\cdot\mathbf{da} = \int_V \rho\, dV \tag{15.3b}$$

$$\oint_{S_o} \mathbf{J}\cdot d\mathbf{a} + \frac{d}{dt}\int_V \rho\, dV = 0 \tag{15.4}$$

Expressed in cylindrical coordinates, the sources are

$$\mathbf{J} = \widehat{\mathbf{z}} J_z \ (r \le R, t)$$

$$I_z(t) = \lim_{R\to 0}\int_0^R 2\pi J_z(r', t)\, dr' = I_o u_{-1}(t)$$

$$\rho = 0$$

where the unit step function, $u_{-1}(t)$, is defined in Appendix B, Section B.1.

From considerations of symmetry, the electric and magnetic fields,

$$\mathbf{E} = \widehat{\mathbf{r}} E_r + \widehat{\phi} E_\phi + \widehat{\mathbf{z}} E_z$$

$$\mathbf{H} = \widehat{\mathbf{r}} H_r + \widehat{\phi} H_\phi + \widehat{\mathbf{z}} H_z$$

satisfy

$$\frac{\partial}{\partial\phi}E_r = \frac{\partial}{\partial z}E_r = 0$$
$$\frac{\partial}{\partial\phi}E_\phi = \frac{\partial}{\partial z}E_\phi = 0$$
$$\frac{\partial}{\partial\phi}E_z = \frac{\partial}{\partial z}E_z = 0$$

$$\frac{\partial}{\partial\phi}H_r = \frac{\partial}{\partial z}H_r = 0$$
$$\frac{\partial}{\partial\phi}H_\phi = \frac{\partial}{\partial z}H_\phi = 0$$
$$\frac{\partial}{\partial\phi}H_z = \frac{\partial}{\partial z}H_z = 0$$

We make use of the cylindrical symmetry by choosing S_o and V to be the surface and volume shown in Figure 15.4 and evaluating Eqs. (15.3a) and (15.3b).

Only the integral over the curved surface contributes (those from the top and bottom surfaces cancel); the result is

$$\varepsilon_o E_r(r) 2\pi r\,\ell = \int_0^r \rho(r') 2\pi\,\ell\, r' dr' = 0$$

$$\mu_o H_r(r) 2\pi r\,\ell = 0$$

Next, we choose C_o and S to be the perimeter and area of the shaded portion of either Figure 15.5 or 15.6.

If Figure 15.5 is applied to Eq. (15.2a) and Figure 15.6 is applied to Eq. (15.2b), the result is

$$2\pi r H_\phi(r, t) = I_o u_{-1}(t) + \frac{d}{dt}\int_0^r 2\pi r' \varepsilon_o E_z(r', t)\, dr' \tag{15.5a}$$

$$[E_z(r, t) - E_z(0, t)]\,\ell = \frac{d}{dt}\int_0^r \mu_o H_\phi(r', t)\, dr' \ell \tag{15.5b}$$

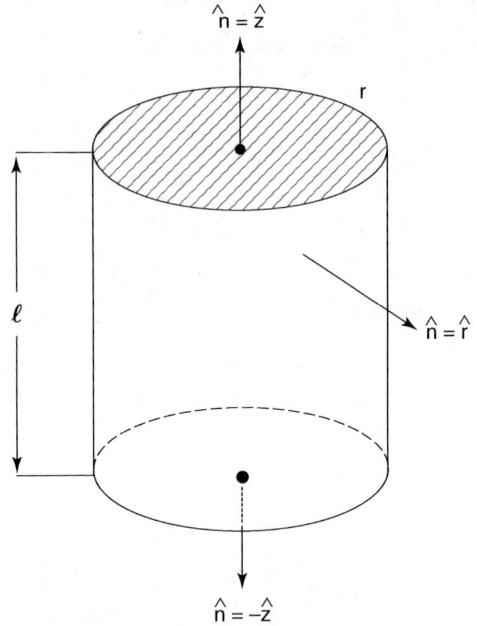

Figure 15.4 Gaussian surface.

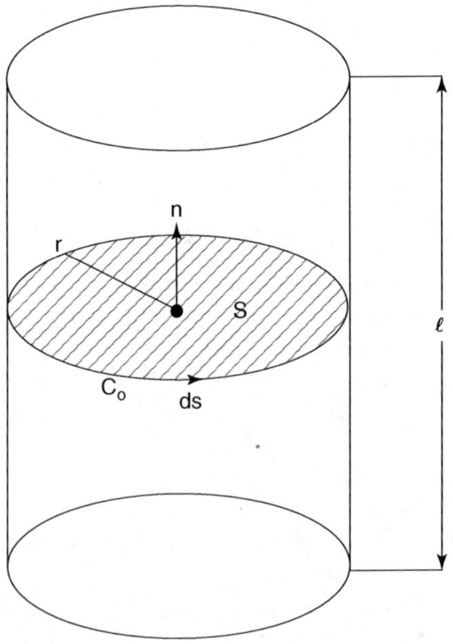

Figure 15.5 Circular integration contour.

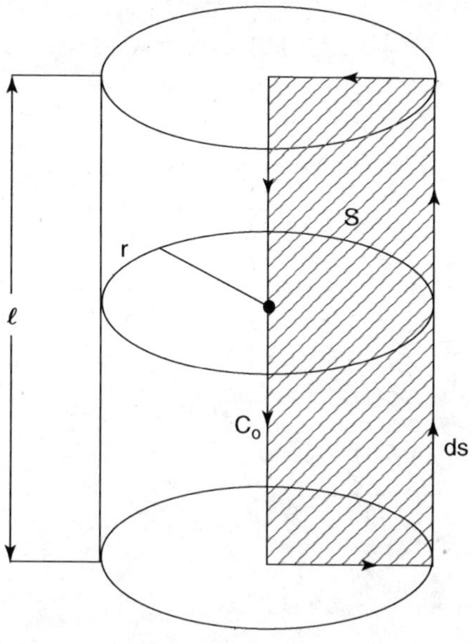

Figure 15.6 Rectangular integration contour.

Notice that the radial contributions to the second integral cancel. When applied to nonzero radii $r_1$and r_2 and the difference of the results is taken:

$$r_2 H_\phi(r_2,t) - r_1 H_\phi(r_1,t) = \frac{d}{dt}\int_{r_1}^{r_2} r'\varepsilon_0 E_z(r',t)\, dr' \qquad (0 < r_1 \le r_2) \tag{15.6a}$$

$$E_z(r_2,t) - E_z(r_1,t) = \frac{d}{dt}\int_{r_1}^{r_2} \mu_0 H_\phi(r',t)\, dr' \qquad (0 < r_1 \le r_2) \tag{15.6b}$$

If, instead, the contour of Figure 15.5 is applied to Eq. (15.2b) and the contour of Figure 15.6 is applied to Eq. (15.2a), the result is

$$2\pi r E_\phi(r,t) = -\frac{d}{dt}\int_0^r 2\pi r' \mu_0 H_z(r',t)\, dr' \tag{15.7a}$$

$$[H_z(r,t) - H_z(0,t)]\,\ell = -\frac{d}{dt}\int_0^r \varepsilon_0 E_\phi(r',t)\, dr'\ell \tag{15.7b}$$

where again the radial contributions to the second integral cancel.

Because $E_\phi(r,t)$ and $H_z(r,t)$ only appear in Eqs. (15.7a) and (15.7b), for which there are no sources, they along with $E_r(r,t)$ and $H_r(r,t)$ vanish; only $E_z(r,t)$ and $H_\phi(r,t)$ are nonzero.

The steady-state solution

Because the current is constant for $t > 0$, one expects that after sufficient time the fields become static. The values

$$\mathbf{H} = \hat{\phi}\frac{I_0}{2\pi r} \tag{15.8a}$$

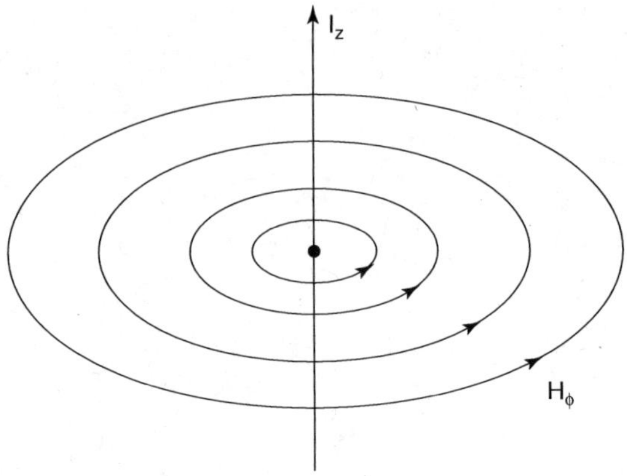

Figure 15.7 Magnetic-field pattern.

$$\mathbf{E} = 0 \tag{15.8b}$$

obviously satisfy all of Maxwell's Equations. The field of Eq. (15.8a) is depicted in Figure 15.7. As for the transient fields, the trial solution,

$$\mathbf{H} = \widehat{\phi}\frac{\mathrm{I_o}}{2\pi r}u_{-1}(t) \tag{15.9a}$$

$$\mathbf{E} = 0 \tag{15.9b}$$

seems a reasonable guess and certainly satisfies Eq. (15.5a). However, when substituted into Eq. (15.6b), an impulse of electric field is generated at $t = 0$ given by

$$\begin{aligned} E_z(r_2,t) - E_z(r_1,t) &= \frac{\mu_o I_o}{2\pi}\ \ln\left(\frac{r_2}{r_1}\right)\frac{d}{dt}u_{-1}(t) \\ &= \frac{\mu_o I_o}{2\pi}u_0(t)\left[\ln(r_2) - \ln(r_1)\right] \end{aligned}$$

or with $r_2 = r$ and $r_1 = R$,

$$E_z(r,t) = \frac{\mu_o \mathrm{I_o}}{2\pi}\ \ln\left(\frac{r}{R}\right)\ u_0(t) \tag{15.10}$$

The constant, R, can be chosen so as to force E_z to be zero at any single radius, but the logarithmic singularity is inescapable. However, it exists at only a single instant of time; for finite R, it occurs at both $r = 0$ and $r = \infty$; for $R \to \infty$, it occurs for *all* finite r. Moreover, when Eq. (15.10) is substituted into Eq. (15.5a), this impulse creates a doublet singularity, $\frac{d}{dt}u_0(t) = u_1(t)$. The magnetic field becomes

$$H_\phi = \frac{\mathrm{I_o}}{2\pi}\left(\frac{1}{r}u_{-1}(t) + \frac{1}{2}\mu_o\varepsilon_o r\left[\ln\left(\frac{r}{R}\right) - \frac{1}{2}\right]u_1(t)\right) \tag{15.11}$$

this in turn adds a term to E_z proportional to $\frac{d}{dt}u_1(t) = u_2(t)$. The feedback occurs in an endless nonconvergent cycle with higher-order singularities being created at each turn. Since these all diverge simultaneously, the trial solution is nonconvergent and obviously false. The inescapable conclusion is that Maxwell's Equations cannot permit the magnetic field to be suddenly generated over all space at any one instant of time, here chosen as $t = 0$. We next explore the possibility that, soon thereafter, H_ϕ might be generated in a limited region close to the line current.

An assumed zone of field generation

Assume that for positive time, the magnetic field is generated within a finite region surrounding the current. From symmetry considerations, the zone (if it exists) must be cylindrical with a radius $r_o(t)$. Since there is no field for negative time, we have $r_o(t \leq 0) = 0$. Consequently,

$$H_\phi = \frac{I_o}{2\pi r}\begin{cases} 0, & t < 0 \\ 1, & 0 < r \leq r_o(t) \\ 0, & r > r_o(t) \end{cases} \tag{15.12}$$

For $t > 0$ and $r \leq r_o(t)$, Eq. (15.5a) is satisfied, provided that

$$\frac{d}{dt}\int_0^{r<r_o(t)} 2\pi r'\varepsilon_o E_z(r', t)\, dr' = 0$$

Consequently, $E_z(r < r_o(t)) = 0$, which is consistent with Eq. (15.9b). However, by hypothesis, there are no fields (electric or magnetic) outside of the zone, so Eq. (15.5a) also requires that for $r > r_o(t)$,

$$0 = I_o + \frac{d}{dt}\int_{r_o(t)-\delta}^{r_o(t)} 2\pi r'\varepsilon_o E_z(r', t)\, dr'$$

This equation is supportable in the limit $\delta \to 0$ only if E_z is infinite at $r = r_o(t)$. It begins to look as if an impulse in the electric field is inescapable, but at least it is now confined to a negligible volume of space! Before taking the limit, we assume that

$$E_z = \begin{cases} 0, & 0 < r < r_o(t) - \delta \\ E_\delta, & r_o(t) - \delta \leq r \leq r_o(t) \\ 0, & r > r_o(t) \end{cases} \tag{15.13}$$

Therefore,

$$I_o = -2\pi\varepsilon_o\delta\frac{d}{dt}\left[r_o(t)E_\delta\right]$$

Because the cylindrical shell of thickness δ is of special significance, we allow for the magnetic field to have a special value, H_δ, within it. Equation (15.12) is therefore amended to

$$H_\phi = \begin{cases} \dfrac{I_o}{2\pi r}, & 0 < r < r_o(t) - \delta \\ H_\delta, & r_o(t) - \delta \leq r \leq r_o(t) \\ 0, & r > r_o(t) \end{cases} \tag{15.14}$$

For $r_o(t) - \delta < r_1 < r_2 < r_o(t)$, Eq. (15.6b) becomes

$$E_\delta(r_2) - E_\delta(r_1) = 0 = \frac{d}{dt}\int_{r_1}^{r_2} \mu_o H_\delta \, dr'$$

$$0 = \mu_o(r_2 - r_1)\frac{dH_\delta}{dt} \tag{15.15}$$

Therefore H_δ is independent of t.

For $r_o(t) - \delta < r_1 < r_o(t)$ and $r_2 > r_o(t)$, Eq. (15.15) becomes

$$-E_\delta = \frac{d}{dt}\int_{r_1}^{r_o(t)} \mu_o H_\phi \, dr'$$

$$= \mu_o H_\delta \frac{dr_o(t)}{dt}$$

and Eq. (15.6a) becomes

$$-r_1 H_\delta = \varepsilon_o \frac{d}{dt}\int_{r_1}^{r_o(t)} r' E_\delta(r', t)\, dr'$$

$$= \varepsilon_o\left(E_\delta\, r_o(t)\frac{dr_o(t)}{dt} + \frac{1}{2}\left[r_o^2(t) - r_1^2\right]\frac{dE_\delta}{dt}\right)$$

When $r_1 \to r_o(t)$, this equation reduces to

$$-H_\delta = \varepsilon_o E_\delta \frac{dr_o(t)}{dt}$$

which remains valid for other values of r_1, provided that $\frac{dE_\delta}{dt} = 0$. Consequently, E_δ, H_δ, and $\frac{dr_o(t)}{dt}$ are all constants. The three relevant equations governing the zone of field generation are summarized below

$$E_\delta = -\mu_o H_\delta \frac{dr_o(t)}{dt} \tag{15.16a}$$

$$H_\delta = -\varepsilon_o E_\delta \frac{dr_o(t)}{dt} \tag{15.16b}$$

$$\mathrm{I}_o = -2\pi\,\varepsilon_o E_\delta\, \delta \frac{dr_o(t)}{dt} \tag{15.16c}$$

The velocity of light

Because E_δ, H_δ, and I_o are nonzero, the product of the first two of Eqs. (15.16) leads to

$$\frac{dr_o(t)}{dt} = \pm c \tag{15.17a}$$

$$c = \frac{1}{\sqrt{\mu_o \varepsilon_o}} \tag{15.17b}$$

and the discovery that the zone expands (or contracts) at the constant velocity, c. We recall that $\mu_o = 4\pi \times 10^{-7}$ Hm^{-1} and $\varepsilon_o \simeq \frac{1}{36\pi} \cdot 10^{-9}$ F m^{-1}; therefore $c \simeq 3 \times 10^{+8}$ ms^{-1}. This is a very good approximation to the measured velocity of light.

The impedance of free-space

The quotient of the same two equations leads to

$$-\frac{E_\delta}{H_\delta} = \pm\eta_o \tag{15.18a}$$

$$\eta_o = c\mu_o = \sqrt{\frac{\mu_o}{\varepsilon_o}} \tag{15.18b}$$

This ratio of the electric to magnetic field has the dimensions of electrical resistance; it is termed the characteristic impedance of free-space, $\eta_o \simeq 120\pi \simeq 377\Omega$. Of course, inside of the zone, $-E_z/H_\phi = 0$.

A modified trial solution

Because $r_o(t) > 0$, the plus sign must be chosen; then $r_o(t) = ct$ and $E_\delta/H_\delta = -\eta_o$. Substituting these values into Eq. (15.16c) leads to

$$H_\delta = \frac{I_o}{2\pi\delta} \tag{15.19}$$

The modified trial solution is therefore

$$H_\phi(r,t) = \begin{cases} \dfrac{I_o}{2\pi r} & (r < ct - \delta) \\ \dfrac{I_o}{2\pi\delta} & (ct - \delta \le r \le ct) \\ 0 & (r > ct) \end{cases} \tag{15.20a}$$

$$E_z(r,t) = \begin{cases} 0 & (r < ct - \delta) \\ \dfrac{-c\mu_o I_o}{2\pi\delta} & (ct - \delta \le r \le ct) \\ 0 & (r > ct) \end{cases} \tag{15.20b}$$

Notice that if the magnetic field implies some sort of distributed inductance (L') through which the line current flows, a voltage per unit length,

$$-E_z = L'\frac{d}{dt}I_o u_{-1}(t) = L'I_o u_0(t) \tag{15.21}$$

should be expected. The impulse in Eq. (15.20b) is therefore at least plausible. Although generated at $r = 0$, $t = 0$, it propagates thereafter as $u_0(t - \frac{r}{c})$, an impulse moving at the speed of light. For any particular radius, r, the steady-state fields emerge as soon as $ct > r$.

The modified trial solution revisited

Because by assumption $\delta \ll r$, the modified trial solution given by Eqs. (15.20) satisfies Eq. (15.6a) for all values of r_1 and r_2 (including cases where one or both values are within the thin shell). On the other hand, they satisfy Eq. (15.6b) for all values of the radii except when $r_1 < ct - \delta$ and $r_2 > ct$. In this case, since E_z and H_ϕ are both zero for $r > ct$, there must be some residual electric field (hopefully small) that we designate as ΔE_z. Evidently,

$$\begin{aligned}\Delta E_z &= -\mu_0 \frac{d}{dt}\left[\int_{r_1}^{ct-\delta} \frac{\mathrm{I_o}}{2\pi r'}\, dr' + H_\delta \delta\right] \\ &= \frac{-c\mu_0 I_o}{2\pi ct} \qquad (r < ct)\end{aligned}$$

Although the magnitude of this field is very small compared to E_δ whenever $ct \gg \delta$, it cannot be ignored because it extends over a much greater range of r. In particular, it contributes importantly to Eq. (15.5a) when $r > ct > 0$.

$$\begin{aligned}0 &= I_o + \frac{d}{dt}\int_0^{ct} 2\pi r' \varepsilon_0 E_z(r', t)\, dr' \\ -I_o &= +\pi \varepsilon_0 \frac{d}{dt}\left[\frac{-c\mu_0 \mathrm{I_o}}{2\pi ct}(ct)^2 + 2ctE_\delta \delta\right] \\ \frac{I_o}{2} &= -2\pi c \varepsilon_0 E_\delta \delta\end{aligned}$$

This last equation replaces Eq. (15.16c) and therefore reduces the strength of both E_δ and H_δ by one-half. A better approximation to the fields is therefore

$$H_\phi(r,t) = \begin{cases} \dfrac{I_o}{2\pi r} & (r < ct - \delta) \\[2ex] \dfrac{I_o}{4\pi \delta} & (ct - \delta \le r \le ct) \\[2ex] 0 & (r > ct) \end{cases} \tag{15.22a}$$

$$E_z(r,t) = \begin{cases} \dfrac{-\mu_0 I_o}{2\pi t} & (r < ct - \delta) \\[2ex] \dfrac{-c\mu_0 I_o}{4\pi \delta} & (ct - \delta \le r \le ct) \\[2ex] 0 & (r > ct) \end{cases} \tag{15.22b}$$

With the changed value of E_z, the value of H_ϕ must also be modified for $r < ct$. Fortunately, the correction to the magnetic field can easily be found, but, in turn, leads to the necessity of again correcting the electric field (and the values of E_δ and H_δ). Next, we do this systematically and show that for any $r < ct$, each successive correction becomes smaller and smaller, as does the amount by which $-E_\delta$ and H_δ are both reduced. Finally, we will be able to sum all of the corrections and find simple analytic expressions that constitute the exact solution of Maxwell's Equations.

A complete series solution

If $\Delta E_z^{(k)}(r,t)$ is the kth correction to the electric field, then $\Delta H_\phi^{(k+1)}(r,t)$, the $(k+1)$th correction to the magnetic field, is given by

$$r\Delta H_\phi^{(k+1)}(r,t) = \varepsilon_0 \frac{d}{dt}\int_0^{r<ct} r'\Delta E_z^{(k)}(r',t)\,dr'$$

Likewise, in terms of $\Delta H_\phi^{(k+1)}(r,t)$, the next correction to the electric field, $\Delta E_z^{(k+2)}(r,t)$, is given by

$$\Delta E_z^{(k+2)}(r,t) = \mu_0 \frac{d}{dt}\int_0^{r<ct} \Delta H_\phi^{(k+1)}(r',t)\,dr'$$

The cycle can be continued as often as desired; the corresponding corrections are then added together and both E_δ and H_δ are adjusted as required from

$$\mathrm{I}_o = -2\pi\varepsilon_0\left(\frac{d}{dt}\left[\int_0^{ct-\delta} r'E_z(r',t)\,dr'\right] + cE_\delta'\delta\right) \tag{15.23}$$

Previously, we calculated the first correction to $E_z(r,t)$, namely

$$\Delta E_z^{(1)} = -\frac{\mu_0 I_o}{2\pi t}$$

Therefore,

$$r\Delta H_\phi^{(2)} = -\mu_0\varepsilon_0\frac{I_o}{2\pi}\frac{d}{dt}\left(\frac{1}{t}\int_0^r r'\,dr'\right)$$

$$\Delta H_\phi^{(2)} = \frac{\mathrm{I}_o}{4\pi}\frac{r}{(ct)^2}$$

Continuing,

$$\Delta E_z^{(3)}(r,t) = \mu_0\frac{d}{dt}\int_0^r \Delta H_\phi^{(2)}(r',t)\,dr'$$

$$\Delta E_z^{(3)} = \frac{-c\mu_0 I_o}{4\pi}\frac{r^2}{(ct)^3}$$

The pattern is apparent; for integers $k = 0, 1, 2, 3, \ldots$, we have

$$\Delta E_z^{(2k+1)} = -\frac{c\mu_0 I_o}{2\pi}C_k\frac{r^{2k}}{(ct)^{2k+1}}$$

$$\Delta H_\phi^{(2k)} = \frac{\mathrm{I}_o}{2\pi}C_k\frac{r^{2k-1}}{(ct)^{2k}}$$

Both share the same coefficients, C_k; these may be calculated from the recursive relation, $C_{k+1} = \frac{2k+1}{2(k+1)}C_k$ (subject to $C_0 = 1$). In closed form we have,

$$C_k = \frac{(2k-1)!}{2^{2k-1}k!(k-1)!}, \qquad k \geq 1$$

However, we started the magnetic-field corrections with $\Delta H_\phi^{(2)}$; observe that $\Delta H_\phi^{(0)} = \frac{I_0}{2\pi r}$ can be included in the series. When combined, the results (for $r \leq ct$) are

$$E_z(r,t) = \sum_{k=0}^{n} \Delta E_z^{(2k+1)} = \frac{-c\mu_0 I_0}{2\pi ct}\left[1 + \frac{1}{2}\left(\frac{r}{ct}\right)^2 + \frac{3}{8}\left(\frac{r}{ct}\right)^4 + \frac{5}{16}\left(\frac{r}{ct}\right)^6 + \cdots\right] \tag{15.24a}$$

$$H_\phi(r,t) = \sum_{k=0}^{n} \Delta H_\phi^{(2k)} = \frac{I_0}{2\pi r}\left[1 + \frac{1}{2}\left(\frac{r}{ct}\right)^2 + \frac{3}{8}\left(\frac{r}{ct}\right)^4 + \frac{5}{16}\left(\frac{r}{ct}\right)^6 + \cdots\right] \tag{15.24b}$$

From Eq. (15.23), one finds that

$$E_\delta(n) = -c\mu_0 H_\delta(n) = \frac{-c\mu_0 I_0}{2\pi\delta} C_{n+1}$$

$$C_{n+1} = 1 - \sum_{k=0}^{n} \frac{C_k}{2(k+1)} = \frac{(2n+1)!}{2^{2n+1}(n+1)!n!}$$

For large n, the factorials can be calculated using the Stirling approximation, $n! \simeq \sqrt{2\pi n}\,(n/e)^n$. As more and more terms are retained, the magnitude of the impulse continues to diminish (rather slowly) because $C_{n+1} \to 1/\sqrt{\pi n}$; it vanishes in the limit $n \to \infty$. (The impulse has been completely transferred to the series.)

The Taylor series that is common to both fields can easily be summed by recognizing that

$$1 + \frac{1}{2}u + \frac{3}{8}u^2 + \frac{5}{16}u^3 + \frac{35}{128}u^4 + \cdots = \frac{1}{\sqrt{1-u}}$$

and substituting $u = (\frac{r}{ct})^2$. The exact closed form of Eqs. (15.24a) and (15.24b) is therefore

$$E_z(r,t) = \frac{-c\mu_0 I_0}{2\pi}\frac{1}{\sqrt{c^2t^2 - r^2}}u_{-1}(ct - r) \tag{15.25a}$$

$$H_\phi(r,t) = \frac{I_0}{2\pi r}\frac{ct}{\sqrt{c^2t^2 - r^2}}u_{-1}(ct - r) \tag{15.25b}$$

Evaluation of Eq. (15.5a) for $r > ct$ and $\delta = 0$ shows that

$$2\pi\varepsilon_0 \frac{d}{dt}\int_0^{ct} E_z(r',t)r'\,dr' = 2\pi\varepsilon_0 \frac{d}{dt}\int_0^{ct} \frac{-c\mu_0 I_0}{2\pi}\frac{r'}{\sqrt{c^2t^2 - r^2}}\,dr' = -I_0$$

which verifies that $E_\delta, H_\delta = 0$ in the limit $n \to \infty$. Of course, the infinity is still present because the series is divergent when $ct = r$.

The ratio of $-E_z$ to H_ϕ is of particular interest; it is given by

$$\frac{-E_z}{H_\phi} = \mu_0\frac{r}{t} = \frac{r}{ct}\eta_0 \tag{15.26}$$

This wave impedance has the value η_0 at the wavefront $r = ct$, but decreases with time at any fixed value of r. For $r \ll ct$, the electric field is negligible in comparison to the magnetic field, and the leading terms of the series are sufficient; we term that region of the fields magnetoquasistatic (*MQS*). Notice that Eq. (15.26) remains valid for the approximate fields, provided that Eqs. (15.24a) and (15.24b) are both truncated at the *same* value of $n \geq 0$.

The electric field is plotted in Figure 15.8 as a function of normalized r when $ct = 12$ for both the approximation $n = 0$, $\delta = .01$ and the exact solution $n = \infty$. Also plotted is the case $ct = 9$ for $n = 2$, $\delta = .01$. The wake of the transient ($r \ll ct$) is well approximated by the zero-order *MQS* magnetic-field and the associated first-order electric field.

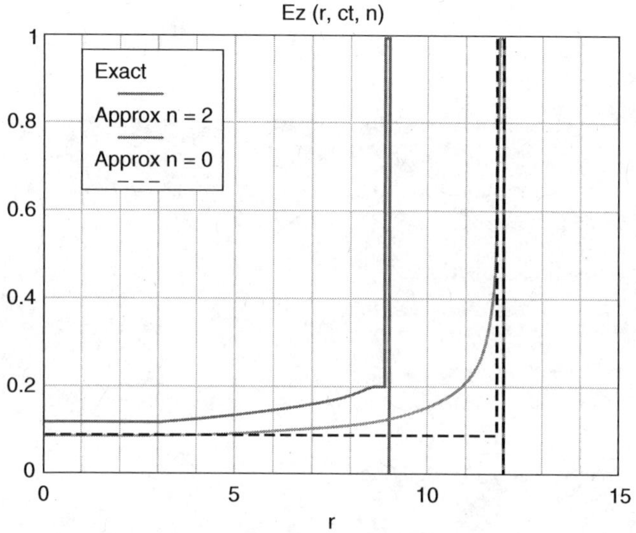

Figure 15.8 $E_z(r, 12, \infty)$ and $E_z(r, 9, 2)$, $E_z(r, 12, 0)$ for $\delta = .01$.

15.2 AXIAL CURRENT STEP (DIFFERENTIAL LAWS)

As in the previous section, a uniform, infinitely long line current is z-directed, of magnitude $I_z(t)$, and surrounded by free-space. This current produces vector and scalar potentials $\mathbf{A} = \widehat{\mathbf{z}}A_z(r,t)$ and (in the Lorenz gauge), $\Phi = 0$; these generate $\mathbf{E} = \widehat{\mathbf{z}}E_z(r,t)$ and $\mathbf{H} = \widehat{\phi}H_\phi(r,t)$.

If $I_z(t) = \int J_z(r,t)\, da = I_0 u_{-1}(t)$ is confined to a circular area of negligible radius, r_0, the cylindrically symmetric two-dimensional impulse response [Part I, Chapter 4, Eq. (4.21)], should be an appropriate solution for Φ or a Cartesian component of one of the fields. Because $\Phi = 0$ and the impulse response vanishes for $t \to \infty$, it cannot be associated with H_x, H_y, or A_z and so we try to match it to E_z. This choice is consistent because differentiating or integrating (with respect to x, y, z, or t) a solution of the homogeneous wave equation produces yet another solution, and the integral of E_z with respect to t generates $-A_z$. The complete solution is found to be

$$A_z(r,t) = \frac{\mu_0 I_0}{2\pi} \cosh^{-1}\left(\frac{ct}{r}\right) u_{-1}(ct - r) \tag{15.27a}$$

$$E_z(r,t) = -\frac{\partial A_z}{\partial t} = \frac{c\mu_0 I_0}{2\pi} \frac{-1}{\sqrt{c^2t^2 - r^2}} u_{-1}(ct - r) \tag{15.27b}$$

$$H_\phi(r,t) = -\frac{1}{\mu_0}\frac{\partial A_z}{\partial r} = \frac{I_0}{2\pi r}\frac{ct}{\sqrt{c^2t^2 - r^2}}u_{-1}(ct - r) \tag{15.27c}$$

As expected, the electric and magnetic fields agree with Eqs. (15.25).

The power density, $-\mathbf{E}\cdot\mathbf{J}$, represents power (per unit volume) extracted from the line current. When integrated over a cylinder of length ℓ and radius $r \geq r_0$ (concentric with the line current), the power delivered by that section of the line source is

$$-\ell\int E_z J_z\, da = -E_z(0,t)\mathrm{I}_0\ell = \frac{\mu_0 \mathrm{I}_0^2 \ell}{2\pi t}u_{-1}(t) \tag{15.28}$$

One portion produces a time rate of change of stored energy within the cylinder, whereas another creates power radiated radially through the surface. We examine these terms in both the Maxwell–Poynting and the Alternate-power representation.

Maxwell–Poynting representation

The Poynting vector, $\mathbf{S} = \mathbf{E}\times\mathbf{H} = \widehat{\mathbf{r}}S_r$, evaluates to

$$S_r = -E_z H_\phi = \frac{c\mu_0 \mathrm{I}_0^2}{4\pi^2 r}\frac{ct}{c^2t^2 - r^2}u_{-1}(ct - r) \tag{15.29}$$

the electric and magnetic energy-densities to

$$W_e = \frac{1}{2}\varepsilon_0 E_z^2 = \frac{\mu_0}{8\pi^2}\mathrm{I}_0^2\frac{1}{c^2t^2 - r^2}u_{-1}(ct - r) \tag{15.30a}$$

$$W_m = \frac{1}{2}\mu_0 H_\phi^2 = \frac{\mu_0 \mathrm{I}_0^2}{8\pi^2 r^2}\frac{c^2t^2}{c^2t^2 - r^2}u_{-1}(ct - r) \tag{15.30b}$$

The power leaving the cylinder is

$$\oint \mathbf{S}\cdot\mathbf{da} = 2\pi r S_r \ell = \frac{c\mu_0 \mathrm{I}_0^2 \ell}{2\pi}\frac{ct}{c^2t^2 - r^2}u_{-1}(ct - r) \tag{15.31}$$

Since there are no localized components, it follows that if $r \to 0$, there can be no stored energy and so Eq. (15.31) must agree with Eq. (15.28). Notice that the radiated power is independent of r whenever $ct \gg r$.

The total electric and magnetic energies contained within the cylindrical volume of length ℓ between radii r_0 and r are (for $ct > r$)

$$\int W_e\, dV = \frac{\mu_0 \mathrm{I}_0^2 \ell}{8\pi}\ln\left[\frac{c^2t^2 - r_0^2}{c^2t^2 - r^2}\right] \tag{15.32a}$$

$$\int W_m\, dV = \frac{\mu_0 \mathrm{I}_0^2 \ell}{8\pi}\ln\left[\frac{r^2\left(c^2t^2 - r_0^2\right)}{r_0^2\left(c^2t^2 - r^2\right)}\right] \tag{15.32b}$$

Whenever $ct \gg r$, there is only time-independent magnetic energy.

Alternate power and energy representation

For $ct > r$, the Alternate-power flux, $\mathbf{S}^{\mathrm{o}} = \widehat{\mathbf{r}}S_r^{\mathrm{o}}$, is expressed as

$$S_r^{\mathrm{o}} = \frac{1}{2\mu_{\mathrm{o}}}\left(A_z\frac{\partial^2 A_z}{\partial r\partial t} - \frac{\partial A_z}{\partial r}\frac{\partial A_z}{\partial t}\right)$$

$$= \frac{c\mu_{\mathrm{o}} I_{\mathrm{o}}^2}{8\pi^2}\frac{\left[r^2\ln\left(\dfrac{\sqrt{c^2t^2-r^2}+ct}{r}\right) + ct\sqrt{c^2t^2-r^2}\right]}{r\left(\sqrt{c^2t^2-r^2}\right)^3} \tag{15.33a}$$

and the Alternate-energy density is

$$W^{\mathrm{o}} = \frac{1}{2}A_z J_z + \frac{1}{2}\varepsilon_{\mathrm{o}}\left[\left(\frac{\partial A_z}{\partial t}\right)^2 - A_z\frac{\partial^2 A_z}{\partial t^2}\right] \tag{15.33b}$$

$$= \frac{1}{2}A_z J_z + \frac{\mu_{\mathrm{o}} \mathrm{I}_{\mathrm{o}}^2}{8\pi^2}\frac{\left[ct\ln\left(\dfrac{\sqrt{c^2t^2-r^2}+ct}{r}\right) + \sqrt{c^2t^2-r^2}\right]}{\left(\sqrt{c^2t^2-r^2}\right)^3}$$

It might be expected that if S_r^{o} is substituted for S_r in Eq. (15.31), the result will be the same (at least whenever $ct \gg r$). Instead,

$$\oint \mathbf{S}^{\mathrm{o}} \cdot \mathbf{da} = 2\pi r S_r^{\mathrm{o}}\ell = P_{\mathrm{rad}}^{\mathrm{o}} \tag{15.34a}$$

$$P_{\mathrm{rad}}^{\mathrm{o}} = \frac{c\mu_{\mathrm{o}}\mathrm{I}_{\mathrm{o}}^2\ell}{4\pi}\frac{\left[r^2\ln\left(\dfrac{\sqrt{c^2t^2-r^2}+ct}{r}\right) + ct\sqrt{c^2t^2-r^2}\right]}{\left(\sqrt{c^2t^2-r^2}\right)^3}u_{-1}(ct-r) \tag{15.34b}$$

Therefore the result (in that limit) is

$$P_{\mathrm{rad}}^{\mathrm{o}} = \frac{\mu_{\mathrm{o}} I_{\mathrm{o}}^2\ell}{4\pi t}u_{-1}(t) \tag{15.35}$$

which is only one-half the Maxwell–Poynting value. As discussed in the previous chapter, the resolution is again found by considering the localized Alternate energy, $\frac{1}{2}A_z J_z$. When integrated over the area of the line current, the result (for $ct \gg r_{\mathrm{o}}$) is

$$U_{\mathrm{localized}} = \frac{1}{2}A_z I_{\mathrm{o}}\ell = \frac{\mu_{\mathrm{o}}\mathrm{I}_{\mathrm{o}}^2\ell}{4\pi}\ln\left(\frac{2ct}{r_{\mathrm{o}}}\right) \tag{15.36}$$

This term accounts for the "missing power" because

$$\frac{d}{dt}U_{\mathrm{localized}} = -\frac{1}{2}E_z \mathrm{I}_{\mathrm{o}}\ell$$

$$= \frac{\mu_{\mathrm{o}}\mathrm{I}_{\mathrm{o}}^2\ell}{4\pi t}u_{-1}(t) = P_{\mathrm{rad}}^{\mathrm{o}} \tag{15.37}$$

As expected, Eq. (15.36) represents stored magnetic energy—but not the same value as given by Eq. (15.32b), which for $ct \gg r$ reduces to $\frac{\mu_0 I_0^2 \ell}{4\pi} \ln(\frac{r}{r_o})$. Added perspective is gained by using circuit-theory concepts and expressing $U_{\text{localized}}$ as $\frac{1}{2}L(t)I_0^2$ with the inductance given by

$$L(t) = \frac{\mu_0 \ell}{2\pi} \ln\left(\frac{2ct}{r_o}\right) \tag{15.38}$$

Despite being evaluated at the origin, this time-varying inductance correctly "understands" that the wave front extends to $r = ct$ even though the volume of integration was limited to $r < ct$. Evidently, the time rate of change of $U_{\text{localized}}$ represents power lost *within* the circuit to a radiation resistance, $\frac{dL(t)}{dt} = \frac{\mu_0 \ell}{2\pi t}$ ohms; the other half of the actual power delivered by the line-current source is accounted for by the Alternate-power flux, S_z^o.

15.3 SUPERPOSITION OF AXIAL LINE CURRENTS

Linear superposition of the previous results for a line current step can be used to analyze a variety of related problems. As examples, we choose three z-directed currents: a uniform volume current density, $J_z(r,t)$, confined to a cylindrical region of infinite length and radius r_o; a uniform surface current density, $K_z(t)$, of infinite length and located at radius R; a uniform surface current density, $K_z(x,t)$, confined to a rectangular strip of infinite length and width w located on the plane $y = 0$. In these cases, I_z is replaced by a differential current, either $J_z dV$ or $K_z dA$, located at x', y'. The differential, dA_z, is then integrated and the resulting vector potential (and $\Phi = 0$) is used to evaluate the electric and magnetic fields. Alternatively, all of the fields can be evaluated be direct superposition.

Uniform J_z within a finite radius, r_o

If within a radius, r_o, the axial current has uniform density J_z that steps from zero to the value J_o at time $t = 0$, Eq. (15.27a) becomes (with $x' = r' \cos\phi'$, $y' = r' \sin\phi'$, and $\eta = \phi' - \phi$)

$$A_z(r,t) = \frac{\mu_0 J_o}{2\pi r_o} \int_0^{r_o} \int_0^{2\pi} \cosh^{-1}\left(\frac{ct}{\sqrt{r^2 + r'^2 - 2rr'\cos\eta}}\right)$$

$$\times\, u_{-1}(ct - \sqrt{r^2 + r'^2 - 2rr'\cos\eta})\, d\eta\, dr' \tag{15.39}$$

Care must be taken in evaluating this integral (and the corresponding field integrals), especially for $r < r_o$.

For the time interval, $0 < t < (r_o - r)/c$, the transient fields at r do not receive information from sources located at $r' \geq r_o$; therefore the fields are the same as if $J_z = J_o u_{-1}(t)$ for *all* r. In that case, symmetry dictates that the fields *everywhere* are given by $\mathbf{H} = 0$ and $J_z + \varepsilon_0 \frac{\partial E_z}{\partial t} = 0$. Consequently, it is advantageous to divide the actual current into regions 1 and 2 that are defined in Figure 15.9. The field integrals are then divided into components that are generated by the regional currents; these (normalized) fields are

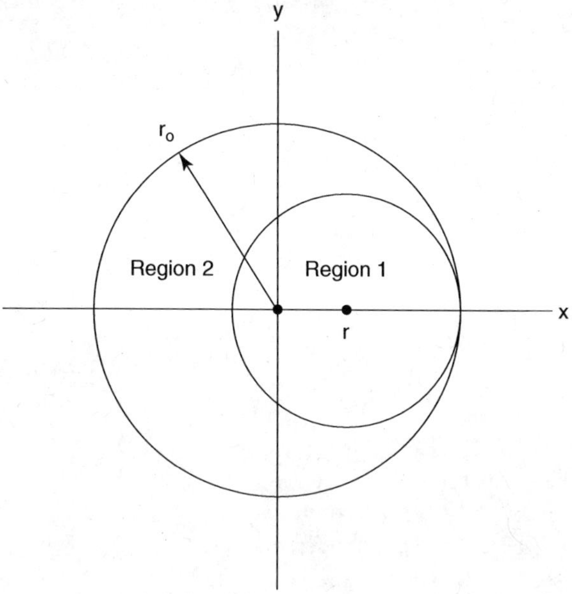

Figure 15.9 Current regions.

labeled $e1, h1, e2$, and $h2$. Region 1 vanishes when the field point p is outside of the current region and $r > r_o$.

It follows that with $I_o = J_o \pi r_o^2$,

$$E_z(r,t) = \frac{c\mu_o I_o}{\pi r_o} u_{-1}(1 - r/r_o)$$
$$[ct/r_o - u_{-1}(ct/r_o - 1 + r/r_o)\sqrt{(ct/r_o)^2 - (1 - r/r_o)^2}]$$
$$+ u_{-1}(ct/r_o - |1 - r/r_o|)E(r,t)$$

where

$$E(r,t) = \frac{c\mu_o I_o}{2\pi r_o} \begin{cases} e1(r,t) & (ct < r_o + r) \\ e2(r,t) & (ct \geq r_o + r) \end{cases}$$

$$e1(r,t) = \frac{2}{\pi} \int_{|1-r/r_o|}^{ct/r_o} \xi \frac{\cos^{-1}\left[\frac{\xi^2 + (r/r_o)^2 - 1}{2\xi r/r_o}\right]}{\sqrt{(ct/r_o)^2 - \xi^2}} d\xi$$

$$e2(r,t) = \frac{2}{\pi} \int_{|1-r/r_o|}^{1+r/r_o} \xi \frac{\cos^{-1}\left[\frac{\xi^2 + (r/r_o)^2 - 1}{2\xi r/r_o}\right]}{\sqrt{(ct/r_o)^2 - \xi^2}} d\xi$$

and

$$H_\phi(r,t) = H(r,t) \begin{cases} 0 & (ct < |r_o - r|) \\ 1 & (ct \geq |r_o - r|) \end{cases}$$

where

$$H(r,t) = \frac{I_o}{2\pi r_o}\begin{cases} h1(r,t) & (ct < r_o + r) \\ h2(r,t) & (ct \geq r_o + r) \end{cases}$$

$$h1(r,t) = \frac{2}{\pi}\int_{|1-r/r_o|}^{ct/r_o} \frac{ct/r_o}{\sqrt{(ct/r_o)^2 - \xi^2}}\sqrt{1 - (\frac{1-(r/r_o)^2 - \xi^2}{2\xi r/r_o})^2}\, d\xi$$

$$h2(r,t) = \frac{2}{\pi}\int_{|1-r/r_o|}^{1+r/r_o} \frac{ct/r_o}{\sqrt{(ct/r_o)^2 - \xi^2}}\sqrt{1 - \left(\frac{1-(r/r_o)^2 - \xi^2}{2\xi r/r_o}\right)^2}\, d\xi$$

It should be noted that because the current density is finite, the amplitudes of both the electric and magnetic fields remain finite for all values of time. For $r \gg r_o$, the wave fronts generated by the current step are spread over $\Delta r = 2r_o$. From Eq. (15.27b) with $r \simeq ct - 2r_o$ a reasonable estimate of the peak fields is

$$E_z(r) \simeq \frac{-c\mu_o I_o}{4\pi\sqrt{r_o r}} \simeq -c\mu_o H_\phi(r)$$

A sequence of plots of the electric and magnetic fields are given in Figure 15.10 for increasing times.

Uniform K_z at constant radius, R

If at radius, R, the axial current has uniform surface density K_z that steps from zero to the value K_o at time $t = 0$, Eq. (15.27a) becomes (with $x' = r'\cos\phi'$, $y' = r'\sin\phi'$, and $\eta = \phi' - \phi$)

$$A_z(r,t) = \frac{\mu_o K_o}{\pi R}\int_0^{2\pi} \cosh^{-1}\left(\frac{ct}{\sqrt{r^2 + R^2 - 2Rr\cos\eta}}\right) \times$$

$$u_{-1}(ct - \sqrt{r^2 + R^2 - 2Rr\cos\eta})\, d\eta \qquad (15.40a)$$

With $I_o = K_o 2\pi R$, the associated fields are

$$E_z(r,t) = \frac{-c\mu_o I_o}{2\pi}\mathrm{Re}\left\{\int_{-R}^{R} \frac{1}{\pi}\frac{1}{\sqrt{R^2 - x^2}}\frac{1}{\sqrt{c^2t^2 - r^2 - R^2 + 2rx}}\, dx\right\} \qquad (15.40b)$$

$$H_\phi(r,t) = \frac{I_o}{(2\pi)^2}\mathrm{Re}\left\{\int_{-R}^{R} \frac{r - x}{r^2 + R^2 - 2rx}\frac{2ct}{\sqrt{R^2 - x^2}}\frac{1}{\sqrt{c^2t^2 - r^2 - R^2 + 2rx}}\, dx\right\}$$

$$(15.40c)$$

where taking the real part is equivalent to introducing the appropriate step functions.

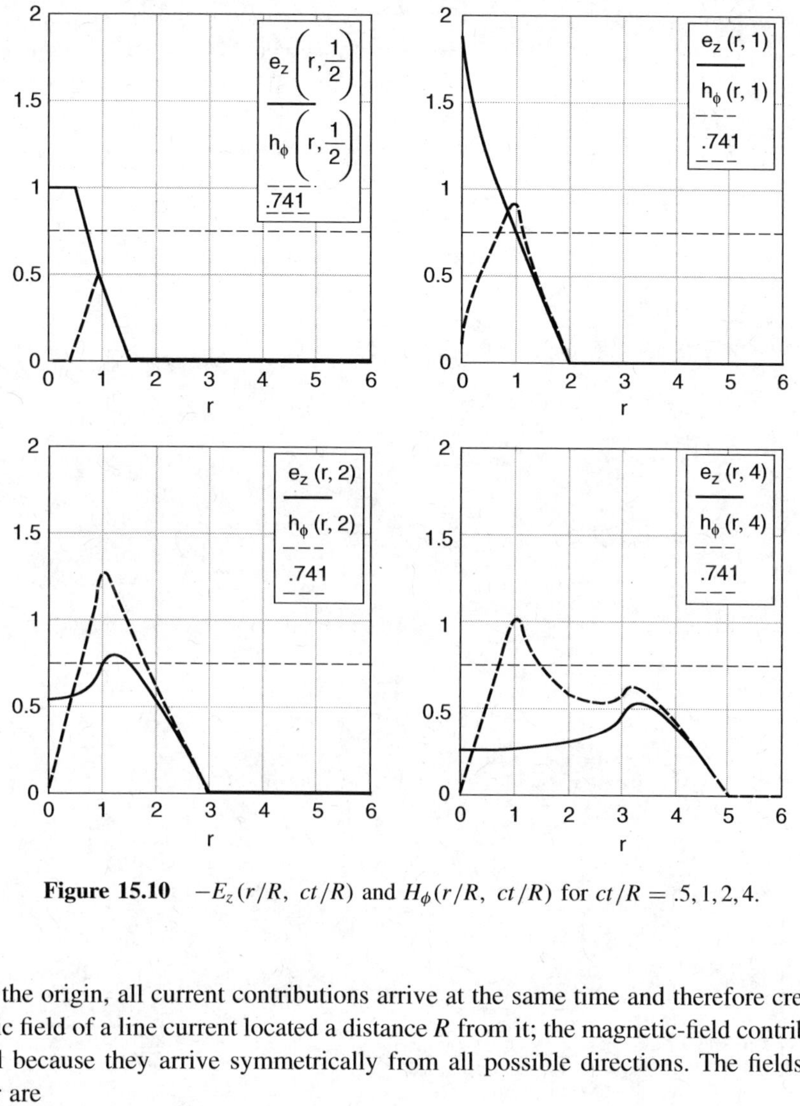

Figure 15.10 $-E_z(r/R,\ ct/R)$ and $H_\phi(r/R,\ ct/R)$ for $ct/R = .5, 1, 2, 4$.

At the origin, all current contributions arrive at the same time and therefore create the electric field of a line current located a distance R from it; the magnetic-field contributions cancel because they arrive symmetrically from all possible directions. The fields at the center are

$$E_z(0,t) = \frac{-c\mu_o I_o}{2\pi}\frac{1}{\sqrt{c^2t^2 - R^2}}u_{-1}(ct - R)$$

$$H_\phi(0,t) = 0$$

Unlike the uniformly distributed current, the peak fields near the wave front are infinite, but, unlike the line-current example, it is a logarithmic infinity. If $n = \frac{c^2t^2 - R^2 - r^2}{2rR}$,

$$E_z(r,t) = -c\mu_o H_\phi(r,t) = \frac{-c\mu_o I_o}{2\pi\sqrt{rR}}F(n)$$

where

$$F(n) = \int_{-1}^{1} \frac{1}{\sqrt{2\pi}}\frac{1}{\sqrt{1-\xi^2}}\frac{1}{\sqrt{n+\xi}}\,d\xi$$

is plotted in Figure 15.11. It is seen that $1/\sqrt{2n}$ is a good approximation except close to the peak. A sequence of plots of the electric and magnetic fields are given in Figure 15.12 for increasing times.

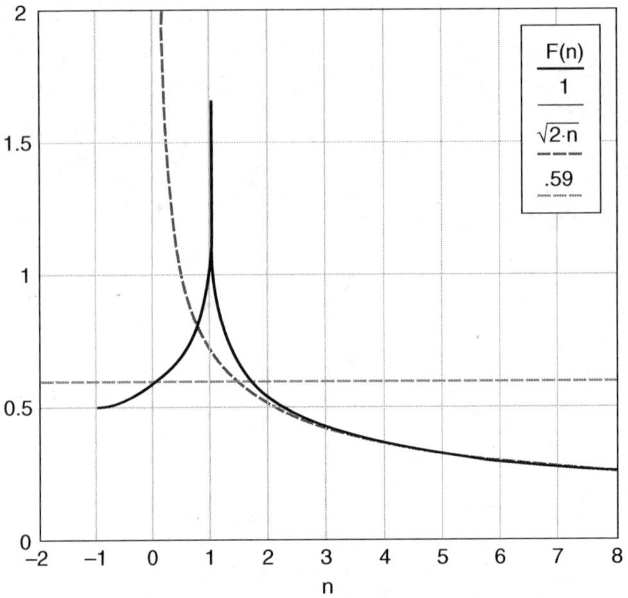

Figure 15.11 Wave-front normalization, $F(n)$.

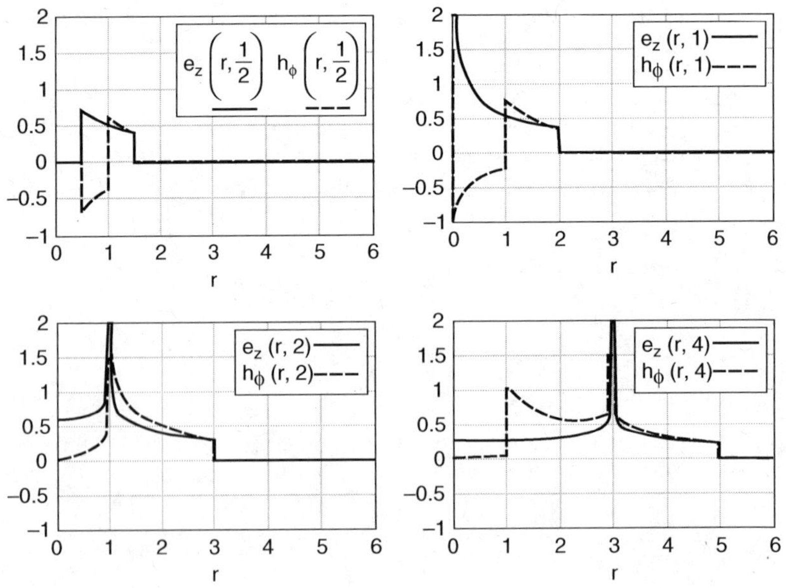

Figure 15.12 $-E_z(r/R,\ ct/R)$ and $H_\phi(r/R,\ ct/R)$ for $ct/R = .5, 1, 2, 4$

Uniform K_z on a strip of constant width

If at $y = 0$, $|x| < w$, the axial current has uniform surface density K_z that steps from zero to the value K_o at time $t = 0$, Eq. (15.27a) becomes with $\xi = x' - x$

$$A_z(x,y,t) = \frac{\mu_o K_o}{2\pi} \int_{-w-x}^{w-x} \cosh^{-1}\left(\frac{ct}{\sqrt{y^2+\xi^2}}\right) u_{-1}(ct - \sqrt{y^2+\xi^2})\, d\xi \qquad (15.41a)$$

$$E_z(x,y,t) = \frac{c\mu_o K_o}{2\pi} \int_{-w-x}^{w-x} \frac{-1}{\sqrt{c^2t^2-y^2-\xi^2}} u_{-1}(ct - \sqrt{y^2+\xi^2})\, d\xi \qquad (15.41b)$$

$$H_x(x,y,t) = \frac{c\mu_o K_o}{2\pi} \int_{-w-x}^{w-x} \frac{-ct}{y^2+\xi^2} \frac{y}{\sqrt{c^2t^2-y^2-\xi^2}} u_{-1}(ct - \sqrt{y^2+\xi^2})\, d\xi \quad (15.41c)$$

$$H_y(x,y,t) = \frac{c\mu_o K_o}{2\pi} \int_{-w-x}^{w-x} \frac{ct}{y^2+\xi^2} \frac{\xi}{\sqrt{c^2t^2-y^2-\xi^2}} u_{-1}(ct - \sqrt{y^2+\xi^2})\, d\xi \quad (15.41d)$$

The total current is $I_o = 2K_o w$. The electric and magnetic fields can be expressed in closed form as

$$E_z(x,y,t) = \frac{-c\mu_o K_o}{2\pi} Re\left\{\left[\sin^{-1}\left(\frac{w+x}{\sqrt{c^2t^2-y^2}}\right) + \sin^{-1}\left(\frac{w-x}{\sqrt{c^2t^2-y^2}}\right)\right]\right\} \qquad (15.42a)$$

$$\begin{aligned} H_x(x,y,t) = \frac{-K_o}{4}\Bigg\{ &\frac{2}{\pi}\tan^{-1}\left(\frac{ct}{y}\frac{w+x}{\sqrt{c^2t^2-y^2-(w+x)^2}}\right) u_{-1}(ct - \sqrt{y^2+(w+x)^2}) \\ &+ \frac{2}{\pi}\tan^{-1}\left(\frac{w-x}{\sqrt{c^2t^2-y^2-(w-x)^2}}\right) u_{-1}(ct - \sqrt{y^2+(w-x)^2}) \\ &+ [u_{-1}(ct - \sqrt{y^2+(w+x)^2})\mathrm{sign}(w+x) \\ &+ u_{-1}(ct - \sqrt{y^2+(w-x)^2})\mathrm{sign}(w-x)]\mathrm{sign}(y)\Bigg\} \end{aligned} \qquad (15.42b)$$

$$\begin{aligned} H_y(x,y,t) = \frac{K_o}{2\pi}\Bigg\{ &\ln\left(\frac{ct+\sqrt{c^2t^2-y^2-(w-x)^2}}{\sqrt{y^2+(w-x)^2}}\right) u_{-1}(ct - \sqrt{y^2+(w-x)^2}) \\ &- \ln\left(\frac{ct+\sqrt{c^2t^2-y^2-(w+x)^2}}{\sqrt{y^2+(w+x)^2}}\right) u_{-1}(ct - \sqrt{y^2+(w+x)^2})\Bigg\} \end{aligned} \qquad (15.42c)$$

Plot sequences of the electric and magnetic fields are given in Figures 15.13 and 15.14 for increasing times. For $ct \gg w$, the fields near the strip are *MQS* in character; the **H** field is static, and the **E** field is nearly isotropic and proportional to $1/t$. The logarithmic singularities in H_y at the strip edges, $|x| = w$, are due to the zero thickness of the current. Directionality of the expanding wavefront is maintained because the high-frequency content of the unit step persists; behind the wave front, the fields with progressively lower-frequency content lose their directionality.

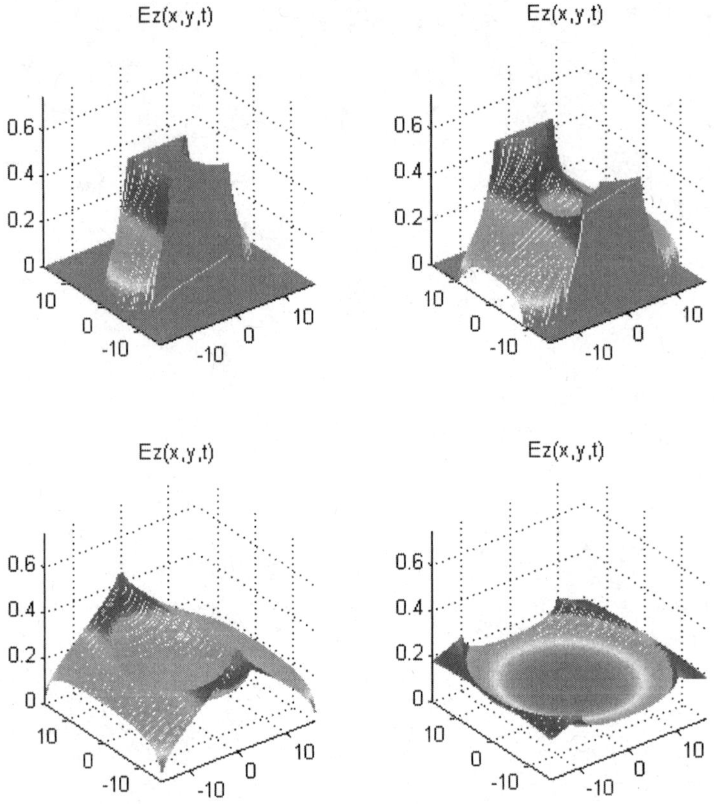

Figure 15.13 $-E_z(x, y, t)$ for $ct/w = 6, 12, 18, 24$.

15.4 AXIAL CURRENT WITH MULTIPLE PULSES

Trapezoidal wave forms

If the line current is of constant magnitude I_o except during the intervals $nt_p < t < nt_p + t_o (n = 0, 1, 2, \ldots, n_o)$ where it has constant slope and reverses polarity at $t = nt_p + \frac{1}{2}t_o$, then

$$I_z(t) = \sum_{n=0}^{n_o} I_n(t) \tag{15.43a}$$

where

$$I_n(t) = (-1)^n \begin{cases} I_o & (t < nt_p) \\ \left(1 + 2\dfrac{nt_p - t}{t_o}\right) I_o & (nt_p < t < nt_p + t_o) \\ -I_o & (t > nt_p + t_o) \end{cases} \tag{15.43b}$$

There are $n_o + 1$ transitions. If n_o is even, the final current is reversed; if n_o is odd, the final current is unchanged. A current transient of this form is shown in Figure 15.15 for $n_o = 2$ when $t_o = .1$ and $t_p = 1$.

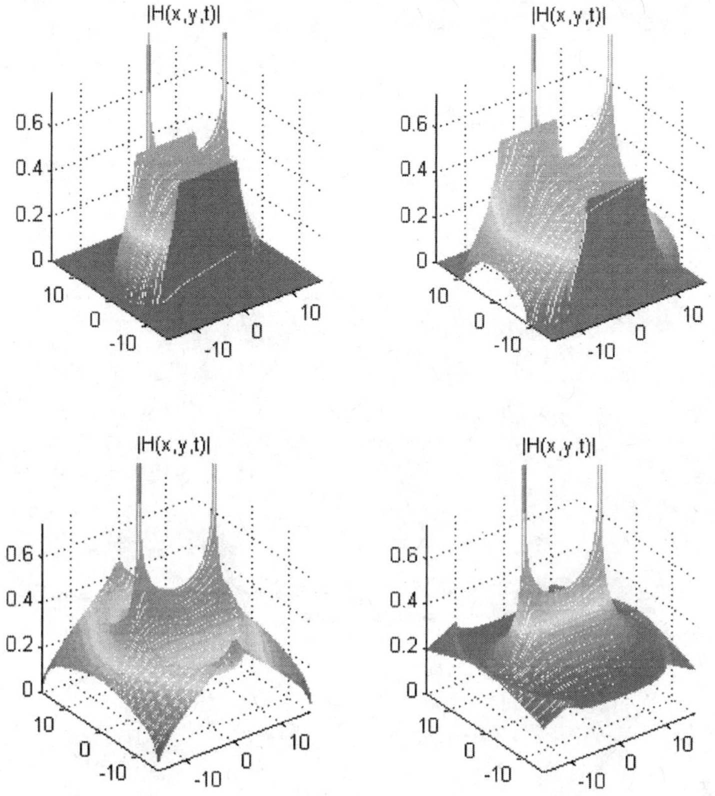

Figure 15.14 $|\mathbf{H}(x, y, t)|$ for $ct/w = 6, 12, 18, 24$.

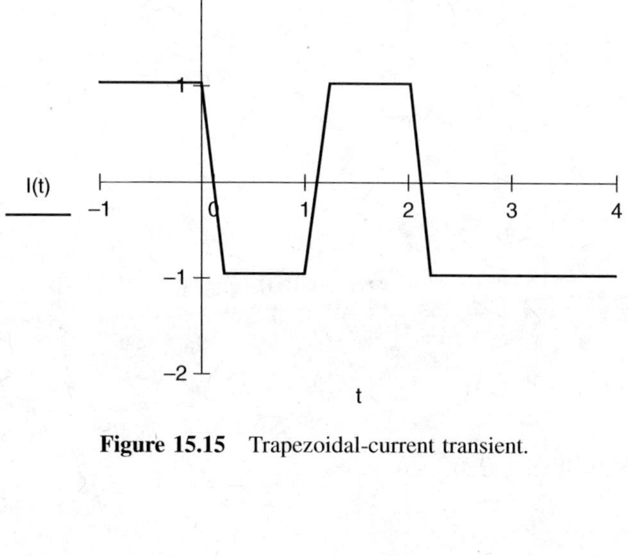

Figure 15.15 Trapezoidal-current transient.

The electric and magnetic fields can be expressed as the superposition of components arising from a single reversal

$$e(r,t,t_o,t_p,n_o) = \sum_{n=0}^{n_o}(-1)^n e_n(r,t,t_o,t_p,n) \tag{15.44a}$$

$$h(r,t,t_o,t_p,n_o) = \sum_{n=0}^{n_o}(-1)^n h_n(r,t,t_o,t_p,n) \tag{15.44b}$$

The Poynting flux is

$$s(r,t,t_o,t_p,n_o) = -e(r,t,t_o,t_p,n_o)\cdot h(r,t,t_o,t_p,n_o)$$

The fields e_n and h_n due to $I_n(t)$ are most easily obtained by differentiating the current. This operation generates a pulse comprised of two current steps. The solution for a single step was studied in Section 15.2; therefore a superposition of such solutions when integrated with respect to time generates the fields for a single value of n. The result is

$$e_n(r,t,t_o,t_p,n) = \begin{cases} e_{n1}, & r \le c(t-nt_p-t_o) \\ e_{n2}, & c(t-nt_p-t_o) \le r \le c(t-nt_p) \\ 0, & c(t-nt_p) \le r \end{cases} \tag{15.45a}$$

$$h_n(r,t,t_o,t_p,n) = \begin{cases} h_{n1}, & r \le c(t-nt_p-t_o) \\ h_{n2}, & c(t-nt_p-t_o) \le r \le c(t-nt_p) \\ h_{n3}, & c(t-nt_p) \le r \end{cases} \tag{15.45b}$$

where

$$e_{n1} = \frac{\mu_o I_o}{\pi t_o}\ln\left(\frac{\sqrt{c^2(t-nt_p)^2-r^2}+c(t-nt_p)}{\sqrt{c^2(t-nt_p-t_o)^2-r^2}+c(t-nt_p-t_o)}\right) \tag{15.45c}$$

$$e_{n2} = \frac{\mu_o I_o}{\pi t_o}\ln\left(\frac{\sqrt{c^2(t-nt_p)^2-r^2}+c(t-nt_p)}{r}\right) \tag{15.45d}$$

$$h_{n1} = \frac{I_o}{\pi}\left(\frac{\frac{r}{ct_o}\left(\sqrt{c^2(t-nt_p-t_o)^2-r^2}-\sqrt{c^2(t-nt_p)^2-r^2}-ct_o\right)}{\left[\sqrt{c^2(t-nt_p-t_o)^2-r^2}+c(t-nt_p-t_o)\right]\left[\sqrt{c^2(t-nt_p)^2-r^2}+c(t-nt_p)\right]} - \frac{1}{2r}\right) \tag{15.45e}$$

$$h_{n2} = \frac{I_o}{\pi r}\left(\frac{1}{2} - \frac{\sqrt{c^2(t-nt_p)^2-r^2}}{ct_o}\right) \tag{15.45f}$$

$$h_{n3} = \frac{I_o}{2\pi r} \tag{15.45g}$$

The electric (solid) and magnetic (dotted) fields are plotted in Figure 15.16(a) as a function of normalized coordinates and amplitudes for $ct = 5$ when $ct_o = 1$, $ct_p = 1$, and there are $n_o + 1 = 3$ reversals; the wake of the transient consists of *MQS* fields. The corresponding Poynting flux is shown in Figure 15.16(b).

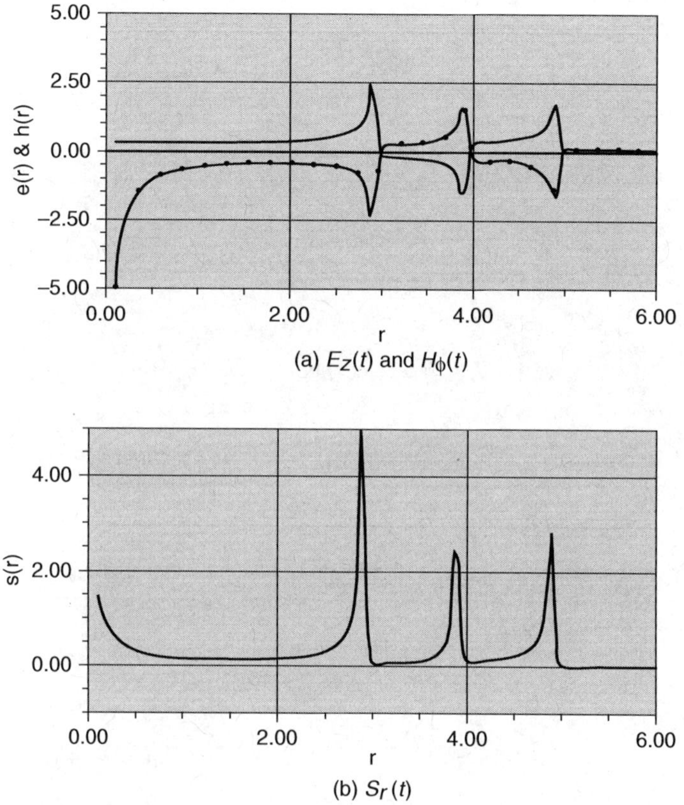

Figure 15.16 Train of current reversals: (a) Fields; (b) Poynting-power flux.

Sine-wave approximation

The trapezoidal approximation to $\sin(x)$ over the interval $0 < x < \pi$ is

$$y(x) = \begin{cases} \beta x/\alpha & (0 < x < \alpha) \\ \beta & (\alpha < x < \pi - \alpha) \\ \beta(\pi - x)/\alpha & (\pi - \alpha < x < \pi) \end{cases}$$

A least-squares fit requires that

$$\alpha = 1.08228$$
$$\beta = .9607$$

and is shown in Figure 15.17.

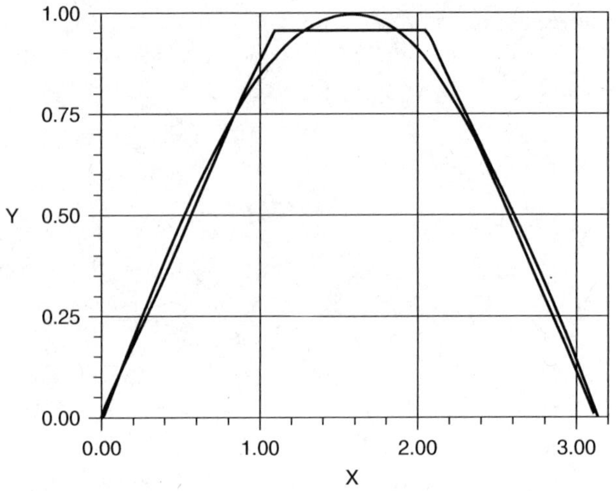

Figure 15.17 Trapezoidal approximation of sin(x).

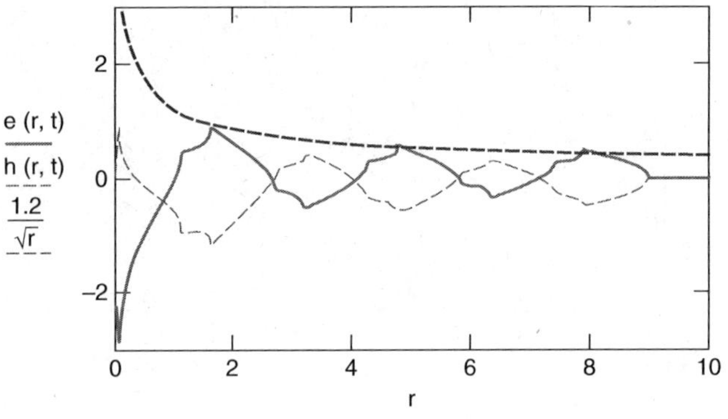

(a) $E_z(t)$ and $H_\varphi(t)$

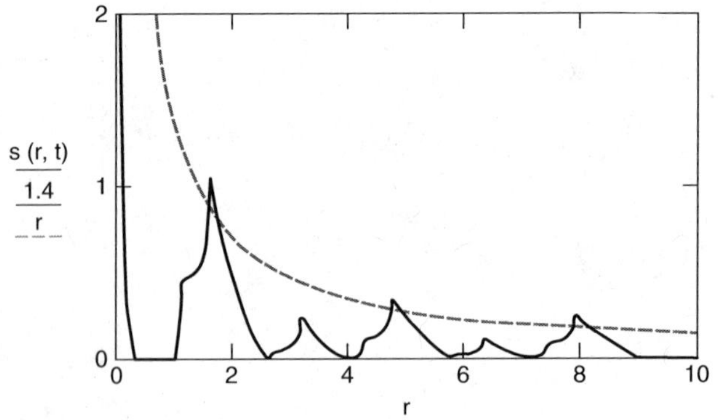

(b) $S_r(t)$

Figure 15.18 Fields and Poynting-power flux of trapezoidal "sine" pulses.

The electric and magnetic fields are plotted as a function of normalized coordinates in Figure 15.18(a) for $ct = 9$ when $ct_o = \alpha$, $ct_p = \pi/2$, and there are $n_o + 1 = 5$ reversals; the corresponding Poynting flux is shown in Figure 15.18(b) The steady-state response to a sinusoidal excitation is well approximated by choosing $n_o \gg 1$, but other methods can be employed. Two powerful approaches are described in the following section.

15.5 FIELDS OF A SINUSOIDAL AXIAL CURRENT

Waves and boundary conditions

If a single axial current is taken as

$$I_z(t) = I_o \cos(\omega t) \tag{15.46}$$

both the electric and magnetic fields are sinusoidal in time. Because the complex Cartesian component, $\underline{E}_z(r)$, satisfies the zero-order Bessel Equation, it follows that a linear combination of $J_0(kr)$ and $Y_0(kr)$ is the general solution. Two boundary conditions are required to find the appropriate coefficients; one is imposed by the source current at $r = 0$, the other by the assumption that there is no reflected wave at $r \to \infty$. The latter condition is met by a Hankel function of zero order and the first kind that represents a wave traveling outward in the radial direction. Making use of Appendix F, Section F.1, one finds

$$-\underline{E}_z(r) = \underline{C}H_0^{(1)}(kr) = \underline{C}[J_0(kr) - jY_0(kr)] \tag{15.47}$$

which for large kr has the asymptotic behavior,

$$-\underline{E}_z(r) \sim \underline{C}\sqrt{\frac{2}{\pi kr}}\exp\left[-j\left(kr - \frac{\pi}{4}\right)\right] \tag{15.48}$$

$\underline{H}_\phi(kr)$ is *not* a Cartesian component satisfying the zero-order Bessel Equation for all kr, rather it satisfies

$$\underline{H}_\phi = \frac{-j}{\omega\mu_o}\frac{\partial \underline{E}_z}{\partial r} \tag{15.49}$$

Because for $kr \to 0$,

$$J_0(kr) \simeq 1$$

$$Y_0(kr) \simeq \frac{2}{\pi}\ln(kr)$$

it follows that

$$I_o = \lim_{r\to 0} 2\pi r \underline{H}_\phi = \frac{4}{\omega\mu_o}\underline{C}$$

and

$$\underline{C} = \tfrac{1}{4}\omega\mu_o I_o = \tfrac{1}{4}\eta_o k I_o$$

For large kr, the asymptotic form of the magnetic field is

$$\underline{H}_\phi(kr) \sim \sqrt{\frac{\varepsilon_o}{\mu_o}}\underline{C}\sqrt{\frac{2}{\pi kr}}\exp\left[-j\left(kr - \frac{\pi}{4}\right)\right] \tag{15.50}$$

and that of the complex Poynting flux is

$$\underline{S}_r = \frac{-1}{2}\underline{E}_z\underline{H}_\phi^* \sim \sqrt{\frac{\varepsilon_o}{\mu_o}}CC^*\frac{1}{\pi kr}$$

The radiated power per-unit axial length is therefore

$$2\pi r\underline{S}_r = \frac{1}{8}\eta_o k I_o^2 = \frac{1}{2}R'_{\text{rad}}I_o^2$$

from which it follows that the radiation resistance per unit length is

$$R'_{\text{rad}} = \frac{1}{4}\eta_o k$$

Each small length, ℓ $(k\ell \ll 1)$, has resistance $\frac{1}{4}\eta_o k\ell$ which should be compared to the radiation resistance of a Hertzian dipole of length ℓ, $\frac{1}{6\pi}\eta_o(k\ell)^2$. The first of these resistances is larger because the remaining axial current causes the length ℓ to radiate more efficiently than it would if it were a solitary element.

Convolution integral

Because we know the electric-field response to a current step of the axial current and the system is linear, we can find the impulse response by simply differentiating the step response. The result (for $I_o = 1$) is

$$h_o(\tau, r) = \frac{c\mu_o}{2\pi}\left[\frac{c^2\tau}{(c^2\tau^2 - r^2)^{\frac{3}{2}}}u_{-1}(\tau - r/c) - \frac{1}{(c^2\tau^2 - r^2)^{\frac{1}{2}}}u_0(\tau - r/c)\right] \tag{15.51}$$

For an arbitrary axial current, $I_z(t)$, the electric field is given by the convolution integral [15, p. 377],

$$E_z(t, r) = \int_{r/c}^{t} h_o(\tau, r)I_z(t - \tau)\, d\tau \tag{15.52}$$

When the current is sinusoidal and in the steady state, the result agrees with that of the previous section. Because the evaluation is nontrivial, especially at the lower limit, we here content ourselves with evaluation at $r \to 0$. Therefore,

$$h_o(\tau, 0) = \frac{\mu_o}{2\pi}\left[\frac{1}{\tau^2}u_{-1}(\tau) - \frac{1}{\tau}u_0(\tau)\right]$$

We approximate $\frac{1}{\tau}u_0(\tau)$ with a rectangular pulse of width δ and height $1/\delta^2$ (where $\delta \sim r_o/c$ is nonzero because of the finite size of the current). Then with $I_z(t) = I_o\cos(\omega t)$,

$$E_z(0, t) = \frac{\mu_o}{2\pi}I_o\left[\int_{\delta}^{t}\frac{1}{\tau^2}\cos[\omega(t - \tau)]\, d\tau - \int_{0}^{\delta}\frac{1}{\delta^2}\cos[\omega(t - \tau)]\, d\tau\right] \tag{15.53a}$$

These integrals are expressed as

$$-E_z(0, t) = -\frac{\omega\mu_o}{4\pi}I_o\left\{[2Ci(\omega t) - 2(\ln(\omega\delta) + \gamma) + 1]\sin(\omega t)\right.$$

$$\left. + 2\text{Si}(\omega t)\cos(\omega t) + \frac{2}{\omega t}\right\} \tag{15.53b}$$

where $\gamma = .577216$ (Euler's constant) and the sine and cosine integrals are defined by

$$\mathrm{Si}(x) = \int_0^x \frac{\sin u}{u}\, du$$

$$\mathrm{Ci}(x) = -\int_x^\infty \frac{\cos u}{u}\, du = \gamma + \ln x + \int_0^x \frac{\cos u - 1}{u}\, du$$

For large values of ωt, $\mathrm{Si}(\omega t) \to \pi/2$, $Ci(\omega t) \to 0$, and

$$-E_z(0,t) \to \frac{\omega\mu_o}{4} I_o \left[\cos(\omega t) - \frac{2}{\pi}\ln\left(\frac{1}{\omega\delta}\right)\sin(\omega t)\right]$$

$$= R'_{\text{rad}} I_z(t) + L' \frac{dI_z(t)}{dt}$$

The input power per unit length is $-E_z(0,t)I_o\cos(\omega t)$; as expected, the average value $\frac{1}{2}R'_{\text{rad}}\mathrm{I}_o^2$ equals the radiated power per unit length calculated in the previous section. The out-of-phase component is proportional to the inductance per unit length,

$$L' = \frac{\mu_o}{2\pi}\ln\left(\frac{1}{\omega\delta}\right)$$

where $\omega\delta \sim kr_o \ll 1$. Notice that, from the complex fields given by Eqs. (15.47) and (15.49), the input impedance (per unit length) can also be evaluated from

$$R'_{\text{rad}} + j\omega L' = \left[\frac{-\underline{E}_z(r)}{2\pi r \underline{H}_\phi(r)}\right]_{r=\delta}$$

CHAPTER 16

PLANE WAVES

16.1 UNIFORM *TEM* PLANE WAVES

Propagation in free-space

Consider two linearly polarized uniform plane waves arising from

$$\mathbf{A} = \widehat{\mathbf{x}}A_o[f_+(z-ct) - f_-(z+ct)]; \quad \Phi = 0 \tag{16.1a}$$

$$\mathbf{E} = -\frac{\partial \mathbf{A}}{\partial t} = \widehat{\mathbf{x}}cA_o\left[\frac{df_+(z-ct)}{d(z-ct)} + \frac{df_-(z+ct)}{d(z+ct)}\right] \tag{16.1b}$$

$$\mu_o\mathbf{H} = \nabla \times \mathbf{A} = \widehat{\mathbf{y}}A_o\left[\frac{df_+(z-ct)}{d(z-ct)} - \frac{df_-(z+ct)}{d(z+ct)}\right] \tag{16.1c}$$

If $f_\mp = 0, \pm E_x/H_y = \sqrt{\mu_o/\varepsilon_o} = \eta_o$ (the free-space characteristic wave impedance),

$$\mathbf{E} \times \mathbf{H} = \widehat{\mathbf{z}}\frac{cA_o^2}{\mu_o}\left(\left[\frac{df_+(z-ct)}{d(z-ct)}\right]^2 - \left[\frac{df_-(z+ct)}{d(z+ct)}\right]^2\right) \tag{16.2}$$

The Power and Beauty of Electromagnetic Fields, First Edition. F. R. Morgenthaler.

and

$$\mathbf{S}^{o} = \widehat{\mathbf{z}}\frac{cA_o^2}{2\mu_o}\left(\begin{array}{c} \left[\frac{df_+(z-ct)}{d(z-ct)}\right]^2 - f_+(z-ct)\frac{d^2 f_+(z-ct)}{d(z-ct)^2} \\ -\left[\frac{df_-(z+ct)}{d(z+ct)}\right]^2 + f_-(z+ct)\frac{df_-(z+ct)}{d(z+ct)} \\ +\frac{\partial}{\partial z}\left[\frac{df_+(z-ct)}{d(z-ct)}f_-(z+ct) - \frac{df_-(z+ct)}{d(z+ct)}f_+(z-ct)\right] \end{array}\right) \tag{16.3}$$

The energy densities for the two representations are

$$W^{\mathrm{em}} = \frac{A_o^2}{\mu_o}\left(\left[\frac{df_+(z-ct)}{d(z-ct)}\right]^2 + \left[\frac{df_-(z+ct)}{d(z+ct)}\right]^2\right) \tag{16.4}$$

and

$$W^{o} = \frac{A_o^2}{2\mu_o}\left(\begin{array}{c} \left[\frac{df_+(z-ct)}{d(z-ct)}\right]^2 - f_+(z-ct)\frac{d^2 f_+(z-ct)}{d(z-ct)^2} \\ +\left[\frac{df_-(z+ct)}{d(z+ct)}\right]^2 - f_-(z+ct)\frac{df_-(z+ct)}{d(z+ct)} \\ +\frac{\partial}{\partial z}\left[\frac{df_+(z-ct)}{d(z-ct)}f_-(z+ct) + \frac{df_-(z+ct)}{d(z+ct)}f_+(z-ct)\right] \end{array}\right) \tag{16.5}$$

The interaction terms between the $f_+(z-ct)$ and $f_-(z+ct)$ waves, which are absent from $\mathbf{S}$ and W^{em}, are present in both $\mathbf{S}^{o}$ and W^{o}. However, when the latter are substituted into the Alternate Poynting Theorem, the resulting terms cancel so that the *net interaction* is zero in both models. The *effective* power flux is therefore of the form $c(W_+ - W_-)$ in both representations. If allowed to remain, these terms are an example of null power and null energy—encountered again in Chapter 18 and discussed more fully in Chapters 19 and 20. Reference is made in Appendix H, Section H.1, to an animation of the overlap interaction when f_+ and f_- are both Gaussian pulses.

When both of the waves are sinusoidal with the same frequency ω, the Poynting flux has both an average value and component varying at the frequency 2ω whereas S_z^o is a constant.[1] As expected, the time-averaged values are equal. These results agree with the sinusoidal steady-state case (with $f_- = 0$) that was considered in Chapter 3, Section 3.5 when $\Phi = 0$.

Propagation in uniform linear materials

If the propagation is within a stationary, uniform, linear material characterized by constant permittivity, ε, and permeability, μ, the Minkowski formulation described in Appendix D, Section D.2 is convenient. Equations (D.7a) and (D.7b) can then be used to calculate

[1]In the sinusoidal steady state, the interaction between f_+ and f_- produces no Alternate power but instead produces time-independent Alternate energy that can be ignored.

$\mathbf{S}^o$ and W^o. If there is no dissipation, $\mathbf{J}_u = 0$, and using the constitutive laws

$$\mathbf{P} = (\varepsilon_o - \varepsilon)\frac{\partial \mathbf{A}}{\partial t}$$

$$\mathbf{M} = \frac{(\mu - \mu_o)}{\mu\mu_o}\nabla \times \mathbf{A}$$

one simply adds the terms $\frac{1}{2}\left(\frac{\partial \mathbf{A}}{\partial t} \times \mathbf{M} - \mathbf{A} \times \frac{\partial \mathbf{M}}{\partial t}\right)$ to Eq. (16.3) and $\frac{1}{2}\left(\mathbf{A} \cdot \frac{\partial \mathbf{P}}{\partial t} - \frac{\partial \mathbf{A}}{\partial t} \cdot \mathbf{P}\right)$ to Eq. (16.5). Because $\nabla\varepsilon = 0$, there is no polarization charge density *inside* the material and $\nabla \cdot \mathbf{A} = 0$. The Alternate power flux and energy density that emerge are the same as the free-space counterparts *except* for the replacement of μ_o and ε_o by μ and ε and the additional power-flux vector,

$$\frac{(\mu - \mu_o)}{2\mu\mu_o}\left[(\mathbf{A} \cdot \nabla)\frac{\partial \mathbf{A}}{\partial t} - \left(\frac{\partial \mathbf{A}}{\partial t} \cdot \nabla\right)\mathbf{A}\right] \tag{16.6}$$

Because Eq. (16.6) has zero divergence, this component can be removed without affecting the Alternate-power theorem. A simpler procedure is to use the free-space formulation *except* for the replacement of μ_o and ε_o by μ and ε in *all* formulas (including $c = 1/\sqrt{\mu_o\varepsilon_o}$ and $\eta_o = \sqrt{\mu_o/\varepsilon_o}$). Both $\mathbf{P}$ and $\mathbf{M}$ then vanish (into the permittivity and permeability). The same can be done for the Minkowski versions of $\mathbf{S}^{em}$ and W^{em}.

16.2 DOPPLER-SHIFTED *TEM* PLANE WAVES

If a uniform *TEM* plane wave of frequency ω (propagating in the $+z$ direction) is normally incident upon a perfectly conducting plane moving along z at a constant velocity, v_o, the reflected wave is Doppler-shifted to the frequency ω'. Assuming that the waves are linearly-polarized with $\mathbf{E}$ parallel to the x axis, the vector potential and scalar potential are, in the Lorenz gauge ($g = 0$), given by

$$\mathbf{A} = \hat{\mathbf{x}}\, A_o\{\cos[\omega(t - z/c) + \theta_+] + \Gamma_A \cos[\omega'(t + z/c) + \theta_-]\} \quad (z \le v_o t) \tag{16.7a}$$

$$\mathbf{A} = 0 \quad (z > v_o t) \tag{16.7b}$$

$$\Phi = 0 \quad (\text{all } z) \tag{16.7c}$$

Assuming that the electric and magnetic fields remain finite, continuity of $\mathbf{A}$ is required. The conducting plane, located at $z = v_o t$, therefore imposes the boundary condition: $A_x = 0$[2]; this is satisfied provided

$$\omega' = \frac{c - v_o}{c + v_o}\omega \tag{16.8a}$$

$$\Gamma_A = -1 \tag{16.8b}$$

$$\theta_+ = \theta_- \tag{16.8c}$$

[2]At the conducting boundary, E_x vanishes only when evaluated in the moving frame of reference; the correct boundary condition, $E_x - v_o\mu_o H_y = 0$, is automatically satisfied by $A_x = 0$.

For simplicity, we choose $\theta_\pm = 0$. Consequently, the electric, magnetic, Poynting vector fields and Poynting energy density (for $z < \mathrm{v_o} t$) are

$$\mathbf{E} = \widehat{\mathbf{x}}\, \mathrm{A_o}\omega \left[\sin\omega(t - z/c) - \frac{c - v_o}{c + v_o} \sin\omega'(t + z/c) \right] \tag{16.9a}$$

$$\mathbf{H} = \widehat{\mathbf{y}} \sqrt{\frac{\varepsilon_o}{\mu_o}} \mathrm{A_o}\omega \left[\sin\omega(t - z/c) + \frac{c - v_o}{c + v_o} \sin\omega'(t + z/c) \right] \tag{16.9b}$$

$$\begin{aligned} \mathbf{S} &= \mathbf{E} \times \mathbf{H} \\ &= \widehat{\mathbf{z}} \sqrt{\frac{\varepsilon_o}{\mu_o}} (\mathrm{A_o}\omega)^2 \left[\sin^2\omega(t - z/c) - \left(\frac{c - v_o}{c + v_o}\right)^2 \sin^2\omega'(t + z/c) \right] \end{aligned} \tag{16.9c}$$

$$\begin{aligned} W^{\mathrm{em}} &= \frac{1}{2}\varepsilon_o \mathbf{E}\cdot\mathbf{E} + \frac{1}{2}\mu_o \mathbf{H}\cdot\mathbf{H} \\ &= \varepsilon_o (\mathrm{A_o}\omega)^2 \left[\sin^2\omega(t - z/c) + \left(\frac{c - v_o}{c + v_o}\right)^2 \sin^2\omega'(t + z/c) \right] \end{aligned} \tag{16.9d}$$

From Eqs. (3.31a) and (3.31b), the Alternate power flux and energy-density are,

$$\mathbf{S}^{\mathrm{o}} = \widehat{\mathbf{z}} \frac{2\, v_o (\mathrm{A_o}\omega)^2 [1 - \cos\omega(t - z/c)\cos\omega'(t + z/c)]}{\mu\, (c + v_o)^2} \tag{16.10a}$$

$$W^{\mathrm{o}} = \varepsilon_o \frac{(\mathrm{A_o}\omega)^2}{(c + v_o)^2} \left(\begin{array}{l} c^2 \left[1 - \cos \dfrac{2\omega(z - v_o t)}{c + v_o} \right] \\ + v_o^2 \left[1 - \cos \dfrac{2\omega(v_o z - c^2 t)}{c(c + v_o)} \right] \end{array} \right) \tag{16.10b}$$

As expected, the time-averaged values of Eqs. (16.9c) and (16.10a) agree; when $v_o = 0$, those values are zero. For nonzero v_o, there are power components in $\mathbf{E} \times \mathbf{H}$ at frequencies 2ω and $2\omega'$and in $\mathbf{S}^{\mathrm{o}}$ at frequencies $\omega + \omega'$ and $\omega - \omega'$. The latter would seem to violate the prohibition of time-dependent nonlocalized Alternate power in harmonic cases, but the theorem only applies to single-frequency situations. It is interesting that in the limit $v_o = c$, where $\omega' \to 0$, the reflected components of $\mathbf{E}$ and $\mathbf{H}$ vanish, but that of $\mathbf{A}$ does not. Consequently, a component of $\mathbf{S}^{\mathrm{o}}$persists at the frequency ω.

The average values of the energy densities, Eqs. (16.9d) and (16.10b), also match. Although the plus and minus waves individually travel with velocities $\pm c$, the energy velocity, v_{E}, of the combination is the same for either representation and is given by

$$v_{\mathrm{E}} = \frac{<S_z(z,t)>}{<W^{\mathrm{em}}(z,t)>} = \frac{<S_z^{\mathrm{o}}(z,t)>}{<W^{\mathrm{o}}(z,t)>} = \frac{2c^2 v_o}{c^2 + v_o^2} \tag{16.11}$$

Notice that, in the limit $\mathrm{v_o} \ll c$, $v_{\mathrm{E}} = 2v_o$; in the limit $v_o \to c$, $v_{\mathrm{E}} = c$.

16.3 NONUNIFORM PLANE WAVES

A planar surface current, $\mathbf{K}$, of magnitude $K_0 \cos(\omega t - \beta z)$ can generate nonuniform plane waves in free-space with evanescent decay that is orthogonal to the propagation direction. The decay constant, α, and propagation constant, β, are related by

$$\alpha^2 = \beta^2 - \omega^2 \mu_o \varepsilon_o \tag{16.12}$$

Evidently, the latter must satisfy the condition, $|\beta| \geq \omega/c$. For convenience, we choose the plane of the current to be $y = 0$.

TE waves

If **K** is parallel to the x axis, there is no divergence of the current and hence no electric surface-charge density. Consequently,

$$\mathbf{K} = \widehat{\mathbf{x}} K_0 \cos(\omega t - \beta z) \tag{16.13a}$$

$$\mathbf{A} = \widehat{\mathbf{x}} \frac{\mu_0 K_0}{2\alpha} e^{-\alpha|y|} \cos(\omega t - \beta z) \tag{16.13b}$$

$$\Phi = 0 \tag{16.13c}$$

The associated electric and magnetic fields are

$$\mathbf{E} = \widehat{\mathbf{x}}\, \frac{\omega \mu_0 K_0}{2\alpha} e^{-\alpha|y|} \sin(\omega t - \beta z) \tag{16.14a}$$

$$\mathbf{H} = \frac{K_0}{2} e^{-\alpha|y|} \left[\widehat{\mathbf{y}} \frac{\beta}{\alpha} \sin(\omega t - \beta z) + \widehat{\mathbf{z}}\, \mathrm{sign}(y) \cos(\omega t - \beta z) \right] \tag{16.14b}$$

Power and energy

Based upon **E** and **H**, the Maxwell–Poynting power flux and energy density are

$$\mathbf{S}^{\mathrm{em}} = \mathbf{E} \times \mathbf{H} = \frac{\omega \mu_0 K_0^2}{8\alpha} e^{-2\alpha|y|} \left[\widehat{\mathbf{y}}\, \mathrm{sign}(y) \sin 2(\omega t - \beta z) + \widehat{\mathbf{z}}\, 2 \frac{\beta}{\alpha} \sin^2(\omega t - \beta z) \right] \tag{16.15a}$$

$$W^{\mathrm{em}} = \frac{\mu_0 K_0^2}{8} e^{-2\alpha|y|} \left[2 \left(\frac{\beta}{\alpha} \right)^2 \sin^2(\omega t - \beta z) + \cos^2(\omega t - \beta z) \right] \tag{16.15b}$$

Based upon **A**, the Alternate power flux and energy density are

$$\mathbf{S}^{\mathrm{o}} = \widehat{\mathbf{z}} \frac{\beta\, \omega \mu_0 K_0^2}{8\alpha^2} e^{-2\alpha|y|} \tag{16.16a}$$

$$W^{\mathrm{o}} = \frac{\omega^2 \mu_0 K_0^2}{8c^2\alpha^2} e^{-2\alpha|y|} + \frac{1}{2} \mathbf{A} \cdot \mathbf{K}\, u_0(y) \tag{16.16b}$$

TM waves

If **K** is parallel to the z axis, there is a nonzero divergence of current which creates both electric surface-charge density, σ_{s}, and electric scalar potential. Consequently,

$$\mathbf{K} = \widehat{\mathbf{z}} K_0 \cos(\omega t - \beta z) \tag{16.17a}$$

$$\sigma_{\mathrm{s}} = \frac{\beta}{\omega} K_0 \cos(\omega t - \beta z) \tag{16.17b}$$

$$\mathbf{A} = \widehat{\mathbf{z}} \frac{\mu_0 K_0}{2\alpha} e^{-\alpha|y|} \cos(\omega t - \beta z) \tag{16.17c}$$

$$\Phi = \frac{\beta K_0}{2\alpha \varepsilon_0 \omega} e^{-\alpha|y|} \cos(\omega t - \beta z) \tag{16.17d}$$

The associated electric and magnetic fields are

$$\mathbf{E} = \frac{K_0}{2\varepsilon_0\omega} e^{-\alpha|y|} \left[\widehat{\mathbf{y}}\, \mathrm{sign}(y)\ \beta \cos(\omega t - \beta z) - \widehat{\mathbf{z}}\, \alpha \sin(\omega t - \beta z) \right] \tag{16.18a}$$

$$\mathbf{H} = -\widehat{\mathbf{x}}\, \mathrm{sign}(y) \frac{K_0}{2} e^{-\alpha|y|} \cos(\omega t - \beta z) \tag{16.18b}$$

Power and energy

Based upon **E** and **H**, the Maxwell–Poynting power flux and energy density are

$$\mathbf{S}^{\mathrm{em}} = \mathbf{E} \times \mathbf{H} = \frac{K_0^2}{8\omega\varepsilon_0} e^{-2\alpha|y|} \left[\widehat{\mathbf{y}}\, \mathrm{sign}(y) \alpha \sin 2(\omega t - \beta z) + \widehat{\mathbf{z}} 2\beta \cos^2(\omega t - \beta z) \right] \tag{16.19a}$$

$$W^{\mathrm{em}} = \frac{K_0^2}{8\omega^2\varepsilon_0} e^{-2\alpha|y|} \left[2\beta^2 \cos^2(\omega t - \beta z) - \alpha^2 \cos^2(\omega t - \beta z) \right] \tag{16.19b}$$

Based upon **A** and Φ, the Alternate power flux and energy density are

$$\mathbf{S}^{\mathrm{o}} = \widehat{\mathbf{z}} \frac{-\beta K_0^2}{8\omega\varepsilon_0} e^{-2\alpha|y|} + \Phi \mathbf{K} u_0(y) \tag{16.20a}$$

$$W^{\mathrm{o}} = \frac{-\mu_o K_0^2}{8} e^{-2\alpha|y|} + \frac{1}{2} (\mathbf{A} \cdot \mathbf{K} + \Phi\, \sigma_{\mathrm{s}}) u_0(y) \tag{16.20b}$$

Energy velocities

For both types of waves, it follows that the energy velocity (group velocity) is z-directed and equal to

$$v_{\mathrm{E}} = \frac{\int_{-\infty}^{+\infty} <S_z> dy}{\int_{-\infty}^{+\infty} <W> dy} = \frac{\omega}{\beta} = c\sqrt{\left(1 - \frac{\alpha^2}{\beta^2}\right)} \tag{16.21}$$

where the power flux, S_z, and energy density, W, are evaluated in either the Maxwell–Poynting or Alternate representation and $<>$ denotes the time average. Notice that for *TE* waves there is no localized component of S_z^{o}; whereas for *TM* waves a nonzero spatial impulse, $\Phi\mathbf{K}$, exists at $y = 0$. Localized components of the Alternate energy density, W^{o}, exist for *both* types of waves because $\frac{1}{2}(\mathbf{A} \cdot \mathbf{K} + \Phi\sigma_{\mathrm{s}})$ is nonzero even when $\Phi = 0$. The contributions from all of these spatial impulses must be included in Eq. (16.21), if the Alternate power-energy representation is used. (Failure to do so produces the erroneous velocity, $c/\sqrt{(1 - \alpha^2/\beta^2)}$.) The exception is for uniform plane waves ($\beta = \omega/\mathrm{c}$, $\alpha = 0$) of finite amplitude, for which $K_0 = 0$.

Because there are no localized terms, the corresponding calculation using the Maxwell–Poynting representation is simpler. In fact, the integrations over y are unnecessary because for both *TE* and *TM* waves (including $\alpha = 0$) we obtain

$$<S_z^{\mathrm{em}}>/<W^{\mathrm{em}}> = \omega/\beta$$

In the free-space regions ($|y| > 0$) we get

$$\mathbf{S}^{\mathrm{o}} = \{\pm\} <\mathbf{S}^{\mathrm{em}}> \text{ and } W^{\mathrm{o}} = \{\pm\}(1 - \alpha^2/\beta^2) <W^{\mathrm{em}}>$$

for $\left\{\begin{smallmatrix} TE \\ TM \end{smallmatrix}\right\}$ waves. These Alternate-energy densities and v_E/c all vanish as $\alpha/|\beta| \rightarrow 1$; such slow waves are quasistatic in character. The *TE* fields are dominated by **H** and become *MQS* waves; the *TM* fields are dominated by **E** and become *EQS* waves. In the Alternate representation, the surface energies dominate for either type of slow wave.

16.4 SKIN-DEPTH-LIMITED CURRENT IN A CONDUCTOR

Assume that the half-space defined by $x \geq 0$ has conductivity, σ, and free-space exists for $x < 0$. Current flows in the $+z$ direction at a frequency, ω, which is high enough to validate the skin-depth regime, where $\delta = \sqrt{\frac{2}{\omega\mu_o\sigma}}$.

The electric field and current density in the conductor are assumed to be

$$\mathbf{E} = \frac{1}{\sigma}\mathbf{J} = \hat{\mathbf{z}}E_o \exp\left(-\frac{x}{\delta}\right)\sin\left(\omega t - \frac{x}{\delta}\right) \tag{16.22}$$

Notice that the integral of $J_z = \sigma E_z$ over the half-space is

$$\int_0^\infty J_z dx = K_z = -\frac{\delta J_o}{\sqrt{2}}\cos\left(\omega t + \frac{\pi}{4}\right)$$

and the current is effectively confined to a region between $x = 0$ and $\delta/\sqrt{2}$. A reasonable approximation is to consider J_z uniform over a skin depth and zero beyond that value. $\delta J_o = \delta\sigma E_o$ can therefore be considered the magnitude of a surface current density, K_z.

Because $\nabla \cdot \mathbf{E} = 0$, there is neither electric charge density nor scalar potential[3]; consequently, $\mathbf{E} = -\frac{\partial \mathbf{A}}{\partial t}$ and the vector potential is

$$\mathbf{A} = \hat{\mathbf{z}}\frac{E_o}{\omega}\exp\left(-\frac{x}{\delta}\right)\cos\left(\omega t - \frac{x}{\delta}\right) \tag{16.23}$$

The magnetic field in the conductor can now be calculated from $\mu_o\mathbf{H} = \nabla \times \mathbf{A}$; the result is

$$\mathbf{H} = -\hat{\mathbf{y}}\frac{1}{\mu_o}\frac{\partial A_z}{\partial x} = \hat{\mathbf{y}}\frac{E_o}{\omega\delta\mu_o}\exp\left(-\frac{x}{\delta}\right)\left[\cos\left(\omega t - \frac{x}{\delta}\right) - \sin\left(\omega t - \frac{x}{\delta}\right)\right] \tag{16.24}$$

where

$$\omega\delta\mu_o = \sqrt{\frac{2\omega\mu_o}{\sigma}}$$

is small enough so that the electric energy densities are negligible. The respective power fluxes and energy densities are (approximately)

$$\mathbf{S} = \hat{\mathbf{x}}\frac{E_o^2}{\omega\delta\mu_o}\exp\left(-\frac{2x}{\delta}\right)\sin\left(\omega t - \frac{x}{\delta}\right)\left[\sin\left(\omega t - \frac{x}{\delta}\right) - \cos\left(\omega t - \frac{x}{\delta}\right)\right] \tag{16.25a}$$

$$W^{em} = \frac{1}{2}\mu_o H_y^2 = \frac{\sigma E_o^2}{4\omega}\exp\left(-\frac{2x}{\delta}\right)\left[1 - \sin 2\left(\omega t - \frac{x}{\delta}\right)\right] \tag{16.25b}$$

$$\mathbf{S}^o = \hat{\mathbf{x}}\frac{E_o^2}{2\omega\delta\mu_o}\exp\left(-\frac{2x}{\delta}\right) \tag{16.25c}$$

$$W^o = \frac{1}{2}A_z J_z = \frac{\sigma E_o^2}{4\omega}\exp\left(-\frac{2x}{\delta}\right)\sin 2\left(\omega t - \frac{x}{\delta}\right) \tag{16.25d}$$

[3]Because $\Phi = 0$, the gauge condition for a conductor, $\nabla \cdot \mathbf{A} + \frac{1}{c^2}\frac{\partial}{\partial t}\Phi = -\mu_o\sigma\Phi$, discussed in Chapter 5, Section 5.3, reduces to the Lorenz gauge (and also the Coulomb gauge).

Interestingly, both average and second-harmonic components of $\mathbf{S}$ and W^{em} are present whereas there is only average $\mathbf{S}^{\text{o}}$ and only second-harmonic W^{o}. However, as expected, both the Poynting and Alternate-power theorems generate $-\sigma E_z^2$.

In this example the electric field is parallel to the surface; consequently there is no electric surface charge. If, instead, a quasi-TEM wave propagates in the air region along z, induced surface charges will generate a scalar potential in both free-space and the conductor. The Alternate-power flux will then have a $\Phi\mathbf{J}$ component that carries z-directed power within a skin-depth layer and associated Alternate-energy components, $\frac{1}{2}(\mathbf{A}\cdot\mathbf{J}+\sigma_s\Phi)$. The volume current can often be approximated as a surface current, $\mathbf{K}=\mathbf{J}\delta$. We shall learn in Chapter 18 how these longitudinal current paths profoundly alter the distribution of the Alternate-power flux.

CHAPTER 17

WAVES INCIDENT AT A MATERIAL INTERFACE

17.1 REFLECTED AND TRANSMITTED PLANE WAVES

When a uniform plane wave is incident upon the interface between two linear and uniform dielectric and/or magnetic materials, some of the wave is transmitted the remainder reflected. If the propagation vector is inclined at an angle θ_i from the normal, the transmitted wave may be diffracted into a different direction, θ_t, with both its amplitude and final state of polarization depending upon both θ_i and the initial polarization. The reflected wave is characterized by the angle, θ_r, and an amplitude and state of polarization. Because of linearity, it is only necessary to solve for the reflected and transmitted waves when the incident wave is linearly polarized in each of two orthogonal directions. The two different media are both assumed to be lossless; each occupies a half-space demarcated by the plane $z = 0$. For $z < 0$, the permittivity and permeability are ε_1 and μ_1; for $z > 0$, they are ε_2 and μ_2. First, the propagation direction of the incident wave is chosen; then (unless it is perpendicular to the surface) it and the normal to the interface define the plane of incidence. Without loss of generality, we then define the remaining Cartesian axes so that the plane of incidence coincides with $y = 0$. This geometry is shown in Figure 17.1. Convenient polarizations that together characterize the problem of reflection and transmission are **E** parallel to $\widehat{\mathbf{y}}$ and **H** parallel to $\widehat{\mathbf{y}}$. The frequency of both incident waves is ω.

The Power and Beauty of Electromagnetic Fields, First Edition. F. R. Morgenthaler.

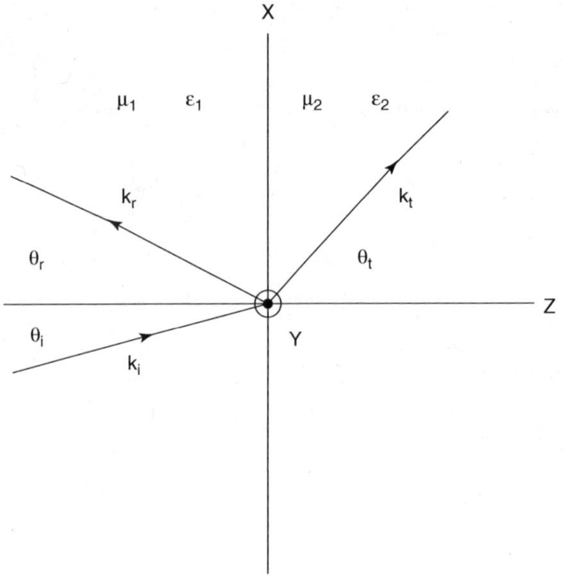

Figure 17.1 Plane of oblique incidence.

The time-honored method of solution is to find the appropriate forms of the complex electric and magnetic fields in both regions and then require the tangential electric and magnetic fields to be continuous at the boundary. It follows that the normal component of the Poynting vector will automatically be continuous and give the fraction of the incident power that is transmitted; the difference is of course the reflected power. We choose instead to focus on the vector potential and Alternate-power flux. The use of complex fields and potentials is highly recommended.

17.2 *TE* POLARIZATION

We already know that it is possible to choose a vector potential that will describe a plane wave without the need for any scalar potential. In the Lorenz gauge, the condition is $\nabla \cdot \underline{\mathbf{A}} = 0$. Because $\underline{\mathbf{E}} = -j\omega\underline{\mathbf{A}}$, we require the incident wave vector potential to be *TE*, y-directed, and of the form

$$\underline{\mathbf{A}}_i = \widehat{\mathbf{y}}\underline{A}_o \exp(-jk_1 \cos\theta_i\, z)\exp(-jk_1 \sin\theta_i\, x) \tag{17.1}$$

$$k_1 = \omega\sqrt{\mu_1\varepsilon_1}$$

where $\underline{A}_o$ and θ_i are the incident-wave amplitude and angle. This potential is presumably generated by y-directed electric currents that are located at $z \to -\infty$.

Both the reflected and transmitted potential waves will also require y-components of the same frequency (in order that the fields are continuous functions of time). These are also uniform plane waves,

$$\underline{\mathbf{A}}_r = \widehat{\mathbf{y}}\underline{\Gamma}_r\underline{A}_o \exp(jk_1 \cos\theta_r z)\exp(-jk_1 \sin\theta_r x) \tag{17.2a}$$

$$\underline{\mathbf{A}}_t = \widehat{\mathbf{y}}\underline{\Gamma}_t\underline{A}_o \exp(-jk_2 \cos\theta_t z)\exp(-jk_2 \sin\theta_t x) \tag{17.2b}$$

$$k_2 = \omega\sqrt{\mu_2\varepsilon_2}$$

where the complex reflection and transmission coefficients, $\underline{\Gamma}_r$ and $\underline{\Gamma}_t$, and the reflected and transmitted angles, θ_r and θ_t, are as yet unknown. The reflection coefficient, $\underline{\Gamma}_r$, applies to both $\underline{\mathbf{A}}$ and $\underline{E}_y$; that associated with $\underline{H}_x$ is $-\underline{\Gamma}_r$. Because, by assumption, the only discontinuity is at $z = 0$, there are no additional reflected waves. The vector potentials of all three waves individually satisfy the appropriate wave equations and are divergence-free. Although it is conceivable that other components of $\underline{\mathbf{A}}_r$ and $\underline{\mathbf{A}}_t$ might be required, we attempt to satisfy the boundary-conditions with these waves and are successful. Any other terms would produce cross-polarized reflected and transmitted waves that would require active sources at the interface; none are present in passive media.

Law of reflection and Snell's Law

We can now apply boundary conditions, by requiring continuity of $\underline{\mathbf{A}}$ at $z = 0$ for all values of x. This requires the phase factors of the complex exponentials to match on both sides of the interface. The result,

$$k_1 \sin\theta_i = k_1 \sin\theta_r = k_2 \sin\theta_t$$

produces two conditions:

$$\theta_r = \theta_i \tag{17.3a}$$

$$\sqrt{\mu_2\varepsilon_2}\,\sin\theta_t = \sqrt{\mu_1\varepsilon_1}\,\sin\theta_i \tag{17.3b}$$

The first is the Law of Reflection (the angle of incidence equals the angle of reflection); the second is Snell's Law. Notice that the smaller angle is in the medium with the slower phase velocity.

Critical angle

If $\sqrt{\frac{\mu_1\varepsilon_1}{\mu_2\varepsilon_2}} > 1$, it follows from Snell's Law that as θ_i is increased, there will come a point where $\sin\theta_t = 1$; beyond that critical angle,

$$\theta_{\text{crit}} = \sin^{-1}\sqrt{\frac{\mu_2\varepsilon_2}{\mu_1\varepsilon_1}} \tag{17.4}$$

the transmitted angle is imaginary and the wave becomes evanescent for $z > 0$. With exponential decay in the $+z$ direction and propagation parallel to the interface, the uniform plane wave is transformed into a nonuniform plane wave of the type encountered in Chapter 16, Section 16.3.

Reflection and transmission coefficients

Continuity of $\underline{\mathbf{A}}$ at the interface, $z = 0$, requires the matching of both phase and amplitude of $\underline{\mathbf{A}}_i + \underline{\mathbf{A}}_r = \underline{\mathbf{A}}_t$; the former has been accomplished, and the latter requires that

$$1 + \underline{\Gamma}_r = \underline{\Gamma}_t$$

This condition is also obtained from continuity of $\underline{E}_y$. To proceed further, we match the normal derivative $\frac{\partial}{\partial z}\left(\frac{\mathbf{A}}{\mu}\right)$ on both sides of $z = 0$; this is equivalent to requiring continuity of $\underline{H}_x$. The result is

$$\frac{k_1 \cos\theta_i}{\mu_1}(1 - \underline{\Gamma}_r) = \frac{k_2 \cos\theta_t}{\mu_2}\underline{\Gamma}_t$$

These two equations are easily solved for the reflection and transmission coefficients:

$$\underline{\Gamma}_r = \frac{\sqrt{\frac{\varepsilon_1}{\mu_1}}\cos\theta_i - \sqrt{\frac{\varepsilon_2}{\mu_2}}\cos\theta_t}{\sqrt{\frac{\varepsilon_1}{\mu_1}}\cos\theta_i + \sqrt{\frac{\varepsilon_2}{\mu_2}}\cos\theta_t} \tag{17.5}$$

$$\underline{\Gamma}_t = \frac{2\sqrt{\frac{\varepsilon_1}{\mu_1}}\cos\theta_i}{\sqrt{\frac{\varepsilon_1}{\mu_1}}\cos\theta_i + \sqrt{\frac{\varepsilon_2}{\mu_2}}\cos\theta_t} \tag{17.6}$$

These coefficients can also be cast in the form expected from transmission-line theory,

$$\underline{\Gamma}_r = \frac{Y_i - Y_t}{Y_i + Y_t}$$

$$\underline{\Gamma}_t = \frac{2Y_i}{Y_i + Y_t}$$

where the admittances, $Y_i = \sqrt{\frac{\varepsilon_1}{\mu_1}}\cos\theta_i$ and $Y_t = \sqrt{\frac{\varepsilon_2}{\mu_2}}\cos\theta_t$, depend on both the material constants and the angle of incidence.

Magnetic Brewster Angle

For certain materials and a special angle of incidence, the reflection coefficient vanishes; this angle (if it exists) is known as the Brewster Angle, θ_B. Evidently, $\sqrt{\frac{\varepsilon_1}{\mu_1}}\cos\theta_B = \sqrt{\frac{\varepsilon_2}{\mu_2}}\cos\theta_t$, but the angle must also satisfy Snell's Law. Both conditions are met if

$$\sin^2\theta_B = \frac{1 - \frac{\mu_1\varepsilon_2}{\mu_2\varepsilon_1}}{1 - \left(\frac{\mu_1}{\mu_2}\right)^2} \le 1 \tag{17.7}$$

Near such angles, the wave is very nearly completely transmitted without reflection. If $\mu_1 = \mu_2$, as is the case when the material is nonmagnetic, there is no Brewster angle. If $\varepsilon_1 = \varepsilon_2$, we have

$$\sin^2\theta_B = \cos^2\theta_t = \frac{1}{1 + \frac{\mu_1}{\mu_2}}$$

and θ_B (sometimes called the magnetic Brewster Angle) exists for all positive values of the permeabilities. At this angle, $\theta_i + \theta_t = \pi/2$ and the reflected wave (if it were present) would be induced, because of the differential permeability, and emerge in a direction parallel to the polarization vector. Because magnetic dipoles do not radiate along their length, the null in that direction explains the absence of the reflection. If $\varepsilon_1 \neq \varepsilon_2$, there are also electric dipoles oriented along y and the combination can shift the angle of the null or can eliminate it all together.

Alternate-power flux

Next, we calculate the complex Alternate power on both sides of the interface from

$$\underline{\mathbf{S}}_1^{\mathrm{o}} = -j\omega\frac{1}{4\mu_1}[(\underline{A}_i + \underline{A}_r)\nabla(\underline{A}_i + \underline{A}_r)^* - (\underline{A}_i + \underline{A}_r)^*\nabla(\underline{A}_i + \underline{A}_r)]$$

$$\underline{\mathbf{S}}_2^{\mathrm{o}} = -j\omega\frac{1}{4\mu_2}(\underline{A}_t\nabla\underline{A}_t^* - \underline{A}_t^*\nabla\underline{A}_t)$$

The z components of the flux evaluate to

$$\underline{S}_{1z}^{\mathrm{o}} = \frac{1}{2\mu_1}\omega k_1\cos\theta_i(1 - \underline{\Gamma}_r\underline{\Gamma}_r^*)\underline{A}_{\mathrm{o}}\underline{A}_{\mathrm{o}}^*$$

$$\underline{S}_{2z}^{\mathrm{o}} = \frac{1}{2\mu_2}\omega k_2\cos\theta_t\underline{\Gamma}_t\underline{\Gamma}_t^*\underline{A}_{\mathrm{o}}\underline{A}_{\mathrm{o}}^*$$

Because $\underline{\Gamma}_r$ and $\underline{\Gamma}_t$ are both real, it is easily verified that these two fluxes are equal.

17.3 *TM* POLARIZATION

Reflection and transmission coefficients

For this *TM* polarization ($\mathbf{H} \parallel \widehat{\mathbf{y}}$) there are both x and z components of $\underline{\mathbf{E}}$ and therefore of $\underline{\mathbf{A}}$. Although these can be found from $\nabla\cdot\underline{\mathbf{A}} = 0$, it is much easier to consider the dual of the fields just evaluated by pretending that magnetic-currents rather than electric currents are the source of the fields. These generate $\underline{\mathbf{A}}_{\mathrm{m}}$ and the appropriate Alternate-power flux is $\underline{\mathbf{S}}^{\mathrm{om}}$. All that remains to do is to swap $\underline{\mathbf{E}}$ and $\underline{\mathbf{H}}$ and μ and ε. Recall that $\underline{\mathbf{H}} = -j\omega\underline{\mathbf{A}}_{\mathrm{m}}$ and $\underline{\mathbf{E}} = \frac{-1}{\varepsilon}\nabla\times\underline{\mathbf{A}}_{\mathrm{m}}$.

$$\underline{\mathbf{A}}_{\mathrm{m}i} = \widehat{\mathbf{y}}\underline{A}_{\mathrm{mo}}\exp(-jk_1\cos\theta_i\, z)\exp(-jk_1\sin\theta_i\, x) \tag{17.8}$$

$$k_1 = \omega\sqrt{\mu_1\varepsilon_1}$$

Both the reflected and transmitted waves will also require y components:

$$\underline{\mathbf{A}}_{\mathrm{m}r} = \widehat{\mathbf{y}}\underline{\Gamma}_r^{\mathrm{m}}\underline{A}_{\mathrm{mo}}\exp(jk_1\cos\theta_r\, z)\exp(-jk_1\sin\theta_r\, x) \tag{17.9a}$$

$$\underline{\mathbf{A}}_{\mathrm{m}t} = \widehat{\mathbf{y}}\underline{\Gamma}_t^{\mathrm{m}}\underline{A}_{\mathrm{mo}}\exp(-jk_2\cos\theta_t\, z)\exp(-jk_2\sin\theta_t\, x) \tag{17.9b}$$

$$k_2 = \omega\sqrt{\mu_2\varepsilon_2}$$

The reflection coefficient, $\underline{\Gamma}_r^{\mathrm{m}}$, applies to both $\underline{\mathbf{A}}_{\mathrm{m}}$ and $\underline{H}_y$; that associated with $\underline{E}_x$ is $-\underline{\Gamma}_r^{\mathrm{m}}$. As before,

$$1 + \underline{\Gamma}_r^{\mathrm{m}} = \underline{\Gamma}_t^{\mathrm{m}}$$

Also, the Law of Reflection, Snell's Law, and the critical angle are unchanged. However, the reflection and transmission coefficients for this polarization are altered to

$$\underline{\Gamma}_r^{\mathrm{m}} = \frac{\sqrt{\frac{\mu_1}{\varepsilon_1}}\cos\theta_i - \sqrt{\frac{\mu_2}{\varepsilon_2}}\cos\theta_t}{\sqrt{\frac{\mu_1}{\varepsilon_1}}\cos\theta_i + \sqrt{\frac{\mu_2}{\varepsilon_2}}\cos\theta_t} \tag{17.10a}$$

$$\underline{\Gamma}_t^{\mathrm{m}} = \frac{2\sqrt{\frac{\mu_1}{\varepsilon_1}}\cos\theta_i}{\sqrt{\frac{\mu_1}{\varepsilon_1}}\cos\theta_i + \sqrt{\frac{\mu_2}{\varepsilon_2}}\cos\theta_t} \tag{17.10b}$$

which, in transmission-line theory form, are written as

$$\underline{\Gamma}_r^{\mathrm{m}} = \frac{Z_i^{\mathrm{m}} - Z_t^{\mathrm{m}}}{Z_i^{\mathrm{m}} + Z_t^{\mathrm{m}}}$$

$$\underline{\Gamma}_t^{\mathrm{m}} = \frac{2Z_i^{\mathrm{m}}}{Z_i^{\mathrm{m}} + Z_t^{\mathrm{m}}}$$

and where the impedances, $Z_i^{\mathrm{m}} = \sqrt{\frac{\mu_1}{\varepsilon_1}}\cos\theta_i$ and $Z_t^{\mathrm{m}} = \sqrt{\frac{\mu_2}{\varepsilon_2}}\cos\theta_t$, depend on both the material constants and the angle of incidence; except for $\theta_i = 0$, these impedances are *not* the reciprocals of Y_i and Y_t.

Brewster Angle

The Brewster Angle is also altered to its dual,

$$\sin^2\theta_B = \frac{1 - \frac{\varepsilon_1\mu_2}{\varepsilon_2\mu_1}}{1 - \left(\frac{\varepsilon_1}{\varepsilon_2}\right)^2} \leq 1 \tag{17.11}$$

Near such angles, the wave is very nearly completely transmitted without reflection. If $\varepsilon_1 = \varepsilon_2$, there is no Brewster Angle. If $\mu_1 = \mu_2$,

$$\sin^2\theta_B = \cos^2\theta_t = \frac{1}{1 + \frac{\varepsilon_1}{\varepsilon_2}}$$

and θ_B exists for all positive values of the permittivities. At this angle, $\theta_i + \theta_t = \pi/2$ and the reflected wave (if it were present) would be induced because of the differential permittivity and emerge in a direction parallel to the polarization vector. Because electric dipoles do not radiate along their length, the null in that direction explains the absence of the reflection. If $\mu_1 \neq \mu_2$, there are also magnetic dipoles oriented along y and the combination can shift the angle of the null or can eliminate it all together.

Dual Alternate-power flux

Next, we calculate the complex dual Alternate-power on both sides of the interface from

$$\underline{\mathbf{S}}_1^{\mathrm{om}} = -j\omega\frac{1}{4\varepsilon_1}[(\underline{A}_{\mathrm{m}i} + \underline{A}_{\mathrm{m}r})\nabla(\underline{A}_{\mathrm{m}i} + \underline{A}_{\mathrm{m}r})^* - (\underline{A}_{\mathrm{m}i} + \underline{A}_{\mathrm{m}r})^*\nabla(\underline{A}_{\mathrm{m}i} + \underline{A}_{\mathrm{m}r})]$$

$$\underline{\mathbf{S}}_2^{\mathrm{om}} = -j\omega\frac{1}{4\varepsilon_2}(\underline{A}_{\mathrm{m}t}\nabla\underline{A}_{\mathrm{m}t}^* - \underline{A}_{\mathrm{m}t}^*\nabla\underline{A}_{\mathrm{m}t})$$

The z components of the flux evaluate to

$$\underline{S}_{1z}^{\mathrm{om}} = \frac{1}{2\varepsilon_1}\omega k_1\cos\theta_i(1 - \underline{\Gamma}_r^{\mathrm{m}}\underline{\Gamma}_r^{\mathrm{m}*})\underline{A}_{\mathrm{o}}\underline{A}_{\mathrm{o}}^*$$

$$\underline{S}_{2z}^{\mathrm{om}} = \frac{1}{2\varepsilon_2}\omega k_2\cos\theta_t\underline{\Gamma}_t^{\mathrm{m}}\underline{\Gamma}_t^{\mathrm{m}*}\underline{A}_{\mathrm{o}}\underline{A}_{\mathrm{o}}^*$$

As with the other polarization, and because $\underline{\Gamma}_r^{\mathrm{m}}$and $\underline{\Gamma}_t^{\mathrm{m}}$ are both real, it is easily verified that these two fluxes are equal.

17.4 ELLIPTICALLY POLARIZED INCIDENT WAVES

An elliptically polarized incident wave can be decomposed into two linearly polarized waves of the types that are analyzed above. The amplitude of each will differ and there will be a phase shift between the two components. If, for example, the polarization is circular, the amplitudes of the components will be equal and the phase shift $\pm\pi/2$, depending upon whether the circular polarization is clockwise or counterclockwise. The reflected and transmitted waves must be calculated for each linearly polarized component and the results summed. The state of polarization may be drastically altered; in the event that θ_i is at the Brewster angle, the net reflection will be linearly polarized because one component will be completely transmitted without reflection.

CHAPTER 18

TEM TRANSMISSION LINES

An idealized uniform transmission line is comprised of two or more perfect conductors that are parallel to one another and extend to infinity without any change in their transverse geometries. In the simplest cases, the conductors are surrounded by free-space and the guided electric and magnetic fields are transverse to the direction of propagation and so denoted *TEM* (transverse electric and magnetic).

18.1 GENERAL TIME-DEPENDENT SOLUTIONS

The free-space vector and scalar potentials, for an arbitrary configuration of perfect conductors (having uniform cross section in every x–y plane) that can support *TEM* modes propagating along z, can be written as

$$A_z = a_o(x,y)\left[f_+(z-ct) - f_-(z+ct)\right] \tag{18.1a}$$

$$\Phi = \phi_o(x,y)\left[f_+(z-ct) + f_-(z+ct)\right] \tag{18.1b}$$

The related electric charge and current distributions (located on the surfaces of the conductors) are

$$J_z = j_o(x,y)\left[f_+(z-ct) - f_-(z+ct)\right] \tag{18.2a}$$

$$\rho = \rho_o(x,y)\left[f_+(z-ct) + f_-(z+ct)\right] \tag{18.2b}$$

The Power and Beauty of Electromagnetic Fields, First Edition. F. R. Morgenthaler.

The conditions $ca_o(x,y) = \phi_o(x,y)$ and $j_o(x,y) = c\rho_o(x,y)$ arise from imposition of the Lorenz gauge and conservation of charge.

The electric and magnetic fields of the mode are

$$\mathbf{E} = \mathbf{e}_o(x,y)\left[f_+(z-ct) + f_-(z+ct)\right]$$

$$\mathbf{H} = \mathbf{h}_o(x,y)\left[f_+(z-ct) - f_-(z+ct)\right]$$

where

$$\mathbf{e}_o(x,y) = -\nabla\phi_o(x,y) = -c\mu_o\hat{\mathbf{z}}\times\mathbf{h}_o(x,y)$$

$$\mu_o\mathbf{h}_o(x,y) = \nabla\times[\hat{\mathbf{z}}a_o(x,y)] = \hat{\mathbf{z}}\times\mathbf{e}_o(x,y)/c$$

Source equivalence

As expected, these fields have only transverse components; let us try to understand why this is so. Of course, the boundary conditions force E_z to vanish on the surface of perfect conductors, but why everywhere else as well? The equivalence principle provides the answer. Because the electric and magnetic fields are both zero inside the conductors, the metal can be removed, *provided that the charges and currents on the surface are maintained.* That is a tall order in practice, but one that can easily be met conceptually. The result is a purely convective current density, $j_o(x,y) = c\rho_o(x,y)$. If $f_- = 0$ and $f_+(z-ct) = u_0(z-ct)$, the associated charge density is the superposition of pairs of opposite polarity, $\pm\rho_o(x,y)dxdydz\ u_0(z-ct)$, traveling in the $+z$ direction at the speed of light. Each of these increments seemingly acts like a point charge and so the fields are those of charges traveling at constant velocity. But, that problem was solved in Chapter 12, Section 12.1, and in the limit $v\to c$ both E_z and H_z vanish. However, in that same limit the transverse fields of a point charge become infinite, whereas the transmission-line fields are finite. Resolution of the paradox is provided by the fact that in a macroscopic theory, the total amount of charge located within any *particular* value of dz is infinitesimal.[1] It is tempting, but incorrect, to conclude that the conductors merely provide a path for the charge distribution to zip along at the speed of light, keeping pace with the generated fields. If it were true, every telecommunication device connected to a phone line would certainly constitute a health hazard because the charges (and their associated mass) would be traveling at relativistic velocities with incredibly large energies. Battery-operated phones could hardly produce such effects, even allowing for finite conductivity that slightly lowers the phase velocity. That conductivity, $\sigma = n_0 q\mu_e$, is the clue to the puzzle because very high values are better modeled by a large carrier concentration, n_0, rather than by a large mobility factor, μ_e. The latter determines the average (drift) velocity of the charges, $v_{\text{drift}} = \mu_e\mathbf{E}$. For copper, the drift velocity corresponding to a current density of 10^6 A/m^2 is $\sim 7\times 10^{-2}$ mm/s—hardly relativistic! Yet somehow the individual charges which (on average move at $\mathbf{v}_{\text{drift}}$) act together to simulate a relativistic current density $c\rho_o(x,y)$. Mobile negative charge which is neutralized by the fixed positive charge of the lattice simply moves very slightly toward or away from the surface of the conductor as the electric field of the *TEM* pulse moves

[1]The discreteness of actual charges would seem to restore the paradox, but in the metal (that we have removed) their exact positions are uncertain and their velocities decidedly nonrelativistic.

by. This creates the surface charges that are needed in order to originate and terminate the electric field. As the pulse moves on (at near the speed of light), those charges move back into the metal and different charges further down the line repeat the action to produce new surface charge as required. Simply put, the *same* charge does not travel with the wave, but the distribution as a whole acts as if it did. The macroscopic Maxwell Equations cannot, of course, distinguish between individual charges. The behavior we have described leads to a prediction: Although the local charge need not move very far as the pulse moves by, finite mobility and carrier concentration together imply that there must be a time lag before **E** disappears; and, in fact, for finite conductivity, this dispersive effect is predicted by theory and borne out by experiment. If n_o (and consequently σ) $\to \infty$, both the required shift of position and the time lag approach zero, even though $\mathbf{v}_{\text{drift}}$ is finite. The result is dispersion-free propagation.

Power energy and stress momentum

Evaluation of Eqs. (3.7) reveals that

$$\mathbf{S}^{\text{em}} = \mathbf{E} \times \mathbf{H} = \widehat{\mathbf{z}}\, c\mu_o \mathbf{h}_o \cdot \mathbf{h}_o \left[f_+^2(z - ct) - f_-^2(z + ct)\right]$$

$$W^{\text{em}} = \mu_o \mathbf{h}_o \cdot \mathbf{h}_o \left[f_+^2(z - ct) + f_-^2(z + ct)\right]$$

$$\overline{\mathbf{T}}^{\text{em}} = \varepsilon_o\, \mathbf{e}_o \mathbf{e}_o \left[f_+(z - ct) + f_-(z + ct)\right]^2 + \mu_o \mathbf{h}_o \mathbf{h}_o \left[f_+(z - ct) - f_-(z + ct)\right]^2 - W^{\text{em}} \overline{\mathbf{I}}$$

$$\mathbf{G}^{\text{em}} = \mathbf{E} \times \mathbf{H}/c^2 = \widehat{\mathbf{z}}\, \mu_o/c\, \mathbf{h}_o \cdot \mathbf{h}_o \left[f_+^2(z - ct) - f_-^2(z + ct)\right]$$

while Eqs. (3.31) can be expressed as

$$S_z^o = \Phi J_z - \frac{c a_o^2}{\mu_o} \frac{\partial}{\partial z} \left[f_+(z - ct) \frac{df_-(z + ct)}{d(z + ct)} - \frac{df_+(z - ct)}{d(z - ct)} f_-(z + ct) \right]$$

$$W^o = \frac{1}{2}(\mathbf{A} \cdot \mathbf{J} + \rho\Phi) + \frac{a_o^2}{\mu_o} \frac{\partial}{\partial z} \left[f_+(z - ct) \frac{df_-(z + ct)}{d(z + ct)} + \frac{df_+(z - ct)}{d(z - ct)} f_-(z + ct) \right]$$

$$T_{xx}^o = \frac{1}{2}(\mathbf{A} \cdot \mathbf{J} - \Phi\rho) - \frac{2}{\mu_o} \left[a_o \frac{\partial^2 a_o}{\partial x^2} - \frac{\partial a_o}{\partial x} \frac{\partial a_o}{\partial x} \right] f_+(z - ct) f_-(z + ct)$$

$$T_{yy}^o = \frac{1}{2}(\mathbf{A} \cdot \mathbf{J} - \Phi\rho) - \frac{2}{\mu_o} \left[a_o \frac{\partial^2 a_o}{\partial y^2} - \frac{\partial a_o}{\partial y} \frac{\partial a_o}{\partial y} \right] f_+(z - ct) f_-(z + ct)$$

$$T_{zz}^o = -W^o + \frac{4}{\mu_o} a_o^2 \frac{df_+(z - ct)}{d(z - ct)} \frac{df_-(z + ct)}{d(z + ct)}$$

$$T_{xy}^o = T_{yx}^o = -\frac{2}{\mu_o} \left[a_o \frac{\partial^2 a_o}{\partial x \partial y} - \frac{\partial a_o}{\partial x} \frac{\partial a_o}{\partial y} \right] f_+(z - ct) f_-(z + ct)$$

$$T_{xz}^o = T_{zx}^o = T_{yz}^o = T_{zy}^o = 0$$

and

$$G_z^{\rm o} = \rho A_z - c\varepsilon_{\rm 0}\, a_{\rm 0}^2 \frac{\partial}{\partial z}\left[f_+(z-ct)\frac{df_-(z+ct)}{d(z+ct)} - \frac{df_+(z-ct)}{d(z-ct)}f_-(z+ct)\right]$$

Whenever *either* $f_+(z-ct)$ or $f_-(z+ct)$ is zero, *only* the localized components of $\mathbf{S}^{\rm o}$, $\mathbf{G}^{\rm o}$, $W^{\rm o}$, and $\overline{\mathbf{T}}^{\rm o}$ persist. In these cases, it follows that

$$\nabla \cdot (\Phi \mathbf{J}) + \frac{\partial}{\partial t}\left[\frac{1}{2}(\mathbf{A}\cdot\mathbf{J} + \rho\Phi)\right] = 0 \tag{18.3a}$$

$$\nabla \cdot \left[-\frac{1}{2}(\mathbf{A}\cdot\mathbf{J} + \rho\Phi)\widehat{\mathbf{z}}\widehat{\mathbf{z}}\right] - \frac{\partial}{\partial t}(\rho\mathbf{A}) = 0 \tag{18.3b}$$

In other circumstances, the localization is not complete, yet Eqs. (18.3) remain perfectly valid because the residuals, involving product functions of $f_+(z-ct)$ and $f_-(z+ct)$, cancel. Therefore, a circuit-theory representation emerges—one in which the wave momentum has also been replaced by the localized circuit momentum, $\rho\mathbf{A}^2$; of course, validation requires that $\mathbf{A}$ and $\mathbf{J}$ have only z-components.[3] For sinusoidal steady-state fields, the reactive power and energy associated with the $\mathbf{E}$ and $\mathbf{H}$ standing waves are, *in the Alternate representation*, located along the electric-current path.

A simple example of a *TEM* transmission line consists of two perfectly conducting plates located at $y = \pm d/2$. In the sinusoidal steady state, the f_+ mode is a uniform plane wave like Eq. (3.43) and if $f_- = 0$, the potentials of Eq. (3.48) are appropriate for the region between the plates. Surface currents, $K_{\rm uz}^s$, and surface charges, $\sigma_{\rm u}^s$, are present on the inner surfaces of the plates; consequently, the Alternate power, energy, and momentum are all localized to the surfaces (even in the limit $d \to \infty$) where in terms of the surface divergence, Eq. (18.3a) becomes

$$\nabla_\Sigma \cdot (\Phi\mathbf{K}^s) + \frac{\partial}{\partial t}\left[\frac{1}{2}(\mathbf{A}\cdot\mathbf{K}^s + \sigma^s\Phi)\right] = 0$$

Notice that each of the densities has the same sign on both plates; therefore one-half of the power and energy are localized to each plate.

18.2 PARALLEL-PLATE TEM LINE IN THE SINUSOIDAL STEADY STATE

Infinite Line

A parallel-plate transmission line consists of two very thin perfectly conducting plates located on the planes $x = \pm d/2$ and surrounded by free-space. A section of the geometry

[2]Notice that, although the Alternate-stress tensor has localized components $-A_iJ_j + \frac{1}{2}(\mathbf{A}\cdot\mathbf{J} - \rho\Phi)\delta_{ij}$, others that are distributed in space must exist—in order to create the forces on the longitudinal conductors that comprise the transmission line. These components (which may cancel some of the localized stresses) vanish when there is a *single* propagating mode, for which $A_z = \pm\Phi/c$ and $\mathbf{A}\cdot\mathbf{J} = \rho\Phi$.

However, when both modes are present, the residual term in $G_z^{\rm o}$ cancels those of $T_{zz}^{\rm o}$ (including the residual of $W^{\rm o}$) and Eq. (18.3b) remains valid.

[3]Strictly speaking, the TEM line continues along the z axis without interruption. Any termination alters this assumption. For example, a resistive sheet placed across the line will cause transverse currents to flow, which in turn create transverse components of $\mathbf{A}$. These, together with modifications to A_z, will seriously alter $\mathbf{S}^{\rm o}$ and $W^{\rm o}$ (as well as $\overline{\mathbf{T}}^{\rm o}$ and $\mathbf{G}^{\rm o}$) in the vicinity of the sheet.

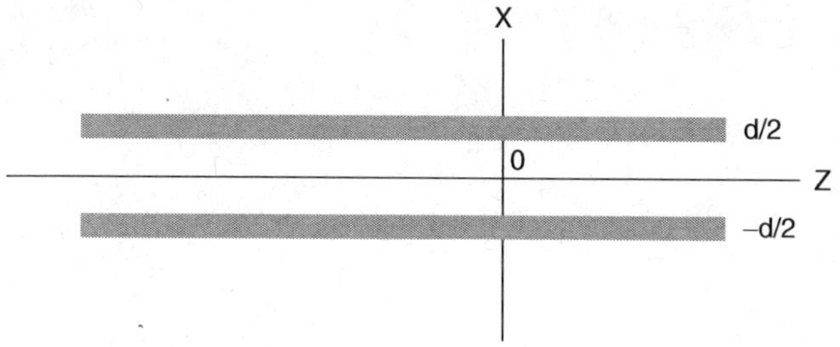

Figure 18.1 Parallel-plate transmission line.

(which is independent of z) is shown in Figure 18.1. The electric and magnetic fields exist only between the plates; they are assumed to be *TEM* modes propagating along the z axis in the sinusoidal steady state. In terms of $\omega = ck$ and the positive constants $B_o^{\pm}$, we have

$$\mathbf{E} = -\widehat{\mathbf{x}}c[B_o^+ \cos(\omega t - kz) - B_o^- \cos(\omega t + kz)] \tag{18.4a}$$

$$\mathbf{B} = \mu_0 \mathbf{H} = -\widehat{\mathbf{y}}[B_o^+ \cos(\omega t - kz) + B_o^- \cos(\omega t + kz)] \tag{18.4b}$$

These fields are terminated on the inner surfaces of the plates, by surface charges,

$$\sigma_s(z,t) = \pm\varepsilon_0 c[B_o^+ \cos(\omega t - kz) - B_o^- \cos(\omega t + kz)] \qquad \left(x = \pm\frac{d}{2}\right)$$

and surface currents,

$$\mathbf{K}(z,t) = \pm\widehat{\mathbf{z}}\left[\frac{B_o^+}{\mu_0}\cos(\omega t - kz) + \frac{B_o^-}{\mu_0}\cos(\omega t + kz)\right] \qquad \left(x = \pm\frac{d}{2}\right)$$

Because the plate currents are z-directed (except for x-directed components due to sources and/or terminations located at $z = \pm\infty$) and we are only interested in the fields for finite z, the Lorenz-gauge vector and scalar potentials are found to be

$$\mathbf{A}(x,z,t) = \widehat{\mathbf{z}}[B_o^+ \cos(\omega t - kz) + B_o^- \cos(\omega t + kz)]\left[u_{-2}\left(x + \frac{d}{2}\right) - u_{-2}\left(x - \frac{d}{2}\right) - \frac{d}{2}\right] \tag{18.5a}$$

$$\Phi(x,z,t) = c[B_o^+ \cos(\omega t - kz) - B_o^- \cos(\omega t + kz)]\left[u_{-2}\left(x + \frac{d}{2}\right) - u_{-2}\left(x - \frac{d}{2}\right) - \frac{d}{2}\right] \tag{18.5b}$$

where $u_{-2}(x) = x\,u_{-1}(x)$ is the unit-ramp function and, as required by symmetry, the potentials vanish at $x = 0$.

The Maxwell–Poynting representations of power energy and stress momentum are (for $|x| \leq d/2$)

$$\mathbf{S} = \mathbf{E} \times \mathbf{H} = \hat{\mathbf{z}} \frac{c}{\mu_o} [(B_o^+)^2 \cos^2(\omega t - kz) - (B_o^-)^2 \cos^2(\omega t + kz)]$$

$$W^{em} = \frac{1}{\mu_o} [(B_o^+)^2 \cos^2(\omega t - kz) + (B_o^-)^2 \cos^2(\omega t + kz)]$$

$$\mathbf{G}^{em} = \frac{\mathbf{S}}{c^2}$$

$$\begin{aligned}\overline{\mathbf{T}}^{em} &= \varepsilon_o \mathbf{E}\,\mathbf{E} + \mu_o \mathbf{H}\,\mathbf{H} - W^{em}\overline{\mathbf{I}} \\ &= 2\frac{B_o^+ B_o^-}{\mu_o} \cos(\omega t - kz)\cos(\omega t + kz)(-\mathbf{x}\,\mathbf{x} + \mathbf{y}\,\mathbf{y}) - W^{em}\,\mathbf{z}\,\mathbf{z}\end{aligned}$$

The Alternate representations of these same quantities are (ignoring the residual components)

$$\mathbf{S}^o = \Phi \mathbf{K} = \hat{\mathbf{z}} \frac{c}{\mu_o}\frac{d}{2} [(B_o^+)^2 \cos^2(\omega t - kz) - (B_o^-)^2 \cos^2(\omega t + kz)] \left[u_0\left(x - \frac{d}{2}\right) + u_0\left(x + \frac{d}{2}\right)\right]$$

$$W^o = \frac{d}{2\mu_o} [(B_o^+)^2 \cos^2(\omega t - kz) + (B_o^-)^2 \cos^2(\omega t + kz)] \left[u_0\left(x - \frac{d}{2}\right) + u_0\left(x + \frac{d}{2}\right)\right]$$

$$\mathbf{G}^o = \rho \mathbf{A} = \frac{\mathbf{S}^o}{c^2}$$

$$T_{ij}^o = \frac{1}{2}(\mathbf{A} \cdot \mathbf{J} - \rho\Phi)\delta_{ij} - A_i J_j + \frac{1}{2\mu_o}\left(\mathbf{A} \cdot \frac{\partial^2 \mathbf{A}}{\partial x_i \partial x_j} - \frac{\partial \mathbf{A}}{\partial x_i} \cdot \frac{\partial \mathbf{A}}{\partial x_j}\right) - \frac{1}{2}\varepsilon_o\left(\Phi \frac{\partial^2 \Phi}{\partial x_i \partial x_j} - \frac{\partial \Phi}{\partial x_i}\frac{\partial \Phi}{\partial x_j}\right)$$

$$\begin{aligned}\overline{\mathbf{T}}^o = \frac{B_o^+ B_o^-}{\mu_o} &\left\{ d\left[u_0\left(x - \frac{d}{2}\right) + u_0\left(x + \frac{d}{2}\right)\right] \widehat{\mathbf{y}}\widehat{\mathbf{y}} \right. \\ &\left. -2\left[u_{-1}\left(x - \frac{d}{2}\right) - u_{-1}\left(x + \frac{d}{2}\right)\right]^2 \widehat{\mathbf{x}}\widehat{\mathbf{x}} \right\} \cos(\omega t - kz)\cos(\omega t + kz) \\ &- W^o \widehat{\mathbf{z}}\widehat{\mathbf{z}}\end{aligned}$$

Because the fields are independent of y, the $\widehat{\mathbf{y}}\widehat{\mathbf{y}}$ stress tensor components in both the Maxwell–Poynting and Alternate formulations can be ignored. As expected, $\mathbf{S}^o$ and $\mathbf{G}^o$ are localized to the surface charges and currents. When $B_o^+ B_o^- = 0$, so too is W^o and $\overline{\mathbf{T}}^o = -W^o \widehat{\mathbf{z}}\widehat{\mathbf{z}}$. At the location of these charges and/or currents, the Lorentz-force volume density is infinite, but only over zero thickness. When integrated over that thickness, the

Lorentz force per unit area acting on the surface is $\mathbf{F}' = \frac{1}{2}(\sigma_s \mathbf{E} + \mathbf{K} \times \mathbf{B})$. (The fields at the surfaces of discontinuity are the average values of those on each side and so the factor of $\frac{1}{2}$ must be inserted.) The boundary condition $\mathbf{n} \cdot (\overline{\mathbf{T}}_{(1)} - \overline{\mathbf{T}}_{(2)}) = \mathbf{F}'$ equates the difference in normal stress on either side of the surface of discontinuity (the positive normal points from side (2) to side (1)) to $\mathbf{F}'$; since the latter is independent of the representation used, we have

$$\mathbf{n} \cdot (\overline{\mathbf{T}}^{\text{o}}_{(1)} - \overline{\mathbf{T}}^{\text{o}}_{(2)}) = \mathbf{n} \cdot (\overline{\mathbf{T}}^{\text{em}}_{(1)} - \overline{\mathbf{T}}^{\text{em}}_{(2)}) = \frac{1}{2}(\sigma_s \mathbf{E} + \mathbf{K} \times \mathbf{B})$$

When applied to the two plates ($x = \pm d/2, \widehat{\mathbf{n}} = \pm\widehat{\mathbf{x}}$), we obtain

$$\mathbf{F}'_{\pm} = \pm\widehat{\mathbf{x}}\frac{2B_{\text{o}}^{+}B_{\text{o}}^{-}}{\mu_{\text{o}}}\cos(\omega t - kz)\cos(\omega t + kz)$$

Notice that if only a single traveling wave is present, the plate forces vanish because the electric and magnetic components balance.

Short-circuit termination at z = 0

Assume now that the top and bottom plates are short-circuited by a perfectly conducting strip of the plane $z = 0$. The boundary condition, $E_x(z = 0) = 0$, is met when $B_{\text{o}}^{-} = B_{\text{o}}^{+}$. Because the source is located at $z = -\infty$, there are no electric or magnetic fields in the region $z > 0$. The potentials $\mathbf{A} = 0$, $\Phi = 0$ certainly produce the correct values of $\mathbf{E}$ and $\mathbf{B}$, but cannot be complete because the current in the short circuit is a source of A_x. Without the necessary fringing potentials, A_z is discontinuous at $z = 0$ and thus violates the Lorenz-gauge condition. Nevertheless, we attempt to calculate the force exerted on the short circuit and find that

$$\widehat{\mathbf{n}} \cdot (\overline{\mathbf{T}}^{\text{o}}_{(1)} - \overline{\mathbf{T}}^{\text{o}}_{(2)}) = -\widehat{\mathbf{z}} \cdot \overline{\mathbf{T}}^{\text{o}}_{(2)} = W^{\text{o}}\widehat{\mathbf{z}}$$

compared with

$$\widehat{\mathbf{n}} \cdot (\overline{\mathbf{T}}^{\text{em}}_{(1)} - \overline{\mathbf{T}}^{\text{em}}_{(2)}) = -\widehat{\mathbf{z}} \cdot \overline{\mathbf{T}}^{\text{em}}_{(2)} = W^{\text{em}}\widehat{\mathbf{z}} = \frac{1}{2}\mathbf{K} \times \mathbf{B} \tag{18.6}$$

Although the Alternate density is incorrect, the *total* force on the short circuit is in agreement. Often, only the total force is required and no further analysis is required; when that is not sufficient, one can calculate it directly from the Lorentz-force density, from the Maxwell-stress tensor, or from $\widehat{\mathbf{n}} \cdot (\overline{\mathbf{T}}^{\text{o}}_{(1)} - \overline{\mathbf{T}}^{\text{o}}_{(2)})$ calculated from either the correct set of vector and scalar potentials or those associated with modified boundary conditions chosen so as not to alter the fields within the transmission line. The latter technique is often very simple as we next demonstrate.

Modified boundary conditions

If we allow the same $\mathbf{E}$ and $\mathbf{B}$ given by Eqs. (18.4a) and (18.4b) to exist both above and below the thin conducting-plates, and extend the short circuit over the entire plane, $z = 0$, there will be surface charges and currents induced on the outer surfaces of the horizontal plates that are of opposite polarity to those on the inner surfaces. In addition, a uniform surface current,

$$K_x = -\frac{2B_{\text{o}}^{+}}{\mu_{\text{o}}}\cos(\omega t)$$

will exist at $z = 0$. Because now $K_z = 0$ (due to the cancellation of currents), only $A_x(z, t)$ is created. In the Lorenz gauge, $\Phi = 0$ because $\nabla \cdot \mathbf{A} = 0$. Since the potentials vanish on the short-circuit plane, none are required for $z > 0$. The complete set of potentials for the modified problem is therefore

$$\mathbf{A}(z, t) = -\widehat{\mathbf{x}}\frac{2B_o^+}{k} \cos(\omega t) \sin(kz)[1 - u_{-1}(z)]$$

$$\Phi = 0$$

Now $-\widehat{\mathbf{z}} \cdot \overline{\mathbf{T}}^{o}_{(2)} = \frac{1}{2}\mathbf{K} \times \mathbf{B} = \widehat{\mathbf{z}}\frac{2(B_o^+)^2}{\mu_o} \cos^2(\omega t)$ is in agreement. Notice, however, that this modified configuration cannot be used to calculate the x-directed plate forces of the original problem.

Exact vector and scalar potentials

If we maintain the original configuration, the current, $K_x(t)$ will be nonzero only between the plates. Although the values of $\mathbf{A}$ and Φ given by Eqs. (18.5a) and (18.5b) generate the correct fields everywhere, the value of A_z is discontinuous at $z = 0$. In a consistent Lorenz gauge, $\nabla \cdot \mathbf{A}$ (and therefore Φ) $\to \infty$, this leads to improper field singularities. These can be removed by adding homogeneous potentials $\mathbf{A}_h = \nabla\Psi$ and $\Phi_h = -\frac{\partial \Psi}{\partial t}$ that create neither $\mathbf{E}$ nor $\mathbf{B}$ yet satisfy the Lorenz gauge everywhere *except* on the plane $z = 0$. This constrains the function, $\Psi(x, z, t)$, to be a solution of the two-dimensional wave equation

$$\nabla^2 \Psi - \frac{1}{c^2}\frac{\partial^2 \Psi}{\partial t^2} = 0 \qquad (\text{except on } z = 0)$$

Because the only source of Ψ is on the plane $z = 0$, we superimpose plane waves propagating away from that plane. The function defined in the region $z > 0$ is denoted by Ψ_+, and that defined in $z < 0$ is denoted by Ψ_-. These are expressed in terms of the wavenumber, k_x, which is allowed to vary continuously from 0 to ∞ so as to generate Fourier integrals:

$$\Psi_\pm = \begin{bmatrix} \int_0^k C_\pm(k_x) \sin(\omega t \mp \sqrt{k^2 - k_x^2} z) \sin(k_x x)\, dk_x \\ + \cos(\omega t) \int_k^\infty C_\pm(k_x) \exp(\mp\sqrt{k_x^2 - k^2} z) \sin(k_x x)\, dk_x \end{bmatrix}$$

In the range $0 < k_x < k = \omega/c$, the waves are uniform plane waves; in the range $k < k_x < \infty$, they are nonuniform plane waves.

We next calculate $\frac{\partial \Psi_\pm}{\partial t}(x, 0, t)$ from

$$\frac{\partial \Psi_\pm}{\partial t} = \begin{bmatrix} \int_0^k \omega C_\pm(k_x) \cos(\omega t \mp \sqrt{k^2 - k_x^2} z) \sin(k_x x)\, dk_x \\ -\omega \sin(\omega t) \int_k^\infty C_\pm(k_x) \exp(\mp\sqrt{k_x^2 - k^2} z) \sin(k_x x)\, dk_x \end{bmatrix}$$

$$\frac{\partial \Psi_\pm}{\partial t}(x, 0, t) = \omega \left[\cos(\omega t) \int_0^k C_\pm(k_x) \sin(k_x x)\, dk_x - \sin(\omega t) \int_k^\infty C_\pm(k_x) \sin(k_x x)\, dk_x \right]$$

and $\frac{\partial \Psi_\pm}{\partial z}(x,0,t)$ from

$$\frac{\partial \Psi_\pm}{\partial z}(x,z,t) = \begin{bmatrix} \mp \int_0^k \sqrt{k^2-k_x^2}C_\pm(k_x)\cos(\omega t \mp \sqrt{k^2-k_x^2}z)\sin(k_x x)\,dk_x \\ \mp\cos(\omega t)\int_k^\infty \sqrt{k_x^2-k^2}C_\pm(k_x)\exp(\mp\sqrt{k_x^2-k^2}z)\sin(k_x x)\,dk_x \end{bmatrix}$$

$$\frac{\partial \Psi_\pm}{\partial z}(x,0,t) = \mp\cos(\omega t)\int_0^\infty \sqrt{|k^2-k_x^2|}C_\pm(k_x)\sin(k_x x)\,dk_x$$

so that we can match the boundary conditions

$$\frac{\partial \Psi_-}{\partial t}(x,0) = \frac{\partial \Psi_+}{\partial t}(x,0) \tag{18.7a}$$

$$A_z(x,0) + \frac{\partial \Psi_-}{\partial z}(x,0) = \frac{\partial \Psi_+}{\partial z}(x,0) \tag{18.7b}$$

Substitution into Eq. (18.7a) (continuity of Φ) reveals that $C_+(k_x) = C_-(k_x) = C(k_x)$. This symmetry also insures continuity of A_x and is consistent with the fact that because Ψ cannot produce fields (except on $z=0$), the existence of the transmission-line plates for $z<0$ is inconsequential (they are, so to speak, "invisible").

Substitution into Eq. (18.7b) (continuity of A_z) leads to

$$A_z(x,0,t) + 2\cos(\omega t)\int_0^\infty \sqrt{|k^2-k_x^2|}C(k_x)\sin(k_x x)\,dk_x = 0$$

It is helpful to differentiate both sides of this equation with respect to x; the result

$$-B_y(x,0,t) + 2\cos(\omega t)\int_0^\infty k_x\sqrt{|k^2-k_x^2|}C(k_x)\cos(k_x x)\,dk_x = 0$$

can be combined with the magnetic flux at the short circuit,

$$B_y(x,0,t) = -2B_\text{o}^+\cos(\omega t)\left[1-u_{-1}\left(|x|-\frac{d}{2}\right)\right]$$

and the Fourier cosine transform of a rectangular-pulse,

$$\left[1-u_{-1}\left(|x|-\frac{d}{2}\right)\right] = \frac{2}{\pi}\int_0^\infty \frac{\sin(k_x\frac{d}{2})}{k_x}\cos(k_x x)\,dk_x$$

to solve for the Fourier-distribution coefficients

$$C(k_x) = -\frac{2B_\text{o}^+}{\pi}\frac{\sin(k_x\frac{d}{2})}{k_x^2\sqrt{|k_x^2-k^2|}}$$

When added to the particular solutions, the homogeneous Ψ produces the necessary Φ_h and $\mathbf{A}_h$ at and near the short circuit. These "fringe" on both sides of the short circuit, but decay as $|z|$ increases. Important contributions to the spectrum extend to $k_x d \sim \pi$; therefore if $kd \ll 1$, most of the plane waves will be of the nonuniform type. In that case, the homogeneous terms decay rapidly. Based upon numerical evaluation of these results, the total vector field $\mathbf{A}$, near the short circuit, is plotted in Figure 18.2 for $t=0$. Symmetry considerations dictate that $A_z(x,0,t)$ is exactly one-half the value of Eq. (18.5a).

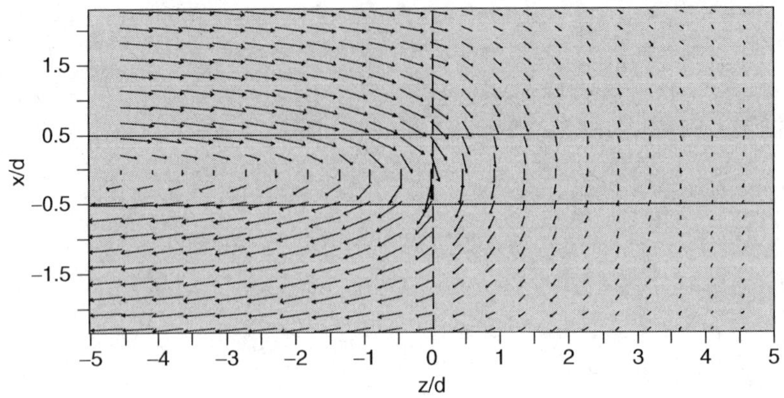

Figure 18.2 Vector field plot of the exact $\mathbf{A}(x/d, z/d, 0)$ near the short circuit.

Because there is Alternate stress on both sides of the short circuit, both $\overline{\mathbf{T}}^{\text{o}}_{(1)}$ and $\overline{\mathbf{T}}^{\text{o}}_{(2)}$ must be included in calculations like Eq. (18.6). It is important to understand that potentials sometimes penetrate perfectly conducting walls and that the consequent diffraction produces "fringing" that severely complicates the Alternate-representation—even for simple configurations. However, as a practical matter, one need never carry out the full analysis.

If, instead of a perfectly conducting plane, a resistive sheet of conductivity σ_{sheet} and thickness δ is placed across the parallel plates at $z = 0$, the transverse surface current can be controlled. As σ_{sheet} increases from zero to infinity, the fringing components of $\underline{\mathbf{A}}$ and $\underline{\Phi}$ (initially zero —assuming the transmission line continues for $z > 0$)[4] approach the values generated by $\Psi_{\pm}$. It follows that the Alternate power-flux, energy-density, momentum-density, and stress-tensor distributions shift continuously from their initial distributions to those required when $\sigma = \infty$. It is $\underline{K}_x$ that produces the required change. Similar behavior is experienced if a resistive sheet is placed across other types of transmission lines or waveguides.

If the space between the parallel plates is loaded with linear, isotropic, dielectric/magnetic material (μ, ε), it is advantageous to pretend that similar material replaces free-space *everywhere*. Then, after the substitutions $\mu_{\text{o}} \rightarrow \mu$ and $\varepsilon_{\text{o}} \rightarrow \varepsilon$ are made, all the previous equations in this chapter remain valid for both the Maxwell–Poynting and Alternate representations. A detailed treatment of the loaded line, without such pretense, is contained in the final section of this chapter.

18.3 TEM TAPERED-PLATE "HORN" TRANSFORMER

Consider once again a pair of thin perfectly conducting parallel plates that are separated by a distance d. They are of infinite width along y and extend from $z = 0$ to $z = \ell$; free-space is between and surrounds them. An electric surface current of frequency ω and magnitude K_0 is located on a plane $z \rightarrow -\infty$ and is oriented parallel to the x axis;

[4]If the plates do not continue for $z > 0$, the open-circuit condition is $B_{\text{o}}^{-} \simeq -B_{\text{o}}^{+}$ rather than $B_{\text{o}}^{-} = 0$.

this generates, throughout all space, a complex vector potential,

$$\underline{A}_x = \frac{\mu_0 \underline{K}_0}{jk} \exp(-jkz)$$

which, in turn, generates a *TEM* uniform plane wave:

$$\underline{E}_x = -c\mu_0 \underline{K}_0 \exp(-jkz)$$

$$\underline{H}_y = -\underline{K}_0 \exp(-jkz)$$

Because the same fields surround each plate, the electric surface charges and surface currents on opposite surfaces of the same plate are of opposing polarity and so the net charge and current on each plate is zero. Consequently, there are no sources that can generate either $\underline{A}_z$ or $\underline{\Phi}$.

The complex Poynting vector or Alternate power flux is everywhere

$$\frac{1}{2}\underline{E}_x \underline{H}_y^* = \underline{S}_z^o = \frac{1}{2} c\mu_0 \underline{K}_0 \underline{K}_0^*$$

and the average power carried between the plates (per unit width along y) is

$$P_z' = \frac{1}{2} c\mu_0 \underline{K}_0 \underline{K}_0^* d$$

Assume now that the plates are both gradually tapered over a length ℓ large compared to the wavelength to provide the horn region shown in Figure 18.3. After the taper, the plates are again parallel, but with reduced spacing; they extend to $z \to \infty$.

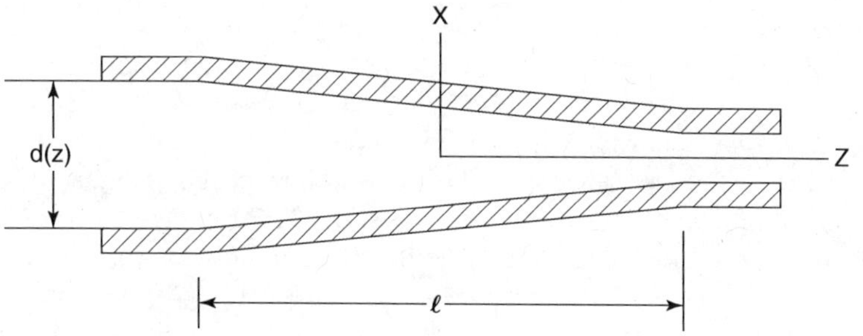

Figure 18.3 Tapered-plate transmission line.

A detailed treatment of tapered transmission lines within the context of distributed circuit theory is contained in Chapter 24, Section 24.3. Here, we simply assume that because $k\ell \gg 1$ and the plate spacing, $d(z)$, decreases very gradually from $d(0)$ to $d(\ell)$, any reflected waves are negligible and the line acts as a transformer. From the Poynting perspective, both $\underline{E}_x$ and $\underline{H}_y$ increase in the region between the plates in order to maintain constant $P_z' \simeq \frac{1}{2}\underline{E}_x \underline{H}_y^* d(z)$; the fields and power fluxes above and below the plates are reduced by negligible amounts because they extend over an infinite range of x. With the fields unbalanced, net charge and current accumulate on the plates. These increase as $d(z)$ decreases, and they provide sources $\underline{K}_z^s(z)$ and $\underline{\sigma}^s(z)$ for both $\underline{A}_z$ and $\underline{\Phi}$. Assuming that $\underline{K}_1(z)$ is the complex magnitude on the top plate and that the horn taper is gradual enough to validate a *TEM* wave solution, the additional potentials between the plates are approximately

$$\underline{A}_z \simeq \mu_0 x \underline{K}_1(z) \exp(-jkz)$$

$$\underline{\Phi} \simeq c\mu_0 x \underline{K}_1(z) \exp(-jkz)$$

where $\underline{K}_1(0) = 0$. These contribute to the electric and magnetic fields, which are now

$$\underline{E}_x \simeq -c\mu_0[\underline{K}_0 + \underline{K}_1(z)] \exp(-jkz) \qquad (0 \leq z \leq \ell)$$

$$\underline{H}_y \simeq -[\underline{K}_0 + \underline{K}_1(z)] \exp(-jkz) \qquad (0 \leq z \leq \ell)$$

Because, despite the taper, the tangential electric field vanishes on the plates, the total Poynting power flowing between them is constant. Consequently,

$$\frac{1}{2}c\mu_0[\underline{K}_0 + \underline{K}_1(z)][\underline{K}_0^* + \underline{K}_1^*(z)]d(z) = \frac{1}{2}c\mu_0\underline{K}_0\underline{K}_0^* d(0)$$

The surface current is generated by the incident magnetic field; therefore $\underline{K}_1$ is in phase with $\underline{K}_0$ and given by

$$\underline{K}_1(z) \simeq \underline{K}_0\left(\sqrt{\frac{d(0)}{d(z)}} - 1\right) \tag{18.8}$$

The Alternate-power flux between and localized on the plates is

$$\underline{S}_z^o \simeq \frac{1}{2}c\mu_0 \left(\underline{K}_0\underline{K}_0^* + \underline{K}_1(z)\underline{K}_1^*(z)\, x \left[u_0(x - d(z)/2) - u_0(x + d(z)/2)\right]\right) \tag{18.9}$$

For $z > \ell$ and if $\sqrt{\frac{d(0)}{d(\ell)}} \gg 1$, the Alternate power has been transferred, nearly entirely, from the volume to the plates.

Notice that although the Poynting flux always vanishes inside perfectly conducting plates, in this case the Alternate flux does not. This fact accounts for the missing Alternate-power cross term which has "leaked" across the plate boundaries to the outside space. Nevertheless, because reflections have been assumed negligible, the total average-power crossing *any* plane of constant z is invariant. Of course, the power delivered to a matched load that may be used to terminate the plate transmission line must agree with the volume integration over $\mathbf{E} \cdot \mathbf{J}$ and is therefore independent of the representation of power and energy that is employed. The Alternate analysis reduces to the integral of $\Phi_{\text{load}}\mathbf{J}_{\text{load}} \cdot \widehat{\mathbf{z}}$ which is, of course, $V_{\text{load}}I_{\text{load}}$.

18.4 TEM LINE WITH PARALLEL PLATES OF HIGH CONDUCTIVITY

In this example, we consider an air-filled transmission line made from parallel plates separated by a distance d. The plates are highly conducting ($\sigma \gg \omega\varepsilon_0$); the inner surfaces are located at $x = \pm d/2$ and are of infinite extent in both the y and z directions. Because the skin depth, δ, is assumed to be very small compared to the plate thickness, the latter can be assumed to be infinite. We assume that a sinusoidal steady-state quasi-*TEM* wave is propagating in the $+z$ direction at the frequency, ω; because of symmetry considerations, we restrict the analysis to the half-space, $x > 0$. The calculations are enormously simplified by working in the complex-frequency domain. In the equations which follow, $\underline{E}_1^\sigma$, $\underline{H}_2^\sigma$, $\underline{E}_3^\sigma$, $\underline{A}_0^\sigma$, $\underline{A}_1^\sigma$, $\underline{A}_3^\sigma$ and $\underline{E}_1$, $\underline{H}_2$, $\underline{E}_3$, $\underline{A}_0$, $\underline{A}_1$, $\underline{A}_3$ are complex amplitudes, and γ_0, γ_1, γ_3 are complex propagation constants.

In the plate, the complex electric and magnetic fields and the associated complex vector and scalar potentials (all with the factor $\exp(j\omega t)$ suppressed) are

$$\underline{\mathbf{E}}^\sigma = (\widehat{\mathbf{x}}\underline{E}_1^\sigma + \widehat{\mathbf{z}}\underline{E}_3^\sigma)\exp\left[-\gamma_0\left(x - \frac{d}{2}\right)\right]\exp(-\gamma_3 z) \tag{18.10a}$$

$$\underline{\mathbf{H}}^\sigma = \widehat{\mathbf{y}}\underline{H}_2^\sigma \exp\left[-\gamma_0\left(x - \frac{d}{2}\right)\right]\exp(-\gamma_3 z) \tag{18.10b}$$

$$\underline{\mathbf{A}}^\sigma = (\widehat{\mathbf{x}}\underline{A}_1^\sigma + \widehat{\mathbf{z}}\underline{A}_3^\sigma)\exp\left[-\gamma_0\left(x - \frac{d}{2}\right)\right]\exp(-\gamma_3 z) \tag{18.10c}$$

$$\underline{\Phi}^\sigma = \underline{A}_0^\sigma \exp\left[-\gamma_0\left(x - \frac{d}{2}\right)\right]\exp(-\gamma_3 z) \tag{18.10d}$$

with complex propagation constants that satisfy the dispersion relation

$$\gamma_0^2 + \gamma_3^2 = j\omega\mu_0(j\omega\varepsilon_0 + \sigma) \simeq j\frac{2}{\delta^2} \tag{18.11}$$

The gauge condition for the conductor, $\nabla \cdot \underline{\mathbf{A}}^\sigma + \mu_0(j\omega\varepsilon_0 + \sigma)\underline{\Phi}^\sigma = 0$, reduces to

$$\gamma_0 A_1^\sigma + \gamma_3 A_3^\sigma \simeq \mu_0\sigma\underline{A}_0^\sigma \tag{18.12}$$

In the free-space region, the corresponding fields and potentials are

$$\underline{\mathbf{E}} = \left[\widehat{\mathbf{x}}\underline{E}_1\cosh(\gamma_1 x) + \widehat{\mathbf{z}}\underline{E}_3\sinh(\gamma_1 x)\right]\exp(-\gamma_3 z) \tag{18.13a}$$

$$\underline{\mathbf{H}} = \widehat{\mathbf{y}}\underline{H}_2\cosh(\gamma_1 x)\exp(-\gamma_3 z) \tag{18.13b}$$

$$\underline{\mathbf{A}} = \left[\widehat{\mathbf{x}}\underline{A}_1\cosh(\gamma_1 x) + \widehat{\mathbf{z}}\underline{A}_3\sinh(\gamma_1 x)\right]\exp(-\gamma_3 z) \tag{18.13c}$$

$$\underline{\Phi} = \underline{A}_0\sinh(\gamma_1 x)\exp(-\gamma_3 z) \tag{18.13d}$$

where

$$\gamma_1^2 + \gamma_3^2 = -\omega^2\mu_0\varepsilon_0 \tag{18.14}$$

The gauge condition for free-space, $\nabla \cdot \underline{\mathbf{A}} + j\omega\mu_0\varepsilon_0\underline{\Phi} = 0$, reduces to

$$-\gamma_1\underline{A}_1 + \gamma_3\underline{A}_3 = j\omega\mu_0\varepsilon_0\underline{A}_0 \tag{18.15}$$

There is no question that working directly with **E** and **H** is the simplest and therefore best way to solve this boundary-value problem. The complex Maxwell's Equations are

$$\nabla \times \underline{\mathbf{E}} = -j\omega\mu_0\underline{\mathbf{H}}$$

$$\nabla \cdot \underline{\mathbf{H}} = 0$$

$$\nabla \times \underline{\mathbf{H}} = (\sigma + j\omega\varepsilon_0)\underline{\mathbf{E}}$$

$$\varepsilon_0\nabla \cdot \underline{\mathbf{E}} = \rho = -\frac{\varepsilon_0}{\sigma}\underline{\mathbf{E}} \cdot \nabla\sigma$$

Except on the surface (where $\nabla\sigma \to \infty$), there is no electric-charge density; when nonzero, $\rho \to \infty$, and represents a surface-charge density that is the source of Φ. In the respective regions, we have

$$\gamma_0 \underline{E}_3^\sigma - \gamma_3 \underline{E}_1^\sigma = -j\omega\mu_0 \underline{H}_2^\sigma$$

$$-\gamma_0 \underline{H}_2^\sigma = (\sigma + j\omega\varepsilon_0)\underline{E}_3^\sigma$$

$$\gamma_3 \underline{H}_2^\sigma = (\sigma + j\omega\varepsilon_0)\underline{E}_1^\sigma$$

$$-\gamma_0 \underline{E}_1^\sigma = \gamma_3 \underline{E}_3^\sigma$$

and

$$\gamma_1 \underline{E}_3 + \gamma_3 \underline{E}_1 = j\omega\mu_0 \underline{H}_2$$

$$\gamma_1 \underline{H}_2 = j\omega\varepsilon_0 \underline{E}_3$$

$$\gamma_3 \underline{H}_2 = j\omega\varepsilon_0 \underline{E}_1$$

$$\gamma_1 \underline{E}_1 = \gamma_3 \underline{E}_3$$

The discontinuity of $\underline{E}_x$ at the plate surface is due to the surface electric-charge density,

$$\underline{\sigma}_s = \varepsilon_0(\underline{E}_1^\sigma - \underline{E}_1)$$

Transverse wave impedance

Because tangential electric and magnetic fields are continuous for all values of x, the ratio $-\underline{E}_z/\underline{H}_y$ is also continuous. Defined as the transverse wave impedance, $\underline{Z}_t(x)$, the values are

$$\underline{Z}_t(x) = \begin{cases} \dfrac{\gamma_1}{j\omega\varepsilon_0}\tanh(\gamma_1 x) & \left(0 < x \le \frac{d}{2}\right) \\ \dfrac{-\gamma_0}{\sigma + j\omega\varepsilon_0} & \left(x \ge \frac{d}{2}\right) \end{cases} \tag{18.16}$$

The boundary condition for this quasi-*TEM* (transverse electric and magnetic) mode, which is actually a *TM* (transverse magnetic) mode, is therefore

$$\frac{\gamma_1}{\gamma_0}\tanh\left(\gamma_1 \frac{d}{2}\right) = \frac{-j\omega\varepsilon_0}{\sigma + j\omega\varepsilon_0} \simeq \frac{-j\omega\varepsilon_0}{\sigma} \tag{18.17}$$

Because σ is assumed to be very large, $\tanh(\gamma_1 \frac{d}{2}) \simeq \gamma_1 \frac{d}{2}$.

Perfectly conducting plates

For a perfect conductor, we have $\sigma \to \infty$, $\gamma_1 \to 0$, $\gamma_0 = (1 + j)/\delta$, $\delta \to 0$, and $\gamma_3 = -j\omega\sqrt{\mu_0\varepsilon_0}$. The mode becomes *TEM* ($\underline{E}_3 = \underline{E}_3^\sigma = \underline{E}_1^\sigma = 0$) with $\underline{E}_1/\underline{H}_2 = \sqrt{\frac{\mu_0}{\varepsilon_0}}$. The complex vector and scalar potentials reduce to

$$\underline{\mathbf{A}} = \left[\widehat{\mathbf{x}}\underline{A}_1 + \widehat{\mathbf{z}}\underline{A}_3\gamma_1 x\right]\exp(-j\omega\sqrt{\mu_0\varepsilon_0}z)$$

$$\underline{\Phi} = \underline{A}_0\gamma_1 x \exp(-j\omega\sqrt{\mu_0\varepsilon_0}z)$$

which generate

$$\underline{\mathbf{E}} = \left[-\widehat{\mathbf{x}}\left(j\omega\underline{A}_1 + \gamma_1\underline{A}_0\right) + \widehat{\mathbf{z}}j\omega\left(-\underline{A}_3 + \sqrt{\mu_0\varepsilon_0}\underline{A}_0\right)\gamma_1 x\right]\exp(-j\omega\sqrt{\mu_0\varepsilon_0}z)$$

$$\mu_0\underline{\mathbf{H}} = -\widehat{\mathbf{y}}(\underline{A}_3\gamma_1 + j\omega\sqrt{\mu_0\varepsilon_0}\underline{A}_1)\exp(-j\omega\sqrt{\mu_0\varepsilon_0}z)$$

Agreement is reached with $\underline{A}_3 = \sqrt{\mu_0\varepsilon_0}\underline{A}_0$ and $\underline{E}_1 = \sqrt{\frac{\mu_0}{\varepsilon_0}}\underline{H}_2 = -(j\ \omega\underline{A}_1 + \gamma_1\underline{A}_0)$. The Lorenz-gauge condition imposes the restriction $\gamma_1\underline{A}_1 = 0$; two possibilities exist: either $\gamma_1 = 0$ and the value of $\underline{A}_0$ is inconsequential or γ_1 is nonzero (but very small) and $\underline{A}_1 = 0$. In the latter case, $\underline{A}_0 \to \infty$ such that $\lim_{\gamma_1\to 0}(\gamma_1\underline{A}_0) = -\underline{E}_1$. Because γ_1 is nonzero when the conductivity is finite, the second possibility is the required choice, consistent with the fact that the currents are purely z-directed. At the interface, the condition $\underline{H}_2^\sigma = \underline{H}_2$ requires that $\underline{A}_3^\sigma = \mu_0\underline{H}_2\frac{\delta}{1+j} \to 0$; also $\underline{A}_1^\sigma = 0$ and $\underline{A}_0^\sigma \to 0$. These results are consistent with the fact that for $x - d/2 \gg \delta$, we have $\underline{\mathbf{E}}^\sigma, \underline{\mathbf{H}}^\sigma \to 0$, but then the potentials are discontinuous. We rectify the problem in this and the general case, by adding homogeneous potentials in the conducting region that generate neither electric or magnetic fields. They are of the form

$$\underline{\mathbf{A}}_h^\sigma = \widehat{\mathbf{z}}\frac{\gamma_3}{j\omega}\underline{A}_h^\sigma\exp(-\gamma_3 z) \tag{18.18a}$$

$$\underline{\Phi}_h^\sigma = \underline{A}_h^\sigma\exp(-\gamma_3 z) \tag{18.18b}$$

Notice that similar terms cannot be added to the free-space region because they would not vanish for $x = 0$ as required by symmetry considerations. For the perfect conductor, $\underline{A}_h^\sigma = -\underline{E}_1\frac{d}{2}$ represents the voltage induced on the plate by the propagating *TEM* mode.[5]

High conductivity revisited (the complete potentials)

We add Eq. (18.18a) to (18.10c) and add Eq. (18.18b) to (18.10d); this adds one more complex variable to the list of unknowns: $\underline{A}_0$, $\underline{A}_1$, $\underline{A}_3$, $\underline{A}_0^\sigma$, $\underline{A}_1^\sigma$, $\underline{A}_3^\sigma$, $\underline{A}_h^\sigma$. The boundary conditions that follow by requiring continuity of $\underline{\mathbf{A}}$ and $\underline{\Phi}$ at $x = d/2$ are

$$\underline{A}_0\sinh\left(\gamma_1\frac{d}{2}\right) = \underline{A}_0^\sigma + \underline{A}_h^\sigma \tag{18.19a}$$

$$\underline{A}_1\cosh\left(\gamma_1\frac{d}{2}\right) = \underline{A}_1^\sigma \tag{18.19b}$$

$$\underline{A}_3\sinh\left(\gamma_1\frac{d}{2}\right) = \underline{A}_3^\sigma + \frac{\gamma_3}{j\omega}\underline{A}_h^\sigma \tag{18.19c}$$

$\nabla\cdot\underline{\mathbf{E}} = \nabla\cdot\underline{\mathbf{E}}^\sigma = 0$ produces

$$\gamma_1\underline{A}_1 - \gamma_3\underline{A}_3 + (\gamma_1^2 + \gamma_3^2)\underline{A}_0 = 0 \tag{18.20a}$$

$$-\gamma_0 A_1^\sigma - \gamma_3 A_3^\sigma + (\gamma_0^2 + \gamma_3^2)\underline{A}_0^\sigma = 0 \tag{18.20b}$$

[5]In the example of Section 16.4, it was assumed that there was no transverse electric field and so no potential was induced. A more accurate model would approximate the current density as a standing wave proportional to $\cos(\omega z/c)$ with the analysis restricted to a range of z centered upon the origin. In that case, a transverse electric field proportional to $\sin(\omega z/c)$ would induce a homogeneous potential.

The results of the current example can be applied to a pair of modes traveling in opposite directions. Using superposition, an exact analysis of the earlier problem can be found.

while the gauge conditions (repeated for convenience) are

$$-\gamma_1\underline{A}_1 + \gamma_3\underline{A}_3 = j\omega\mu_o\varepsilon_o\underline{A}_0 \tag{18.21a}$$

$$\gamma_0\underline{A}_1^\sigma + \gamma_3\underline{A}_3^\sigma = \mu_o(\sigma + j\omega\varepsilon_o)\,\underline{A}_0^\sigma \tag{18.21b}$$

Continuity of $\underline{E}_z$ and $\underline{H}_y$ produces

$$(-j\omega\underline{A}_3 + \gamma_3\underline{A}_0)\sinh\left(\gamma_1\frac{d}{2}\right) = -j\omega\underline{A}_3^\sigma + \gamma_3\underline{A}_0^\sigma \tag{18.22a}$$

$$(\gamma_1\underline{A}_3 + \gamma_3\underline{A}_1)\cosh\left(\gamma_1\frac{d}{2}\right) = -\gamma_0\underline{A}_3^\sigma + \gamma_3\underline{A}_1^\sigma \tag{18.22b}$$

But, substitution of Eqs. (18.19a) and (18.19c) reveals that the first of these equations is redundant. Notice also that the addition of Eqs. (18.20a) and (18.21a) and Eqs. (18.20b) and (18.21b) verify the dispersion relations for the conducting and free-space regions, Eqs. (18.11) and (18.14). The solution of these equations reestablishes the validity of Eq. (18.17) and provides the complex mode coefficients:

$$\underline{A}_0 = \frac{\underline{A}_h^\sigma + \underline{A}_0^\sigma}{\sinh(\gamma_1\frac{d}{2})}$$

$$\underline{A}_1 = \frac{(j\gamma_1^2\underline{A}_h^\sigma + \omega\mu_o\sigma\underline{A}_0^\sigma)}{\omega\gamma_0\cosh(\gamma_1\frac{d}{2})}\,\frac{(\sigma + j\omega\varepsilon_o)}{\sigma}$$

$$\underline{A}_3 = \frac{\varepsilon_o(\gamma_1^2 + j\omega\mu_o\sigma)\,\underline{A}_h^\sigma}{\sigma\gamma_3\sinh(\gamma_1\frac{d}{2})}$$

$$\underline{A}_1^\sigma = \frac{(j\gamma_1^2\underline{A}_h^\sigma + \omega\mu_o\sigma\underline{A}_0^\sigma)}{\omega\gamma_0}\,\frac{(\sigma + j\omega\varepsilon_o)}{\sigma}$$

$$\underline{A}_3^\sigma = \frac{-j\gamma_1^2\underline{A}_h^\sigma}{\omega\gamma_3}\,\frac{(\sigma + j\omega\varepsilon_o)}{\sigma}$$

In order to prevent the gauge in the conductor from being infinite when $\sigma \to \infty$, we set $\underline{A}_0^\sigma = 0$[6]. This has the effect of maintaining the Lorenz gauge in a perfect conductor.

Because $\sigma \gg \omega\varepsilon_o$, and assuming that $d \gg \delta$, the propagation constants are well approximated by

$$\gamma_0 \simeq \frac{1+j}{\delta} \tag{18.23}$$

$$\gamma_1 \simeq \alpha_1 + j\beta_1 = \omega\sqrt{\mu_o\varepsilon_o}\sqrt{\frac{\delta}{2d}}\left(\sqrt{(\sqrt{2}+1)} - j\sqrt{(\sqrt{2}-1)}\right) \tag{18.24}$$

[6]In the skin-depth limit, $\underline{A}_0^\sigma \to (1+j)\sqrt{\frac{\omega}{2\mu_o\sigma}}\underline{A}_1 - \frac{j\gamma_1^2}{\omega\mu_o\sigma}\underline{A}_h^\sigma$; therefore it vanishes in the limit $\sigma \to \infty$ (assuming $\underline{A}_1$ and $\underline{A}_h^\sigma$ are finite).

$$\gamma_3 \simeq \alpha_3 + j\beta_3 = \omega\sqrt{\mu_o\varepsilon_o}\left(\frac{\delta}{2d} + j\right) \tag{18.25}$$

As expected, as $\delta \to 0$, the mode becomes lossless. Notice, however, that if the plate spacing is reduced until it becomes comparable to δ, the real part of γ_3 (the attenuation factor of the mode) becomes large. In the Maxwell–Poynting representation, this is due to the reduced area for $\mathbf{E} \times \mathbf{H}$ power flow—in comparison to that available for current flow. In the Alternate representation, reduced plate-spacing merely reduces the induced voltage and thus the mode power associated with a given current density that maintains constant dissipation in the plates. In both circuit theory and transmission-line parlance, the plates have become a low-impedance device.

Application of the complex power theorems

Because we are working with complex fields, it is appropriate to use the complex Poynting Theorem and the complex Alternate Theorem. For convenience, we repeat the results of Chapter 6, Section 6.2.

The complex Poynting Theorem is

$$\nabla \cdot \underline{\mathbf{S}} + j2\omega\left(\frac{1}{4}\mu_o\,\underline{\mathbf{H}}\cdot\,\underline{\mathbf{H}}^* - \frac{1}{4}\varepsilon_o\underline{\mathbf{E}}\cdot\underline{\mathbf{E}}^*\right) = -\frac{1}{2}\underline{\mathbf{E}}\cdot\underline{\mathbf{J}}^* \tag{18.26}$$

where the complex Poynting vector is defined by $\underline{\mathbf{S}} = \frac{1}{2}\underline{\mathbf{E}} \times \underline{\mathbf{H}}^*$. For the propagating mode, the time-averaged electric and magnetic energy densities are nearly equal. The reactive power and energy are negligible and the principal flux is

$$\underline{\mathbf{S}} \simeq \hat{\mathbf{z}}\frac{1}{2}\sqrt{\frac{\varepsilon_o}{\mu_o}}|\underline{E}_1|^2\exp(-2\alpha_3 z) \tag{18.27}$$

There is also a very small x-directed flux which increases nearly linearly with x; it is required to feed the losses in the plates. Since we have all of the coefficients and propagation constants, it is straightforward to carry out the compete evaluations if more exact analysis is required.

As derived in Section 6.2, the corresponding theorem for complex-power flux in the Alternate (localized) representation is

$$\nabla \cdot \underline{\mathbf{S}}^o + j2\omega\left[\frac{1}{8}\left(\underline{\mathbf{A}}\cdot\,\underline{\mathbf{J}}^* + \underline{\mathbf{A}}^*\cdot\,\underline{\mathbf{J}}\right) - \frac{1}{8}\left(\underline{\Phi}\,\underline{\rho}^* + \underline{\Phi}^*\,\underline{\rho}\right)\right] = -\frac{1}{2}\underline{\mathbf{E}}\cdot\underline{\mathbf{J}}^* \tag{18.28}$$

with the Alternate-complex-power flux defined by

$$\begin{aligned}\underline{\mathbf{S}}^o = \frac{1}{2}\,\underline{\Phi}\underline{\mathbf{J}}^* - j\omega\frac{1}{4\mu_o}\left(\underline{A}_k\nabla\underline{A}_k^* - \underline{A}_k^*\nabla\underline{A}_k\right)\\ + j\omega\frac{1}{4}\varepsilon_o\left(\underline{\Phi}\nabla\underline{\Phi}^* - \underline{\Phi}^*\nabla\underline{\Phi}\right)\\ + j\omega\frac{1}{4\mu_o}\left(\underline{\mathbf{A}}\,\underline{g}^* - \underline{\mathbf{A}}^*\underline{g}\right)\end{aligned} \tag{18.29}$$

where summation over $k = 1, 2, 3$ is assumed and

$$\underline{g} = \nabla \cdot \underline{\mathbf{A}} + j\omega\mu_o\varepsilon_o\underline{\Phi} \tag{18.30}$$

is the complex gauge. Because we chose $\underline{A}_0^\sigma = 0$, we obtain $\underline{g} = \frac{j\gamma_1^2}{\omega}\underline{A}_h^\sigma \exp(-\gamma_3 z) \simeq 0$ within the conductor (in the high σ limit) – as well as in the region between the plates.

Because the electric-charge density is zero except on the surface of the conductor where it forms a surface charge, the electric stored energy density is as localized as possible. The magnetic stored energy density is distributed over a few skin depths in the same form as the current density. As in the Poynting representation, the average energies very nearly balance[7] and thus prevent any appreciable reactive terms in either the energy or power. The Alternate power flux is confined to the current density and thus within a skin depth of the surface is approximately

$$\underline{\mathbf{S}}^{\mathrm{o}} \simeq \frac{1}{2}\,\underline{\Phi\mathbf{J}}^* \simeq \widehat{\mathbf{z}}\frac{1}{2}\sqrt{\frac{\varepsilon_{\mathrm{o}}}{\mu_{\mathrm{o}}}}|\underline{E}_1|^2\frac{d}{2\delta}\exp(-2\alpha_3 z)$$

In this picture, there is negligible Alternate-power flux in the space between the plates. Finally, we recall that there are two plates and so the total power flows and the dissipation are doubled.

Simple estimate of the attenuation factor for low-loss propagation

The real part of Eq. (18.28) yields

$$\nabla \cdot Re\underline{\mathbf{S}}^{\mathrm{o}} = -\frac{1}{2}Re\underline{\mathbf{E}} \cdot \underline{\mathbf{J}}^* \tag{18.31}$$

and since the losses are assumed to be very small, $\underline{\mathbf{S}}^{\mathrm{o}} \simeq \widehat{\mathbf{z}}\frac{1}{2}\,\underline{\Phi J}_z^* \sim \exp(-2\alpha_3 z)$, where α_3 is unknown. It follows from Eq. (18.31) that $\frac{\partial}{\partial z}(Re\underline{\Phi J}_z^*) = Re 2\alpha_3 \underline{\Phi J}_z^* \simeq Re\underline{E}_z\underline{J}_z^*$ which requires

$$2\alpha_3\underline{\Phi} \simeq \underline{E}_z = \frac{\underline{J}_z\delta}{\sigma\delta} \simeq \frac{\underline{K}_z}{\sigma\delta}$$

But, $\underline{\Phi} = -\underline{E}_1 d/2$ and $-\underline{K}_z = \underline{H}_2 = \sqrt{\frac{\varepsilon_{\mathrm{o}}}{\mu_{\mathrm{o}}}}E_1$; therefore, it follows that

$$\alpha_3 \simeq \frac{\omega\sqrt{\mu_{\mathrm{o}}\varepsilon_{\mathrm{o}}}\delta}{2d} = \beta_3\frac{\delta}{2d}$$

which agrees exactly with Eq. (18.25).

In the Maxwell–Poynting representation, the dissipation is in the walls of the conductors while the stored energy is in the space between them. Nevertheless energy and power arguments still allow the attenuation to be found from a simple perturbation calculation, because the *total* energy density (when integrated over the cross-section of the mode) is related to the dissipation power density by

$$2\alpha_3\iint \langle W^{em}\rangle c\; dx\, dy = \iint \tfrac{1}{2}Re\underline{\mathbf{E}} \cdot \underline{\mathbf{J}}^* dx\, dy$$

Since the fields are independent of y, the second integration can be taken over an arbitrary width.

With the Alternate representation, a similar formula can be employed; however, with the approach just followed, no integrations are required nor is it necessary to calculate the total energy density.

[7] The balance is exact for a perfect conductor.

18.5 PARALLEL-PLATE TEM LINE LOADED WITH LINEAR MATERIAL

In the final example of this chapter, we consider a simple parallel-plate transmission line made from perfectly conducting ground planes of infinite extent and located at $x = \pm d/2$. The space between the plates is assumed filled with a lossless linear material characterized by permeability μ and permittivity ε; that outside the plates with free-space.[8] A *TEM* mode of frequency ω is propagating in the +z direction. The electric, magnetic, polarization, and magnetization fields in the time domain are

$$\left.\begin{aligned} E_x &= E_o \cos(\omega t - kz) \\ B_y &= \mu H_y = B_o \cos(\omega t - kz) \\ P_x &= (\varepsilon - \varepsilon_o) E_o \cos(\omega t - kz) \\ M_y &= \frac{\mu - \mu_o}{\mu \mu_o} B_o \cos(\omega t - kz) \end{aligned}\right\} \left(|x| \le \frac{d}{2}\right) \tag{18.32}$$

where $\frac{B_o}{E_o} = \frac{k}{\omega} = \sqrt{\mu\varepsilon}$. All four fields vanish in the regions $|x| \ge d/2$.

The free-charge and free-current surface densities (located on the inner plate surfaces at $x = \pm d/2$) are

$$\sigma_{su} = \mp \varepsilon E_o \cos(\omega t - kz)$$

$$K_{uz} = \mp \frac{B_o}{\mu} \cos(\omega t - kz)$$

The vector and scalar potentials for this geometry are

$$\left.\begin{aligned} \mathbf{A} = {} & \widehat{\mathbf{x}}[C_1 \cosh(\alpha x) + C_2] \sin(\omega t - kz) \\ & + \widehat{\mathbf{z}} C_3 \sinh(\alpha x) \cos(\omega t - kz) \\ \Phi = {} & C_0 \sinh(\alpha x) \cos(\omega t - kz) \end{aligned}\right\} \left(|x| \le \frac{d}{2}\right) \tag{18.33a}$$

$$\left.\begin{aligned} \mathbf{A} &= \begin{bmatrix} \widehat{\mathbf{x}} C_1' \sin(\omega t - kz) \\ \pm \widehat{\mathbf{z}} C_3' \cos(\omega t - kz) \end{bmatrix} \exp(-\alpha(|x| - \tfrac{d}{2})) \\ \Phi &= \pm C_0' \exp(-\alpha(|x| - \tfrac{d}{2})) \cos(\omega t - kz) \end{aligned}\right\} \left(|x| \ge \frac{d}{2}\right) \tag{18.33b}$$

In order that both **E** and **B** vanish outside the plates and are *TEM* inside them,

$$\alpha C_3 = -kC_1$$

$$\alpha C_3' = kC_1'$$

$$\omega C_3 = kC_0$$

$$\omega C_3' = kC_0'$$

[8] Since there are no electric or magnetic fields outside the plates, we could pretend that the linear material inside the plates is present everywhere. This artifice would allow free-space analyses to be substituted, provided that the replacements $\varepsilon_o \to \varepsilon$ and $\mu_o \to \mu$ were made in the final results. Because our goal is to gain deeper understanding of the effects of the material on the Alternate-power and Alternate-energy distributions, we forego that simplification.

The Lorenz-gauge condition requires that

$$\omega C_0 - c^2\alpha C_1 - c^2 k C_3 = 0$$

$$\omega C_0' + c^2\alpha C_1' - c^2 k C_3' = 0$$

and from Eqs. (1.9) and (1.10) that $E_o = -\omega C_2$ and $B_o = -kC_2$. Continuity of **A** and Φ at $|x| = d/2$ requires that

$$C_0 = \frac{-E_o \exp(-\alpha \frac{d}{2})}{\alpha}, \qquad C_0' = C_0 \sinh(\alpha \tfrac{d}{2})$$

$$C_1 = \frac{E_o \exp(-\alpha \frac{d}{2})}{\omega}, \qquad C_1' = -C_1 \sinh(\alpha \tfrac{d}{2})$$

$$C_3 = \frac{-E_o k \exp(-\alpha \frac{d}{2})}{\alpha\omega}, \qquad C_3' = C_3 \sinh(\alpha \tfrac{d}{2})$$

with

$$\alpha = \omega\sqrt{\mu\varepsilon - \mu_o\varepsilon_o} = k\sqrt{1 - \frac{\mu_o}{\mu}\frac{\varepsilon_o}{\varepsilon}} \tag{18.34}$$

It is interesting (but complicating) that **A** and Φ have transverse spatial dependences that are frequency-dependent—despite the fact that **E** and **B** do not. This is a consequence of using the free-space Lorenz gauge.

The Maxwell–Poynting–Minkowski and Alternate-Minkowski representations are obvious choices to use. Both are derived in Part II, Chapter 9, Section 9.3 and summarized in Appendix D, Section D.2. The modified Alternate power flux and energy density (expressed in terms of **P** and **M**) are found by substituting the results given above into Eqs. (9.11b) and (9.11c). $\mathbf{S}^{ou}$ and W^{ou} can then be integrated over x and a unit width along y to find the total power and the energy per unit length along z. The result of the power flux integration is

$$P_1 = \iint_{\text{plates}} \Phi K_{uz}\, dz dy = \sqrt{\frac{\varepsilon}{\mu}} E_o^2 \frac{2\sinh(\alpha\frac{d}{2})\exp(-\alpha\frac{d}{2})}{\alpha}\cos^2(\omega t - kz)$$

$$P_2 = \iiint_{\text{material}} S_z^{ou} dx\, dy\, dz$$

$$= \sqrt{\frac{\varepsilon}{\mu}} E_o^2 \left[\frac{d}{2} - \frac{\sinh(\alpha\frac{d}{2})\exp(-\alpha\frac{d}{2})}{\alpha} - \frac{\frac{\mu}{\mu_o}\sinh^2(\alpha\frac{d}{2})\exp(-\alpha d)}{\alpha}\right]$$

$$P_3 = \iiint_{\text{air}} S_z^{ou} dx\, dy\, dz = \sqrt{\frac{\varepsilon}{\mu}} E_o^2 \frac{\frac{\mu}{\mu_o}\sinh^2(\alpha\frac{d}{2})\exp(-\alpha d)}{\alpha}$$

When added, the total time-averaged power is $\frac{1}{2}\sqrt{\frac{\varepsilon}{\mu}} E_o^2 d$ as expected. The energy contributions corresponding to these powers are found to be

$$E_1 = \iint_{\text{plates}} \frac{1}{2}(\Phi\sigma_{su} + A_z K_{uz})\, dydz$$

$$= \varepsilon E_o^2 \frac{2\sinh(\alpha\frac{d}{2})\exp(-\alpha\frac{d}{2})}{\alpha}\cos^2(\omega t - kz)$$

$$E_{2a} = \varepsilon_o E_o^2 \left[\frac{d}{2} - \frac{\sinh(\alpha \frac{d}{2}) \exp(-\alpha \frac{d}{2})}{\alpha} - \frac{\sinh^2(\alpha \frac{d}{2}) \exp(-\alpha d)}{\alpha} \right]$$

$$E_{2b} = \iiint_{\text{material}} \frac{1}{2} \left(\mathbf{A} \cdot \frac{\partial \mathbf{P}^o}{\partial t} - \frac{\partial \mathbf{A}}{\partial t} \cdot \mathbf{P}^o \right) dx\, dy\, dz$$

$$= (\varepsilon - \varepsilon_o) E_o^2 \left[\frac{d}{2} - \frac{\sinh(\alpha \frac{d}{2}) \exp(-\alpha \frac{d}{2})}{\alpha} \right]$$

$$E_3 = \varepsilon_o E_o^2 \frac{\sinh^2(\alpha \frac{d}{2}) \exp(-\alpha d)}{\alpha}$$

The sum of the time-averaged terms gives $\frac{1}{2}\varepsilon E_o^2 d$, again the expected value.

These results are clarified somewhat by plotting normalized energies and powers as a function of kd; the results are shown in Figure 18.4.

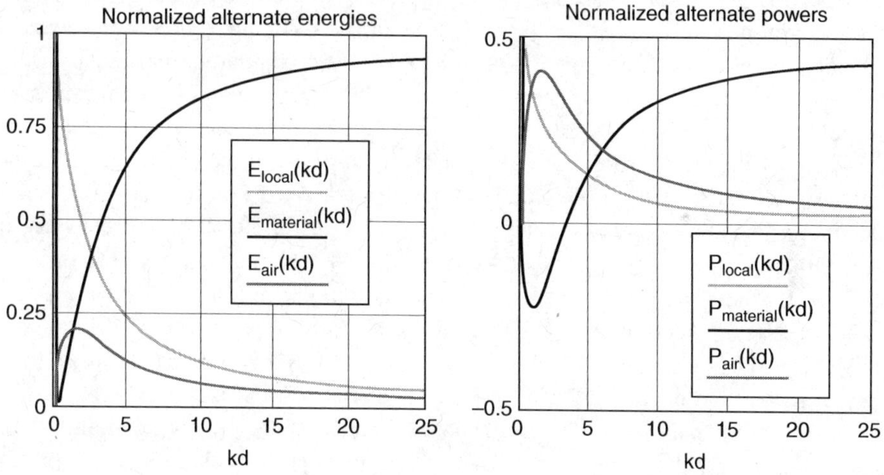

Figure 18.4 Parallel-plate transmission line ($\varepsilon = 9\varepsilon_o$, $\mu = 2\mu_o$).

Despite providing the correct results, the complexity of this representation—especially in comparison with the standard Poynting model—would certainly seem to discourage its use. However, consider the case where $\mu = \mu_o$ and $\varepsilon = \varepsilon_o$. Then $\alpha = 0$, $P_l = c\, E_l = \sqrt{\frac{\varepsilon_o}{\mu_o}} E_o^2 d \cos^2(\omega t - kz)$, and $P_2 = P_3 = 0$. In addition, $E_{2a} = E_{2b} = E_3 = 0$ and localization of power and energy to the surface currents and charges is complete. This is in accord with the general analysis of free-space TEM lines given earlier. Now restore the material, but assume that $0 < \alpha d \ll 2$. In this case,

$$P_l = \frac{E_l}{\sqrt{\mu\varepsilon}} \approx \left(1 - \frac{\alpha d}{2}\right) \sqrt{\frac{\varepsilon}{\mu}} E_o^2 d \cos^2(\omega t - kz) \tag{18.35}$$

and negligible contributions are distributed within the material and *outside* of the plates. In such situations, the complexity is avoided since $\Phi \mathbf{J}_u$ and $\frac{1}{2}(\Phi \rho_u + \mathbf{A} \cdot \mathbf{J}_u)$are easily calculated and by themselves give a very satisfactory approximation.

Although the small αd limit is the usual case, additional understanding is gained by considering large values. Then, for $\alpha d \gg 2$ we have

$$P_1 = \frac{E_1}{\sqrt{\mu\varepsilon}} = \frac{1}{\alpha}\sqrt{\frac{\varepsilon}{\mu}}E_o^2\cos^2(\omega t - kz)$$

$$P_2 = \frac{1}{2}\sqrt{\frac{\varepsilon}{\mu}}E_o^2\left(d - \frac{1}{\alpha}(1 + \frac{\mu}{2\mu_o})\right)$$

$$P_3 = \frac{1}{4\alpha}\frac{\mu}{\mu_o}\sqrt{\frac{\varepsilon}{\mu}}E_o^2$$

$$E_{2a} + E_{2b} = \frac{1}{2}\varepsilon E_o^2\left(d - \frac{1}{\alpha}\right) - \frac{1}{4\alpha}\varepsilon_o E_o^2$$

$$E_3 = \frac{1}{4\alpha}\varepsilon_o E_o^2$$

and it is seen that the circuit power P_1(including the second-harmonic power) vanishes as $1/\alpha$ along with P_3, the component *outside* of the plates. In contrast, P_2, the component associated with the material loading, eventually increases[9] until all of the power resides in the material polarization and/or magnetization currents. Similar effects occur with the corresponding energies. The extinction of P_1 and E_1 mirrors that of Φ. In summary, when αd is very small, the circuit power and energy components are maximum on the plates and negligible everywhere else. As αd increases, the energy components within the material and *outside* of the plate region both increase at the expense of the circuit term. Further increase of αd causes the outside contributions to peak (at $\alpha d = .628$) and then begin to decline. Eventually, the entire energy has shifted to the material region between the plates. Although $\frac{P_1}{E_1} = \frac{1}{\sqrt{\mu\,\varepsilon}}$ and $\frac{P_3}{E_3} = \frac{\sqrt{\mu\,\varepsilon}}{\mu_o\varepsilon_o}$, the material power and energy have different frequency dependencies; when $\mu \gg \mu_o$, $P_2(\alpha d)$ is initially negative for small αd, and then it reverses sign and finally approaches the positive asymptote where $\frac{P_2}{E_2} \rightarrow \frac{1}{\sqrt{\mu\,\varepsilon}}$.

If free-space replaces the material loading, there are no transverse currents to "attract" Alternate power and energy and $\alpha = 0$. When material loading is present, competition occurs between the polarization/magnetization currents and the unpaired current, with **A** and Φ being scattered by the former. The surface to effective volume ratio favors the unpaired or circuit current when *kd* (hence αd) is small; this is generally true in the quasistatic regime. Finally, we again observe that because there are no **E** or **H** fields for $|x| > d/2$, the material parameters outside of the plate region are inconsequential and would normally be chosen so as to trivialize the power calculations. (If the artifice of assuming that the same μ and ε exists outside the plates is made, and the gauge is altered to $\nabla \cdot \mathbf{A} + \mu\varepsilon\frac{\partial\Phi}{\partial t} = 0$, then $\alpha = 0$ and $P_1 = \frac{1}{\sqrt{\mu\,\varepsilon}}E_1 = \sqrt{\frac{\varepsilon}{\mu}}E_o^2 d\cos^2(\omega t - kz)$ are rigorous. In that case, there is no Alternate power or Alternate energy inside the material (or outside the plates) regardless of the magnitude of *kd*. This is clearly the analysis strategy of choice. Notice that this result is consistent with the complex Alternate-power theorem derived in Part I, Chapter 6, Eq. (6.10) when $\underline{g} = 0$ in the material region but free-space (or arbitrary μ and ε) exists outside the plate region. The polarization and magnetization currents have disappeared into the linear permittivity and permeability.

[9]For $\mu \gg \mu_o$, $P_2(\alpha d)$ is initially negative for small αd, and then it reverses sign and finally approaches the positive asymptote as noted. With respect to $P_3(\alpha d)$, the functional dependence is not altered but the amplitude is scaled by the factor μ/μ_o.

CHAPTER 19

RECTANGULAR WAVEGUIDE MODES

19.1 INTRODUCTION

The analysis of waveguide modes in a uniform rectangular guide (shown in Figure 19.1), with thin perfectly conducting walls located at $0 \leq x \leq a$, $y = 0, b$ and $0 \leq y \leq b$, $x = 0, a$ can be accomplished by the artifice that employs the superposition of uniform plane waves throughout all space to produce appropriate standing wave periodicity in both the x and y directions and simple propagation along the waveguide axis, z.

If the nulls of the standing wave pattern are placed properly (where both the tangential E fields and normal H fields vanish), very thin perfectly conducting planes can be positioned at $x = 0, \pm a, \pm 2a, \ldots$ and at $y = 0, \pm b, \pm 2b, \ldots$ without altering the fields. Although there are now electric surface currents, **K**, and surface charges, σ_s, of opposite polarity on the two sides of each conducting plane (and thus no net current or charge), the **E** and **H** fields inside each rectangular waveguide cell of cross-sectional area $a \times b$ are isolated from all others. It follows that if all currents and charges are extinguished, except those residing on the inner walls of the guide, then **E** and **H** outside the guide will also vanish. Although $\mathbf{A} = 0$ and $\Phi = 0$ are Lorenz-gauge potentials that appear to be consistent with these outer fields, the required continuity of **A** through the boundary may create non-periodic fringing of the potentials (including Φ)—both inside and outside the waveguide. These components cannot modify $\mathbf{E} \times \mathbf{H}$, but can alter the distribution of the

The Power and Beauty of Electromagnetic Fields, First Edition. F. R. Morgenthaler.

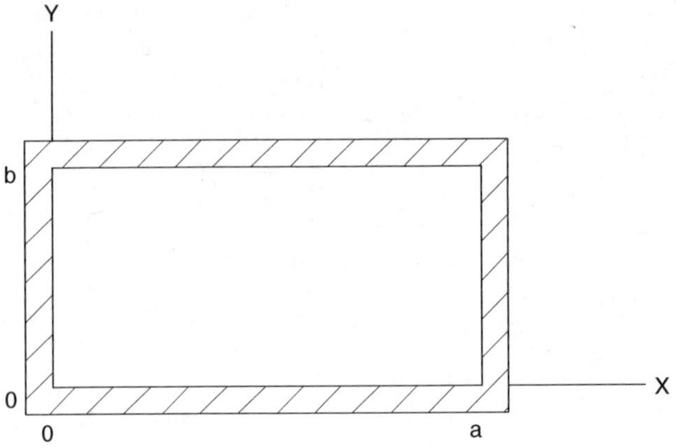

Figure 19.1 Rectangular waveguide geometry.

Alternate-power flux, $\mathbf{S}^{\mathrm{o}}$. In addition, nonzero values of Φ on the boundary may result in localized Alternate-power flow, $\Phi\mathbf{K}$. The spatial average over the waveguide may be affected by the modifications and components of Alternate power may exist *outside* the waveguide. Nevertheless, when integrated over the entire transverse plane, the total Alternate power must agree with the time average of the integrated Poynting-flux.

The surface currents and charges of any particular mode can easily be calculated; they provide the source terms of the inhomogeneous wave equations that govern the complete $\mathbf{A}$ and Φ. For one class of modes (*TM*), the outer potentials are zero and there is no fringing; for another class (*TE*) this is not the case. Because of the rectangular shape, the calculation of the nonzero fringing components for a single waveguide is difficult and will not be attempted here (large arrays of identical waveguides are easily treated and discussed later in this chapter); those for a circular waveguide are carried out in Chapter 20 and fully illustrate the issues involved.

19.2 PERIODIC POTENTIALS AND FIELDS

The electric and magnetic fields of a rectangular waveguide are confined to the waveguide and are the superposition of *TE* (transverse electric, $E_z = 0$) and *TM* (transverse magnetic, $H_z = 0$) modes. Transverse-periodic vector and scalar potentials are suitable for generating these fields. Inside of the guide, the potentials that govern either type of mode are of the form

$$\mathbf{A} = \left(\begin{bmatrix} \hat{\mathbf{x}} A_1 \cos(k_x x)\sin(k_y y) \\ +\hat{\mathbf{y}} A_2 \sin(k_x x)\cos(k_y y) \end{bmatrix} \cos(\omega t - k_z z) \\ + \hat{\mathbf{z}} A_3 \sin(k_x x)\sin(k_y y)\sin(\omega t - k_z z) \right) \tag{19.1a}$$

$$\Phi = A_0 \sin(k_x x)\sin(k_y y)\sin(\omega t - k_z z) \tag{19.1b}$$

where $\omega^2 = c^2(k_x^2 + k_y^2 + k_z^2)$. The Lorenz gauge imposes the condition

$$\frac{\omega}{c^2} A_0 = k_x A_1 + k_y A_2 + k_z A_3 \tag{19.2}$$

The associated electric and magnetic fields are

$$\mathbf{E} = \left(\begin{bmatrix} \widehat{\mathbf{x}}(\omega A_1 - k_x A_0)\cos(k_x x)\sin(k_y y) \\ +\widehat{\mathbf{y}}(\omega A_2 - k_y A_0)\sin(k_x x)\cos(k_y y) \end{bmatrix} \sin(\omega t - k_z z) \atop - \widehat{\mathbf{z}}(\omega A_3 - k_z A_0)\sin(k_x x)\sin(k_y y)\cos(\omega t - k_z z) \right) \tag{19.3a}$$

$$\mu_0 \mathbf{H} = \left(\begin{bmatrix} \widehat{\mathbf{x}}(k_y A_3 - k_z A_2)\sin(k_x x)\cos(k_y y) \\ +\widehat{\mathbf{y}}(k_z A_1 - k_x A_3)\cos(k_x x)\sin(k_y y) \end{bmatrix} \sin(\omega t - k_z z) \atop + \widehat{\mathbf{z}}(k_x A_2 - k_y A_1)\cos(k_x x)\cos(k_y y)\cos(\omega t - k_z z) \right) \tag{19.3b}$$

Since these fields are periodic in both x and y, it follows that, provided that we have $k_x = n\pi/a$ and $k_y = m\pi/b$ with n, m positive integers, the tangential **E** fields and normal **H** fields vanish where required. Notice that the spatially periodic components of both the scalar potential and the tangential vector potential vanish along the boundary. The modes are labeled as TE_{nm} or TM_{nm}.

19.3 WAVEGUIDE DISPERSION

Because

$$\omega^2 = c^2(k_x^2 + k_y^2 + k_z^2) \tag{19.4a}$$

all rectangular waveguide modes have radian frequencies, $\omega = 2\pi f = ck = 2\pi c/\lambda_0$, that are related to the guide wavelength, $\lambda_g = 2\pi/k_z$, by

$$\lambda_g = \frac{\lambda_0}{\sqrt{1 - (\frac{\lambda_0}{\lambda_c})^2}} \tag{19.4b}$$

where it is convenient to define a cutoff wavelength, λ_c, by

$$\left(\frac{1}{\lambda_c}\right)^2 = \left(\frac{n}{2a}\right)^2 + \left(\frac{m}{2b}\right)^2 \tag{19.4c}$$

and the corresponding cutoff frequency by $f_c = c/\lambda_c$. For $f > f_c$, the value of λ_g is real and the mode in question propagates; otherwise the mode is below cutoff (λ_g is imaginary) and the evanescent fields decay exponentially, as expected from the analysis of nonuniform plane waves. Although average power cannot be carried by a single cutoff mode, it can be carried by a pair of such modes having (imaginary) k_z values of opposite sign. The propagating mode with the smallest value of f_c is termed the dominant mode; if it is nondegenerate, there is a frequency range between $(f_c)_{\min}$ and the next highest cutoff frequency where only a single mode can propagate. Both *TE* and *TM* modes must be considered. The superscript □ is sometimes used to label rectangular waveguide modes.

The phase and group velocities of a given mode are both z-directed and easily calculated from

$$v_{\text{phase}} = \frac{\omega}{k_z} = \frac{c}{\sqrt{1 - (\frac{\lambda_0}{\lambda_c})^2}} \tag{19.5a}$$

and

$$v_{\text{group}} = \frac{\partial \omega}{\partial k_z} = c^2 \frac{k_z}{\omega} = c\sqrt{1 - \left(\frac{\lambda_o}{\lambda_c}\right)^2} \tag{19.5b}$$

It follows that for any mode,

$$v_{\text{phase}} v_{\text{group}} = c^2 \tag{19.5c}$$

Although the phase velocity of any propagating mode is greater than c, special relativity is not contradicted because signals carrying information propagate at the group velocity; the magnitude of the latter is always less than or equal to the velocity of light.

19.4 TE_{NM} MODES

For TE modes ($E_z = 0$), for which $n \geq 1$, $m \geq 1$, the constants satisfy

$$A_3 = \frac{k_x A_1 + k_y A_2}{k_x^2 + k_y^2} k_z \tag{19.6a}$$

$$A_0 = \frac{k_x A_1 + k_y A_2}{k_x^2 + k_y^2} \omega \tag{19.6b}$$

When $n = 0$, $m \geq 1$ (or $n \geq 1$, $m = 0$), the amplitudes of **E** and **H** depend only on A_1 (or A_2); in either case, the remaining constants satisfy Eq. (19.2). The dominant mode is either $TE^{\square}_{01}$ with $\lambda_c = 2b$ or $TE^{\square}_{10}$ with $\lambda_c = 2a$; if $a = 2b$, the $TE^{\square}_{10}$ mode is dominant and the next modes $TE^{\square}_{20}$ and $TE^{\square}_{01}$ both have the same cutoff wavelength, $\lambda_c = a = 2b$.

The electric (solid lines) and the magnetic (dashed lines) fields of the $TE^{\square}_{10}$ mode are sketched in Figure 19.2[1]; the electric surface currents in Figure 19.3. The fields of the $TE^{\square}_{20}$ and $TE^{\square}_{01}$ modes are shown in Figure 19.4.

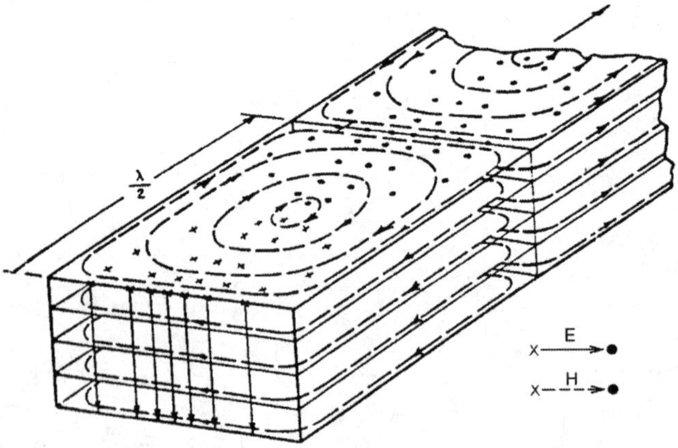

Figure 19.2 $TE^{\square}_{10}$ mode: electric and magnetic fields.

[1]The author is grateful to Prof. Frank Reintjes for permission to use figures of the electric and magnetic fields of both rectangular and circular waveguide modes taken from **Principles of Radar** by Reintjes and Coate (McGraw-Hill, 1994) (based upon notes used at the M.I.T. Radar School during World War II). Although drawn by hand, they convey the essential features of TE_{nm} and TM_{nm} modes very clearly, are of historic interest, and equal or surpass many published illustrations produced by current computer graphics programs.

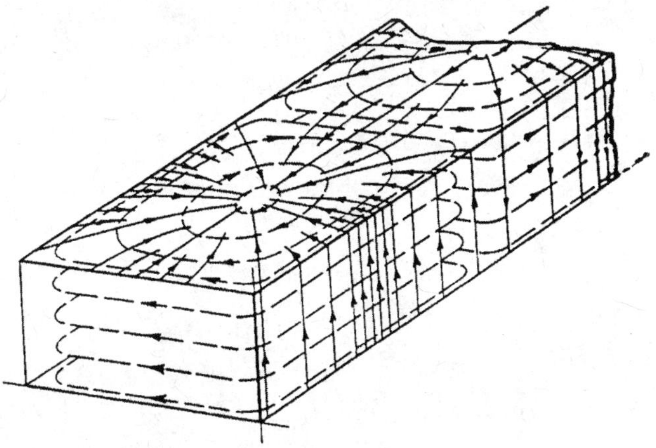

Figure 19.3 $TE_{10}^{\square}$ mode: electric surface currents.

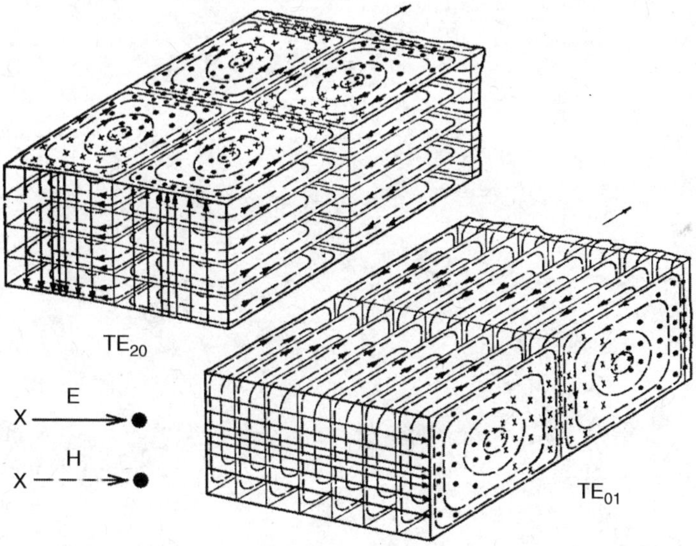

Figure 19.4 $TE_{20}^{\square}$ and $TE_{01}^{\square}$ modes: electric and magnetic fields.

TE power fluxes

The time-independent power flux, $\mathbf{S}^{o}$, has only a z component with expected standing wave character along x and y.

$$S_z^{o} = \frac{k_z \omega}{2\mu_o(k_x^2 + k_y^2)} \left(\begin{array}{c} (k_x^2 + k_y^2) \begin{bmatrix} A_1^2 \cos^2(k_x x)\sin^2(k_y y) \\ +A_2^2 \sin^2(k_x x)\cos^2(k_y y) \end{bmatrix} \\ -\ (A_1 k_x + A_2 k_y)^2 \sin^2(k_x x)\sin^2(k_y y) \end{array} \right) = W^{o} c \frac{k_z}{k} \qquad (19.7a)$$

where W^{o} is the Alternate energy density.

The corresponding component of $\mathbf{E} \times \mathbf{H}$ is

$$S_z = \frac{k_z \omega (A_1 k_y - A_2 k_x)^2}{\mu_o (k_x^2 + k_y^2)^2} \begin{bmatrix} k_x^2 \sin^2(k_x x) \cos^2(k_y y) \\ + k_y^2 \cos^2 k_x x \sin^2 k_y y \end{bmatrix} \sin^2(\omega t - k_z z) \neq W^{em} c \frac{k_z}{k} \quad (19.7b)$$

When integrated over the waveguide cross section, the time-averaged Maxwell–Poynting densities satisfy $\int <S_z> da = ck_z/k \int <W^{em}> da$.

Without specification of A_1 and A_2, the spatial distribution of S_z^o is not unique. Nevertheless, when the integral is taken over the guide cross section, the resulting time-averaged waveguide power,

$$TE \text{ power} = \frac{k_z \omega (A_1 k_y - A_2 k_x)^2}{8\mu_o (k_x^2 + k_y^2)} ab \quad (19.8)$$

is uniquely expressed in terms of the amplitude of the longitudinal magnetic field,

$$\mu_o H_z = (A_2 k_x - A_1 k_y) \cos(k_x x) \cos(k_y y) \cos(\omega t - k_z z) \quad (19.9)$$

19.5 *TM*$_{NM}$ MODES

For *TM* modes ($H_z = 0$), the constants satisfy

$$A_2 = \frac{k_y}{k_x} A_1 \quad (19.10a)$$

$$A_0 = \frac{c^2 \left[A_1 (k_x^2 + k_y^2) + A_3 k_x k_z \right]}{k_x \omega} \quad (19.10b)$$

and $n \geq 1$, $m \geq 1$. The dominant *TM* mode is $TM_{11}^{\square}$ with $\lambda_c = \frac{2ab}{\sqrt{a^2+b^2}}$ (If $a = 2b$, the value is $\frac{2a}{\sqrt{5}}$.) However, both the $TE_{10}^{\square}$ and $TE_{01}^{\square}$ modes have lower cutoff frequencies. The electric and the magnetic fields of the $TE_{11}^{\square}$ and $TM_{11}^{\square}$ modes are compared in Figure 19.5.

TM power fluxes

The time-independent power flux, $\mathbf{S}^o$, also has only a z-component with expected standing wave character along x and y. With $k^2 = k_x^2 + k_y^2 + k_z^2$, we have

$$S_z^o = \frac{c^2 k_z}{2\omega \mu_o k_x^2} \begin{pmatrix} A_1^2 k^2 \left[k_x^2 \cos^2(k_x x) \sin^2(k_y y) + k_y^2 \sin^2(k_x x) \cos^2(k_y y) \right] \\ + (k_x^2 + k_y^2) \left[(A_1 k_z - A_3 k_x)^2 - A_1^2 k^2 \right] \sin^2(k_x x) \sin^2(k_y y) \end{pmatrix}$$

$$= W^o c \frac{k_z}{k} \quad (19.11a)$$

where W^o is the Alternate-energy density. The corresponding component of $\mathbf{E} \times \mathbf{H}$ is

$$S_z = \frac{c^2 k_z (A_1 k_z - A_3 k_x)^2}{\omega \mu_o k_x^2} \begin{bmatrix} k_x^2 \cos^2(k_x x) \sin^2(k_y y) \\ + k_y^2 \sin^2(k_x x) \cos^2(k_y y) \end{bmatrix} \sin^2(\omega t - k_z z) \neq W^{em} c \frac{k_z}{k} \quad (19.11b)$$

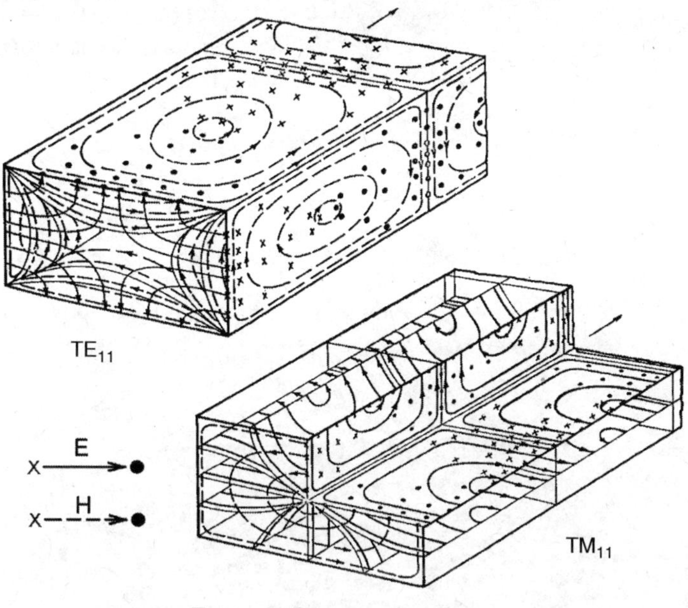

Figure 19.5 $TE^{\square}_{11}$ and $TM^{\square}_{11}$ modes: electric and magnetic fields.

As before, the integrals of the Maxwell–Poynting densities satisfy

$$\int <S_z> da = ck_z/k \int <W^{\text{em}}> da.$$

As in the case of *TE* modes, without further specification of either A_1 or A_3, the spatial distribution of S_z^{o} is not unique. Again, when the integral is taken over the guide cross section, the resulting time-averaged waveguide power,

$$TM\,\text{Power} = \frac{c^2(k_x^2 + k_y^2)k_z(A_1k_z - A_3k_x)^2}{8\mu_0\omega\, k_x^2}ab \qquad (19.12)$$

is uniquely expressed in terms of the amplitude of the longitudinal electric field,

$$E_z = \frac{c^2(k_x^2 + k_y^2)(A_1k_z - A_3k_x)}{\omega k_x}\sin(k_x x)\sin(k_y y)\cos(\omega t - k_z z) \qquad (19.13)$$

If, as is often the case, calculation of the fields and the time-averaged power is all that is required, the analysis can end here; however, we seek a deeper understanding and thus continue.

19.6 NULL ALTERNATE-POWER AND ALTERNATE-ENERGY DISTRIBUTIONS

Consider *TE* modes with $A_1k_y = A_2k_x$ (or *TM* modes with $A_1k_z = A_3k_x$). Then **E**, **H**, **S** and the surface charges and currents on the walls *all* vanish[2] so that the distinction

[2]Null modes can be generated by choosing $\mathcal{A} = \square\Psi$, where $\square^2\Psi = 0$. In our example, $\Psi = (\phi_0/\omega)\sin(k_x x)\sin(k_y y)\cos(\omega t - k_z z)$.

between *TE* and *TM* is lost. Interestingly, unless $k_x k_y = 0$, the Alternate-flux vector, $\mathbf{S}^{\text{o}}$, *persists*. The result is both a null power flux, $\mathbf{S}^{\text{o}}_{\text{null}}$, and a null energy density, $W^{\text{o}}_{\text{null}}$, given by

$$\mathbf{S}^{\text{o}}_{\text{null}} = \hat{\mathbf{z}}\frac{ck_z}{k} W^{\text{o}}_{\text{null}} \tag{19.14a}$$

$$W^{\text{o}}_{\text{null}} = \frac{A_1 A_2 k^2}{2\mu_{\text{o}} k_x k_y}[k_x^2 \cos(2k_x x)\sin^2(k_y y) + k_y^2 \sin^2(k_x x)\cos(2k_y y)] \tag{19.14b}$$

Evidently, null energy travels at the group velocity to produce null power. A portion of this function is shown in Figure 19.6 for the case, $k_x = k_y$.

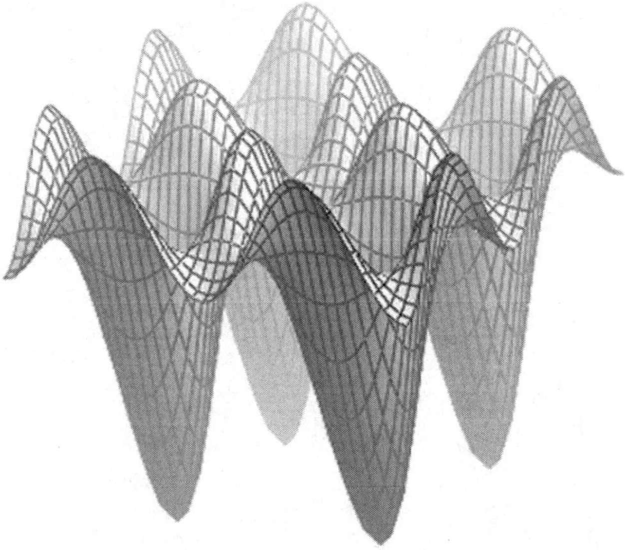

Figure 19.6 $\text{Null}^{\square}_{11}$ mode ($a = b$): $S^{\text{o}}_z(x, y)$ or $W^{\text{o}}(x, y)$.

Even though the integral of $W^{\text{o}}_{\text{null}}$ over the guide cross section vanishes so that the *total* null power and energy have the correct values of zero, there must be sources, generating **A** and Φ, located somewhere *outside* of the guide. If the potentials of Eqs. (19.1a) and (19.1b) are assumed to extend throughout all space, those sources are hidden at infinity.

19.7 UNIQUENESS RESOLVED

If the *complete* **A** and Φ for a particular $TE^{\square}_{nm}$ or $TM^{\square}_{nm}$ mode of unit average power were available, the four constants (A_1, A_2, A_3, and A_0) that determine the periodic components would be known, along with any fringing components. This is trivial in the case of $TM^{\square}_{nm}$ modes because the set of constants

$$A_1 = A_2 = 0 \tag{19.15a}$$

$$A_0 = \frac{c^2 k_z}{\omega} A_3 \tag{19.15b}$$

is consistent with the external potentials: $\mathbf{A} = 0$, $\Phi = 0$. Although it is tedious to calculate these constants for the $TE_{nm}^{\square}$ modes of a single rectangular waveguide, it is easy to do so for certain large arrays of identical waveguides.

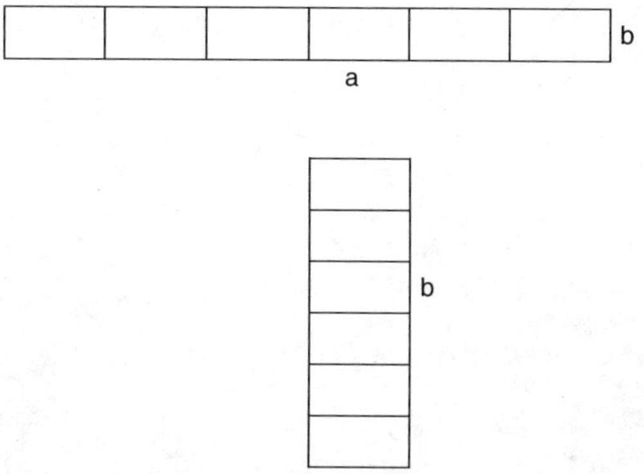

Figure 19.7 Horizontally and vertically stacked waveguides.

As shown schematically in Figure 19.7, consider $2N$ identical guides placed either side by side along the x axis so as to occupy the space $-Na \leq x \leq Na$ and $0 \leq y \leq b$ or stacked vertically along the y axis so as to occupy the space $0 \leq x \leq a$ and $-Nb \leq y \leq Nb$. Further assume that $N \gg 1$ and that the fields within each waveguide are all coherent versions of the same mode, are of equal amplitude, and are phased so that Eqs. (19.1a) and (19.1b) apply. In the first configuration (except for the end walls at $x = \pm Na$), all the currents parallel to the y axis cancel, whereas those parallel to either the x or z axes do not; there are sources for A_x and A_z, but none for A_y. In this case, when $m > 0$, $A_2 = 0$[3] and Eq. (19.7a) reduces to

$$S_z^{\text{o}}(TE_{nm}) = \frac{A_1^2 k_z \omega}{2\mu_{\text{o}}} \sin^2\left(\frac{m\pi}{b}y\right)\left[\cos^2\left(\frac{n\pi}{a}x\right) - \frac{(nb)^2}{(nb)^2 + (ma)^2}\sin^2\left(\frac{n\pi}{a}x\right)\right] \tag{19.16a}$$

In the second configuration (except for the end walls at $y = \pm Nb$), all the currents parallel to the x axis cancel, whereas those parallel to either the y or z axes do not. There are sources for A_y and A_z, but none for A_x. In this case, when $n > 0$, $A_1 = 0$ and Eq. (19.7a) reduces to

$$S_z^{\text{o}}(TE_{nm}) = \frac{A_2^2 k_z \omega}{2\mu_{\text{o}}} \sin^2\left(\frac{n\pi}{a}x\right)\left[\cos^2\left(\frac{m\pi}{b}y\right) - \frac{(ma)^2}{(nb)^2 + (ma)^2}\sin^2\left(\frac{m\pi}{b}y\right)\right] \tag{19.16b}$$

In order that these configurations carry equal total powers, the values of A_1 and A_2 must satisfy $A_1 am = A_2 bn$.

The distribution (but not the total) of Alternate TE power within a single waveguide that is part of the array is therefore seen to be affected by its neighbors! This occurs (even when $a = b$) because, unlike the electric and magnetic fields, $\mathbf{A}$ and Φ are not shielded by the conducting walls.

[3]For the horizontally stacked configuration, the neglect of end currents, K_y and setting $A_y \to 0$, is a reasonable approximation only if $m > 0$; then $E_y = -\frac{\partial \Phi}{\partial y}$ has a source. For the vertically stacked configuration, neglect of the end currents, K_x and setting $A_x \to 0$, requires $n > 0$.

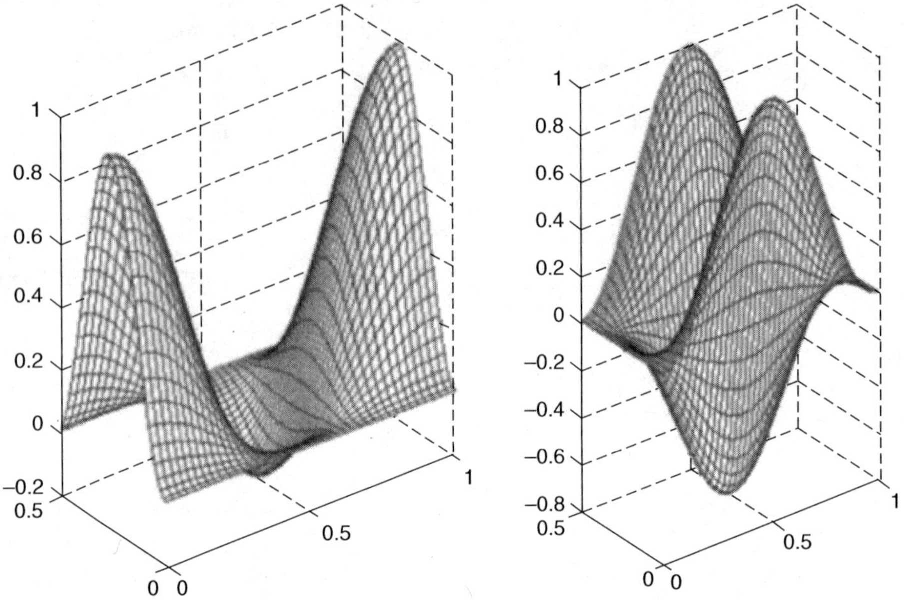

(a) Horizonatally stacked (b) Vertically stacked

Figure 19.8 $TE_{11}^{\square}$ mode Alternate flux, S_z^{o}, in stacked waveguides.

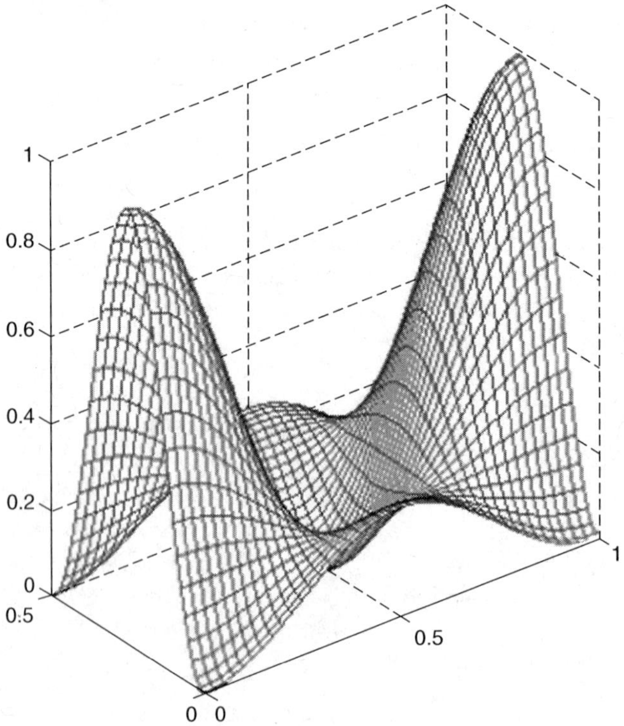

Figure 19.9 $TE_{11}^{\square}$ mode Poynting flux, $< S_z >$ (independent of stacking).

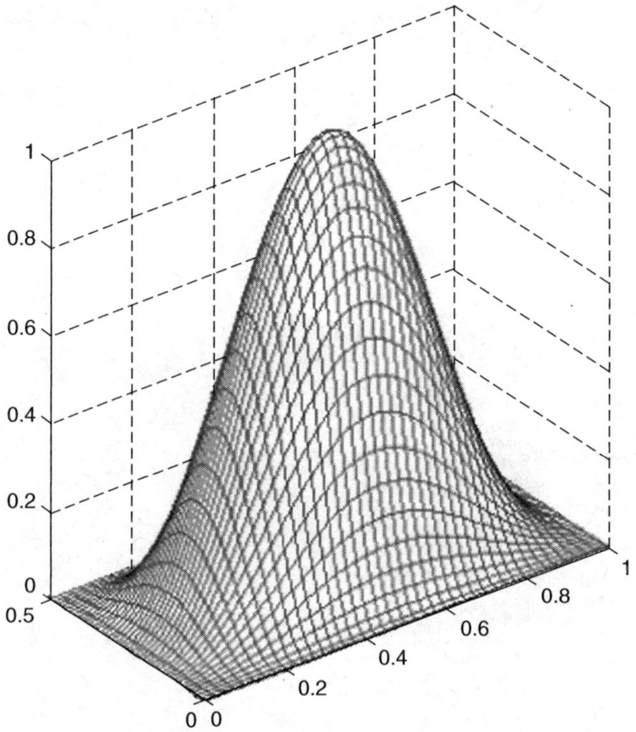

Figure 19.10 $TM_{11}^{\square}$ mode Alternate flux, $< S_z^o >$ (single waveguide).

Because for $TM_{nm}^{\square}$ modes, $A_1 = A_2 = 0$, Eq. (19.11a) is identical for both configurations (and a single waveguide):

$$S_z^o(TM_{nm}) = \frac{A_3^2 c^2 \pi^2 k_z}{2\omega\mu_o}\left[\left(\frac{n}{a}\right)^2 + \left(\frac{m}{b}\right)^2\right]\sin^2\left(\frac{n\pi}{a}x\right)\sin^2\left(\frac{m\pi}{b}y\right) \tag{19.17}$$

The Alternate and Poynting power-flux distributions are plotted and compared in Figures 19.8 and 19.9 for the TE_{11} modes and in Figures 19.10 and 19.11 for the TM_{11} modes. In all cases, $a = 2b$.

Because the potentials for these *TE* and *TM* modes vanish on the boundaries, $\mathbf{A} = 0$, $\Phi = 0$ outside of the guides satisfy the requirements of continuity. It follows that there is neither fringing nor localized Alternate power anywhere. Calculation of the Alternate-energy density for both configurations reveals that the energy velocity (group velocity) of any mode is $v_E = S_z^o / W^o = ck_z/k$.

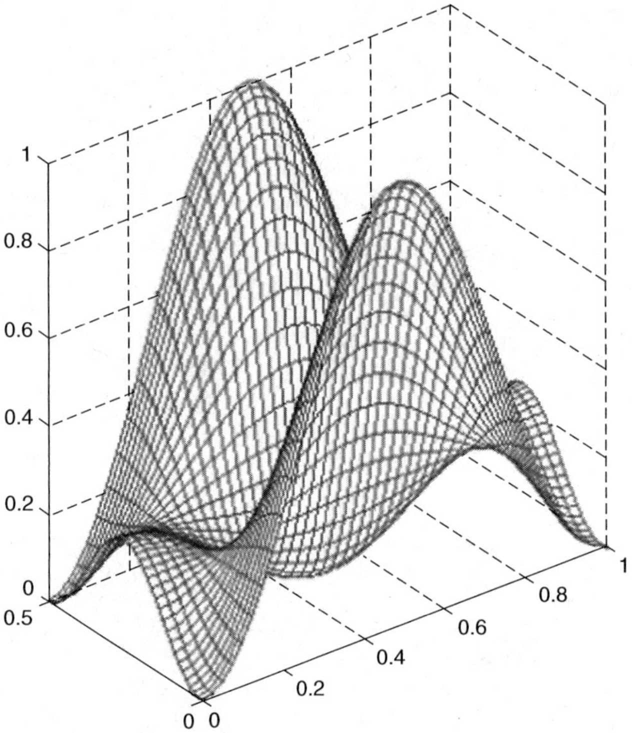

Figure 19.11 $TM_{11}^{\square}$ mode Poynting flux, $< S_z >$ (single waveguide).

Notice, too, that for either stacked configuration, those walls with no net charge and current can be removed, producing a single pair of plates separated either vertically by the distance b or horizontally by the distance a. The analysis is therefore applicable to parallel-plate waveguides.[4]

[4]Because the end walls of the resulting structure are still present even when $N \to \infty$, only *TE* and *TM* plate modes are included in this analysis. *TEM* plate modes are considered separately in Section 18.2.

CHAPTER 20

CIRCULAR WAVEGUIDE MODES

20.1 INTRODUCTION

The electric and magnetic fields of a uniform circular waveguide of radius R (shown in Figure 20.1) share many of the properties of those of a rectangular waveguide (Chapter 19). In general, a superposition of *TE* and *TM* modes, that satisfy the cylindrical wave equation, is confined within the guide; some are propagating modes, some are evanescent (cutoff). The analysis of these modes can be accomplished by first finding the magnetic vector and scalar electric potentials. Inside of the guide, the potentials that govern *either* a TE_{nm} or TM_{nm} mode (sometimes labeled with the superscript o) are now expressed in terms of cylinder (rather than trigonometric) functions of the form *TE* or *TM*:

$$\mathbf{A}_{\text{inner}} = \left(\begin{bmatrix} +\hat{\mathbf{r}} \left(C_1 \frac{n}{r} J_n(k_c r) + C_2 \frac{d}{dr} J_n(k_c r) \right) \left\{ \begin{smallmatrix} \sin \\ \cos \end{smallmatrix} n\phi \right\} \\ +\hat{\phi} \left(C_1 \frac{d}{dr} J_n(k_c r) + C_2 \frac{n}{r} J_n(k_c r) \right) \left\{ \begin{smallmatrix} \cos \\ -\sin \end{smallmatrix} n\phi \right\} \end{bmatrix} \cos(\omega t - k_z z) \atop +\hat{\mathbf{z}} \left(C_3 k_z J_n(k_c r) \right) \left\{ \begin{smallmatrix} \sin \\ \cos \end{smallmatrix} n\phi \right\} \sin(\omega t - k_z z) \right) \tag{20.1a}$$

$$\Phi_{\text{inner}} = C_0 \omega J_n(k_c r) \left\{ \begin{smallmatrix} \sin \\ \cos \end{smallmatrix} n\phi \right\} \sin(\omega t - k_z z) \tag{20.1b}$$

The Power and Beauty of Electromagnetic Fields, First Edition. F. R. Morgenthaler.

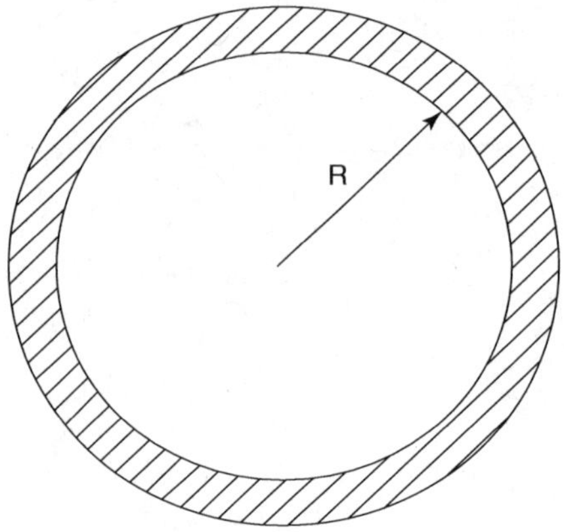

Figure 20.1 Circular waveguide geometry.

when the ϕ variation assumes either of two orthogonal standing wave patterns or

$$\mathbf{A}_{\text{inner}} = \begin{pmatrix} \hat{\mathbf{r}} \left[C_1 \dfrac{n}{r} J_n(k_c r) + C_2 \dfrac{d}{dr} J_n(k_c r) \right] \cos(\omega t - k_z z - n\phi) \\ + \hat{\phi} \left[C_1 \dfrac{d}{dr} J_n(k_c r) + C_2 \dfrac{n}{r} J_n(k_c r) \right] \sin(\omega t - k_z z - n\phi) \\ + \hat{\mathbf{z}}\, C_3 k_z J_n(k_c r) \sin(\omega t - k_z z - n\phi) \end{pmatrix} \tag{20.2a}$$

$$\Phi_{\text{inner}} = C_0 \omega J_n(k_c r) \sin(\omega t - k_z z - n\phi) \tag{20.2b}$$

when the ϕ variation is a wave circulating in an azimuthal direction (a linear combination of standing waves) that depends upon the sign of the integer n. Here, J_n is the Bessel function (of the first kind) of order n, $\omega^2 = c^2(k_c^2 + k_z^2)$, and $k_c R$ is a constant that depends upon both mode integers: n, m. Properties of these functions are reviewed in Appendix F, Section F.1. Although misleading because the direction of the transverse fields is actually a function of r, we shall refer to standing ϕ-wave modes with stationary nulls as "linearly polarized" and refer to the circulating ϕ-wave modes without nulls as "circularly polarized." Each type is a linear superposition of the other although, as with uniform plane waves, the general superposition produces "elliptically polarized" modes.

The Lorenz-gauge condition requires that

$$k_c^2 C_2 + k_z^2 C_3 = (k_c^2 + k_z^2) C_0 \tag{20.3}$$

The boundary conditions at the perfectly conducting wall require the tangential electric fields, E_ϕ and E_z, and the normal magnetic field, H_r, to vanish at $r = R$. In a consistent gauge, the scalar potential, Φ, and all components of $\mathbf{A}$ must be continuous.

Waveguide dispersion

Because

$$\omega^2 = c^2(k_c^2 + k_z^2) = c^2 k^2 \tag{20.4a}$$

all circular waveguide modes have radian frequencies, $\omega = 2\pi f = 2\pi c/\lambda_o$, that are related to the guide wavelength, $\lambda_g = 2\pi/k_z$, by

$$\lambda_g = \frac{\lambda_o}{\sqrt{1 - (\frac{\lambda_o}{\lambda_c})^2}} \tag{20.4b}$$

the same relation as for rectangular waveguide modes. The cutoff wavelength is defined by $\lambda_c = 2\pi/k_c$ where $k_c R$ is a constant that depends upon the specific mode. The corresponding cutoff frequency is $f_c = c/\lambda_c$. For $f > f_c$, the value of λ_g is real and the mode in question propagates; otherwise the mode is below cutoff (λ_g is imaginary) and the evanescent fields decay exponentially. As in a rectangular guide, that mode with the smallest value of f_c is termed the dominant mode; if it is non-degenerate, there is a frequency range between $(f_c)_{\min}$ and the next highest cutoff frequency where only a single mode can propagate. Both *TE* and *TM* modes must be considered.

The phase and group velocities of a given mode are both z-directed and easily calculated from Eqs. (19.5a) and (19.5b); their product by Eq. (19.5c).

20.2 *TM*$_{nm}$ MODES

For *TM* modes, the boundary conditions require that $J_n(k_c R) = 0$ and $C_1 = C_2 = 0$; the positive values of $k_c R$ that satisfy this equation are labeled u_{nm} starting with the smallest value, u_{n1}. Values for $n, m \leq 3$ are tabulated below.

u_{nm}	$m = 1$	$m = 2$	$m = 3$
u_{0m}	2.405	5.520	8.654
u_{1m}	3.832	7.016	10.174
u_{2m}	5.136	8.417	11.620
u_{3m}	6.380	9.761	13.015

These constraints force Φ and all components of $\mathbf{A}$ to vanish at $r = R$; therefore, assuming that the only currents are on the cylindrical wall, the potentials outside of the guide are everywhere zero and the complete solution is known. From Eqs. (1.9) and (1.10), the electric and magnetic fields of the Eqs. (20.1) modes are

$$\mathbf{E} = -\omega C_0 \left(\begin{bmatrix} \hat{\mathbf{r}} \dfrac{d}{dr} J_n(k_c r) \left\{ \begin{smallmatrix} \sin \\ \cos \end{smallmatrix} n\phi \right\} \\ +\hat{\phi} \dfrac{n}{r} J_n(k_c r) \left\{ \begin{smallmatrix} \cos \\ -\sin \end{smallmatrix} n\phi \right\} \end{bmatrix} \sin(\omega t - k_z z) \atop + \hat{\mathbf{z}} \dfrac{k_c^2}{k_z} J_n(k_c r) \left\{ \begin{smallmatrix} \sin \\ \cos \end{smallmatrix} n\phi \right\} \cos(\omega t - k_z z) \right) \tag{20.5a}$$

and

$$\mathbf{H} = \frac{k_z C_3}{\mu_o} \begin{bmatrix} \hat{\mathbf{r}} \dfrac{n}{r} J_n(k_c r) \left\{ \begin{smallmatrix} \cos \\ -\sin \end{smallmatrix} n\phi \right\} \\ -\hat{\phi} \dfrac{d}{dr} J_n(k_c r) \left\{ \begin{smallmatrix} \sin \\ \cos \end{smallmatrix} n\phi \right\} \end{bmatrix} \sin(\omega t - k_z z) \tag{20.5b}$$

The surface charge, σ_s, and surface current, **K**, on the inner wall are found by applying the boundary conditions $\sigma_s = -\widehat{\mathbf{r}} \cdot \varepsilon_0 \mathbf{E}$ and $\mathbf{K} = -\widehat{\mathbf{r}} \times \mathbf{H}$. The result is

$$\sigma_s = \varepsilon_0 \omega C_0 \left.\frac{d}{dr} J_n(k_c r)\right|_{r=R} \left\{\begin{matrix}\sin\\ \cos\end{matrix}\, n\phi\right\} \sin(\omega t - k_z z) \tag{20.6a}$$

$$\mathbf{K} = -\widehat{\mathbf{z}} \frac{k_z C_3}{\mu_0} \left.\frac{d}{dr} J_n(k_c r)\right|_{r=R} \left\{\begin{matrix}\sin\\ \cos\end{matrix}\, n\phi\right\} \sin(\omega t - k_z z) \tag{20.6b}$$

In terms of these quantities and the unit-impulse function, the volume charge and current densities can be written as

$$\rho = \sigma_s u_0(r - R) \tag{20.7a}$$

$$\mathbf{J} = \mathbf{K}\; u_0(r - R) \tag{20.7b}$$

The dominant TM mode is $TM_{01}{}^{\mathrm{O}}$ with $k_c R = 2.405$; the electric (solid lines) and the magnetic (dashed lines) fields are sketched in Figure 20.2.

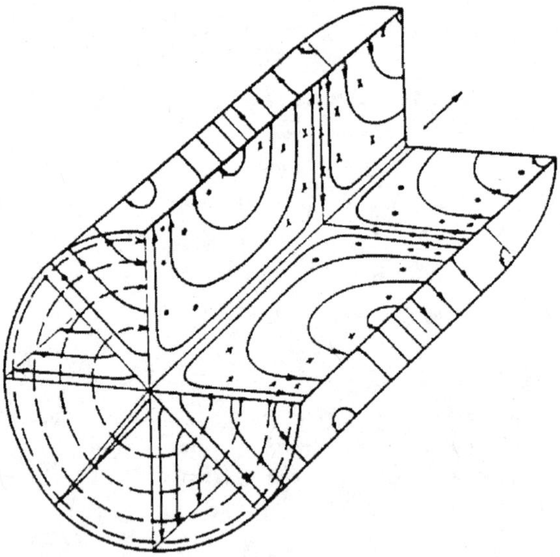

Figure 20.2 $TM_{01}{}^{\mathrm{O}}$ mode ($k_c R = 2.405$): electric and magnetic fields.

The mode index n is both the order of the Bessel function and a measure of the ϕ-dependence; it corresponds to the maximum number of thin perfectly conducting axial planes that could be inserted (at appropriate positions) without affecting the waveguide fields. The mode index m is the maximum number of thin perfectly conducting cylinders (including the guide itself) that could be inserted (at appropriate radii) without affecting the waveguide fields. The case of $n = m = 2$ is depicted in Figure 20.3: The division of the guide is shown in (a); the transverse electric-field pattern of the $TE_{22}{}^{\mathrm{O}}$ mode is shown in (b); the transverse magnetic-field pattern of the $TM_{22}{}^{\mathrm{O}}$ mode is shown in (c).

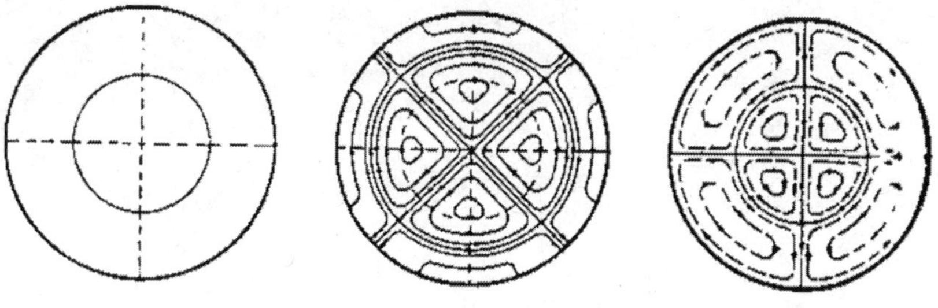

(a) (b) (c)

Figure 20.3 Circular-mode divisions ($n = m = 2$).

TM power fluxes

The Poynting flux and Alternate-power flux of these modes are

$$\mathbf{S} = -E_z H_\phi \widehat{\mathbf{r}} + E_z H_r \widehat{\phi} + \left(E_r H_\phi - E_\phi H_r\right) \widehat{\mathbf{z}}$$

$$\mathbf{S}^{\text{o}} = S_z^{\text{o}} \widehat{\mathbf{z}}$$

where the z components are

$$S_z = \frac{\omega k_z C_0 C_3}{\mu_{\text{o}}} \begin{pmatrix} \left[\dfrac{d}{dr} J_n(k_c r)\right]^2 \left\{ {\sin \atop \cos} n\phi \right\}^2 \\ + \left[\dfrac{n}{r} J_n(k_c r)\right]^2 \left\{ {\cos \atop \sin} n\phi \right\}^2 \end{pmatrix} \sin^2(\omega t - k_z z) \tag{20.8a}$$

and

$$S_z^{\text{o}} = W^{\text{o}} \frac{c k_z}{k} = \frac{\omega k_z C_0 C_3}{2\mu_{\text{o}}} k_c^2 J_n^2(k_c r) \left\{ {\sin \atop \cos} n\phi \right\}^2 \tag{20.8b}$$

For the general mode, the Poynting flux includes components of reactive power in all three directions; the time-averaged flux is, of course, purely z-directed. The Alternate-power flux is independent of time and has only an axial-component. The electromagnetic energy density, W^{em}, includes reactive energy; the Alternate-energy density, W^{o}, does not. Notice that because both **A** and Φ are zero on the waveguide wall, no localized Alternate-power or Alternate-energy contribution is present. The group velocity of any mode is simply $S_z^{\text{o}}/W^{\text{o}} = c k_z / k$.

Application of Appendix F, Eq. (F.12) reveals that the integrals of both the Alternate and the time-averaged Poynting flux, taken over the guide cross section, are equal to

$$P_z^{TM} = (1 + \delta_{n0}) \frac{\pi \omega k_z C_0 C_3}{4\mu_{\text{o}}} u_{nm}^2 J_{n+1}^2(u_{nm}) \tag{20.9}$$

Although the powers are equal, the spatial distributions are not. The Alternate-power flux, S_z^{o}, of the TM_{01}^{O} mode plotted in Figure 20.4 and the time-averaged Poynting flux, $<S_z>$, in Figure 20.5 make this apparent. Similar plots are provided for the TM_{11}^{O} and TM_{02}^{O} modes.

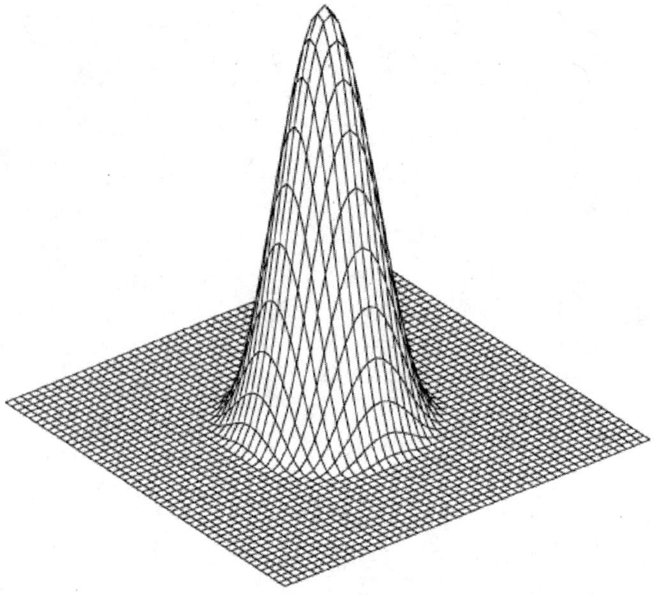

Figure 20.4 $TM_{01}{}^{\text{O}}$ mode ($k_c R = 2.405$): Alternate-power flux, $S_z^{\text{o}}(x, y)$.

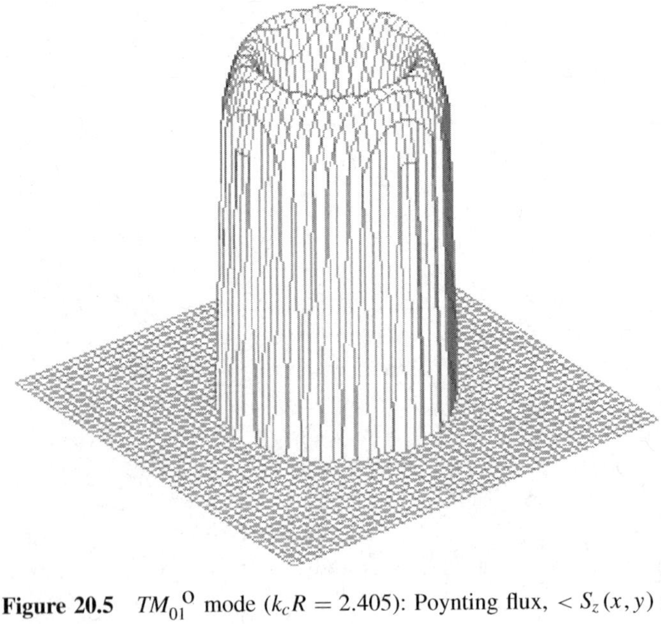

Figure 20.5 $TM_{01}{}^{\text{O}}$ mode ($k_c R = 2.405$): Poynting flux, $< S_z(x, y) >$.

20.3 TE_{nm} MODES

For *TE* modes, the boundary conditions require that $\frac{d}{dr} J_n(k_c r)\big|_{r=R} = 0$; the positive values of $k_c R$ that satisfy this equation are labeled u'_{nm} starting with the smallest value,

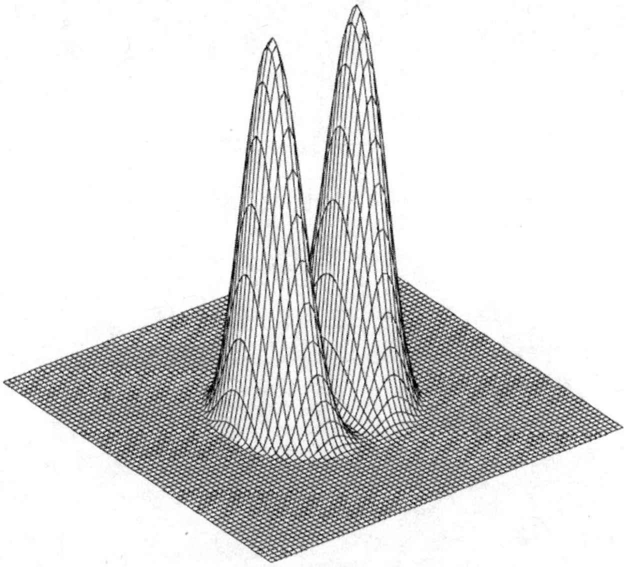

Figure 20.6 TM_{11}^{o} mode ($k_c R = 3.832$): Alternate-power flux, $S_z^{\text{o}}(x, y)$.

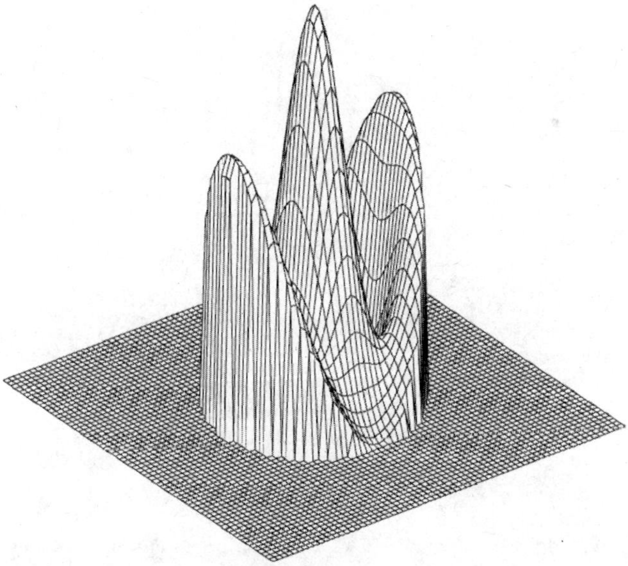

Figure 20.7 TM_{11}^{o} mode ($k_c R = 3.832$): Poynting flux, $S_z(x, y)$.

u'_{n1}. Values for $n, m \leq 3$ are tabulated below.

u'_{nm}	$m = 1$	$m = 2$	$m = 3$
u'_{0m}	3.832	7.016	10.174
u'_{1m}	1.841	5.331	8.536
u'_{2m}	3.054	6.706	9.969
u'_{3m}	4.201	8.015	11.346

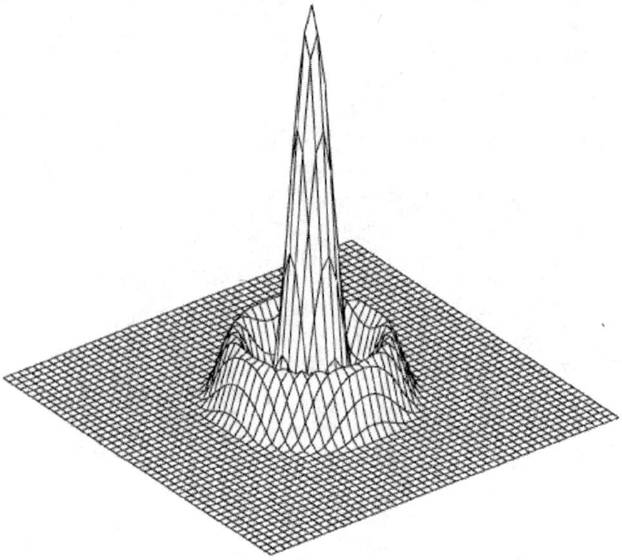

Figure 20.8 $TM_{02}{}^{\mathrm{O}}$ mode ($k_cR = 5.520$): Alternate-power flux, $S_z^{\mathrm{o}}(x, y)$.

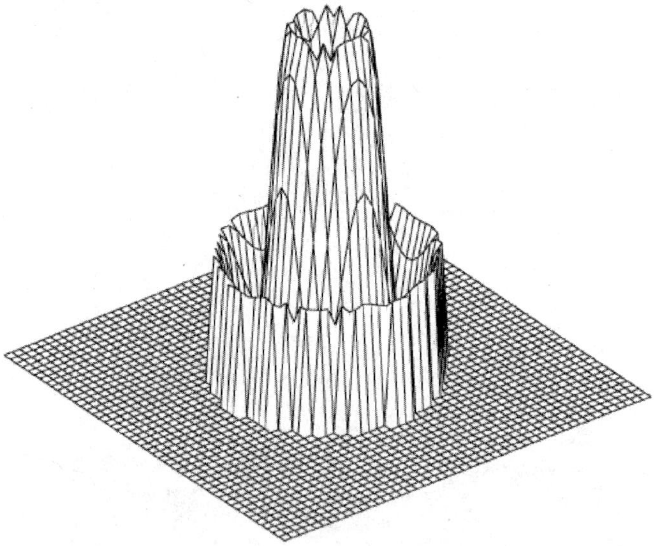

Figure 20.9 $TM_{02}{}^{\mathrm{O}}$ mode ($k_cR = 5.520$): Poynting flux, $S_z(x, y)$.

If only the **E** and **H** fields are required, it is permissible to set $C_0 = C_2 = C_3 = 0$ even though continuity of A_r at $r = R$ requires fringing potentials outside of the guide that may require additional potentials inside. By themselves, these extra terms cannot create either **E** or **H** fields. Consequently, the electric and magnetic fields of the Eqs. (20.1) modes are

$$\mathbf{E} = \omega C_1 \begin{bmatrix} \widehat{\mathbf{r}}\ \dfrac{n}{r} J_n(k_c r) \left\{ {\sin \atop \cos} n\phi \right\} \\ + \widehat{\phi}\ \dfrac{d}{dr} J_n(k_c r) \left\{ {\cos \atop -\sin} n\phi \right\} \end{bmatrix} \sin(\omega t - k_z z) \tag{20.10a}$$

and

$$\mathbf{H} = \frac{-C_1}{\mu_o}\left(\begin{bmatrix} \hat{\mathbf{r}}\, k_z \dfrac{d}{dr} J_n(k_c r)\left\{\begin{smallmatrix}\cos\\ -\sin\end{smallmatrix} n\phi\right\} \\ +\hat{\phi}\, k_z \dfrac{n}{r} J_n(k_c r)\left\{\begin{smallmatrix}\sin\\ \cos\end{smallmatrix} n\phi\right\} \end{bmatrix} \sin(\omega t - k_z z) \\ +\, \hat{\mathbf{z}}\, k_c^2 J_n(k_c r)\left\{\begin{smallmatrix}\cos\\ -\sin\end{smallmatrix} n\phi\right\} \cos(\omega t - k_z z)\right) \tag{20.10b}$$

The surface charge, σ_s, and surface current, $\mathbf{K}$, are again found by applying the boundary conditions: $\sigma_s = -\hat{\mathbf{r}} \cdot \varepsilon_o \mathbf{E}$ and $\mathbf{K} = -\hat{\mathbf{r}} \times \mathbf{H}$. The result is

$$\sigma_s = -\varepsilon_o \omega C_1 \frac{n}{R} J_n(k_c R)\left\{\begin{smallmatrix}\sin\\ \cos\end{smallmatrix} n\phi\right\} \sin(\omega t - k_z z) \tag{20.11a}$$

$$\mathbf{K} = \frac{C_1}{\mu_o} J_n(k_c R)\begin{bmatrix} \hat{\mathbf{z}}\, k_z \dfrac{n}{R}\left\{\begin{smallmatrix}\sin\\ \cos\end{smallmatrix} n\phi\right\} \sin(\omega t - k_z z) \\ -\hat{\phi}\, k_c^2 \left\{\begin{smallmatrix}\cos\\ -\sin\end{smallmatrix} n\phi\right\} \cos(\omega t - k_z z) \end{bmatrix} \tag{20.11b}$$

Notice that both σ_s and K_z vanish for all $n = 0$ modes.

The dominant mode is TE_{11}^{o} with $k_c R = 1.841$; the electric and the magnetic fields are sketched in Figure 20.10 Notice that the field patterns of the $TE_{10}^{\square}$ and TE_{11}^{o}modes near the center of their respective waveguides are both essentially the same—that of a linearly polarized plane wave. The difference in m values is due to the mode designation conventions. For comparison, the fields of the $TE_{01}^{o}$and TE_{02}^{o} modes are sketched in Figures 20.11 and 20.12.

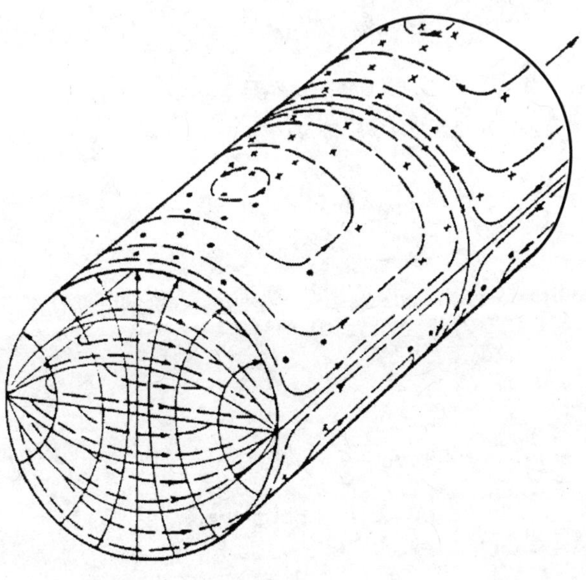

Figure 20.10 TE_{11}^{o} mode ($k_c R = 1.841$): electric and magnetic fields.

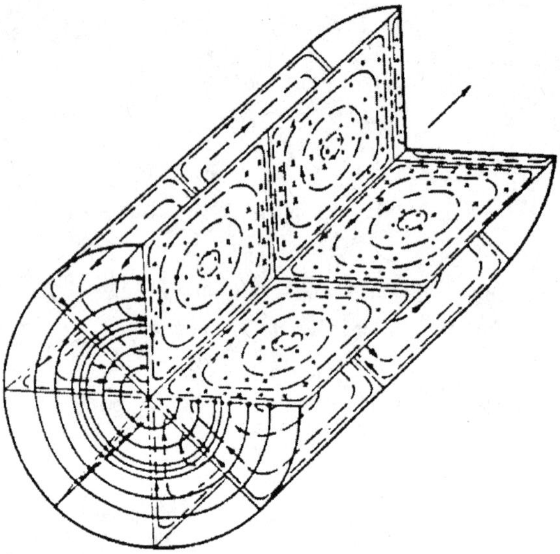

Figure 20.11 TE_{01}^{O} mode ($k_cR = 3.832$): electric and magnetic fields.

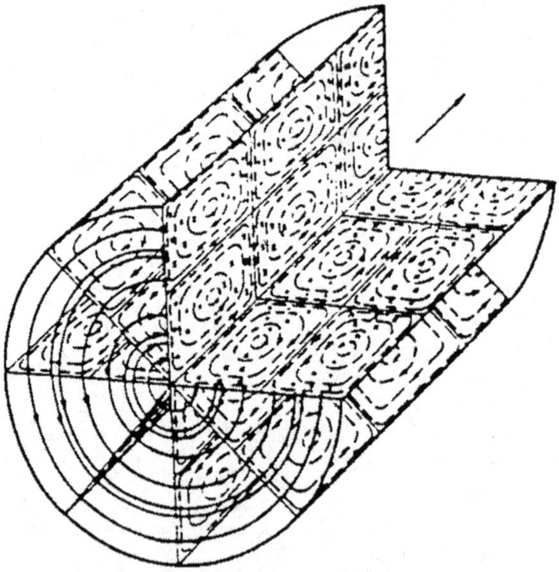

Figure 20.12 TE_{02}^{O} mode ($k_cR = 7.016$): electric and magnetic fields.

TE power fluxes

The z component of the Poynting flux of these modes is

$$S_z = \frac{\omega k_z}{\mu_o} C_1^2 \left(\begin{array}{l} \left[\frac{d}{dr} J_n(k_c r) \right]^2 \left\{ \begin{smallmatrix} \cos \\ \sin \end{smallmatrix} n\phi \right\}^2 \\ + \left[\frac{n}{r} J_n(k_c r) \right]^2 \left\{ \begin{smallmatrix} \sin \\ \cos \end{smallmatrix} n\phi \right\}^2 \end{array} \right) \sin^2(\omega t - k_z z) \qquad (20.12)$$

However, before the correct Alternate flux is calculated, it is necessary to find additional potentials both outside and inside the guide that provide continuity yet do not produce either **E** or **H**. These potentials satisfy $\mathbf{A}_\mathrm{h} = \nabla\psi$ and $\Phi_\mathrm{h} = -\frac{\partial\psi}{\partial t}$ with ψ a solution of the homogeneous wave equation. For $r \geq R$, one must include Bessel functions of the second kind, $Y_n(k_c r)$. Properties of these functions are also reviewed in Appendix F, Section F.1. With

$$\psi_{\text{inner}} = 0 \tag{20.13a}$$

$$\psi_{\text{outer}} = [C_4 J_n(k_c r) + C_5 Y_n(k_c r)] \left\{{\sin \atop \cos} n\phi\right\} \cos(\omega t - k_z z) \tag{20.13b}$$

it is possible to provide continuity of **A** and Φ at $r = R$ by choosing

$$C_0 = C_2 = C_3 = 0$$

$$C_4 = -C_1 \frac{n Y_n(k_c R)}{R \left.\dfrac{d}{dr} Y_n(k_c r)\right|_{r=R}}$$

$$C_5 = C_1 \frac{n J_n(k_c R)}{R \left.\dfrac{d}{dr} Y_n(k_c r)\right|_{r=R}}$$

Nevertheless, the boundary condition at $r \to \infty$ is not properly accounted for (unless $n = 0$) since only an outgoing wave can emanate from currents and charges located at $r = R$ and, without further modification, cylindrical standing waves are present that imply sources at $r \to \infty$. This is corrected by *everywhere* adding

$$\psi = -J_n(k_c r)[C_4 \cos(\omega t - k_z z) + C_5 \sin(\omega t - k_z z)] \left\{{\sin \atop \cos} n\phi\right\}$$

so that the final values are

$$\psi = \begin{cases} -J_n(k_c r)[C_4 \cos(\omega t - k_z z) + C_5 \sin(\omega t - k_z z)] \left\{{\sin \atop \cos} n\phi\right\} & (r < R) \\ \\ C_5[Y_n(k_c r) \cos(\omega t - k_z z) - J_n(k_c r) \sin(\omega t - k_z z)] \left\{{\sin \atop \cos} n\phi\right\} & (r > R) \end{cases} \tag{20.14}$$

For large r, the asymptotic limit of $J_n(k_c r) \sin(\omega t - k_z z) - Y_n(k_c r) \cos(\omega t - k_z z)$ (a Hankel function of the first kind) is $\sqrt{\frac{2}{\pi k_c r}} \sin(\omega t - k_z z - k_c r + \dfrac{\pi}{4} + \dfrac{n\pi}{2})$.

On the inner wall of the waveguide the values of $\mathbf{A} - \widehat{\mathbf{r}} A_r$, and Φ can now be calculated: as

$$\mathbf{A} - \widehat{\mathbf{r}} A_r = J_n(k_c R) \begin{pmatrix} -\widehat{\phi}\, \dfrac{n}{R} \begin{bmatrix} C_4 \cos(\omega t - k_z z) \\ +C_5 \sin(\omega t - k_z z) \end{bmatrix} \left\{{\cos \atop -\sin} n\phi\right\} \\ +\widehat{\mathbf{z}}\, k_z \begin{bmatrix} C_5 \cos(\omega t - k_z z) \\ -C_4 \sin(\omega t - k_z z) \end{bmatrix} \left\{{\sin \atop \cos} n\phi\right\} \end{pmatrix} \tag{20.15a}$$

(20.15b)

$$\Phi = \omega J_n(k_c R) \begin{bmatrix} C_5 \cos(\omega t - k_z z) \\ -C_4 \sin(\omega t - k_z z) \end{bmatrix} \left\{ {}^{\sin}_{\cos}\, n\phi \right\} \tag{20.15c}$$

The Alternate-power flux and energy density have localized surface contributions, respectively, $\Phi\mathbf{K}\ u_0(r - R)$ and $\frac{1}{2}\left(\mathbf{A}\cdot\mathbf{K}+\Phi\sigma_s\right) u_0(r - R)$. Because Φ and $\mathbf{A}$ are not zero at the inner wall (where the surface currents and charges are located), these include both time-averaged and reactive components.

$$\Phi\mathbf{K} = \frac{\omega C_1 C_5}{\mu_0} J_n(k_c R) \begin{pmatrix} [J_n(k_c R)\cos(\omega t - k_z z) + Y_n(k_c R)\sin(\omega t - k_z z)] \\ \times \begin{bmatrix} \hat{\boldsymbol{\phi}}\ k_c^2\{\mp\}\sin(2n\phi)\cos(\omega t - k_z z) \\ +\hat{\mathbf{z}}\ k_z \frac{n}{R} \left\{ {}^{\sin}_{\cos}\, n\phi \right\}^2 \sin(\omega t - k_z z) \end{bmatrix} \end{pmatrix} \tag{20.16a}$$

$$\frac{1}{2}\mathbf{A}\cdot\mathbf{K} = \frac{-C_1 C_5}{2\mu_0}\frac{n}{R} J_n(k_c R) \begin{pmatrix} k_c^2 \begin{bmatrix} Y_n(k_c R)\cos^2(\omega t - k_z z) \\ -J_n(k_c R)\sin(\omega t - k_z z)\cos(\omega t - k_z z) \end{bmatrix} \left\{ {}^{\cos}_{\sin}\, n\phi \right\}^2 \\ +k_z^2 \begin{bmatrix} J_n(k_c R)\sin(\omega t - k_z z)\cos(\omega t - k_z z) \\ +Y_n(k_c R)\sin^2(\omega t - k_z z) \end{bmatrix} \left\{ {}^{\sin}_{\cos}\, n\phi \right\}^2 \end{pmatrix} \tag{20.16b}$$

$$\frac{1}{2}\Phi\sigma_s = \frac{-C_1 C_5}{2\mu_0}\frac{n}{R} J_n(k_c R)\, k^2 \begin{bmatrix} J_n(k_c R)\sin(\omega t - k_z z)\cos(\omega t - k_z z) \\ +Y_n(k_c R)\sin^2(\omega t - k_z z) \end{bmatrix} \left\{ {}^{\sin}_{\cos}\, n\phi \right\}^2 \tag{20.16c}$$

Notice that all of the localized terms vanish for $n = 0$.

In the free-space interior, $r < R$, the Alternate flux has time-independent components

$$S_r^{\mathrm{o}} = \frac{C_1 C_5}{2\mu_0} k_c^2 \omega n \frac{J_{n-1}(k_c r) J_{n+1}(k_c r) - J_n^2(k_c r)}{r} \{\pm\}\cos(2n\phi) \tag{20.17a}$$

$$S_\phi^{\mathrm{o}} = \frac{C_1 C_5}{2\mu_0} k_c^2 \omega \frac{J_{n-1}(k_c r) J_{n+1}(k_c r)}{r} \{\pm\}\sin(2n\phi) \tag{20.17b}$$

$$S_z^{\mathrm{o}} = \frac{\omega k_z}{2\mu_o} \begin{pmatrix} \begin{bmatrix} C_1^2\left(\frac{n}{r}J_n\right)^2 - 2C_1 C_4 \frac{n}{r} J_n J_n' \\ +\left(C_4^2 + C_5^2\right)\left(J_n'^2 - k_c^2 J_n^2\right) \end{bmatrix} \left\{ {}^{\sin}_{\cos}\, n\phi \right\}^2 \\ + \begin{bmatrix} C_1^2 J_n'^2 - 2C_1 C_4 \frac{n}{r} J_n J_n' \\ +\left(C_4^2 + C_5^2\right)\left(\frac{n}{r}J_n\right)^2 \end{bmatrix} \left\{ {}^{\cos}_{\sin}\, n\phi \right\}^2 \end{pmatrix} \tag{20.17c}$$

where $J_n' = \frac{d}{dr} J_n(k_c r)$. Outside the guide, $r > R$, the Alternate flux is

$$S_r^{\mathrm{o}} = \frac{\omega k_c n^2 C_5^2}{2\mu_0} \frac{\{\pm\}\cos(2n\phi)}{r^2} [Y_{n-1}(k_c r) J_n(k_c r) - J_{n-1}(k_c r) Y_n(k_c r)] \tag{20.18a}$$

$$S^{\text{o}}_{\phi} = \frac{\omega k_c n C_5^2}{2\mu_{\text{o}}} \frac{\{\pm\} \sin(2n\phi)}{r^2} [Y_{n-1}(k_c r) J_n(k_c r) - J_{n-1}(k_c r) Y_n(k_c r)] \tag{20.18b}$$

$$S^{\text{o}}_{z} = \frac{\omega k_z C_5^2}{4\mu_{\text{o}}} \nabla^2 [(J_n^2 + Y_n^2) \left\{ {\sin \atop \cos} \right\}^2 (n\phi)] \tag{20.18c}$$

Substitution of the Wronskian, $Y_{n-1}(x)J_n(x) - J_{n-1}(x)Y_n(x) = \frac{2}{\pi x}$, simplifies the transverse components to

$$S^{\text{o}}_{r} = \frac{\omega n^2 C_5^2}{\pi \mu_{\text{o}}} \frac{\{\pm\} \cos(2n\phi)}{r^3} \tag{20.19a}$$

$$S^{\text{o}}_{\phi} = \frac{\omega n C_5^2}{\pi \mu_{\text{o}}} \frac{\{\pm\} \sin(2n\phi)}{r^3} \tag{20.19b}$$

whereas the longitudinal flux evaluates to

$$S^{\text{o}}_{z} = \frac{\omega k_z C_5^2}{2\mu_{\text{o}}} \left[\begin{array}{l} \left(J_n'^2 + Y_n'^2 - k_c^2 (J_n^2 + Y_n^2)\right) \left\{ {\sin \atop \cos} n\phi \right\}^2 \\ +(\frac{n}{r})^2 (J_n^2 + Y_n^2) \left\{ {\cos \atop \sin} n\phi \right\}^2 \end{array} \right] \tag{20.20a}$$

$$S^{\text{o}}_{z} \to \frac{\omega n^2 k_z C_5^2}{4\mu_{\text{o}} k_c r^3} \qquad (k_c r \gg 1) \tag{20.20b}$$

Notice that for $n = 0$, the Alternate flux components external to the guide (as well as the localized components) vanish and $S^{\text{o}}_{z}(x, y) = \langle S_z(x, y) \rangle$. For $n > 0$, all components are nonzero but independent of z and t; therefore the divergences of the transverse and longitudinal components separately vanish. Evaluation of the nonlocalized Alternate-energy density reveals that for TE^{o}_{nm} (as well as TM^{o}_{nm}) modes we have

$$S^{\text{o}}_{z} = W^{\text{o}} \frac{ck_z}{k} = W^{\text{o}} \, \mathrm{v}_{\text{group}} \tag{20.21}$$

The localized power flux and the surface energy density evaluate to

$$\Phi \mathbf{K} = \left(\begin{array}{l} \widehat{\phi} \left[\begin{array}{l} \frac{\omega k_c^2 C_1 C_5}{2\mu_{\text{o}}} J_n(k_c R) \cos(\omega t - k_z z) [Y_n(k_c R) \sin(\omega t - k_z z) \\ + J_n(k_c R) \cos(\omega t - k_z z)] \{\pm\} \sin(2n\phi) \end{array} \right] \\ +\widehat{\mathbf{z}} \left[\begin{array}{l} \frac{-\omega k_z C_1 C_5}{\mu_{\text{o}}} \frac{n}{R} J_n(k_c R) \sin(\omega t - k_z z) [Y_n(k_c R) \sin(\omega t - k_z z) \\ + J_n(k_c R) \cos(\omega t - k_z z)] \left\{ {\sin \atop \cos} n\phi \right\}^2 \end{array} \right] \end{array} \right) \tag{20.22a}$$

and

$$\frac{1}{2}(\mathbf{A}\cdot\mathbf{K}+\sigma_s\Phi) = \frac{C_1C_5\,n}{2\mu_o\ R}J_n(k_cR)\times \left[\begin{array}{l} (k_c^2+2k_z^2)Y_n(k_cR)\sin^2(\omega t-k_zz)\left\{{\sin \atop \cos}n\phi\right\}^2 \\ +k_c^2Y_n(k_cR)\cos^2(\omega t-k_zz)\left\{{\cos \atop \sin}n\phi\right\}^2 \\ +J_n(k_cR)\left[k_z^2\left\{{\sin \atop \cos}n\phi\right\}^2-\frac{1}{2}\{\pm\}k_c^2\cos(2n\phi)\right]\sin 2(\omega t-k_zz) \end{array}\right] \tag{20.22b}$$

As expected, they include reactive components. When the surface power and energy are averaged over both t and ϕ, the z component of power is found to be related to the Alternate energy by

$$\begin{aligned} <\Phi K_z> &= <\frac{1}{2}(\mathbf{A}\cdot\mathbf{K}+\sigma_s\Phi)>\frac{ck_z}{k} \\ &= -\frac{C_1C_5}{4\mu_0}\frac{n\omega k_z}{R}J_n(k_cR)Y_n(k_cR) \end{aligned} \tag{20.23}$$

The total average power must be calculated by integrating $<S_z^o>$ over an entire $x--y$ plane and is the sum of the inner, outer, and surface contributions; the result of using Appendix F, Eqs. (F.13a)–(F.14) is

$$2\pi R<\Phi K_z> = <P_z^o>_{\text{outer}} = \frac{\pi n^2\omega k_z C_1^2}{2\mu_0}\frac{-Y_n(u'_{nm})}{R\left.\dfrac{d}{dr}Y_n(u'_{nm}\dfrac{r}{R})\right|_{r=R}}J_n^2(u'_{nm})$$

$$<P_z^o>_{\text{inner}} = (1+\delta_{n0})\frac{\pi\omega k_z C_1^2}{4\mu_0}[(u'_{nm})^2-n^2]J_n^2(u'_{nm})-2<P_z^o>_{\text{outer}}$$

which agrees with that obtained by integrating the time-averaged Eq. (20.12) over the waveguide cross section; the well-known result is

$$<P_z^o>_{\text{total}} = (1+\delta_{n0})\frac{\pi\omega k_z C_1^2}{4\mu_0}[(u'_{nm})^2-n^2]J_n^2(u'_{nm}) \tag{20.24}$$

The ratio of the average surface power to the total average power is equal to

$$\left[\frac{<P_z^o>_{\text{outer}}}{<P_z^o>_{\text{total}}}\right]_{nm} = \frac{2n^2Y_n(u'_{nm})}{[(u'_{nm})Y_{n+1}(u'_{nm})-nY_n(u'_{nm})][(u'_{nm})^2-n^2]} \tag{20.25}$$

$$\left[\frac{<P_z^o>_{\text{outer}}}{<P_z^o>_{\text{total}}}\right]_{nm}$$

n	$m=1$	$m=2$	$m=3$
0	0	0	0
1	.152	.0013	.00019
2	.157	.0025	.00045
3	.158	.0033	.00067

Notice that the localized components of the Alternate flux vanish for all $TE_{0m}^{\ o}$ modes and that $S_z^o(x,y) = <S_z(x,y)>$.

Alternate-power fluxes of the dominant $TE_{11}^{\ o}$ mode are plotted below as a function of x and y; the longitudinal nonlocalized component in Figure 20.13(a), the surface-power $<\Phi K_z>$ in Figure 20.13(b), and the transverse component (shown as a vector-field plot)

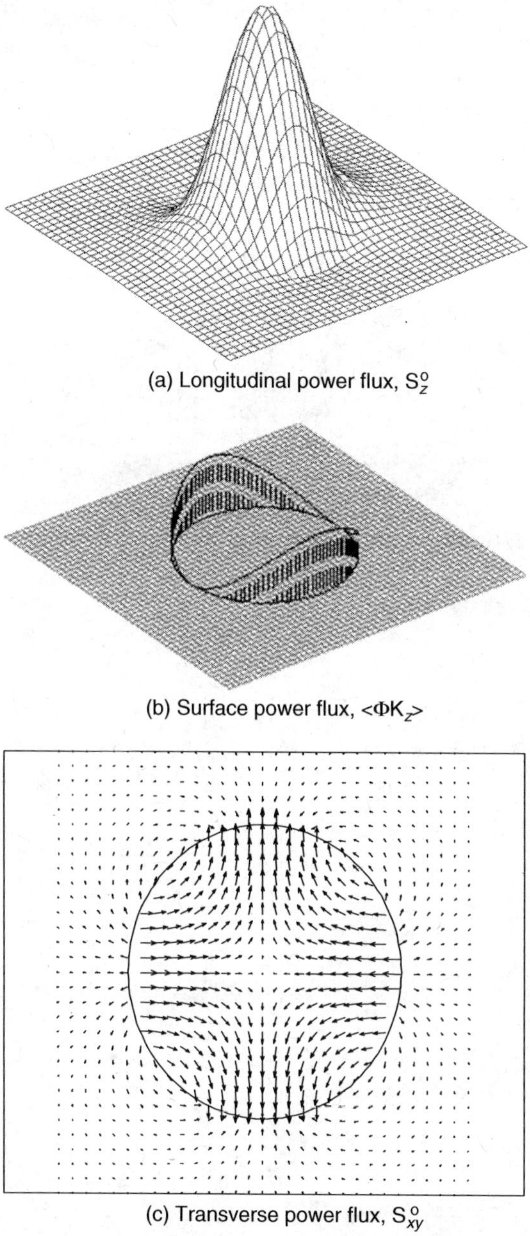

(a) Longitudinal power flux, S_z^o

(b) Surface power flux, $<\Phi K_z>$

(c) Transverse power flux, S_{xy}^o

Figure 20.13 $TE_{11}{}^{O}$ mode ($k_c R = 1.841$): Alternate-power fluxes.

in Figure 20.13(c). The time-averaged Poynting flux, $<S_z>$, of the same mode is plotted in Figure 20.14. In both cases, the polarization of the mode plotted is described by the trigonometric functions of $n\phi$ (or sign) listed on the top line between the curly braces. Plots of the orthogonal polarization are obtained by rotating these plots through 90 degrees.

For the $TE_{11}{}^{O}$ mode, 15.2% of the total z-directed Alternate power is localized on the surface current with an equal amount located *outside* of the waveguide; the Alternate

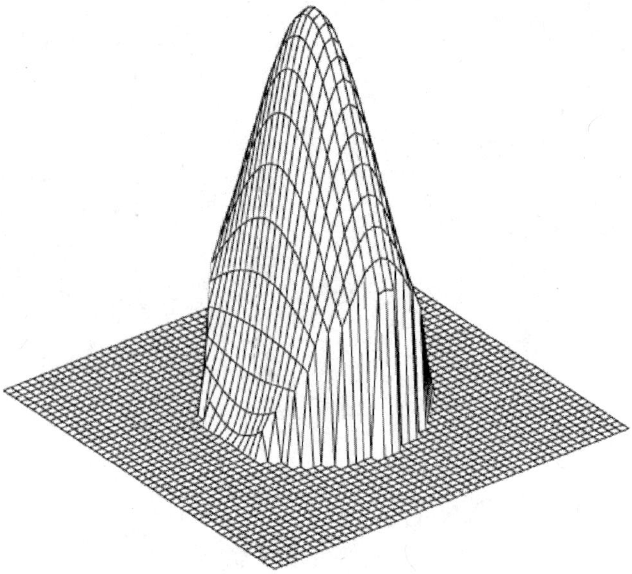

Figure 20.14 TE_{11}^{O} mode ($k_c R = 1.841$): Poynting flux, $< S_z(x, y) >$.

power inside the guide is reduced accordingly. These percentages are maintained (to within 1%) for *all* TE_{n1}^{O} modes ($n > 0$), but very rapidly decrease for modes with $m > 2$; the transverse fluxes S_{xy}^{O} are reduced accordingly as shown in Figure 20.15.

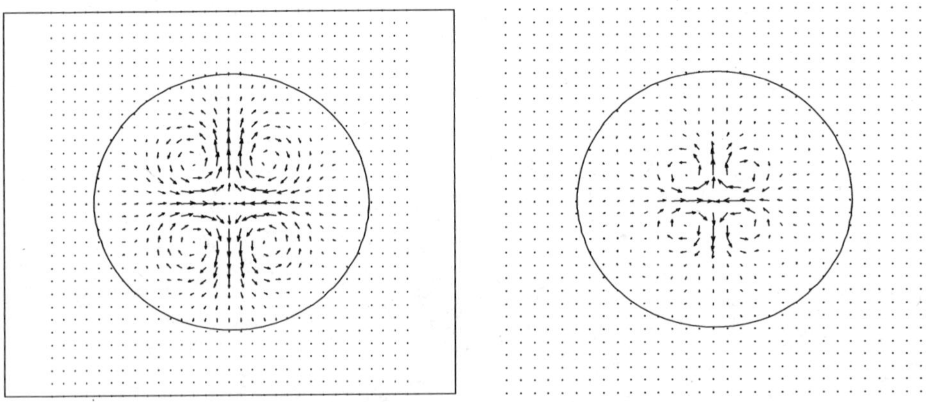

Figure 20.15 *TE* Alternate-power fluxes: $S_{xy}^{\text{O}}(x, y)$.

If the fringing potentials, both inside and outside the waveguide, are ignored (by setting $C_4 = C_5 = 0$), the localized power and energy distributions vanish[1] (as do the components *outside* of the waveguide). Although the Alternate-flux distribution is then

[1]If the TE_{nm}^{O} mode fields are assumed to exist (continuously) outside the guide, oppositely directed surface currents and surface charges cancel K_z and σ_s. In that event the localized power and energy vanish, along with the homogeneous potentials. Such an artifice has clear computational advantages.

in error, the total average power is not. In this instance, the modified S_z^{o} exactly coincides with the time-averaged value of S_z and thus obviously yields the correct total average-power when integrated over the waveguide cross section.

Similar plots of power fluxes for the TE_{21}^{o} and TE_{01}^{o} modes are also provided in Figures 20.16 through 20.18.

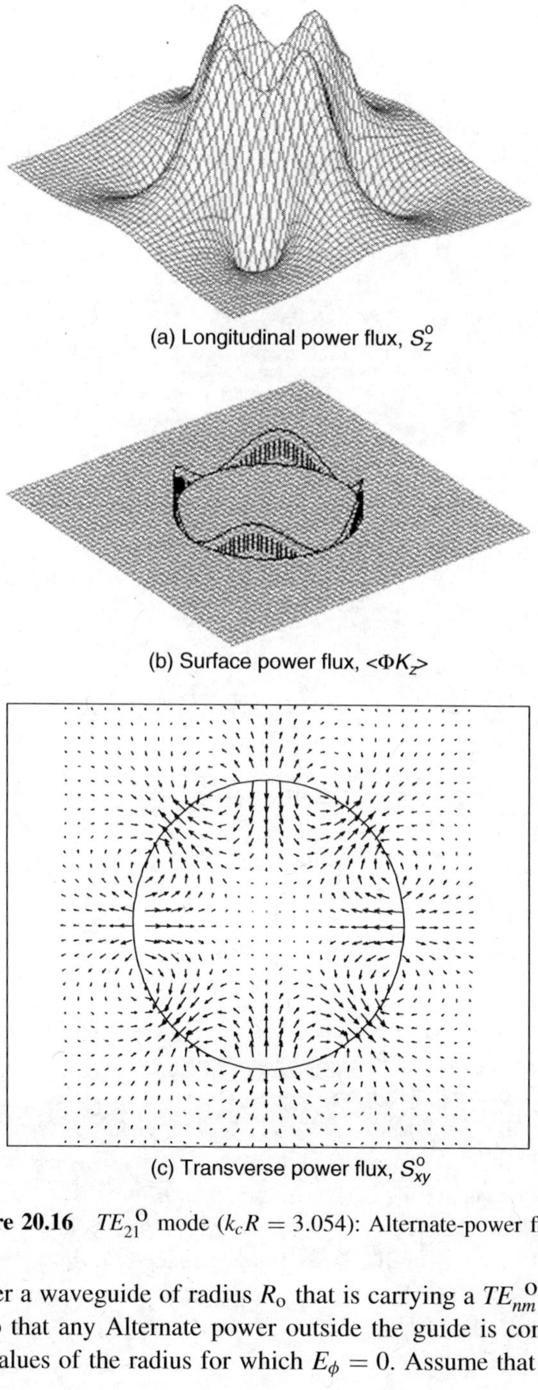

(a) Longitudinal power flux, S_z^{o}

(b) Surface power flux, $\langle \Phi K_z \rangle$

(c) Transverse power flux, S_{xy}^{o}

Figure 20.16 TE_{21}^{o} mode ($k_c R = 3.054$): Alternate-power fluxes.

Finally, consider a waveguide of radius R_{o} that is carrying a TE_{nm}^{o} mode and where m is large enough so that any Alternate power outside the guide is completely negligible. There will be m values of the radius for which $E_\phi = 0$. Assume that the first of these is

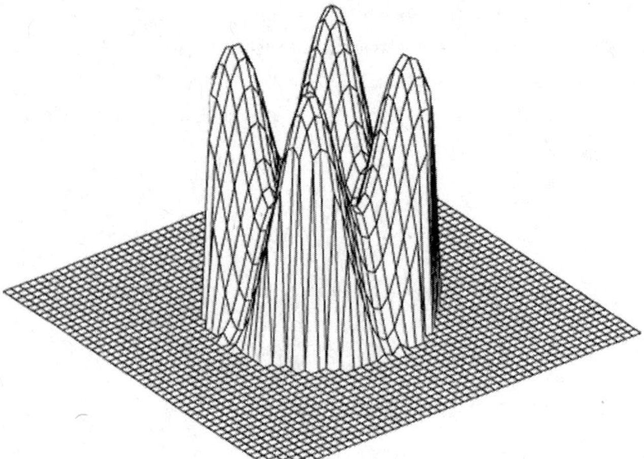

Figure 20.17 $TE_{21}^{\,\mathrm{O}}$ mode ($k_cR = 3.054$): Poynting flux, $< S_z(x, y) >$.

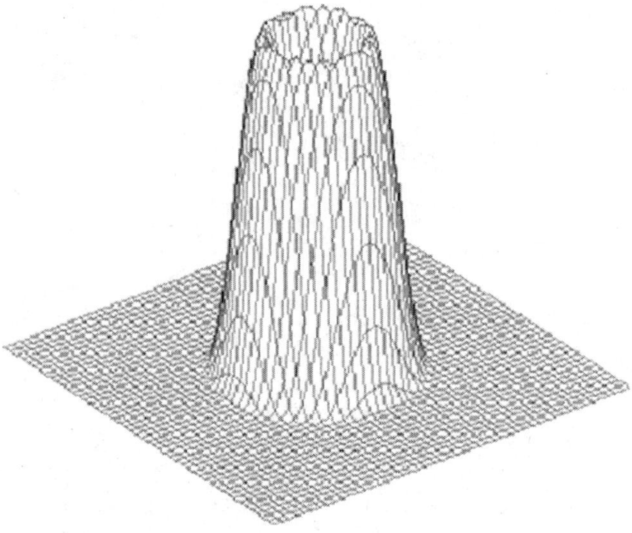

Figure 20.18 $TE_{01}^{\,\mathrm{O}}$ mode ($k_cR = 3.832$): $S_z^{\mathrm{o}}(x, y) =< S_z(x, y) >$.

R and at that radius a very thin perfectly conducting cylindrical wall is inserted without disturbing the fields on either side; it follows that $R = \frac{u'_{n1}}{u'_{nm}} R_{\mathrm{o}}$ and the inner fields are those of a $TE_{n1}^{\,\mathrm{O}}$ mode. Because the currents and charges on opposite surfaces of the wall cancel, there is no localized Alternate power or Alternate energy *anywhere* and the distributions inside of the wall are the same as the Maxwell–Poynting values for the $TE_{n1}^{\,\mathrm{O}}$ mode. Notice, however, that if the fields between R and R_{o} are extinguished, localized Alternate power appears that will affect the overall distribution! Remarkably, the presence of a coherent mode for $r > R$ alters the distribution inside. A similar effect occurred with the periodic array of rectangular waveguides (stacked horizontally or vertically) that was analyzed in Chapter 19, Section 19.7.

20.4 NULL ALTERNATE POWER AND ENERGY DISTRIBUTIONS

In the discussion of rectangular waveguides in Chapter 19, Section 19.6, certain vector and scalar potentials, periodic in both x and y directions, were found that produce neither **E** nor **H** field yet do create Alternate power. A similar situation exists in cylindrical coordinates.

Consider modes described by Eqs. (20.1) with $C_1 = 0$, $C_0 = C_2 = C_3$. Then **E**, **H** and the surface charges and currents on the walls *all* vanish.[2] Interestingly, unless $k_c = 0$, the Alternate flux vector, $\mathbf{S}^{\text{o}}$, *persists* (although $\mathbf{E} \times \mathbf{H} = 0$). The result is both a null power flux, $\mathbf{S}^{\text{o}}_{\text{null}}$, and a null energy density, $W^{\text{o}}_{\text{null}}$, related by

$$\mathbf{S}^{\text{o}}_{\text{null}} = \hat{\mathbf{z}}\frac{k_z}{k}cW^{\text{o}}_{\text{null}} \tag{20.26a}$$

With $C_0 = A_{\text{null}}/k_c$, and making use of Appendix F, Eqs. (F.8a)–(F.9b), the standing ϕ-wave null-mode energy densities are given by

$$W^{\text{o}}_{\text{null}} = \frac{\varepsilon_0 \omega^2 A^2_{\text{null}}}{4k_c^2 r^2}\left(\begin{array}{l} k_c^2 r^2 [J^2_{n-1}(k_c r) + J^2_{n+1}(k_c r) - 2J^2_n(k_c r)] \left\{ {}^{\sin}_{\cos}\, n\phi \right\}^2 \\ +2n^2 J^2_n(k_c r)\{\pm\}\cos(2n\phi) \end{array}\right) \tag{20.26b}$$

Circulating null-mode energy densities can be calculated from the superposition of orthogonal standing wave modes (with equal amplitudes). Then the ϕ-dependence vanishes and

$$W^{\text{o}}_{\text{null}} = \frac{1}{4}\varepsilon_0 \omega^2 A^2_{\text{null}} \left[J^2_{n-1}(k_c r) + J^2_{n+1}(k_c r) - 2J^2_n(k_c r)\right] \tag{20.27a}$$

However, for modes described by Eqs. (20.2) with $C_1 = 0$, $C_0 = C_2 = C_3$, there is also ϕ-directed power and

$$\mathbf{S}^{\text{o}}_{\text{null}} = (\hat{\phi}\frac{n}{kr} + \hat{\mathbf{z}}\frac{k_z}{k})cW^{\text{o}}_{\text{null}} \tag{20.27b}$$

This result is consistent with the standing ϕ-wave power because equal contributions of $\pm n$ modes cancel the azimuthal component.

For any null mode (with a value of k_c that corresponds to either a TE_{nm}^{o} or TM_{nm}^{o} mode of a guide with radius, R), the integral of S_z^{o} over the guide cross section vanishes so that the average power has the correct value of zero. As with the rectangular modes, null energy travels at the group velocity to produce null power.[3] S_z^{o} is shown in Figure 20.19 and Figure 20.20 for the TE_{11}^{o} and TM_{01}^{o} standing ϕ-wave null modes.

20.5 ALTERNATE ENERGY MOMENTUM AND PHOTONS

When a uniform planewave is circularly polarized, the associated spin angular momentum is nonzero. The same is true for waveguide modes with circular ϕ variation, but, in addition, orbital angular momentum is present.

[2] As in the case of rectangular waveguide, null modes can be generated by choosing $\mathcal{A} = \Box\Psi$ where $\Box^2\Psi = 0$. In our example, $\Psi = C_0 J_n(k_c r)\{{}^{\sin}_{\cos}\, n\phi\}\cos(\omega t - k_z z)$.

[3] A spin 1 particle has three states $(+1, 0, -1)$, yet a photon has only two (corresponding to the ± 1 circularly polarized transverse waves—the longitudinal polarization is missing). Nevertheless, it is tempting, and probably harmless, to associate null power and energy with the "missing" state.

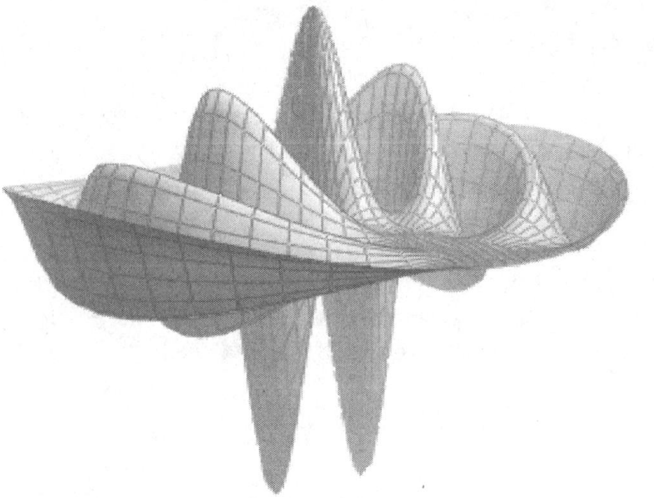

Figure 20.19 $TE_{11}{}^{\mathrm{o}}$ null mode ($k_c R = 1.8412$): $S_z^{\mathrm{o}}(x, y)$ or $W^{\mathrm{o}}(x, y)$.

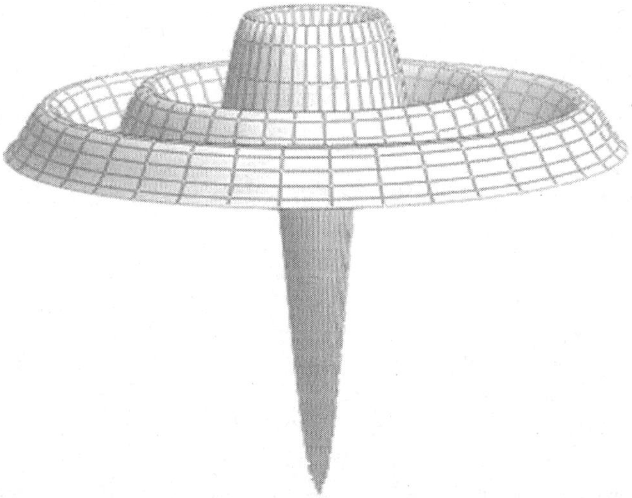

Figure 20.20 $TM_{01}{}^{\mathrm{o}}$ null mode ($k_c R = 2.405$): $S_z^{\mathrm{o}}(x, y)$ or $W^{\mathrm{o}}(x, y)$.

TE_{nm}^{o} "circularly polarized" modes

Based upon Eqs. (20.2), the Lorenz-gauge potentials (of the particular solution) are of the form

$$\mathbf{A} = \frac{A_{\mathrm{o}}}{k_c}\begin{pmatrix} \hat{\mathbf{r}}\,\dfrac{n}{r}J_n(k_c r)[\cos(\omega t - k_z z - n\phi) + \Gamma_\phi \cos(\omega t - k_z z + n\phi)] \\ +\hat{\phi}\,\dfrac{d}{dr}J_n(k_c r)[\sin(\omega t - k_z z - n\phi) - \Gamma_\phi \sin(\omega t - k_z z + n\phi)] \end{pmatrix} \tag{20.28a}$$

$$\Phi = 0 \tag{20.28b}$$

where n is an integer of either sign and the azimuthal reflection coefficient Γ_ϕ is in the range $-1 \le \Gamma_\phi \le +1$. In what follows, we assume $\Gamma_\phi = 0$, but the results are easily generalized to arbitrary "elliptically polarized" modes.

With $\mathbf{r} = r\widehat{\mathbf{r}} + z\widehat{\mathbf{z}}$, the spin and orbital angular-momentum densities defined in Appendix D, Section D.3 are expressed as

$$\mathbf{L}^{\text{spin}} = \varepsilon_0 \mathbf{E} \times \mathbf{A} = \widehat{\mathbf{z}} \frac{\omega \varepsilon_0}{(u'_{nm})^2} A_0^2 R^2 \left[\frac{n}{r} J_n(k_c r) \frac{d}{dr} J_n(k_c r) \right] \tag{20.29a}$$

$$\mathbf{L}^{\text{orbit}} = \varepsilon_0 \left[\mathbf{E} \cdot \frac{\partial \mathbf{A}}{\partial \phi} (\widehat{\mathbf{z}} - \frac{z}{r} \widehat{\mathbf{r}}) - r\, \mathbf{E} \cdot \frac{\partial \mathbf{A}}{\partial z} \widehat{\boldsymbol{\phi}} \right] \tag{20.29b}$$

$$L_z^{\text{orbit}} = \frac{\omega n \varepsilon_0}{(u'_{nm})^2} A_0^2 R^2 \left[\begin{array}{l} \left(\frac{n}{r}\right)^2 J_n^2(k_c r) \sin^2(\omega t - k_z z - n\phi) \\ + \left[\frac{d}{dr} J_n(k_c r)\right]^2 \cos^2(\omega t - k_z z - n\phi) - \frac{1}{r} J_n(k_c r) \frac{d}{dr} J_n(k_c r) \end{array} \right] \tag{20.29c}$$

$$\mathbf{L}^{\text{charge}} = \mathbf{r} \times \rho\, \mathbf{A} = \mathbf{r} \times \sigma_s u_0(r - R)\, \mathbf{A} \tag{20.29d}$$

Notice that L_z^{orbit} vanishes at the center of the guide for all modes, but $\mathbf{L}^{\text{spin}}$ is uniform near $r = 0$ for $n = 1$. When integrated over the area of the guide, the axial components are

$$\frac{L_{z\text{ total per unit length}}^{\text{spin}}}{P_{\text{ave}}} = \frac{n}{c^2 k_z} \frac{2}{(u'_{nm})^2 - n^2}$$

$$\frac{<L_{z\text{ total per unit length}}^{\text{orbit}}>}{P_{\text{ave}}} = \frac{n}{c^2 k_z} \frac{(u'_{nm})^2 - (n^2 + 2)}{(u'_{nm})^2 - n^2}$$

$$\frac{<L_{z\text{ total per unit length}}^{\text{spin + orbit}}>}{P_{\text{ave}}} = \frac{n}{c^2 k_z}$$

where P_{ave} is twice the value of Eq. (20.24) (which applies to standing ϕ-wave modes). Notice that, as expected, these ratios become very large as the cutoff condition, $k_z \to 0$, is approached.

Since A_ϕ^{p} and A_z^{p} vanish at $r = R$, $\mathbf{L}_{\text{total per unit length}}^{\text{charge}} = 0$ and $\mathbf{L}_{\text{total per unit length}}^{\text{spin +orbit}}$ should and do agree with the integration of $\varepsilon_0 \mathbf{r} \times (\mathbf{E} \times \mathbf{B})$ over the guide. We have used the particular value $\mathbf{A}_{\text{p}}$ rather than the continuous potentials $\mathbf{A}_{\text{p}}+\mathbf{A}_{\text{h}}$ and Φ_{h} which satisfy the Lorenz gauge everywhere. Consequently, the results for $\mathbf{L}^{\text{spin}}$ $\mathbf{L}^{\text{orbit}}$ and $\mathbf{L}^{\text{charge}}$ *individually* are in doubt. Results based upon $\mathbf{A}_{\text{h}} = \nabla\psi$ and $\Phi_{\text{h}} = -\frac{\partial \psi}{\partial t}$ when Eq. (20.14) is replaced by

$$\psi = \begin{cases} -J_n(k_c r)[C'_4 \cos(\omega t - k_z z - n\phi) + C'_5 \sin(\omega t - k_z z - n\phi)] & (r < R) \\ C'_5[Y_n(k_c r) \cos(\omega t - k_z z - n\phi) - J_n(k_c r) \sin(\omega t - k_z z - n\phi)] & (r > R) \end{cases}$$

(chosen to make **A** and Φ continuous) reveal that for the TE_{11}^{o} mode:

TE_{11}^{o} mode	$\mathbf{A} = \mathbf{A}_\text{p}$	$\mathbf{A} = \mathbf{A}_\text{p}+\mathbf{A}_\text{h}$
$<L_{z\text{ total per unit length}}^{\text{spin}}>/P_{\text{ave}}$	$\frac{.837}{c^2 k_z}$	$\frac{.655}{c^2 k_z}$
$<L_{z\text{ total per unit length}}^{\text{orbit}}>/P_{\text{ave}}$	$\frac{.163}{c^2 k_z}$	$\frac{.193}{c^2 k_z}$
$<L_{z\text{ total per unit length}}^{\text{charge}}>/P_{\text{ave}}$	0	$\frac{.152}{c^2 k_z}$

The totals are the same, $\frac{1}{c^2 k_z}$. As expected, this metric favors low frequencies—especially those near cutoff.

For a single TE_{nm}^{o} mode propagating in the $+z$ direction, the free-space values of W^{o}, $\mathbf{S}^{\text{o}} = c^2\mathbf{G}^{\text{o}}$, T_{zz}^{o}, $T_{\phi z}^{\text{o}}$, $T_{z\phi}^{\text{o}}$, and $\mathbf{L}^{\text{spin}} = \varepsilon_0 \mathbf{E} \times \mathbf{A}$ are all independent of time. These and the time-averaged $\mathbf{L}^{\text{orbit}}$ can be expressed in terms of the photon density, $n_{\text{photon}}(n,r) = n_+(n,r) + n_-(n,r)$, the net spin-density, $n_{\text{spin}}(n,r) = n_+(n,r) - n_-(n,r)$, and, because the waveguide mode is "circularly polarized," a propagation vector $\mathbf{k} = -\nabla\Psi = \frac{n}{r}\widehat{\phi} + k_z\widehat{\mathbf{z}}$ that corresponds to the phase of the fields, $\Psi = \omega t - k_z z - n\phi$. The results, based upon $\mathbf{A}_\text{p}$, $\Phi_\text{p} = 0$, which place in evidence the particle aspects, are

$$
\begin{aligned}
W^{\text{o}} &= n_{\text{photon}}\hbar\omega \\
\mathbf{G}^{\text{o}} &= \frac{1}{c^2}\mathbf{S}^{\text{o}} = n_{\text{photon}}\hbar\mathbf{k} - n_{\text{spin}}\hbar\frac{1}{r}\widehat{\phi} \\
n_{\pm}\hbar &= \tfrac{1}{2}(n_{\text{photon}} \pm n_{\text{spin}})\hbar = \tfrac{1}{4}\varepsilon_0\omega A_0^2 J_{n\mp 1}^2(k_c r) \\
S_{\phi}^{\text{o}} &= c^2 G_{\phi}^{\text{o}} = c^2 n_{\text{orbit}}\hbar\frac{n}{r} \\
n_{\text{orbit}} &= n_{\text{photon}} - \frac{n_{\text{spin}}}{n} = \left(1 - \tfrac{1}{n}\right)n_+ + \left(1 + \tfrac{1}{n}\right)n_- \\
<S_{\phi}> &= c^2 <G_{\phi}> = c^2 n_{\text{em}}\hbar\frac{n}{r} \\
n_{\text{em}}\hbar &= \tfrac{1}{2}\varepsilon_0\omega A_0^2 J_n^2(k_c r) \\
T_{zz}^{\text{o}} &= -G_z^{\text{o}} v_{\text{group }z} \\
T_{\phi z}^{\text{o}} &= -G_{\phi}^{\text{o}} v_{\text{group }z} \\
T_{z\phi}^{\text{o}} &= -G_z^{\text{o}} v_{\phi} \\
\mathbf{v}_{\text{group}} &= \frac{\partial\omega[k_c(n,m),k_z]}{\partial k_z}\widehat{\mathbf{z}} + \frac{\partial\omega[k_c(n,m),k_z]}{\partial(n/r)}\widehat{\phi} \\
&= c\left[\frac{k_z}{k}\widehat{\mathbf{z}} + \frac{k_c}{k}\frac{r}{R}\frac{\partial u'_{nm}}{\partial n}(1-\delta_{n0})\widehat{\phi}\right] \quad (r \le R) \\
<\mathbf{L}^{\text{orbit}}> &= \mathbf{r} \times \mathbf{G}^{\text{o}} = n_{\text{orbit}}\, n\hbar\, (\widehat{\mathbf{z}} - \frac{z}{r}\widehat{\mathbf{r}}) - n_{\text{photon}}\hbar\, k_z r\,\widehat{\phi} \\
\mathbf{L}^{\text{spin}} &= n_{\text{spin}}\hbar\widehat{\mathbf{z}}
\end{aligned}
$$

The free-space Alternate stress tensor is symmetric; therefore,

$$
v_{\phi}(n,r) = \frac{n_{\text{orbit}}}{n_{\text{photon}}}\frac{nc}{kr}
$$

which is consistent with $v_{\text{groupz}}\widehat{\mathbf{z}} + v_\phi\widehat{\phi} = \mathbf{S}^{\text{o}}/W^{\text{o}}$. The azimuthal speed v_ϕ is *not* the ϕ-component of the group velocity, $r\omega_{\text{energy}}$, but is related to the frequency of Alternate-energy circulation by

$$\omega^{\text{o}}_{\text{energy}} = \frac{\int_0^R S^{\text{o}}_\phi\, dr}{\int_0^R r\, W^{\text{o}} dr} = \frac{\int_0^R n_{\text{photon}} v_\phi\, dr}{\int_0^R n_{\text{photon}} r\, dr} = \frac{\int_0^R n_{\text{orbit}} \frac{nc}{kr} dr}{\int_0^R n_{\text{photon}} r\, dr} \tag{20.30a}$$

(where the integrations include the localized surface contributions and exterior region in case the homogeneous potentials are included). This frequency is *not* identical to that of the Poynting-energy circulation,

$$\omega_{\text{energy}} = \frac{\int_0^R <S_\phi>\, dr}{\int_0^R r <W^{\text{em}}>\, dr} = \frac{\int_0^R n_{\text{em}} \frac{nc}{kr}\, dr}{\int_0^R n_{\text{photon}} r\, dr} = c\,\frac{k_c}{k}\frac{\partial k_c(n,m)}{\partial n}(1-\delta_{n0}) \tag{20.30b}$$

because the axial spin angular momentum is contained in rG_ϕ, but *not* in rG^{o}_ϕ. Notice that both denominator integrals are equal, even though $<W^{\text{em}}> \neq W^{\text{o}}$. This follows directly from

$$<W^{\text{em}}> = \left[\frac{(2k_z^2/k_c^2+1)n_{\text{photon}} + n_{\text{em}}}{2(k_z^2/k_c^2+1)}\right]\hbar\omega \tag{20.31}$$

and Appendix F, Eqs. (F.13). For the dominant $TE_{11}{}^{\text{O}}$ mode, the average Maxwell–Poynting energy density is depicted in Figure 20.21, as a function of k_z/k_c. Only when $|k_z| \gg k_c$ does $<W^{\text{em}}> = W^{\text{o}} = n_{\text{photon}}\hbar\omega$; in this limit, $v_{\text{groupz}} = c$.

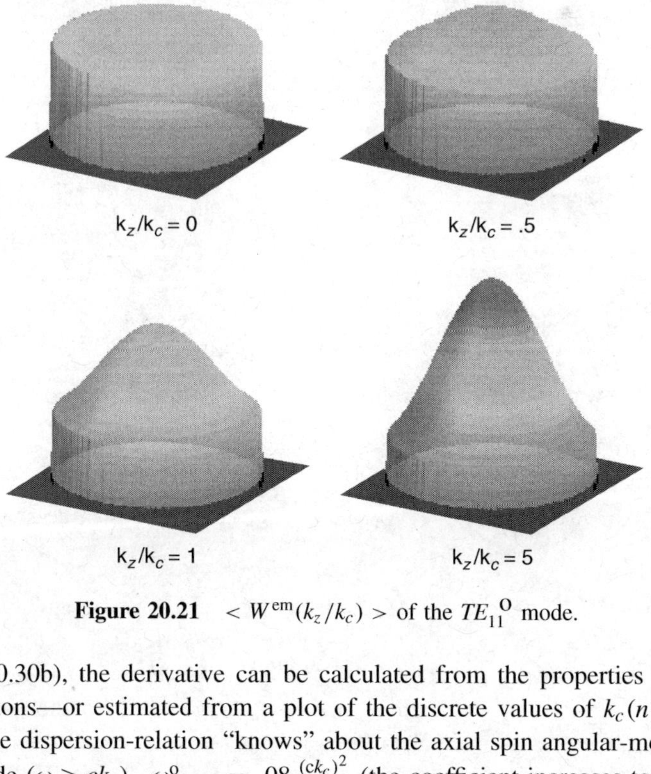

Figure 20.21 $< W^{\text{em}}(k_z/k_c) >$ of the $TE_{11}{}^{\text{O}}$ mode.

In Eq. (20.30b), the derivative can be calculated from the properties of noninteger Bessel functions—or estimated from a plot of the discrete values of $k_c(n)$ for fixed m. Evidently, the dispersion-relation "knows" about the axial spin angular-momentum. For the $TE_{11}{}^{\text{O}}$mode ($\omega \geq ck_c$), $\omega^{\text{o}}_{\text{energy}} = .08\,\frac{(ck_c)^2}{\omega}$ (the coefficient increases to .11 when the homogeneous potentials are added) whereas $\omega_{\text{energy}} = .69\,\frac{(ck_c)^2}{\omega}$. Their ratio is reasonable, given that $<\mathbf{L}^{\text{orbit}}_{\text{total per unit length}}>/<\mathbf{L}^{\text{total}}_{\text{total per unit length}}> = .16$ or $.19$. The maximum velocities occur for $\omega = ck_c$ and $r = R$; then $\omega^{\text{o}}_{\text{energy}}R = .14\text{c}$ and $\omega_{\text{energy}}R = 1.27\text{c}$.

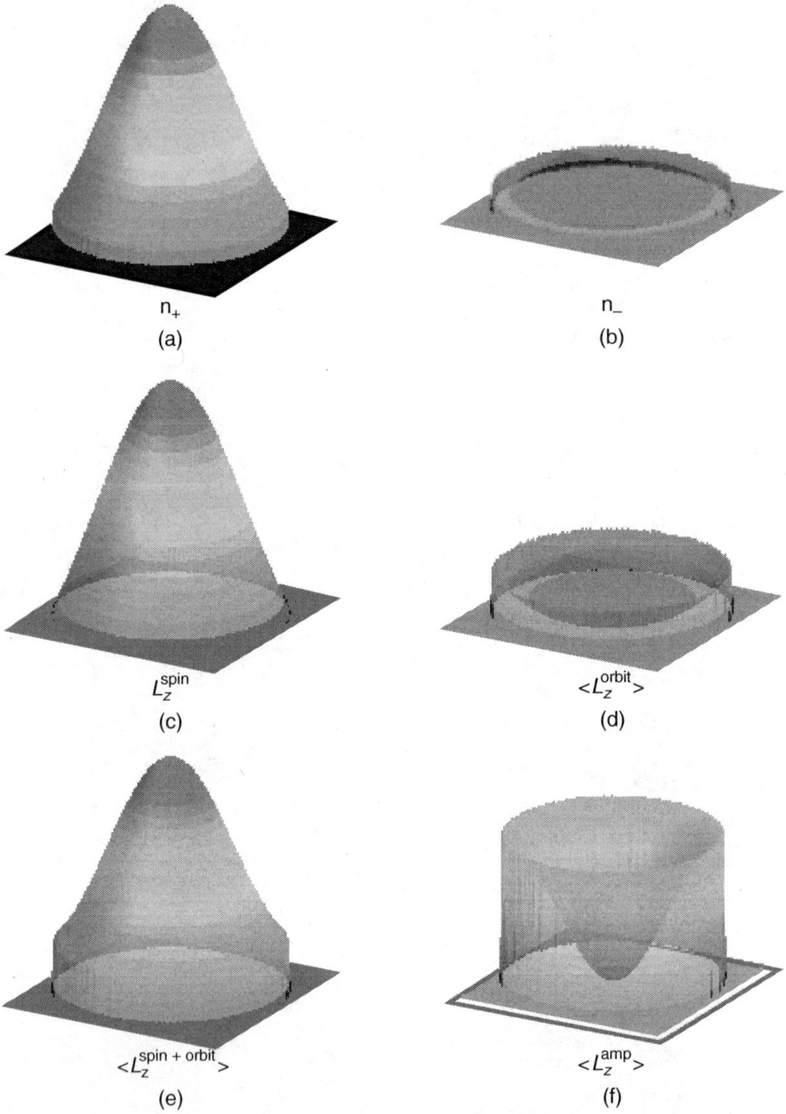

Figure 20.22 "Circularly polarized" TE_{11}^{O} mode: surface plots of spin densities and angular momenta.

Contour plots of the densities $n_\pm(r)$ and the spin and orbital angular momenta are shown in Figure 20.22 for a "circularly polarized" TE_{11}^{O} mode. The total axial densities, $<L_z^{\text{spin + orbit}}> = n(n_+ + n_-)\hbar$ and $<L_z^{\text{amp}}> = [\mathbf{r} \times (\varepsilon_0\mathbf{E} \times \mathbf{B})] \cdot \widehat{\mathbf{z}}$, are also plotted. Although when integrated over the cross section of the guide, the photon and Maxwell–Poynting representations are equivalent, the distributions are very different. Notice that $L_z^{\text{spin}} = 0$ at $r = R$. For TE_{nm}^{O} modes with standing wave ϕ variation ($\Gamma_\phi^2 = 1$), there are

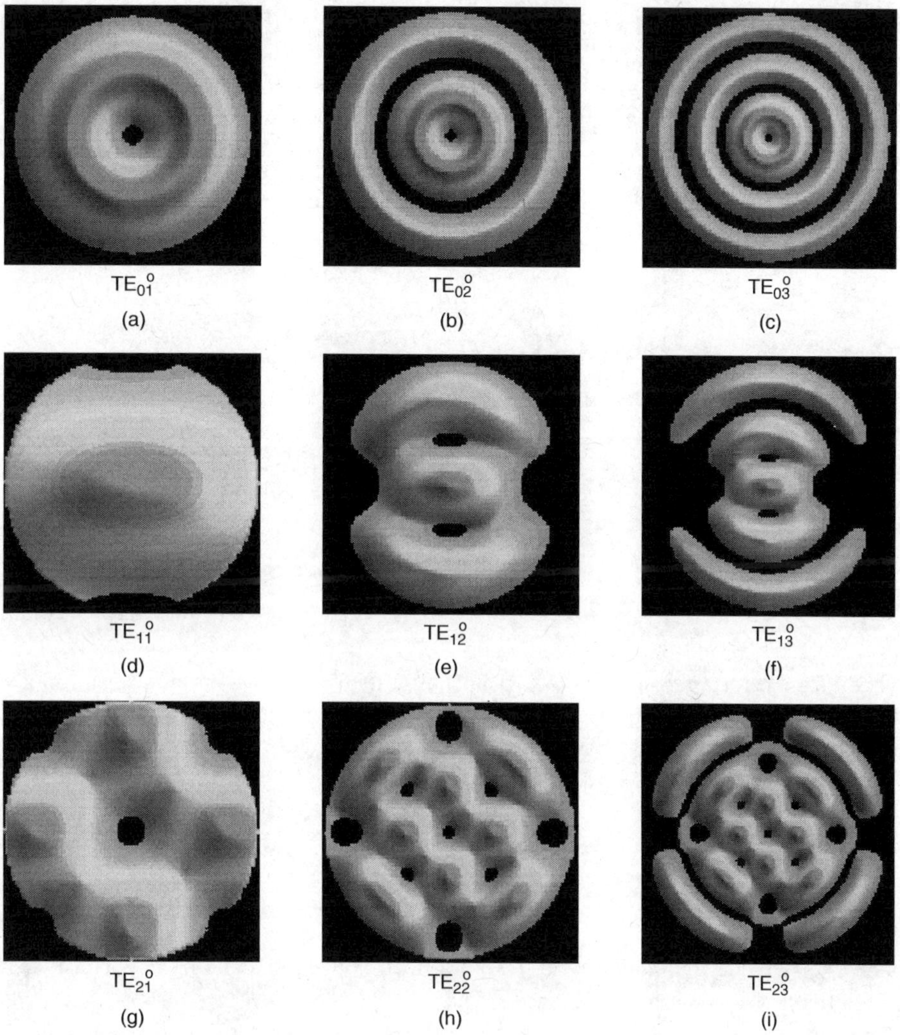

Figure 20.23 Spin densities, $n_+(r) = n_-(r)$, of TE_{nm}^{O} standing ϕ-wave modes.

equal densities of spin $+1$ and spin -1 photons and $\mathbf{L}^{\mathrm{spin}} = <L_z^{\mathrm{orbit}}> = 0$. Contour plots of the densities $n_+(r) = n_-(r)$ are shown in Figure 20.23 for TE_{nm}^{O} modes when $\Gamma_\phi = 1$ and $n = 0, 1, 2$; $m = 1, 2, 3$. These plots are rotated by 90 degrees when $\Gamma_\phi = -1$.

TM_{nm}^{o} "circularly polarized" modes

Lorenz gauge

Based upon Eqs. (20.2), the Lorenz-gauge potentials that vanish for $r \geq R$ are of the form

$$\mathbf{A} = \widehat{\mathbf{z}}\, A_{\mathrm{o}} J_n(k_c r)[\cos(\omega t - k_z z - n\phi) + \Gamma_\phi \cos(\omega t - k_z z + n\phi)] \tag{20.32a}$$

$$\Phi = c\frac{k_z}{k}\, A_{\mathrm{o}} J_n(k_c r)[\cos(\omega t - k_z z - n\phi) + \Gamma_\phi \cos(\omega t - k_z z + n\phi)] \tag{20.32b}$$

where again n is an integer of either sign and the azimuthal reflection coefficient Γ_ϕ is in the range $-1 \le \Gamma_\phi \le +1$ In what follows, we assume $\Gamma_\phi = 0$, but the results are easily generalized to arbitrary "elliptically polarized" modes.

With $\mathbf{r} = r\widehat{\mathbf{r}} + z\widehat{\mathbf{z}}$, the spin, axial-orbital, and charge angular-momentum densities are expressed as

$$\mathbf{L}^{\text{spin}} = \varepsilon_o \mathbf{E} \times \mathbf{A} \tag{20.33a}$$

$$= \varepsilon_o c \frac{k_z}{k} A_o^2 \begin{bmatrix} -\widehat{\mathbf{r}}\frac{n}{r} J_n^2(k_c r)\sin(\omega t - k_z z - n\phi)\cos(\omega t - k_z z - n\phi) \\ +\widehat{\phi} J_n(k_c r)\frac{d}{dr} J_n(k_c r)\cos^2(\omega t - k_z z - n\phi) \end{bmatrix}$$

$$\mathbf{L}^{\text{orbit}} = \varepsilon_o \left[\mathbf{E} \cdot \frac{\partial \mathbf{A}}{\partial \phi}\left(\widehat{\mathbf{z}} - \frac{z}{r}\widehat{\mathbf{r}}\right) - r\,\mathbf{E} \cdot \frac{\partial \mathbf{A}}{\partial z}\widehat{\phi} \right] \tag{20.33b}$$

$$= \varepsilon_o c \frac{k_c^2}{k} A_o^2 J_n^2(k_c r) \left(n\widehat{\mathbf{z}} - k_z r\,\widehat{\phi} - \frac{n}{r} z\widehat{\mathbf{r}} \right) \sin^2(\omega t - k_z z - n\phi)$$

$$\mathbf{L}^{\text{charge}} = 0 \tag{20.33c}$$

As is the case with TE_{nm}^{O} modes, L_z^{orbit} vanishes at the center of the guide for all modes. When integrated over the area of the guide, the axial components are

$$\frac{L_{z\text{ total per unit length}}^{\text{spin}}}{P_{\text{ave}}} = 0$$

$$\frac{<L_{z\text{ total per unit length}}^{\text{orbit}}>}{P_{\text{ave}}} = \frac{n}{c^2 k_z}$$

where P_{ave} is twice the value of Eq. (20.9) (which applies to standing ϕ-wave modes). The total is the same as for TE_{nm}^{O} modes.

The particle representation of the TE_{nm}^{O} modes can be used for TM_{nm}^{O} modes with the same definitions of $\mathbf{k}$, $\mathbf{v}_{\text{group}}$, $\overline{\mathbf{T}}^{\text{o}}$, and W^{o}, but with the following changes:

$$\mathbf{G}^{\text{o}} = \frac{1}{c^2}\mathbf{S}^{\text{o}} = n_{\text{photon}}\hbar\mathbf{k}$$

$$S_\phi^{\text{o}} = <S_\phi> = c^2 <G_\phi> = c^2 n_{\text{orbit}}\hbar\frac{n}{r}$$

$$n_{\text{photon}}\hbar = \tfrac{1}{2}\varepsilon_o c A_o^2 \frac{k_c^2}{k} J_n^2(k_c r)$$

$$n_{\text{spin}}\hbar = \tfrac{1}{2}\varepsilon_o c A_o^2 \frac{k_z}{k} J_n(k_c r)\frac{d}{dr} J_n(k_c r)$$

$$n_{\text{orbit}} = n_{\text{em}} = n_{\text{photon}}$$

$$<\mathbf{L}^{\text{orbit}}> = \mathbf{r} \times \mathbf{G}^{\text{o}} = n_{\text{orbit}}\,\hbar(n\,\widehat{\mathbf{z}} - k_z r\,\widehat{\phi} - \frac{n}{r} z\widehat{\mathbf{r}})$$

$$<\mathbf{L}^{\text{spin}}> = n_{\text{spin}}\hbar\,\widehat{\phi}$$

$$v_\phi = \frac{nc}{kr} \qquad (k \geq k_c)$$

$$\mathbf{v}_{\text{group}} = c[\frac{k_z}{k}\widehat{\mathbf{z}} + \frac{k_c}{k}\frac{r}{R}\frac{\partial u_{nm}}{\partial n}(1 - \delta_{n0})\widehat{\phi}] \quad (r \leq R)$$

It is interesting, to express these particle densities in the form

$$n_{\text{photon}} = n_+ + n_- - n_{\text{null}}$$

$$n_{\text{spin}} = (n_+ - n_-)\frac{k_z}{2n/r}$$

$$n_\pm \hbar = \tfrac{1}{4}\varepsilon_0 \omega A_o^2 \frac{k_c^2}{k^2} J_{n\mp1}^2(k_c r)$$

$$n_{\text{null}} \hbar = \tfrac{1}{4}\varepsilon_0 \omega A_o^2 \frac{k_c^2}{k^2}[J_{n-1}^2(k_c r) + J_{n+1}^2(k_c r) - 2J_n^2(k_c r)]$$

The null energy, $n_{\text{null}}\hbar\omega$, and associated null-power flux, $c^2 n_{\text{null}}\hbar\mathbf{k}$, are seen to correspond exactly with the forms of Eqs. (20.27a) and (20.27b). Although they disappear when integrated over the waveguide cross section, n_{null} is required in order to satisfy the Lorenz-gauge boundary conditions of the $TM_{nm}^{\ \text{O}}$ modes.

Although the value of $|v_\phi|$ can exceed c and, in fact, diverges at $r = 0$, this velocity is *not* the azimuthal group velocity—even though $S_\phi^{\text{o}} = <(\mathbf{E} \times \mathbf{H})_\phi> = W^{\text{o}} v_\phi$. Consequently, S_ϕ^{o} is *not* a signal circulating at speeds above c. In fact, $S_\phi^{\text{o}} \sim \frac{nk_c}{k}(k_c r)^{2n-1} \to 0$ at the center of the guide. Equations (20.30) are still applicable except $k_c R = u_{nm}$. Because $n_{\text{em}} = n_{\text{orbit}}$, we have $\omega_{\text{energy}} = \omega^{\text{o}}_{\text{energy}}$. (For the $TM_{11}^{\ \text{O}}$mode, $\omega_{\text{energy}} = .35\,\frac{(ck_c)^2}{\omega}$.)

For $n \neq 0$, $\int_0^R n_{\text{spin}} dr = 0$, and along any radius there are equal numbers of photons with ± 1 spin. For these, and also for $n = 0$ photons, the time-averaged spin density is aligned with either the $\pm\widehat{\phi}$ direction. These orientations do not depend upon the sign of n, but they reverse when k_z changes sign. Although for $|k_z| > 0$, there is spin angular-momentum, the z component is always zero. For $TM_{nm}^{\ \text{O}}$ modes with standing wave ϕ variation ($\Gamma_\phi^2 = 1$), we have $<L_z^{\text{orbit}}> = 0$.

As examples, the ϕ-directed vector fields $<\mathbf{L}^{\text{spin}}>$ of the $TM_{01}^{\ \text{O}}$ and $TM_{11}^{\ \text{O}}$ modes are shown in Figure 20.24, assuming $k_z > 0$.

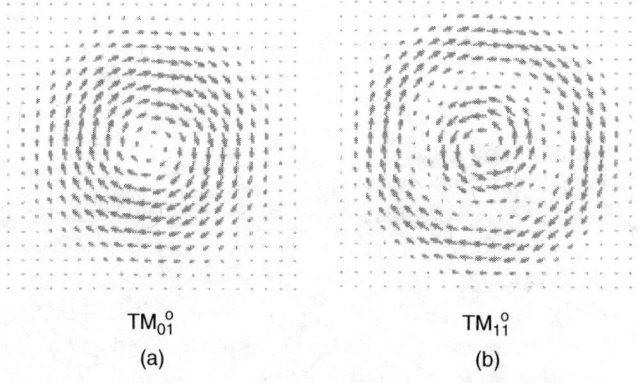

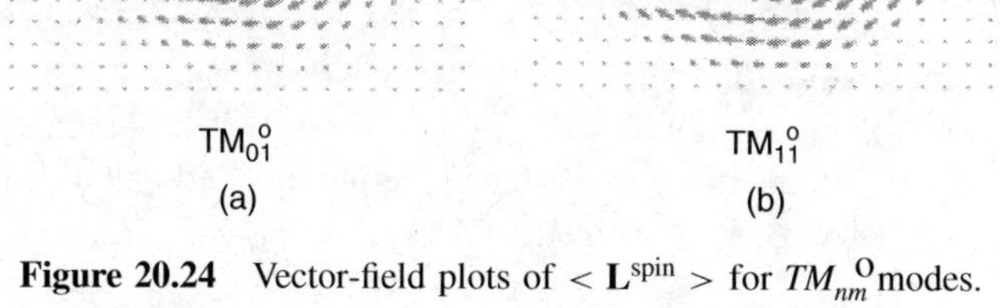

TM_{01}^{o}
(a)

TM_{11}^{o}
(b)

Figure 20.24 Vector-field plots of $< \mathbf{L}^{\text{spin}} >$ for $TM_{nm}^{\ \text{O}}$modes.

Coulomb gauge

If the special form of the Coulomb gauge ($\nabla \cdot \mathbf{A} = 0$, $\Phi = 0$) is used *without regard to matching the potentials exterior to the waveguide*, the particle representation is altered. Such a solution, based upon Eqs. (20.2) with $C_1 = C_0 = 0$ and $C_3 = -(k_c/k_z)^2 C_2$, is exact, only if the TM_{nm}^{O} mode continues uninterrupted for $r \geq R$. (We followed a similar strategy when we ignored the homogeneous potentials of the TE_{nm}^{O}modes.) After substituting $C_2 = \frac{1}{k}A_{\text{o}} = \frac{1}{k_c}A'_{\text{o}}$ (so that the total z-directed Alternate power within the waveguide is unchanged), the results are

$$W^{\text{o}} = n'_{\text{photon}}\hbar\omega$$

$$c^2\mathbf{G}^{\text{o}} = \mathbf{S}^{\text{o}} = W^{\text{o}}\left(c\frac{k_z}{k}\widehat{\mathbf{z}} + v'_\phi\widehat{\phi}\right)$$

$$\mathbf{L}^{\text{spin}} = n'_{\text{spin}}\hbar\left(\widehat{\mathbf{z}} + \frac{k_c^2}{k_z n/r}\widehat{\phi}\right)$$

$$<\mathbf{L}^{\text{orbit}}> = \mathbf{r}\times\mathbf{G}^{\text{o}} = nn'_{\text{orbit}}\hbar(\widehat{\mathbf{z}} - z/r\,\widehat{\mathbf{r}}) - n'_{\text{photon}}\hbar k_z r\,\widehat{\phi}$$

$$n'_{\text{photon}} = n'_+ + n'_- - \frac{k_c^2}{k^2}n'_{\text{null}}$$

$$n'_{\text{spin}} = \frac{k_z^2}{k^2}(n'_+ - n'_-)$$

$$n'_{\text{orbit}} = n'_{\text{photon}} - n'_{\text{spin}}/n$$

$$n'_\pm\hbar = \tfrac{1}{4}\varepsilon_{\text{o}}\omega A'^2_{\text{o}} J^2_{n\mp1}(k_c r)$$

$$n'_{\text{null}}\hbar = \tfrac{1}{4}\varepsilon_{\text{o}}\omega A'^2_{\text{o}}[J^2_{n-1}(k_c r) + J^2_{n+1}(k_c r) - 2J^2_n(k_c r)]$$

$$v'_\phi = \frac{nn'_{\text{orbit}}}{n'_{\text{photon}}}\frac{c}{kr} \qquad (k \geq k_c)$$

Now, $\mathbf{L}^{\text{spin}}$ is time-independent, with an axial component. At cutoff ($k_z = 0$), $n'_{\text{photon}}$is identical to n_{photon} but changes as k_z increases. The quantities n'_{orbit}, v'_ϕ, and $\omega'^{\text{o}}_{\text{energy}}$ are also modified. (For the TM_{11}^{O}mode, the ratio $\omega\,\omega'^{\text{o}}_{\text{energy}}/\,(ck_c)^2$ varies from .14 when $k_z/k \to 0$ to .35 when $k_z/k \to 1$.) Despite the changes, $<L_{z\text{ total per unit length}}^{\text{spin + orbit}}>/P_{\text{ave}}$ remains the same. So, too, does the form of $<L_\phi^{\text{spin}}>$; therefore, Figure 20.24 still applies.

Modes of a square waveguide

In case a rectangular guide has a square cross section, the $TE_{nm}^{\square}$ and $TE_{mn}^{\square}$ modes share identical frequencies and are therefore termed degenerate. This means that a linear combination of phase quadrature modes is "elliptically polarized" in the sense used above and "circularly polarized" if the amplitudes are equal. When carried out, analyses similar to those of the previous sections place in evidence the photon spin and angular momentum densities. Naturally, the same is true for the $TM_{nm}^{\square}$ and $TM_{mn}^{\square}$ modes.

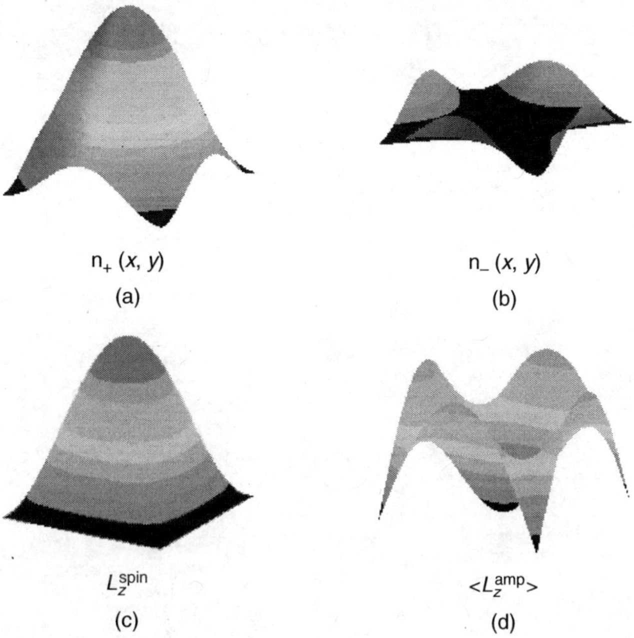

Figure 20.25 Spin and angular-momentum densities of a "circularly polarized" $TE_{01}^{\square}$–$TE_{10}^{\square}$ mode.

Contour plots of $n_+(x, y)$, $n_-(x, y)$, $L^{spin}(x, y) = [n_+(x, y) - n_-(x, y)]\hbar$, and $L_z^{amp}(x, y)$ for a "circularly polarized" $TE_{01}^{\square}$–$TE_{10}^{\square}$ mode are shown in Figure 20.25. The latter is calculated from $\mathbf{L}^{amp} = \mathbf{r} \times (\varepsilon_0 \mathbf{E} \times \mathbf{B})$ (with the origin at the center of the guide). Because $<L_z^{orbit}> = 0$, the spatial averages of $L_z^{spin}(x, y)$ and $L_z^{amp}(x, y)$ are equal. These plots should be compared with Figure 20.22.

CHAPTER 21

DIELECTRIC WAVEGUIDES

21.1 INTRODUCTION

Dielectric waveguides can be fabricated in many different geometries. Here we discuss a uniform parallel-plate waveguide partially loaded with a rectangular slab of lossless dielectric material. The cross-sectional geometry and coordinate system is shown in Figure 21.1. For simplicity we restrict the analysis to modes that have complex fields without y variation, which are traveling in the $+z$ direction as $\exp(-jk_z z)$. Both uniform and nonuniform plane waves are present. (These have already been discussed in Chapter 16.) In the dielectric slab, the x variations are either $\sin(k_x x)$ or $\cos(k_x x)$; in the air regions, they are linear combinations of $\exp(\pm\alpha_x x)$. To simplify the analysis, we neglect fringing fields near the edges and outside the plate region. The symmetry of the guide requires the Cartesian components of both the electric and magnetic field to be either an even or odd function of x. We define the mode as even or odd, depending upon the symmetry of $\underline{E}_y(x)$; the fields of each mode (which must be *TE* because, without y variation, $\underline{E}_x = \underline{E}_z = 0$) can be written down by inspection.

The Power and Beauty of Electromagnetic Fields, First Edition. F. R. Morgenthaler.

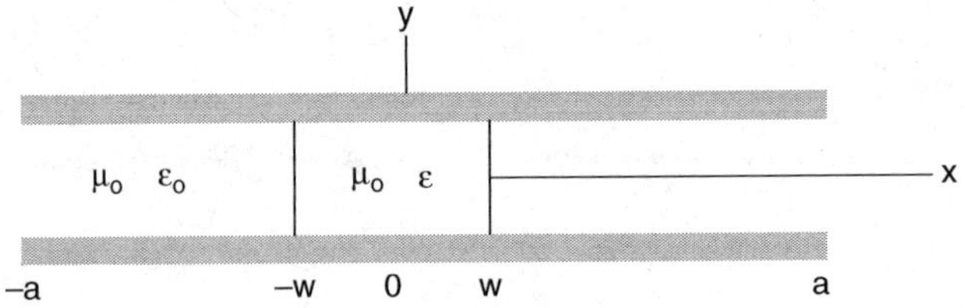

Figure 21.1 Dielectric waveguide geometry.

21.2 SYMMETRIC *TE* MODES

$$\underline{E}_y^d = \underline{E}_0 \cos(k_x x) \exp(-jk_z z) \tag{21.1a}$$

$$\underline{H}_x^d = -\frac{k_z}{\omega\mu_o} \underline{E}_0 \cos(k_x x) \exp(-jk_z z) \tag{21.1b}$$

$$\underline{H}_z^d = -j\frac{k_x}{\omega\mu_o} \underline{E}_0 \sin(k_x x) \exp(-jk_z z) \tag{21.1c}$$

$$\underline{E}_y^a = [\underline{E}_1 \exp(\alpha_x |x|) + \underline{E}_2 \exp(-\alpha_x |x|)] \exp(-jk_z z) \tag{21.2a}$$

$$\underline{H}_x^a = -\frac{k_z}{\omega\mu_o} [\underline{E}_1 \exp(\alpha_x |x|) + \underline{E}_2 \exp(-\alpha_x |x|)] \exp(-jk_z z) \tag{21.2b}$$

$$\underline{H}_z^a = j\frac{x}{|x|} \frac{\alpha_x}{\omega\mu_o} [\underline{E}_1 \exp(\alpha_x |x|) - \underline{E}_2 \exp(-\alpha_x |x|)] \exp(-jk_z z) \tag{21.2c}$$

The boundary conditions at the interfaces $x = \pm w$ require continuity of $\underline{E}_y$ and $\underline{H}_z$. ($\underline{H}_x$ is then automatically continuous.)

$$\underline{E}_0 \cos(k_x w) = \underline{E}_1 \exp(\alpha_x w) + \underline{E}_2 \exp(-\alpha_x w)$$

$$-k_x \underline{E}_0 \sin(k_x w) = \alpha_x [\underline{E}_1 \exp(\alpha_x w) - \underline{E}_2 \exp(-\alpha_x w)]$$

The open-circuit conditions at $x = \pm a$ require that $\underline{K}_x = 0$. This condition is met when $\underline{H}_z = 0$ or

$$\underline{E}_1 \exp(\alpha_x a) + \underline{E}_2 \exp(-\alpha_x a) = 0$$

One relationship between a_x and k_x is found by solving the three boundary-condition equations; the result is

$$\alpha_x = k_x \tan(k_x w) \frac{\exp[2\alpha_x (a - w)] + 1}{\exp[2\alpha_x (a - w)] - 1} \tag{21.3}$$

21.3 ANTISYMMETRIC *TE* MODES

$$\underline{E}_y^d = \underline{E}_{20} \sin(k_x x) \exp(-jk_z z) \tag{21.4a}$$

$$\underline{H}_x^d = \underline{H}_{10} \sin(k_x x) \exp(-jk_z z) \tag{21.4b}$$

$$\underline{H}_z^d = -j\frac{k_x}{k_z} \underline{H}_{10} \cos(k_x x) \exp(-jk_z z) \tag{21.4c}$$

$$\underline{E}_y^a = \frac{x}{|x|}[\underline{E}_1 \exp(\alpha_x |x|) - \underline{E}_2 \exp(-\alpha_x |x|)] \exp(-jk_z z) \tag{21.5a}$$

$$\underline{H}_x^a = -\frac{x}{|x|}\frac{k_z}{\omega\mu_o}[\underline{E}_1 \exp(\alpha_x |x|) - \underline{E}_2 \exp(-\alpha_x |x|)] \exp(-jk_z z) \tag{21.5b}$$

$$\underline{H}_z^a = j\frac{\alpha_x}{\omega\mu_o}[\underline{E}_1 \exp(\alpha_x |x|) + \underline{E}_2 \exp(-\alpha_x |x|)] \exp(-jk_z z) \tag{21.5c}$$

As with the even modes, the boundary conditions at the interfaces $x = \pm w$ require continuity of $\underline{E}_y$ and $\underline{H}_z$.

$$\underline{E}_0 \sin(k_x w) = \underline{E}_1 \exp(\alpha_x w) - \underline{E}_2 \exp(-\alpha_x w)$$

$$k_x \underline{E}_0 \cos(k_x w) = \alpha_x [\underline{E}_1 \exp(\alpha_x w) + \underline{E}_2 \exp(-\alpha_x w)]$$

Again, the open-circuit conditions at $x = \pm a$ require that $\underline{K}_x = 0$. This condition is met when $\underline{H}_z = 0$ or

$$\underline{E}_1 \exp(\alpha_x a) - \underline{E}_2 \exp(-\alpha_x a) = 0$$

One relationship between a_x and k_x is found by solving the three boundary-condition equations; the result is

$$\alpha_x = -k_x \cot(k_x w)\frac{\exp[2\alpha_x(a - w)] - 1}{\exp[2\alpha_x(a - w)] + 1} \tag{21.6}$$

21.4 DISPERSION RELATIONS

In the dielectric slab, the uniform plane waves require that $k_x^2 + k_z^2 = \omega^2\mu_o\varepsilon$; in the air regions, the nonuniform plane waves require that $-\alpha_x^2 + k_z^2 = \omega^2\mu_o\varepsilon_o$. The difference of these equations generates the equation of a circle,

$$\alpha_x^2 + k_x^2 = \omega^2\mu_o(\varepsilon - \varepsilon_o) \tag{21.7}$$

This relationship between a_x and k_z is common to both sets of modes and together with either Eq. (21.3) or Eq. (21.6) allows α_x and k_x to be found as a function of ω. This is conveniently done by plotting $\alpha_x(k_x)$ for either the even or odd modes on the circle plot. The intersections produce a graphical solution to the equations that then determines the dispersion relation $k_z(\omega)$. This is shown in Figure 21.2 for the case $a \to \infty$. The solid curve is for even modes, and the dotted curve is for odd modes. The circles are drawn for two different frequencies; for the lower frequency, only one even mode can propagate; for the higher frequency, two even and one odd mode can propagate.

The group velocities can then be found from $\frac{\partial\omega}{\partial k_z}$. As with waveguide modes, a particular mode may not propagate below some critical frequency. In general, the even and odd modes have different cutoff frequencies, but the lowest even mode can always propagate; in the low-frequency limit, it becomes a quasi-*TEM* mode subject to general transmission line analysis. For other modes and depending upon the frequency, the value of α_x may be large; in such cases the Poynting energy is stored mainly in the dielectric. Here the confinement is one-dimensional, but it is also possible to localize the energy in a dielectric in two or three dimensions. The latter forms the basis of dielectric resonators discussed further in Chapter 24, Section 24.9.

The graphical analysis is illuminating and (it or its numerical counterpart) is necessary when a is finite. However, for the case $a \to \infty$, analytic expressions can easily be found for $\omega(k_x)$, $k_z(k_x)$, $v_{\text{phase}}(k_x)$, and $v_{\text{group}}(k_x)$. For the two types of modes the results are:

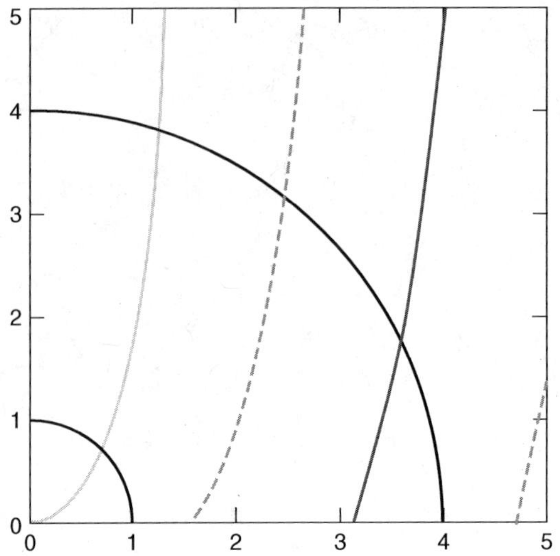

Figure 21.2 $\alpha_x w$ versus $k_x w$.

Symmetric modes

$$\alpha_x = k_x \tan(k_x w) \geq 0 \tag{21.8a}$$

$$\omega = \frac{ck_x\sqrt{\tan^2 k_x w + 1}}{\sqrt{\varepsilon/\varepsilon_0 - 1}} \tag{21.8b}$$

$$k_z = \frac{k_x\sqrt{\varepsilon/\varepsilon_0 \tan^2 k_x w + 1}}{\sqrt{\varepsilon/\varepsilon_0 - 1}} \tag{21.8c}$$

$$v_{\text{phase}} = \frac{\omega}{k_z} = \frac{c}{\sqrt{\varepsilon/\varepsilon_0 + (1 - \varepsilon/\varepsilon_0)\cos^2 k_x w}} \tag{21.8d}$$

$$v_{\text{group}} = \frac{\partial\omega}{\partial k_z} = \frac{c(k_x w \sin k_x w + \cos k_x w)\sqrt{1 + (\varepsilon/\varepsilon_0 - 1)\sin^2 k_x w}}{\varepsilon/\varepsilon_0(k_x w \sin k_x w + \cos k_x w) + (1 - \varepsilon/\varepsilon_0)\cos^3 k_x w} \tag{21.8e}$$

When $k_x w \ll 1$, the dielectric slab has negligible effect, $\omega \simeq ck_z$ and $v_{\text{phase}} \simeq v_{\text{group}} \simeq c$; when $k_x w \to \pi/2$, there is complete confinement and $v_{\text{phase}} \simeq v_{\text{group}} \simeq c/\sqrt{\varepsilon/\varepsilon_0}$.

Antisymmetric modes

$$\alpha_x = -k_x \cot(k_x w) \geq 0 \tag{21.9a}$$

$$\omega = \frac{ck_x\sqrt{\cot^2 k_x w + 1}}{\sqrt{\varepsilon/\varepsilon_0 - 1}} \tag{21.9b}$$

$$k_z = \frac{k_x\sqrt{\varepsilon/\varepsilon_0 \cot^2 k_x w + 1}}{\sqrt{\varepsilon/\varepsilon_0 - 1}} \tag{21.9c}$$

$$v_{\text{phase}} = \frac{\omega}{k_z} = \frac{c}{\sqrt{1 + (\varepsilon/\varepsilon_0 - 1)\cos^2 k_x w}} \tag{21.9d}$$

$$v_{\text{group}} = \frac{\partial \omega}{\partial k_z} = \frac{c(k_x w \cos k_x w - \sin k_x w)\sqrt{1 + (\varepsilon/\varepsilon_0 - 1)\cos^2 k_x w}}{\varepsilon/\varepsilon_0(k_x w \cos k_x w - \sin k_x w) + (\varepsilon/\varepsilon_0 - 1)\sin^3 k_x w} \tag{21.9e}$$

When $k_x w \gtrapprox \pi/2$, the dielectric slab has negligible effect, $\omega \simeq ck_z$ and $v_{\text{phase}} \simeq v_{\text{group}} \simeq c$; when $k_x w \to \pi$, there is complete confinement and $v_{\text{phase}} \simeq v_{\text{group}} \simeq c/\sqrt{\varepsilon/\varepsilon_0}$.

CHAPTER 22

ANTENNAS AND DIFFRACTION

22.1 INTRODUCTION

The Hertzian dipoles (electric and magnetic), which have been discussed in Chapter 13, form the basis for understanding all antenna configurations which can be analyzed as linear superpositions of these important elements. Although this is fairly obvious in the case of thin-wire antennas (considered individually or in arrays), it is also true for aperture antennas, such as large paraboloid satellite dishes. This is because, by the equivalence principle, the aperture fields can be replaced with surface currents (electric and/or magnetic); consequently, each differential element of area acts as an elementary dipole. Interactions of electromagnetic waves with apertures of all sorts—and with reflecting surfaces (planar or curved)—are topics of diffraction theory that employ similar analytical techniques. The examples analyzed in this chapter provide an introduction to antenna and diffraction theory.[1] The receiving circuit of a thin-wire dipole antenna is discussed in Chapter 24.

[1]For many years, Lan Chu taught a legendary graduate subject on antennas. Because he never wrote a textbook or issued class notes, there was concern, after his untimely death in 1973, that some of his unique perspectives might be lost. Fortunately, Dr. Allan C. Schell, his former student and an antenna expert in his own right, agreed to return to M.I.T. for a semester to both teach the subject and prepare course notes. Allan was the ideal person to do so both because he had great familiarity with the course and enjoyed Chu's full confidence. (Allan had previously been selected by Chu to substitute for him during several academic leaves.) Subsequently, the present author took over teaching the course and for two decades used those notes (and his own memories of the subject) to try and keep the Chu approach to antennas alive. It is a pleasure to acknowledge that the preparation of this chapter benefited from both influences.

The Power and Beauty of Electromagnetic Fields, First Edition. F. R. Morgenthaler.

22.2 HALF-WAVE DIPOLES

Wire antenna

The analysis of the half-wave wire dipole is similar to that of the short Hertzian electric dipole, except that the current is nonuniform and the phase differences between waves originating at different points along its length must be taken into account. We assume that the dipole is fed by a current source inserted in a small gap at the midpoint of the antenna, and that the current distribution along the wire is a sinusoidal standing wave with nodes at both ends. As the first step, we evaluate the complex vector potential from

$$\underline{A}_z = \frac{\mu_0}{4\pi r_p} \exp(-jkr_p) \int_{-\ell/2}^{+\ell/2} \underline{I}_z(z_q) \exp\left(jk \frac{\mathbf{r}_q \cdot \mathbf{r}_p}{r_p}\right) dz_q \tag{22.1}$$

where $\ell = \lambda/2$, $\underline{I}_z(z_q) = \underline{\mathrm{I}}_0 \cos(\pi z_q/\ell)$, and $r_p \gg \ell$. Because the current is an even function, the integral simplifies to

$$\underline{A}_z = \frac{\mu_0 \underline{\mathrm{I}}_0}{2\pi k r_p} \exp(-jkr_p) \int_0^{\pi/2} \cos(kz_q) \cos(kz_q \cos\theta_p)\, d(kz_q) \tag{22.2}$$

Making use of the trigonometric identity,

$$2\cos(kz_q)\cos(kz_q\cos\theta_p) = \cos[(1+\cos\theta_p)kz_q] + \cos[(1-\cos\theta_p)kz_q]$$

and dropping the subscript p, we obtain

$$\underline{A}_z = \frac{\mu_0 \underline{\mathrm{I}}_0}{2\pi kr} \exp(-jkr) \frac{\cos(\pi/2\cos\theta)}{\sin^2\theta} \tag{22.3}$$

From the Lorenz-gauge condition, we have

$$\underline{\Phi} = \frac{1}{-j\omega\mu_0\varepsilon_0} \frac{\partial \underline{A}_z}{\partial z}$$

In the far field, this is expressed as

$$\underline{\Phi} = \frac{1}{\omega\varepsilon_0} \frac{\underline{\mathrm{I}}_0}{2\pi r} \left(1 + \frac{1}{jkr}\right) \exp(-jkr) \cos\theta \frac{\cos(\pi/2\cos\theta)}{\sin^2\theta} \tag{22.4}$$

but only the $1/r$ term need be retained.

The complex Alternate-power flux is then easily evaluated from

$$\underline{\mathbf{S}}^o = \frac{-j\omega}{4}\left[\frac{1}{\mu_0}(\underline{A}_z \nabla \underline{A}_z^* - \underline{A}_z^* \nabla \underline{A}_z) - \varepsilon_0(\underline{\Phi}\nabla\underline{\Phi}^* - \underline{\Phi}^*\nabla\underline{\Phi})\right]$$

and results in

$$\underline{\mathbf{S}}^o = \hat{\mathbf{r}} \sqrt{\frac{\mu_0}{\varepsilon_0}} \frac{|\underline{\mathrm{I}}_0|^2}{8\pi^2 r^2} \left[\frac{\cos(\pi/2\cos\theta)}{\sin^2\theta}\right]^2 (1 - \cos^2\theta)$$

where, as in the case of the Hertzian dipole, the factor $(1 - \cos^2\theta)$ arises from the difference between the vector and scalar potential terms and is the dipole element factor. When absorbed into the final formula, we get

$$\underline{\mathbf{S}}^o = \hat{\mathbf{r}} \sqrt{\frac{\mu_0}{\varepsilon_0}} \frac{|\underline{\mathrm{I}}_0|^2}{8\pi^2 r^2} \left[\frac{\cos(\pi/2\cos\theta)}{\sin\theta}\right]^2 \tag{22.5}$$

Of course, the same formula can be obtained by solving for $\underline{\mathbf{E}}$ and $\underline{\mathbf{H}}$ and calculating $\frac{1}{2}\underline{\mathbf{E}} \times \underline{\mathbf{H}}^*$, but there is no need to do so.

Directivity and radiation resistance

The total average power radiated by the half-wave dipole is found by integrating $\underline{S}^{\mathrm{o}}$ over the surface of a sphere of radius r; the result is

$$<P_{\mathrm{rad}}> = \sqrt{\frac{\mu_{\mathrm{o}}}{\varepsilon_{\mathrm{o}}}} \frac{|\underline{\mathrm{I}}_{\mathrm{o}}|^2}{4\pi} \int_0^{\pi} \frac{\cos^2(\pi/2 \cos\theta)}{\sin\theta} \, d\theta \qquad (22.6)$$

$$= \frac{1}{2}\sqrt{\frac{\mu_{\mathrm{o}}}{\varepsilon_{\mathrm{o}}}} \frac{|\underline{\mathrm{I}}_{\mathrm{o}}|^2}{2\pi} 1.219$$

which, when equated to $<P_{\mathrm{rad}}> = \frac{1}{2} R_{\mathrm{rad}} |\underline{\mathrm{I}}_{\mathrm{o}}|^2$, determines the radiation resistance,

$$R_{\mathrm{rad}} = 73.14 \text{ ohms}$$

Because this value is large compared to copper losses in the wire, and the dipole is self-resonant, high-efficiency operation is possible.

The directivity or gain follows from the basic definition and is expressed as

$$D(\theta) = \frac{4\pi r^2 \underline{S}^{\mathrm{o}}}{<P_{\mathrm{rad}}>} = 1.641 \left[\frac{\cos(\pi/2 \cos\theta)}{\sin\theta} \right]^2 \qquad (22.7)$$

The peak directivity, 1.641, is only slightly greater than the 1.5 of a Hertzian dipole. The power pattern is plotted in Figure 22.1 and compared to that of a simple dipole.

Gain-resistance product

For any symmetrically fed wire antenna, the peak directivity occurs when $\theta = \pi/2$; for this direction, the retarded phase factor $k\mathbf{r}_q \cdot \widehat{\mathbf{r}}_p = 0$ and the integral of the line current

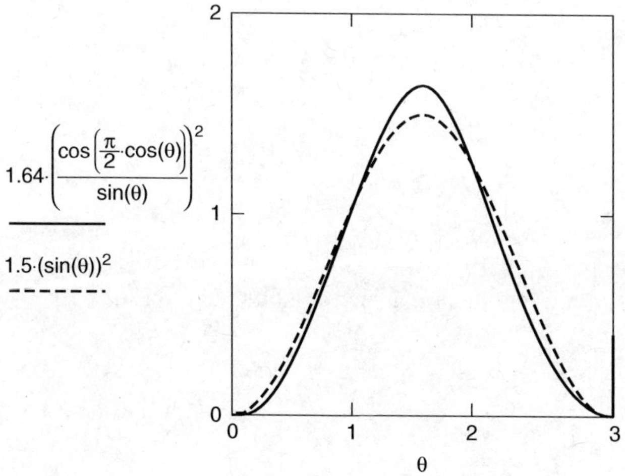

Figure 22.1 Directivity (gain) of a half-wave dipole.

over the length is simply the average current times the length. This before can be written as $<I_z(z)> \ell = I_0 \ell_{eff}$, where I_0 is the peak current. It follows that for any (symmetric) antenna, the product of the peak directivity and the radiation resistance is given by

$$D_o R_{rad} = \frac{1}{4\pi}\sqrt{\frac{\mu_0}{\varepsilon_0}}(k\ell_{eff})^2 \simeq 30(k\ell_{eff})^2 \tag{22.8}$$

This agrees with the analysis just completed since for the cosine dependence of the current, $\ell_{eff} = 2/\pi\ \ell$ and $k\ell_{eff} = 2$.

Thin slot in a ground plane

Although we have modeled the current as a line current of zero radius, the theory also applies to current flow on conductors of uniform but arbitrary cross section, as long as the transverse dimensions are all very small compared to a wavelength. In particular, consider a very thin rectangular strip made from a perfect conductor of width a, which is $\lambda/2$ in length. Assume that it lies in the $x - z$ plane. If a small gap is made at $z = 0$ (the midpoint of its length) and an ideal voltage source is inserted, the far fields will be identical to the half-wave dipole we have just analyzed, provided that we can express the gap current in terms of the voltage. But the complementary structure of the narrow rectangular strip is an infinite ground plane ($y = 0$) with a rectangular slot cut out of it (of width a and length $\lambda/2$). Naturally, any practical ground plane is of finite size, but the analysis is still approximately valid as long as the dimensions of the plane are large compared to a wavelength. The far fields generated by this thin slot are *nearly* the dual fields of the strip dipole, provided that the dual of the current source connected across the small gap is an ideal voltage source that connects the edges of the slot at mid-length. The difference between the two sets of fields lies in the type of plane symmetry that characterizes them. In the nomenclature of Chapter 6, Section 6.9, the strip-antenna fields, $\mathbf{E}^{strip}$ and $\mathbf{H}^{strip}$, are Type I fields; the slot-antenna fields, $\mathbf{E}^{slot}$ and $\mathbf{H}^{slot}$, are Type II. Therefore on *either side* of the ground plane (but not on both sides at the same time), the fields and potentials are related by

$$\mathbf{E}^{slot} = \sqrt{\frac{\mu_0}{\varepsilon_0}}\mathbf{H}^{strip} \tag{22.9a}$$

$$\mathbf{H}^{slot} = -\sqrt{\frac{\varepsilon_0}{\mu_0}}\mathbf{E}^{strip} \tag{22.9b}$$

$$\mathbf{A}_m^{slot} = -\sqrt{\frac{\varepsilon_0}{\mu_0}}\mathbf{A}^{strip} \tag{22.9c}$$

$$\Phi_m^{slot} = -\sqrt{\frac{\varepsilon_0}{\mu_0}}\Phi^{strip} \tag{22.9d}$$

The z-directed current on the strip antenna and the dual quantity (the slot voltage) are given by

$$\underline{I}_z(z) = 2\int_{-a/2}^{a/2} \underline{H}_x^{strip}(x, 0_+, z)\, dz = \underline{\mathrm{I}}_0^{strip}\cos\left(\frac{\pi}{\ell}z\right)$$

$$\underline{V}_x(z) = \int_{-a/2}^{a/2} \underline{E}_x^{slot}(x, 0, z)\, dz = \underline{\mathrm{V}}_0^{slot}\cos\left(\frac{\pi}{\ell}z\right)$$

The voltage and current appearing across the respective sources, each of which occupy a gap of size g, are

$$\underline{V}^{\text{strip}} = \int_{-g/2}^{g/2} \underline{E}_z^{\text{strip}}(x, 0, z)\, dz$$

$$\underline{I}^{\text{slot}} = 2\int_{-g/2}^{g/2} \underline{H}_z^{\text{slot}}(x, 0_+, z)\, dz$$

The factors of two arises because, for Type I symmetry, $\underline{H}_x^{\text{strip}}(x, 0_+, z) = -\underline{H}_x^{\text{strip}}(x, 0_-, z)$ and for Type II symmetry, $\underline{H}_z^{\text{slot}}(x, 0_+, z) = -\underline{H}_z^{\text{slot}}(x, 0_-, z)$. It follows that the dual fields are equal when

$$2\underline{V}_{\text{o}}^{\text{slot}} = \sqrt{\frac{\mu_{\text{o}}}{\varepsilon_{\text{o}}}}\,\underline{I}^{\text{strip}}$$

$$2\underline{V}^{\text{strip}} = \sqrt{\frac{\mu_{\text{o}}}{\varepsilon_{\text{o}}}}\,\underline{I}^{\text{slot}}$$

From the product of these two equations, it follows that

$$\frac{\underline{V}_{\text{o}}^{\text{slot}}}{\underline{I}^{\text{slot}}}\,\frac{\underline{V}^{\text{strip}}}{\underline{I}_{\text{o}}^{\text{strip}}} = \frac{1}{4}\frac{\mu_{\text{o}}}{\varepsilon_{\text{o}}}$$

and, in fact, the product of the impedances of *any* planar antenna and its complement is

$$\underline{Z}\,\underline{Z}' = \frac{1}{4}\eta_{\text{o}}^2 = (60\pi)^2 \text{ ohms} \tag{22.10}$$

Because the power patterns of the antenna and its complement are the same, the directivities of the two antennas are equal, but the radiation resistances are not. For the slot dipole, $R_{\text{rad}}^{\text{slot}} = 485.8$ ohms.

22.3 SELF-COMPLEMENTARY PLANAR ANTENNAS

There is a class of planar antennas (made by selectively removing sections of a large ground plane) that are approximately self-complementary. In other words, the antenna and its complement are essentially identical (except for rotation and/or lateral displacement). A simple example of such a pattern is the chessboard, where the black squares are thin conducting sheets attached at their corners and the white squares are open spaces. The attachment between two centrally located black squares is broken to provide a small gap for feeding power into the structure. Other, more practical geometries are bow-tie or pin-wheel-like patterns such as those depicted in Figure 22.2.

Since $\underline{Z} = \underline{Z}'$, it follows that $\underline{Z} = 188.5$ ohms; this is a real impedance that is constant over a wide frequency range. However, the analysis is approximately valid only as long as the dimensions of the structure are large compared to a wavelength. This restriction places a low-frequency cutoff on this type of antenna, a characteristic shared by all antennas of finite size.

22.4 TRAVELING-WAVE WIRE ANTENNAS

Assume now that the complex current on the antenna is a traveling wave of the form

$$\underline{I}_z(z_q) = \underline{I}_{\text{o}} \exp(-jk_z z_q)$$

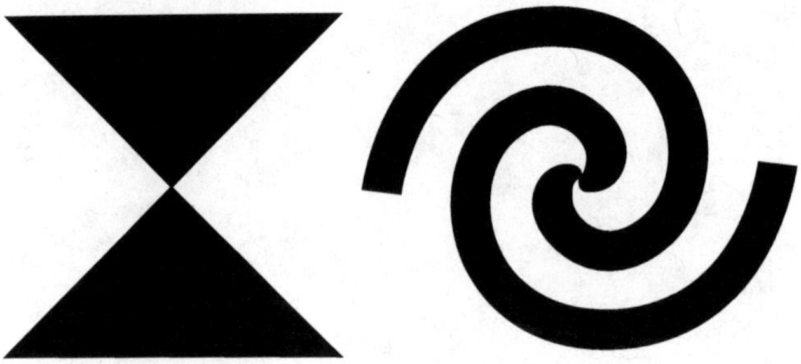

Figure 22.2 Self-complementary antennas.

where k_z is a constant not necessarily equal to the free-space wavenumber, k.

$$\underline{A}_z = \frac{\mu_o}{4\pi r_p}\exp(-jkr_p)\int_{-\ell/2}^{+\ell/2} \underline{I}_z(z_q)\exp(jkz_q\cos\theta)\, dz_q \tag{22.11}$$

Then

$$\underline{A}_z = \frac{\mu_o \underline{I}_o}{4\pi r_p}\exp(-jkr_p)\int_{-\ell/2}^{+\ell/2} \exp[j(k\cos\theta - k_z)z_q]\, dz_q$$

$$= \frac{\mu_o \underline{I}_o \ell}{4\pi r_p}\exp(-jkr_p)\frac{\sin[(k\cos\theta - k_z)\ell/2]}{(k\cos\theta - k_z)\ell/2}$$

and from the Lorenz-gauge condition, the far-field potential is

$$\underline{\Phi} = \frac{-1}{j\omega\mu_o\varepsilon_o}\frac{\partial \underline{A}_z}{\partial z} \to \mathrm{c}\,\underline{A}_z\cos\theta$$

With this value, the complex Alternate-power flux is easily evaluated in the far-field from

$$\underline{\mathbf{S}}^{\mathrm{o}} \to \frac{-j\omega}{4}\frac{1}{\mu_o}(\underline{A}_z\nabla\underline{A}_z^* - \underline{A}_z^*\nabla\underline{A}_z)(1-\cos^2\theta)$$

and is found to be

$$\underline{\mathbf{S}}^{\mathrm{o}} = \hat{\mathbf{r}}\frac{\mathrm{c}\mu_o}{2}\frac{|\underline{I}_o|^2}{16\pi^2 r^2}k^2\ell^2\left\{\frac{\sin[(k\cos\theta - k_z)\ell/2]}{(k\cos\theta - k_z)\ell/2}\right\}^2\sin^2\theta \tag{22.12}$$

Because the Alternate power flux and energy density are independent of time, this formula must be valid for all $r > \ell$. As expected, the flux is proportional to the product of the array and element factors. For large values of $k\ell$, the $(\sin x)/x$ function is sharply peaked at $x = 0$. Then, unless the peak occurs near $\theta = 0$, the dipole element factor can be considered constant over the main beam which occurs for $\theta = \theta_o$ where

$$\cos\theta_o = k_z/k \le 1$$

The beam width, $\Delta\theta$, defined by the first set of nulls, is approximately given by

$$\Delta\theta = \frac{4}{\pi k\ell\sin\theta_o}$$

provided that $\Delta\theta \ll \theta_o$.

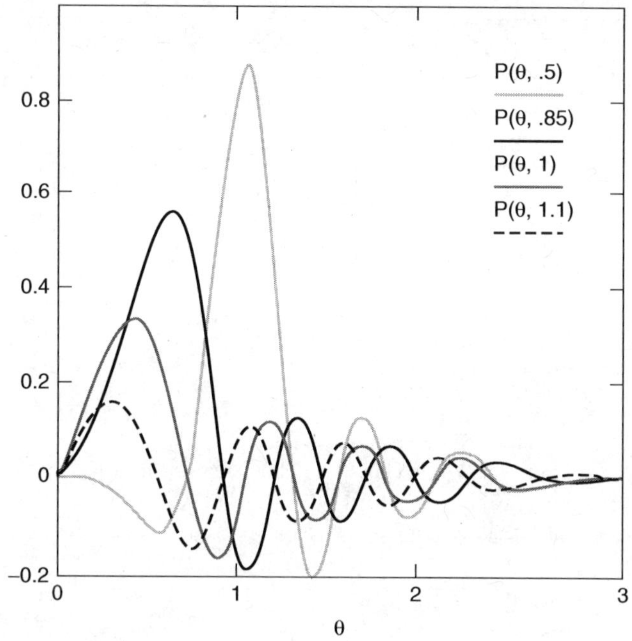

Figure 22.3 Pattern function, $P(\theta, k_z/k)$.

If $k_z = k$, the position of the beam, $\theta = 0$ coincides with the null of the dipole. In this case, there is no radiation in the end-fire direction. But what of values of $k_z/k > 1$ where the peak of $(\sin x)/x$ is shifted to the "invisible region" of imaginary θ? In Figure 22.3, the pattern function,

$$P(\theta, k_z/k) = \sin\theta \frac{\sin[(k\cos\theta - k_z)\ell/2]}{(k\cos\theta - k_z)\ell/2}$$

is plotted for $k\ell = 4\pi$ and $k_z/k = .5, .85, 1, 1.1$.

Super-gain and end-fire antennas

Although our example is a very poor end-fire antenna, let us for the moment neglect the element factor and plot, in Figure 22.4, the array power pattern

$$\left\{\frac{\sin[(k\cos\theta - k_z)\ell/2]}{(k\cos\theta - k_z)\ell/2}\right\}^2$$

(normalized to unity), for $k\ell = 4\pi$ and $k_z/k = .5, .85, 1, 1.1125, 1.2$.

In Figure 22.5, also for $k\ell = 4\pi$, the peak directivity (gain) is plotted as a function of k_z/k.

It is seen that as k_z/k increases above unity, the directivity rises sharply to generate a "super-gain" situation. However, after approximately doubling, the gain falls and then oscillates. The peak occurs when $(k_z - k)\ell \simeq \pi/2$; this modest increase in phase shift is termed the Hansen–Woodyard condition and is commonly used in the design of end-fire arrays. In this case, $D_o \simeq 30$ and the beam width is approximately $\theta_o \simeq \sqrt{\frac{\pi}{kl}} = \frac{1}{2}$ rad. (28°). Of course, we have somehow to overcome the $\sin^2\theta$ element factor.

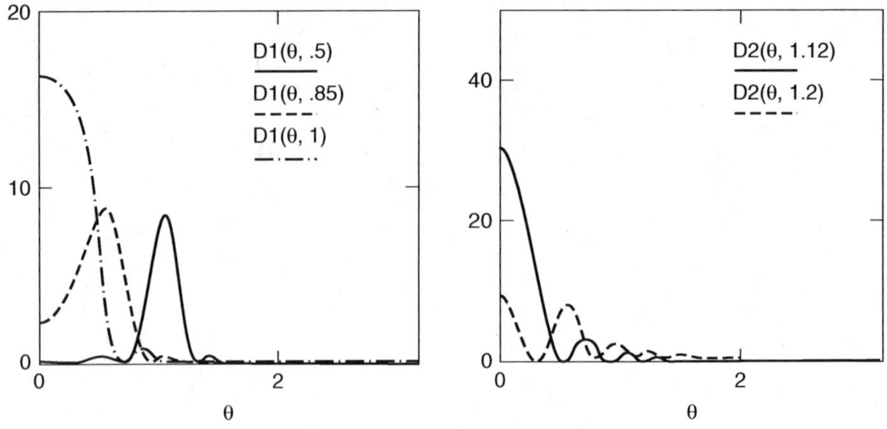

Figure 22.4 Directivity (without element factor).

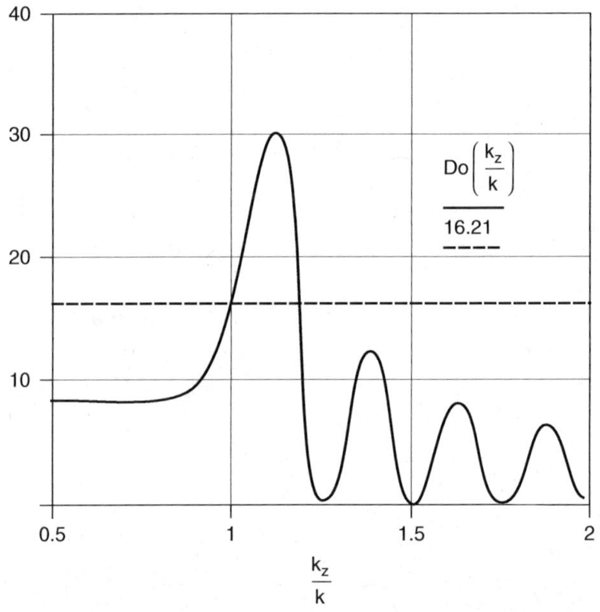

Figure 22.5 Peak directivity, $D_0(k_z/k)$.

The dipoles which comprise this traveling-wave array are the incremental current elements, $I_z\,dz$. A discrete array with very similar characteristics can be made with identical uniformly spaced finite dipoles that are fed with currents of equal amplitude and appropriate progressive phase shifts. If the dipoles are closely spaced, the antenna approximates a continuous array. Such an antenna is more flexible because the dipoles need not be oriented parallel to the common z-axis, but can be oriented at an inclined angle, often chosen to be 90°. With that orientation, the element factor switches to $\cos^2\theta$ and there is no difficulty in achieving end-fire radiation with $k_z \geq k$. Also the dipoles can be made larger. In practice, the element spacing is often set equal to $\lambda/2$ to form a discrete array.

Because higher gain is achieved with greater phase shift between elements, it was first believed that the "super-gain" principle might be applied to standing-wave arrays and other antennas. However, increasing the phase shift between (discrete or continuous) elements generates nonuniform plane waves of the type discussed in Chapter 16, Section 16.3. Such waves are evanescent in directions orthogonal to their propagation direction and so decay exponentially along the outward radial direction. Although they are not present in the far field, the energy that they store becomes simply enormous as larger gains are attempted. This in turn requires huge element currents. In addition to the energy issue and neglecting the inevitable I^2R losses, the phasing of the array elements becomes so critical and the bandwidth of operation so small as to be totally impractical. A super-gain antenna with zero efficiency is not very desirable. Any hopes that some new (as yet undiscovered) antenna configuration would surmount these difficulties were put to rest when the fundamental limits governing antennas of a given physical size were set out by Lan Chu in 1948 [23].

22.5 THE THEORY OF SIMPLE ARRAYS

We have already encountered the concept of continuous arrays when we superimposed differential-dipole elements on both standing-wave and traveling-wave wire antennas. The alternative practice of using a series of discrete radiators driven at the same frequency by excitations, each with controlled amplitude and phase, was also briefly discussed. In this section, we develop the theory of simple arrays made from a set of N Hertzian dipoles that are identical in size (length ℓ) and orientation (parallel to the z axis). They are located at discrete locations, q_n (from the common origin of spherical coordinates) given by the set of vector distances

$$\mathbf{r}_{q_n} = \mathbf{d}_n, \qquad n = 1, 2, \ldots, N$$

and are fed by the set of complex currents,

$$\underline{I}_n = \mathrm{I}_n \ \exp(j\psi_n)$$

$$\mathrm{I}_n = \mathrm{I}_{\max} \ \mathrm{w}_n$$

Because it is simpler, we choose to analyze the radiation from this dipole array using the complex Alternate-power flux which in free-space is

$$\underline{\mathbf{S}}^{\mathrm{o}} = \frac{-j\omega}{4\mu_{\mathrm{o}}}[(\underline{A}_z \nabla \underline{A}_z^* - \underline{A}_z^* \nabla \underline{A}_z) - \frac{1}{c^2}(\underline{\Phi}\nabla\underline{\Phi}^* - \underline{\Phi}^*\nabla\underline{\Phi})]$$

From Eq. (1.32), the far-field complex vector potential and the (Lorenz-gauge) scalar potential are

$$\underline{A}_z \simeq \frac{\mu_{\mathrm{o}}}{4\pi r}\exp(-jkr)\sum_{n=1}^{N} \underline{I}_n \exp(jk\mathbf{d}_n \cdot \widehat{\mathbf{r}})$$

$$\underline{\Phi} \simeq \mathrm{c}\underline{A}_z \cos\theta$$

These can be used to evaluate the free-space complex Alternate-power flux,

$$\underline{\mathbf{S}}^{\mathrm{o}} \simeq \widehat{\mathbf{r}}\frac{\omega k}{2\mu_{\mathrm{o}}}\underline{A}_z\underline{A}_z^* \sin^2\theta$$
$$\simeq \widehat{\mathbf{r}}\tfrac{1}{2}\eta_{\mathrm{o}}\mathrm{I}_{\max}^2\frac{(k\ell)^2}{4\pi r^2}\sin^2\theta\left|\sum_{n=1}^{N}\left(\frac{\underline{I}_n}{\mathrm{I}_{\max}}\right)\exp(jk\mathbf{d}_n\cdot\widehat{\mathbf{r}})\right|^2$$

where $\widehat{\mathbf{r}}$ is the unit vector in the radial direction (from the coordinate-system origin). The angular dependence is the product of the dipole element factor, $\sin^2\theta$, and the array factor,

$$F(\theta,\phi) = \left|\sum_{n=1}^{N} \mathrm{w}_n \exp[j(k\mathbf{d}_n\cdot\widehat{\mathbf{r}} + \psi_n)]\right|^2 \qquad (22.13)$$

where $\mathrm{w}_n = I_n/I_{\max} \leq 1$ are normalized coefficients. The $\mathbf{d}_n$ vectors describe element positions that, in general, do not lie on a straight line and are unequally spaced. Nevertheless, it is common practice to space the elements equally and arrange them along a straight line, a circle, or some other symmetric configuration.

Directivity (Gain)

The integral of $\underline{S}_r^{\mathrm{o}}$ over the surface of any sphere of radius r yields the total average radiated power $<P_{\mathrm{rad}}>$ and the directivity (gain) is defined by

$$D(\theta,\phi) = \frac{4\pi r^2 \underline{S}_r^{\mathrm{o}}}{< P_{\mathrm{rad}} >} = KF(\theta,\phi)\sin^2\theta \qquad (22.14a)$$

$$K = \frac{4\pi}{\int_{\phi=0}^{2\pi}\int_{\theta=0}^{\pi}[F(\theta,\phi)\sin^2\theta]\sin\theta\, d\theta d\phi} \qquad (22.14b)$$

For moderate to large values of N, the directivity depends only on the array factor; therefore array theory can be applied to non-Hertzian elements (such as half-wave dipoles) as long as the element gain is small compared to D_{o}.

We next consider linear arrays with uniform spacing.

Uniform linear arrays

Because the element factor is maximized for $\theta = \pi/2$, we want the geometry to allow constructive interference to be maximized in the $x-y$ plane; this can be accomplished by making the element phases equal ($\psi_n = 0$). In addition, the peak directivity is largest when all element currents are equal. With these choices, there are two interesting cases: one with equal spacing along the z axis, the second with equal spacing along the x axis (or any axis orthogonal to z). The first case makes $F(\theta,\phi)$ independent of the azimuthal angle, ϕ.

Uniform spacing along the z axis

When calculating the electric and magnetic fields, it is convenient to place the midpoint of the array at the origin and thus with the spacing between dipoles equal to d:

$$\mathbf{d}_n = \widehat{\mathbf{z}} \left(n - \frac{N+1}{2}\right) d, \qquad n = 1, 2, \ldots, N$$

$$\mathrm{w}_n = 1$$

$$\psi_n = 0$$

$$\widehat{\mathbf{z}} \cdot \widehat{\mathbf{r}} = \cos\theta$$

However, the constant shift added to $\mathbf{d}_n$ does not affect the array power factor, which is

$$F(\theta) = \left| \sum_{n=1}^{N} \exp[j(kd \cos\theta\ )n] \right|^2 = \left[\frac{\sin\left(\frac{N}{2} kd \cos\theta\ \right)}{\sin\left(\frac{1}{2} kd \cos\theta\ \right)} \right]^2 \qquad (22.15)$$

Because the sum is a geometric series, it can be expressed in closed form. Notice that as expected, for $\theta = \pi/2$, all dipoles add in phase and $F(\pi/2) = N^2$. Nulls in the pattern occur wherever $Nkd|\cos\theta| = 2m\pi$, $m = 1, 2, \ldots$. For large N, the beamwidth defined by the $m = 1$ zeros at $\theta = \pi/2 \pm \Delta\theta/2$ is

$$\Delta\theta \simeq \frac{4\pi}{Nkd} \ll 1$$

Uniform spacing along the x axis

It is again convenient to place the midpoint of the array at the origin and thus with the spacing between dipoles equal to d:

$$\mathbf{d}_n = \widehat{\mathbf{x}} \left(n - \frac{N+1}{2}\right) d, \qquad n = 1, 2, \ldots, N$$

$$\mathrm{w}_n = 1$$

$$\theta_n = 0$$

$$\widehat{\mathbf{x}} \cdot \widehat{\mathbf{r}} = \sin\theta \cos\phi$$

However, the constant shift added to $\mathbf{d}_n$ does not affect the array power factor, which is

$$F(\theta, \phi) = \left| \sum_{n=1}^{N} \exp[j(kd \sin\theta \cos\phi)n] \right|^2 = \left[\frac{\sin\left(\frac{N}{2} kd \sin\theta \cos\phi\right)}{\sin\left(\frac{1}{2} kd \sin\theta \cos\phi\right)} \right]^2 \qquad (22.16)$$

Notice that for $\theta = \phi = \pi/2$, all dipoles add in phase and again $F(\pi/2, \pi/2) = N^2$. Nulls in the pattern occur wherever $Nkd|\sin\theta \cos\phi| = 2m\pi$, $m = 1, 2, \ldots$. For large N and $\theta = \pi/2$, the azimuthal beam width defined by the $m = 1$ zeros at $\phi = \pi/2 \pm \Delta\phi/2$ is

$$\Delta\phi \simeq \frac{4\pi}{Nkd} \ll 1$$

Directivity as a function of *N*, *kd*, and angle

For a plane in which $F(\theta, \phi)$ has the maximum value, both Eqs. (22.15) and (22.16) are of the form

$$F(\vartheta) = \left[\frac{\sin(\frac{N}{2}kd \cos \vartheta)}{\sin(\frac{1}{2}kd \cos \vartheta)} \right]^2$$

where $\vartheta = \theta$ or $\vartheta = \phi$ $(\theta = \pi/2)$. It is instructive to plot this power-pattern function as a function of angle for different values of N and kd. In Figure 22.6 a series of normalized plots of $F(\vartheta)/N^2$ are presented for $N = 2$ and $kd = \pi/2, \pi, 3\pi/2, 2\pi, 5\pi/2, 3\pi$; in Figure 22.7 a similar series is presented for $N = 4$.

The width of the central peak decreases and D_o increases with increasing kd, which might seem to be an advantage, but notice that the secondary peaks (side lobes) increase for $kd > \pi$; this is the reason that $d = \lambda/2$ is commonly employed. The normalized directivity (gain) in the chosen plane is the same as $F(\vartheta)/N^2$ except that the dipole element factor must be included when the array is oriented parallel to the z axis. For this case, the two series of plots are repeated with the normalized gain, $F(\theta) \sin^2 \theta / N^2$, and presented in Figures 22.8 and 22.9.

Finally, the $N = 4$ series of curves are presented in polar form ($x - z$ plane) in Figures 22.10 and 22.11.

The width of the central peak again decreases with increasing kd, but notice that the side lobes are partially suppressed by the element factor. Nevertheless, $d = \lambda/2$ is still commonly employed especially for larger values of N where (with $kd = \pi$) the peak

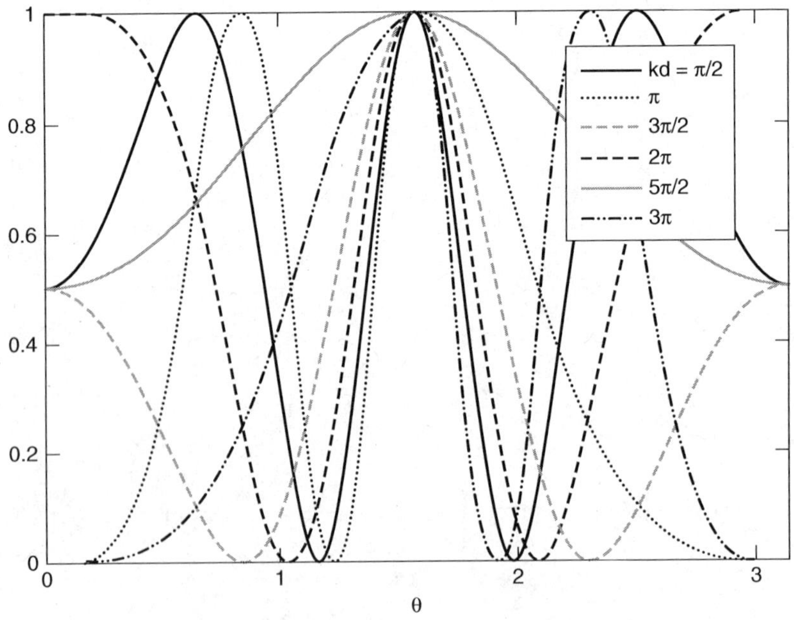

Figure 22.6 Normalized $F(\vartheta, kd)$ for $N = 2$.

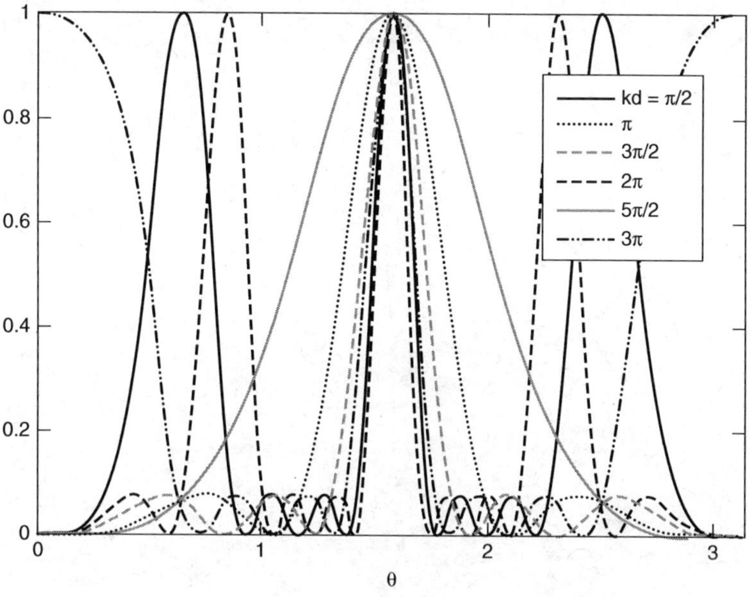

Figure 22.7 Normalized $F(\vartheta, kd)$ for $N = 4$.

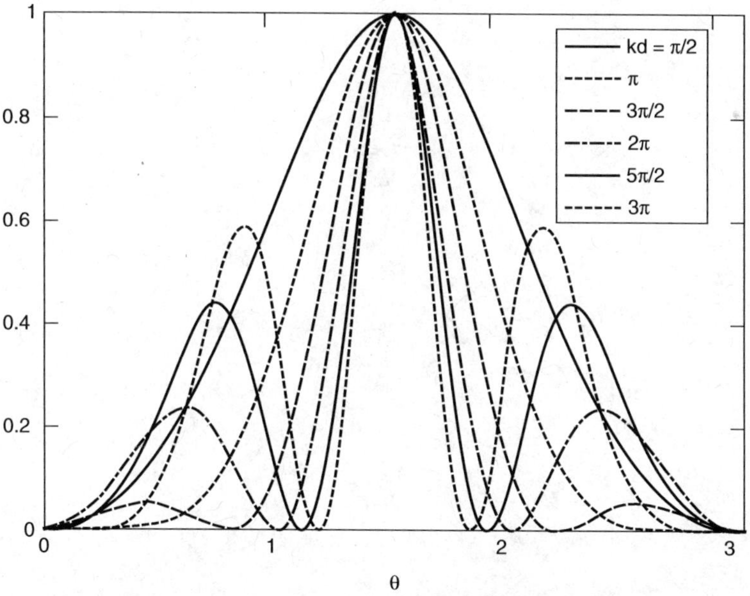

Figure 22.8 Normalized $F(\theta, kd)\sin^2\theta$ for $N = 2$.

directivity (gain) D_o is related to the beam width $\Delta\vartheta$ by

$$D_o \simeq \frac{2}{\Delta\vartheta} \simeq \frac{N}{2} \gg 1$$

This result assumes that $F(\vartheta) \simeq N^2$ for $|\vartheta - \pi/2| \leq \Delta\vartheta/2$ and zero elsewhere; the element factor of the dipole can be ignored.

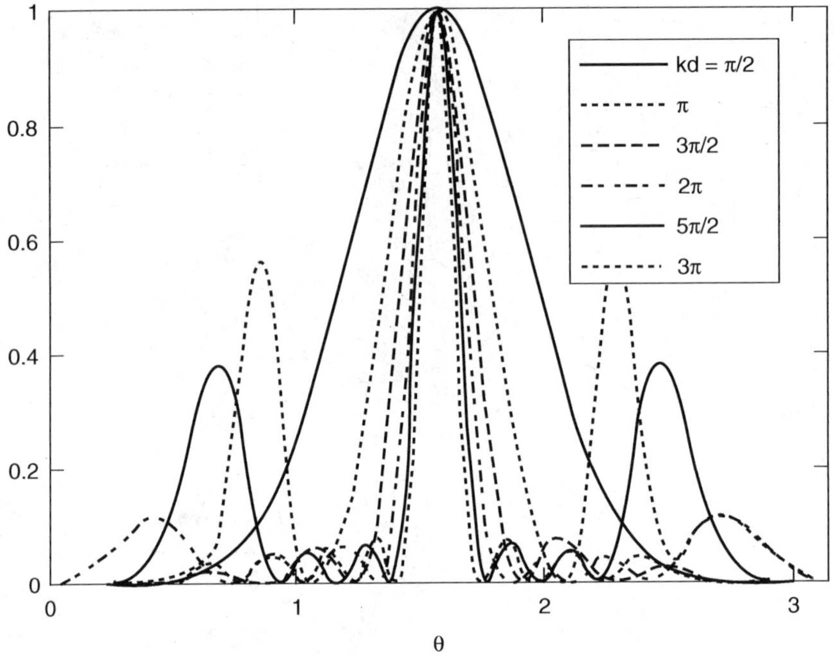

Figure 22.9 Normalized $F(\theta, kd)\sin^2\theta$ for $N = 4$.

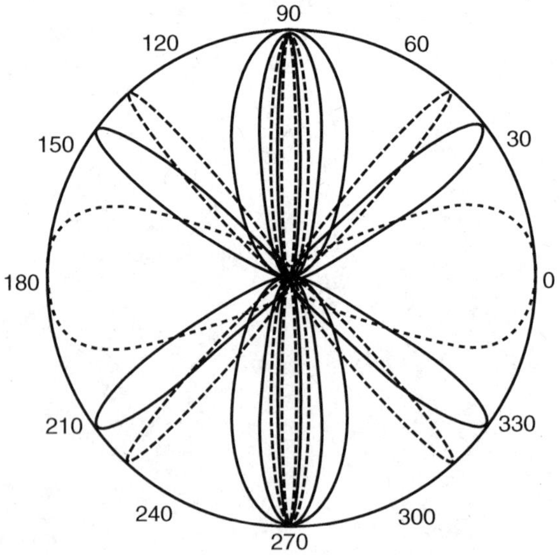

Figure 22.10 Normalized $F(\vartheta, kd)$ (polar plots) for $N = 4$.

Control of the main beam-width and side-lobe levels requires an understanding of how these parameters interact. Generally, a narrow beam (high directivity) requires a longer array and lower side-lobes require the amplitude of excitation to be reduced near the ends. The element factor produces an effective taper for the z-oriented examples (but not

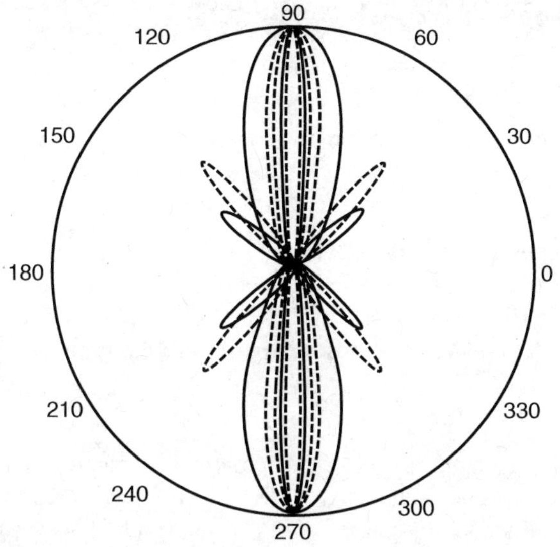

Figure 22.11 Normalized $F(\theta, kd)\sin^2\theta$ (polar plots) for $N = 4$.

when N is large); similar effects can be produced by tapering the array coefficients, w_n. Phase control of the aperture is afforded by adjusting ψ_n. Additional insight is provided by observing that the complex array function,

$$s(z) = \sum_{n=0}^{N} \underline{b}_n z^n = B(z - z_1)(z - z_2)\cdots(z - z_N)$$

$$z = \exp(jkd\cos\vartheta)$$

which determines the directivity $|s(z)|^2$, has zeros that all lie on the unit circle of the complex plane. Notice that as ϑ varies from 0 to π, z varies from $\exp(jkd)$ to $\exp(-jkd)$ and thus moves completely around the unit circle when $kd = \pi$. That value is the smallest one that permits all of the zeros to be accessed; their position can be controlled by adjusting both kd and the complex array coefficients, $w_n \exp(j\psi_n)$. Synthesis of the directivity of a uniform linear array is then carried out by appropriately positioning the complex zeros. This technique was developed by Schelkunoff [24] and initially led some workers to the believe that using $kd > \pi$, in combination with certain sets of complex array coefficients, would allow the production of miniaturized very high-gain antennas. However, as discussed previously, such "super-gain" antennas were proved to be completely impractical.[2] Nevertheless, array theory has led to many clever and practical array designs; among the best is the often-used Dolph–Chebyshev synthesis procedure [25], which generates the lowest side-lobe level for a prescribed maximum antenna directivity.

[2]For a uniform array 50 λ in length ($N = 100$), a super-gain antenna of half that physical length would require the reactive power to be 10^{59} times that of the radiated power!

22.6 DIFFRACTION BY A RECTANGULAR SLIT

Maxwell–Poynting analysis

The problem we wish to analyze consists of a uniform linearly polarized plane wave with electric field,

$$E_x^{\text{incident}}(x, z, t) = E_{\text{o}} \cos(\omega t - kz)$$

normally incident upon a perfectly conducting plane, located at $z = 0$. The plane has a slit cut from it that is of width d in the x-direction and independent of y; the midpoint is located at $x = 0$. The width is assumed large compared to a wavelength; therefore $kd \gg 1$ and the aperture electric field is approximately the same as that of the incident wave,

$$E_x(x, 0, t) = E_{\text{o}} \cos(\omega t)[1 - u_{-1}(|x| - d/2)] \tag{22.17}$$

Each Cartesian component of both the electric and magnetic field satisfies the free-space wave equation; we choose to focus on E_x. Because the sources for the fields in the region $z > 0$ are on the aperture plane, we superimpose plane waves propagating away from that plane. These are expressed in terms of the wavenumber, k_x, which is allowed to vary continuously from 0 to ∞ so as to generate the Fourier integral

$$\begin{aligned} E_x(x, z, t) = &\int_0^k C(k_x) \cos\left(\omega t - \sqrt{k^2 - k_x^2}\, z\right) \cos(k_x x)\, dk_x \\ &+ \cos(\omega t) \int_k^\infty C(k_x) \exp\left(-\sqrt{k_x^2 - k^2}\, z\right) \cos(k_x x)\, dk_x \end{aligned} \tag{22.18}$$

In the range $0 < k_x < k = \omega/c$, the waves are uniform plane waves; in the range $k < k_x < \infty$, they are nonuniform plane waves. For $z = 0$,

$$E_x(x, 0, t) = \cos(\omega t) \int_0^\infty C(k_x) \cos(k_x x)\, dk_x$$

which, in view of Eq. (22.17), has the Fourier transform

$$\int_0^\infty \int_0^\infty C(k_x) \cos(k'x) \cos(k_x x)\, dk_x dx = E_{\text{o}} \int_0^{d/2} \cos(k'x)\, dx$$

or equivalently

$$\lim_{L \to \infty} \int_0^\infty \frac{1}{2} \left[\frac{\sin L(k_x - k')}{(k_x - k')} + \frac{\sin L(k_x + k')}{(k_x + k')} \right] C(k_x)\, dk_x = E_{\text{o}} \frac{\sin(k'd/2)}{k'}$$

The integral has contributions only from $k_x = \pm k'$ and so

$$C(k') = \frac{E_{\text{o}} d}{\pi} \frac{\sin(k'd/2)}{k'd/2}$$

Substituting $C(k_x)$ into Eq. (22.18) leads to

$$\begin{aligned} E_x(x, z, t) = \frac{E_{\text{o}} d}{\pi} \Bigg[&\int_0^k \frac{\sin(k_x d/2)}{k_x d/2} \cos\left(\omega t - \sqrt{k^2 - k_x^2}\, z\right) \cos(k_x x)\, dk_x \\ &+ \cos(\omega t) \int_k^\infty \frac{\sin(k'd/2)}{k'd/2} \exp\left(-\sqrt{k_x^2 - k^2}\, z\right) \cos(k_x x)\, dk_x \Bigg] \end{aligned}$$

or in complex notation

$$\underline{E}_x(x,z) = \frac{E_o d}{\pi}\left[\int_0^k \frac{\sin(k_x d/2)}{k_x d/2}\cos(k_x x)\exp\left(-j\sqrt{k^2-k_x^2}z\right)dk_x\right.$$
$$\left.+\int_k^\infty \frac{\sin(k_x d/2)}{k_x d/2}\exp\left(-\sqrt{k_x^2-k^2}z\right)\cos(k_x x)\,dk_x\right]$$

For very small values of z, large values of k_x are required to produce the rectangular spatial pulse; these nonuniform waves decay rapidly with increasing z and so, with respect to k_x, the aperture acts like a low-pass filter. Only $k_x < k$ can propagate away from the slit.

Because $kd \gg \pi$, it is reasonable to approximate

$$\frac{\sin(k_x d/2)}{k_x d/2} \simeq \begin{cases} 1 & \left(k_x < \dfrac{\pi}{d}\right) \\ 0 & \left(k_x > \dfrac{\pi}{d}\right) \end{cases}$$

The direction of the propagating waves is therefore largely confined within the angles (from the z axis) $\theta = \pm\theta_o$, where $2\theta_o \simeq \frac{2\pi}{kd} = \lambda/d$ defines the beam width.

If $z^3 > x^4/\lambda$, the phase errors incurred by approximating

$$\sqrt{k^2-k_x^2}z \simeq \left(k - \frac{k_x^2}{2k}\right)z$$

are negligible and

$$\underline{E}_x(x,z) \simeq \frac{E_o d}{2\pi}\exp(-jkz)\int_0^{\pi/d<k}[\exp(jk_x x)+\exp(-jk_x x)]\exp\left(j\frac{k_x^2}{2k}z\right)dk_x \quad (22.19)$$

which, with the substitutions,

$$u = \sqrt{\frac{z}{\pi k}}k_x \quad \text{and} \quad u_o = x\sqrt{\frac{k}{\pi z}}$$

can be written as

$$\underline{E}_x(x,z) \simeq \frac{E_o d}{2\pi}\sqrt{\frac{\pi k}{z}}\exp\left[-jk\left(z+\tfrac{x^2}{2z}\right)\right]\int_0^{\sqrt{\frac{\pi z}{kd^2}}}\{[\exp[j\tfrac{\pi}{2}(u-u_o)^2]+\exp[j\tfrac{\pi}{2}(u+u_o)^2]\}\,du \quad (22.20)$$

Equation (22.20) can be evaluated in terms of the Fresnel (cosine and sine) integrals,

$$C(x) = \int_0^x \cos(\tfrac{\pi}{2}u^2)\,du$$
$$S(x) = \int_0^x \sin(\tfrac{\pi}{2}u^2)\,du$$

which are plotted in Figure 22.12.

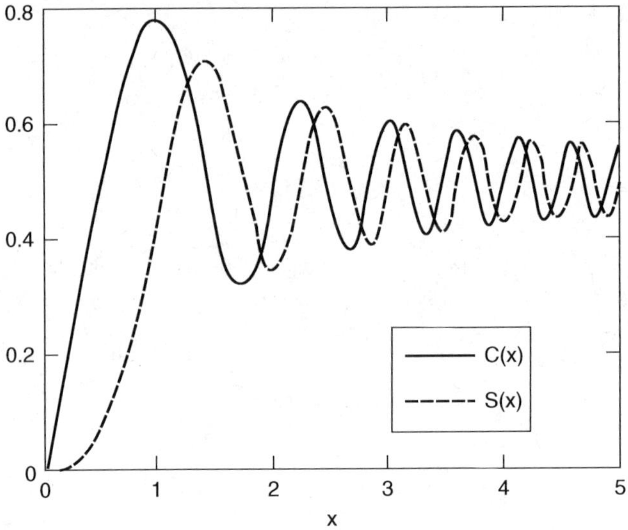

Figure 22.12 Fresnel cosine and sine integrals.

The result is

$$\underline{E}_x(x,z) \simeq \frac{E_o d}{2\pi}\sqrt{\frac{\pi k}{z}} \exp\left[-jk\left(z+\frac{x^2}{2z}\right)\right]\left\{\left[C\left(\sqrt{\frac{\pi z}{kd^2}}-x\sqrt{\frac{k}{\pi z}}\right)\right.\right.$$
$$+C\left(\sqrt{\frac{\pi z}{kd^2}}+x\sqrt{\frac{k}{\pi z}}\right)\Bigg]+j\Bigg[S\left(\sqrt{\frac{\pi z}{kd^2}}-x\sqrt{\frac{k}{\pi z}}\right)$$
$$\left.\left.+S\left(\sqrt{\frac{\pi z}{kd^2}}+x\sqrt{\frac{k}{\pi z}}\right)\right]\right\} \tag{22.21}$$

There are Fresnel ripples in the x and z variations of E_x; the former depend upon the value of z and mediate the transition from the main beam to the shadow region where the field intensity approaches zero.

The on-axis field simplifies to

$$\underline{E}_x(0,z) \simeq E_o d\sqrt{\frac{k}{\pi z}} \exp(-jkz)\left[C\left(\sqrt{\frac{\pi z}{kd^2}}\right)+jS\left(\sqrt{\frac{\pi z}{kd^2}}\right)\right] \tag{22.22}$$

The ripples are revealed in Figure 22.13, a plot of the normalized magnitude as a function of $w = \sqrt{\frac{\pi}{kd}}\sqrt{\frac{z}{d}}$.

Equation (22.22) approaches the limit (for large z),

$$\underline{E}_x(0,z) \rightarrow \frac{E_o d}{2}\sqrt{\frac{k}{\pi z}} \exp(-jkz)(1+j)$$

Because the slit is infinite in the y direction, the "far field" does *not* decline as $1/z$. In this region, the on-axis complex Poynting flux is

$$\underline{S}_z = \frac{1}{2}\sqrt{\frac{\varepsilon_o}{\mu_o}}|\underline{E}_x(0,z)|^2 \rightarrow \frac{1}{2}\sqrt{\frac{\varepsilon_o}{\mu_o}}E_o^2 d\frac{kd}{2\pi z} \tag{22.23}$$

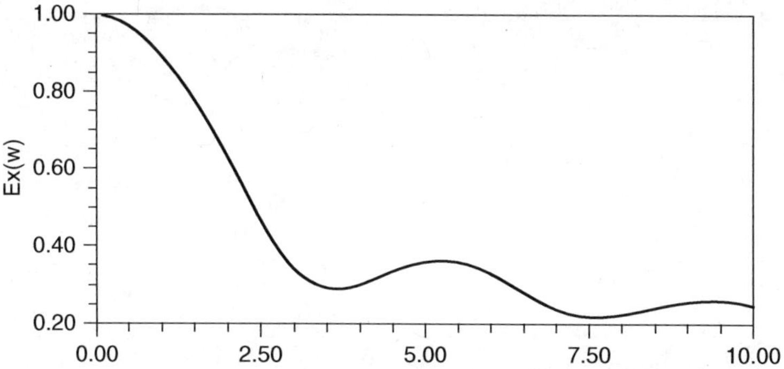

Figure 22.13 Normalized on-axis Fresnel ripples.

from which the peak directivity is seen to be

$$D_o = kd$$

The preceding analysis represents the Maxwell–Poynting representation carried out directly in terms of the Cartesian component of the electric field. We turn next to the dual Alternate-power representation.

Alternate-representation analysis

In the free-space region $z > 0$, the equivalence principle allows the fields to be expressed either as

$$\mu_o \underline{\mathbf{H}} = \nabla \times \underline{\mathbf{A}}$$
$$\underline{\mathbf{E}} = -j\omega \underline{\mathbf{A}} - \nabla \underline{\Phi}$$

or

$$\varepsilon_o \underline{\mathbf{E}} = -\nabla \times \underline{\mathbf{A}}_m$$
$$\underline{\mathbf{H}} = -j\omega \underline{\mathbf{A}}_m - \nabla \underline{\Phi}_m$$

where $\underline{\mathbf{A}}$ is generated by $\underline{\mathbf{K}} = \widehat{\mathbf{x}}\underline{K}_x$, or $\underline{\mathbf{A}}_m$ by $\underline{\mathbf{K}}_m = \widehat{\mathbf{y}}\underline{K}_{my}$. We choose the latter representation because (without y variation) $\nabla \cdot \underline{\mathbf{A}}_m = 0$ and in the Lorenz gauge, $\underline{\Phi}_m = 0$; this simplifies the calculation of both $\underline{\mathbf{H}}$ and the dual Alternate-power flux, $\underline{\mathbf{S}}^{om}$.

We approximate $\underline{E}_x(x,z)$ by Eq. (22.21), which includes only uniform plane wave components with $k_z \simeq k$. This is reasonable except for values of z close to the aperture where the nonuniform plane waves are important. Therefore,

$$\frac{\partial \underline{A}_{my}}{\partial z} \simeq -jk\underline{A}_{my} \simeq \varepsilon_o \underline{E}_x(x,z)$$

$$\underline{A}_{my} \simeq j\varepsilon_o \frac{E_o d}{2\pi k}\sqrt{\frac{\pi k}{z}} \exp\left[-jk\left(z + \frac{x^2}{2z}\right)\right] \left\{ \left[C\left(\sqrt{\frac{\pi z}{kd^2}} - x\sqrt{\frac{k}{\pi z}}\right) \right.\right.$$
$$\left.\left. + C\left(\sqrt{\frac{\pi z}{kd^2}} + x\sqrt{\frac{k}{\pi z}}\right)\right] + j\left[S\left(\sqrt{\frac{\pi z}{kd^2}} - x\sqrt{\frac{k}{\pi z}}\right) + S\left(\sqrt{\frac{\pi z}{kd^2}} + x\sqrt{\frac{k}{\pi z}}\right)\right]\right\}$$

The dual Alternate-power flux,

$$\underline{\mathbf{S}}^{\mathrm{om}} = \frac{-j\omega}{4\varepsilon_{\mathrm{o}}}[\underline{A}_{\mathrm{my}}\nabla\underline{A}^*_{\mathrm{my}} - \underline{A}^*_{\mathrm{my}}\nabla\underline{A}_{\mathrm{my}}]$$

is real and has the principal component

$$\begin{aligned}\underline{S}_z^{\mathrm{om}} &\simeq \frac{\omega k}{2\varepsilon_{\mathrm{o}}}\underline{A}_{\mathrm{my}}\underline{A}^*_{\mathrm{my}} \\ &\simeq \frac{kd}{8\pi}\sqrt{\frac{\varepsilon_{\mathrm{o}}}{\mu_{\mathrm{o}}}}E_{\mathrm{o}}^2\frac{d}{z}\left\{\left[C\left(\sqrt{\frac{\pi z}{kd^2}}+x\sqrt{\frac{k}{\pi z}}\right)+C\left(\sqrt{\frac{\pi z}{kd^2}}-x\sqrt{\frac{k}{\pi z}}\right)\right]^2\right. \\ &\left.+\left[S\left(\sqrt{\frac{\pi z}{kd^2}}+x\sqrt{\frac{k}{\pi z}}\right)+S\left(\sqrt{\frac{\pi z}{kd^2}}-x\sqrt{\frac{k}{\pi z}}\right)\right]^2\right\}\end{aligned}$$

which agrees with the "far-field" on-axis Poynting flux, Eq. (22.23). Normalized plots of $\underline{S}_z^{\mathrm{om}}(x/d, z/d = 1, 5, 10)$ and $\underline{S}_z^{\mathrm{om}}(x/d = 0, .5, 1, z/d)$ are presented in Figures 22.14 and 22.15 for the case $kd = 10$; both sets of traces are in order of decreasing amplitude.

22.7 DIFFRACTION BY A LARGE CIRCULAR APERTURE

We assume that a uniform plane wave of frequency ω is normally incident upon an infinite perfectly conducting screen located on the plane $z = 0$, which contains a circular hole of radius R (centered at $x = y = 0$). Free-space exists on both sides of the screen. The incident wave is linearly polarized and given by

$$\underline{\mathbf{E}}^{\mathrm{i}} = \hat{\mathbf{x}}\underline{E}_{\mathrm{o}}\exp(-jkz) \tag{22.24a}$$

$$\underline{\mathbf{H}}^{\mathrm{i}} = \hat{\mathbf{y}}\sqrt{\frac{\varepsilon_{\mathrm{o}}}{\mu_{\mathrm{o}}}}\underline{E}_{\mathrm{o}}\exp(-jkz) \tag{22.24b}$$

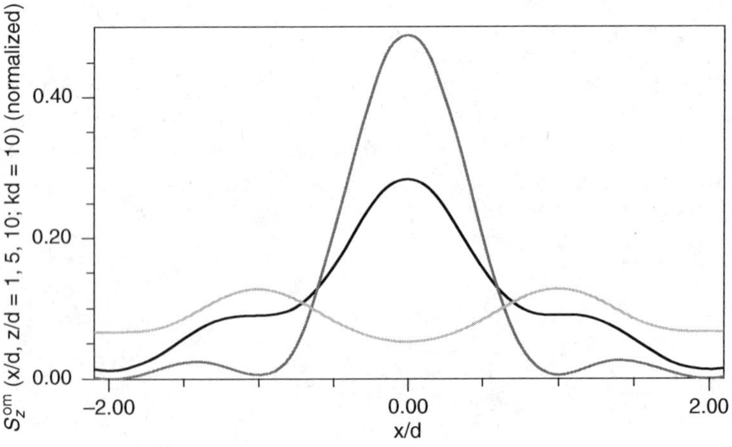

Figure 22.14 $S_z^{\mathrm{om}}(x/d)/(\frac{1}{2}\sqrt{\frac{\varepsilon_{\mathrm{o}}}{\mu_{\mathrm{o}}}}E_{\mathrm{o}}^2)kd = 10.$

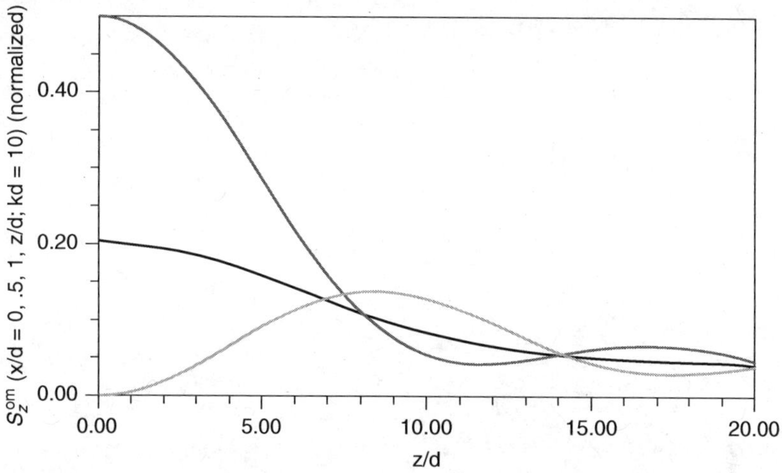

Figure 22.15 $S_z^{\mathrm{om}}(z/d)/(\frac{1}{2}\sqrt{\frac{\varepsilon_o}{\mu_o}}E_o^2)kd = 10.$

Assuming that $kR \gg 1$, we choose to analyze this diffraction problem by using the induction theorem described in Chapter 6, Section 6.8. The electric and magnetic surface currents, $\underline{\mathbf{K}}_e^s$ and $\underline{\mathbf{K}}_m^s$, mounted at $z = \pm\delta$ replace the incident wave and allow us to focus on the scattered fields on either side of the screen; we are interested primarily in the fields for $+z$. As previously taught, when $\delta \to 0$, the electric surface current cancels everywhere, and the magnetic current is doubled where there is conductor and cancels in the aperture. We are left with a source distribution

$$\underline{\mathbf{K}}_m^s = \begin{cases} 0 & (r \le R) \\ \widehat{\mathbf{z}} \times 2\,\underline{\mathbf{E}}\big|_{z=0} = \widehat{\mathbf{y}}2\underline{E} & (r > R) \end{cases}$$

that will generate the scattered fields. Although we could proceed now to find the dual vector potential, it is advantageous to superpose a second current, $\underline{\mathbf{K}}_m^s = -\widehat{\mathbf{y}}2\underline{E}_o$, that exists *everywhere* on the plane. By itself, it generates the *negative* of the scattered field when there is no opening in the screen. Since there is obviously no total field for $z > 0$, the scattered fields would be $-\underline{\mathbf{E}}^i$ and $-\underline{\mathbf{H}}^i$, but their negatives are the incident fields themselves. It follows that the excitation

$$\underline{\mathbf{K}}_m^s = \begin{cases} -\widehat{\mathbf{y}}2\underline{E}_o & (r \le R) \\ 0 & (r > R) \end{cases}$$

generates the total fields (incident plus scattered). Now we can integrate over the finite aperture rather than the infinite plane. Although $\underline{\mathbf{K}}_m^s$ by itself allows calculation of $\underline{\mathbf{A}}_m$ and hence $\underline{\mathbf{E}}$, we also need $\underline{\Phi}_m$ in order to calculate $\underline{\mathbf{H}}$. The source of the dual potential is magnetic charge that must be present at the edge of the hole in order to terminate the magnetic current. Conservation of magnetic charge demands that the boundary condition is

$$-\widehat{\mathbf{r}} \cdot \underline{\mathbf{K}}_m^s + j\omega\underline{Q}'_m = 0$$

where $\underline{Q}'_m$ is the line-charge density at $r = R$. With $\mathbf{r}_q$ the vector from the origin to point q on the aperture and $\mathbf{r}_p$ the vector from the origin to the field point p and $r_{qp}^2 = |\mathbf{r}_p -$

$\mathbf{r}_q|^2$, the required dual potentials satisfy

$$\underline{\mathbf{A}}_{\mathrm{m}p} = \widehat{\mathbf{y}} \int\int \frac{\varepsilon_{\mathrm{o}} \underline{K}^s_{\mathrm{m}q}}{4\pi r_{qp}} \exp(-jkr_{qp})\, da_q \tag{22.25a}$$

$$\underline{K}^s_{\mathrm{m}q} = -2\underline{E}_{\mathrm{o}}$$

$$\underline{\Phi}_{\mathrm{m}p} = \int \frac{\underline{Q}'_{\mathrm{m}q}}{4\pi \mu_{\mathrm{o}} r_{qp}} \exp(-jkr_{qp}) R\, d\phi_q \tag{22.25b}$$

$$j\omega \underline{Q}'_{\mathrm{m}q} = \underline{K}^s_{\mathrm{m}q} \sin\phi_q$$

and generate the fields

$$\varepsilon_{\mathrm{o}} \underline{\mathbf{E}} = -\nabla \times \underline{\mathbf{A}}_{\mathrm{m}}$$

$$\underline{\mathbf{H}} = -j\omega \underline{\mathbf{A}}_{\mathrm{m}} - \nabla \underline{\Phi}_{\mathrm{m}}$$

On-axis fields and power

One might expect that if the point p is restricted to lie on the z axis, symmetry would make evaluation of the integrals easier and in fact closed-form solutions are possible. We therefore restrict our present discussion to the on-axis case, for which

$$\underline{\mathbf{A}}_{\mathrm{my}} = \int_0^R \frac{-\varepsilon_{\mathrm{o}} 2\underline{E}_{\mathrm{o}}}{4\pi r_{qp}} \exp(-jkr_{qp}) 2\pi r_q dr_q$$

$$\underline{\Phi}_{\mathrm{m}} = \frac{-2\underline{E}_{\mathrm{o}} R}{j\omega\mu_{\mathrm{o}} 4\pi \sqrt{z^2+R^2}} \exp(-jk\sqrt{z^2+R^2}) \int_0^{2\pi} \sin\phi_q d\phi_q = 0$$

Because $r_{qp}^2 = r_q^2 + z^2$, it follows that $r_q dr_q = r_{qp} dr_{qp}$; consequently, the factor r_{qp} cancels and

$$\underline{\mathbf{A}}_{\mathrm{my}} = \int_z^{\sqrt{z^2+R^2}} -\varepsilon_{\mathrm{o}} \underline{E}_{\mathrm{o}} \exp(-jkr_{qp})\, dr_{qp}$$

which is trivially evaluated as

$$\underline{A}_{\mathrm{my}} = \varepsilon_{\mathrm{o}} \underline{E}_{\mathrm{o}} \frac{1}{jk} \left[\exp\left(-jk\sqrt{z^2+R^2}\right) - \exp(-jkz) \right]$$

This on-axis function is sufficient to allow calculation of the curl and therefore the electric field

$$\underline{E}_x = \frac{1}{\varepsilon_{\mathrm{o}}} \frac{\partial \underline{A}_{\mathrm{my}}}{\partial z} = \underline{E}_{\mathrm{o}} \left[\exp(-jkz) - \frac{z}{\sqrt{z^2+R^2}} \exp(-jk\sqrt{z^2+R^2}) \right]$$

Because $\underline{H}_y = -j\omega \underline{A}_{\mathrm{my}} - \frac{\partial}{\partial y} \underline{\Phi}_{\mathrm{m}}$ the on-axis $\underline{A}_{\mathrm{my}}$ is still sufficient. Although the on-axis value of $\underline{\Phi}_{\mathrm{m}} = 0$, it is necessary (and complicating) to calculate off-axis values before evaluating $\nabla \underline{\Phi}_{\mathrm{m}}$. With $r_q = R$ and $r_{qp}^2 = R^2 + z^2 + r_p^2 - 2Rr_p \cos(\phi_p - \phi_q)$, one can choose $r_p \ll R$ so that

$$r_{qp} \simeq \sqrt{R^2+z^2} - \frac{Rr_p \cos(\phi_p - \phi_q)}{\sqrt{R^2+z^2}}$$

and

$$\underline{\Phi}_{\mathrm{m}} = \frac{\underline{K}^s_{\mathrm{m}q} R \exp(-jk\sqrt{R^2+z^2})}{j\omega 4\pi \mu_o \sqrt{R^2+z^2}} \int_0^{2\pi} \frac{\exp\left(jk\dfrac{Rr_p\cos(\phi_p-\phi_q)}{\sqrt{R^2+z^2}}\right)}{\left[1-\dfrac{Rr_p\cos(\phi_p-\phi_q)}{R^2+z^2}\right]} \sin\phi_q d\phi_q$$

the integrand can be further expanded in a Taylor series. The only terms that survive the integration are (with $r_p \sin\phi_p = y$)

$$\underline{\Phi}_{\mathrm{m}} = \frac{2\underline{E}_{\mathrm{o}} R \exp(-jk\sqrt{R^2+z^2})}{j\omega 4\mu_o \sqrt{R^2+z^2}} \frac{(jk\sqrt{R^2+z^2}+1)}{R^2+z^2} Ry$$

Finally,

$$\underline{H}_y = \sqrt{\frac{\varepsilon_o}{\mu_o}}\underline{E}_{\mathrm{o}}\left[\exp(-jkz) + \left[\frac{R^2}{2(z^2+R^2)}\left(1+\frac{1}{jk\sqrt{z^2+R^2}}\right) - 1\right]\exp(-jk\sqrt{z^2+R^2})\right]$$

Let us check these results by allowing $R \to \infty$ and thus removing the screen. Then for finite z, we have

$$\underline{E}_x = \underline{E}_{\mathrm{o}} \exp(-jkz)$$

$$\underline{H}_y = \sqrt{\frac{\varepsilon_o}{\mu_o}}\underline{E}_{\mathrm{o}}\left[\exp(-jkz) - \frac{1}{2}\exp(-jkR)\right]$$

This is *almost* the incident field, but the term involving $\exp(-jkR)$ is puzzling. It represents a wave traveling from the edge of the aperture and arrives at the axis with arbitrary phase; it is an artifact of the diffraction integral mathematics which are really Fourier type integrals. Off-axis, one should be prepared for the Gibbs Phenomenon [26]. Notice that if the tiniest amount of dissipation is added so that $k \to k' - jk''$, the offending term will decay to zero.

Far fields, power, and directivity

With the screen replaced, it is of interest to examine the fields for large values of z. Then,

$$\underline{E}_x = \sqrt{\frac{\mu_o}{\varepsilon_o}}\underline{H}_y \simeq \underline{E}_{\mathrm{o}}[\exp(-jkz) - \exp(-jkz\sqrt{1+R^2/z^2})]$$

(Notice that there are no far-field components of $\underline{H}_y$ due to $\nabla\underline{\Phi}_{\mathrm{m}}$.) But, $\sqrt{1+R^2/z^2} \simeq 1 + \frac{1}{2}R^2/z^2$; therefore

$$\underline{E}_x = \sqrt{\frac{\mu_o}{\varepsilon_o}}\underline{H}_y \to jk\frac{R^2}{2z}\underline{E}_{\mathrm{o}}\exp(-jkz)$$

and the z-directed complex Poynting flux is expressed as

$$\underline{S}_z = \frac{1}{2}\underline{E}_x\underline{H}_y^* \to \frac{k^2R^2}{4\pi z^2}\left(\frac{1}{2}\sqrt{\frac{\varepsilon_o}{\mu_o}}\underline{E}_{\mathrm{o}}\underline{E}_{\mathrm{o}}^*\pi R^2\right)$$

The term in parentheses is the total average power passing through the aperture.

In general, the far-field radial flux can be written as

$$\underline{S}_r = \frac{<P_{\rm total}>}{4\pi r^2} D(\theta, \phi)$$

where $D(\theta, \phi)$ is defined as the directivity[3] (gain) of the radiator. (It follows from the reciprocity theorem that the angular pattern of the radiator is the same for both the transmitting and receiving cases.) If $D(\theta, \phi) = 1$, the power is radiated uniformly in all directions. For the case at hand, the z axis coincides with the maximum directivity, $D_{\rm o}$, and

$$D_{\rm o} = k^2 R^2 = \frac{4\pi}{\lambda_{\rm o}^2}(\pi R^2)$$

This is an example of the very basic antenna-theory formula,

$$D_{\rm o} = \frac{4\pi}{\lambda_{\rm o}^2} A_{\rm eff}$$

that relates maximum directivity to the effective area of the antenna through which (by the reciprocity theorem) it transmits or receives radiated power. For large-aperture antennas ($kR \gg 1$) the effective area equals that of the geometric aperture.[4]

Alternate power

The on-axis complex Alternate-power flux, based upon the dual vector and scalar potentials that are generated by the fictitious magnetic currents and magnetic charges, is

$$\underline{S}_z^{\rm m} = -j\omega\frac{1}{4}\left[\frac{1}{\varepsilon_{\rm o}}\left(\underline{A}_{\rm my}\frac{\partial}{\partial z}\underline{A}^*_{\rm my} - \underline{A}^*_{\rm my}\frac{\partial}{\partial z}\underline{A}_{\rm my}\right) - \mu_{\rm o}\left(\underline{\Phi}_{\rm m}\frac{\partial}{\partial z}\underline{\Phi}^*_{\rm m} - \underline{\Phi}^*_{\rm m}\frac{\partial}{\partial z}\underline{\Phi}_{\rm m}\right)\right]$$

Evaluation is very straightforward; because $\underline{\Phi}_{\rm m} = 0$, there is no need to calculate the off-axis potential. Also there is no reactive term

$$\underline{S}_z^{\rm m} = \frac{1}{2}\sqrt{\frac{\varepsilon_{\rm o}}{\mu_{\rm o}}}\underline{E}_{\rm o}\underline{E}^*_{\rm o}\left(1 + \frac{z}{\sqrt{z^2+R^2}}\right)\left(1 - \cos\left[k\left(\sqrt{z^2+R^2} - z\right)\right]\right) \qquad (22.26)$$

The far-field limit agrees exactly with that of $\underline{S}_z$.

Fresnel region

Examination of the factor $(1 - \cos[k(\sqrt{z^2+R^2} - z)]$ reveals that the cosine argument starts at kR for $z = 0$ and decreases monotonically as z increases. Because kR has been assumed to be large, the factor itself rapidly oscillates between 0 and 2 with a frequency that decreases with increasing z. These oscillations are known as Fresnel ripples and occur for values of z known as the Fresnel region. (Although we will not calculate them, ripples can also occur as a function of radius when the point p is moved off-axis at constant z; this was seen in the diffraction of the rectangular slit.) This behavior is shown in Figure 22.16 for three different values of R $(\lambda_{\rm o}, 2\lambda_{\rm o}, 3\lambda_{\rm o})$; the z coordinate

[3]If $<P_{\rm total}>$ is the input power to an antenna, the power gain, $G(\theta, \phi)$, must take into account dissipation in the antenna as well as any propagation loss; if, instead, $<P_{\rm total}>$ is the actual power radiated, and propagation loss is neglected, the power gain and directivity are equal.

[4]For any type of Hertzian dipole, $D_{\rm o} = 3/2$ and $A_{\rm eff} = \frac{3}{8\pi}\lambda_{\rm o}^2$ independent of the physical size!

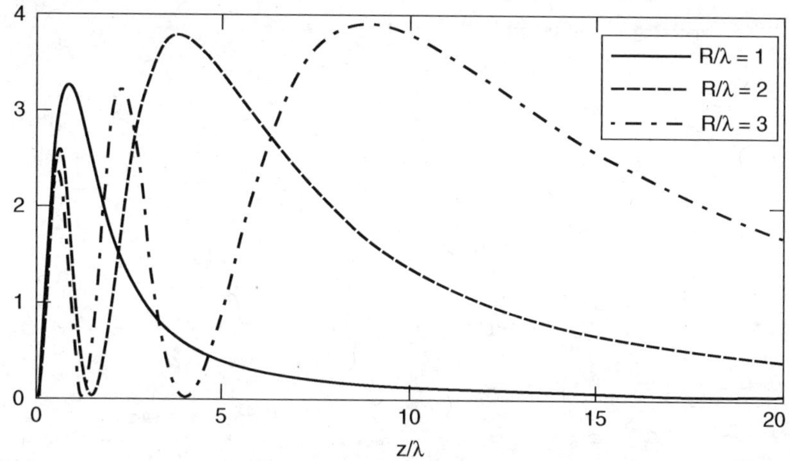

Figure 22.16 Fresnel ripples (on-axis).

is normalized to the wavelength, λ_o. Of particular interest is the location of the so-called "last axial-maximum" that occurs when the cosine phase has been reduced to π. The corresponding value of z is $R^2/\lambda_o - \lambda_o/4$. For larger values, the phase difference decreases until at some point it can be considered negligible. (For that and larger values of z, all waves traveling from all points in the aperture arrive in phase.) The phase that *defines* the onset of the far field is commonly set at $\pi/8$; this translates to the far field beginning at

$$z \geq \frac{2(2R)^2}{\lambda_o}$$

An aperture that forms an annular ring between radii, R_a and R_b, can be analyzed by means of linear superposition. Simply take the difference between the circular-aperture solutions for the larger and smaller radii. In the event that $R_a = R$ and $R_b \to \infty$, the Babinet solution for the complementary screen is recovered.

Fresnel zones

Suppose that we stand on-axis at a fixed value of z and observe the power flux as R is increased from a small value to infinity. The amplitude plotted as a function of R is shown in Figure 22.17 for $z = 5\lambda$; it oscillates with a nonuniform but increasing frequency.

The values of R that define the boundaries between the regions of positive and negative slope are labeled R_n, where

$$R_n = \sqrt{n\lambda\left(z + \frac{n}{4}\lambda\right)}, \qquad n = 0, 1, 2, 3, \ldots$$

The area between R_{n-1} and R_n is known as the nth Fresnel zone and is given by

$$A_n = \pi\lambda\left(z + \frac{2n-1}{4}\lambda\right)$$

All but the first are annular rings. For $z \gg \lambda$, all zones have the same area, $\pi\lambda z$. The first eight values of R_n/λ are plotted in Figure 22.18 for $z = 5\lambda$. Only zones with $R_n \leq R$ are active.

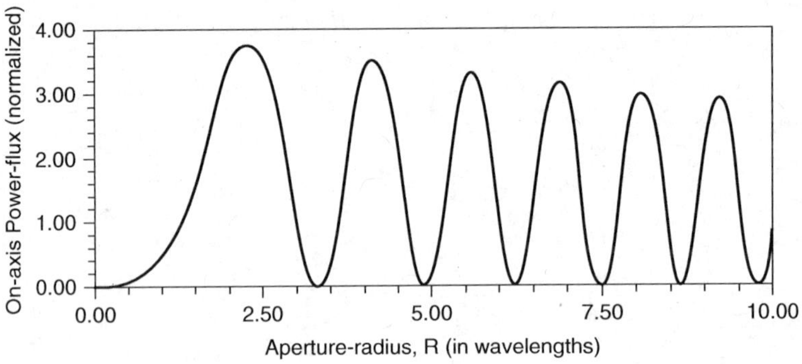

Figure 22.17 $S_z^o\ (R/\lambda, z = 5\lambda)$ (normalized).

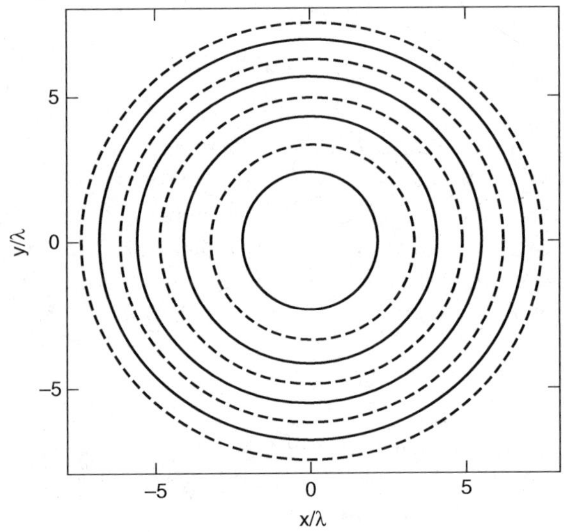

Figure 22.18 Fresnel zones (R_n/λ, $z/\lambda = 5$).

The waves arriving at z from all even zones are basically in phase with one another, but out of phase with the waves arriving from all the odd zones; the converse is also true. To simplify calculations, the phase of an entire zone is often quantized at either 0 or π. In general, the highest-numbered active zone has only a fraction of its area within the aperture and so its contribution is weakened. As z increases, the number of active zones decreases until $R_1 = R$ and the last axial maximum is reached; beyond that point, only the area fraction, $\pi R^2/A_1 \rightarrow R^2/\lambda z$, contributes to the field. This is the Fraunhofer, far-field region. Notice that if $R < \lambda/2$, the maximum occurs at $z = 0$ and there are no on-axis Fresnel ripples. Because each zone (but the first) is an annular ring, the superposition method described previously can provide exact on-axis solutions.

Off-axis zones

To analyze off-axis diffraction at the point (r, z) by means of Fresnel zones, one simply calculates $R_n(z)$ and then translates their common center by the shift r. Each zone contributes the fraction of its area that is within the aperture and is summed according to its phase. For values of z where phase oscillations occur, Fresnel ripples can also be expected in the radial direction until $r/R > 1$ is well within the shadow region. Before the advent of the digital computer, graph paper, planimeters, and quantized phases allowed many diffraction problems to be solved graphically—even those which can be approximated by the Fresnel-integral functions. However, the real importance of the zone concept is as an aide to the development of intuitive-thinking that can (and did) lead to technical innovation. Numerical analysis—greatly facilitated by excellent software—can provide the solution of many well-posed electromagnetic problems. But now, as in the past, the ability to understand the solution is what really counts. The invention of the Fresnel lens is a wonderful example.

Fresnel lens

The result plotted in Figure 22.17 reveals that the on-axis power flux, at *any* prespecified (finite) value of z, can be increased over its aperture value provided R is chosen properly. This is a surprising result because from naive intuition, we would expect the aperture power to spread out (with a consequent monotonic decrease in the flux) as it propagates away from the aperture. Although this picture is accurate for large ratios of z/R, it is certainly not the whole story; instead the circular aperture (if large enough) focuses the power at a series of axial positions. In this respect the aperture acts like a lens with multiple focal lengths and thus has a limited ability to concentrate the power. Nevertheless, the result is intriguing and we can understand why attempts were made to correct the deficiencies.

Returning to the figure, we observe that if our goal is to maximize the field at z, increasing the aperture beyond the first maximum is actually counterproductive and one sort of engineer would leave it at that. But more thoughtful minds would (and did) observe that by using a large aperture and blocking out all even zones *or* all odd zones, the result (a Fresnel lens) will increase the power at the focus by approximately N^2, where N is the number of unblocked zones, each of which contributes equally. Better still, use all of the zones but correct the phases[5] so that they add constructively and increase the focal power by another factor of four. In this case, none of the aperture area is being wasted. Since all of the aperture is focused at the same point, the aberrations are also greatly reduced.

Off-axis far-field radiation pattern

In the far-field region, the off-axis dual vector potential,

$$\underline{\mathbf{A}}_{mp} = -\widehat{\mathbf{y}} \int\int \frac{2\varepsilon_0 \underline{E}_o}{4\pi r_{qp}} \exp(-jkr_{qp})\, da_q$$

can be approximated by

$$\underline{\mathbf{A}}_{mp} = -\widehat{\mathbf{y}}\varepsilon_0 \underline{E}_o \frac{\exp(-jkr)}{r} \sin\theta \int_0^R [\frac{1}{2\pi} \int_0^{2\pi} \exp(jkr_q \sin\theta \cos\psi)\, d\psi] r_q dr_q$$

[5]If the phase correction within the aperture can be made a continuous function of radius, the maximum focusing possible from a plane aperture is achieved.

where $\psi = \phi_q - \phi_p$. The inner integral is recognizable as the real Bessel function (Appendix F, Eq. (F.2a))

$$J_0(kr_q \sin\theta) = \frac{1}{\pi}\int_0^{\pi} \cos(kr_q \sin\theta \cos\psi)\, d\psi$$

and because of the recurrence relation (Appendix F, Eq. (F.8a))

$$xJ_0(x) = \frac{d}{dx}[xJ_1(x)]$$

the outer integral can be replaced by a first-order Bessel function; thus

$$\underline{\mathbf{A}}_{\rm m} = -\widehat{\mathbf{y}}\varepsilon_{\rm 0}\underline{E}_{\rm o}R\frac{\exp(-jkr)}{kr}\frac{J_1(kR\sin\theta)}{\sin\theta} \tag{22.27}$$

Rather than integrate Eq. (22.25b) to find the far-field dual scalar potential, it is simpler to use the Lorenz-gauge condition,

$$j\omega\mu_{\rm 0}\varepsilon_{\rm 0}\underline{\Phi}_{\rm m} = -\frac{\partial \underline{A}_{\rm my}}{\partial y} \rightarrow jk\frac{y}{r}\underline{A}_{\rm my}$$

from which

$$\underline{\Phi}_{\rm m} = -c\varepsilon_{\rm 0}\underline{E}_{\rm o}R\frac{\exp(-jkr)}{kr}\frac{y}{r}\frac{J_1(kR\sin\theta)}{\sin\theta}$$

The fields are then well-approximated by

$$\underline{E}_x \rightarrow j\underline{E}_{\rm o}R\frac{\exp(-jkr)}{r}\frac{J_1(kR\sin\theta)}{\sin\theta}\frac{z}{r} \tag{22.28a}$$

$$\underline{E}_z \rightarrow -j\underline{E}_{\rm o}R\frac{\exp(-jkr)}{r}\frac{J_1(kR\sin\theta)}{\sin\theta}\frac{x}{r} \tag{22.28b}$$

$$\underline{H}_x \rightarrow -jc\varepsilon_{\rm 0}\underline{E}_{\rm o}R\frac{\exp(-jkr)}{r}\frac{J_1(kR\sin\theta)}{\sin\theta}\frac{xy}{r^2} \tag{22.28c}$$

$$\underline{H}_y \rightarrow jc\varepsilon_{\rm 0}\underline{E}_{\rm o}R\frac{\exp(-jkr)}{r}\frac{J_1(kR\sin\theta)}{\sin\theta}\left[1-\left(\frac{y}{r}\right)^2\right] \tag{22.28d}$$

$$\underline{H}_z \rightarrow -jc\varepsilon_{\rm 0}\underline{E}_{\rm o}R\frac{\exp(-jkr)}{r}\frac{J_1(kR\sin\theta)}{\sin\theta}\frac{yz}{r^2} \tag{22.28e}$$

Notice that for $z = 0$, both $\underline{E}_x$ and $\underline{H}_z$ vanish as required by the ground-plane boundary conditions. The peak fields are located near $\theta = 0$, where $\lim_{x\to 0} J_1(x)/x = 1/2$. Consequently,

$$\underline{E}_x = \sqrt{\frac{\mu_{\rm 0}}{\varepsilon_{\rm 0}}}\underline{H}_y \rightarrow j\frac{\underline{E}_{\rm o}R}{2}\frac{\exp(-jkr)}{r}kR$$

which agrees with the on-axis result since $r = z$.

It follows (as expected) that the far-field complex Poynting vector is radially directed and given by

$$\underline{S}_r = \frac{\frac{1}{2}\sqrt{\frac{\varepsilon_{\rm 0}}{\mu_{\rm 0}}}\underline{E}_{\rm o}\underline{E}_{\rm o}^*\pi R^2}{4\pi r^2}(kR)^2\left[\frac{2J_1(kR\sin\theta)}{kR\sin\theta}\right]^2(1-\sin^2\theta\sin^2\phi) \tag{22.29}$$

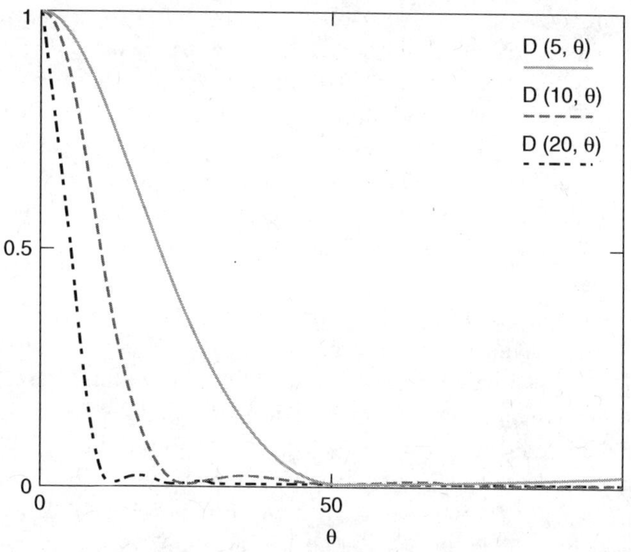

Figure 22.19 Normalized directivity.

The directivity is seen to be

$$D(\theta,\phi) = (kR)^2 \left[\frac{2J_1(kR\sin\theta)}{kR\sin\theta}\right]^2 (1-\sin^2\theta\sin^2\phi)$$

which is the product of two pattern functions: the element factor of a single dipole oriented along the y axis and the array factor which arises from the interference of the continuous distribution of dipoles that are equivalent to the aperture fields. Because by assumption $kR \gg 1$, the directivity is concentrated near $\theta = 0$ and the element-factor term in parentheses can be replaced by unity. Plots of the normalized directivity are plotted in Figure 22.19 for the cases $kR = 5$ (solid), 10 (dotted), 20 (dot-dash). The beam width of the power can be defined by the first-pattern null or at half-height (−3 dB); for $kR \gg 1$, the latter choice leads to

$$\theta_o = \frac{3.2}{kR} \simeq \frac{\lambda}{2R} \ll 1$$

The dual Alternate-power flux, calculated from

$$\underline{\mathbf{S}}^{\mathrm{om}} = -j\omega\frac{1}{4}\left[\frac{1}{\varepsilon_o}(\underline{A}_{\mathrm{my}}\nabla\underline{A}^*_{\mathrm{my}} - \underline{A}^*_{\mathrm{my}}\nabla\underline{A}_{\mathrm{my}}) - \mu_o(\underline{\Phi}_{\mathrm{m}}\nabla\underline{\Phi}^*_{\mathrm{m}} - \underline{\Phi}^*_{\mathrm{m}}\nabla\underline{\Phi}_{\mathrm{m}})\right]$$

leads to the same far-field result as Eq. (22.29); as with the on-axis case, the calculation is very much simpler.

22.8 DIFFRACTION BY A SMALL CIRCULAR APERTURE

When the hole in the perfectly conducting plane is reduced in size so that kR becomes comparable to and then smaller than unity, the analysis of the previous section fails – for reasons that are important to understand. Before trying to do so, we attempt to use the large hole limit when kR is reduced. The directivity drops dramatically as it spreads out over the entire half-space; when $kR \ll 1, D(\theta,\phi) \to (kR)^2(1-\sin^2\theta\sin^2\phi)$. Eventually, $D \to 0$ (which is certainly the correct limit for $R = 0$) with a pattern that is simply the dipole element factor. It is either isotropic ($\phi = 0$) or proportional to $\cos^2\theta$ ($\phi = \pi/2$). Although these predictions are qualitatively correct, the assumption that the value of $\underline{E}_o$remains independent of kR is not. This is evident because the aperture electric field must originate at the rim from positive charges and terminate on negative ones. But, in the limit $k = 0$, the entire aperture plane must be an equipotential that forces $E_o = 0$; it is the time variation of the tangential magnetic field, $\underline{H}_o$, that induces rim charges of both polarities. In addition, as the area of the hole decreases, the bending of the aperture-plane electric-field lines (so that they are purely radial at the rim) occurs over a larger portion of the area, and the assumption of field uniformity throughout the aperture is lost.

The aperture fields constitute an *MQS* problem with zero-order $\underline{\mathbf{H}}$ and $\underline{\mathbf{K}}_e^s$ generated by the low-frequency incident plane wave (and modified by the hole). The first-order $\underline{\mathbf{E}}$ in and near the aperture satisfies the quasistatic equations; it provides an effective value for $\underline{E}_o$ that can be used in the previous theory. Before deriving the *MQS* solution, we first content ourselves with a simple *estimate*: $\underline{E}_o \simeq j\omega\mu_o R\underline{H}_o = j\sqrt{\frac{\mu_o}{\varepsilon_o}}kR\underline{H}_o$. This is the value of the standing-wave electric field a distance R in front of the screen; it is approximately uniform because the dipole-like fields caused by the hole are no longer appreciable. Because there is little magnetic flux within the hole and for a distance comparable to R in front of it, integration over an "aperture" displaced by R is approximately the same as integrating the actual nonuniform field over the true aperture. Then,

$$\underline{S}_r = \frac{\sqrt{\frac{\mu_o}{\varepsilon_o}}\underline{H}_o\underline{H}_o^*}{12r^2}k^4R^6\frac{3}{2}(1-\sin^2\theta\sin^2\phi) \tag{22.30}$$

which matches the power flux of a magnetic dipole, $\frac{\mu_o c}{72\pi^2 r^2}\underline{\mathrm{mm}}^* k^4\frac{3}{2}(1-\sin^2\theta\sin^2\phi)$, provided that the magnetic moment is equated to

$$\underline{\mathrm{m}} = \underline{H}_o\sqrt{6\pi}R^3$$

and the directivity is equated to

$$D(\theta,\phi) = \frac{3}{2}(1-\sin^2\theta\sin^2\phi)$$

Because its physical size is small compared to a wavelength, this dipole radiates symmetrically into both half-spaces. A key point is that the magnetic dipole strength is proportional to R^3 and therefore the total power diffracted in the small kR regime is proportional to R^6 rather than R^2. This is characteristic of Rayleigh scattering from any small object. Small apertures of different shapes can be treated using *MQS* analysis to evaluate $\underline{\mathrm{m}}$. Here we consider a small circular hole. Because at very low frequencies there is little charge at the rim of the aperture, the true electric surface current (*not* the induction-theorem sources) must alter its uniform x-directed flow to avoid the hole; this implies that the zero-order *MQS* magnetic field in the hole is zero. Therefore, the hole can be filled with a lossless nondielectric linear magnetic material, without changing either of the aperture fields. Because the ground plane is very thin, assume a thickness, $\Delta \ll R$, that can later be reduced to zero. We now are dealing with a small disk of magnetic material that is placed in an applied magnetic field $\mathbf{\underline{H}}^{\text{applied}} = \widehat{\mathbf{y}}\underline{H}_o$; it responds by generating a magnetization vector, $\underline{\mathbf{M}} = \widehat{\mathbf{y}}\underline{M}$. From the results of Appendix C, Section C.6, $\underline{H}^{\text{inside}} = \underline{H}_o - N_t\underline{M} = 0$ and $N_t \to \frac{\pi}{8}\frac{\Delta}{R}$. The total magnetic moment is $\underline{M}V$, where the volume of the disk[6] is $\pi R^2\Delta$. It follows that

$$\underline{\mathrm{m}} = \frac{\underline{H}_o}{N_t}\pi R^2\Delta = 8\underline{H}_o R^3$$

independent of the thickness. This result replaces the estimate of $\sqrt{6\pi} = 7.7$ with the factor 8. However, the closeness of these results must be considered fortuitous.

22.9 DIFFRACTION BY THE COMPLEMENTARY SCREEN

Use of Babinet's Principle (Part I, Chapter 6, Section 6.9) allows one to use all of the preceding results to analyze diffraction through a complementary screen, namely a perfectly conducting disk of radius R surrounded by free-space.

For $kR \gg 1$, Eqs.(22.18)–(22.28e) are the total fields (incident and scattered) for $z > 0$. Because

$$\mathbf{E}^i + \mathbf{E}^s \leftrightarrow \sqrt{\frac{\mu_o}{\varepsilon_o}}\mathbf{H}^{s\prime}$$

the components scattered by the complementary screen are

$$\underline{H}_x^{s\prime} \to j\sqrt{\frac{\varepsilon_o}{\mu_o}}\underline{E}_o R\frac{\exp(-jkr)}{r}\frac{J_1(kR\sin\theta)}{\sin\theta}\frac{z}{r}$$

$$\underline{H}_z^{s\prime} \to -j\sqrt{\frac{\varepsilon_o}{\mu_o}}\underline{E}_o R\frac{\exp(-jkr)}{r}\frac{J_1(kR\sin\theta)}{\sin\theta}\frac{x}{r}$$

$$\underline{E}_x^{s\prime} \to -j\underline{E}_o R\frac{\exp(-jkr)}{r}\frac{J_1(kR\sin\theta)}{\sin\theta}\frac{xy}{r^2}$$

$$\underline{E}_y^{s\prime} \to j\underline{E}_o R\frac{\exp(-jkr)}{r}\frac{J_1(kR\sin\theta)}{\sin\theta}\left[1 - \left(\frac{y}{r}\right)^2\right]$$

$$\underline{E}_z^{s\prime} \to -j\underline{E}_o R\frac{\exp(-jkr)}{r}\frac{J_1(kR\sin\theta)}{\sin\theta}\frac{yz}{r^2}$$

For the limit $kR \ll 1$, the scattered fields can be estimated from these formulas after approximating $J_1(x) \simeq \frac{1}{2}x$ and reducing $\underline{E}_o$ by the factor jkR. The result is

$$\underline{H}_x^{s\prime} \to -k^2R^3\sqrt{\frac{\varepsilon_o}{\mu_o}}\underline{E}_o\frac{\exp(-jkr)}{2r}\frac{z}{r}$$

$$\underline{H}_z^{s\prime} \to k^2R^3\sqrt{\frac{\varepsilon_o}{\mu_o}}\underline{E}_o\frac{\exp(-jkr)}{2r}\frac{x}{r}$$

$$\underline{E}_x^{s\prime} \to \underline{E}_o k^2R^3\frac{\exp(-jkr)}{2r}\frac{xy}{r^2}\,.$$

[6]The volume of the very oblate spheroid being used to model the disk and calculate N_t has only $\frac{2}{3}$ as much volume, so there will be some uncertainty in our final result.

$$\underline{E}_y^{s\prime} \rightarrow -\underline{E}_o k^2 R^3 \frac{\exp(-jkr)}{2r}\left[1-\left(\frac{y}{r}\right)^2\right]$$

$$\underline{E}_z^{s\prime} \rightarrow \underline{E}_o k^2 R^3 \frac{\exp(-jkr)}{2r}\frac{yz}{r^2}$$

The k^2R^3 dependence of the fields is what is expected.

22.10 PARAXIAL WAVE EQUATION

Introduction

In the preceding sections, we have seen that in the far field of an antenna or diffraction screen, the fields are often confined to a narrow cone surrounding a single axis. This fact suggests that it might be possible to simplify the analysis of such situations by making that assumption at the outset.

The free-space wave equation governing the complex field, $\underline{\Psi}$, of frequency ω is the Helmholtz Equation, Eq. (4.9),.

$$\nabla^2\underline{\Psi} + \omega^2\mu_0\varepsilon_0\underline{\Psi} = 0$$

where $\underline{\Psi}$ can be the scalar potential $\underline{\Phi}$ or any Cartesian component of $\underline{\mathbf{A}}$, $\underline{\mathbf{E}}$, or $\underline{\mathbf{H}}$.

If the field is propagating at angles close to the z axis, it is reasonable to write the field in the form

$$\underline{\Psi} = \underline{u}(x,y,z)\exp(-jkz)$$

where $k = \omega\sqrt{\mu_0\varepsilon_0}$. Then,

$$\frac{\partial^2\underline{u}}{\partial x^2} + \frac{\partial^2\underline{u}}{\partial y^2} - j2k\frac{\partial\underline{u}}{\partial z} + \frac{\partial^2\underline{u}}{\partial z^2} = 0$$

In the so-called paraxial approximation, k is assumed large enough that one can neglect the last term, and the wave equation reduces to

$$\frac{\partial^2\underline{u}}{\partial x^2} + \frac{\partial^2\underline{u}}{\partial y^2} - j2k\frac{\partial\underline{u}}{\partial z} = 0 \qquad (22.31)$$

Consistent with this approximation, the gradient operation becomes

$$\nabla\underline{\Psi} \simeq \left(\hat{\mathbf{x}}\frac{\partial\underline{u}}{\partial x} + \hat{\mathbf{y}}\frac{\partial\underline{u}}{\partial y} - jk\hat{\mathbf{z}}\underline{u}\right)\exp(-jkz)$$

and the divergence and curl operations on the complex vector $\underline{\mathbf{A}} = \underline{\mathbf{a}}\exp(-jkz)$ become

$$\nabla\cdot\underline{\mathbf{A}} \simeq \left(\frac{\partial\underline{a}_x}{\partial x} + \frac{\partial\underline{a}_y}{\partial y} - jk\underline{a}_z\right)\exp(-jkz)$$

and

$$\nabla\times\underline{\mathbf{A}} \simeq \left[\hat{\mathbf{x}}\left(\frac{\partial\underline{a}_z}{\partial y} + jk\underline{a}_y\right) + \hat{\mathbf{y}}\left(-jk\underline{a}_x - \frac{\partial\underline{a}_z}{\partial x}\right) + \hat{\mathbf{z}}\left(\frac{\partial\underline{a}_y}{\partial x} - \frac{\partial\underline{a}_x}{\partial y}\right)\right]\exp(-jkz)$$

Naturally, there are solutions to the paraxial wave equation in one, two, and three dimensions. In one dimension (z), $\frac{\partial \underline{u}}{\partial z} = 0$ and the solution is both trivial and exact. In two dimensions (x, z),

$$\frac{\partial^2 \underline{u}}{\partial x^2} - j2k\frac{\partial \underline{u}}{\partial z} = 0$$

Solutions of the form

$$\underline{u}(x,z) = \exp\left[j\left(\underline{p}(z) + \frac{x^2}{\underline{q}(z)}\right)\right] \tag{22.32}$$

are valid, provided that

$$\frac{d\underline{p}}{dz} = \frac{-j}{k\underline{q}}$$

$$\frac{d\underline{q}}{dz} = \frac{-2}{k}$$

In three dimensions (r, ϕ, z), but assuming cylindrical symmetry, Eq. (22.31) is again two-dimensional, namely

$$\frac{1}{r}\frac{\partial}{\partial r}(r\frac{\partial \underline{u}}{\partial r}) - j2k\frac{\partial \underline{u}}{\partial z} = 0 \tag{22.33}$$

In this case, solutions of the form

$$\underline{u}(r,z) = \exp\left[j\left(\underline{p}(z) + \frac{r^2}{\underline{q}(z)}\right)\right] \tag{22.34}$$

are valid, provided that

$$\frac{d\underline{p}}{dz} = \frac{-j2}{k\underline{q}}$$

$$\frac{d\underline{q}}{dz} = \frac{-2}{k}$$

In both the two- and three-dimensional cases, we have

$$\underline{q}(z) = \frac{-2}{k}z + \underline{c}_1$$

$$\underline{p}(z) = j\frac{1}{s}\ln|2z - k\underline{c}_1| + \underline{c}_2 \qquad (s = 2 \text{ or } 1)$$

where $\underline{c}_1$ and $\underline{c}_2$ are complex integration constants.

Gaussian-beam solutions

One-dimensional confinement

With $s = 2$ and appropriate integration and normalization constants, Eq. (22.32) becomes

$$\underline{u} = C\sqrt{\frac{jkW^2}{2z + jkW^2}}\exp\left(-\frac{jkx^2}{2z + jkW^2}\right) \tag{22.35}$$

with density

$$\underline{u}\underline{u}^* = \underline{C}\underline{C}^* \frac{|k|W^2}{\sqrt{4z^2 + k^2W^4}} \exp\left(\frac{-2k^2W^2x^2}{4z^2 + k^2W^4}\right)$$

Because

$$\frac{1}{2}\int_{-\infty}^{+\infty} \underline{u}\,\underline{u}^* dx = \frac{1}{2}\underline{C}\underline{C}^* \sqrt{\frac{\pi}{2}} W$$

total wave power is conserved and maximum confinement of the strip beam at the focal plane ($z = 0$) is approximately within $|x| < \sqrt{\frac{\pi}{8}} W$.

Two-dimensional confinement

We again look for a family of complex solutions that embrace both a simple unconfined plane wave ($\frac{\partial u}{\partial r} = \frac{\partial u}{\partial z} = 0$) and one confined within some radius, $R(z)$. With $s = 1$, $R(0) = R$, and appropriate integration constants, Eq. (22.34) becomes

$$\underline{u} = \underline{C} \frac{\exp\left[\frac{-(r/R)^2}{1 + (2z/kR^2)^2}\right]}{\sqrt{1 + (2z/kR^2)^2}} \exp\left[-j\left(\frac{2kz(r/R)^2}{(2z/R)^2 + (kR)^2} - \tan^{-1}\left(\frac{2z}{kR^2}\right)\right)\right] \quad (22.36)$$

Notice that the integral of $\frac{1}{2}\underline{u}\underline{u}^*$ over any infinite cross-sectional plane is $\frac{1}{4}\underline{C}\underline{C}^* \pi R^2$ so that (as with one-dimensional confinement) the total wave power propagating along z is conserved. Notice, too, that the amplitude of $\underline{u}$ is a symmetric function of z with $z = 0$ the focal plane, where maximum confinement is within $r < R/\sqrt{2}$. An optical Gaussian beam can be refocused to multiple positions by the use of lenses spaced periodically along z. In such cases the solution, Eq. (22.36), is valid between neighboring lenses.

In the far field, $z \gg kR^2$, $\tan^{-1}(\frac{2z}{kR^2}) \to \pi/2$ (and on axis because of the paraxial approximation),

$$\underline{u}\exp(-jkz) \to j\underline{C}\frac{kR^2}{2z}\exp(-jkz)$$

As expected, this result agrees with the on-axis analysis of diffraction from a large circular hole of radius R carried out in Section 22.7. However, the two diffraction problems differ in that the Gaussian aperture distribution at $z = 0$ provides a gradual taper that completely eliminates side lobes in the far-field. This is consistent with the fact that the Fourier transform of a Gaussian function is itself a Gaussian function.

Higher-order solutions

Because the partial derivative of Eq. (22.34) with respect to x or y is also a solution of Eq. (22.31), higher-order solutions can easily be generated. Together these form a complete set in terms of which the complex field $\underline{u}$ can be expanded. The nth derivative with respect to x and mth derivative with respect to y is itself a solution that can be expressed as

$$\frac{\partial^{n+m}\underline{u}(r,z)}{\partial x^n \partial y^m} = \frac{(-1)^{n+m}}{[iq(z)]^{(n+m)/2}} H_n\left(\frac{x}{\sqrt{iq(z)}}\right) H_m\left(\frac{y}{\sqrt{iq(z)}}\right) \underline{u}(r,z)$$

where $H_p(w)$ is a Hermite polynomial of integer order p and complex argument w. These polynomials and their properties are defined in Appendix F, Section F.3. The expansion of Eq. (22.34) (with $r^2 = x^2 + y^2$) follows immediately from Eq. (F.25a). However, it should be kept in mind that as n and m increase, these higher-order solutions become poorer and poorer paraxial approximations. Expansion in terms of Bessel functions (circular-waveguide modes) is an alternate strategy.

A similar expansion can be made for one-dimensionally confined beams ($s = 2$); in these cases $m = 0$ and $\underline{u}(x, z)$ replaces $\underline{u}(r, z)$ for confinement along x and $n = 0$ and $\underline{u}(y, z)$ replaces $\underline{u}(r, z)$ for confinement along y.

CHAPTER 23

WAVES AND RESONANCES IN FERRITES

23.1 INTRODUCTION

In certain magnetic materials, electromagnetic waves can propagate in extraordinary ways that permit the construction of devices of great technical importance. In such cases, mechanical as well as electromagnetic components of the wave exist and cause what first seems to be counterintuitive behavior. All of this is possible because, at an atomic level, an individual electron that carries a magnetic moment also carries angular momentum. In most solids the net moment and momentum of an individual atom is zero, but in those rare cases when nonzero magnetic moments in a solid exist and are strongly aligned parallel to one another, organized angular momentum also exists and is the basis of magnetic resonance effects [27]. Because a precessing mechanical top is inherently nonreciprocal, so too are angular momentum (spin) waves and thus associated electromagnetic waves when the mechanical component is sufficiently strong. This fact is the underlying reason why *nonreciprocal* ferrite devices have proved to be possible.[1]

The desired characteristics of a magnetic material for use at microwave frequencies are:

[1]Early (initially forgotten) work by Lord Rayleigh proved that nonreciprocal optical devices were possible, but impractical using materials then available. Later, Tellegen, independently proposed a different nonreciprocal device; it too was impractical, but had the effect of motivating others to develop practical alternatives.

The Power and Beauty of Electromagnetic Fields, First Edition. F. R. Morgenthaler.

1. Magnetic Order A relatively high density of magnetic ions is necessary in order that the magnetic interactions be large. Strong nearest-neighbor coupling leads to a spontaneous magnetization below some critical temperature, T_c, which should be high enough to allow practical applications at and above room temperature.

2. High Resistivity A large skin depth is required to allow penetration of electromagnetic energy inside the material so that the rf magnetic field can interact with the magnetic moments.

3. Low Magnetic Loss To reduce magnetic-resonance damping, the magnetic ions should be weakly coupled to the lattice; therefore, spin–orbit coupling effects should be small. For a given material, losses also depend upon the configuration of the magnetization and can be minimized by applying a large enough DC magnetic field to remove all domain walls (and hence all wall resonances). It is desirable that the material saturate in a reasonable field.

23.2 FERRITES

Fortunately, a number of actual materials combine these properties and belong to the class termed ferrimagnetic. These materials, commonly called ferrites, are oxides of the ferromagnetic metals which may also contain ions of one or more nonmagnetic atoms. Examples are Fe_3O_4(magnetite), $BaFe_2O_4$, $MnFe_2O_4$, $CoFe_2O_4$, and $Y_3Fe_5O_{12}$(yttrium iron garnet) The first of these (also known as lodestone) was the first magnetic material discovered by man, but is of limited importance for microwave applications because it suffers from fairly low resistivity. The last (commonly abbreviated YIG) is of great practical importance because of its unusually low magnetic loss.

Angular momentum and magnetic moments

The bound electrons of a many-electron atom or ion give rise to a total angular momentum vector $\mathbf{J}$,[2] made up of both spin and orbital contributions. Associated with each $\mathbf{J}$ there is a magnetic moment vector given by

$$\mu_{\mathrm{J}} = \gamma_{\mathrm{g}}\mathbf{J} \tag{23.1}$$

where

$$\gamma_{\mathrm{g}} = -g'\frac{e}{2m_{\mathrm{e}}}$$

is the gyromagnetic ratio (negative) with e/m_{e} the ratio of electron charge to mass and g' the dimensionless Landé g-factor,

$$g' = \frac{3}{2} + \frac{S(S+1) - L(L+1)}{2J(J+1)} \tag{23.2}$$

where L is the orbital (integer) value, S the spin (multiple half-integer) value, and J the total that takes on the values $|L-S|, \ldots, |L+S|$. (The ground-state value is either

[2]This is a quantum vector that (in units of $\hbar$—Planck's constant divided by 2π) has a magnitude $\sqrt{J(J+1)}$ but a maximum value J in any given direction.

the minimum or maximum.) In most practical ferrites the orbital moment is quenched; therefore $L = 0$, $J = S$, and $g' = 2$.

In addition, the individual magnetic moments are strongly coupled to their nearest neighbors and below some ordering temperature (termed the Curie temperature T_c) give rise to a spontaneous macroscopic magnetization vector where

$$\mathbf{M} = n_o \mu_J = n_o \mathbf{J}\ \mu_B$$

n_o is the volume density of the moments and $\mu_B = \frac{e}{2m_e}\hbar$ (the Bohr magneton). According to Eq. (23.1), there must also be a macroscopic angular momentum density given by

$$\mathfrak{J} = n_o \mathbf{J} = \frac{1}{\gamma_g}\mathbf{M}$$

In the absence of an applied magnetic field, a ferrite specimen will subdivide into many small regions or domains in which the direction of magnetization alternates. The transition regions between oppositely magnetized domains can oscillate at microwave frequencies, thereby absorbing power. Such losses can be prevented by applying a DC magnetic field strong enough to wipe out the domain pattern, and we assume this has been done. In the magnetically saturated sample the magnitude of $\mathbf{M}$ (and hence also $\mathfrak{J}$) is a constant at every point within the material.

Constitutive relations

The assumption that the ferrite is magnetically saturated implies that both the magnetization, $\mathbf{M}$, and associated angular-momentum density, $\mathbf{M}/\gamma_g$, have constant magnitudes. The energy density, $-\mu_o \mathbf{M} \cdot \mathbf{H}$, is minimized when $\mathbf{M}$ and $\mathbf{H}$ are parallel; when they are not an electromagnetic torque density, $\mu_o \mathbf{M} \times \mathbf{H}$ is created. The total torque density acting upon $\mathbf{M}$ must be orthogonal to it and so Newton's Law takes the form

$$\frac{1}{\gamma_g}\frac{\partial \mathbf{M}}{\partial t} = \mu_o \mathbf{M} \times (\mathbf{H} + \mathbf{H}_{\text{material}}) \tag{23.3}$$

where the *effective* magnetic field, $\mathbf{H}_{\text{material}}$, accounts for any material torques in excess of the electromagnetic component; it is not a true magnetic field in the Maxwellian sense. This constitutive law forces $\mathbf{M} \cdot \frac{\partial \mathbf{M}}{\partial t} = 0$, and therefore is consistent with the saturation condition, $|\mathbf{M}| = M_s$. Although torques due to dissipation, crystalline anisotropy, magnetoelasticity, and quantum-mechanical interactions must be accounted for in a more complete theory, for simplicity, we here assume $\mathbf{H}_{\text{material}} = 0$. In other respects, the ideal ferrite is a lossless linear dielectric with isotropic permittivity, ε.

Magnetic resonance

If a static magnetic field, $\mathbf{H} = \widehat{\mathbf{z}} H_z$, exists inside the ferrite and there are no other components, the solution of Eq. (23.3) is

$$\mathbf{M} = M_s(\widehat{\mathbf{z}}\cos\theta + \sin\theta[\widehat{\mathbf{x}}\cos(\omega_o t - \psi) + \widehat{\mathbf{y}}\sin(\omega_o t - \psi)]) \tag{23.4}$$

$$\omega_o = -\gamma_g \mu_o H_z = g' \frac{e}{2m_e}\mu_o H_z$$

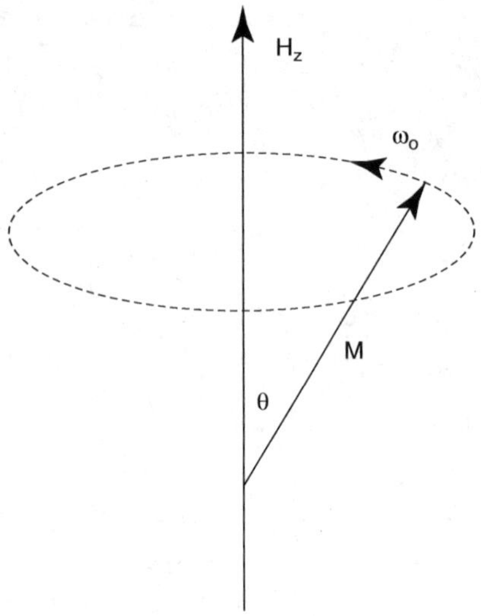

Figure 23.1 Magnetic resonance precession.

where the angles θ and ψ are arbitrary constants. The magnetization is precessing about the DC field at the magnetic-resonance frequency, ω_o, while maintaining a constant cone angle, θ. The geometry is indicated in Figure 23.1.

Because it appears that the angle ψ can be an arbitrary function of position without invalidating Eq. (23.4), the prospects for interesting sorts of wave propagation motivate further study. Because (for $g' = 2$) the frequency increases at the rate of 2.8 MHz per oersted (1 Oe $= 250/\pi \; Am^{-1}$), a value of 1000 Oe produces resonance at the microwave frequency of 2.8 GHz. Fields of this magnitude can easily be produced with either permanent magnets or electromagnets and so ferrites are of special interest to microwave engineers. The motion of **M** is similar to that of a child's top, which is mechanically nonreciprocal (the direction of precession is not arbitrary, but depends upon the spin direction), and so nonreciprocal wave propagation in a ferrite can be expected; this proves to be true and, following the initial success by C. L. Hogan, led to the development of nonreciprocal microwave devices such as gyrators, isolators, and circulators [28]. When very low-loss ferrites such as yttrium iron garnet (YIG) became available, high-Q filters with tunable resonances were also developed.

23.3 LARGE-SIGNAL EQUATIONS

Maxwell's Equations for a nonconducting ferrite can be written, in the Chu formulation, as

$$\nabla \times \mathbf{H} - \varepsilon \frac{\partial \mathbf{E}}{\partial t} = 0$$

$$\nabla \times \mathbf{E} + \mu_o \frac{\partial \mathbf{H}}{\partial t} = -\mathbf{J}_m$$

$$\nabla \cdot \varepsilon \mathbf{E} = 0$$

$$\nabla \cdot \mu_o \mathbf{H} = \rho_m$$

arranged so that the sources appear to be magnetic currents and magnetic charges,

$$\mathbf{J}_{\mathrm{m}} = \mu_{\mathrm{o}} \frac{\partial \mathbf{M}}{\partial t}$$

$$\rho_{\mathrm{m}} = -\mu_{\mathrm{o}} \nabla \cdot \mathbf{M}$$

Because $J_{\mathrm{u}} = 0$, $\rho_{\mathrm{u}} = 0$, it is possible and convenient to employ dual vector and scalar potentials, $\mathbf{A}_{\mathrm{m}}$ and Φ_{m}, the Lorenz gauge, and the dual Alternate-power flux and Alternate-energy density, $\mathbf{S}^{\mathrm{om}}$ and W^{om}[3].

The large-signal Poynting Theorem,

$$\nabla \cdot (\mathbf{E} \times \mathbf{H}) + \frac{\partial}{\partial t}(\frac{1}{2}\varepsilon \mathbf{E} \cdot \mathbf{E} + \mu_{\mathrm{o}} \mathbf{H} \cdot \mathbf{H}) = -\mathbf{H} \cdot \mathbf{J}_{\mathrm{m}} \quad (23.5)$$

is a conservation law because (neglecting any material torques)

$$-\mathbf{H} \cdot \mathbf{J}_{\mathrm{m}} = -\gamma_{\mathrm{g}} \mu_{\mathrm{o}}^2 (\mathbf{H} \times \mathbf{M}) \cdot \mathbf{H}_{\mathrm{material}} = 0$$

23.4 LINEARIZED (SMALL-SIGNAL) EQUATIONS

Time-dependent equations

Although the energy-density term, $-\mu_{\mathrm{o}} \mathbf{M} \cdot \mathbf{H} = -\mu_{\mathrm{o}} M_s H_z \cos\theta$, does not appear in Eq. (23.5), it is responsible for creating the $\mu_{\mathrm{o}} \mathbf{M} \times \mathbf{H}$ torque density. Although the cone angle of precession in Eq. (23.4) is arbitrary, the minimum energy occurs for $\theta = 0$. Since this is the stable ground-state energy, it is expected that, normally, only small deviations will be permitted.[4] Under these small-signal conditions, $\sin\theta \simeq \theta$, $\cos\theta \simeq 1$, and

$$\mathbf{M} \simeq \widehat{\mathbf{z}} M_s + \mathbf{m}, |\mathbf{m}| \ll M_s$$

$$\mathbf{m} \cdot \widehat{\mathbf{z}} = 0$$

$$\mathbf{H} \simeq \widehat{\mathbf{z}} H_z + \mathbf{h}, |\mathbf{h}| \ll H_z$$

where $\theta = \frac{|\mathbf{m}|}{M_s}$ and we now include a small deviation $\mathbf{h}$ from the applied DC magnetic field. If the torque $\mu_{\mathrm{o}} \mathbf{m} \times \mathbf{h}$ is neglected, the linearized forms of Eq. (23.3) and the Maxwell Equations result in

$$\frac{\partial \mathbf{m}}{\partial t} = \gamma_{\mathrm{g}} \mu_{\mathrm{o}} M_s \widehat{\mathbf{z}} \times \left(\mathbf{h} - \frac{H_z}{M_s} \mathbf{m} \right) \quad (23.6a)$$

$$\nabla \times \mathbf{h} - \varepsilon \frac{\partial \mathbf{e}}{\partial t} = 0 \quad (23.6b)$$

$$\nabla \times \mathbf{e} + \mu_{\mathrm{o}} \frac{\partial \mathbf{h}}{\partial t} = -\mathbf{j}_{\mathrm{m}} \quad (23.6c)$$

[3]When $\rho_{\mathrm{u}} \neq 0$, a superposition of both dual and normal vector potentials is required (as detailed in Chapter 6, Section 6.5). Fortunately, in the nonconducting case, we need only $\mathbf{A}_{\mathrm{m}}$. Alternatively, one can switch from the Chu to Amperian formulation and use only $\mathbf{A}$ (generated by $\mathbf{J}_{\mathrm{u}} + \frac{\partial}{\partial t}\mathbf{P} + \nabla \times \mathbf{M}$).

[4]Even when $\theta \ll 1$, nonlinear effects can occur (due to the parametric excitation of spin waves) that limit the linear operating range of ferrite devices. These effects, first analyzed by H. Suhl [32], are beyond the scope of this text but are discussed in Morgenthaler [33].

$$\nabla \cdot \varepsilon \mathbf{e} = 0 \tag{23.6d}$$

$$\nabla \cdot \mu_0 \mathbf{h} = \rho_m \tag{23.6e}$$

$$\mathbf{j}_m = \mu_0 \frac{\partial \mathbf{m}}{\partial t}, \qquad \rho_m = -\mu_0 \nabla \cdot \mathbf{m} \tag{23.6f}$$

The form of these equations allows the electric and magnetic fields to be expressed as

$$\varepsilon \mathbf{e} = -\nabla \times \mathbf{a}_m$$

$$\mathbf{h} = -\frac{\partial \mathbf{a}_m}{\partial t} - \nabla \phi_m$$

when the linearized dual-potentials are the solutions of

$$\nabla^2 \mathbf{a}_m - \mu_0 \varepsilon \frac{\partial^2 \mathbf{a}_m}{\partial t^2} = -\varepsilon\, \mathbf{j}_m$$

$$\nabla^2 \phi_m - \mu_0 \varepsilon \frac{\partial^2 \phi_m}{\partial t^2} = -\frac{1}{\mu_0} \rho_m$$

$$\nabla \cdot \mathbf{a}_m + \mu_0 \varepsilon \frac{\partial \phi_m}{\partial t} = 0$$

Complex Polder susceptibility and permeability tensors

For sinusoidal steady-state fields of frequency ω, these equations can be cast in complex form. When this done for Eq. (23.6a), the result in Cartesian-component form becomes

$$j\omega \underline{m}_x = -\gamma_g \mu_0 M_s \left(\underline{h}_y - \frac{H_z}{M_s} \underline{m}_y \right)$$

$$j\omega \underline{m}_y = \gamma_g \mu_0 M_s \left(\underline{h}_x - \frac{H_z}{M_s} \underline{m}_x \right)$$

These two equations can be solved for $\underline{m}_x$ and $\underline{m}_y$; the result expressed as a matrix equation is

$$\begin{bmatrix} \underline{m}_x \\ \underline{m}_y \end{bmatrix} = \begin{bmatrix} \chi(\omega) & -j\kappa(\omega) \\ j\kappa(\omega) & \chi(\omega) \end{bmatrix} \cdot \begin{bmatrix} \underline{h}_x \\ \underline{h}_y \end{bmatrix} \tag{23.7}$$

where, with $\omega_M = -\gamma_g \mu_0 M_s$ and $\omega_Z = -\gamma_g \mu_0 H_z$,

$$\chi(\omega) = \frac{\omega_M \omega_Z}{\omega_Z^2 - \omega^2}$$

$$\kappa(\omega) = \frac{-\omega_M \omega}{\omega_Z^2 - \omega^2}$$

are the elements of the Polder susceptibility tensor. It is very important to notice that this tensor is *nonreciprocal*.

It is often convenient to work with $\underline{\mathbf{b}}$ instead of $\underline{\mathbf{m}}$; in these cases one employs the Polder permeability tensor defined by

$$\begin{bmatrix} \underline{b}_x \\ \underline{b}_y \end{bmatrix} = \mu_0 \begin{bmatrix} 1+\chi(\omega) & -j\kappa(\omega) \\ j\kappa(\omega) & 1+\chi(\omega) \end{bmatrix} \cdot \begin{bmatrix} \underline{h}_x \\ \underline{h}_y \end{bmatrix} \tag{23.8a}$$

which is often written as

$$\begin{bmatrix} \underline{b}_x \\ \underline{b}_y \end{bmatrix} = \begin{bmatrix} \mu & -j\kappa \\ j\kappa & \mu \end{bmatrix} \cdot \begin{bmatrix} \underline{h}_x \\ \underline{h}_y \end{bmatrix} \tag{23.8b}$$

with $\mu = \mu_o(1+\chi)$ and (inconsistently) $\kappa = \mu_o\kappa$. Either form of the Polder tensor can be inserted into the complex Maxwell Equations; except for the reciprocity theorem, all of the complex theorems of Chapter 6 apply.

When the frequency is such that $\underline{\mathbf{m}}$ is not resonant, the fields are mainly electromagnetic in character and it is advantageous to focus on $\underline{\mathbf{e}}$ and/or $\underline{\mathbf{h}}$. When, instead the frequency is at or close to magnetic resonance, the fields are dominated by the mechanical aspects and it is advantageous to focus on $\underline{\mathbf{m}}$. We follow the latter technique in the following sections.

Dissipation caused by small or moderate magnetic losses can be included in the small-signal model by allowing ω_z to become the complex frequency, $\omega_z + j\omega_{\Delta H}$. Here, $\omega_{\Delta H} = -\gamma_g\mu_o\Delta H$ is a phenomenological constant that, in general, is frequency- and wavenumber-dependent; ΔH is defined as the half-linewidth of the resonance. For simplicity, we set $\Delta H \to 0$ in what follows.

23.5 UNIFORM PRECESSION IN A SMALL ELLIPSOID

As detailed in Appendix C, Section C.6, if a uniform static magnetic field, $\hat{\mathbf{z}}H_o$, is applied parallel to one of the principal axes of a small ferrite ellipsoid, the DC field inside the sample will also be uniform, but of reduced strength. Provided that H_o is large enough to saturate the sample, the internal field is found to be

$$\mathbf{H} = \hat{\mathbf{z}}(H_o - N_z M_s) \quad > 0$$

where N_z is the appropriate demagnetizing factor. The magnetization vector consists of uniform static and small-signal (time-varying) components,

$$\mathbf{M} = \hat{\mathbf{z}}M_s + \hat{\mathbf{x}}m_x + \hat{\mathbf{y}}m_y$$

where x and y are also principal axes. Because it is uniform, there is only a surface divergence of $\mathbf{m}$, and the small-signal magnetic field is created by rotating magnetic surface charges. In the *MQS* limit, the time delay due to electromagnetic retardation is negligible as long as the ellipsoid dimensions are small compared to the free-space wavelength. Assuming this is the case, we have

$$\mathbf{h} = -\hat{\mathbf{x}}N_x m_x - \hat{\mathbf{y}}N_y m_y$$

which is the field that would exist if $\mathbf{m}$ were also a static vector. In this quasistatic context, N_x and N_y are sometimes referred to as "rf-demagnetizing" factors; the three factors sum to unity. When these fields are substituted into the linearized torque-equation, the Cartesian components become

$$\frac{\partial m_x}{\partial t} = \gamma_g\mu_o(H_o - N_z M_s + N_y M_s)m_y$$

$$\frac{\partial m_y}{\partial t} = -\gamma_g\mu_o(H_o - N_z M_s + N_x M_s)m_x$$

Differentiation of either equation with respect to time, followed by substitution of the other equation, produces a second-order equation that can easily be solved. The solution known as the Kittel uniform-precession resonance is [29]

$$\begin{aligned} m_x &= m_a \cos(\omega_0 t - \psi) \\ m_y &= m_b \sin(\omega_0 t - \psi) \end{aligned} \tag{23.9}$$

where ψ is an arbitrary phase constant and

$$\omega_0 = -\gamma_g \mu_0 \sqrt{(H_0 - N_z M_s + N_x M_s)(H_0 - N_z M_s + N_y M_s)} \tag{23.10a}$$

$$\frac{m_a}{m_b} = \sqrt{\frac{(H_0 - N_z M_s + N_y M_s)}{(H_0 - N_z M_s + N_x M_s)}} \tag{23.10b}$$

In general, the small-signal magnetization vector traces out an ellipse and, because $\gamma_g < 0$, the precession follows the right-hand rule: with thumb pointing in the $+z$ direction, the curled fingers point in the direction of the motion. Because $H_z = H_0 - N_z M_s \geq 0$, the *minimum* value of ω_0 ($-\gamma_g \mu_0 \sqrt{N_x N_y} M_s$) is both shape- and material-dependent.

In case of an ellipsoid of revolution (spheroid), $N_x = N_y = N_t = (1 - N_z)/2$ with N_z given by Eqs. (C.31) or (C.33). The precession is circular with $\omega_0 = -\gamma_g \mu_0 [H_0 - (1 - 3N_t) M_s]$. For a sphere, $N_t = 1/3$ and $\omega_0 = -\gamma_g \mu_0 H_0$; for a long cylinder, $N_t = 1/2$ and $\omega_0 = -\gamma_g \mu_0 (H_0 + \frac{1}{2} M_s)$; for a thin disk, $N_t = 0$ and $\omega_0 = -\gamma_g \mu_0 (H_0 - M_s)$. The formulas can also be applied to a thin film magnetized in a direction either perpendicular to its plane ($N_t = 0$) or parallel to it ($N_x = 1$ or $N_y = 1$) for which $\omega_0 = -\gamma_g \mu_0 \sqrt{H_0 (H_0 + M_s)}$.

It might be expected that nonuniform magnetostatic modes (those with nonzero volume divergence of **m**) exist and they do. The analysis of all modes of a spheroid was first carried out by L. R. Walker [30] and are known as Walker modes. Naturally, the Kittel resonance is one of them.

23.6 PLANE WAVE SOLUTIONS

We assume that for plane wave propagation, the small-signal version of Eq. (23.9) is

$$\mathbf{m} = \widehat{\mathbf{x}} m_a \cos(\omega t - \mathbf{k} \cdot \mathbf{r}) + \widehat{\mathbf{y}} m_b \sin(\omega t - \mathbf{k} \cdot \mathbf{r}) \tag{23.11}$$

We allow for $m_a \neq m_b$; if true, the magnetization is elliptically polarized and the directions $\widehat{\mathbf{x}}$ and $\widehat{\mathbf{y}}$ take on special significance based upon $\mathbf{k}$. It follows that

$$\mathbf{j}_m = \omega \mu_0 [-\widehat{\mathbf{x}} m_a \sin(\omega t - \mathbf{k} \cdot \mathbf{r}) + \widehat{\mathbf{y}} m_b \cos(\omega t - \mathbf{k} \cdot \mathbf{r})]$$

$$\rho_m = \mu_0 [-k_x m_a \sin(\omega t - \mathbf{k} \cdot \mathbf{r}) + k_y m_b \cos(\omega t - \mathbf{k} \cdot \mathbf{r})]$$

and, with $k_0^2 = \omega^2 \mu_0 \varepsilon$, the dual potentials and the fields they generate are

$$\mathbf{a}_m = \omega \mu_0 \varepsilon \frac{-\widehat{\mathbf{x}} m_a \sin(\omega t - \mathbf{k} \cdot \mathbf{r}) + \widehat{\mathbf{y}} m_b \cos(\omega t - \mathbf{k} \cdot \mathbf{r})}{\mathbf{k} \cdot \mathbf{k} - k_0^2} \tag{23.12a}$$

$$\phi_m = \frac{-k_x m_a \sin(\omega t - \mathbf{k} \cdot \mathbf{r}) + k_y m_b \cos(\omega t - \mathbf{k} \cdot \mathbf{r})}{\mathbf{k} \cdot \mathbf{k} - k_0^2} \tag{23.12b}$$

$$\mathbf{e} = \frac{\omega\mu_o \mathbf{m} \times \mathbf{k}}{\mathbf{k}\cdot\mathbf{k} - k_o^2} \tag{23.12c}$$

$$\mathbf{h} = \frac{-\mathbf{k}(\mathbf{m}\cdot\mathbf{k}) + k_o^2\mathbf{m}}{\mathbf{k}\cdot\mathbf{k} - k_o^2} \tag{23.12d}$$

Notice that the complex Amperian current density,

$$\nabla \times \underline{\mathbf{m}} = j\,\underline{\mathbf{m}} \times \mathbf{k}$$

produces small-signal potentials,

$$\underline{\mathbf{a}} = \frac{j\mu_o \underline{\mathbf{m}} \times \mathbf{k}}{\mathbf{k}\cdot\mathbf{k} - k_o^2}$$

$$\underline{\phi} = 0$$

that generate the correct fields, $\underline{\mathbf{e}}$ and $\underline{\mathbf{b}} = \underline{\mathbf{h}} + \underline{\mathbf{m}}$.

Finally, substitution of Eq. (23.12d) into Eq. (23.6a) reveals that the assumed form of **m** is consistent only when

$$k_x k_y = 0 \tag{23.13a}$$

$$\omega^2(\mathbf{k}) = \left(\omega_Z + \frac{k_x^2 - k_o^2}{\mathbf{k}\cdot\mathbf{k} - k_o^2}\omega_M\right)\left(\omega_Z + \frac{k_y^2 - k_o^2}{\mathbf{k}\cdot\mathbf{k} - k_o^2}\omega_M\right) \tag{23.13b}$$

$$\left(\frac{m_b}{m_a}\right)^2 = \frac{\left(\omega_Z + \dfrac{k_x^2 - k_o^2}{\mathbf{k}\cdot\mathbf{k} - k_o^2}\omega_M\right)}{\left(\omega_Z + \dfrac{k_y^2 - k_o^2}{\mathbf{k}\cdot\mathbf{k} - k_o^2}\omega_M\right)} \tag{23.13c}$$

where, as before, we have made use of the definitions:

$$\omega_Z = -\gamma_g \mu_o H_z$$

$$\omega_M = -\gamma_g \mu_o M_s$$

The transverse wave vector must be parallel to either the major or minor axis of the precession ellipse; when **k** is parallel to the z axis, the precession is circularly polarized as expected from symmetry considerations.

The dispersion relation, $\omega(\mathbf{k})$, normalized to ω_M, is shown in Figure 23.2 for $\mathbf{k} = \hat{\mathbf{z}}k_z$ (the straight and curved solid lines respectively represent positive and negative circular polarization) and in Figure 23.3 for $\mathbf{k} = \hat{\mathbf{x}}k_x$ (they respectively represent $\mathbf{h}\cdot\hat{\mathbf{z}} = 0$ and $\mathbf{e}\cdot\hat{\mathbf{z}} = 0$). In both cases, $H_z = M_s$; the dotted curves are the magnetostatic-wave limits (normalized values of 1 and $\sqrt{2}$).

Notice that for any propagation angle defined by k_x/k_z, the frequency approaches a limit that is independent of $|\mathbf{k}|$. This is consistent with the earlier speculation that, in Eq. (23.4), the precession phase angle, ψ, can be an arbitrary function of position; it holds for the magnetostatic-wave limits that are discussed below in greater detail.

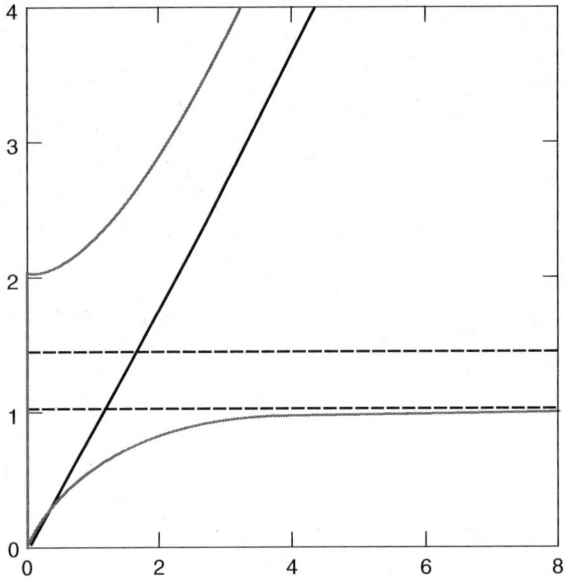

Figure 23.2 Normalized dispersion: ω/ω_M vs. $k_z/(\omega_M\sqrt{\mu_0\varepsilon})$.

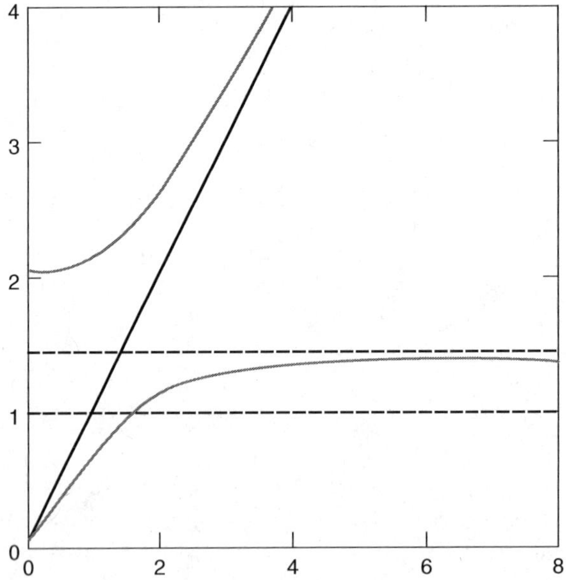

Figure 23.3 Normalized dispersion: ω/ω_M vs. $k_x/(\omega_M\sqrt{\mu_0\varepsilon})$.

Electromagnetic waves

When $\mathbf{k}\cdot\mathbf{k} = k^2 \simeq k_o^2$, the waves are basically electromagnetic, but may have important magnetic attributes, even though $\mathbf{m}$ is nonresonant. We investigate propagation when $\mathbf{k}$ is either parallel or perpendicular to $\widehat{\mathbf{z}}$.

k parallel to $\hat{\mathbf{x}}$ ($\mathbf{e} \times \hat{\mathbf{z}} = 0$)

$$k_x^2 = k_o^2 = \omega^2 \mu_o \varepsilon$$

$$m_a = m_b = 0$$

This case occurs because **h** is linearly polarized parallel to $\hat{\mathbf{z}}$. With $m_z = 0$, the magnetic properties of the ferrite are those of free-space.

k *parallel to* $\hat{\mathbf{x}}$ ($\mathbf{h} \times \hat{\mathbf{z}} = 0$)

$$\left[\omega^2 = \gamma_g \mu_o^2 (H_z + M_s)\left(H_z - \frac{k_o^2}{k_x^2 - k_o^2} M_s\right); \left(\frac{m_b}{m_a}\right)^2 = (H_z + M_s)/\left(H_z - \frac{k_o^2}{k_x^2 - k_o^2} M_s\right)\right]$$

This case occurs because **e** is linearly polarized parallel to $\hat{\mathbf{z}}$ and $\mathbf{b} \cdot \mathbf{k} = 0$ ($b_x = 0$). It can be expressed as

$$k_x^2 = \omega^2 \mu_o \varepsilon \frac{(\omega_z + \omega_M)^2 - \omega^2}{\omega_z(\omega_z + \omega_M) - \omega^2}$$

$$\left|\frac{m_b}{m_a}\right| = \left|\frac{\omega_z + \omega_M}{\omega}\right|$$

The propagation is reciprocal. Notice that there are frequencies for which $k_x \to 0$ and $k_x \to \infty$; these are magnetostatic wave regions.

k *parallel to* $\hat{\mathbf{z}}$

$$\left[\omega = -\gamma_g \mu_o \left(H_z - \frac{k_o^2}{k_z^2 - k_o^2} M_s\right); m_a^2 = m_b^2\right]$$

This result can be rewritten in the form

$$k_z^2 = \omega^2 \mu_o \varepsilon \left(1 + \frac{\omega_M}{\omega_z - \omega}\right)$$

$$m_a = \frac{\omega}{|\omega|} m_b$$

Circularly polarized waves with opposite rotations (depending upon the sign of ω) propagate with different values of $k_z = k_\pm$. It follows that a linearly polarized wave (the superposition of left and right circularly polarized waves of equal amplitude) will have its plane of polarization rotated by the Faraday angle, $\theta_F = \frac{1}{2}(k_+ - k_-)\ell$, in a distance ℓ. This is a nonreciprocal effect, because the rotation angle is independent of the direction of propagation. Because $k_z \to \infty$ when $\omega \to \omega_z$, this is also a magnetostatic wave limit.

Magnetostatic waves

There are two quasistatic regions $k \ll k_o$ and $k \gg k_o$. We investigate each in turn:

$k \ll k_o$

In the extreme limit, $k \to 0$, $\mathbf{h} = -\mathbf{m}$, and with the flux expelled ($\mathbf{b} = 0$), $\mathbf{e} = 0$. Both the power flux and group velocity approach zero. The limiting frequency of the wave is

$$\omega \to -\gamma_g \mu_o (H_z + M_s)$$

$k \gg k_o$

In this region, $\mathbf{h} \to -\mathbf{k}(\mathbf{m}\cdot\mathbf{k})/k^2$, and $\mathbf{e} = \omega\mu_0(\mathbf{m}\times\mathbf{k})/k^2$. The small-signal magnetic field depends upon the orientation of the wavevector, but *not* its magnitude[5]; because $\mathbf{e} \to 0$, $\nabla \times \mathbf{h} \to 0$, validating the designation "magnetostatic wave." Again, the power-flux and group-velocity approach zero. This regime is one in which as the wavelength gets smaller, the quasistatic approximation gets better. The limiting frequency of these waves is (with $k_y = 0$),

$$\omega(\mathbf{k}) = -\gamma_g\mu_0\sqrt{H_z(H_z + \frac{k_x^2}{k^2}M_s)}$$

Notice that for $k_x = 0$, $\omega(k_z) \to \omega_z$ in agreement with an earlier observation. Because k can be very large, it is possible for it to equal that of an elastic wave of the same frequency; in these cases, in single crystals of ferrite with very small losses, magnetoelastic coupling may lead to the excitation of elastic and magnetoelastic waves at microwave frequencies. At still larger values of k, the phase shift between neighboring magnetic moments brings into play quantum-mechanical exchange coupling. This spin-wave region of the spectrum is also magnetostatic, but characterized by frequencies that increase as the wavelength becomes smaller; an additional channel for exchange-power flow and the boundary conditions governing it become important in this region. These topics require an appropriate $\mathbf{H}_{\text{material}}$ to be included in Eq. (23.3), but are beyond the scope of this text. For those interested, surveys written by Morgenthaler [31], [33] and the more recent text on the theory of magnetostatic waves by Stancil [34] can be consulted.

Although these solutions are all uniform plane waves, nonuniform magnetostatic waves generated at interfaces between a ferrite and a normal dielectric are also permitted. If the applied magnetic field is parallel to a planar ferrite–air interface and the propagation is mutually perpendicular to the field and the interface normal, the nonreciprocal Damon–Eschbach mode [35] propagates at the frequency $-\gamma_g\mu_0(H_o + \frac{1}{2}M_s)$ with exponential decay normal to the interface. Propagation is allowed in only one direction. This is the same frequency as that of the uniform precession in a long cylinder.

23.7 SMALL-SIGNAL POWER AND ENERGY

Maxwell–Poynting representation

From the linearized torque equation and the linearized Maxwell Equations, the small-signal Poynting Theorem for a dissipationless material with uniform parameters (ε, M_s, H_z, γ_g) is (in the Minkowski form—summarized in Appendix D, Section D.2)

$$\nabla\cdot(\mathbf{e}\times\mathbf{h}) + \frac{\partial}{\partial t}\left[\frac{1}{2}\varepsilon\mathbf{e}\cdot\mathbf{e} + \frac{1}{2}\mu_0\mathbf{h}\cdot(\mathbf{h}+\mathbf{m})\right] = \frac{1}{2}\mu_0\left(\mathbf{m}\cdot\frac{\partial\mathbf{h}}{\partial t} - \mathbf{h}\cdot\frac{\partial\mathbf{m}}{\partial t}\right) \quad (23.14a)$$

[5] In the event that $\mathbf{k}$ is parallel to $\widehat{\mathbf{z}}$, $\mathbf{m}\cdot\mathbf{k} = 0$ and $\mathbf{h} \to 0$. In this special case the electric field, although itself weak, may be larger. This is certainly unusual for a magnetostatic field, but it simply means that $\mathbf{m}$ is the dominant field and w_m is the dominant energy.

or equivalently (in the Chu form)

$$\nabla \cdot (\mathbf{e} \times \mathbf{h}) + \frac{\partial}{\partial t}\left(\frac{1}{2}\varepsilon \mathbf{e} \cdot \mathbf{e} + \frac{1}{2}\mu_0 \mathbf{h} \cdot \mathbf{h}\right) = -\mu_0 \mathbf{h} \cdot \frac{\partial \mathbf{m}}{\partial t} \tag{23.14b}$$

Unlike its large-signal counterpart, the right-hand side of the latter does not vanish. Rather it can be written as

$$-\mu_0 \left(\mathbf{h} - \frac{H_z}{M_s}\mathbf{m}\right) \cdot \frac{\partial \mathbf{m}}{\partial t} - \mu_0 \frac{H_z}{M_s}\mathbf{m} \cdot \frac{\partial \mathbf{m}}{\partial t} = -\frac{\partial}{\partial t}\left(\frac{1}{2}\mu_0 \frac{H_z}{M_s}\mathbf{m} \cdot \mathbf{m}\right)$$

(The first term on the left-hand side vanishes.) This represents the transfer into the energy of a subsystem. It follows that

$$\nabla \cdot (\mathbf{e} \times \mathbf{h}) + \frac{\partial}{\partial t}\left(\frac{1}{2}\varepsilon \mathbf{e} \cdot \mathbf{e} + \frac{1}{2}\mu_0 \mathbf{h} \cdot \mathbf{h} + \frac{1}{2}\mu_0 \frac{H_z}{M_s}\mathbf{m} \cdot \mathbf{m}\right) = 0 \tag{23.15}$$

constitutes a conservation law when applied to all three small-signal energy densities. The physical origin of the new term is evidently $-\mu_0 \mathbf{M} \cdot \mathbf{H} = -\mu_0 M_s H_z \cos\theta \simeq -\mu_0 M_s H_z \left(1 - \frac{1}{2}\theta^2\right)$. Because $\theta = \frac{|\mathbf{m}|}{M_s}$, the increase in this energy density above the ground-state value is exactly $\frac{1}{2}\mu_0 \frac{H_z}{M_s}\mathbf{m} \cdot \mathbf{m}$. This is sometimes referred to as the Zeeman energy density and given the symbol w_Z. It must be emphasized that because small-signal theorems are based upon linearized equations, the small-signal power flux and energy density may not always be exactly equivalent to the second-order terms that result from an expansion of the respective large-signal quantities. This was a controversial point when L. J. Chu developed the small-signal kinetic-power theorem [36] for electron beams – a theorem of great utility when applied to traveling-wave amplifiers. In this instance, because there is no electric field in the absence of a small-signal excitation, the large-signal Maxwell Equations are (correct to second-order)

$$\nabla \times \mathbf{e} = -\mu_0 \frac{\partial}{\partial t}\left\{\widehat{\mathbf{z}}\left[H_z + \left(M_s - \frac{1}{2}\frac{\mathbf{m} \cdot \mathbf{m}}{M_s}\right)\right] + \mathbf{h} + \mathbf{m}\right\}$$

$$\nabla \times (\widehat{\mathbf{z}} H_z + \mathbf{h}) = \varepsilon \frac{\partial \mathbf{e}}{\partial t}$$

Therefore, because $\mathbf{m} \cdot \widehat{\mathbf{z}} = 0$, the large-signal Poynting theorem,

$$\nabla \cdot (\mathbf{e} \times \mathbf{H}) + \frac{\partial}{\partial t}\left(\frac{1}{2}\epsilon \mathbf{e} \cdot \mathbf{e} + \frac{1}{2}\mu_0 \mathbf{H} \cdot \mathbf{H}\right) = 0$$

is equivalent to

$$H_z \mu_0 \frac{\partial}{\partial t}\left(H_z + M_s - \frac{1}{2}\frac{\mathbf{m} \cdot \mathbf{m}}{M_s}\right) + \mathbf{h} \cdot \nabla \times \mathbf{e} - \mathbf{e} \cdot \nabla \times (\widehat{\mathbf{z}} H_z + \mathbf{h})$$
$$+ \frac{\partial}{\partial t}\left(\frac{1}{2}\epsilon \mathbf{e} \cdot \mathbf{e} + \frac{1}{2}\mu_0 H_z^2 + \frac{1}{2}\mu_0 \mathbf{h} \cdot \mathbf{h}\right) = 0$$

But H_z and M_s are assumed to be independent of position and time, therefore the result of this second-order expansion is identical to Eq. (23.15). We have learned that the small-signal Zeeman energy is hiding in $\nabla \cdot (\mathbf{e} \times \mathbf{H})$.

Substitute Zeeman energy density

Manipulation of the linearized torque equation produces a very interesting and useful result. We start by cross-multiplying Eq. (23.6a) with **m**; the result

$$\mathbf{m} \times \frac{\partial \mathbf{m}}{\partial t} = \gamma_g \mu_0 M_s (\mathbf{m} \cdot \mathbf{h} - \frac{H_z}{M_s} \mathbf{m} \cdot \mathbf{m}) \widehat{\mathbf{z}}$$

is then dot multiplied by $\widehat{\mathbf{z}}$ and divided by $2\gamma_g M_s$. After rearrangement,

$$\frac{1}{2} \mu_0 \frac{H_z}{M_s} \mathbf{m} \cdot \mathbf{m} = \frac{1}{2} \mu_0 \mathbf{m} \cdot \mathbf{h} + \frac{1}{2\gamma_g M_s} \left(\frac{\partial \mathbf{m}}{\partial t} \times \mathbf{m} \right) \cdot \widehat{\mathbf{z}}$$

This substitute for w_Z when inserted into Eq. (23.15) produces

$$\nabla \cdot (\mathbf{e} \times \mathbf{h}) + \frac{\partial}{\partial t} \left[\frac{1}{2} \varepsilon \mathbf{e} \cdot \mathbf{e} + \frac{1}{2} \mu_0 \mathbf{h} \cdot (\mathbf{h} + \mathbf{m}) + \frac{1}{2\gamma_g M_s} \left(\frac{\partial \mathbf{m}}{\partial t} \times \mathbf{m} \right) \cdot \widehat{\mathbf{z}} \right] = 0 \quad (23.16a)$$

In addition to the Minkowski form of the small-signal energy density, there is a term that can be interpreted as excess kinetic energy associated with the precession of the angular momentum. We define the latter as

$$w_m = \frac{1}{2\gamma_g M_s} \left(\frac{\partial \mathbf{m}}{\partial t} \times \mathbf{m} \right) \cdot \widehat{\mathbf{z}} = \frac{\mu_0}{2\omega_M} \left(\mathbf{m} \times \frac{\partial \mathbf{m}}{\partial t} \right) \cdot \widehat{\mathbf{z}} \quad (23.16b)$$

and use an italicized subscript to differentiate it from the usual magnetic energy density.

For the uniform plane waves and using this substitute form of the energy, we obtain

$$\mathbf{e} \times \mathbf{h} = \frac{\omega \mu_0}{\mathbf{k} \cdot \mathbf{k} - k_0^2} \left[\frac{k_0^2 \mathbf{m} \cdot \mathbf{m} - (\mathbf{m} \cdot \mathbf{k})^2}{\mathbf{k} \cdot \mathbf{k} - k_0^2} \mathbf{k} + (\mathbf{m} \cdot \mathbf{k})\, \mathbf{m} \right]$$

$$w_{\text{total}} = \mu_0 \left[k_0^2 \frac{k^2 \mathbf{m} \cdot \mathbf{m} - (\mathbf{m} \cdot \mathbf{k})^2}{(\mathbf{k} \cdot \mathbf{k} - k_0^2)^2} + \frac{1}{2} \frac{\omega}{\omega_M} m_a m_b \right]$$

Dual Alternate representation

We turn now to the Minkowski forms of the Alternate representation (also summarized in Appendix D, Section D.2). Because, by assumption, $\mathbf{J}_u$, $\rho_u = 0$, the dual Alternate Minkowski representation is a convenient choice.[6] Based upon Eqs. (D.8), the small-signal Poynting theorem becomes

$$\nabla \cdot \mathbf{s}^{om} + \frac{\partial}{\partial t} \left(w^{om} + \frac{1}{2} \mu_0 \frac{H_z}{M_s} \mathbf{m} \cdot \mathbf{m} \right) = 0$$

$$\nabla \cdot \mathbf{s}^{om} + \frac{\partial}{\partial t} \left[w^{om} + \frac{1}{2} \mu_0 \mathbf{m} \cdot \mathbf{h} + \frac{1}{2\gamma_g M_s} \left(\frac{\partial \mathbf{m}}{\partial t} \times \mathbf{m} \right) \cdot \widehat{\mathbf{z}} \right] = 0$$

[6]When $\mathbf{J}_u$, $\rho_u \neq 0$, the Alternate-Minkowski representation based upon the small-signal version of Eqs. (D.7) can be used.

$$\mathbf{s}^{\text{om}} = \frac{1}{2\varepsilon}\left(a_{\text{m}x}\nabla\frac{\partial a_{\text{m}x}}{\partial t} - \frac{\partial a_{\text{m}x}}{\partial t}\nabla a_{\text{m}x} + a_{\text{m}y}\nabla\frac{\partial a_{\text{m}y}}{\partial t} - \frac{\partial a_{\text{m}y}}{\partial t}\nabla a_{\text{m}y}\right)$$
$$- \frac{1}{2}\mu_{\text{o}}\left(\phi_{\text{m}}\nabla\frac{\partial \phi_{\text{m}}}{\partial t} - \frac{\partial \phi_{\text{m}}}{\partial t}\nabla\phi_{\text{m}}\right) + \frac{1}{2}\mu_{\text{o}}\left(\frac{\partial \phi_{\text{m}}}{\partial t}\mathbf{m} - \phi_{\text{m}}\frac{\partial \mathbf{m}}{\partial t}\right)$$

$$\text{w}^{\text{om}} = -\frac{1}{2}\mu_{\text{o}}\left[a_{\text{m}x}\frac{\partial^2 a_{\text{m}x}}{\partial t^2} - \left(\frac{\partial a_{\text{m}x}}{\partial t}\right)^2 + a_{\text{m}y}\frac{\partial^2 a_{\text{m}y}}{\partial t^2} - \left(\frac{\partial a_{\text{m}y}}{\partial t}\right)^2\right]$$
$$+ \frac{1}{2}\mu_{\text{o}}^2\varepsilon\left[\phi_{\text{m}}\frac{\partial^2 \phi_{\text{m}}}{\partial t^2} - \left(\frac{\partial \phi_{\text{m}}}{\partial t}\right)^2\right] + \frac{1}{2}\mu_{\text{o}}\left(\mathbf{a}_{\text{m}}\cdot\frac{\partial \mathbf{m}}{\partial t} - \frac{\partial \mathbf{a}_{\text{m}}}{\partial t}\cdot\mathbf{m}\right)$$

For the uniform plane waves, with $k_y = 0$,

$$\mathbf{s}^{\text{om}} = \frac{\omega\mu_{\text{o}}}{2(\mathbf{k}\cdot\mathbf{k} - k_{\text{o}}^2)^2}\begin{bmatrix} \widehat{\mathbf{x}}k_x(\mathbf{k}\cdot\mathbf{k} - k_{\text{o}}^2)m_a^2 \\ +\,\mathbf{k}\,[k_{\text{o}}^2(m_a^2 + m_b^2) - k_x^2 m_a^2] \end{bmatrix}$$

The average power flux and energy density is the same in both the Maxwell–Poynting and dual Alternate representations. When the total energy density is included, we obtain

$$\frac{<\mathbf{s}^{\text{em}}>}{<\text{w}_{\text{total}}>} = \frac{\partial\omega}{\partial\mathbf{k}} = \mathbf{v}_{\text{group}} \tag{23.17a}$$

$$\mathbf{v}_{\text{group}} = \frac{\omega<[k_{\text{o}}^2\mathbf{m}\cdot\mathbf{m} - (\mathbf{m}\cdot\mathbf{k})^2]\mathbf{k} + (\mathbf{k}\cdot\mathbf{k} - k_{\text{o}}^2)(\mathbf{m}\cdot\mathbf{k})\,\mathbf{m}>}{k_{\text{o}}^2<k^2\mathbf{m}\cdot\mathbf{m} - (\mathbf{m}\cdot\mathbf{k})^2> + \frac{1}{2}(\mathbf{k}\cdot\mathbf{k} - k_{\text{o}}^2)^2\dfrac{\omega}{\omega_{\text{M}}}m_a m_b} \tag{23.17b}$$

In general, the group velocity is not parallel to **k**, a phenomenon termed beam steering. Evaluation requires the use of Eqs. (23.13b) and (23.13c).

23.8 SMALL-SIGNAL STRESS AND MOMENTUM

Maxwell–Poynting representation

The small-signal Minkowski version of the force density is, for a ferrite with uniform parameters,

$$\nabla\cdot\{\varepsilon\mathbf{e}\,\mathbf{e} + \mu_{\text{o}}\mathbf{h}\,(\mathbf{h}+\mathbf{m}) - \frac{1}{2}[\varepsilon\mathbf{e}\cdot\mathbf{e} + \mu_{\text{o}}\mathbf{h}\cdot(\mathbf{h}+\mathbf{m})]\bar{\mathbf{I}}\}$$
$$-\frac{\partial}{\partial t}[\mu_{\text{o}}\varepsilon\mathbf{e}\times(\mathbf{h}+\mathbf{m})] = \frac{1}{2}\mu_{\text{o}}(m_k\nabla h_k - h_k\nabla m_k) \tag{23.18}$$

where summation over $k = 1, 2, 3$ is assumed.

When this small-signal force density is compared to the small-signal power conversion density, $\frac{1}{2}\mu_{\text{o}}(\mathbf{m}\cdot\frac{\partial\mathbf{h}}{\partial t} - \mu_{\text{o}}\mathbf{h}\cdot\frac{\partial\mathbf{m}}{\partial t})$, it is seen that (as expected) they are components of a small-signal four-vector

$$\left[\frac{1}{2}\mu_{\text{o}}(m_k\Box h_k - h_k\Box m_k)\right]$$

This suggests that the energy density w_m which is associated with the fourth component, should have a small-signal momentum counterpart $\mathbf{g}_m$ that is connected with the first three. Because w_m can be expressed as $\frac{-ic}{-2\gamma_g M_s}(\frac{\partial \mathbf{m}}{\partial x_4} \times \mathbf{m}) \cdot \widehat{\mathbf{z}}$, we examine the consequences of setting $-icg_{m\,i} = \frac{-ic}{-2\gamma_g M_s}(\frac{\partial \mathbf{m}}{\partial x_i} \times \mathbf{m}) \cdot \widehat{\mathbf{z}}$, which leads to

$$g_{m\,i} = \frac{\mu_0}{2\omega_M}(\frac{\partial \mathbf{m}}{\partial x_i} \times \mathbf{m}) \cdot \widehat{\mathbf{z}} \tag{23.19}$$

The test comes when the time derivative is evaluated and Eq. (23.6a) is substituted. The result vindicates our choice because

$$\begin{aligned}\frac{\partial g_{m\,i}}{\partial t} &= \frac{\mu_0}{2\omega_M}\left(\frac{\partial^2 \mathbf{m}}{\partial x_i \partial t} \times \mathbf{m} + \frac{\partial \mathbf{m}}{\partial x_i} \times \frac{\partial \mathbf{m}}{\partial t}\right) \cdot \widehat{\mathbf{z}} \\ &= \frac{1}{2}\mu_0\left(\mathbf{m} \cdot \frac{\partial \mathbf{h}}{\partial x_i} - \frac{\partial \mathbf{m}}{\partial x_i} \cdot \mathbf{h}\right)\end{aligned}$$

Now small-signal stress momentum also satisfies a conservation law,

$$\begin{aligned}&\nabla \cdot \{\varepsilon \mathbf{e}\mathbf{e} + \mu_0 \mathbf{h}(\mathbf{h} + \mathbf{m}) - \frac{1}{2}[\varepsilon \mathbf{e} \cdot \mathbf{e} + \mu_0 \mathbf{h} \cdot (\mathbf{h} + \mathbf{m})]\bar{\mathbf{I}}\} \\ &\quad - \frac{\partial}{\partial t}[\mu_0 \varepsilon \mathbf{e} \times (\mathbf{h} + \mathbf{m}) + \mathbf{g}_m] = 0\end{aligned} \tag{23.20}$$

If either the Alternate representation of Eq. (23.18), based upon the small-signal equivalent of Eqs. (D.7), or the dual form based upon Eqs. (D.8) is used, equivalent conservation laws require the same $\mathbf{g}_m$. Notice that, in the Maxwell–Poynting–Minkowski representation, the small-signal momentum density can be written as $\mathbf{d} \times \mathbf{b} + \mathbf{g}_m$.

For plane waves of frequency ω, the time-averaged quantities (in either representation) satisfy

$$<\bar{\mathbf{t}}_{\text{total}}> = -<\mathbf{g}_{\text{total}}>\mathbf{v}_{\text{group}}$$

as well as Eq. (23.17a),

$$<\mathbf{S}_{\text{total}}> = <\mathrm{W}_{\text{total}}>\mathbf{v}_{\text{group}}$$

The average value of $\mathbf{g}_{\text{total}}$ is consistent with that obtained from the complex momentum theorems derived in Part I, Chapter 6, Section 6.4 – provided that the scalar $\mu(\omega)$ is generalized to the Polder tensor. But note that when $\mathbf{k}$ is parallel to the dc field, the waves are circularly polarized and the tensor reduces to a scalar. In that case, Eq. (6.23) can be used with $\mu = \mu^{\pm}(\omega) = \mu_0\big(1 + \frac{\omega_M}{\omega_z \mp |\omega|}\big)$.

The large-signal model of a ferrite that is locally saturated (but may be spatially nonuniform) has been discussed by the author [17]. That model includes a generalization of Eq. (23.19) taken to be

$$G_{m\,i} = \frac{1}{-2\gamma_g \mathbf{M} \cdot \widehat{\mathbf{u}}}\left(\frac{\partial \mathbf{M}}{\partial x_i} \times \mathbf{M}\right) \cdot \widehat{\mathbf{u}} \qquad (i = 1, 2, 3) \tag{23.21}$$

where $\widehat{\mathbf{u}}$ is the unit vector defining the equilibrium direction of the mesoscopic[7] magnetization vector that is locally saturated, but may be spatially nonuniform.

[7]Mesoscopic dimensions are on the order of magnetic domain-wall thicknesses. They are intermediate between the microscopic (atomic) and macroscopic distance scales.

23.9 QUASIPARTICLE INTERPRETATION (MAGNONS)

We have already learned that if plane waves propagate in a ferrite medium with weak electromagnetic characteristics, they are magnetostatic waves. In this and other magnetostatic cases, the quasistatic electric field makes negligible contributions to the energy and momentum densities and they are very well approximated by

$$\mathrm{w}_{\text{total}} \simeq \frac{1}{2}\mathbf{h}\cdot\mathbf{b} + \mathrm{w}_m$$

$$\mathbf{g}_{\text{total}} \simeq \mathbf{g}_m$$

But in the magnetostatic approximation, $\mathbf{h} = -\nabla\phi_{\text{m}}$ because $\nabla \times \mathbf{h} = 0$; therefore, $\mathbf{h}\cdot\mathbf{b} = \nabla\cdot(-\phi_{\text{m}}\mathbf{b})$ and, using the divergence theorem,

$$\int_{V_{\text{o}}} \mathrm{w}_{\text{total}} dV \simeq \frac{1}{2}\oint_{S_{\text{o}}} -\phi_{\text{m}}\mathbf{b}\cdot\hat{\mathbf{n}}\, da + \int_{V_{\text{o}}} \mathrm{w}_m\, dV$$

Consistent with the magnetostatic approximation is the neglect of any far-field radiation; therefore, if the closed surface S_{o} is allowed to recede to infinity, the surface integral vanishes ($\phi_{\text{m}} \to 0$ at least as fast as r^{-2}) and since it provides the proper total, w_m is an acceptable alternative. This *effective* energy density vanishes outside the ferrite; nevertheless, any energy due to fields outside the ferrite is properly accounted for!

Returning to the plane wave, we have

$$\mathrm{w}_m = \frac{1}{2}\mu_{\text{o}}\frac{\omega}{\omega_{\text{M}}}m_a m_b \tag{23.22a}$$

$$\mathbf{g}_m = \frac{1}{2}\mu_{\text{o}}\frac{\mathbf{k}}{\omega_{\text{M}}}m_a m_b \tag{23.22b}$$

If this energy density is written as $n_m\hbar\omega$, the momentum density becomes $\mathbf{g}_m = n_m\hbar\mathbf{k}$ and the particle interpretation is alive and well. These quasiparticles are called "magnons" (which explains the choice of the italicized subscript); the density,

$$n_m = \frac{1}{2}\mu_{\text{o}}\frac{m_a m_b}{\hbar\omega_{\text{M}}} \tag{23.23}$$

is seen to depend upon the area of the ellipse ($\pi m_a m_b$) swept out by the precession of $\mathbf{m}$.

It should be noted that these results apply to the Kittel uniform-precession mode of the small ellipsoid (for which $\mathbf{k} = 0$) as well as to both magnetostatic regions: $k \ll k_{\text{o}}$ and $k \gg k_{\text{o}}$. For the large k region, the small-signal momentum increases as the quasistatic approximation becomes even better; this is very different from the electromagnetic component, $\mathbf{d}\times\mathbf{b}$, which is completely negligible in the quasistatic limit.

Finally, there is a simple interpretation of n_m that can be made when the magnetic moments are due to spin ($g' = 2$) and the precession is both uniform and circular; then with $m_a = m_b = M_s\theta$ and V the volume of the ferrite, the total number of magnons $n_m V$ is

$$N_m = \frac{M_s^2}{\hbar\omega_{\text{M}}}V\frac{1}{2}\theta^2$$

where $\hbar\omega_{\text{M}} = g'\mu_B\mu_{\text{o}}M_s$. When all of the magnetic moments in the volume, N_{o}, are completely aligned, they create the saturation magnetization and a total magnetic moment,

$M_s V = N_0 \mu_B$ (expressed in Bohr magnetons, $\mu_B = \frac{e}{2m_e}\hbar$). Consequently, $N_m = N_0 \frac{1}{4}\theta^2$. Now consider that a small number of the spins are reversed (flipped) so as to point in the opposite direction. Because each reversed spin lowers the total magnetic moment by $2\mu_B$, it follows that N_{flipped} of them (spread out uniformly throughout the volume) will cause M_z to decrease slightly from M_s to $M_s \cos\theta$ according to

$$(N_0 - 2N_{\text{flipped}})\mu_B = M_s V \cos\theta$$

or

$$N_{\text{flipped}} = N_0 \frac{1}{2}(1 - \cos\theta)$$

Because the decrease is assumed to be small, we have $N_{\text{flipped}} \ll N_0$ and $\cos\theta \simeq 1 - \frac{1}{2}\theta^2$.

$$N_m = N_{\text{flipped}} = N_0 \frac{1}{4}\theta^2$$

On a per unit volume basis, the magnon density is simply equal to that of the flipped spins.

CHAPTER 24

EQUIVALENT CIRCUITS

24.1 RECEIVING CIRCUIT OF A DIPOLE

Two dipoles labeled #1 and #2 are made of conducting wires of small radii (r_w) that have lengths d_1 and d_2; they are surrounded by free-space and separated by a distance r_{12}. Ideal current sources $I_1(0)$ and $I_2(0)$ of frequency ω are applied to the small gaps that exist at the midpoint of each antenna. The lengths are such that for a given frequency ω, each dipole is in the far field of the other. This condition implies that

$$kr_{12} \gg (kd_i)^2, \qquad i = 1, 2$$

Because of linearity, the complex fields are the superposition of those that are due to the separate excitations.

$$\underline{\mathbf{E}} = \underline{\mathbf{E}}_1 + \underline{\mathbf{E}}_2$$
$$\underline{\mathbf{H}} = \underline{\mathbf{H}}_1 + \underline{\mathbf{H}}_2$$

but because we cannot superimpose material, both antennas must be present when either set of fields is calculated.

The Power and Beauty of Electromagnetic Fields, First Edition. F. R. Morgenthaler.

In the Maxwell–Poynting representation, we have

$$\underline{\mathbf{S}} = \tfrac{1}{2}(\underline{\mathbf{E}}_1 \times \underline{\mathbf{H}}_1^* + \underline{\mathbf{E}}_2 \times \underline{\mathbf{H}}_2^* + \underline{\mathbf{E}}_1 \times \underline{\mathbf{H}}_2^* + \underline{\mathbf{E}}_2 \times \underline{\mathbf{H}}_1^*)$$

Consider now that a closed surface S_2 located in the far fields of *both* dipoles surrounds dipole #2 (but not dipole #1). Then because $\underline{\mathbf{E}}_1$ and $\underline{\mathbf{H}}_1 = \sqrt{\frac{\varepsilon_0}{\mu_0}}\mathbf{k} \times \underline{\mathbf{E}}_1$ are (approximately) uniform plane waves with linear polarizations of constant orientation, the net power fluxes $\underline{\mathbf{E}}_1 \times \underline{\mathbf{H}}_1^*$ and $\underline{\mathbf{E}}_2 \times \underline{\mathbf{H}}_1^*$ separately vanish when integrated over S_2 and

$$\oint_{S_2} \underline{\mathbf{S}} \cdot \widehat{\mathbf{n}}\, da = \oint_{S_2} \tfrac{1}{2}(\underline{\mathbf{E}}_2 \times \underline{\mathbf{H}}_2^* + \underline{\mathbf{E}}_1 \times \underline{\mathbf{H}}_2^*) \cdot \widehat{\mathbf{n}}\, da \tag{24.1}$$

The first term (conveniently evaluated in the far field) is the average radiated power that is independent of the choice of the surface as long as S_2 completely surrounds dipole #2. If the surface is moved closer to the dipole, the imaginary component of the power increases as the reactive electric and magnetic fields and the associated stored Poynting energies are encountered. However, the reactive power arises from the sum of volume and surface contributions due, respectively, to the reactive energy and radiated/dissipated power. As long as we calculate the correct total, we can alter S_2 as we please.

The second term of Eq. (24.1) is unaltered by interchanging the dot and cross; the result

$$\oint_{S_2} \tfrac{1}{2}\, \underline{\mathbf{E}}_1 \times \underline{\mathbf{H}}_2^* \cdot \widehat{\mathbf{n}}\, da = \tfrac{1}{2} \oint_{S_2} \underline{\mathbf{E}}_1 \cdot \underline{\mathbf{H}}_2^* \times \widehat{\mathbf{n}}\, da$$

can be evaluated by choosing the z axis parallel to d_2 and shrinking S_2 to a concentric cylinder of length d_2 and cylindrical radius, r_c. If the radius were reduced so that $r_c \leq r_w$, then we have $\underline{E}_{z1} \simeq 0$ (because of the induced surface charges that are necessary to cancel the incident tangential field) except in the gap where it is greatly increased. If, instead, $r_c \gg r_w$ yet $kr_c \ll 1$, $\underline{E}_{z1}$ *will approximate the value that would exist if there were no wire*. Then because

$$\underline{\mathbf{H}}_2^* = \widehat{\phi}\frac{\underline{I}_2(z)}{2\pi r_c}$$

$$\tfrac{1}{2}\underline{\mathbf{E}}_1 \cdot \oint_{S_2} \underline{\mathbf{H}}_2^* \times \widehat{\mathbf{n}}\, da = -\tfrac{1}{2}\underline{E}_{z1}\underline{I}_2^*(0) \int_{-d_2/2}^{+d_2/2} \frac{\underline{I}_2^*(z)}{\underline{I}_2^*(0)}\, dz$$

$$\underline{P}_{21} = -\tfrac{1}{2}\underline{\mathbf{E}}_1 \cdot \mathbf{d}_{\text{eff}2}\underline{I}_2^*(0)$$

The complex-power theorem applied to the gap terminals of dipole #2 yields

$$\tfrac{1}{2}\underline{V}_{g2}\underline{I}_2^*(0) = \underline{P}_{21} + \langle P_{\text{rad}2}\rangle + \langle P_{\text{dissip}2}\rangle + j2\omega(\langle E_{\text{m}2}\rangle - \langle E_{\text{e}2}\rangle)$$

where the radiated power,

$$\langle P_{\text{rad}2}\rangle = \oint_{S_2} \frac{1}{2}\underline{\mathbf{E}}_2 \times \underline{\mathbf{H}}_2^* \cdot \widehat{\mathbf{n}}\, da$$

the dissipation in the wires (copper losses),

$$\langle P_{\text{dissip}2}\rangle = \tfrac{1}{2}\int_{-d_2/2}^{+d_2/2} R'(\omega)\underline{I}_2(z)\underline{I}_2^*(z)\, dz$$

$$= \tfrac{1}{2}R_{\sigma 2}\underline{I}_2(0)\underline{I}_2^*(0)$$

and

$$X_2 = \frac{4\omega}{|\underline{I}_2(0)|^2}(<E_{\mathrm{m}}> \; - \; <E_{\mathrm{e}}>)$$

lead to

$$\underline{V}_{g2} = -\underline{\mathbf{E}}_1 \cdot \mathbf{d}_{\mathrm{eff2}} + (R_{\mathrm{rad2}} + R_{\sigma 2} + jX_2)\underline{I}_2^*(0)$$

A similar analysis can be made for dipole #1; the result is

$$\underline{V}_{g1} = -\underline{\mathbf{E}}_2 \cdot \mathbf{d}_{\mathrm{eff1}} + (R_{\mathrm{rad1}} + R_{\sigma 1} + jX_1)\underline{I}_1^*(0)$$

These equations are of the form

$$\underline{V}_{g1} = \underline{Z}_{11}\underline{I}_1(0) + \underline{Z}_{12}\underline{I}_2(0)$$

$$\underline{V}_{g2} = \underline{Z}_{21}\underline{I}_1(0) + \underline{Z}_{22}\underline{I}_2(0)$$

where

$$\underline{Z}_{11}(j\omega) = R_{\mathrm{rad1}} + R_{\sigma 1} + jX_1$$

$$\underline{Z}_{12}(j\omega) = \frac{-\underline{\mathbf{E}}_2 \cdot \mathbf{d}_{\mathrm{eff1}}}{\underline{I}_2(0)}$$

$$\underline{Z}_{21}(j\omega) = \frac{-\underline{\mathbf{E}}_1 \cdot \mathbf{d}_{\mathrm{eff2}}}{\underline{I}_1(0)}$$

$$\underline{Z}_{22}(j\omega) = R_{\mathrm{rad2}} + R_{\sigma 2} + jX_2$$

$\underline{Z}_{11}(j\omega)$ and $\underline{Z}_{22}(j\omega)$ are the input impedances of the two dipole antennas (assuming the gap capacitances are negligible). For Hertzian dipoles with triangular current distributions, $\mathrm{d}_{\mathrm{eff}i} = \mathrm{d}_i/2$, $(k\mathbf{d}_i \ll 1)$, and

$$R_{\mathrm{rad}i} = \frac{\eta_{\mathrm{o}}}{6\pi}(kd_{\mathrm{eff}i})^2$$

$$R_{\sigma i} = R' d_i$$

$$X_i = \frac{-1}{\omega C_i}$$

where the value of R' depends upon r_w/δ (δ is the skin depth) and the reactance is capacitive. If *EQS* fields are used to estimate the value, the result is $C_i \simeq \dfrac{\pi \varepsilon_{\mathrm{o}} d_{\mathrm{eff}i}}{\ln(d_{\mathrm{eff}i}/r_w)}$. (Refer to Chapter 25, Problem 25.1-28.) Because of the reciprocity theorem (Part I, Chapter 6, Section 6.10), we have $\underline{Z}_{12}(j\omega) = \underline{Z}_{21}(j\omega)$.

The equivalent receiving circuit of either dipole is shown in Figure 24.1 when the source is a Norton equivalent and in Figure 24.2 when the source is a Thevenin equivalent. The gap capacitance reduces the actual current flowing on the antenna.

Finally, it should be noted that the same final result can be obtained using the Alternate representation of complex power and energy calculated in terms of

$$\underline{\mathbf{A}} = \underline{\mathbf{A}}_1 + \underline{\mathbf{A}}_2$$

$$\underline{\Phi} = \underline{\Phi}_1 + \underline{\Phi}_2$$

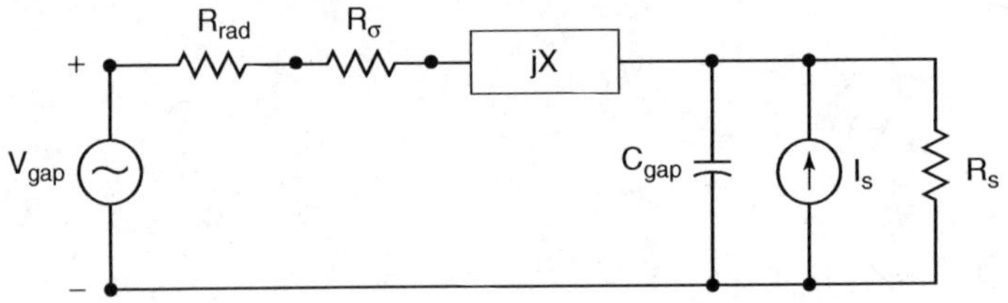

Figure 24.1 Dipole equivalent circuit with Norton source.

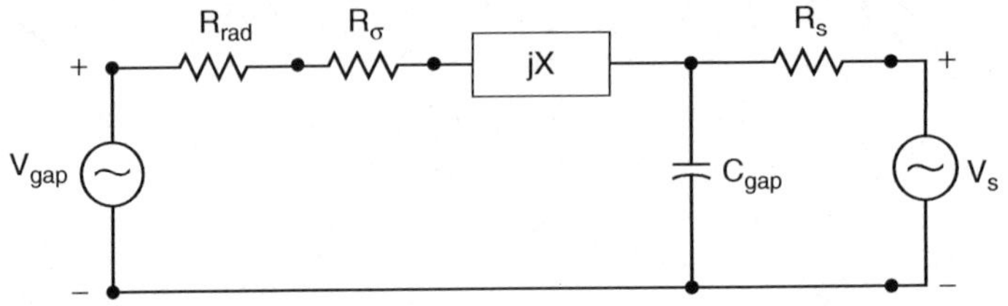

Figure 24.2 Dipole equivalent circuit with Thevenin source.

The banishment of Alternate reactive power and energy in the free-space region applies to the cross-terms and so it is the real part of $\underline{P}_{21}$ that travels through space between dipole #1 and dipole #2; the imaginary part is generated by the localized Alternate energy. For a short dipole, this is primarily capacitive with density, $-j\omega\frac{1}{4}(\underline{\Phi}_1\underline{\rho}_2^* + \underline{\Phi}_1^*\underline{\rho}_2)$. The circuit power $\frac{1}{2}\underline{V}_1\underline{I}_2^*$ arises from the localized flux, $\frac{1}{2}(\underline{\Phi}_1 + \underline{\Phi}_2)\underline{\mathbf{J}}_2^*$.

24.2 *TEM* TRANSMISSION LINES

Basic equations

A distributed-circuit model of a lossless two-conductor transmission line is shown in Figure 24.3. It consists of an endless repetition of differential sections of series inductance and shunt capacitance.[1] For a balanced structure, half of the inductance, $L'dz$, is on each side of the line; in order to emphasize the symmetry, the capacitance, $C'dz$, is divided into two equal components connected in series. The node between them coincides with the midline and, in agreement with the scalar potential in the physical structure, can be set at ground potential. This choice allows one to connect a short-circuit ground between all similar nodes and work with one-half of the circuit, now an unbalanced circuit; to avoid the factors of two, it is convenient to double both the half-voltage and half-inductance and half the doubled capacitance. The generalization that includes dissipation adds series resistance and shunt conductance. The result is shown in Figure 24.4.

[1]If each section is made from lumped elements and occupies a physical length ℓ, the differential equation is replaced by a difference equation that links each section with its nearest neighbors. Provided that $k\ell \ll 1$, an infinite chain of identical lumped circuits can simulate a continuous transmission line. This result is of practical benefit.

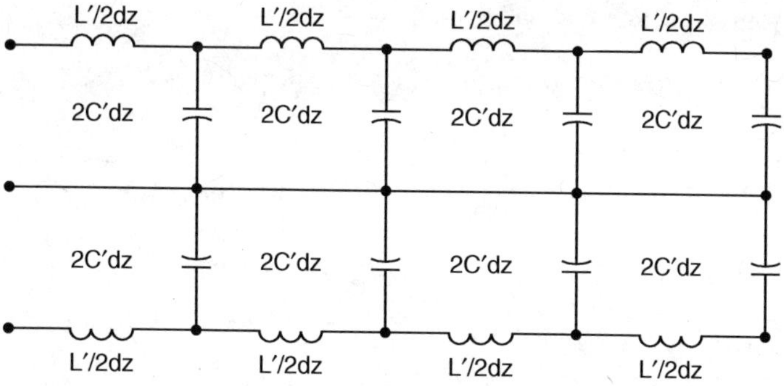

Figure 24.3 Balanced incremental circuit.

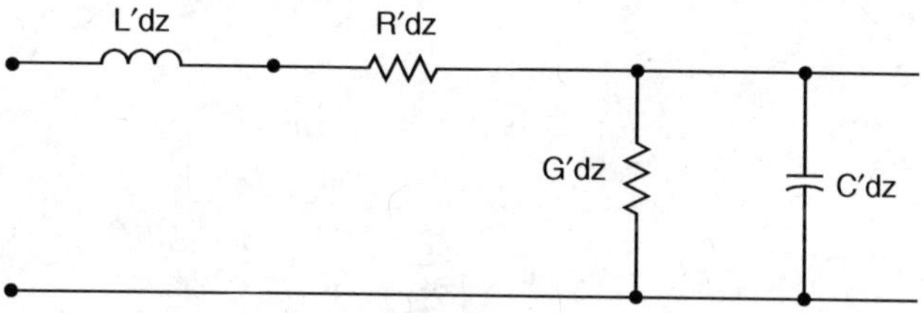

Figure 24.4 Unbalanced incremental circuit.

When Kirchhoff's voltage and current laws are applied, the result is a pair of coupled first-order differential equations,

$$-\frac{d\underline{V}(z)}{dz} = (R' + j\omega L')\underline{I}(z) = \underline{Z}'_s\underline{I}(z) \tag{24.2a}$$

$$-\frac{d\underline{I}(z)}{dz} = (G' + j\omega C')\underline{V}(z) = \underline{Y}'_p\underline{V}(z) \tag{24.2b}$$

that govern the complex voltage and current. In the past, these were known as the Telegrapher's Equations. On a per unit length basis, the inductance L' and capacitance C' are already familiar; the series resistance R' is mainly due to copper losses in the conductors[2] and the shunt conductance G' to conductivity of dielectric materials between the conductors. The series elements comprise the impedance, $\underline{Z}'_s(j\omega) = R' + j\omega L'$; the shunt or parallel elements comprise the admittance, $\underline{Y}'_p(j\omega) = G' + j\omega C'$. When one or the other of the variables is eliminated, a pair of second-order equations results:

$$\frac{d^2\underline{V}(z)}{dz^2} - \underline{Z}'_s\underline{Y}'_p\underline{V}(z) = 0 \tag{24.3a}$$

$$\frac{d^2\underline{I}(z)}{dz^2} - \underline{Z}'_s\underline{Y}'_p\underline{I}(z) = 0 \tag{24.3b}$$

[2]Copper losses in highly conducting ground planes that are thick compared to the skin depth is considered in Chapters 16 (Section 16.4) and 18 (Section 18.4). In such cases, the parameters R' and the (often negligible) component of L' due to the magnetic field inside the conductor are both frequency-dependent.

The general solution of these equations is

$$\underline{V}(z) = \underline{V}^{+}(0)\exp(-\sqrt{\underline{Z}'_s\underline{Y}'_p}z) + \underline{V}^{-}(0)\exp(\sqrt{\underline{Z}'_s\underline{Y}'_p}z) \tag{24.4a}$$

$$\underline{I}(z) = \sqrt{\frac{\underline{Y}'_p}{\underline{Z}'_s}}[\underline{V}^{+}(0)\exp(-\sqrt{\underline{Z}'_s\underline{Y}'_p}z) - \underline{V}^{-}(0)\exp(\sqrt{\underline{Z}'_s\underline{Y}'_p}z)] \tag{24.4b}$$

Because of dissipation in the circuit, the propagation constant,

$$\sqrt{\underline{Z}'_s\underline{Y}'_p} = \sqrt{(R' + j\omega L')(G' + j\omega C')} = a + j\beta \tag{24.5}$$

is complex and leads to attenuation of the waves. Some authors prefer to define the complex propagation constant as $\underline{k} = k' - jk''$, where $\beta = k'$ and $\alpha = k''$.

When a single wave is propagating, $\underline{V}^{+}\underline{V}^{-} = 0$, and the ratio of the complex voltage to current is

$$\frac{\underline{V}(z)}{\underline{I}(z)} = \pm\underline{Z}_o(j\omega) \tag{24.6a}$$

$$\underline{Z}_o(j\omega) = \frac{1}{\underline{Y}_o(j\omega)} = \sqrt{\frac{\underline{Z}'_s}{\underline{Y}'_p}} = \sqrt{\frac{R' + j\omega L'}{G' + j\omega C'}} \tag{24.6b}$$

where the plus sign is for the case $\underline{V}^{-} = 0$ and $\underline{Z}_o(j\omega)$ is defined as the characteristic impedance of the transmission line; its reciprocal $\underline{Y}_o(j\omega)$ as the characteristic admittance.

Power and energy

Complex power associated with a transmission line has always been expressed in terms of voltage and current and so is really Alternate power. Multiplying Eq. (24.2a) by $\frac{1}{2}\underline{I}^*(z)$, the conjugate of Eq. (24.2b) by $\frac{1}{2}\underline{V}(z)$, and adding results in

$$\begin{aligned}-\frac{d}{dz}[\frac{1}{2}\underline{V}(z)\underline{I}^*(z)] &= \frac{1}{2}[R'\underline{I}(z)\underline{I}^*(z) + G'\underline{V}(z)\underline{V}^*(z)]\\ &\quad + j2\omega[\frac{1}{4}L'\underline{I}(z)\underline{I}^*(z) - \frac{1}{4}C')\underline{V}(z)\underline{V}^*(z)]\end{aligned} \tag{24.7}$$

The average dissipated power and the difference between the average magnetic and electric stored energies (all per unit length) are present exactly as one should expect.

Lossless, low-loss, and distortionless lines

If there is no dissipation,

$$Z_o = \sqrt{\frac{L'}{C'}}$$

$$\beta = \omega\sqrt{L'C'}$$

$$\alpha = 0$$

Both C' and L' can be calculated from static-field evaluations of the electric charge per unit length, $Q' = C'V$, and the magnetic flux per unit length, $\Psi' = L'I$. Since the voltage and current travel at the same phase velocity as that governing the electric and magnetic field, it follows that

$$L'C' = \mu\varepsilon$$

and only one calculation is required. In addition, conformal transformations can often be used to map the static Laplacian fields of one two-conductor geometry into those of another. The resulting formulas of Z_o for several important line geometries are included in Appendix E, Section E.4.

If both $R' \ll \omega L'$ and $G' \ll \omega C'$, the low-loss approximation is

$$Z_o \simeq \sqrt{\frac{L'}{C'}}$$

$$\beta \simeq \omega\sqrt{L'C'} \simeq \omega\sqrt{\mu\varepsilon}$$

$$\alpha \simeq \frac{1}{2}\left(\frac{R'}{Z_o} + \frac{G'}{Y_o}\right)$$

If $R'C' = G'L'$, the values

$$Z_o = \sqrt{\frac{L'}{C'}}$$

$$\beta = \omega\sqrt{L'C'} = \omega\sqrt{\mu\varepsilon}$$

$$\alpha = \frac{R'}{Z_o} = \frac{G'}{Y_o}$$

are exact. Therefore, independent of the amount of loss, all frequencies propagate at the same phase velocity and with the same attenuation factor. Consequently, an arbitrary traveling wave will propagate with attenuation, but undistorted shape. Such a transmission line is termed a "distortionless line."[3] Note that although any low-loss line is approximately distortionless, transmission over very long distances can accumulate phase errors that produce unacceptable waveform distortion. This is the reason that long-distance communication systems often require equalization filters to restore waveforms. In many cases, Z_oG' is small compared to R'/Z_o, and therefore α can be reduced by an increase in the characteristic impedance. This is the rationale behind the invention patented by Michael Pupin (but proposed earlier by Oliver Heaviside) of loading coils that were used to raise the average L' of long-distance telephone lines. A drawback associated with this method is the lower propagation velocity, hence longer delay, that results.

Reflection coefficient and line impedance

Except for the amplitudes which depend upon the magnitude(s) of the source(s), the voltage and current patterns are determined by the ratio $\underline{V}^-(z)/\underline{V}^+(z)$. In the usual case a source at the left end of the line generates a wave that is reflected by a load at the right end and so

$$\underline{\Gamma}(z) = \frac{\underline{V}^-(z)}{\underline{V}^+(z)} = -\frac{\underline{I}^-(z)}{\underline{I}^+(z)} \tag{24.8}$$

[3]Because as noted, R' is actually a function of frequency, the line can only approximate a distortionless line over a limited bandwidth.

is defined as the voltage reflection coefficient; it is the negative of the current reflection coefficient. The ratio of Eqs. (24.4a) and (24.4b) can be expressed as

$$\frac{\underline{V}(z)}{\underline{I}(z)} = \sqrt{\frac{\underline{Z}'_s}{\underline{Y}'_p}} \frac{[1 + \underline{\Gamma}(0)\exp(2\sqrt{\underline{Z}'_s \underline{Y}'_p} z)]}{[1 - \underline{\Gamma}(0)\exp(2\sqrt{\underline{Z}'_s \underline{Y}'_p} z)]} \tag{24.9a}$$

or, since $\underline{\Gamma}(z) = \underline{\Gamma}(0)\exp(2\sqrt{\underline{Z}'_s \underline{Y}'_p} z)$, as

$$\frac{\underline{V}(z)}{\underline{I}(z)} = \underline{Z}_t(z) = \underline{Z}_o \frac{1 + \underline{\Gamma}(z)}{1 - \underline{\Gamma}(z)} \tag{24.9b}$$

For a lossless line, it is convenient to normalize the line impedance to Z_o; then (switching notation from β to k)

$$\underline{Z}_n(z) = \frac{1 + \underline{\Gamma}(z)}{1 - \underline{\Gamma}(z)} = \frac{1 + \underline{\Gamma}(0)\exp(j2kz)}{1 - \underline{\Gamma}(0)\exp(j2kz)} \tag{24.10}$$

It is seen that $\underline{\Gamma}(z) = \underline{\Gamma}(0)\exp(j2kz) = u + jv$ is formed by a simple rotation of $\underline{\Gamma}(0)$. Since for passive loads $|\underline{\Gamma}(z)| \leq 1$, we need only be concerned with the unit circle and its interior. The complex $\underline{\Gamma}$ plane is a very useful graphical representation of a transmission line.

With reference to Figure 24.5, pretend that the vectors $\underline{\Gamma}(z)$ and $-\underline{\Gamma}(z)$ are the locations of the pedals of a stationary bicycle, except that the radius, $\underline{\Gamma}(z)$, is adjustable from zero to unity. Now attach a rubber band between a fixed point on the frame ($u = -1$, $v = 0$) and each of the pedals. As one pedals in the clockwise direction, z decreases in the direction of the source; in the counterclockwise direction, z increases in the direction of the load. One full revolution moves the point z exactly $\lambda/2$ from its starting point. The vector $1 + \underline{\Gamma}(z)$ is a measure of the complex voltage; the vector $1 - \underline{\Gamma}(z)$ is a measure of the complex current. As the pedals rotate, the rubber bands stretch or contract as first

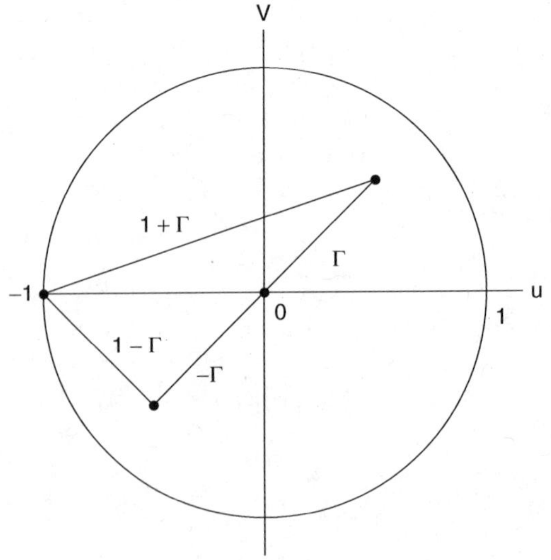

Figure 24.5 Γ plane.

$\underline{V}(z)$ then $Z_o\underline{I}(z)$ becomes larger. The angle between the rubber bands is the angle of $\underline{Z}_n(z)$; when both pedals are at the same height ($v = 0$) the principal angle is either zero or π and the impedance is real. The magnitude of the voltage cycles (every $\lambda/4$) between a maximum proportional to $1 + |\underline{\Gamma}|$ and a minimum proportional to $1 - |\underline{\Gamma}|$; these points correspond respectively to impedance maxima and minima. The current is minimum at a voltage maximum and vice versa. The ratio of a voltage maximum to minimum is termed the voltage standing-wave ratio (*VSWR*), given by

$$VSWR = \frac{1 + |\underline{\Gamma}|}{1 - |\underline{\Gamma}|}$$

The inverse formulas are

$$\underline{\Gamma} = \frac{\underline{Z}_n(z) - 1}{\underline{Z}_n(z) + 1}$$

and

$$|\underline{\Gamma}| = \frac{VSWR - 1}{VSWR + 1}$$

Smith Chart

The normalized impedance can be written as $r + jx = \dfrac{1 + u + jv}{1 - u - jv}$. Although this is a straightforward transformation easily handled by a computer, that was not true in the 1930s and 1940s when radar antennas and many microwave devices were being invented. A researcher at the (then) Bell Telephone Laboratories, P. H. Smith came to the rescue by inventing a graphical aide that became known as the Smith Chart [37]. Like many elegant solutions, once it was proposed it seemed obvious. The loci of constant r and x in the $\underline{\Gamma}$ plane are governed by a conformal transformation that maps straight lines into circles:

$$\left(u - \frac{r}{r+1}\right)^2 + v^2 = \frac{1}{(1+r)^2}$$

$$(u-1)^2 + \left(v - \frac{1}{x}\right)^2 = \frac{1}{x^2}$$

This means that a graph can be prepared once and for all that provides a graphical solution. The graph can include the entire unit circle or an expansion of the central portion so as to provide higher resolution. In Figure 24.6 are plotted the loci of constant r (for $r = 0, .5, 1, 2$) and constant x (for $x = 0, \pm.5, \pm 1, \pm 2$); both sets are in order of decreasing radius. Notice that $xv \geq 0$, and all curves are orthogonal and pass through the point ($u = 1$, $v = 0$).

Now imagine that the chart has been printed on a sheet of transparent plastic. It can be placed on top of the $\underline{\Gamma}$ plane whenever the transformation is needed. A Smith Chart is simply a labeled $\underline{\Gamma}$ plane; an actual chart is shown in Figure 24.7; they were available in the form of inexpensive paper pads; measurements taken by hand could be plotted directly on them. The scales displayed near the circumference provide conversion of angles to distance in wavelengths and remind the user which direction of rotation moves z toward the load. In this era of automated test equipment and data processing, there is no longer a need for the pads, but it is still useful to have the computer generate a Smith Chart with all of the data plotted on it. Fortunately, some test equipment provides that option; however, a few lines of code can add it to any data-analysis software.

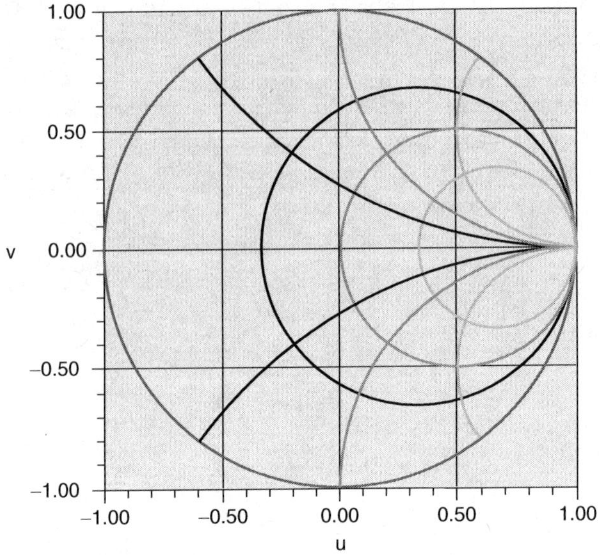

Figure 24.6 Simplified Smith Chart.

For some problems it is more convenient to work with the normalized admittance $g + jb = \frac{1 - u - jv}{1 + u + jv}$. There is no need for a new graph, simply rotate a Smith Chart by 180°, but keep the underlying $\underline{\Gamma}$ plane in its usual position. One must keep in mind that it is only the plastic overlay (real or imagined) that is rotated.

The line impedance at any position along a transmission line can also be found by integrating the first-order nonlinear differential equation that governs $\underline{Z}_\ell(z) = \underline{V}/\underline{I}$. From

$$\frac{d}{dz}\underline{Z}_\ell(z) = \frac{1}{\underline{I}}\frac{d\underline{V}}{dz} - \underline{Z}_\ell^2(z)\frac{1}{\underline{V}}\frac{d\underline{I}}{dz}$$

and substitution of Eqs. (24.2a) and (24.2b), one finds that

$$\frac{d}{dz}\underline{Z}_\ell(z) = -\underline{Z}_s' + \underline{Y}_p'\underline{Z}_\ell^2(z)$$

or in normalized form

$$\frac{d}{d(\sqrt{\underline{Z}_s'\underline{Y}_p'}z)}\left[\frac{\underline{Z}_\ell(z)}{\sqrt{\underline{Z}_s'/\underline{Y}_p'}}\right] = \left[\frac{\underline{Z}_\ell(z)}{\sqrt{\underline{Z}_s'/\underline{Y}_p'}}\right]^2 - 1 \qquad (24.11)$$

The general solution of Eq. (24.11) is Eq. (24.9a) or, in the event that there is no dissipation, Eq. (24.10).

Impedance matching

Complex loads

If we wish to match a complex load impedance, $Z_L = R_L + jX_L$, to a transmission line of real characteristic impedance, Z_o, there are several possibilities. One method is to connect either a short-circuited or open-circuited length of transmission line across the

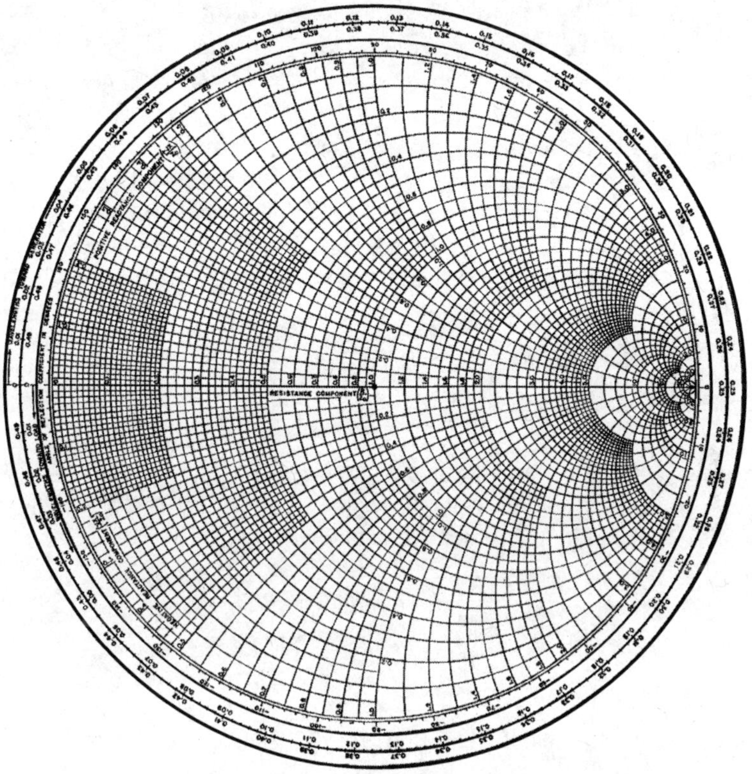

Figure 24.7 Smith Chart.

input line at the closest point to the load where the real part of the line admittance is $1/Z_o$; the minimum length is chosen that makes the stub susceptance the negative of the line susceptance at that point. The resulting parallel combination of admittances is real and of value $1/Z_o$. The "single-stub tuner" works with any load for which $R_L > 0$, but both the length of the stub and its line position must be made adjustable in order to accommodate different loads. Another technique employs two adjustable stubs separated by a fixed distance (normally $\frac{3}{8}\lambda$) with one of these located either at the position of the load or $\frac{1}{4}\lambda$ in front of it. It is left as an exercise for the student to show that the "double-stub tuner" can match all possible loads for at least one of these positions.

Resistive loads

If $X_L = 0$, the previous methods are still applicable, but there are several additional possibilities that do not require the use of stubs. One method makes use of the fact that the normalized line-impedance cycles between a maximum that is real and a minimum that is its reciprocal as the point z is moved in either direction by $\lambda/4$. In between, the impedance is complex. It follows that if a one-quarter wavelength section of a transmission line with characteristic impedance Z_o' is placed between the load, the input resistance will be $Z_o'^2/R_L$. If that value is made equal to Z_o, a perfect match results. The condition,

$$Z_o' = \sqrt{Z_o R_L}$$

completes the design of the "quarter-wave transformer." Notice that if the length of the transformer is doubled, the input impedance is R_L *independent* of the value of Z_0'. One complete rotation around the Smith Chart occurs for every $\lambda/2$ of translation along the line.

A second method of achieving a match is to use a section of tapered line to provide a gradual transition between the impedances. Very low reflections can be achieved over wide bandwidths if the taper length is at least several wavelengths. Although it operates over a wider band width than the quarter-wave transformer, it occupies more space—especially at lower frequencies. The analysis of lossless tapered lines or "horns" is the subject of the next section.

24.3 LOSSLESS TAPERED LINES

When the inductance and/or capacitance per unit length are functions of z, a transmission line is said to be tapered. Provided that the variations are reasonably gradual, the *TEM* approximation validates the first-order equations

$$\frac{d\underline{V}(z)}{dz} = -j\omega L'(z)\underline{I}(z) \tag{24.12a}$$

$$\frac{d\underline{I}(z)}{dz} = -j\omega C'(z)\underline{V}(z) \tag{24.12b}$$

which lead to modified Helmholtz equations for the complex voltage and current:

$$\frac{d^2\underline{V}(z)}{dz^2} - \frac{1}{L'(z)}\frac{dL'(z)}{dz}\frac{d\underline{V}(z)}{dz} + \omega^2 L'(z)C'(z)\underline{V}(z) = 0 \tag{24.13a}$$

$$\frac{d^2\underline{I}(z)}{dz^2} - \frac{1}{C'(z)}\frac{dC'(z)}{dz}\frac{d\underline{I}(z)}{dz} + \omega^2 L'(z)C'(z)\underline{I}(z) = 0 \tag{24.13b}$$

The inductance and/or capacitance per unit length can be tapered when any of μ, ε, or the geometry of the transmission line is made a function of z. If only the geometry varies, $L'(z) = \mu g(z)$, $C'(z) = \varepsilon/g(z)$, and $\omega^2 L'(z)C'(z) = \omega^2\mu\varepsilon = k^2$; if the material parameters vary, the wavenumber becomes $k(z)$.

As the first example, consider a tapered line that is a parallel-plate structure. If the ratio of the plate spacing, d, and width, w, is small, then, neglecting fringing fields, $g(z) = d(z)/w(z)$. For a coaxial line with inner and outer radii, $a(z)$ and $b(z)$, $g(z) = \frac{1}{2\pi}\ln[b(z)/a(z)]$. In these and other cases, either or both of the dimensions may be tapered.

Liouville–Green (WKB) Approximation

In practice a taper provides a smooth transition from one size line to another and, if space permits, is usually made gradual enough to prevent appreciable reflections from being generated. In such cases, the Liouville–Green approximation [38], rediscovered nearly a century later by Wentzel, Kramers, and Brillouin (WKB), is suitable and, when applied, leads to

$$\underline{V}(z) \simeq \sqrt{\frac{L'(z)}{k(z)}}\left[\underline{A}^+ \exp\left(-j\int k(z)\,dz\right) + \underline{A}^- \exp\left(j\int k(z)\,dz\right)\right] \tag{24.14}$$

provided that

$$k^2(z) \gg \left| \frac{1}{2L'} \frac{d^2L'}{dz^2} - \frac{3}{4} \left(\frac{1}{L'} \frac{dL'}{dz} \right)^2 \right|$$

and

$$\underline{I}(z) \simeq \sqrt{\frac{C'(z)}{k(z)}} \left[\underline{A}^+ \exp\left(-j \int k(z)\, dz \right) - \underline{A}^- \exp\left(j \int k(z)\, dz \right) \right] \qquad (24.15)$$

provided that

$$k^2(z) \gg \left| \frac{1}{2C'} \frac{d^2C'}{dz^2} - \frac{3}{4} \left(\frac{1}{C'} \frac{dC'}{dz} \right)^2 \right|$$

If $\underline{A}^- \underline{A}^+ = 0$, we obtain

$$\frac{\underline{V}(z)}{\underline{I}(z)} \simeq \pm Z_o(z)$$

$$Z_o(z) \simeq \sqrt{\frac{L'(z)}{C'(z)}}$$

where the plus sign goes with $\underline{A}^- = 0$.

A constant impedance taper is useful for providing a transition between lines that are of different size, but share the same characteristic impedance. This is the principle behind the design of horn antennas and acoustic megaphones.

Exponential taper

If $k_o = \omega\sqrt{\mu\varepsilon}$ is constant, the family of exponential tapers,

$$L'(z) = L'(0) \exp(-\alpha z)$$

$$C'(z) = C'(0) \exp(+\alpha z)$$

(where α can be any positive or negative constant), forces both Eqs. (24.13a) and (24.13b) to be linear differential equations with constant coefficients:

$$\frac{d^2\underline{V}(z)}{dz^2} + \alpha \frac{d\underline{V}(z)}{dz} + k_o^2 \underline{V}(z) = 0 \qquad (24.16a)$$

$$\frac{d^2\underline{I}(z)}{dz^2} - \alpha \frac{d\underline{I}(z)}{dz} + k_o^2 \underline{I}(z) = 0 \qquad (24.16b)$$

Exact solutions are available and can provide helpful insights. This form of taper is of technical importance in its own right; in addition an exponential can approximate a linear taper.

If a solution of the form $\exp(pz)$ is assumed for $\underline{V}(z)$, it follows that

$$p = -\frac{1}{2}\alpha \pm j\sqrt{k_o^2 - \frac{1}{4}\alpha^2}$$

Therefore in terms of $k = \sqrt{k_o^2 - \frac{1}{4}\alpha^2}$, we have

$$\underline{V}(z) = \exp\left(-\frac{1}{2}\alpha z\right)[\underline{V}^+ \exp(-jkz) + \underline{V}^- \exp(jkz)] \tag{24.17a}$$

$$\underline{I}(z) = \frac{1}{\omega L'(0)} \exp\left(\frac{1}{2}\alpha z\right)\left[\left(k - j\frac{\alpha}{2}\right)\underline{V}^+ \exp(-jkz) - (k + j\frac{\alpha}{2})\underline{V}^- \exp(jkz)\right] \tag{24.17b}$$

where use has been made of

$$\underline{V}(z) = \frac{j}{\omega C'(0)} \exp(-\alpha z)\frac{d\underline{I}(z)}{dz}$$

$$\underline{I}(z) = \frac{j}{\omega L'(0)} \exp(\alpha z)\frac{d\underline{V}(z)}{dz}$$

If $\underline{V}^- = 0$, we obtain

$$\frac{\underline{V}(z)}{\underline{I}(z)} = \sqrt{\frac{L'(0)}{C'(0)}} \exp(-\alpha z)\frac{k_o}{\sqrt{k_o^2 - \frac{1}{4}\alpha^2} - j\frac{1}{2}\alpha}$$

If $\underline{V}^+ = 0$, we have

$$\frac{\underline{V}(z)}{\underline{I}(z)} = -\sqrt{\frac{L'(0)}{C'(0)}} \exp(-\alpha z)\frac{k_o}{\sqrt{k_o^2 - \frac{1}{4}\alpha^2} + j\frac{1}{2}\alpha}$$

These solutions are valid within the taper, which is assumed to extend between $z = 0$ and $z = \ell$. If one attempts to greatly reduce ℓ by increasing α, k may become imaginary; this leads to imaginary characteristic impedances that make it impossible to transmit average power through the taper without reflections. The quasi-*TEM* assumption may also be invalidated.

24.4 TRANSIENTS ON TRANSMISSION LINES

The voltage and current on a nondispersive *TEM* transmission line satisfy the one-dimensional wave equation and in the time domain are of the form

$$v(z,t) = f_+(t - z/c) + f_-(t + z/c) \tag{24.18a}$$

$$Z_o\, i(z,t) = f_+(t - z/c) - f_-(t + z/c) \tag{24.18b}$$

The sum of Eqs. (24.18a) and (24.18b) produces

$$2f_+(t - z/c) = v(z,t) + Z_o\, i(z,t)$$

which at any point along the line can be interpreted as the Thevenin equivalent circuit shown in Figure 24.8.

It is important to realize that this circuit replaces everything that exists to the left of a chosen line position, z.

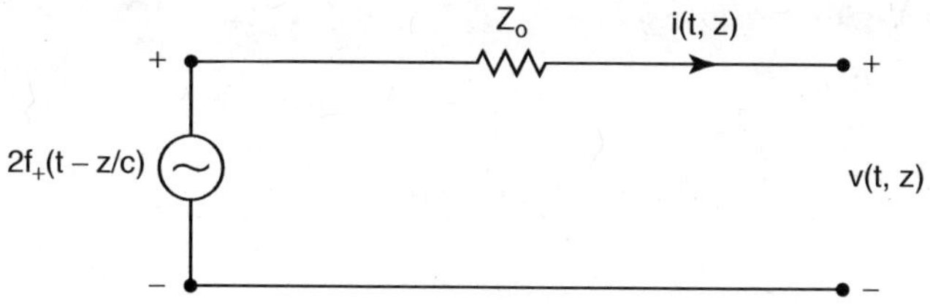

Figure 24.8 Thevenin equivalent circuit with positive-wave voltage source.

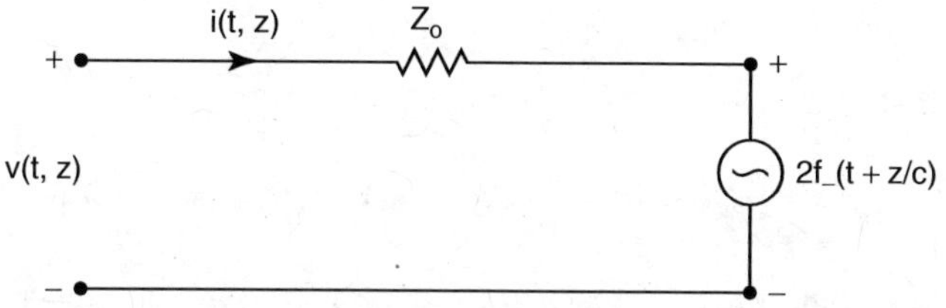

Figure 24.9 Thevenin equivalent circuit with negative-wave voltage source.

The difference of Eqs. (24.18a) and (24.18b) produces

$$2f_-(t + z/c) = v(z, t) - Z_o\, i(z, t)$$

which at any point along the line can be interpreted as the Thevenin equivalent circuit shown in Figure 24.9.

It is important to realize that this circuit replaces everything that exists to the right of a chosen line position, z.

It is helpful to work through an example that illustrates how these equivalent circuits can be applied. Assume that the transmission line is of length ℓ, terminated at $z = 0$ by a load resistance R_L and excited at $z = -\ell$ by a Thevenin equivalent circuit (a voltage source, $v_s(t)$ in series with resistance, R_s). Assume further that $v_s(t) = V_s u_{-1}(t)$ and the line voltage and current are both zero for all time $t < 0$. It is helpful to define $T = \ell/c$, the time required for a signal to travel the length of the line. Applying the negative-wave circuit at the input, $z = -\ell$ produces the circuit equation,

$$V_s u_{-1}(t) - (Z_o + R_s) i(-\ell, t) = 2f_-(t - T)$$

But, no reflected wave, f_-, exists anywhere along the line for $t < T$ (and not at the input for $t < 2T$); therefore, $i(-\ell, t < 2T) = f_+(t + T)/Z_o$ and

$$f_+(t + T) = \frac{Z_o}{(Z_o + R_s)} V_s u_{-1}(t)$$

At the input, the line acts like a simple resistor with value Z_o. Elsewhere along the line (as long as $f_- = 0$), we have

$$f_+(t - z/c) = \frac{Z_o}{(Z_o + R_s)} V_s u_{-1}(t - z/c - T)$$

At $z = 0$, we apply the positive-wave circuit to produce the circuit equations,

$$2f_+(t) - Z_o\, i(0,t) = v(0,t) = R_L i(0,t)$$

$$v(0,t) = f_+(t) + f_-(t)$$

$$Z_o\, i(0,t) = f_+(t) - f_-(t)$$

As expected, the solution introduces the voltage reflection coefficient (at the load), $\Gamma_L = \dfrac{R_L - Z_o}{R_L + Z_o}$, and

$$f_-(t) = \Gamma_L f_+(t)$$

Elsewhere along the line (as long as f_+ has not changed), we have

$$f_-(t + z/c) = \frac{R_L - Z_o}{R_L + Z_o} \frac{Z_o}{(Z_o + R_s)} V_s u_{-1}(t + z/c - T)$$

If $R_s = Z_o$, the input circuit is matched (because of the source reflection coefficient, $\Gamma_s = \dfrac{R_s - Z_o}{R_s + Z_o} = 0$). In that event, neither f_+ nor f_- will ever change and we have the complete solution of the transient. The steady state is reached everywhere along the line for $t > 2T$, and the final line voltage and current are those of a simple voltage divider:

$$v(z) = i(z)R_s = \frac{R_L}{R_L + R_s} V_s \tag{24.19}$$

If $R_s \neq Z_o$, there will be a reflection at the input starting at time $t = 2T$ that will change f_+; this in turn will travel down the line and generate a change in f_- at $t = 3T$. This process will continue for all even values of T at the input and all odd values at the load. When either or both values of R_s and R_L are finite, the dissipation will cause the successive reflections to become smaller and smaller. In the limit of very large t/T, the series of reflections can be summed and the same steady-state values of Eq. (24.19) will be reached. On the other hand, if there is no loss ($R_L = R_s = 0$), we have $\Gamma_s = \Gamma_L = -1$, and the series of reflections produces a current that grows to an unbounded value. It is expected that the short-circuit current is infinite, but it takes an infinite time to reach the limit.

When the load impedance is not a resistance, but rather a single reactive element or a complex lumped circuit, the value of $f_-(t)$ is related to $f_+(t)$ by a linear differential equation that must be solved before one can proceed. In simple cases such as those of a single capacitor or inductor, the step response can be found by inspection. For these and more complicated loads, the method of Laplace transforms can be used. In terms of the complex frequency s, one introduces

$$f_\pm(t \mp z/c) \leftrightarrow \exp(\pm sz/c) F_\pm(s)$$

and the complex reflection coefficient, $\Gamma_L(s) = \dfrac{Z_L(s) - Z_o}{Z_L(s) + Z_o}$. Then at $z = 0$,

$$F_-(s) = \frac{Z_L(s) - Z_o}{Z_L(s) + Z_o} F_+(s)$$

The inverse transform can then be used to evaluate $f_-(t + z/c)$.

24.5 PLANE WAVES (OBLIQUE INCIDENCE)

Plane waves striking the interface between dielectric/magnetic materials at oblique angles of incidence are discussed in Chapter 17. Here we make simple equivalent circuits that encourage the solution of many problems by the use of transmission-line analogies. The wave and interface geometry and the Cartesian coordinates used to analyze them is the same as before.

TE waves

Because the electric field is perpendicular to the plane of incidence, the introduction of thin perfectly conducting sheets that are normal to $\underline{\mathbf{E}} = \widehat{\mathbf{y}}\,\underline{E}_y(x,z)$ will have no effect on the wave propagation. Assume that there are two such planes spaced by a distance d; both are, of course, parallel to the x–z plane of incidence. The field originates and terminates on electric surface charges of opposite polarity, $\underline{\sigma}_{\mathrm{u}}^s = \pm\varepsilon\underline{E}_y(x,z)$, but identical charges of opposite polarity, on the other side of each sheet, render each of them electrically neutral.

Now consider the electric surface-current density,

$$\underline{\mathbf{K}}^s = \pm[\widehat{\mathbf{z}}\underline{H}_x(x,z) - \widehat{\mathbf{x}}\underline{H}_z(x,z)]$$

that flows on the inner surfaces. As with the charge, oppositely directed currents on the outer surfaces produce zero net current flow on each sheet. It is just this cancellation of both charge and current that permits the sheets to be inserted without disturbing the fields.

Since plane waves are infinite in extent, the power they carry is also infinite. In order to make our equivalent circuit manageable, we focus on the finite cross-sectional area defined by the plate-spacing, d, and an arbitrary width w in the x direction. The voltage,

$$\underline{V}(x,z) = -\underline{E}_y(x,z)d$$

between the plates and the current,

$$\underline{I}(x,z) = \underline{K}_z^s w = \underline{H}_x(x,z)w$$

flowing in the z direction on the top sheet both appear suitable because it is the fields transverse to the $z = 0$ interface that must be continuous. The voltage takes care of $\underline{E}_y$, the current $\underline{H}_x$. The current flowing parallel to x is of little concern because we did not cut the conducting sheets into strips of width w! The x dependence of the voltage and current is a little disconcerting until we recall that the x and z dependencies are separable. The factor $\exp(-jk_x x)$ is set by the incident wave and is common to *all* of the fields; it is this commonality that is the basis of Snell's Law. Consequently,

$$\underline{V}(x,z) = \underline{V}(z)\exp(-jk_x x)$$

$$\underline{I}(x,z) = \underline{I}(z)\exp(-jk_x x)$$

and their ratio, which is the line impedance, depends only upon z. If we remember to invoke Snell's Law, the x dependence can be ignored; the result is a complex voltage and current that depends only upon z.

Transmission-line parameters

The line impedance is

$$\frac{\underline{V}(z)}{\underline{I}(z)} = -\frac{\underline{E}_y(z)d}{\underline{H}_x(z)w}$$

Since d and w are arbitrary, we might as well set them equal (simply for convenience). Then, for a single wave traveling with positive k_z, we have $\underline{H}_x = -\sqrt{\frac{\varepsilon}{\mu}}\underline{E}_y \cos\theta$ and

$$Z_0^{te} = \frac{1}{Y_0^{te}} = \frac{\sqrt{\frac{\mu}{\varepsilon}}}{\cos\theta}$$

$$k_z = \omega\sqrt{\mu\varepsilon}\cos\theta$$

TM waves

Because it is the magnetic field that is perpendicular to the plane of incidence, the introduction of the thin perfectly conducting sheets that are normal to $\underline{\mathbf{H}} = \widehat{\mathbf{y}}\ \underline{E}_y(x,z)$ would have a catastrophic effect on the wave propagation. But, since this polarization is the dual of the *TE* wave, we can instead imagine using thin sheets of perfect magnetically conducting material. Magnetic surface charges and magnetic surface currents replace their electric counterparts.

In order to make our equivalent circuit manageable, we again focus on the finite cross-sectional area defined by the plate-spacing, d and an arbitrary width w in the x direction. The "magnetic voltage,"

$$\underline{V}^m(x,z) = -\underline{H}_y(x,z)d$$

between the plates and the "magnetic current,"

$$\underline{I}^m(x,z) = \underline{K}_z^{sm} w = -\underline{E}_x(x,z)w$$

flowing in the z direction on the top sheet together carry the dual power. As in the *TE* wave, we have

$$\underline{V}^m(x,z) = \underline{V}^m(z)\exp(-jk_x x)$$

$$\underline{I}^m(x,z) = \underline{I}^m(z)\exp(-jk_x x)$$

and their ratio, which is the "magnetic line impedance," depends only upon z. If we again remember to invoke Snell's Law, the x dependence can be ignored. Finally, we can shift back to electric voltage and electric current (so that our equivalent circuit can be connected to ordinary sources and loads) by realizing that $\underline{V}^m(z) \to \underline{I}^e(z)$ and $\underline{I}^m(z) \to \underline{V}^e(z)$.

Transmission-line parameters

The line admittance ("magnetic impedance") is therefore

$$\frac{\underline{I}^e(z)}{\underline{V}^e(z)} = \frac{\underline{H}_y(z)d}{\underline{E}_x(z)w}$$

We again set $d = w$. Then, for a single wave traveling with positive k_z, we have $\underline{E}_x = \sqrt{\frac{\mu}{\varepsilon}}\underline{H}_y \cos\theta$ and

$$Y_o^{tm} = \frac{1}{Z_o^{tm}} = \frac{\sqrt{\frac{\varepsilon}{\mu}}}{\cos\theta}$$

$$k_z = \omega\sqrt{\mu\varepsilon}\cos\theta$$

Summary of parameters

Only two parameters characterize a lossless uniform *TEM* transmission line: L' and C'; or, equivalently, we have $Z_o = \sqrt{L'/C'}$ and the phase velocity, $v_p = \omega/k_z = 1/\sqrt{L'C'}$. Continuity of k_x (Snell's Law) provides the angle, θ. A single differential section of the equivalent (unbalanced) line for both *TE* or *TM* waves is depicted in Figure 24.10. The required values of L' and C' are

TE:	$Z_o^{te} = \dfrac{\sqrt{\frac{\mu}{\varepsilon}}}{\cos\theta}$	$L' = \mu$	$C' = \varepsilon\cos^2\theta$
TM:	$Z_o^{tm} = \sqrt{\frac{\mu}{\varepsilon}}\cos\theta$	$L' = \mu\cos^2\theta$	$C' = \varepsilon$

$$k_z = \omega\sqrt{\mu\varepsilon}\cos\theta$$

$$k_x = \omega\sqrt{\mu\varepsilon}\sin\theta$$

A series cascade of these circuits can be used to model multiple parallel layers of dielectric/magnetic material of different thicknesses and parameters when the incident-wave angle and polarization is specified (Linearity and superposition allow any polarization to be resolved into its *TE* and *TM* components.) Then the $\underline{\Gamma}$ plane and Smith Chart provide the tools required for the analysis of the reflection from and transmission through the structure.

24.6 WAVEGUIDES

Waveguides made from hollow conductors of uniform cross-section can support electromagnetic modes of propagation provided the frequency exceeds threshold values. The

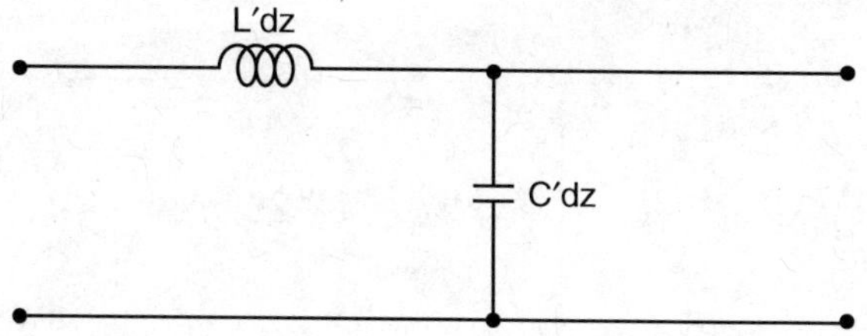

Figure 24.10 Equivalent circuit (*TE* and *TM* waves).

dispersion relation governing a specific mode, that may be either *TE* or *TM* (but *not TEM*) in character, is

$$k_z^2 = k^2 - k_c^2$$

or equivalently,

$$\lambda_g = \frac{\lambda_o}{\sqrt{1 - (\frac{\lambda_o}{\lambda_c})^2}}$$

where λ_g is the guide wavelength, λ_o the free-space wavelength, and λ_c the cutoff wavelength that depends upon the waveguide dimensions and the specific mode. They are defined by

$$k_z = \frac{2\pi}{\lambda_g}$$

$$k = \frac{2\pi}{\lambda_o}$$

$$k_c = \frac{2\pi}{\lambda_c} = \frac{\omega_c}{c}$$

The distributed circuit that represents a nondispersive *TEM* transmission line can be modified so that it will simulate the waveguide dispersion of a particular mode. Two versions (duals of one another) that are appropriate are shown in Figures 24.11 and 24.12. Either one adds capacitance in series with the inductance, or inductance in parallel with the shunt capacitance. Equations (24.3a) and (24.3b) are still valid.

For circuit 1 they become

$$\underline{Z}'_s(j\omega) = j\omega L'_1 + \frac{S'}{j\omega}$$

$$\underline{Y}'_p(j\omega) = j\omega C'_1$$

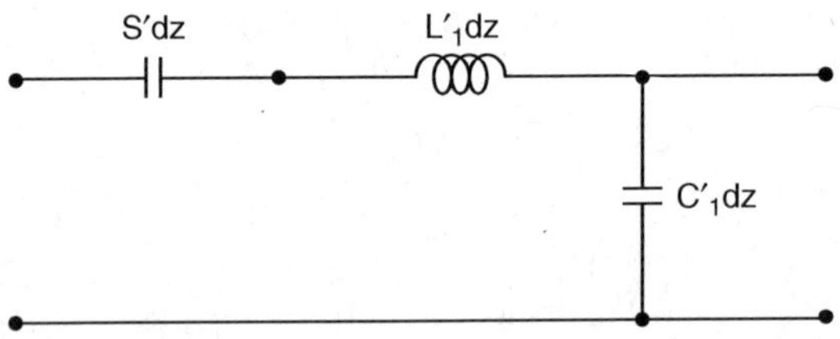

Figure 24.11 Equivalent circuit 1.

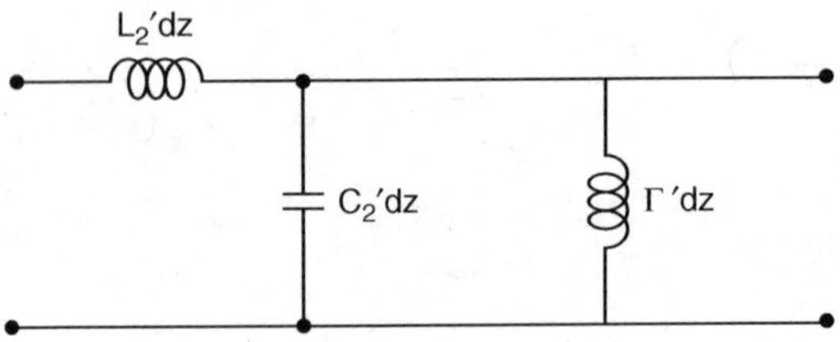

Figure 24.12 Equivalent circuit 2.

where S' is the reciprocal capacitance per unit length

$$k_z^2 = \omega^2 L_1' C_1' - S' C_1'$$

$$Z_o = \sqrt{\frac{\underline{Z}_s'(j\omega)}{\underline{Y}_p'(j\omega)}} = \sqrt{\frac{L_1'}{C_1'}} \frac{\sqrt{\omega^2 - S'/L_1'}}{\omega}$$

For circuit 2 they become

$$\underline{Z}_s'(j\omega) = j\omega L_2'$$

$$\underline{Y}_p'(j\omega) = j\omega C_2' + \frac{\Gamma'}{j\omega}$$

where Γ' is the reciprocal inductance per unit length.

$$k_z^2 = \omega^2 L_2' C_2' - \Gamma' L_2'$$

$$Z_o = \sqrt{\frac{\underline{Z}_s'(j\omega)}{\underline{Y}_p'(j\omega)}} = \sqrt{\frac{L_2'}{C_2'}} \frac{\omega}{\sqrt{\omega^2 - \Gamma'/C_2'}}$$

Below cutoff, both circuits have imaginary values of both k_z and Z_o; circuit 1 is capacitive, circuit 2 is inductive. In this same frequency range, the principal type of energy stored in *TM* and *TE* waveguide modes is, respectively, electric and magnetic. It follows that circuit 1 can model *TM* modes and circuit 2 can model *TE* modes.

TM modes

Circuit 1 can model a *specific TM* mode, provided that we have

$$L_1' C_1' = \mu\varepsilon$$

$$k_c = \sqrt{S' C_1'} = \omega_c \sqrt{\mu\varepsilon}$$

so that the phase velocities and cutoff frequencies match. In general, the value of S' is different for each mode. The frequency dependence of the characteristic admittance must also match

$$Y_o = \sqrt{\frac{C_1'}{L_1'}} \frac{\dfrac{\omega}{\omega_c}}{\sqrt{\left(\dfrac{\omega}{\omega_c}\right)^2 - 1}}$$

$$\sqrt{\frac{C_1'}{L_1'}} = g^{tm} \sqrt{\frac{\varepsilon}{\mu}}$$

but the geometric scale factor, g^{tm}, is arbitrary provided the circuit and waveguide powers are equal. For a single traveling-wave mode, we require that the circuit 1 complex voltage and current satisfy

$$\frac{1}{2} Y_o \underline{V}_o \underline{V}_o^* = P_{TM}$$

$$\underline{I}_o = Y_o \underline{V}_o$$

with

$$L_1' = \mu, \qquad C_1' = g^{tm}\varepsilon, \qquad S' = k_c^2/C_1'$$

Rectangular guides

For *TM* rectangular waveguide modes, the pertinent results of Chapter 19 can be summarized as

$$\underline{E}_z = E_o \sin(k_x x)\sin(k_y y)\exp(-jk_z z)$$

$$k_c^2 = \left(\frac{n\pi}{a}\right)^2 + \left(\frac{m\pi}{b}\right)^2$$

where $n \geq 1$ and $m \geq 1$ are integers. The average power, expressed in terms of the longitudinal electric field, is

$$\mathrm{P}_{TM}^{\text{rectangular}} = \frac{\varepsilon_o \omega k_z E_o^2}{8k_c^2}ab$$

Therefore,

$$\underline{V}_o = \frac{E_o}{2}\sqrt{\left[\left(\frac{\omega}{\omega_c}\right)^2 - 1\right]\frac{ab}{g^{tm}}}$$

Although the factor g^{tm} is completely arbitrary, some have argued (sometimes vociferously) for a specific choice. *If used consistently*, any value is appropriate; perhaps the simplest is $g^{tm} = 1$.

Circular guides

For *TM* circular waveguide modes, the pertinent results of Chapter 20 can be summarized as

$$\underline{E}_z = E_o J_n(k_c r)\left\{{}^{\sin}_{\cos} n\phi\right\}\exp(-jk_z z)$$

$$k_c R = u_{nm}$$

where n and m are integers and the roots, $J_n(u_{nm}) = 0$, are tabulated in Appendix F. The average power, expressed in terms of the longitudinal electric field, is

$$\mathrm{P}_{TM}^{\text{circular}} = (1 + \delta_{n0})\varepsilon\frac{\pi\omega k_z}{4k_c^2}R^2 J_{n+1}^2(u_{nm})\, E_o^2$$

Therefore,

$$\underline{V}_o = E_o R\, J_{n+1}(u_{nm})\sqrt{\frac{(1+\delta_{n0})}{2g^{tm}}\left[\left(\frac{\omega}{\omega_c}\right)^2 - 1\right]}$$

As with rectangular waveguide, the choice of the factor g^{tm} is completely arbitrary. Again, perhaps the simplest choice is $g^{tm} = 1$.

TE modes

Circuit 2 can model a *specific TE* mode, provided that we have

$$L_2' C_2' = \mu\varepsilon$$

$$k_c = \sqrt{L_2'\Gamma'} = \omega_c\sqrt{\mu\varepsilon}$$

so that the phase velocities and cutoff frequencies match. In general, the value of Γ' is different for each mode. The frequency dependence of the characteristic impedance must also be equal, which requires that

$$Z_o = \sqrt{\frac{L_2'}{C_2'}} \frac{\dfrac{\omega}{\omega_c}}{\sqrt{\left(\dfrac{\omega}{\omega_c}\right)^2 - 1}}$$

$$\sqrt{\frac{L_2'}{C_2'}} = g^{te}\sqrt{\frac{\mu}{\varepsilon}}$$

but the geometric scale factor, g^{te}, is arbitrary, provided that the circuit and waveguide powers are equal. For a single traveling-wave mode, we require that the circuit 2 complex voltage and current satisfy

$$\frac{1}{2} Z_o \underline{I}_o \underline{I}_o^* = \mathrm{P}_{TE}$$

$$\underline{V}_o = Z_o \underline{I}_o$$

with

$$L_2' = g^{te}\mu, \qquad C_2' = \varepsilon, \qquad \Gamma' = k_c^2 / L_2'$$

Rectangular guides

For *TE* rectangular waveguide modes, the pertinent results of Chapter 19 can be summarized as

$$\underline{H}_z = H_o \cos(k_x x)\cos(k_y y)\exp(-jk_z z)$$

$$k_c^2 = \left(\frac{n\pi}{a}\right)^2 + \left(\frac{m\pi}{b}\right)^2$$

where $n \geq 0$ and $m \geq 0$ are integers (but *not* $n = m = 0$). The average power, expressed in terms of the longitudinal magnetic field, is

$$P_{TE}^{\text{rectangular}} = \frac{\mu\omega k_z H_o^2}{8k_c^2} ab$$

Therefore,

$$\underline{I}_o = \frac{H_o}{2}\sqrt{\left[\left(\frac{\omega}{\omega_c}\right)^2 - 1\right]\frac{ab}{g^{te}}}$$

Here, too, the choice of the factor g^{te} is completely arbitrary, but consistency may well dictate $g^{te} = g^{tm}$. Again, perhaps the simplest value is $g^{te} = 1$.

Circular guides

For *TE* circular waveguide modes, the pertinent results of Chapter 20 can be summarized as

$$\underline{H}_z = H_o \frac{J_n(k_c r)}{J_n(u'_{nm})} \{ {}^{\cos}_{-\sin} n\phi \} \exp(-jk_z z)$$

$$k_c R = u'_{nm}$$

where n and m are integers and the roots of $\frac{d}{dx} J_n(x) = 0$, u'_{nm} are tabulated in Appendix F. The average power, expressed in terms of the longitudinal magnetic field, is

$$P_{TE}^{\text{circular}} = (1 + \delta_{n0})\mu \frac{\pi \omega k_z}{4k_c^2} \left[1 - \left(\frac{n}{u'_{nm}} \right)^2 \right] H_o^2 R^2$$

Therefore,

$$\underline{I}_o = \frac{H_o R}{2} \sqrt{\frac{[(\frac{\omega}{\omega_c})^2 - 1]}{g^{te}}} \sqrt{1 - \left(\frac{n}{u'_{nm}} \right)^2}$$

As with rectangular waveguide, the choice of the factor g^{te} is completely arbitrary. Again, perhaps the simplest value is $g^{te} = 1$.

24.7 THE SCATTERING MATRIX

Single-port

Consider a lossless *TEM* transmission line of characteristic impedance, Z_o, connected to a load at $z = 0$. In the sinusoidal steady state, at frequency ω, the average complex power in the line can be expressed as

$$Re \frac{1}{2} \underline{V}(z) \underline{I}^*(z) = \frac{1}{2Z_o} [\underline{V}_+(z) \underline{V}_+^*(z) - \underline{V}_-(z) \underline{V}_-^*(z)] = \underline{a}(z) \underline{a}(z)^* - \underline{b}(z) \underline{b}(z)^*$$

where $\underline{V}_+$ and $\underline{V}_-$ or the normalized variables $\underline{a}(z)$ and $\underline{b}(z)$ are traveling waves (in opposite directions) that obey the wave equation. Evidently,

$$\underline{a}(z) = \frac{1}{\sqrt{8Z_o}} [\underline{V}(z) + Z_o \underline{I}(z)] \tag{24.20a}$$

$$\underline{b}(z) = \frac{1}{\sqrt{8Z_o}} [\underline{V}(z) - Z_o \underline{I}(z)] \tag{24.20b}$$

is a satisfactory choice. At the load, the ratio $\underline{b}(0)/\underline{a}(0)$ is defined as the complex scattering coefficient which (because $\underline{V}(0) = Z_L \underline{I}(0)$) is identical to the voltage reflection coefficient,

$$\frac{\underline{b}(0)}{\underline{a}(0)} = \frac{\underline{V}_-(0)}{\underline{V}_+(0)} = \frac{Z_L - Z_o}{Z_L + Z_o} = \underline{\Gamma}(0)$$

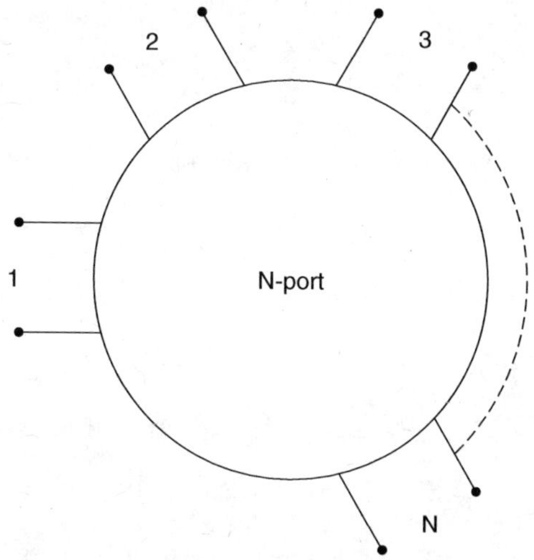

Figure 24.13 N-port junction.

N-port junction

If, instead of a single line terminated by a simple lumped-circuit load, there are N identical transmission lines each connected to one port of an N-port junction (shown in Figure 24.13), there will be positive and negative traveling waves in each line.

Following Eq. (24.20), we define $\underline{a}_i$ and $\underline{b}_i$ (at each input plane of the junction) as the wave amplitudes entering and leaving port i.

$$\underline{a}_i = \frac{1}{\sqrt{8Z_o}}[\underline{V}_i + Z_o\underline{I}_i] \tag{24.21a}$$

$$\underline{b}_i = \frac{1}{\sqrt{8Z_o}}[\underline{V}_i - Z_o\underline{I}_i] \tag{24.21b}$$

The waves exiting any one port will be affected by all of the other waves. Because the system is linear,

$$\underline{b}_i = \sum_{j=1}^{N} \underline{S}_{ij}\underline{a}_j, \quad i = 1, 2, \ldots, N \tag{24.22}$$

where the array of complex coefficients, $\underline{S}_{ij}$, forms the square scattering matrix. From linear circuit theory,

$$\underline{V}_i = \sum_{j=1}^{N} \underline{Z}_{ij}\underline{I}_j, \quad i = 1, 2, \ldots, N \tag{24.23}$$

Therefore, after substitution of Eqs. (24.23) into Eqs. (24.22), use of the Kronecker delta, δ_{ij}, and rearrangement, we have

$$\sum_{j=1}^{N}[\underline{Z}_{ij}\underline{I}_j - Z_o\delta_{ij}\underline{I}_j] = \sum_{k=1}^{N}\underline{S}_{ik}\sum_{j=1}^{N}[\underline{Z}_{kj}\underline{I}_j + Z_o\delta_{kj}\underline{I}_j]$$

which implies

$$[\underline{Z}_{ij} - Z_o\delta_{ij}] = \sum_{k=1}^{N} \underline{S}_{ik}[\underline{Z}_{kj} + Z_o\delta_{kj}]$$

Reversing the last equation and switching to nonindex tensor notation, we obtain

$$\overline{\underline{\mathbf{S}}} \cdot [\overline{\underline{\mathbf{Z}}} + Z_o\overline{\mathbf{I}}] = [\overline{\underline{\mathbf{Z}}} - Z_o\overline{\mathbf{I}}]$$

or, equivalently,

$$\overline{\underline{\mathbf{S}}} \cdot [\overline{\underline{\mathbf{Z}}} + Z_o\overline{\mathbf{I}}] \cdot [\overline{\underline{\mathbf{Z}}} + Z_o\overline{\mathbf{I}}]^{-1} = [\overline{\underline{\mathbf{Z}}} - Z_o\overline{\mathbf{I}}] \cdot [\overline{\underline{\mathbf{Z}}} + Z_o\overline{\mathbf{I}}]^{-1}$$

Finally, since $[\overline{\underline{\mathbf{Z}}} + Z_o\overline{\mathbf{I}}] \cdot [\overline{\underline{\mathbf{Z}}} + Z_o\overline{\mathbf{I}}]^{-1} = \overline{\mathbf{I}}$, the scattering matrix is found to be

$$\overline{\underline{\mathbf{S}}} = [\overline{\underline{\mathbf{Z}}} - Z_o\overline{\mathbf{I}}] \cdot [\overline{\underline{\mathbf{Z}}} + Z_o\overline{\mathbf{I}}]^{-1} \tag{24.24}$$

This formula is the generalization of the reflection coefficient. Often, the value $Z_o = 50$ ohms is used to standardize the definition of the scattering matrix.

In the event the transmission lines connected to the ports have different values of Z_o, Eqs. (24.21) become

$$\underline{a}_i = \frac{1}{\sqrt{8Z_{o(i)}}}[\underline{V}_i + Z_{o(i)}\underline{I}_i] \tag{24.25a}$$

$$\underline{b}_i = \frac{1}{\sqrt{8Z_{o(i)}}}[\underline{V}_i - Z_{o(i)}\underline{I}_i] \tag{24.25b}$$

and the matrix equation that must be solved is

$$\sum_{k=1}^{N} \underline{S}_{ik}\left[\frac{\underline{Z}_{kj}}{\sqrt{Z_{o(k)}}} + \sqrt{Z_{o(k)}}\delta_{kj}\right] = \left[\frac{\underline{Z}_{ij}}{\sqrt{Z_{o(i)}}} + \sqrt{Z_{o(i)}}\delta_{ij}\right]$$

Lossless junctions

In the event that the junction is lossless, the average powers entering and leaving are equal and $\sum_{i=1}^{N}(\underline{a}_i\underline{a}_i^* - \underline{b}_i\underline{b}_i^*) = 0$. It follows from Eq. (24.22) that

$$\sum_{i=1}^{N} \underline{S}_{ij}\underline{S}_{ik}^* = \delta_{jk}$$

and therefore the scattering matrix is unitary, satisfying

$$\underline{S}_{ij} = \underline{S}_{ji}^* = \underline{S}_{ij}^{-1} \tag{24.26a}$$

$$|\det \overline{\underline{\mathbf{S}}}| = 1 \tag{24.26b}$$

When every port $j \neq i$ is terminated by a resistor equal to Z_{oj} (so that $\underline{a}_j = 0$), the junction is said to be "perfectly matched" if (when $\underline{a}_i = 1$) $\underline{b}_i = 0$ for all values of i. Physical structures that are lossless, symmetric, and reciprocal and satisfy these matched conditions are technically important. The cases $N = 1$ and $N = 2$ are trivial and the reader should verify that there are no solutions for $N = 3$ by proving that for a symmetric junction, $|\underline{S}_{11}| \geq 1/3$.[4] The case $N = 4$ has solutions that include directional couplers with equal power division, so-called "magic Ts." Both waveguide[5] and coaxial transmission-line versions were developed during World War II by the M.I.T. Radiation Laboratory; the contributions made by R. L. Kyhl are specifically acknowledged in Montgomery et al. [39]. If the reference planes for the four ports of a "magic T" are appropriately defined, the scattering matrix can be shown to have one of the following forms:

$$\underline{\overline{\mathbf{S}}} = \frac{\exp(j\delta)}{\sqrt{2}} \begin{pmatrix} 0 & 0 & 1 & -1 \\ 0 & 0 & 1 & 1 \\ 1 & 1 & 0 & 0 \\ -1 & 1 & 0 & 0 \end{pmatrix}$$

or

$$\underline{\overline{\mathbf{S}}} = \frac{\exp(j\delta)}{\sqrt{2}} \begin{pmatrix} 0 & 0 & 1 & j \\ 0 & 0 & j & 1 \\ 1 & j & 0 & 0 \\ j & 1 & 0 & 0 \end{pmatrix}$$

These devices have also been developed in combined waveguide-coaxial versions; the low-frequency analog is the hybrid coil originally developed for telephone repeater circuits. The "magic T" is a versatile device that has been applied to impedance bridges, balanced mixers, balanced duplexers, and microwave discriminators.

24.8 DIRECTIONAL COUPLERS

Directional couplers are devices that consist of two lossless transmission lines or waveguides that are coupled in a manner that allows a portion of the power to transfer from one line or guide to the other with that portion splitting unequally between the positive and negative directions of propagation in line #2, depending upon the direction of the power in line #1. Unless ferrite loading is used, the device forms a reciprocal junction with four ports. An ideal coupler will transfer power from line #1 to only one of the wave directions in line #2 without causing a reflected wave in line #1. The two ports of line #2 each sample wave power traveling in different directions in line #1. Any lossless directional coupler is a matched four-port junction. If half of the power is transferred, the device is a "magic T."

When identical rectangular waveguides are used, they are often placed side by side and share a common wall that contains small holes, slots or other coupling elements; the common wall can be either the broad or narrow side.

24.9 RESONATORS

Introduction

A resonator is a device that can both confine and allow oscillatory transfer between at least two forms of energy when the oscillation is close to a characteristic (resonant)

[4] When structures that employ ferrite materials are considered, nonreciprocal solutions that are perfectly matched can be found; these are known as three-port circulators.

[5] If they support orthogonal modes, each waveguide may act as two ports of the junction.

frequency, ω_0. A mass and spring or a pendulum provide mechanical examples involving kinetic energy and potential energy of either elastic or gravitational character. Here we are concerned primarily with electric and magnetic energies in transmission-line and dielectric and waveguide resonators, but the combination of magnetostatic and angular-kinetic energies make ferrite resonators possible and they also are considered. No matter how constructed, all resonators share certain characteristics which we enumerate at the outset.

Quality factors

An ideal lossless resonator must not only have no internal dissipation due to friction, finite conductivity, or other dissipative property, it must be completely shielded from its external environment so that no coupling (including radiation) can occur. In that state, the total energy (assuming that somehow it was introduced and is associated with a single mode) will remain constant forever and the oscillation frequency will have a precise value. But, to be useful, there must be some coupling to the external world so that energy can be exchanged and, in practice, both internal and external dissipation is inevitable. These effects taken together determine the value of a dimensionless parameter, Q, which together with ω_0 characterizes the resonator. It is defined by

$$Q = \frac{\omega_o \langle W_{\text{total}} \rangle}{\langle P_{\text{dissipation}} \rangle}$$

where $\langle W_{\text{total}} \rangle$ is the total average energy and $\langle P_{\text{dissipation}} \rangle$ the average dissipated power at the resonator frequency. It follows that Q is 2π times the ratio of the average energy to the energy dissipated per cycle and we will soon learn that the width of the resonance is $\Delta\omega = \omega_0/Q$. Frequency selectivity in high-Q resonators explains their use in filter applications such as spectrum analyzers.

It is convenient to divide the power dissipation into internal and external components, P_{int} and P_{ext}. Then

$$\frac{1}{Q} = \frac{\langle P_{\text{int}} \rangle}{\omega_o \langle W_{\text{total}} \rangle} + \frac{\langle P_{\text{ext}} \rangle}{\omega_o \langle W_{\text{total}} \rangle}$$

It is useful to define separate values of Q for the internal, external, and combined dissipation

$$Q_u = \frac{\omega_o \langle W_{\text{total}} \rangle}{\langle P_{\text{int}} \rangle}$$

$$Q_x = \frac{\omega_o \langle W_{\text{total}} \rangle}{\langle P_{\text{ext}} \rangle}$$

$$Q_\ell = \frac{\omega_o \langle W_{\text{total}} \rangle}{\langle P_{\text{int}} + P_{\text{ext}} \rangle}$$

It follows that the loaded, unloaded, and external Q factors[6] are related by

$$\frac{1}{Q_\ell} = \frac{1}{Q_u} + \frac{1}{Q_x} \tag{24.27}$$

[6] Alternate notations are prevalent. Some authors use Q_i (internal) instead of Q_u and Q_r (radiation) instead of Q_x.

(They combine like parallel resistors.) Although both P_{int} and P_{ext} will alter W_{total}, the changes are negligible for high-Q resonators; in such cases, the total energy can be calculated assuming the dissipation is zero.

Transmission-line resonator

A transmission-line resonator, which serves as a load impedance, consists of a lossless line of characteristic impedance, Z_o, and length, ℓ; the internal losses due to R', the small distributed resistance per unit length in the wires, can be modeled by a lumped resistor $R_\ell = \frac{1}{2}R'\ell$ at $z = 0$. (The factor $1/2$ is due to the spatial average of $I_z^2(z)$ over the length.) As shown in Figure 24.14, a Thevenin equivalent circuit is coupled to the line at $z = -\ell + x$.

The coupling is equivalent to part (a) of Figure 24.15; if the stub of length x is electrically short, $k(x) \ll 1$, it can be replaced by a lumped inductor $L_o = L'x = xZ_o/c$ as shown in part (b).

The reflection coefficient in the resonator line evaluated at $z = -\ell + x - \delta$ (in the limit $\delta \to 0$) and the corresponding admittance are

$$\Gamma_1 = \frac{R_\ell - Z_o}{R_\ell + Z_o}\exp[j2k(\ell - x)]$$

$$Y_1 = Y_o\frac{1 - \Gamma_1}{1 + \Gamma_1}$$

The values in the stub line evaluated at $z = -\ell + x + \delta$ (in the limit $\delta \to 0$) are

$$\Gamma_2 = -\exp(j2kx)$$

$$Y_2 = Y_o\frac{1 - \Gamma_2}{1 + \Gamma_2} = \frac{1}{jZ_o \tan kx}$$

The admittance seen by the Thevenin source is the parallel combination, $Y_1 + Y_2 = \frac{1}{Z_{\text{in}}(j\omega)}$. The input impedance, $R + jX$, can also be evaluated from the complex Poynting theorem (Part I, Chapter 6, Section 6.2) as

$$Z_{\text{in}}(s) = \frac{2}{|\underline{I}|^2}[<P_{\text{int}}> + 2\sigma_\omega(<W_m> + <W_e>) + j2\omega(<W_m> - <W_e>)] \quad (24.28)$$

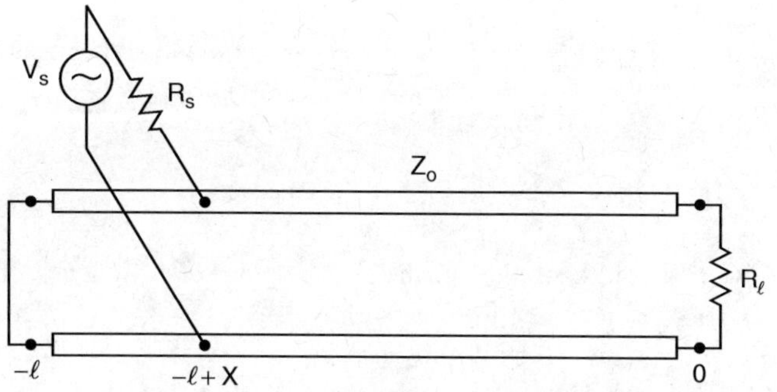

Figure 24.14 Transmission-line resonator.

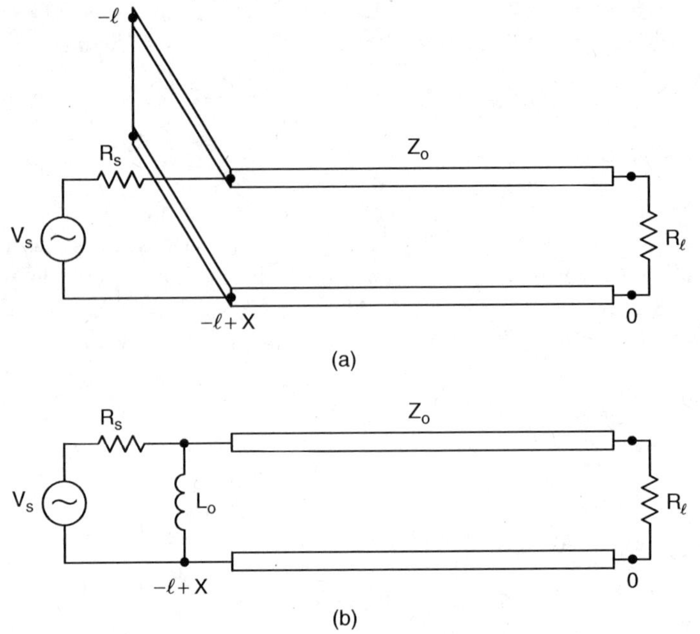

Figure 24.15 Equivalent coupling: (a) stub and (b) lumped inductor.

The average power delivered to the resonator is

$$<P_R> = \frac{V_s V_s^*}{8R_s} \frac{4RR_s}{(R+R_s)^2 + X^2} \tag{24.29}$$

and is maximized when $X = 0$, $R = R_s$. The value $\frac{V_s V_s^*}{8R_s}$ is the maximum available power that can be extracted from the source by a passive load. The bandwidth of the filter depends upon R_s as well as R.

Although the analysis using the stub equivalent provides an exact solution for all frequencies, our interest is mainly in the vicinity of a resonance where $X = 0$ and the average stored electric and magnetic energies are equal, $<W_m> = <W_e> = \frac{1}{2}<W_{\text{total}}>$. This occurs when

$$k\ell = p\pi \qquad \left(\ell = p\frac{\lambda_o}{2}\right), \qquad p = 1, 2, 3, \ldots$$

The position of the s-plane zeroes, where $Z_{\text{in}}(-\alpha \pm j\omega_o) = 0$, occur at

$$\omega = \pm\omega_o$$

$$-\sigma_\omega = \alpha = \frac{1}{2}\frac{<P_{\text{int}}>}{<W_{\text{total}}>}$$

and determine the bandwidth of the intrinsic resonance. For high Q_u, the half-power frequencies of $Z_{\text{in}}(j\omega)$ occur when $\omega - \omega_o = \pm\alpha$ at which $X = \pm R$. Near a resonance, Eq. (24.28) can approximated (for $\sigma_\omega = 0$) as

$$Z_{\text{in}}(j\omega) = \frac{2}{|\underline{I}|^2}[<P_{\text{int}}>_o + j(\omega - \omega_o)(<W_m> + <W_e>)_o] = R + jX$$

with the average dissipated power and the average total energy evaluated at resonance—as denoted by the subscript o.

The true bandwidth of the filter depends upon R_s as well as R. It follows that

$$R + R_s = \frac{2}{|\underline{I}|^2}(<P_{\text{int}}> + <P_{\text{ext}}>)$$

$$X = \frac{2}{|\underline{I}|^2}(\omega - \omega_{\text{o}})<W_{\text{total}}> = \frac{(\omega - \omega_{\text{o}})}{\omega_{\text{o}}} Q_\ell (R + R_s)$$

and that Eq. (24.29) can be approximated as

$$\frac{<P_R>}{\left(\dfrac{V_s^2}{8R_s}\right)} = \frac{4\xi}{(\xi + 1)^2} \frac{1}{\left[1 + \left(\dfrac{\omega - \omega_{\text{o}}}{\Delta\omega}\right)^2\right]}$$

where the filter bandwidth is defined by

$$\Delta\omega = \frac{\omega_{\text{o}}}{Q_\ell}$$

and the coupling factor is defined by

$$\xi = \frac{Q_u}{Q_x}$$

The unloaded Q (calculated assuming $L_{\text{o}} = 0$, or equivalently $R_s \to \infty$) is defined by

$$Q_u = \frac{\omega_{\text{o}}<W>_{\text{o}}}{<P_{\text{int}}>_{\text{o}}} = \frac{\omega_{\text{o}}\frac{1}{4}L' I_{\text{o}}^2 \ell}{\frac{1}{2}R_\ell I_{\text{o}}^2} = \frac{p\pi Z_{\text{o}}}{2R_\ell}$$

If the stub is electrically short, it can be replaced by the shunt inductance,

$$L_{\text{o}} = L' x \ll \frac{Z_{\text{o}}}{\omega_{\text{o}}}$$

The external Q (calculated assuming $R_\ell = 0$, $V_s = 0$, $R_s \gg \omega_{\text{o}} L_{\text{o}}$) depends upon

$$R_x = Re \frac{R_s j\omega L_{\text{o}}}{R_s + j\omega L_{\text{o}}} = \frac{R_s(\omega L_{\text{o}})^2}{R_s^2 + (\omega L_{\text{o}})^2} = \frac{(\omega L_{\text{o}})^2}{R_s}$$

and is defined by

$$Q_x = \frac{\omega_{\text{o}}<W>_{\text{o}}}{<P_{\text{ext}}>_{\text{o}}} = \frac{\omega_{\text{o}}\frac{1}{4}L' I_{\text{o}}^2 \ell}{\frac{1}{2}R_x I_{\text{o}}^2}$$
$$= \frac{p\pi Z_{\text{o}} R_s}{2(\omega_{\text{o}} L_{\text{o}})^2}$$

At resonance,

$$\frac{<P_R>}{\left(\dfrac{V_s^2}{8R_s}\right)} = \frac{4\xi}{(\xi + 1)^2}$$

where

$$\xi = \frac{Q_u}{Q_x} = \frac{(\omega_{\text{o}} L_{\text{o}})^2}{R_\ell R_s}$$

When $\omega_{\text{o}} L_{\text{o}} = \sqrt{R_\ell R_s}$, the average power delivered to the resonator is the maximum that is available.

Coupling coefficient

The three regions of coupling are defined by

$\xi < 1$ undercoupled

$\xi = 1$ critically coupled $\qquad Q_\ell = \frac{1}{2}Q_u = \frac{1}{2}Q_x$

$\xi > 1$ overcoupled

The average power delivered to any resonator is maximized at resonance when $\xi = 1$; this is the condition of critical coupling. Both under- and overcoupling results in less power being delivered to the resonator.

Transmission resonator

A transmission-line resonator, which serves to transmit power from a Thevenin source to a separate load impedance, has an output as well as an input circuit coupled to the line. For simplicity we consider both coupling elements to be shunt inductances[7] as shown in Figure 24.16.

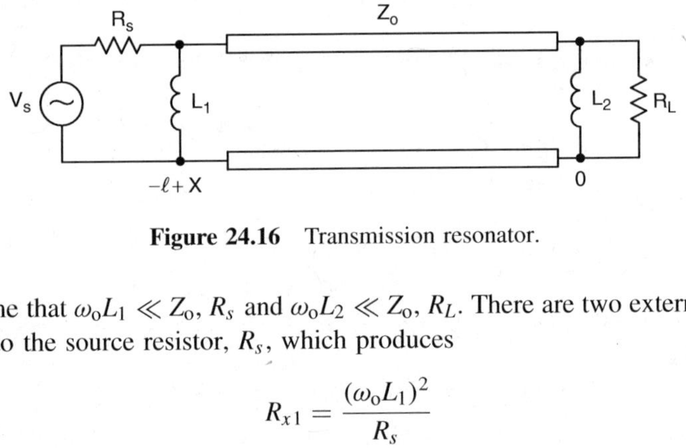

Figure 24.16 Transmission resonator.

We assume that $\omega_o L_1 \ll Z_o$, R_s and $\omega_o L_2 \ll Z_o$, R_L. There are two external Q factors: One is due to the source resistor, R_s, which produces

$$R_{x1} = \frac{(\omega_o L_1)^2}{R_s}$$

while the other is due to the load resistor, R_L, which produces

$$R_{x2} = \frac{(\omega_o L_2)^2}{R_L}$$

The latter can be combined with the internal losses to produce an effective Q_u as seen by the source. Often the internal losses are negligible compared to the load and

$$Q_u^{\text{eff}} = Q_{x2} = \frac{p\pi Z_o R_L}{2(\omega_o L_2)^2}$$

As before,

$$Q_x = Q_{x1} = \frac{p\pi Z_o R_s}{2(\omega_o L_1)^2}$$

[7]Shunt capacitors can also be used, but the high electric fields generated in a high-Q resonator may produce electrical breakdown; this is not a problem when shunt inductors are used.

The input coupling coefficient is

$$\xi = \frac{Q_x}{Q_u^{\text{eff}}} = \frac{R_s}{R_L}\frac{L_2^2}{L_1^2}$$

It follows that for equal coupling ($L_1 = L_2$ and $R_L = R_s$), the transmission resonator is critically coupled at resonance. This is a consequence of the symmetry of the circuit, which automatically produces $Q_{x1} = Q_{x2}$. Notice that when internal resonator losses are included, one can never deliver all of the available power to the load resistor even though it is of course possible to perfectly match the resonator to the input circuit.

If the output resistor is replaced by a second identical resonator with a resistive output, the bandwidth of the combined circuit will be decreased. This process can be continued to provide a periodic cascade of transmission resonators with even higher frequency selectivity.

Waveguide resonators

Rectangular or circular waveguides can be used instead of the *TEM* transmission line to produce either single-ended or transmission cavity resonators. Instead of using lumped-circuit inductances to provide the coupling, a common practice is to place thin, perfectly conducting (short-circuit) planes across the waveguide cross section at $z = 0$ and $z = -\ell$. One or both of these planes is an iris with an opening that is often a small circular hole. Although any TE_{nm} or TM_{nm} mode can be used, it is often the dominant $TE_{10}^{\square}$ or TE_{11}^{o} mode that is used. If the hole that forms an iris is very small, the length of the cavity satisfies $k_z\ell = p\pi, p = 1, 2, \ldots$, where k_z satisfies the waveguide mode dispersion. Equivalent circuits for the waveguide modes can be used; these were given in Section 24.6. Equivalent circuits for inductive and capacitive irises of many varieties were developed at the M.I.T. Radiation Laboratory and are included in the *Waveguide Handbook* edited by N. Marcuvitz [40]; these topics are developed in the text by Collin [41].

Irises for TE$_{10}$ rectangular waveguide

A hole of radius, r_o, centered in the rectangular cross-sectional plane of a rectangular waveguide of width, a, and height, b, can be shown to be inductive with a normalized susceptance given by

$$\frac{B}{Y_o} \simeq \frac{3}{16\pi}\frac{ab\lambda_g}{r_o^3} \qquad (r_o \ll b,\ a > b) \tag{24.30}$$

where $\lambda_g = 2\pi/k_z$ and provided that $2a > \lambda_o > 2a/3$ and the $TE_{11}^{\square}$ mode is not propagating.

Irises for TE$_{11}$ circular waveguide

A hole of radius r_o, centered in the rectangular cross-sectional plane of a circular waveguide of radius R_o, can be shown to be inductive with a normalized susceptance given by

$$\frac{B}{Y_o} \simeq \frac{\lambda_g}{4R}\left[\frac{6R^3}{8.4r_o^3} - 2.344\right] \qquad (r_o \ll R) \tag{24.31}$$

provided that $2.61R < \lambda_o < 3.41R$.

Plane wave diffraction from a small circular hole

Because the dominant mode waveguide fields incident upon the small hole are essentially linearly polarized, and $r_o \ll R$, the scattering analyzed in Chapter 22, Section 22.8, is approximately valid. The power radiated through the hole (in terms of the transverse magnetic field) is then

$$\int_0^{2\pi}\int_0^{\pi/2} r^2 \underline{S}_r \sin\theta \, d\theta d\phi = \frac{\sqrt{\frac{\mu_o}{\varepsilon_o}}\underline{H}_o\underline{H}_o^*}{4} k^4 r_o^6 \pi \int_0^{\pi/2} (1 - \frac{1}{2}\sin^2\theta)\sin\theta \, d\theta$$

$$<P_r> = \frac{1}{2}\sqrt{\frac{\mu_o}{\varepsilon_o}}\underline{H}_o\underline{H}_o^* \pi r_o^2 \left(\frac{1}{3}k^4 r_o^4\right)$$

$$k_z \ell = p\pi$$

We apply this result and those of Chapters 19 and 20 to both rectangular and circular waveguide transmission-cavity resonators with symmetric irises. It is advantageous to use the Alternate power and Alternate energy formulation. In all cases, the losses due to finite wall conductivity are neglected and the spacing between the irises is assumed large enough that interactions between them can be ignored. The results, based upon standing-wave fields, are summarized below.

TE_{10} mode in rectangular waveguide of dimensions a x b

$$0 \le x \le a\,, \quad 0 \le y \le b\,, \quad 0 \le z \le \ell$$

$$k_x a = \pi\,, \qquad k_z \ell = \pi$$

$$A_y = -A_o \sin\left(\frac{\pi}{a}x\right)\sin(k_z z)\cos\omega t$$

$$\mu_o H_x = -\frac{\partial A_y}{\partial z} = A_o k_z \sin(\frac{\pi}{a}x)\cos(k_z z)\cos\omega t$$

$$W^o = \tfrac{1}{2}\varepsilon_o A_o^2 \omega^2 \sin^2\left(\frac{\pi}{a}x\right)\sin^2(k_z z)$$

$$(Energy)^o = \tfrac{1}{8}\varepsilon_o A_o^2 \omega^2 ab\ell$$

$$<P_r> = \tfrac{1}{2} c\mu_o \underline{H}_o\underline{H}_o^* \pi r_o^2 (\tfrac{1}{3}k^4 r_o^4)$$

$$|\underline{H}_o| = \frac{A_o k_z}{\mu_o} \qquad \text{(evaluated at } x = \frac{a}{2},\ z = 0, \ell)$$

$$Q_\ell = \frac{\omega(Energy)^o}{2<P_r>} = \frac{\tfrac{1}{8}\varepsilon_o A_o^2 \omega^2 ab\ell \ ck}{c\mu_o \underline{H}_o\underline{H}_o^* \pi r_o^2 (\tfrac{1}{3}k^4 r_o^4)}$$

$$Q_\ell = \frac{\tfrac{3}{8}ab}{k_z^3 k r_o^6}$$

TE*$_{11}$ *mode in circular waveguide of radius R

$$r \le R, \quad 0 \le z \le \ell$$

$$k_c R = u'_{11} = 1.84\,, \quad k_z \ell = \pi$$

$$A_r = C_1 \tfrac{1}{r} J_1(k_c r)\sin(k_z z)\{{}^{\sin}_{\cos}\,\phi\}\cos\omega t$$

$$A_\phi = C_1 \tfrac{d}{dr} J_1(k_c r)\sin(k_z z)\{{}^{\cos}_{-\sin}\,\phi\}\cos\omega t$$

$$(Energy)^{o} = \frac{\pi \omega k C_1^2}{8 c \mu_o}[(u'_{11})^2 - 1] J_1^2(u'_{11})\ell$$

$$<P_r> = \tfrac{1}{2} c \mu_o \underline{H}_o \underline{H}_o^* \pi r_o^2 (\tfrac{1}{3} k^4 r_o^4)$$

$$\mu_o |\underline{H}_o| = \tfrac{1}{2} C_1 k_z k_c \quad \text{(evaluated at } r = 0,\ z = 0, \ell)$$

$$Q_\ell = \frac{\omega (Energy)^{o}}{2<P_r>}$$

$$Q_\ell = \frac{\frac{3}{2}[1 - \frac{1}{(u'_{11})^2}] J_1^2(u'_{11}) \pi R^2}{k_z^3 k r_o^6} = \frac{.36\ \pi R^2}{k_z^3 k r_o^6}$$

Within appropriate ranges of the parameters, the simplified analyses agree reasonably well with the more exact treatments that use waveguide equivalent circuits and Eqs. (24.30) and (24.31).

Dielectric resonators

Small disks, cylinders, or spheres of low-loss, high-dielectric-constant material can be used to confine electromagnetic fields without the necessity of employing conducting walls; the high dielectric constant reduces the wavelength and therefore the physical size of the resonator. Confinement occurs when waves inside the dielectric produce total internal reflections at the surfaces. (Refer to Chapter 17, Section 17.2.2 and Chapter 21.) The fields inside are superpositions of uniform plane waves; those outside are superpositions of nonuniform (evanescent) plane waves. The filter elements are often arranged in periodic arrays to increase the frequency selectivity; coupling to and between them is provided by the evanescent fields. A brief review and additional references are given in Ramo et al. [11, pp. 521–525].

YIG sphere filter

In Chapter 23, we learned that magnetostatic resonances in ferrite material are determined by the DC magnetic field and not the physical dimensions; in particular the frequency of the uniform precession (Kittel resonance) of a small sphere is magnetically tunable and independent of the radius. This is of great technical importance because size-dependent cavity resonators of necessity become larger at lower frequencies (a giant organ pipe is required for low pedal notes) and require mechanical adjustment for tuning (like a trombone). In addition, the nonreciprocity of ferrites can offer additional advantages.

A simple geometry will suffice to illustrate the basic principle of operation. Consider two identical single-turn loops made of perfectly conducting wire that are oriented at right angles to each other as shown in Figure 24.17. The gaps that form terminals 1 and

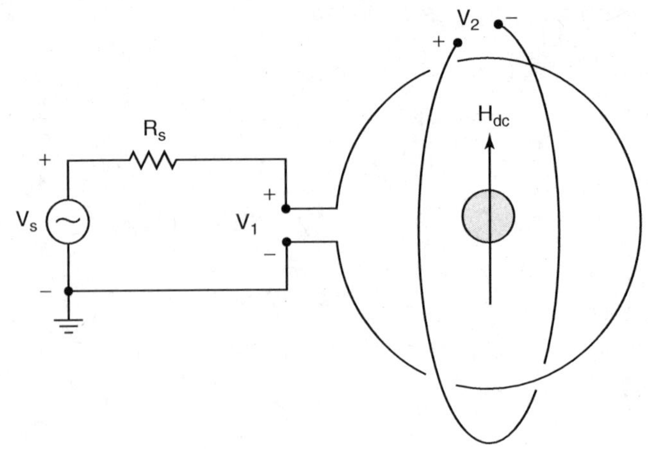

Figure 24.17 YIG sphere filter.

2 are very small. A small single-crystal sphere of yttrium iron garnet (YIG) is placed at the common center of the loops, in a DC magnetic field that is spatially uniform over its volume and large enough to saturate the DC magnetization. The surface of the sphere is highly polished so as to minimize the resonance line width. The uniform precession resonance is adjusted to the desired filter frequency,

$$\omega_o = \frac{e}{m_e}\mu_o H_{dc}$$

(The proportionality factor is 2.8 MHz/Oe.) A Thevenin equivalent circuit with a voltage source, $V_s \cos(\omega t)$, of microwave frequency ω and resistance R_s is connected to loop 1; the output loop 2 is open-circuited. The free-space wavelength is assumed to be large compared to the sphere and loop radii. Unless the frequency is close to ω_o, the amplitude of the small-signal magnetization precessing at the uniform precession is very small and the rf-magnetic field generated by loop 1 is orthogonal to loop 2; consequently there is no flux linkage and $V_2 = 0$. Near resonance, the circularly polarized precession generates an rf magnetic field component parallel to the plane of loop 1; the time variation of the flux linking loop 2 generates a voltage, $V_2(t)$. We wish to calculate the input and transfer impedances of the filter and, for the voltage gain $\frac{V_2}{V_s}(j\omega)$, determine the center frequency, maximum value, and the 3-dB bandwidth.

To proceed further with the analysis, we assume the following parameters and conditions:

- CGS units are commonly used to specify magnetic materials and fields; helpful conversions are 1 Oe $= 250/\pi\ Am^{-1}$ and $B(G) = H(\text{Oe}) + 4\pi M(G)$.
- The YIG sphere has a saturation magnetization, $4\pi M_s = 1750\ G$; a diameter of $2R = 1\ mm$; and a resonance linewidth, $2\Delta H_o = .5$ Oe. For simplicity, magnetic anisotropy is neglected.
- The applied magnetic field is $H_{dc} = 1000$ Oe.
- The diameter of the wire loops is $2R_o = 6\ mm.$; the diameter of the circular cross section of the wire is $2r = 1/2\ mm$.

- Any shift in the resonance frequency of the sphere caused by the conducting loops is negligible.
- The Thevenin source resistance is $R_s = 50\ \Omega$.

Static fields

If the z axis is chosen parallel to H_{dc}, the magnetic field inside the YIG sphere is (refer to Appendix C, Section C.6)

$$\mathbf{H} = \widehat{\mathbf{z}}(H_{\mathrm{dc}} - \tfrac{1}{3}M_s) \qquad (r \le R)$$

where, in order that the sphere is saturated,

$$H_{\mathrm{dc}} > \tfrac{1}{3}M_s$$

Outside the sphere, the field is the superposition of a uniform and dipole component,

$$\mathbf{H} = \widehat{\mathbf{z}}H_{\mathrm{dc}} + \tfrac{1}{3}M_s \left(\frac{R}{r}\right)^3 (\widehat{\mathbf{r}}\ 2\cos\theta + \widehat{\theta}\ \sin\theta) \qquad (r \ge R)$$

Rf fields

If loop 1 lies in the y–z plane (with the $+x$ axis pointing out of the paper) and the current $i_1(t)$ due to the voltage source is counterclockwise, the rf magnetic field near the center (with the sphere removed) is approximately uniform and from the Biot–Savart Law [Part I, Chapter 1, Section 1.5, Eq. (1.23b)] is given by

$$\mathbf{h}^{\mathrm{a}} = \widehat{\mathbf{x}}\, h_x^{\mathrm{a}}$$

$$h_x^{\mathrm{a}} = \frac{i_1(t)}{2R_{\mathrm{o}}}$$

With the sphere replaced, the *MQS* rf magnetic field inside is

$$h_x = h_x^{\mathrm{a}} - \frac{1}{3}m_x$$

$$h_y = -\frac{1}{3}m_y$$

The rf field outside the sphere is that of a magnetic dipole, $\frac{4}{3}\pi R^3(\widehat{\mathbf{x}}m_x + \widehat{\mathbf{y}}m_y)$.

The complex rf fields are related by the Polder susceptibility tensor; therefore the complex small-signal magnetization is

$$\underline{m}_x = \chi \underline{h}_x - j\kappa \underline{h}_y$$

$$\underline{m}_y = j\kappa \underline{h}_x + \chi \underline{h}_y$$

which, in terms of the applied rf field, $\underline{h}_x^{\mathrm{a}}$, is

$$\underline{m}_x = \chi^{\mathrm{e}} \underline{h}_x^{\mathrm{a}}$$

$$\underline{m}_y = j\kappa^{\mathrm{e}} \underline{h}_x^{\mathrm{a}}$$

where

$$\chi = \frac{\omega_M \omega_z}{\omega_z^2 - \omega^2}, \qquad \kappa = \frac{-\omega_M \omega}{\omega_z^2 - \omega^2}$$

$$\chi^e = \frac{\omega_M \omega_o}{\omega_o^2 - \omega^2}, \qquad \kappa^e = \frac{-\omega_M \omega}{\omega_o^2 - \omega^2}$$

and with $-\gamma_g = \frac{e}{m_e} > 0$,

$$\omega_z = -\gamma_g \mu_o (H_{dc} - \tfrac{1}{3} M_s)$$

$$\omega_M = -\gamma_g \mu_o M_s$$

$$\omega_o = -\gamma_g \mu_o H_{dc} > \tfrac{1}{3} \omega_M$$

Notice that, as expected, ω_o is the Kittel uniform precession resonance which for YIG must be a frequency greater than 1.63 GHz (in order that the ferrite be saturated).

For $H_{dc} = 1000$ Oe, $f_o = 2800$ MHz $= 2.8$ GHz.

Magnetic-flux coupling to the loops

The magnetic flux that threads loop 1 is due to the applied rf field, $h_x^a(t)$, produced by $i_1(t)$ and the dipole field produced by $m_x(t)$. Although the applied field in the plane $x = 0$ is not uniform everywhere inside the loop, the flux linking it can be expressed as $Li_1(t)$, where it is well known that

$$L \simeq \mu_o R_o [\ln(8R_o/r) - 2] = 9.7 \times 10^{-9} \ H$$

is the self-inductance of a single loop of radius R_o made from circular wire of radius r. It follows that $\omega_o L \simeq 170 \ \Omega$.

The flux linking loop 1 that is caused by m_x is composed of the flux inside the sphere, $\mu_o m_x \frac{2}{3}\pi R^2$, and that in the air between the sphere and the loop, $-\mu_o m_x \frac{2}{3}\pi R^3(\frac{1}{R} - \frac{1}{R_o})$. The sum of all three terms produces a complex flux,

$$\underline{\mathfrak{F}}_x = L\underline{I}_1 + \mu_o \underline{m}_x \tfrac{2}{3}\pi \left(\frac{R^3}{R_o}\right)$$

The complex voltage is $\underline{V}_1 = j\omega \underline{\mathfrak{F}}_x = \underline{V}_s - R_s \underline{I}_1$; therefore

$$\underline{Z}_{in} = \frac{\underline{V}_s}{\underline{I}_1} = R_s + j\omega \left[L + \mu_o \tfrac{1}{3}\pi \left(\frac{R^3}{R_o^2}\right) \frac{\omega_M \omega_o}{\omega_o^2 - \omega^2} \right]$$

Because loop 2 is open-circuited, the complex magnetic flux that threads it is due to the dipole field produced by $\underline{m}_y$. The result is

$$\underline{\mathfrak{F}}_y = \mu_o \underline{m}_y \tfrac{2}{3}\pi \left(\frac{R^3}{R_o}\right) = -j\mu_o \tfrac{1}{3}\pi \left(\frac{R^3}{R_o^2}\right) \frac{\omega_M \omega}{\omega_o^2 - \omega^2} \underline{I}_1$$

The complex open-circuit voltage, $\underline{V}_2 = j\omega \underline{\mathfrak{F}}_y$, allows calculation of the transfer impedance,

$$Z_{21} = \frac{\underline{V}_2}{\underline{I}_1} = \omega \mu_o \tfrac{1}{3}\pi \left(\frac{R^3}{R_o^2}\right) \frac{\omega_M \omega}{\omega_o^2 - \omega^2}$$

and the voltage gain,

$$\frac{\underline{V}_2}{\underline{V}_s} = \frac{\omega\mu_0\frac{1}{3}\pi(R^3/R_o^2)\dfrac{\omega_M\omega}{\omega_0^2-\omega^2}}{R_s + j\omega\left[L + \mu_0\frac{1}{3}\pi(R^3/R_o^2)\dfrac{\omega_M\omega_0}{\omega_0^2-\omega^2}\right]}$$

Magnetic loss

To add magnetic resonance dissipation, we simply make the substitution $H_{dc} \to H_{dc} + j\Delta H_0$. The half-power points of the resonance are $\omega = -\gamma_g\mu_0(H_{dc} \pm \Delta H_0)$. The unloaded Q is therefore

$$Q_u = \frac{H_{dc}}{2\Delta H_0} = 2000$$

At resonance, even though $R = .5$ mm is very small, the terms proportional to ω_M dominate and

$$\frac{\underline{V}_2}{\underline{V}_s}(j\omega_0) \simeq -j$$

Neglecting ΔH_0, and replacing $\omega \to \omega_0$ except $\omega_0^2 - \omega^2 \to 2\omega_0(\omega_0 - \omega)$, the voltage ratio near resonance is

$$\frac{\underline{V}_2}{\underline{V}_s} \simeq \frac{1}{\dfrac{2\frac{(\omega_0-\omega)}{\omega_0}(R_s + j\omega_0 L)}{\omega_M\mu_0\frac{1}{3}\pi R^3/R_o^2} + j}$$

The approximate half-power fractional bandwidth (the range of ω over which $|\underline{V}_2/\underline{V}_s| \geq 1/\sqrt{2}$) is given by

$$\frac{\Delta\omega}{\omega_0} = \omega_M\mu_0\frac{1}{3}\pi\left(\frac{R^3}{R_o^2}\right)\frac{\sqrt{R_s^2 + 2(\omega_0 L)^2}}{R_s^2 + (\omega_0 L)^2} = \frac{1}{Q_x}$$

The numerical values of Q_x, Q_ℓ, and the actual half-power bandwidth are therefore

$$Q_x \simeq 227$$

$$Q_\ell = \frac{Q_x\, Q_u}{Q_x + Q_u} \simeq 204$$

$$\Delta f = \frac{f_0}{Q_\ell} \simeq 14 \text{ MHz}$$

The YIG filter is very compact, is magnetically tunable, and can combine high Q with low insertion loss. The ratio R/R_0 allows for adjustment of the Q. For a small sphere, ω_0 does not depend upon the saturation magnetization; therefore, although M_s is a function of temperature, the filter frequency is not. Nevertheless, filters based upon other shapes such as disks and thin films have been constructed.

CHAPTER 25

PRACTICE PROBLEMS

25.1 STATICS

Problem 25.1-1

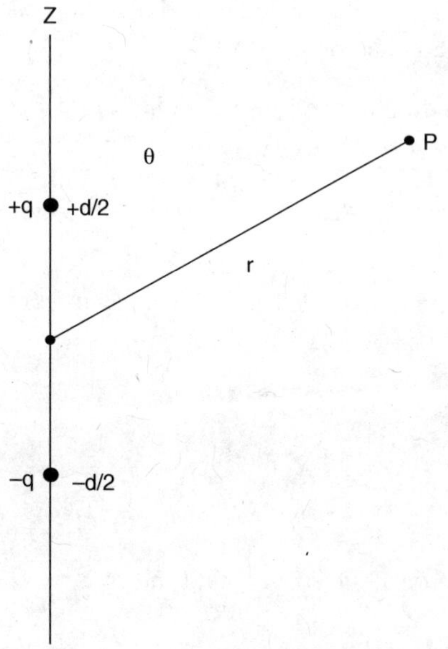

The Power and Beauty of Electromagnetic Fields, First Edition. F. R. Morgenthaler.

A pair of electric charges, $\pm q$, is located on the z axis at locations $z = \pm d/2$. The magnitude of the electric dipole moment is defined as $p = qd$.

a) In terms of p, find the electric potential at the point P expressed in both Cartesian coordinates (x, y, z) and spherical coordinates (r, θ, ϕ).

Use the approximation $\frac{1}{\sqrt{1+\Delta}} \simeq 1 - \frac{1}{2}\Delta$ when $|\Delta| \ll 1$, to simplify the results of part(a) when

b) $r \gg d$

c) $r \ll d$

d) For parts (b) and (c), find separate approximate expressions for the electric field.

Problem 25.1-2

$\mathbf{F}_1, \mathbf{F}_2$ and $\mathbf{F}_3$ are vector fields defined by

$$\mathbf{F}_1 = \widehat{\mathbf{x}}\,3z^2 + \widehat{\mathbf{y}}\,2yz + \widehat{\mathbf{z}}\left(6xz + y^2\right)$$

$$\mathbf{F}_2 = \widehat{\mathbf{x}}\,y + \widehat{\mathbf{y}}\,x$$

$$\mathbf{F}_3 = \widehat{\mathbf{x}}\cos y + \widehat{\mathbf{y}}\sin x$$

a) Evaluate the curl of each field.

b) Evaluate the divergence of each field.

c) Which (if any) could be a static electric field in a free-space region, $x^2 + y^2 + z^2 < R^2$, that might contain electric charge?

d) Which (if any) could be a static magnetic-field in a free-space region, $x^2 + y^2 + z^2 < R^2$, that might contain electric current?

e) Which (if any) could be either a static electric or magnetic field in a free-space region, $x^2 + y^2 + z^2 < R^2$.

f) If any, find the electric charge density, ρ, and electric current density, $\mathbf{J}$, associated with parts (c) and (d).

Problem 25.1-3

Consider a vector field, $\mathbf{F} = \widehat{\mathbf{x}}F_o \sin(by)$ where F_o and b are constants.

a) Could $\mathbf{F}$ be an electric field? If yes, find the electric charge that is its source; if no, explain why not.

b) Could $\mathbf{F}$ be a magnetic field? If yes, find the electric-current density that is its source; if no, explain why not.

c) Evaluate $\oint_C \mathbf{F} \cdot \mathbf{ds}$ directly—when the closed contour C is a square of side a with center at $x = y = z = 0$ that lies in the plane:

(i) $x = 0$

(ii) $z = 0.$

d) Repeat part (c) this time using Stokes Law.

Problem 25.1-4

The electric potential, Φ in free-space that may contain electric charges is known to be

$$\Phi = C_o \exp(-|y|) \sin |x|$$

a) Find the charge distribution.

b) Find the electric-field.

Problem 25.1-5

In cylindrical coordinates (r, ϕ, z) the electric charge density, ρ in an otherwise free-space region is

$$\rho = \rho_o e^{-\alpha r^2}$$

where ρ_o and α are constants.

a) Find the electric field, $\mathbf{E}$.

b) Find the electric potential, Φ, such that $\mathbf{E} = -\nabla\Phi$.

c) Find the total charge per unit length along the z axis.

Problem 25.1-6

a) If $\mathbf{f}$ is the Lorentz force-density acting upon charge and current densities, show that

$$\mathbf{E} \cdot \mathbf{J} = \mathbf{f} \cdot \mathbf{v} + (\mathbf{E} + \mathbf{v} \times \mu_o \mathbf{H}) \cdot (\mathbf{J} - \rho \mathbf{v})$$

b) Show that if the velocity $\mathbf{v}$ is nonrelativistic, the second term is $\mathbf{E}' \cdot \mathbf{J}'$ where the primed fields are those evaluated in the moving frame of reference.

HINT: Galilean relativity applies.

Problem 25.1-7

The Lorentz force acting in free-space upon a positive charge, q, with mass m causes the charge to follow the trajectory:

$$x = r_o \cos \omega t$$

$$y = r_0 \sin \omega t$$

$$z = \frac{1}{2} a_0 t^2$$

for times $t > 0$ such that $a_0 t$ and and $\omega r_0$are both small compared to the velocity of light, $c = \frac{1}{\sqrt{\mu_0 \varepsilon_0}} \approx 3 \times 10^8\ ms^{-1}$.

a) Find the time-independent electric and magnetic fields, which act upon the charge.

b) Prove that only the electric field does work on the charge. How much work is done in 1 second (after $t = 0$)?

Problem 25.1-8

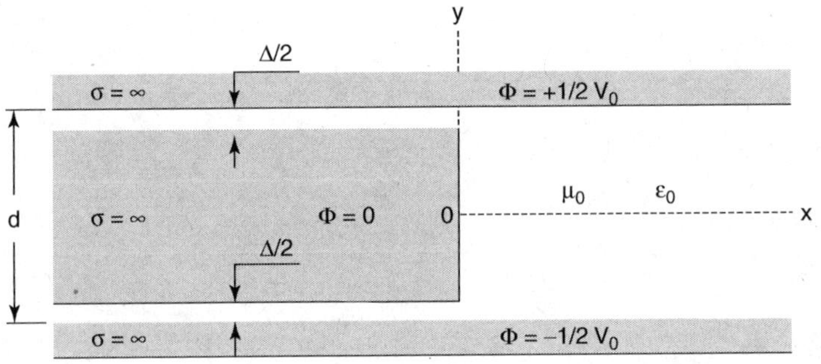

The structure shown is independent of z. The top and bottom plates and the inner slab are made of perfect conductors; the air gaps are very small ($\Delta \ll d$). The potentials of the three conductors are $\pm V_0/2$ and 0 as shown.

a) Use reasonable approximations to find expressions for the potential in the air regions for $|x| < 2d$, given that the plates extend for at least $|x| = 10d$.

b) Is there a net electric force exerted on the center conductor due to the fields in the region $|x| < 2d$? If yes, in what direction?

c) Is there a net vertical electric force exerted on the center conductor? Explain.

d) Is the configuration stable?

Problem 25.1-9

In order to accelerate electrons along the $+z$ axis in an otherwise free-space region while at the same time focusing them, it is decided to employ the cylindrically symmetric **E** field:

$$\mathbf{E} = E_0 \frac{1}{L} (\hat{\mathbf{z}} r - \hat{\mathbf{r}} 2z)$$

(Neglect the fields generated by the electrons themselves.) The electrons start from rest at ($t = 0$, $z = 0$, $r = r_0 = \frac{L}{10}$) and reach the point ($z = L$, $r = 0$) at time T.

a) What sign must the constant E_o have?

b) Find the electric potential associated with **E**.

c) Find the shape of two electrodes (each of which is an equipotential surface), with one passing through the point $z = 0$, $r = 0$ while the other passes through $z = L$, $r = 0$. Truncate them so that both electrodes fit within a cylindrical volume of length L and radius $R = 3L$.

d) If $L = .1m$ and $T = 2 \times 10^{-8}s$, what is the minimum voltage that should be applied across the electrodes.

e) Calculate the maximum velocity of an electron and verify that is nonrelativistic ($\frac{e}{m} = 1.8 \times 10^{11} Ckg^{-1}$).

Problem 25.1-10

A point charge q is located a distance d above the surface of a nonmagnetic dielectric half-space $z \leq 0$ characterized by uniform linear permittivity ε; free-space exists for $z > 0$ and there are no other free charges.

a) Find the static electric field in all space.

b) Find the polarization charge distribution everywhere.

c) Find the total electrical force exerted on the charge.

HINT: Use the method of images.

Problem 25.1-11

A line current I is located parallel to and a distance d above the surface of a magnetic half-space $z \leq 0$ characterized by zero polarization and uniform linear permeability μ; free-space exists for $z > 0$ and there are no other free currents.

a) Find the static magnetic field in all space.

b) Find the magnetization current distribution everywhere.

c) Find the total magnetic force exerted on a unit length of the current.

HINT: Use the method of images.

Problem 25.1-12

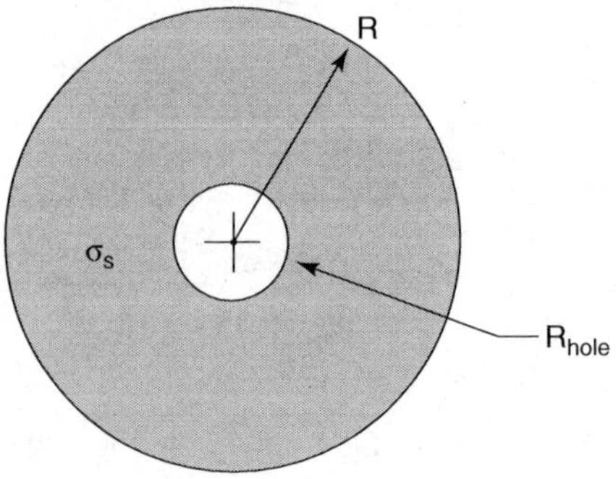

A very thin disk of radius R is made from a perfect insulator with free-space values μ_o, ε_o and lies in the x–y plane; the surface of the disk is coated with uniform electric surface charge density, σ_s $[Cm^{-2}]$.

a) Find the electric field and the electric potential everywhere along the z axis that passes through the center of the disk.

b) Repeat part (a) when a hole of radius $R_{\text{hole}} < R$ is now cut in the disk and the associated charge removed.

HINT: Use superposition.

Problem 25.1-13

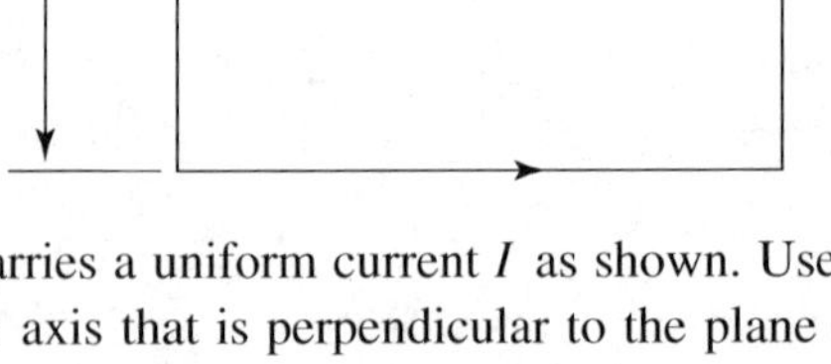

A square loop of wire carries a uniform current I as shown. Use the Biot–Savart Law to find the **H** field on the z axis that is perpendicular to the plane of the loop at its center.

Problem 25.1-14

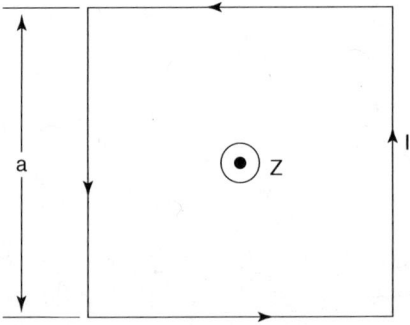

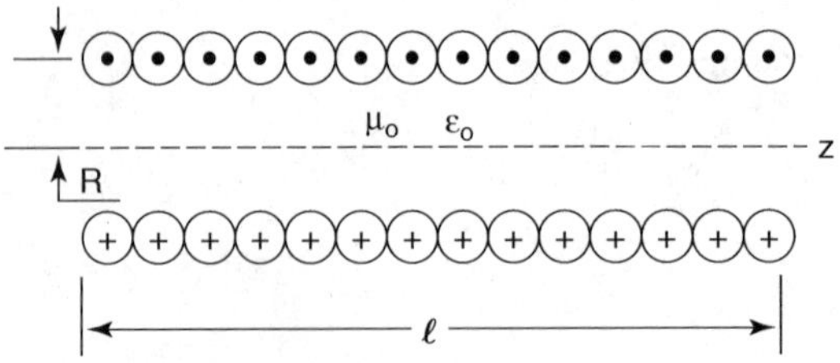

N closely spaced circular turns, each of radius R, form a solenoid of length ℓ that carries a DC current I_o.

a) Find the magnetic field along the z axis:

(i) using the Biot–Savart Law

(ii) using the scalar magnetic potential

When $\ell \gg R$, find:

b) the approximate **H** field inside the solenoid.

c) the approximate inductance.

Problem 25.1-15

A ferrite circular cylinder of length ℓ and radius R is magnetized to saturation by a large DC magnetic field applied parallel to the z axis of symmetry; the magnetization vector, $\mathbf{M} = \hat{\mathbf{z}}M_s$, is assumed to be uniform throughout the volume of the cylinder (the Sommerfeld approximation).

a) Assume that the source of **M** is a magnetic charge density, $\rho_m = -\mu_o \nabla \cdot \mathbf{M}$, and integrate over it in order to find inside the cylinder the <u>on-axis</u> **H** field due to M_s(the demagnetizing contribution to the total field).

b) Assume that the source of **M** is an Amperian electric current density, $\mathbf{J}_e = \nabla \times \mathbf{M}$, and integrate over it in order to find inside the cylinder the <u>on-axis</u> **B** field and **H** field.

Problem 25.1-16

A positive point charge q is mounted between two perfectly conducting planes in the $x - y$ plane ($z = 0$) as shown below.

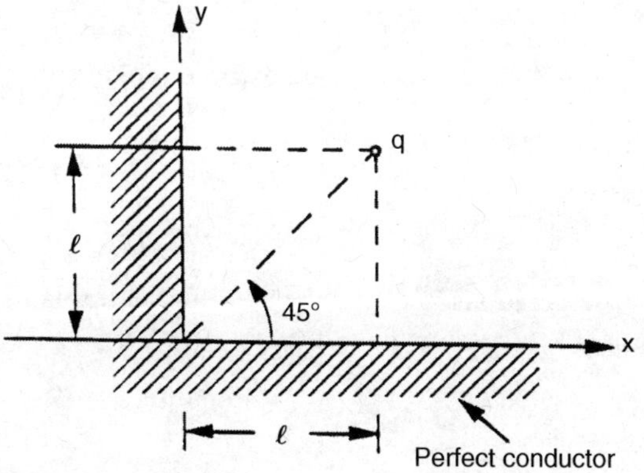

a) What are the boundary conditions at the perfectly conducting surfaces?

b) Place images so that the boundary conditions are satisfied.

c) Find the surface charge density σ_s, along x at $y = z = 0$.

Problem 25.1-17

In a very long cylindrical region of free-space there are two electric charge distributions, $+\rho_0$ and $-\rho_0$, where ρ_0 is a constant. There is no charge outside of the cylinder. The positive charges are stationary; the negative charges all move at constant velocity, v parallel to the axis of the cylinder (z axis). Neglect end effects to find inside and outside the cylindrical region of radius R:

a) The electric field

b) The magnetic field

Problem 25.1-18

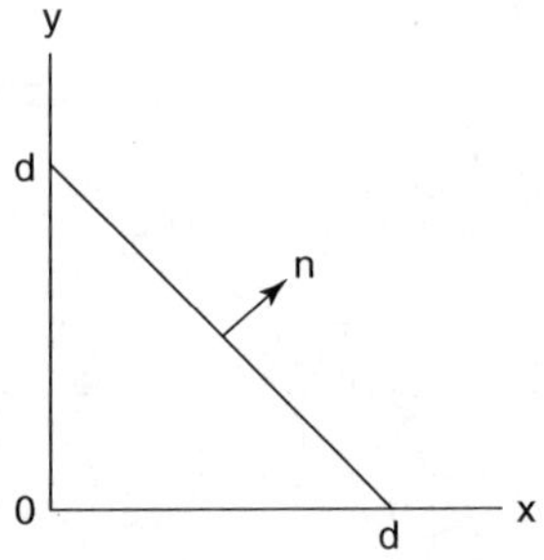

A triangular region bounded by the x axis, y axis, and line $x + y = d$, contains an electric-potential, $\Phi(x, y)$. There is no electric-charge within the region. The boundary conditions on the perimeter of the region are known to be

$$E_x(0, y) = E_o \frac{y}{d}$$

$$\Phi(x, 0) = 0$$

$$\mathbf{E} \cdot \widehat{\mathbf{n}} = 0$$

a) Find the electric potential inside the region.

b) If the plane $y = 0$ is a perfectly conducting ground-plane, find any surface charges on the boundary $0 \leq x \leq d$, $y = 0+$.

Problem 25.1-19

Three-dimensional representations of four potential functions, $\Phi(x, y)$, are plotted below; the vertical dimension (z) represents the potential.

a) Which (if any) could not possibly be solutions of Laplace's Equation? Explain briefly.

b) Which potentials could **not** be associated with magnetic fields?

c) The functions not selected in either part a or part b are solutions of Laplace's equation. Sketch two-dimensional contour plots of equipotentials for each of those Laplacian cases. Also sketch the field lines.

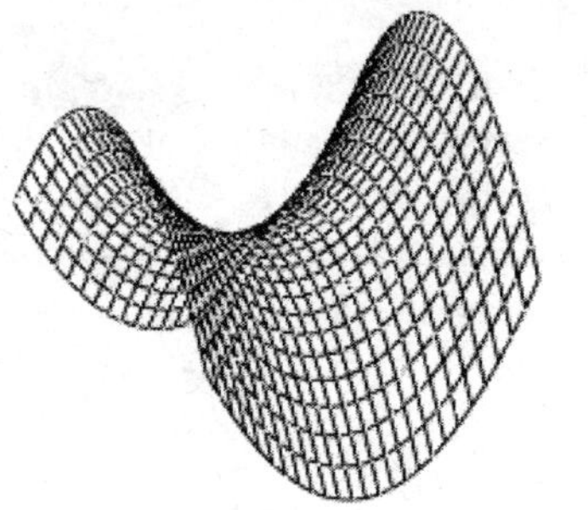

M

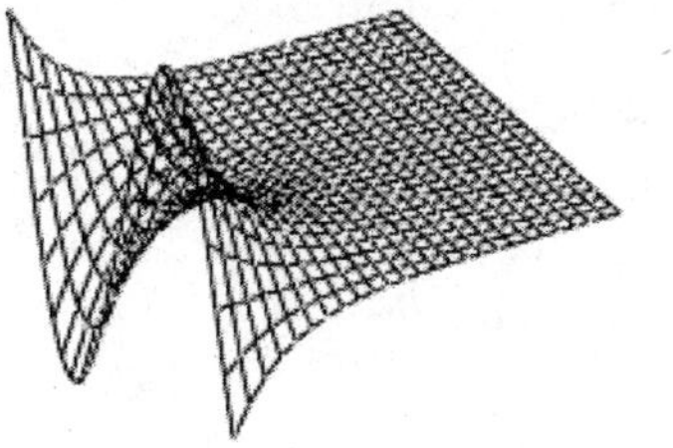

S

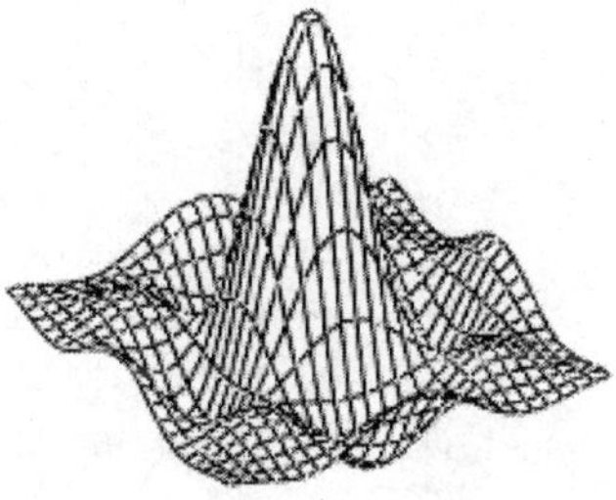

P

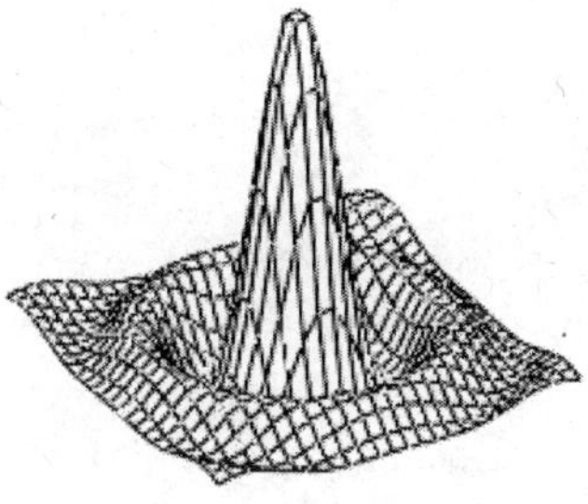

N

Problem 25.1-20

It is required that a static electric field be created by sources that are outside a cylindrical region of free-space of length L and radius $R = L/10$. No charges or currents are permitted inside the cylinder. An additional requirement is that the field maintain cylindrical symmetry with respect to the z axis and that the on-axis be of the form

$$\mathbf{E}(r=0,z) = \widehat{\mathbf{z}}\left(E_0 + E_1\frac{z}{L}\right) \qquad (0 < z < L)$$

where both E_0 and E_1 are positive constants.

a) Prove that it is **not** possible to achieve this result with a field that has only a z component within the cylindrical region.

Assume that the design goal can be accomplished with a potential $\Phi(x,y,z)$ that is of the form

$$\Phi(x,y,z) = A_0 + A_1 z + A_2 z^2 + B_2(x^2+y^2)$$

b) Find approximate values for the constants A_0, A_1, A_2, and B_2.

c) Choose suitable equipotentials and design two or more electrodes that when properly charged will provide the required field, when $E_1 = 2E_0$.

d) Are the choices made in (c) unique?

e) Qualitatively, what will happen to a positive charge, initially at rest, if it is placed near (but not on) the z axis? What happens if it is a negative charge?

Problem 25.1-21

An M.I.T. student is the victim of a computer virus that has deleted some of the laboratory measurements stored in a data file that had not been backed up. The DC measurements were made by probing voltages at periodic intervals along the x and y surface coordinates of a thin semiconductor wafer. Six of the twenty-five values that had been measured in a square area are now missing; the remaining voltages (with respect to the zero reference) in mV are

−2	−.75	1	3.25	6
−.75	.5	2.25	4.5	7.25
0	x	x	x	8
.25	x	x	x	8.25
0	1.25	3	5.25	8

There is not enough time to repeat the measurements before the lab report is due, but, because theory predicts that the voltages are samples of a solution of Laplace's Equation and because the spacing between data points is not too great, the student's knowledge of electromagnetic theory allows the missing values to be inferred from what has remained.

a) In the original data set, what was the maximum voltage? The minimum voltage?

b) Use your best estimate to restore the missing six values.

NOTE: Later, the campus police apprehended a suspect—someone always considered a potential thief. But, he was not charged, because a wily lawyer provided a defense well-grounded in the law.

Problem 25.1-22

The space between large, flat, perfectly conducting capacitor plates is filled with a lossless dielectric fluid that is characterized by a linear dielectric constant, $K_d = \varepsilon/\varepsilon_o > 1$. A single spherical air bubble exists as shown; the radius R is small compared to the distance to the nearest plate. The electric-field breakdown strength of the fluid is $E_{\text{fluid}}^{\max}$; that of air is $E_{\text{air}}^{\max} < E_{\text{fluid}}^{\max}$ (neglect fringing).

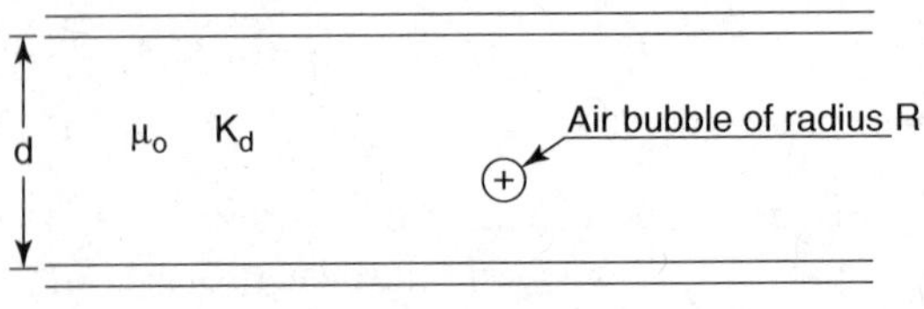

a) What is the maximum voltage that can be applied to the capacitor without causing breakdown?

b) If the voltage exceeds the part (a) value, where will breakdown be initiated?

Problem 25.1-23

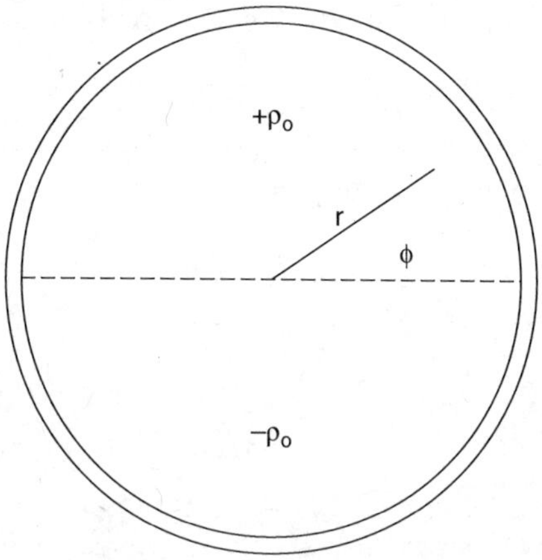

A long, perfectly conducting, thin-walled cylinder is filled with uniform electric charge density $\pm\rho_0$ in the upper/lower halves as shown; the tube is at ground potential.

a) Find a particular solution of Poisson's Equation for each region. (They need not satisfy all (or any) of the boundary conditions.)

Any failure of the part (a) solution to satisfy the boundary conditions of the problem must be compensated for by the addition of a homogeneous (Laplacian) solution.

b) Find the total potential in terms of an appropriate Fourier series.

c) Find and sketch the electric field.

d) Find the electric surface charges on the inner wall of the cylinder.

Problem 25.1-24

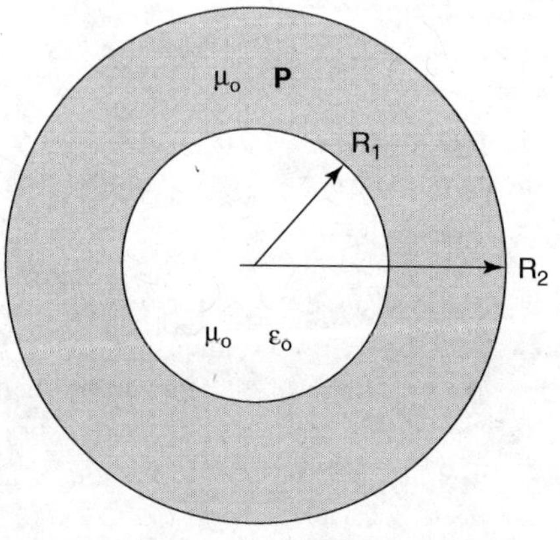

A spherical shell bounded by concentric spheres of radius R_1 and R_2 is made from a permanently polarized dielectric having a polarization vector given by

$$\mathbf{P} = \hat{\mathbf{r}} P_0 \exp(-br) \qquad (R_1 \leq r \leq R_2)$$

where P_o and b are positive constants.

a) Find the polarization charge densities (volume and surface) everywhere.

b) Find the **E** field everywhere. (Assume there is no unpaired "free" electric charge.)

Dust containing unpaired charge of one polarity settles uniformly over the outer surface of the sphere until the total surface charge at $r = R_2$ is zero.

c) What is the polarity of the charge?

d) Find the modified **E** field everywhere.

Problem 25.1-25

An infinitely long strip of electric surface charge of constant uniform density, σ_o, is located on the plane $z = 0$. It extends from $-\infty < x < +\infty$ and $0 \le y \le W$ and is surrounded by free-space.

a) Find an expression for the electric potential, $\Phi(x, y, z)$ everywhere.

b) Find an expression for the electric-field component, $E_z(x, y, z)$, for finite values of x and y, in the limit $W \to \infty$.

Problem 25.1-26

A large thin sheet of uniform conductivity σ and thickness Δ is surrounded by free-space. Two perfectly conducting electrodes are painted on the same side of the sheet in a pattern made of solid areas that are far from the edges of the sheet and separated from each other by a distance large compared to Δ. The DC electrical-resistance measured between the electrodes is R.

A three-dimensional (3-D) air-filled structure is made from two perfectly conducting electrodes that are parallel to one another along z. The uniform cross section is similar in geometry to that of the painted electrodes described above except that there is a linear scale factor, sf, between them (1 inch along x or y of the conducting sheet corresponds to sf inches along x or y of the 3-D structure). The length ℓ (along z) is large compared to the spacing between the parallel electrodes.

a) The 3-D structure is employed as a capacitor by applying a DC voltage between the open-circuited electrodes. Set up an analogy between the conduction problem of the sheet and the electric field of the capacitor that is generated by a DC voltage. Which fields are in correspondence? Which boundary conditions? How can a two-dimensional field (sheet) be analogous to a 3-D field (capacitor)?

b) Calculate the approximate capacitance of the structure in terms of σ, Δ, ε_o, sf, ℓ, and R.

c) The 3-D structure is now employed as an inductor by placing a short circuit at one end and applying a DC current which flows in the $+z$ direction along one electrode and in the $-z$ direction along the other one. Set up an analogy between the conduction problem of the sheet and the magnetic field of the inductor that is generated by the DC current. Which fields are in correspondence? Which boundary conditions? How can a 2-D field (sheet) be analogous to a 3-D field (inductor)?

d) Calculate the approximate inductance of the structure in terms of σ, Δ, μ_o, sf, ℓ, and R.

Problem 25.1-27

A sphere of radius R is made of uniform lossless dielectric with isotropic scalar permittivity ε. The sphere is surrounded by free-space and a concentric electric surface-charge density, $\sigma_s(r = R_0, \theta) = -3\varepsilon_o E_o \cos\theta$, where $R_0 > R$.

a) In the absence of the dielectric sphere, show that σ_s produces a uniform constant electric field, $\mathbf{E} = \hat{\mathbf{z}} E_o$ for $r \leq R_0$.

b) Find the electric field everywhere when the dielectric sphere is present.

c) Find the change in the electric energy that occurs when the sphere is removed, but the value of E_o remains the same.

d) Repeat part (**c**) when the sphere has finite conductivity, σ. Separately, consider the case $\sigma = \infty$.

Problem 25.1-28

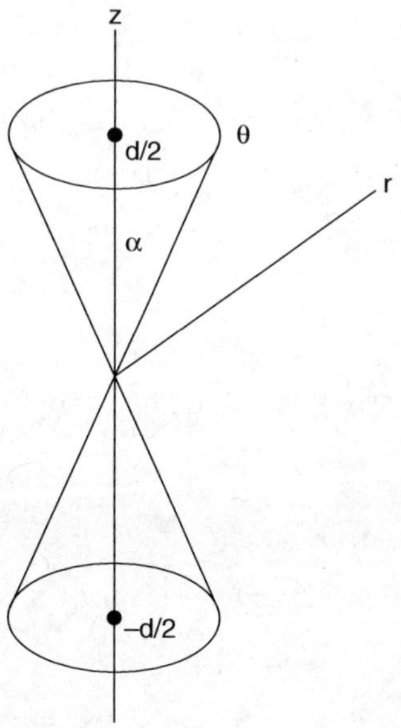

A dipole antenna is made from perfectly conducting cylindrical wire of length d and radius r_w. A gap of length g exists at the center ($r_w < g \ll d$) and the structure is surrounded by free-space. The voltage applied across the gap is V. This antenna can be approximated by the biconical structure shown in the figure.

a) If the bicones extend to $\pm\infty$, and there is no gap between them, verify that the exact electrostatic potential is

$$\Phi(r,\theta) = \frac{V}{2}\frac{\tanh(\cos\theta)}{\tanh(\cos\alpha)} \qquad (\alpha \le \theta \le \pi - \alpha)$$

Assume that the part (**a**) potential remains valid for the truncated structure within the spherical radius, $r \le d/2$.

b) Find the cone angle, $\alpha \ll 1$, that makes the combined surface area of the cones equal to the curved surface area of the cylindrical wire ($2\pi r_w d$).

c) Calculate the electric surface charge on the cones.

d) Use the result of part (**c**) to calculate the capacitance between the cones. (This value neglects the gap capacitance—negligible because $r_w < g$.)

25.2 QUASISTATICS

Problem 25.2-1

A single electric charge, $+q$, moves slowly in free-space along the z axis; at time t it is located at

$$z_o = v_o t + \frac{1}{2}at^2$$

where v_o and a are constants.

a) Find the approximate *EQS* electric field expressed in spherical coordinates assuming $r \gg |z_o|$.

b) Find the approximate magnetic field due to the motion (also expressed in spherical coordinates) assuming $r \gg |z_o|$.

c) Is the *EQS* approximation valid for very larger? Explain, briefly.

Problem 25.2-2

A long cylindrical conductor (μ_o, ε, σ) of circular cross section (radius, R) is surrounded by free-space; it is parallel to and centered on the z axis. Far from the cylinder, the *EQS* electric field is

$$\mathbf{E} = \hat{\mathbf{x}}E_o(t)$$

Neglect any z variation in the fields and find:

a) The complete *EQS* field inside and outside the conductor.

b) The *EQS* current density, $\mathbf{J}$, everywhere.

c) The *EQS* electric charge densities (volume and surface; both unpaired and polarization).

d) The magnetic field due to time variation of $E_o(t)$.

Problem 25.2-3

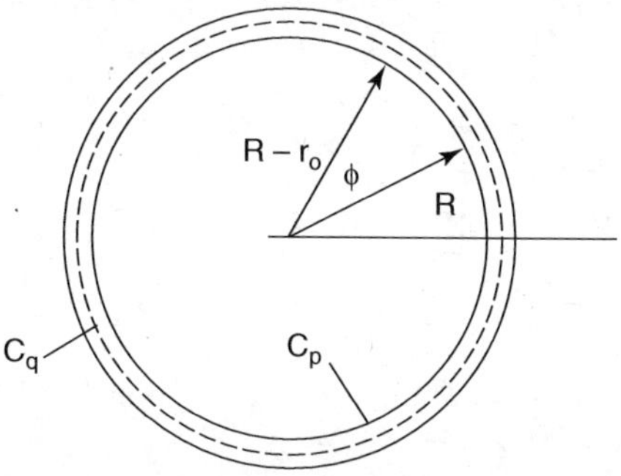

Loop geometry

A perfectly conducting wire is formed into a single-turn circular-loop of radius R as shown in the figure. The wire is surrounded by free-space, has a circular cross section of radius $r_o \ll R$, and carries a low-frequency current $I(t)$. Although the current is an electric surface current, its effect can be approximated by a line-current image that is positioned at the center of the cross section (the dotted circle).

a) Show that the total *MQS* magnetic flux, $Flux = LI(t)$, can be approximated as

$$Flux = \frac{\mu_o I(t)}{4\pi} \oint_{C_p} \oint_{C_q} \frac{\mathbf{ds}_q \cdot \mathbf{ds}_p}{r_{qp}}$$

$$r_{qp} = \sqrt{R^2 + (R - r_o)^2 - 2R(R - r_o)\cos(\phi_p - \phi_q)}$$

b) Show that the inductance, L, is approximately given by the elliptic integral,

$$L = \mu_o \int_0^{\pi} \frac{R(R - r_o)\cos\phi \; d\phi}{\sqrt{r_o^2 + 2R(R - r_o)(1 - \cos\phi)}}$$

where $\phi = \phi_p - \phi_q$. Because $r_o \ll R$, this integral can be divided into two intervals: $0 \leq \phi \leq \phi_o$, in which $\cos\phi \simeq 1 - \frac{1}{2}\phi^2$, and $\phi_o \leq \phi \leq \pi$, in which it is permissible to set $r_o = 0$. A reasonable value is $\phi_o = 1/2$; then

$$L \simeq \mu_o [\int_0^{1/2} \frac{R^2(1 - \frac{1}{2}\phi^2)\; d\phi}{\sqrt{r_o^2 + R^2\phi^2}} + \int_{1/2}^{\pi} \frac{R\cos\phi \; d\phi}{\sqrt{2(1 - \cos\phi)}}]$$

c) Evaluate the integrals in part (b) and show that, because $\sinh^{-1} x \simeq \ln(2x)$ for large x,

$$L \simeq \mu_o R[\ln(\frac{R}{r_o}) + .07389]$$

d) Compare the part (c) approximation with the more exact result,

$$L \simeq \mu_o R\left[\ln\left(\frac{8R}{r_o}\right) - 2\right]$$

that can be derived from the elliptic integral when $r_o \ll R$.

Problem 25.2-4

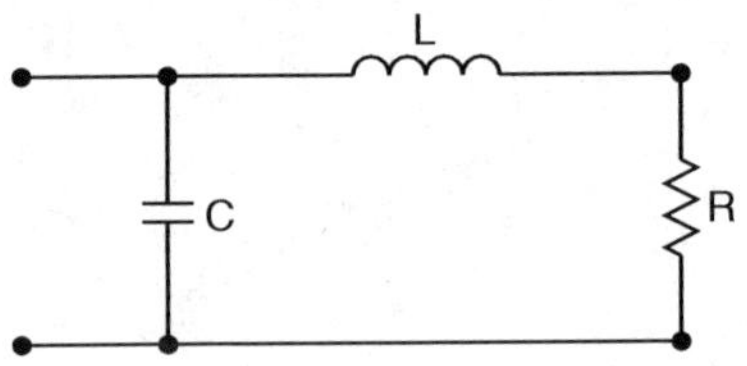

The voltage across the terminals is $V_s = V_o \cos \omega t$. In the sinusoidal steady state, all of the voltages across and currents through the circuit branches are sinusoids of the same frequency that can be expressed as

$$v(t) = Re\left\{\underline{V}\, e^{j\omega t}\right\}$$

$$i(t) = Re\left\{\underline{I}\, e^{j\omega t}\right\}$$

a) Calculate the instantaneous current through R and the instantaneous power delivered to the circuit by the voltage source.

b) Find both the time-averaged electric and magnetic energies ($<W_e>$ and $<W_m>$), each as a function of R, L, C, ω and V_o.

c) Find the time-averaged dissipated power, $<P_d>$, in the circuit as a function of the same parameters used in part (b).

d) Use the results of parts (a) and (b) to evaluate the quantity $\left[<P_d> + j2\omega\,(<W_m> - <W_e>)\right]$.

e) Calculate the complex power, $\underline{P} = \frac{1}{2}\, \underline{V}\underline{I}^*$, delivered by the voltage source and compare it with the result of part (c).

f) Use the results of parts (d) and (e) to evaluate either the input impedance $Z(j\omega)$ or input admittance, $Y(j\omega)$, of the circuit. Verify that your answer is correct.

g) If electric energy predominates, the associated electric fields are governed by electroquasistatics (*EQS*); if magnetic energy predominates, the associated magnetic fields are governed by magnetoquasistatics (*MQS*). For a fixed frequency ω, find the conditions in terms of R, L, C and ω such that

(i) *EQS* applies.

(ii) *MQS* applies.

The circuit is considered low-loss if the energy that is lost in one cycle of the sinusoidal excitation is small compared to the total energy stored in the circuit during the same cycle.

h) Find the conditions on R, L, C and ω such that the circuit is *not* low loss.

Problem 25.2-5

Parallel plates of area A and spacing d are filled with a linear material characterized by

$$\sigma(x) = \sigma_0 \frac{d}{x}$$

$$\varepsilon(x) = \varepsilon_0 \exp(x/d)$$

where the perfectly conducting plates are located at $x = 0$ and $x = d$. The plate dimensions are large compared to the spacing and so the no fringing approximation can be made. The voltage between the plates is a slowly varying $V_0(t)$.

a) Find the quasistatic electric field between the plates.

b) Find the volume and surface unpaired (free) electric-charge distributions.

c) Find the volume and surface polarization-charge distributions.

d) Find the lumped equivalent circuit which models the low-frequency behavior of the device.

Problem 25.2-6

A positive charge, $+q$, of mass m is, at the time $t = 0$, located at $x = y = z = 0$ and moving with nonrelativistic velocity $v_x = v_0$, $v_y = v_z = 0$. The only force acting upon the charge is that due to a static uniform magnetic field, $\mathbf{H} = \hat{\mathbf{z}} H_0$

a) Find the trajectory of the charge for $t > 0$.

b) Repeat part (a) when a uniform static electric field, $\mathbf{E} = \hat{\mathbf{z}} E_0$, is also present.

Problem 25.2-7

In a region of free-space defined by

$$0 < x < \mathrm{d}$$

$$0 < y < \mathrm{w}$$

$$-\ell < z < 0$$

the electric and magnetic fields are *either*

$$\mathbf{E} = \hat{\mathbf{x}} E_a \cos(\omega t)\ \sin(\tfrac{\omega}{c} z)$$

$$\mathbf{H} = \hat{\mathbf{y}} H_a \sin(\omega t)\ \cos(\tfrac{\omega}{c} z)$$

or

$$\mathbf{E} = \hat{\mathbf{x}} E_b \cos(\omega t)\ \cos(\tfrac{\omega}{c} z)$$

$$\mathbf{H} = \hat{\mathbf{y}} H_b \sin(\omega t)\ \sin(\tfrac{\omega}{c} z)$$

a) Verify that each pair of fields can separately satisfy Maxwell's Equations in free-space without charge or current, provided that the ratios, E_a/H_a and E_b/H_b, are appropriate.

b) Find those ratios

If $\frac{\omega}{c}|z| \ll 1$, the z-dependence of the fields can be approximated by the leading terms of the Taylor-series expansion of $\sin\left(\frac{\omega}{c}z\right)$ and $\cos\left(\frac{\omega}{c}z\right)$.

$$\sin u \simeq u - \frac{1}{3!}u^3 + \cdots$$

$$\cos u \simeq 1 - \frac{1}{2}u^2 + \cdots$$

c) Keep only the leading term in each field component and verify that one pair of fields is *EQS*, the other *MQS*. Which is which?

d) How high can the frequency ω become before the quasistatic approximations break down?

Problem 25.2-8

A circular cylinder of radius R and infinite length (along z) is made from material of uniform conductivity σ, permittivity ε, and permeability μ_o; it is surrounded by free-space and placed in a uniform magnetic field,

$$\mathbf{H} = \hat{\mathbf{z}} H_o \cos \omega t$$

The frequency ω is such that $\omega\varepsilon \gg \sigma$.

If the cylinder is solid and uniform:

a) Find the *MQS* electric field (correct to first-order) inside and outside the cylinder.

b) Find the average power (per-unit length) dissipated in the cylinder.

c) How large can the radius be before the quasistatic analysis is invalidated?

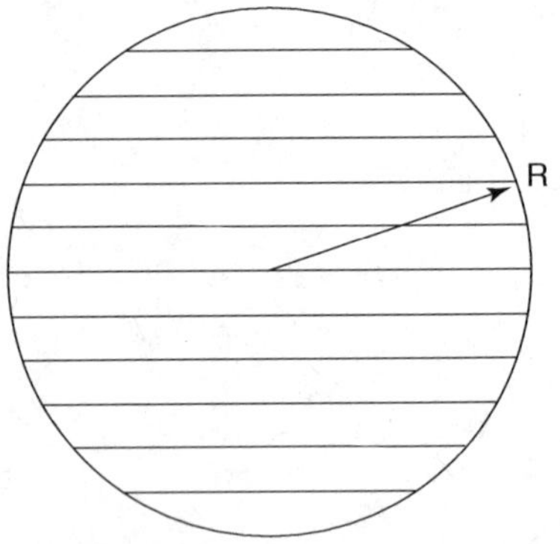

Laminated conducting cylinder.

If the cylinder is divided into a stack of N laminations (sheets of conductor spaced by *very thin* insulating layers of permittivity, ε_i and permeability μ_o) as shown, the eddy currents induced by the time-varying magnetic field will be modified. If the curvature at the ends is neglected, each sheet may be approximated to have a rectangular cross section of thickness, $2R/N$, that is very small compared to its width.

d) Use Cartesian coordinates to find the *MQS* electric field (correct to first order) inside a single (typical) lamination and the adjacent insulating layers. (Neglect end effects and apply boundary conditions to the planar interfaces.)

e) Show that, when $N \gg 1$, the laminations reduce the power dissipated in the solid cylinder by $1/N$.

f) What critical dimension determines how high ω can be raised before the quasistatic analysis is invalidated? What changes occur when $\sigma \gg \omega\varepsilon$.

g) Use these results to explain why transformer cores made from iron or steel are often laminated.

Problem 25.2-9

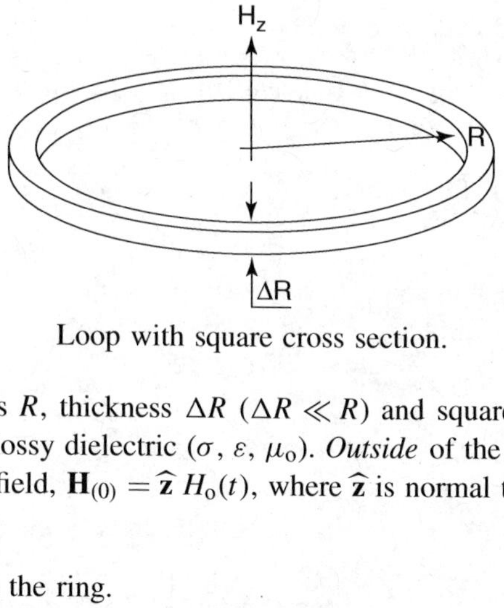

Loop with square cross section.

A hollow ring of radius R, thickness ΔR ($\Delta R \ll R$) and square cross section is made from a linear isotropic lossy dielectric (σ, ε, μ_o). *Outside* of the ring exists a zero-order time varying magnetic field, $\mathbf{H}_{(0)} = \hat{\mathbf{z}}\, H_o(t)$, where $\hat{\mathbf{z}}$ is normal to the plane of the ring. $\mathbf{E}_{(0)} = 0$.

a) Find $H_o(t)$ inside the ring.

b) Find the first-order electric field everywhere.

c) Find the approximate first-order magnetic field everywhere along the symmetry (z) axis.

d) Find the approximate first-order magnetic field everywhere far from the ring ($r \gg R$).

e) Calculate, correct to second order, the average power dissipated in the ring, if $H_o(t) = H_o \cos \omega t$.

Problem 25.2-10

A solid nonmagnetic conducting sphere (σ, μ_o) of radius R is surrounded by free-space. Far from the sphere is an essentially uniform slowly varying zero-order magnetic field,

$$\mathbf{H}_{(0)} = \hat{\mathbf{z}}\, H_o \cos \omega t$$

$$\mathbf{E}_{(0)} = 0$$

where, for parts (a)–(e), $\sqrt{\omega \mu_o \sigma} R \ll 1$.

a) Find the complete zero-order magnetic field inside and outside the sphere.

b) Find the first-order electric field inside and outside the sphere.

c) Find the first-order magnetic field inside and outside the sphere.

HINT: Find a particular $\mathbf{H}_{(1)}$ inside the sphere that is of the form

$$r^n [\hat{\mathbf{r}}\, A(t) \cos \theta + \hat{\theta}\, B(t) \sin \theta]$$

and then add a homogeneous field such that the resultant field satisfies the boundary conditions.

d) Find, correct to second order, the power dissipated in the sphere.

e) Check the relative strengths of $\mathbf{H}_{(0)}$ and $\mathbf{H}_{(1)}$ to get an indication of the convergence of the quasistatic expansion.

f) For $\sqrt{\omega \mu_o \sigma} R \gg 1$, find a reasonable approximation to the magnetic field outside of the sphere.

g) Qualitatively, sketch the total power dissipated in the sphere as a function of σ for fixed ω and also as a function of ω for fixed σ.

Problem 25.2-11

Perfectly conducting parallel plates of width w (along x), spacing d (along y), and length ℓ (along z) are terminated at $z = 0$ with a resistance R. The region between the plates $(0 < y < d)$ is filled with a linear isotropic material of permittivity ε and permeability μ. Because $\ell, w \gg d$, fringing fields can be neglected. A source of frequency ω is connected to the input at $z = -\ell$. Quasistatics applies because $\omega\sqrt{\mu\varepsilon}\ell \ll 1$.

a) If $R = \infty$, and the source is an ideal voltage source, $V = V_o \cos \omega t$, find the zero-order and first-order fields inside the structure. Find a lumped equivalent circuit as seen by the source.

b) If $R = 0$, and the source is an ideal current source, $I = I_o \cos \omega t$, find the zero-order and first-order fields inside the structure. Find a lumped equivalent circuit as seen by the source.

Problem 25.2-12

A simple linear model of fields in a uniform superconductor, due to F. London, assumes convective current flow of electrons

$$\mathbf{J} = -ne\mathbf{v}$$

where n is the electron density, $-e$ the electron charge, and $\mathbf{v}$ the nonrelativistic continuum velocity. Newton's second Law and the linearized Lorentz force (without damping) yields

$$m_e \frac{\partial \mathbf{v}}{\partial t} = -e\mathbf{E}$$

where m_e is the effective mass of the electron. When combined with Faraday's Law, the result is

$$\nabla \times \mathbf{E} = \frac{\partial}{\partial t} \nabla \times \left(\frac{m_e}{-e} \mathbf{v} \right) = -\frac{\partial \mathbf{B}}{\partial t}$$

After integrating with respect to time, and setting the integration constant to zero, the result is

$$\nabla \times \left(\frac{m_e}{ne^2} \mathbf{J} \right) = -\mathbf{B}$$

This equation (known as the London Equation) predicts the Meissner Effect ($\mathbf{B} = 0$ when $\mathbf{J} = 0$).

a) Show that the London Equation also follows from the assumption that the total momentum associated with each electron, $m_e\mathbf{v} - e\mathbf{A}$, vanishes.

b) Combine the London Equation with the magnetoquasistatic Ampere's Law

$$\nabla \times \mathbf{B} = \mu_0 \mathbf{J}$$

to obtain the partial differential equation governing $\mathbf{J}$.

c) Show that the quantity $\delta_L = \sqrt{\frac{m_e}{ne^2 \mu_0}}$ plays a role that is similar to the skin depth in a normal conductor.

d) Assume m_e is the free electron mass and numerically evaluate the London penetration depth for $n = 10^{30} m^{-3}$.

e) Find expressions for the Maxwell–Poynting and Alternate-energy densities that apply within a London superconductor.

Problem 25.2-13

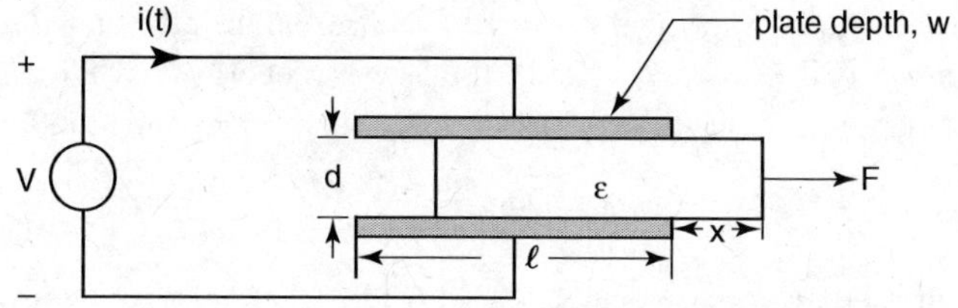

An ideal constant voltage source, V is connected to a parallel-plate capacitor in which a moveable dielectric slab of linear isotropic permittivity ε is inserted as shown in the figure. The length and depth of the plates are ℓ and w; the plate separation is $d \ll \ell, w$. The position x is initially zero. A mechanical force F acts upon the slab and *slowly* withdraws it halfway from between the plates ($x = \ell/2$). Neglecting kinetic energy as well as electrical and mechanical dissipation, conservation of quasistatic energy demands that

$$Vi\ dt = dW_{\text{electric}} + Fdx$$

where i is the first-order current associated with the *EQS* fields.

a) Prove that the force (due to the fringing electric field) is given by

$$F = -\left(\frac{\partial W_{\text{electric}}}{\partial x}\right)_{d\text{ charge} = 0}$$

where W_{electric} is the total zero-order energy that *neglects* fringing. In terms of the geometry:

b) Find the value of F required to initiate movement of the dielectric slab.

c) Find the value of F required to maintain the position $x = \ell/2$.

d) Find the mechanical work done by the force.

e) Find the electrical energy supplied (or absorbed) by the voltage source.

Problem 25.2-14

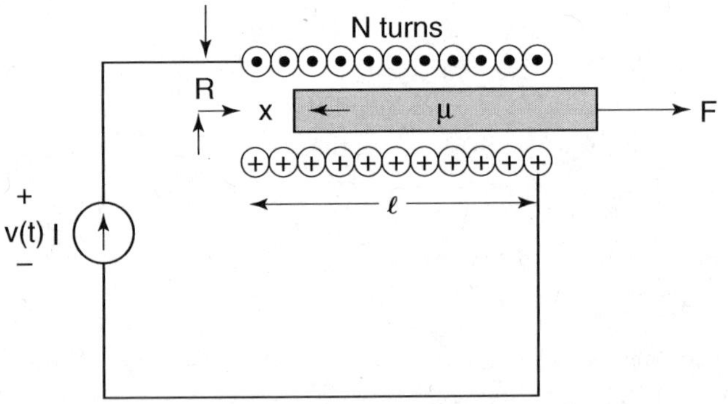

An ideal constant current source, I, is connected to a solenoid inductor in which a moveable magnetic core of linear isotropic permeability μ is inserted as shown in the figure. The radius of the N circular turns is R; the length of the coil is $\ell \gg R$. The position x is initially zero. A mechanical force F acts upon the core and *slowly* withdraws it halfway from the coil ($x = \ell/2$). Neglecting kinetic energy as well as electrical and mechanical dissipation, conservation of quasistatic energy demands that

$$I v\ dt = dW_{\text{magnetic}} + Fdx$$

where v is the first-order voltage associated with the *MQS* fields.

a) Prove that the force (due to the fringing magnetic field) is given by

$$F = -\left(\frac{\partial W_{\text{magnetic}}}{\partial x}\right)_{d\,\text{flux}\,=\,0}$$

where W_{magnetic} is the total zero-order energy that *neglects* fringing.

In terms of the geometry:

b) Find the value of F required to initiate movement of the magnetic core.

c) Find the value of F required to maintain the position $x = \ell/2$.

d) Find the total mechanical work done by the force.

e) Find the electrical energy supplied (or absorbed) by the current source.

Problem 25.2-15

Show that, in the Lorenz gauge, the Alternate-power theorem can be split into two parts,

$$\nabla \cdot (\Phi \mathbf{J} + \mathbf{S}^{\text{o}}_{\Phi}) + \frac{\partial}{\partial t}\left(\frac{1}{2}\rho\Phi + W^{\text{o}}_{\Phi}\right) = \nabla\Phi \cdot \mathbf{J}$$

$$\nabla \cdot \mathbf{S}^{\text{o}}_{A} + \frac{\partial}{\partial t}\left(\frac{1}{2}\mathbf{A}\cdot\mathbf{J} + W^{\text{o}}_{A}\right) = \frac{\partial \mathbf{A}}{\partial t} \cdot \mathbf{J}$$

a) Find $\mathbf{S}^{\text{o}}_{\Phi}$, W^{o}_{Φ} and $\mathbf{S}^{\text{o}}_{A}$, W^{o}_{A}.

b) Compare these two theorems to the approximate quasistatic *EQS* and *MQS* power theorems derived in Chapter 3.

Show that, in the Lorenz gauge, the Alternate-stress theorem can be split into two parts ($\bar{\mathbf{I}}$ is the identity tensor),

$$\nabla \cdot \left(-\frac{1}{2}\rho\Phi\,\bar{\mathbf{I}} + \overline{\mathbf{T}}^{\text{o}}_{\Phi}\right) - \frac{\partial}{\partial t}\mathbf{G}^{\text{o}}_{\Phi} = -\rho\nabla\Phi$$

$$\nabla \cdot \left(-\mathbf{A}\,\mathbf{J} + \frac{1}{2}\mathbf{A}\cdot\mathbf{J}\,\bar{\mathbf{I}} + \overline{\mathbf{T}}^{\text{o}}_{A}\right) - \frac{\partial}{\partial t}(\rho\mathbf{A} + \mathbf{G}^{\text{o}}_{A}) = -\rho\frac{\partial \mathbf{A}}{\partial t} + \mathbf{J}\times(\nabla\times\mathbf{A})$$

c) Find $\overline{\mathbf{T}}^{\text{o}}_{\Phi}$, $\mathbf{G}^{\text{o}}_{\Phi}$, and $\overline{\mathbf{T}}^{\text{o}}_{A}$, $\mathbf{G}^{\text{o}}_{A}$. These form one basis for quasistatic *EQS* and *MQS* Lorentz-force theorems.

Show that similar splits of the Alternate-power and Alternate-stress theorems can be made in the Coulomb gauge.

d) Repeat parts (**a**), (**b**), and (**c**).

25.3 PLANE WAVES

Problem 25.3-1

In free-space, an elliptically polarized uniform plane wave is propagating in the z direction with frequency ω. The tip of the **E** field traces out an ellipse with major and minor axes: E_{max} and E_{min}.

a) Find both the time-dependent and complex free-space electric and magnetic fields.

b) Evaluate the Poynting vector ($\mathbf{E} \times \mathbf{H}$) and the associated electric and magnetic energy densities, W_e and W_m.

c) Assume that $\Phi = 0$ and evaluate the Alternate power-flux vector ($\mathbf{S}^o$) and the associated energy density, W^o.

In the absence of an electric scalar potential, is "electric energy" contained in W^o? Explain.

d) Compare parts (b) and (c). Specifically consider the cases of linear and circular polarization.

Problem 25.3-2

An electromagnetic uniform plane wave of frequency ω propagates through an electrically neutral cold plasma existing in free-space. The plasma consists of equal numbers of free electrons and positive ions and, in a linear model, is characterized by

$$\rho = 0$$

$$\mathbf{J} = -n_o e \mathbf{v}$$

$$m \frac{\partial \mathbf{v}}{\partial t} = -e\mathbf{E}$$

where n_o is the number density (assumed uniform) of both the mobile electrons (of charge $-e$ and mass m) and the heavy positive ions whose motion is neglected. Because the velocity, $\mathbf{v}$, is first order in the electric and magnetic field strengths, the Lorentz force due to the magnetic field is absent in the linearized equation of motion.

a) Show that in the sinusoidal steady state, the plasma can be characterized as a dielectric where $\underline{\mathbf{D}} = \varepsilon(\omega)\underline{\mathbf{E}}$. Find $\varepsilon(\omega)$ in terms of the physical parameters and show that at a particular frequency defined as the plasma frequency $\varepsilon(\omega_p) = 0$.

The electric field in the plasma is known to be

$$\mathbf{E} = \widehat{\mathbf{x}}\mathbf{E}_o \cos(\omega t - k_z z)$$

b) Find k_z and the magnetic field, **H**.

c) Find the vector and scalar potentials, **A** and Φ.

d) Find the Poynting-power flux, **S**, and the associated-energy density, W^{em}.

e) Find the Alternate-power flux, $\mathbf{S}^o$, and the Alternate-energy density, W^o.

Problem 25.3-3

Consider a thin planar sheet of linear material characterized by μ_o, ε_o, finite conductivity σ, and thickness, Δ. The sheet (surrounded by free-space) is of infinite extent along x and y and located at $z = z_o(t)$. A uniform plane wave of frequency ω traveling in the $+z$ direction is incident upon the moving sheet. The wave is linearly polarized with the **E** field parallel to the x axis; the amplitude of the incident wave is E_o. Assume that the current density inside the sheet is uniform

a) For $z_o = vt$, find the steady-state **E** and **H** fields for all z. Consider both positive and negative values of v.

b) Is there a reflected wave when $v = +c$? A transmitted wave? Explain.

c) Find the potentials **A** and Φ (Lorenz gauge) for all z, assuming that the former has a single component (parallel to the surface current induced on the moving sheet).

d) Find both the Poynting- and Alternate-power fluxes and energy densities.

e) Evaluate $\mathbf{E} \cdot \mathbf{J}$ (as a function of z and t) from both the Poynting- and Alternate-power theorems.

f) Find the mechanical pressure that must be exerted on the plate so as to maintain the velocity.

g) Repeat (a)–(f) when $\sigma = \infty$.

Problem 25.3-4

Consider a linearly polarized uniform plane wave propagating in the $+z$ direction in free-space when the scalar-potential, $\Phi = 0$.

a) Find and compare $\mathbf{S} = \mathbf{E} \times \mathbf{H}$ with the Alternate flux, $\mathbf{S}^o$.

b) Compare the Maxwell–Poynting energy densities, W_e and W_m with the Alternate densities, W_e^o and W_m^o.

Problem 25.3-5

A uniform traveling plane wave of frequency ω propagates in a region of free-space in the $+z$ direction. The Poynting vector, $\mathbf{S} = \mathbf{E} \times \mathbf{H}$, is given by

$$\mathbf{S}(x, y, z, t) = \widehat{\mathbf{z}} S_o$$

where S_o is a positive constant.

For $z = 0$ and $t = 0$, we obtain

$$\mathbf{E}(x, y, 0, 0) = \widehat{\mathbf{x}} E_o$$

where E_o is a positive constant.

Find $\mathbf{E}(r, t)$ and $\mathbf{H}(r, t)$ in terms of S_o

HINT: The wave may be the superposition of two or more linearly polarized waves.

Problem 25.3-6

In the Lorenz gauge, the free-space vector potential is known to be

$$\mathbf{A} = \widehat{\mathbf{y}} A_{\mathrm{o}} \sin\left[\omega_1\left(t - \frac{|z|}{c}\right) + \frac{1}{2}\frac{(\omega_2 - \omega_1)}{T}\left(t - \frac{|z|}{c}\right)^2\right] u_{-1}\left(t - \frac{|z|}{c}\right) u_{-1}\left(T - t + \frac{|z|}{c}\right)$$

where A_{o}, ω_1, ω_2, and T are constants; $(\omega_1 + \omega_2)T = n\pi$ and n is an integer.

a) Find the scalar potential, Φ.

b) Find the electric currents and/or electric charges that produce these potentials.

c) Find and compare $\mathbf{S} = \mathbf{E} \times \mathbf{H}$ with the Alternate flux, $\mathbf{S}^{\mathrm{o}}$.

d) Compare the Maxwell–Poynting energy densities, W_{e} and W_{m}, with the Alternate densities $W_{\mathrm{e}}^{\mathrm{o}}$ and $W_{\mathrm{m}}^{\mathrm{o}}$.

e) When $\omega_2 \gg \omega_1$and $n \gg 1$, interpret the results in terms of photon energy, number density, and group velocity.

Problem 25.3-7

A very thin uniform conducting disk of radius R, thickness d, conductivity σ, and mass m is located on the plane $z = 0$ and surrounded by free-space. The disk is free to rotate about a frictionless pivot located at its center (but restrained from moving along z). A uniform plane wave of circular polarization and frequency ω is normally incident upon the disk. The incident electric field is

$$\mathbf{E} = E_{\mathrm{o}}[\widehat{\mathbf{x}} \cos(\omega t - kz) + \widehat{\mathbf{y}} \sin(\omega t - kz)]$$

where $kR \gg 1$. The dielectric and magnetic properties of the disk are the same as free-space.

a) Find the reflected and transmitted waves (both electric and magnetic fields).

b) Find the effective surface current density ($\mathbf{J}\,d$) and electric line-charge density (at the edge of the disk).

c) Find the pressure exerted on the disk by evaluating the z component of the Lorentz force. What value of σd maximizes the pressure?

d) Find the net torque (if any) on the disk by evaluating the $\mathbf{r} \times \mathbf{f}$ Lorentz torque. Find the angular acceleration. What value of σd maximizes the angular acceleration?

e) From the Maxwell stress tensor, evaluate $\widehat{\mathbf{z}} \cdot (\overline{\overline{\mathbf{T}}}^{\mathrm{em}} \cdot \widehat{\mathbf{z}})$ and $\widehat{\mathbf{z}} \cdot [\mathbf{r} \times (\overline{\overline{\mathbf{T}}}^{\mathrm{em}} \cdot \widehat{\mathbf{z}})]$ on both sides of the disk and integrate the differences over the disk area to confirm the results of parts (c) and (d).

f) Repeat part (e) using the Alternate stress-tensor $\overline{\overline{\mathbf{T}}}^{\mathrm{o}}$ and confirm the results of part (c), but *not* part (d).

g) Integrate the localized torque density $\mathbf{J} \times \mathbf{A}$ over the volume of the disk and reconcile the apparent discrepancy of part (f). Demonstrate that the result does not require the evaluation of $\widehat{\mathbf{z}} \cdot [\mathbf{r} \times (\overline{\mathbf{T}}^{\text{o}} \cdot \widehat{\mathbf{z}})]$.

h) Evaluate separately the spin and orbital angular-momenta in the fields for both $z < 0$ and $z > 0$.

HINT: Use the results of Appendix D, Section D.3.

Problem 25.3-8

An elliptically polarized uniform plane wave of frequency ω is propagating in free-space. The macroscopic electric field in the Lorenz gauge is known to be

$$\mathbf{E} = \widehat{\mathbf{x}} E_1 \cos(\omega t - kz) + \widehat{\mathbf{y}} E_2 \sin(\omega t - kz + \psi)]$$

and arises from a large number of photons (each with energy $\hbar\omega$).

In parts (a)–(c), the phase angle $\psi = 0$.

a) Find the number density n_p of the photons.

b) Find spin angular-momentum density, $\mathbf{L}^{\text{spin}}$ of the wave and the *average* value per photon in terms of the ratio E_1/E_2. What is the range of possible values for that average?

c) Repeat part (b) for the orbital angular-momentum density, $\mathbf{L}^{\text{orbit}}$.

In general, $n_p = n_+ + n_-$ and $n_s = n_+ - n_-$, where $n_\pm$ are the densities of photons with spin ± 1 (in units of $\hbar$) and the net spin angular-momentum density is $L_z^{\text{spin}} = n_s \hbar$.

d) Use the Alternate representation in the Lorenz gauge to find n_+ and n_- and the ratio n_+/n_- in terms of E_1, E_2, and ψ.

e) Evaluate the ratio n_+/n_- for the special cases: $E_1 \neq E_2$, $\psi = 0$ and $E_1 = E_2$, $\psi \neq 0$.

HINT: Use the results of Appendix D, Section D.3.

Problem 25.3-9

A layer of lossless dielectric material (characterized by permittivity ε, permeability μ_{o}, and thickness d) is located between the planes $z = 0$ and $z = d$. Free-space exists on either side. A linearly polarized uniform plane wave of frequency ω is obliquely incident upon the surface $z = 0$ at the angle θ (measured from the normal). When $\theta = 60^{\text{o}}$, there is no reflected wave regardless of the frequency.

a) Find the value of $\varepsilon/\varepsilon_{\text{o}}$ and the wave polarization (*TE* or *TM*).

b) For the same value of ε, find all conditions (values of θ and ω) for which the reflected wave is zero.

c) Repeat part (b) for a wave with the opposite polarization.

The configuration is altered. Now the front surfaces of layers of lossless dielectric/magnetic material (characterized by permittivity ε, permeability μ, and thickness d) are located on the *infinite* set of planes $z = 0, 2d, 4d, \ldots$, etc. Free-space exists between the layers and for all $z < 0$. A linearly polarized uniform plane wave of frequency ω is normally incident ($\theta = 0$) upon the surface $z = 0$.

d) For $\varepsilon = 4\varepsilon_0$, $\mu = \mu_0$, find as a function of ω, the reflected wave present in the half-space, $z < 0$.

e) Under what conditions (if any) is the reflection zero? The reflection total?

f) Repeat parts (d) and (e) when $\varepsilon = \varepsilon_0$, $\mu = 4\mu_0$.

g) Repeat parts (d) and (e) when $\varepsilon = 2\varepsilon_0$, $\mu = 2\mu_0$.

HINT: Use equivalent transmission-line circuits for *TE*, *TM*, and *TEM* waves. Make use of the fact that the second configuration is periodic.

25.4 RADIATION AND DIFFRACTION

Problem 25.4-1

A Hertzian electric dipole of effective length d is placed at the center of a perfectly conducting spherical cavity of radius R, where $kR \gg 1$. The interior of the cavity is a free-space region.

a) Find the complex electric and magnetic fields inside of the cavity for $r \gg d$.

For $r < R$, but $kr \gg 1$:

b) Evaluate the time-dependent Poynting vector $(\mathbf{E} \times \mathbf{H})$ and the associated electric and magnetic energy densities.

c) Evaluate the Alternate-power flux vector $(\mathbf{S}^o)$ and the associated energy density, W^o.

d) Compare parts (b) and (c).

Problem 25.4-2

A Hertzian magnetic-dipole antenna is fed from the gap in the thin wire that forms a small circular loop of radius R; the frequency is such that $kR \ll 1$.

a) Use reasonable approximations to develop an equivalent receiving circuit that is analogous to that derived for a Hertzian electric-dipole antenna in Section 24.1.

b) Derive the effective capture area, A_c, of a small lossless circular loop of radius R and compare it to that of a short dipole of length ℓ ($k\ell \ll 1$). The capture area is defined by $\langle S \rangle A_c = \langle P_{max} \rangle$ when the dipole is oriented to optimally intercept a uniform plane wave of time-averaged power flux $\langle S \rangle$ and the load impedance is adjusted to maximize the average power that is received.

Problem 25.4-3

The loop described in Problem 25.2-3 is excited by a sinusoidal current of frequency ω. Assume that the radius is small enough that $kR \ll 1$; therefore the current in the loop is uniform. Including the phase retardation factor, the complex flux linking the loop is

$$\underline{Flux} = \frac{\mu_0 \underline{I}}{4\pi} \oint_{C_p} \oint_{C_q} \frac{\exp(-jkr_{qp})}{r_{qp}} \mathbf{ds}_q \cdot \mathbf{ds}_p$$

$$r_{qp} = \sqrt{R^2 + (R - r_0)^2 - 2R(R - r_0)\cos(\phi_p - \phi_q)}$$

a) If the complex voltage across a small gap in the loop is $\underline{V} = j\omega \underline{Flux}$, find the real part of the complex input impedance and verify that it is identical to the radiation resistance of the small loop as calculated from consideration of the far-field.

b) How is the imaginary part of the impedance modified from the low-frequency limit, $j\omega L$?

Problem 25.4-4

Consider a thin-wire antenna of effective electrical length, $kd = .05$, and wire radius, $r_w = .01d$. Assume that there are no copper losses in the dipoles.

a) Find what passive load impedance placed across the terminals of the Hertzian dipole will extract the maximum power from a uniform plane wave that is incident upon it.

b) If the load is a pure resistance, R, what value will cause the *maximum* power to be scattered from the dipole? How does the maximum value compare with the maximum load power of part (a)?

c) What value of R will cause the *minimum* power to be scattered?

d) Repeat parts (a)–(c) when the antenna is a half-wave dipole.

Problem 25.4-5

A small wire loop of diameter d carries a uniform circulating current $I_0 \cos \omega t$. A short wire antenna of length ℓ is located at the center of and oriented perpendicular to the plane of the loop; it carries an identical uniform current. The frequency is such that both kd, $k\ell \ll 1$. The antennas are surrounded by free-space; *both* are excited at the same time.

a) Find the electric and magnetic far fields.

b) Find the vector and scalar potentials in the far-field region.

c) Find the total radiated power.

d) If the antenna terminals (small gaps) are connected in series with each other and both wires are perfectly conducting, calculate the real part of the input impedance.

e) What ratio ℓ/d will result in the loop-delivering half of that power?

Problem 25.4-6

Consider a thin-wire dipole antenna of length ℓ that is surrounded by free-space and is vertically mounted above an infinite ground plane as shown.

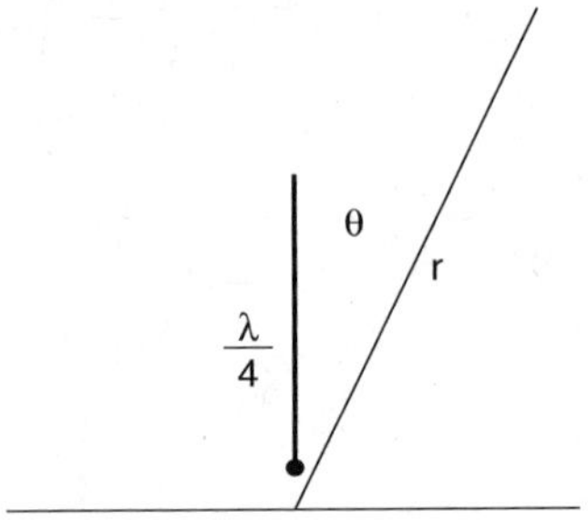

a) Calculate the directivity (lossless gain) and radiation resistance when $\ell = \lambda_0/4$.

b) Qualitatively, what is the effect of making the ground plane a circular disk of finite radius R when the antenna is located at the center. Estimate how large R must be before the part (a) values are reasonable.

Problem 25.4-7

The complex vector potential, generated by sinusoidal currents of frequency ω, is assumed known in the source free region(s) of free-space. In the Lorenz gauge, do the following:

a) Prove that the complex Alternate-power flux can be expressed as

$$\underline{\mathbf{S}}^{\text{o}} = \frac{-j\omega}{4\mu_0}\left\{\left[\sum_{i=1}^{3} \underline{A}_i \nabla \underline{A}_i^* - \frac{1}{k^2}(\nabla\cdot\underline{\mathbf{A}})\nabla(\nabla\cdot\underline{\mathbf{A}}^*)\right] - c.c.\right\}$$

where $c.c.$ is the complex conjugate.

b) If there is only one Cartesian component of $\underline{\mathbf{A}}$ (which we choose to be $\underline{A}_z$), show that the far-field value of $\underline{\mathbf{S}}^{\text{o}}$ reduces to

$$\underline{\mathbf{S}}^{\text{o}} \to \frac{-j\omega}{4\mu_0}[\underline{A}_z \nabla \underline{A}_z^* - \underline{A}_z^* \nabla \underline{A}_z]\sin^2\theta$$

where $\cos\theta = z/r$.

Problem 25.4-8

A computer company is designing a new model to replace their current version; they plan to substitute computer chips with a clock rate three times faster than those now in use. Although they would like to continue to use the same metal case as before, they realize that electromagnetic radiation leaking out of the small holes used to provide ventilation may increase to unacceptable levels because of the higher frequencies. If they decrease the hole size, the ventilation suffers unless they also increase the number of holes, but that may also increase the radiation. The company's package designer proposes a solution. First, scale the hole radius r so that r/λ is kept constant; second, increase the number of holes N so that the total open area $N\pi r^2$ is at the *minimum* kept constant. (More area will probably be required because of increased power dissipation in the faster chips.) In the present design, all of the holes make up a square array uniformly spaced in both the x and y directions—defined along the sides, $L \ll \lambda$, of the square.

You have been brought in as a consultant to examine this solution and make recommendations.

a) First, consider a single hole and a simple model that replaces the box with an infinite ground plane to estimate the level of power that radiates as a function of frequency and hole radius.

b) Second, consider a square array of N identical holes (excited equally); estimate the total radiated power when $L \ll \lambda$.

c) What do you think of the proposed new design? Can you improve upon it?

Problem 25.4-9

A uniform plane wave of frequency ω traveling in the $+z$ direction is normally incident upon a large perfectly conducting screen located on the plane $z = 0$. The thin screen contains a square hole with edges L parallel to the x and y axes; The wave is linearly polarized with the electric field (of amplitude E_o) parallel to the x axis and $kL \gg 1$.

a) Within the Fresnel region, find the average power flux on the $+z$ axis.

b) Find the on axis electric and magnetic fields in the far-field region.

c) Calculate the maximum directivity of the aperture.

HINT: It is easier to calculate the Alternate power.

Problem 25.4-10

The array power factor, $F(\theta) = \underline{s}(\underline{z})\underline{s}^*(\underline{z})$, of a linear array with N elements uniformly spaced (by $\lambda_o/2$) along the z axis, can be synthesized using Chebyshev polynomials of the first kind (see Appendix F, Section F.2). In terms of $\underline{z} = \exp(j\pi\cos\theta)$,

$$\underline{s}(\underline{z}) = \tfrac{1}{2}\sum_{n=0}^{N} a_n \underline{z}^n$$

where a_n are the real array coefficients. These can be relabeled $a'_{\pm 1}, a'_{\pm 2}, \ldots, a'_{\pm(N+1)/2}$ if N is odd or $2a'_0, a'_{\pm 1}, a'_{\pm 2}, \cdots, a'_{\pm N/2}$ if N is even. If the array coefficients have even symmetry with respect to the center of the array, $a'_n = a'_{-n}$. Then $\underline{s}(\underline{z})$ is real and, in terms of $z = Re\underline{z}$, equal to

$$s(z) = \begin{cases} \sum_{n=1}^{(N+1)/2} a_n' \cos(n \cos^{-1} z), & N \text{ odd} \\ \sum_{n=0}^{N/2} a_n' \cos(n \cos^{-1} z), & N \text{ even} \end{cases}$$

$$z = \cos(\pi \cos\theta)$$

Although a complete set of functions allows *any* $s(z)$ to be synthesized, this expansion in terms of the partial set of orthogonal Chebyshev polynomials will suffice if N is large enough. Instead of dealing with an arbitrary $s(z)$, Dolph chose a particular form of array function that is analytic, is flexible in that either beam-width or side-lobe levels can be specified, and can be expanded *without error* in terms of either $N' = (N+1)/2$ or $N/2 + 1$ values of a_n'. The desired form of $s(z)$ is the continuous function

$$s(z) = T_{N'}(\alpha z) = \begin{cases} \cos(N' \cos^{-1} \alpha z) & (|z| \le 1/\alpha) \\ \cosh(N' \cosh^{-1} \alpha z) & (|z| > 1/\alpha) \end{cases}$$

All of the pattern nulls are confined to the range $|z| < 1/\alpha$ and all side lobes have the same peak value of unity. The peak of the main beam occurs at $z = 1$ as expected.

a) Determine the beam width of $T_{N'}(\alpha z)$ (defined between the set of pattern nulls closest to $\theta = \pi/2$) as a function of N' and α.

b) Determine the side-lobe level of $T_{N'}(\alpha z)$ (the ratio between subsidiary peak and the main beam maxima) as a function of N' and α.

c) For $N = 5$ ($N' = 3$) and a desired side-lobe level of .05 (−26 dB), find the beam-width.

d) Compare the part (c) value with those of a uniform array (all $a_n = 1$) of the same size.

e) Find the a_n' coefficients (given $a_1' = 1$).

f) For a desired beam width of 5° and side-lobe level of −20 dB, what value of N' is required?

g) Design an $N = 5$ array that has all equal lobes (no main beam).

Problem 25.4-11

In the geometrical-optics limit, the complex vector potential can be expressed as

$$\underline{\mathbf{A}} = \underline{\mathbf{A}}_o(x, y, z) \exp[-j\phi(x, y, z)]$$

where, because the wavelength is very small, the phase variation is normally large compared to the amplitude variations (focal regions are excepted). In such cases, $\frac{\partial \phi}{\partial x_i}$ serves as k_i and the dispersion relation is

$$\left(\frac{\partial \phi}{\partial x}\right)^2 + \left(\frac{\partial \phi}{\partial x}\right)^2 + \left(\frac{\partial \phi}{\partial x}\right)^2 = \omega^2 \mu \varepsilon$$

This "eikonal" equation is still valid when the linear isotropic permittivity and/or permeability is nonuniform, provided that the variations are slowly varying compared to the wavelength. Such a medium can serve as a lens; an important example is the Luneburg lens that we here consider. It consists of a sphere of radius R centered at the origin of Cartesian coordinates and is made of linear material with $\mu = \mu_o$ and $\varepsilon = \varepsilon_o(2 - \frac{x^2+y^2+z^2}{R^2})$. In practice, the lens is approximated by a series of concentric dielectric shells, each with a constant dielectric constant that together provide a step approximation to the continuous function. The radius is large compared to the free-space wavelength ($\omega\sqrt{\mu_o \varepsilon_o} R \gg 1$).

A uniform plane wave propagating in the $+z$ direction has a frequency ω and a constant phase of zero at the plane $z = -R$. As discussed in Part I, Chapter 5, Section 5.3, the wave can be described as particles (photons) subject to Hamilton's Equations.

a) Show that the differential of the "wavevector" $\mathbf{k}$ is related to the differential phase by $d\mathbf{k} = \frac{1}{2}\frac{\nabla\varepsilon}{\varepsilon} d\phi$.

b) Find the on-axis phase for $-R \leq z \leq R$.

c) Find the phase everywhere on the surface of the sphere at $z = 0$.

d) Show that the values of $\mathbf{k}$ located at the part (c) positions are bent by $\nabla\varepsilon$ and follow the surface of the sphere.

e) Show that the phases of the part (b) and part (d) rays arrive at the point $x = y = 0$, $z = R$ with the same phase.

f) Show that *all* other rays that intercept and are bent by the sphere arrive at the same (focal) point with the same phase found in part (e).

g) Find the trajectory of a ray that lies in the x–z plane, starts at $x = y = 0$, $z = -R$ with an initial angle θ_o with respect to the $+z$ direction and ends at $z = +R$. Check the result for the values $\theta_o = 0$ and $\pi/2$.

h) Describe what happens if the incident wave is tilted by an angle θ with respect to the z axis.

The lens can be used either to transmit or receive a plane wave. The figure depicts the rays when the feed is located at $z = -R$.

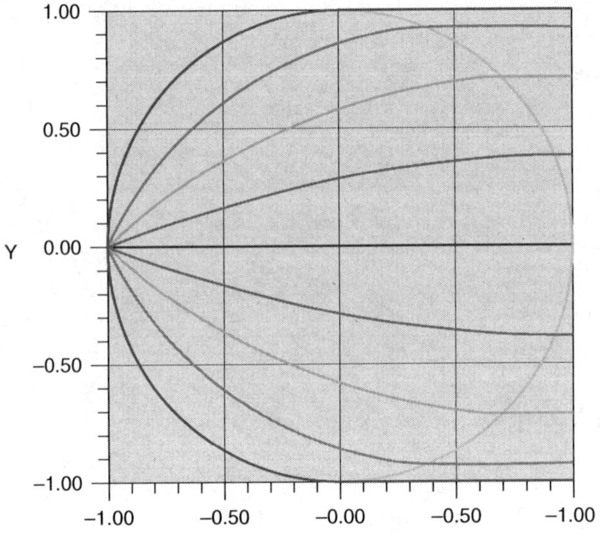

Luneburg lens

Problem 25.4-12

The results of the preceding problem show that when the linear permittivity and permeability are uniform, the wave vectors are not bent, but travel in straight lines until they reach a discontinuity created by an interface or boundary that causes reflection and refraction (as governed by the Law of Reflection and Snell's Law). Ray tracing in the geometrical optics limit is therefore a simpler alternative to more exact solutions and can serve to analyze (and design) a variety of antennas and lenses.

a) Use geometrical optics to prove that a perfectly conducting parabolic dish [formed by truncating a paraboloid, $z = C(x^2 + y^2)$] will focus a plane wave (with constant phase across its aperture), at the focal point.

b) Show that a portion of a perfectly conducting sphere of radius R can approximate a paraboloid provided the area of the aperture is not too large compared to πR^2. Calculate the position of the approximate focal point.

Problem 25.4-13

The antenna of a radar system has a maximum directivity (gain) G_o. The signal transmitted is a waveform that is a rectangular pulse of duration τ modulated by a sine wave of frequency ω; the peak power is P_{peak}. After transmission, the same antenna is connected to a receiver with bandwidth $\Delta f = 1/\tau$ and noise power $kT\Delta f$. The radar target is a large, perfectly conducting sphere of radius R, where $kR \gg 1$.

a) Use the induction theorem to show that the scattering cross section of the sphere is πR^2.

b) Assuming that the antenna is oriented properly, find (as a function of the peak power P_{peak}) the maximum range (the distance between antenna and sphere) that produces a signal-to-noise ratio of unity in the receiver. The result is the radar-range equation.

c) What is the minimum range that can be determined?

Problem 25.4-14

In free-space, two charges, $\pm q$, are symmetrically located at $z = \pm\frac{1}{2}d(t)$; together they form an electric-dipole moment, $qd(t)$, oriented in the $+z$ direction. The velocities, $v = \pm\frac{d}{dt}d(t)$, are nonrelativistic.

a) Use the approximations developed in Appendix E, Eqs. (E.9a) and (E.9b), to calculate the potentials of this dipole.

b) Show that in the far-field region,

$$S_r^{o} = \frac{1}{2}\frac{\mu_0 q^2}{(4\pi r)^2}\left(v^* \frac{\partial^2 v^*}{\partial r \partial t} - \frac{\partial v^*}{\partial r}\frac{\partial v^*}{\partial t}\right)\sin^2\theta$$

$$W^{o} = \frac{-1}{2}\frac{\mu_0 q^2}{(4\pi c r)^2}\left[v^* \frac{\partial^2 \mathrm{v}^*}{\partial t^2} - \left(\frac{\partial v^*}{\partial t}\right)^2\right]\sin^2\theta$$

where v^* is the velocity evaluated at the retarded time, $t^* = t - r/c$.

c) For $d(t) = d_0 \cos(\omega t)$, show that the result is the same as that calculated for a Hertzian dipole and that nonrelativistic velocities imply $kd \ll 1$.

Problem 25.4-15

a) Prove that, in the Lorenz gauge,

$$\nabla\cdot\left\{\mathbf{S}^{o} - \mathbf{E}\times\mathbf{H} + \frac{1}{2}\frac{\partial}{\partial t}\left[\varepsilon_0\left(\Phi\mathbf{E} + \frac{\partial\Phi}{\partial t}\mathbf{A} - \Phi\frac{\partial\mathbf{A}}{\partial t}\right) - \mathbf{A}\times\mathbf{H}\right]\right\} = 0$$

where $\mathbf{S}^{o}$ is the Alternate-power flux given by Eq. (3.31a).

b) Show that the divergence-free vector of part (a) can be expressed as

$$\mathbf{S}^{\text{beauty}} = \frac{1}{2\mu_0}\nabla\times\left(\mathbf{A}\times\frac{\partial\mathbf{A}}{\partial t}\right)$$

Problem 25.4-16

Within the paraxial approximation (see Chapter 22, Section 22.10), a Gaussian beam of frequency ω propagating in free-space (with focal plane $z = 0$) is elliptically polarized such that

$$\underline{\mathbf{E}} = (\widehat{\mathbf{x}}E_a - j\widehat{\mathbf{y}}E_b)\frac{\exp\left[\dfrac{-(r/R)^2}{1+(2z/kR^2)^2}\right]}{\sqrt{1+(2z/kR^2)^2}}\exp\left[-j\left(kz + \frac{2kz(r/R)^2}{(2z/R)^2+(kR)^2} - \tan^{-1}\left(\frac{2z}{kR^2}\right)\right)\right]$$

where E_a, E_b, and R are constants and $k = \omega/c$.

a) Calculate the total average power in the beam.

b) Calculate the total average angular momentum per unit length (along z) in the beam from integrating $\mathbf{r} \times (\mathbf{E} \times \mathbf{H})/c^2$ over the x–y plane. How does the result depend upon the state of polarization? Contrast linear polarization with left- and right-hand circular polarization.

c) When $\underline{\Phi} = 0$, separately calculate the spin and orbital contributions [defined in Appendix D, Section D.3, Eqs. (D.13a) and (D.13b)] by integrating them individually. Compare the sum with the result of part (b) and show that there is a discrepancy.

d) Show that the discrepancy in part (c) is due to the paraxial approximation which leads to $\nabla \cdot \mathbf{E} \neq 0$ in violation of the free-space assumption. But, this implies that electric-charge density, $\rho = \varepsilon_0 \nabla \cdot \mathbf{E}$, is present. Integrate the integral of $\mathbf{r} \times \rho\mathbf{A}$ over the x–y plane and show that the discrepancy has been resolved.

e) From parts (a) and (b), calculate the ratio of the total average angular-momentum per unit length to the total average power in the beam. How is it possible that the result is independent of R when in the limit $R \to \infty$ the beam becomes an unconfined plane wave and $\mathbf{z} \cdot [\mathbf{r} \times (\mathbf{E} \times \mathbf{H})/c^2] = 0$?

f) Evaluate the time-dependent value of $\mathbf{S}^{\text{beauty}}$ (defined in the previous problem).

Problem 25.4-17

A rectangular aperture of width $2d$ is located in the $z = 0$ plane. Between $-d < x < +d$ there exists an equivalent surface-current distribution,

$$\mathbf{K} = \widehat{\mathbf{y}} \sum_{n=0}^{N} K_n \cos\left(\frac{n\pi}{2d}x\right) \cos \omega t$$

where n is an integer and the currents extend over all y.

a) Find the radiation polar pattern of the far field (not the power pattern) in terms of $\sin\theta = x/r$.

b) Show that the pattern nulls are equally spaced along the coordinate $\sin\theta$ and that at each of these locations the amplitude of the pattern depends upon a single value of K_n.

c) Use the results of part (b) to find the values of K_n that synthesize a specified pattern $P(\theta)$ at N sample points. This method, originally due to Woodward, is a simple alternative to finding the Fourier coefficients by integration over the specified pattern. It can easily be extended to apertures with two-dimensional variation.

d) Apply this method to approximate the cosec pattern defined by

$$P(\theta) = \begin{cases} \theta/5 & (0 \leq \theta < 5^{\text{o}}) \\ 1 & (5^{\text{o}} \leq \theta \leq 20^{\text{o}}) \\ \sin 20^{\text{o}}/\sin\theta & (20^{\text{o}} \leq \theta \leq 90^{\text{o}}) \end{cases}$$

with an $N = 12$ Woodward synthesis.

Problem 25.4-18

In this problem, the Hertzian electric-dipole transient analyzed in Chapter 13, Section 13.1, includes linear dissipation at the source. Consequently, the current step is now

$$I(t) = I_0[\sin(\omega t + \alpha) - \sin(\alpha)\exp(-t/\tau)]u_{-1}(t).$$

The line current is assumed uniform along the dipole; therefore, electric charges $\pm Q(t)$ are generated at the dipole ends ($z = \pm d/2$).

a) Find $Q(t)$ and the angle α (in terms of ω and the time constant τ) that satisfies the condition $Q(0) = 0$.

b) Find **A** and Φ generated by the current transient and verify that both potentials vanish on the wave front $r = ct$. If $t \gg \tau$, what range of r corresponds to the sinusoidal steady state?

c) Find the power fluxes in both the Maxwell–Poynting and Alternate representations when $\omega\tau \gg 1$.

Problem 25.4-19

Consider, once again, the Hertzian electric dipole of length d and uniform current, $I_0 \cos\omega t$. The dipole is oriented parallel to the z axis, surrounded by free-space, and $kd \ll 1$. The Lorenz-gauge potentials, **A** and Φ, are known. Consider also a gauge transformation: $\mathbf{A}' = \mathbf{A} + \nabla\Psi$, $\Phi' = \Phi - \partial\Psi/\partial t$ that produces $\nabla \cdot \mathbf{A}' = 0$ (the Coulomb gauge—also known as the radiation gauge). Assume that $\mathbf{A}'$ simultaneously satisfies the Lorenz gauge.

a) Explain why, and where, this "dual" gauge fails. Where it does not fail, and what equation does Ψ satisfy?

b) Find Ψ, such that the *same* dipole fields are generated for $r \gg d$.

c) In both the Lorenz and "dual" gauges, calculate and compare the free-space values of: $\mathbf{S}^o$, W^o, and $\mathbf{L}^{spin} = \varepsilon_0\, \mathbf{E} \times \mathbf{A}$.

Problem 25.4-20

A uniform planar surface-current of infinite extent is located at $z = 0$ and surrounded by free-space. The x-directed current is of the form $K_x = K_s(t)\, u_{-1}(t)$.

a) Prove that in Cartesian coordinates, the vector potential can be evaluated from the integral

$$A_x(z,t) = \frac{c\mu_0}{2} \int_{|z|/c}^{t} K_s(t - t')\, dt'$$

b) What is the physical significance of t'?

c) Find $A_x(z,t)$ and the electric and magnetic fields when $K_s(t) = K_0$.

d) Repeat part (c) when $K_s(t) = K_0 \cos\omega t$.

Problem 25.4-21

A uniform axial line current of infinite extent is located at $r = 0$ and surrounded by free-space. The z-directed current is of the form $I_z = I_s(t)\, u_{-1}(t)$.

a) Prove that in cylindrical coordinates, the vector potential can be evaluated from the integral

$$A_z(r,t) = \frac{\mu_0}{2\pi} \int_r^{ct} \frac{I_s(t - r'/c)}{\sqrt{(r')^2 - r^2}}\, dr'$$

b) What is the physical significance of r'?

c) Find $A_z(r,t)$ and the electric and magnetic fields when $I_s(t) = I_0$.

Problem 25.4-22

Verify that

$$\frac{\partial}{\partial t} \nabla \cdot \frac{1}{2} \left[\mathbf{A} \times \mathbf{H} - \varepsilon_0 \left(\Phi \mathbf{E} + \Phi \frac{\partial \mathbf{A}}{\partial t} - \frac{\partial \Phi}{\partial t} \mathbf{A} \right) \right] = \frac{\partial^2 \beta_{ij}}{\partial x_i \partial x_j}$$

where the Lorenz gauge is assumed, repeated indices are summed, and

$$\beta_{ij} = \frac{1}{\mu_0} \left[E_i A_j - \frac{1}{2} \mathbf{E} \cdot \mathbf{A}\, \delta_{ij} - \frac{1}{2} \left(\Phi \frac{\partial A_j}{\partial x_i} - \frac{\partial \Phi}{\partial x_i} A_j \right) \right]$$

HINT: Use Macsyma/Maxima or an equivalent math program (*4d-em.mac* or *4d-em-wxm.mac* is not required).

25.5 TRANSMISSION LINES

Problem 25.5-1

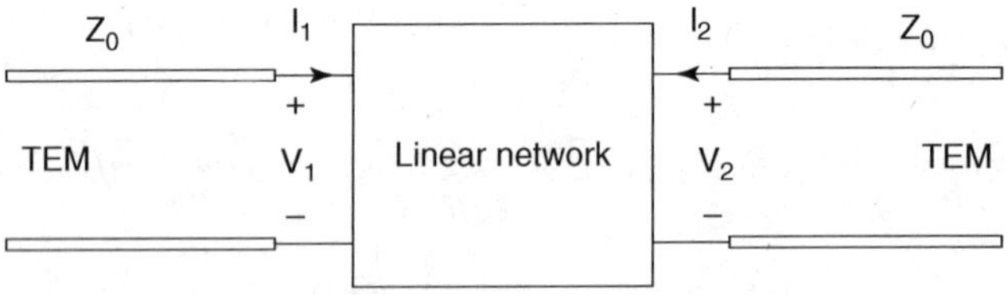

A two-terminal lumped linear passive circuit is characterized in the frequency domain by

$$\begin{bmatrix} V_1 \\ V_2 \end{bmatrix} = \begin{bmatrix} Z_{11} & Z_{12} \\ Z_{21} & Z_{22} \end{bmatrix} \cdot \begin{bmatrix} I_1 \\ I_2 \end{bmatrix}$$

or alternatively by

$$\begin{bmatrix} V_1 \\ I_1 \end{bmatrix} = \begin{bmatrix} A & -B \\ C & -D \end{bmatrix} \cdot \begin{bmatrix} V_2 \\ I_2 \end{bmatrix}$$

Identical lossless TEM transmission lines of characteristic impedance, Z_o, are connected to the terminals as shown. In the general case of sinusoidal steady-state excitation at frequency, ω, traveling waves of amplitude a_1 and a_2 bring power into the network while reflected waves b_1 and b_2 carry power away from it.

a) Find A, B, C, and D in terms of Z_{11}, Z_{12}, Z_{21}, Z_{22}.

With suitable normalization, the net time-averaged power entering the network through terminal pair k is $\underline{a}_k\underline{a}_k^\star - \underline{b}_k\underline{b}_k^\star$.

b) Find $\underline{a}_k$ and $\underline{b}_k$ in terms of the complex $\underline{V}_k$ and $\underline{I}_k$.

The scattering matrix, $\overline{\overline{\underline{S}}}$, is defined by

$$\begin{bmatrix} \underline{b}_1 \\ \underline{b}_2 \end{bmatrix} = \begin{bmatrix} \underline{S}_{11} & \underline{S}_{12} \\ \underline{S}_{21} & \underline{S}_{22} \end{bmatrix} \cdot \begin{bmatrix} \underline{a}_1 \\ \underline{a}_2 \end{bmatrix}$$

c) Find $\overline{\overline{\underline{S}}}$ in terms of $\overline{\overline{\underline{Z}}}$.

d) What properties does $\overline{\overline{\underline{S}}}$ have when the network is

(i) lossless

(ii) reciprocal

(iii) symmetric

e) For the resistive network shown below and $Z_o = 50\ \Omega$, find A, B, C, D, $\underline{Z}_{11}$, $\underline{Z}_{12}$, $\underline{Z}_{21}$, $\underline{Z}_{22}$, and $\underline{S}_{11}$, $\underline{S}_{12}$, $\underline{S}_{21}$, $\underline{S}_{22}$.

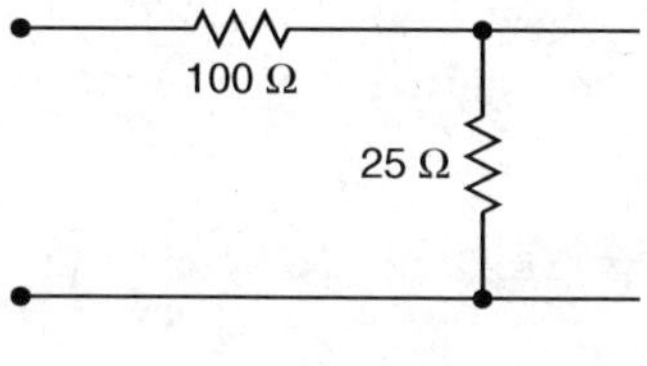

Problem 25.5-2

An infinite cascade of identical two-terminal lumped networks forms a periodic chain. Assume that each network is reciprocal. The voltage across the input to the kth network is defined as V_k, where k is an integer; that across the output of the same network is V_{k+1}.

a) Find the (second-order) difference equation that governs V_k.

b) Assume a solution of the form $V_k = C\alpha^k$ and verify that there are two independent values of α. By superposition, the general solution is of the form $V_k = C_1\alpha_1^k + C_2\alpha_2^k$. Find α_1 and α_2.

c) Find the associated current, I_k.

d) If a finite cascade of N identical networks is terminated by a shunt impedance Z_i that is chosen properly, the impedance V_k/I_k at every pair of terminals will be Z_i.

Find this so-called "image impedance" for the general case and then for symmetric lossless networks.

d) Use the results of parts (b) and (c) to find all of the voltages, V_k, for the four-section resistive network shown below, given that $V_0 = 1\ V$. Verify your results by direct circuit theory calculations.

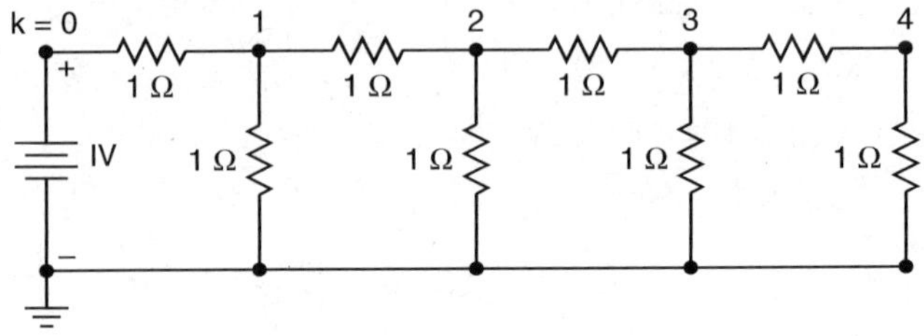

Problem 25.5-3

Consider an arbitrary length, ℓ, of air-filled *TEM* transmission line of characteristic impedance, Z_o^ℓ. Terminal pairs are defined at either end.

a) Find the frequency-dependent impedance matrix elements: $\underline{Z}_{11}, \underline{Z}_{12}, \underline{Z}_{21}, \underline{Z}_{22}$.

b) Find the frequency-dependent scattering matrix elements $\underline{S}_{11}, \underline{S}_{12}, \underline{S}_{21}, \underline{S}_{22}$ defined with respect to an arbitrary value of both Z_o and Z_o^ℓ.

c) Assume $\omega\ell/c \ll 1$, and use these "lumped circuits" to form a periodic cascade as in the previous problem. Find α_1, α_2 and the image impedances, Z_i. Show that these "lumped circuits" are equivalent to a series inductor ($L = L'\ell$) and a shunt capacitance ($C = C'\ell$).

d) Compare the results of part (c) with the exact results obtained from transmission-line theory.

Problem 25.5-4

Consider a long wire of length ℓ and circular cross section surrounded by free-space that carries an electric current, $I_o \cos(\omega t)$. The radius of the conductor is a and the uniform conductivity is σ. Assume that the permittivity and permeability are that of free-space and that the current is uniformly distributed over the cross section.

a) Find an upper bound on the frequency, ω.

b) Find approximate expressions for the electric and magnetic fields (both inside and outside the wire). Neglect the fields beyond a radius of $\ell/2$. (Explain why this is reasonable.)

c) Find approximate expressions for the vector and scalar potentials (both inside and outside the wire) that are consistent with part (b).

d) Evaluate (in the time domain) the power flux, **S**, and the total energy density, W, in both the Maxwell–Poynting and Alternate (circuit) representations. Evaluate, in both representations,

$$\nabla \cdot \mathbf{S} + \frac{\partial}{\partial t} W$$

e) Evaluate (in the frequency domain) the complex power flux, $\underline{\mathbf{S}}$, and the time-averaged electric and magnetic energy densities in both the Maxwell–Poynting and Alternate (circuit) representations. Evaluate the complex Poynting theorem in both representations.

Problem 25.5-5

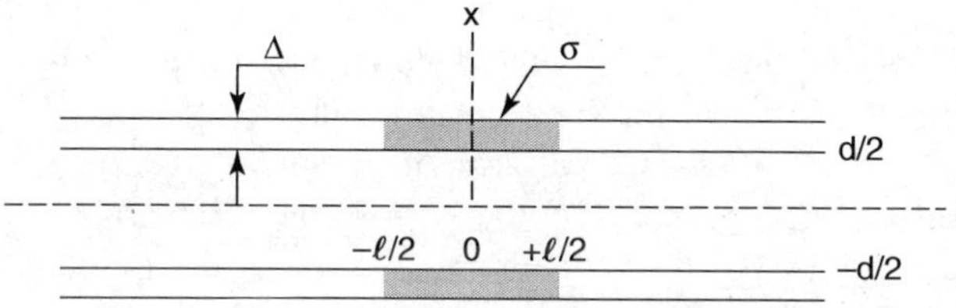

Except for the resistive sections of length ℓ, an air-filled parallel-plate transmission line is made of two perfectly conducting plates each of thickness, Δ; they are of width w (into the paper) and are spaced by a distance d. The resistive sections are characterized by μ_o and conductivity $\sigma \gg \omega\varepsilon_o$. The geometry and frequency of operation, $\omega = ck$ is such that

$$d \ll w$$

$$\ell \gg d$$

$$k\ell \ll 1$$

$$\Delta \ll \delta = \sqrt{\frac{2}{\omega\mu_o\sigma}}$$

where δ is the skin depth of the resistive sections.

A *TEM* wave with electric field

$$\mathbf{E} = \widehat{\mathbf{x}}E_o \cos(\omega t - kz)$$

is incident from the left. Assume that there is no magnetic field outside of the transmission line and that the longitudinal electric field components generated by the resistor rapidly become negligible for $|z| > \ell/2$. The transmission line is matched with an appropriate resistor at its right-hand end so as to eliminate any reflected wave for positive z; the position of the load can be taken at $+\infty$.

a) If $w = 10d$, what is the value (in ohms) of the matching load resistor?

b) Find the *approximate* **E** and **H** fields for $|z| > \ell$.

c) Find the *approximate* **E**, **H**, **A** and Φ in the region, $|z| \leq \ell/2$. (Note that quasistatics applies.)

d) *Sketch* the $\mathbf{E} \times \mathbf{H}$ power flux in the region $|z| \leq \ell$.

e) Evaluate and *sketch* the Alternate-power flux in the region $|z| \leq \ell$.

Problem 25.5-6

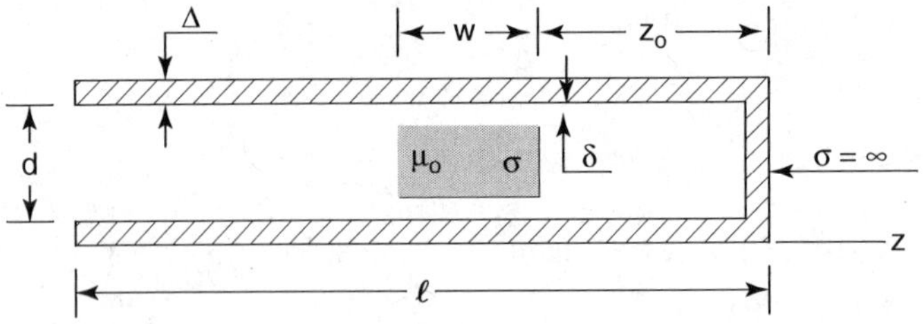

A *TEM* parallel-plate transmission line of length ℓ is made from perfectly conducting plates terminated by a short circuit. The plates, surrounded by free-space, have a uniform width (into the paper) that is large compared to the spacing d. A block of high-conductivity material of length w (the same width as the plates) is placed a distance z_o from the short circuit as shown; there are small symmetric air gaps δ. The parameters are: $\ell = 100\ cm$, $w = 10\ cm$, $z_o = 20cm$, $d = 5cm$, and $\sigma = 10^8 Sm^{-1}$.

A voltage step, $V_s u_{-1}(t)$, is applied at the input of the plates, $z = -\ell$, from a source that is matched to the line. For $t < 0$, there are no fields in the structure.

When $\delta = 0$,

a) *Estimate* the time at which appreciable current starts to flow in the short circuit.

b) Find the steady-state value of the short-circuit current.

When $\delta = \frac{1}{4}$mm,

c) Find (approximately) both the short-circuit current, $i(0, t)$, and the voltage at the input of the block, $V(-w - z_o, t)$.

Problem 25.5-7

Consider an air-filled coaxial transmission line of circular cross section made of perfect conductors. *TEM* fields exist within the space $r_1 \le r \le r_2$. It is known that $r_2 = 1\ cm$.

a) What value of load resistance will terminate the line with minimum reflections.

b) If the electric breakdown strength of air is 30,000 Vcm^{-1}, what value of r_1 will allow maximum power to be transmitted along the line (assuming part (a) conditions hold).

c) Find the numerical values of both peak and average power for part (b) when the transmitted signal is sinusoidal with frequency ω.

d) Repeat part (c) when the line length is greater than $\lambda/2$ and the load resistor has a value that is

(i) twice the value of part (a).

(ii) half the value of part (a).

Problem 25.5-8

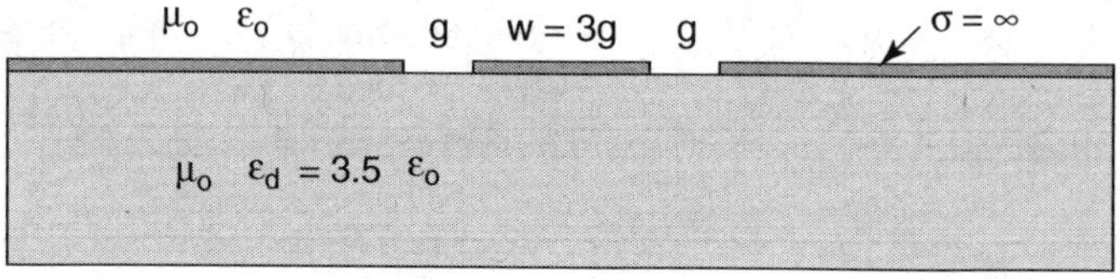

The cross section of a quasi-*TEM* mode transmission line is shown above. Identical air gaps of width g are cut in a thin, perfectly conducting ground plane; the width of the center section is $w = 3g$. Free-space exists above the plane; below is a lossless uniform linear isotropic dielectric with permittivity $\varepsilon_d = 3.5\varepsilon_o$ and permeability μ_o. Assume that the ground plane is of infinite extent, with both outer sections held at ground potential, and that the dielectric occupies the entire lower half-space.

Find the numerical values of:

a) the inductance per unit length, L'/μ_o

b) the characteristic impedance, Z_o

c) the phase velocity of the transmission line

HINT: Make use of the principle of complementarity.

Problem 25.5-9

A lossless *TEM* transmission line of characteristic impedance Z_o and propagation velocity c is terminated at $z = 0$ by a resistance R_L. It is excited at the input end ($z = -cT$) by a Thevenin voltage source $V_s u_{-1}(t)$ through a resistance R_s.

a) Find the positive and negative voltage waves, $f_+(t)$ and $f_-(t)$, as a series of reflections involving Γ_L and Γ_S, the voltage reflection coefficients at the load and source ends of the line.

b) Find and sketch the voltage and current at the midpoint of the line ($z = -\frac{1}{2}cT$) as a function of time for $0 < t < 4T$.

c) Find and sketch the voltage and current at the time $t = \frac{3}{2}T$ as a function of position z.

d) Find the steady-state voltage and current distributions along the line by summing the series found in part (a) and compare it with that found by physical reasoning.

e) Repeat parts (a)–(c) when $R_s = R_L = 0$.

f) Compare the input current with that found by replacing the line with a lumped inductor of value L. What value gives the best match?

Problem 25.5-10

A lossless *TEM* transmission line of characteristic impedance Z_o and propagation velocity c is terminated at $z = 0$ by a capacitance C. It is excited at the input end ($z = -cT$) by a Thevenin voltage source, $V_s u_{-1}(t)$, through a resistance $R_s = Z_o$.

a) Find the positive and negative voltage waves, $f_+(t)$ and $f_-(t)$, for all time.

b) Find and sketch the voltage and current at the midpoint of the line ($z = -\frac{1}{2}cT$) as a function of time.

c) Find and sketch the voltage and current at the time $t = \frac{3}{2}T$ as a function of position z.

d) Repeat parts (a)–(c) when the capacitance is replaced by an inductance L.

Problem 25.5-11

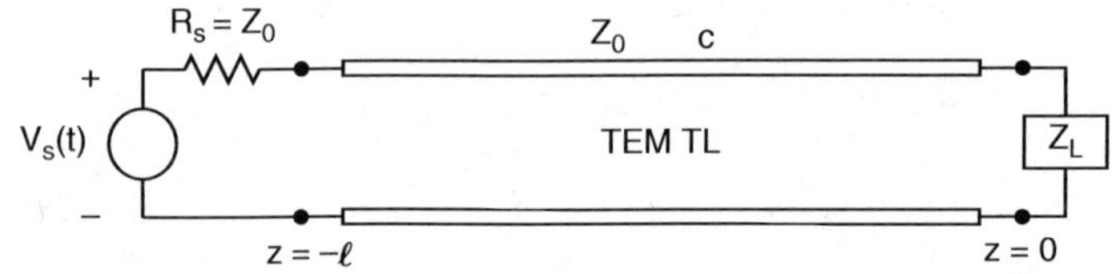

A lossless air-filled *TEM* transmission line of length ℓ and characteristic impedance $Z_o = 75\ \Omega$ is terminated by a load impedance Z_L. An ideal voltage source, $v_S(t) = V_o \cos(\omega_o t)$ in series with a 75-Ω resistor is connected across the other end. In the steady state, the voltage minimum (V_{min}) closest to the load is located 1 *cm* from it. The current minimum (I_{min}) closest to the load is located 5 *cm* from the load. For $V_o = 200\ V$, the time-averaged power delivered to the load is 50 *W*.

a) Find the frequency, ω_o.

b) Find $Z_L(\omega_o)$.

c) What is the maximum peak value of the voltage that occurs along the line?

d) If the transmission-line length, ℓ, is reduced to zero, but V_o and Z_L remain the same, what average power is dissipated in the load?

Problem 25.5-12

A perfectly conducting cylinder of infinite length and radius a is surrounded by free-space. A line charge of magnitude $+q'$ is placed parallel to the cylinder at a radial distance r from its center. An equal, but opposite polarity, surface charge is induced upon the surface.

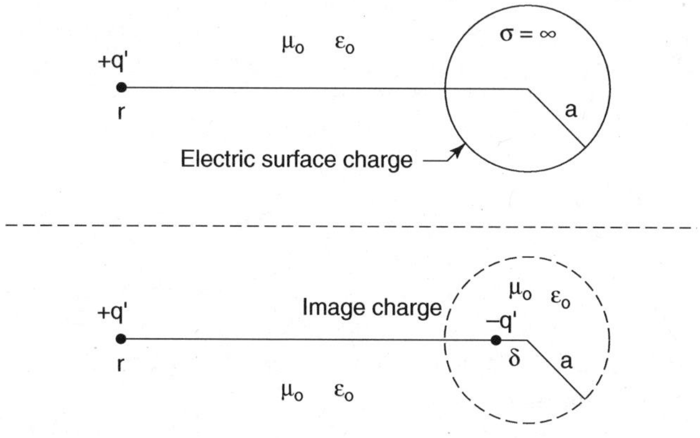

a) Find the location of a fictitious line charge $-q'$ that together with $+q'$ produces the same electric field **outside** of the cylindrical region (now considered to be free-space). The image charge is thus seen to be equivalent to the actual surface charge distribution—with respect to the exterior space.

b) Find the limiting values of δ for the cases $r/a \gg 1$ and $r/a \rightarrow 1$.

c) Find the static electric field everywhere.

d) Find the actual charge distribution on the cylinder.

Problem 25.5-13

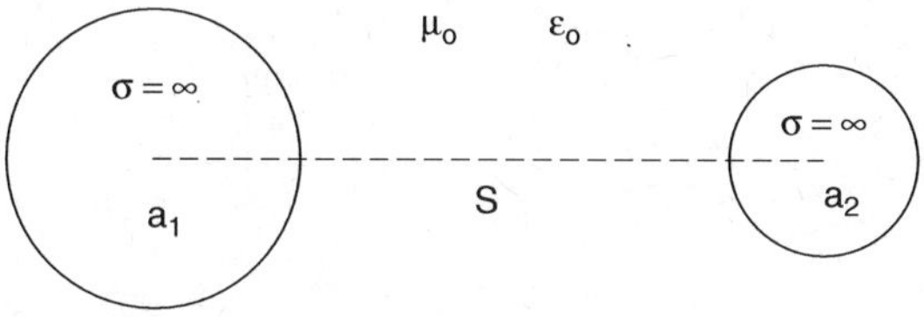

Consider a uniform *TEM* transmission line with the cross section shown above. The center-to-center spacing between the two cylindrical conductors is s; the radii are a_1 and a_2.

a) Use the results of the previous problem to calculate the characteristic impedance, Z_o, of the line.

b) Calculate the inductance per unit length and capacitance per unit length of the line.

c) For a symmetric line ($a_1 = a_2 = a$), find the s/a ratio required to design a 300-Ω line.

d) When $s/a_{1,2} \gg 1$, the surface charges on the cylinders can be approximated as uniform. (Why?) Use this fact to derive an approximate formula for Z_o and compare it with the exact result obtained in part (a).

e) What function of the complex variable, $x + iy$ can be used to generate the Laplacian potential that is equivalent to this image charge problem?

Problem 25.5-14

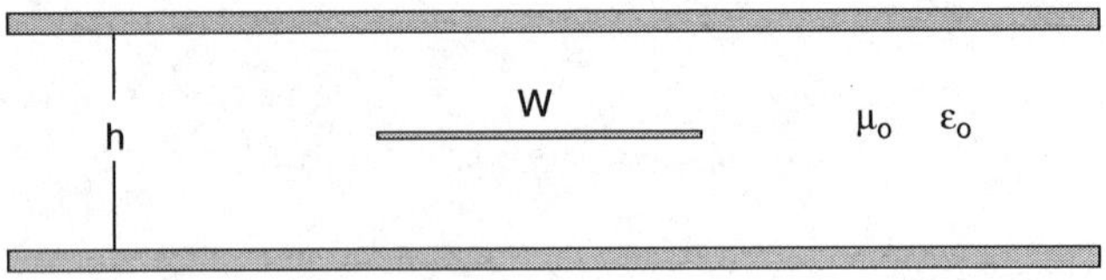

The symmetric stripline made from perfect conductors has a plate spacing, h, a center strip of width, w, and negligible thickness. The outer conductors are both at ground potential and can be assumed to be of infinite width. The cross section shown above is uniform in the z direction (into the paper).

a) Find the characteristic impedance of the stripline as a function of the geometry.

HINT: Find an appropriate conformal mapping in the complex plane.

b) Design a 50-ohm line.

Problem 25.5-15

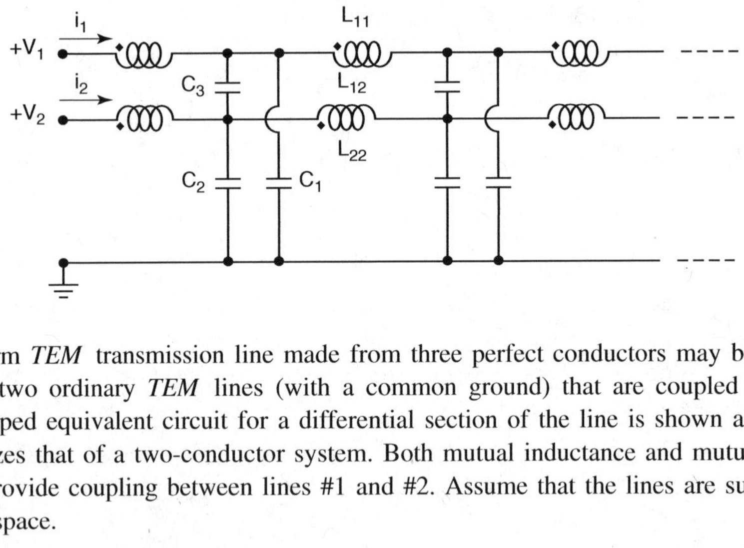

A uniform *TEM* transmission line made from three perfect conductors may be considered as two ordinary *TEM* lines (with a common ground) that are coupled together. The lumped equivalent circuit for a differential section of the line is shown above and generalizes that of a two-conductor system. Both mutual inductance and mutual capacitance provide coupling between lines #1 and #2. Assume that the lines are surrounded by free-space.

a) Generalize the Telegrapher's Equations to express the line voltages, $v_1(z, t)$ and $v_2(z, t)$, in terms of $i_1(z, t)$ and $i_2(z, t)$ and the inductance and capacitance matrices, L'_{ij} and C'_{ij} (per unit length).

b) Express the C'_{ij} in terms of C'_1, C'_2, and C'_3.

c) What relationship(s) exist among L'_{ij} and C'_{ij}.

d) Find the general form of the line voltages and currents in both the time and frequency domains.

Problem 25.5-16

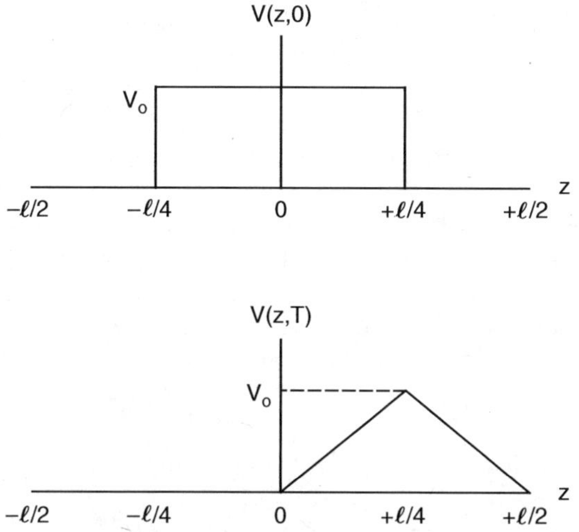

TEM transmission-line voltages

A uniform air-filled *TEM* transmission line of length ℓ is open-circuited at the input, $z = -\ell/2$, and short-circuited at the output, $z = +\ell/2$. The characteristic impedance of the line is Z_o; the transit time is defined as $T = \ell/c$. The voltage along the line, $V(z, t)$, is known and plotted at times $t = 0$ and $t = T$.

a) Find and plot the line voltage along the line at $t = T/2$ and $t = 3T/2$.

b) Find and plot the line current along the line at $t = 0$ and $t = T$.

c) Find and plot $V(-\ell/2,t)$ for $0 \leq t \leq 4T$.

d) Find in terms of V_o, Z_o, and ℓ, the total electromagnetic energy stored in the line at $t = T$.

25.6 WAVEGUIDES

Problem 25.6-1

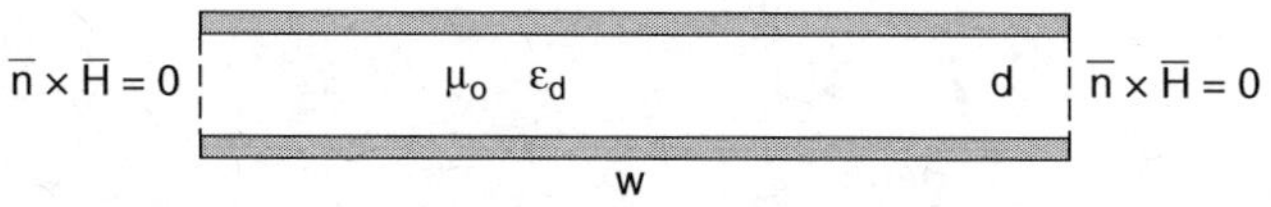

A parallel-plate transmission line made of perfectly conducting plates of width w and separation d is filled with a lossless linear isotropic dielectric of permittivity ε_d. Because $w \gg d$, the no-fringing approximation (no fields outside the structure) is equivalent to a boundary condition of $\mathbf{n} \times \mathbf{H} = 0$ as shown in the figure of the guide cross section.

a) Find all modes (*TEM*, *TE*, *TM*) within the structure. Give both the **E** and **H** fields and the dispersion relation, $\omega(k_z)$.

b) Find an equivalent transmission-line circuit for each mode.

Problem 25.6-2

An air-filled uniform rectangular waveguide with transverse dimensions $a = 2\ cm$ and $b = 1\ cm$ is propagating a traveling mode at a frequency of 10 GHz.

a) What is the maximum average power that can propagate without the peak electric field exceeding the breakdown strength of air ($30,000\ Vcm^{-1}$)?

b) If the metallic walls of the waveguide are nonmagnetic and have an electrical conductivity, $\sigma = 2 \times 10^8\ Sm^{-1}$. (mhos/m), what is the skin depth?

c) For the conditions of part (b), use reasonable approximations to derive the attenuation constant, α of the mode. How far can the mode travel before experiencing 3 dB of loss?

Problem 25.6-3

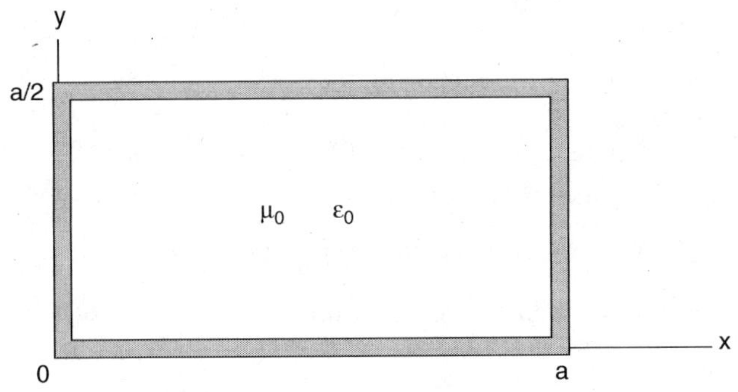

Consider a uniform rectangular waveguide with perfectly conducting walls that is filled with free-space; the width is $a = 3\ cm$, the height is $a/2$. For sinusoidal steady-state excitation at the frequency ω, the complex vector potential (Lorenz gauge) inside the guide is of the form

$$\underline{A}_x = A_0 \sin(2\pi \tfrac{y}{a})\, f_1(z)$$

$$\underline{A}_y = A_0 \sin(\pi \tfrac{x}{a})\, f_2(z)$$

$$\underline{A}_z = A_0 \sin(\pi \tfrac{x}{a}) \sin(2\pi \tfrac{y}{a})\, f_3(z)$$

where $f_1(z)$, $f_2(z)$, and $f_3(z)$ (each of magnitude unity) all represent traveling waves in the $+z$ direction.

a) For a frequency of 15 GHz, find the time-dependent electric and magnetic fields inside the waveguide. What mode(s) are present?

b) Find the surface electric charges and currents on the waveguide walls.

c) Find (in terms of A_0) the average power propagating in the waveguide when the frequency is reduced to 7.5 GHz.

Problem 25.6-4

Consider an air-filled rectangular waveguide of width a and height b propagating the dominant TE_{10} mode at frequency ω.

a) Find the locations at which the magnetic field is circularly polarized.

b) Describe the sense of polarization and the dependence upon the direction of propagation.

Problem 25.6-5

Consider a uniform cylindrical waveguide of radius R in which the TM_{01} mode at frequency ω is below cutoff ($k_z^2 = -\alpha^2 < 0$). Assume that in the vicinity of $z = 0$,

$$\Phi = R(r)\left[C_1 \exp(-\alpha z)\cos\omega t + C_2 \exp(\alpha z)\sin\omega t\right]$$

a) Find expressions for the normal and Alternate-energy densities.

b) Find expressions for the normal and Alternate-power flux.

c) find the average total power for $C_2 \leq C_1$ and $C_2 = 0$.

d) Find ω, if $R = 5\ cm$ and $\alpha = 1/10\ cm^{-1}$.

Problem 25.6-6

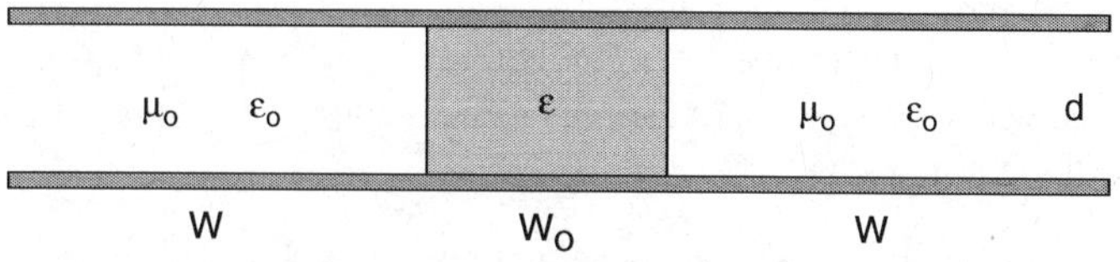

A linear isotropic dielectric of permittivity $\varepsilon = 3\varepsilon_0$ and width w_o is placed between perfectly conducting ground planes spaced a distance d.

a) Find the approximate dispersion relation $k_z(\omega)$ for the lowest-order mode when $\omega \to 0$ and $k_x w \ll 1$. This is the quasi-*TEM* mode.

b) Find the characteristic impedance and the phase and group velocities of the quasi-*TEM* mode (when $w = 2w_o$ and $d = .5w_o$) by calculating the static capacitance and inductance (per unit length). Show that the results are in agreement with part (a). Over what frequency range are these approximate values valid?

c) As the frequency is increased, the fields in the air regions become more and more nonuniform until at some value, ω_o, they are found to decay nearly exponentially between their values at the dielectric-air interface to $e^{-\pi} = .04$ of those values at the edge of the plate region. If $w_o = 2.5\ cm$, what is the approximate value of ω_o? How is the mode characterized?

d) Evaluate the numerical values of both the phase and group velocities for the frequency, ω_o, of part (c). Compare the results with the values found in part (b).

Problem 25.6-7

Waveguide modes with azimuthal circulation propagating without reflections in a perfectly conducting circular waveguide of radius R (enclosing free-space) have complex potentials inside the guide of the form

$$\underline{\mathbf{A}}_{\text{inner}} = \begin{pmatrix} \widehat{\mathbf{r}}\left[\underline{C}_1 \frac{n}{r} J_n(k_c r) + \underline{C}_2 \frac{d}{dr} J_n(k_c r)\right] \\ +j\widehat{\boldsymbol{\phi}}\left[\underline{C}_1 \frac{d}{dr} J_n(k_c r) + \underline{C}_2 \frac{n}{r} J_n(k_c r)\right] \\ -j\widehat{\mathbf{z}}\,\underline{C}_3 k_z J_n(k_c r)) \end{pmatrix} \exp[-j(k_z z - n\phi)]$$

$$\underline{\Phi}_{\text{inner}} = -j\underline{C}_0 \omega J_n(k_c r)\exp[-j(k_z z - n\phi)]$$

Assume the potentials satisfy the Lorenz gauge. In the case of $TE_{nm}^{\ \ \text{O}}$ modes, use the particular solution, $\underline{\Phi}_{\text{inner}} = 0$.

a) For both $TM_{nm}^{\ \ \text{O}}$ and $TE_{nm}^{\ \ \text{O}}$ modes, find the spin and orbital components of the time-averaged angular momentum density (defined in Appendix D, Section D.3).
For a given mode propagating above cutoff:

b) Compare the total average angular momentum per unit length to the average power propagating in the guide.

c) Find the net average torque per unit length created by a $TE_{11}^{\ \ \text{O}}$ mode when the waveguide walls are perfectly conducting.

d) Find the net average torque per unit length created by a $TE_{11}^{\ \ \text{O}}$ mode when the waveguide walls are highly conducting. What happens as the frequency is lowered close to the cutoff value?

NOTE: When the guide is highly conducting, the wall currents produce tangential electric fields that are approximately given by $(\sigma\delta)\mathbf{E} = \mathbf{K}_s$ where δ is the skin depth (assumed to be very small compared to R) and $\mathbf{K}_s$ is the surface-current density calculated under the assumption that $\sigma = \infty$.

Problem 25.6-8

A waveguide of square cross-section (with edge dimensions a) can support degenerate TE_{n0} and TE_{0n} modes of the same frequency ω and propagation constant k_z. Assume that n is an odd integer and that the amplitudes of E_x and E_y are such that the electric field at the center of the guide ($x = a/2$, $y = a/2$) is circularly polarized.

a) Find the time-dependent electric and magnetic fields (assuming a propagating mode).

b) Find the approximate fields near the axis of the guide, $x = a/2 + r\cos\phi$, $y = a/2 + r\sin\phi$, where $r \ll a/2$.

c) Find the average power propagating in the waveguide.

d) Find both spin and orbital components of the approximate average angular-momentum density (defined in Appendix D, Section D.3) near the waveguide axis.

A thin circular rod of length ℓ and radius r_o ($r_o \ll \ell$, $r_o \ll a$) is made of a lossy dielectric with $\varepsilon \simeq \varepsilon_o$ and conductivity σ. It is placed in the waveguide along and concentric with the z axis. Because $\sigma \ll \omega\varepsilon_o$, the rod only weakly disturbs the *TE* fields.

e) Find the net torque acting upon the rod. Express the answer in terms of the average power found in part (c). What value of n maximizes the torque?

HINT: Use the Alternate formulation of the angular-momentum theorem.

25.7 JUNCTIONS AND COUPLERS

Problem 25.7-1

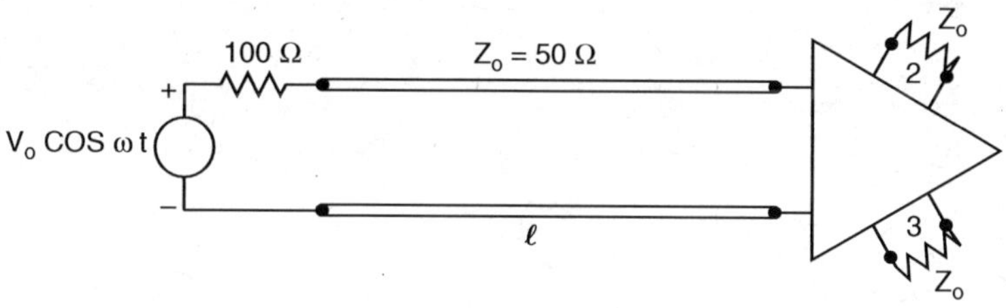

A **lossless, reciprocal**, 3-port junction is **completely symmetrical**. It has been designed to **minimize** the reflected power in the input line—when matched loads terminate the other two ports. The scattering matrix of the junction is defined with respect to $Z_o = 50\ \Omega$. At the frequency ω, the value of S_{11} is real.

A lossless *TEM* line ($Z_o = 50\ \Omega$) is connected to port #1 as shown; matched loads ($50\ \Omega$) are used to terminate ports #2 and #3. The ideal voltage source ($V_o = 80\ V$) is connected to the line through a 100-Ω resistance. When the line length, ℓ, is made equal to $\lambda/4$, the total power delivered to the matched loads is found to be a **maximum**.

a) Find the coefficients of the scattering matrix.

b) Find the power delivered to each load when:

(i) $\ell = \lambda/4$

(ii) $\ell = \lambda/2$

Problem 25.7-2

Consider the scattering matrix for a lossless three-port junction that *may* be nonreciprocal. All three ports have lossless TEM mode transmission lines attached with identical values of Z_o. When any two of the ports are terminated with matched loads, Z_o, the reflected power in the third line is zero. The ports are labeled 1, 2, and 3 in clockwise order.

a) Find the general form of the scattering matrix. Can the junction be reciprocal?

b) If the port labels are permuted, so that $1 \to 2$, $2 \to 3$, and $3 \to 1$ and the new scattering matrix is unchanged, what form must it have? Is the matrix unique?

c) How does the power incident upon port #1 divide between the matched loads used to terminate ports #2 and #3? Is such a device useful?

Problem 25.7-3

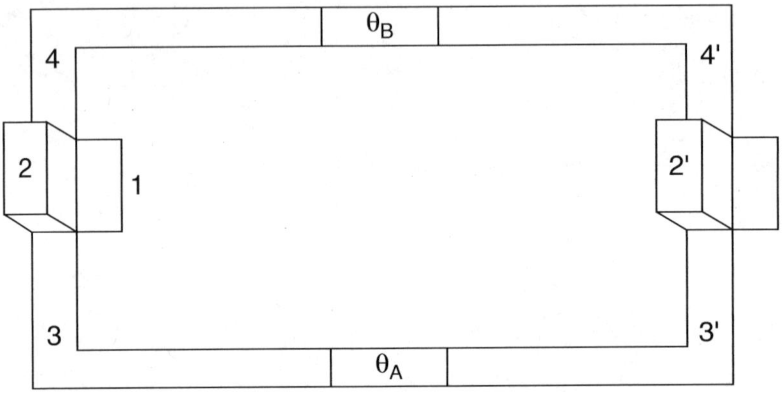

Two "magic Ts" are connected as shown above with terminals defined so that

$$\begin{bmatrix} b_1 \\ b_2 \\ b_3 \\ b_4 \end{bmatrix} = \begin{bmatrix} 0 & 0 & 1/\sqrt{2} & -1/\sqrt{2} \\ 0 & 0 & 1/\sqrt{2} & 1/\sqrt{2} \\ 1/\sqrt{2} & 1/\sqrt{2} & 0 & 0 \\ -1/\sqrt{2} & 1/\sqrt{2} & 0 & 0 \end{bmatrix} \cdot \begin{bmatrix} a_1 \\ a_2 \\ a_3 \\ a_4 \end{bmatrix}$$

The boxes marked θ_A and θ_B are linear phase shifters (matched in both directions so as not to introduce reflections) that have the following characteristics: For waves traveling from 3 to 3′ and 4 to 4′, the phase shifts are equal $\overrightarrow{\theta}_A = \overrightarrow{\theta}_B = \theta_L$. For waves traveling from 3′ to 3, $\overleftarrow{\theta}_A = \theta_L$; from 4′ to 4, $\overleftarrow{\theta}_B = \theta_R$. The differential phase shift, $\theta_L - \theta_R$ is given by

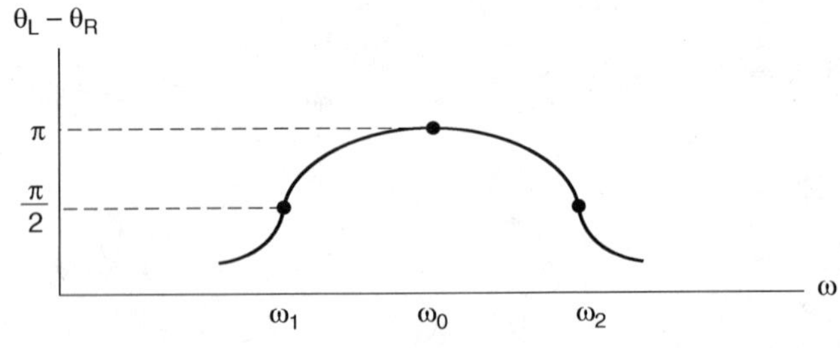

For $\omega_1 \le \omega \le \omega_2$, the shape of the curve is approximately **parabolic** with $\omega_o = \dfrac{\omega_1 + \omega_2}{2}$.

Except for the phase shifts θ_A and θ_B, the path length between 3 and 3′ equals that between 4 and 4′.

a) Find the scattering matrix of the four-port device (1, 2, 1′, 2′) as a function of frequency for $\omega_1 \le \omega \le \omega_2$.

b) Over what frequency range does the device approximate an ideal circulator in that no unwanted port power is greater than 10% of the main signal power?

Problem 25.7-4

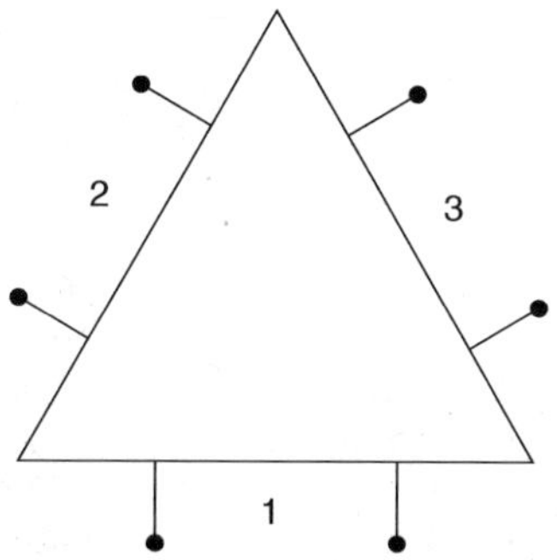

A **lossless, reciprocal**, 3-port junction is **completely symmetrical** and fitted with coaxial connectors ($Z_o = 90\ \Omega$). Matched loads (90 Ω) are used to terminate ports #2 and #3. A lossless *TEM* line is connected to port #1; that coaxial line is completely filled with lossless dielectric of relative permittivity (dielectric constant) $\varepsilon_d/\varepsilon_o = 9$ so that the resultant impedance is $Z_o = 30\ \Omega$. Power traveling in line #1 at frequency ω_o is equally split between ports #2 and #3 **without reflection**. The phase delay between ports is less than 90°.

a) Find the scattering matrix of the junction (defined with respect to $Z_o = 90\ \Omega$ at all three ports) for the frequency ω_o.

b) If the lossless, reciprocal junction is replaced with one of improved design (still with 90-Ω connectors), the value of $\varepsilon_d/\varepsilon_o$ required to prevent power from being reflected in the input coaxial line can be reduced. What is the **minimum** value of the dielectric constant that would be required?

Problem 25.7-5

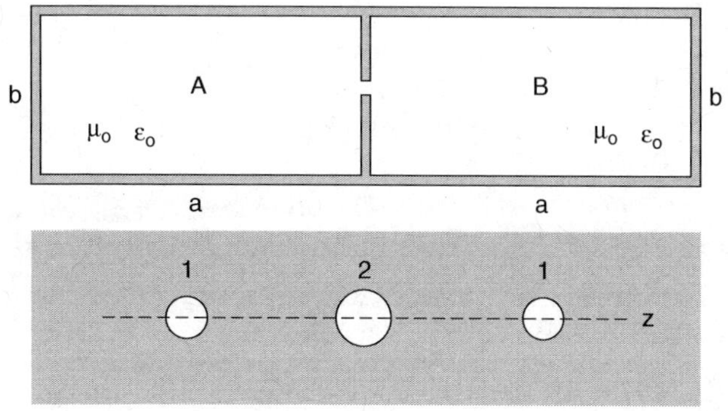

A directional coupler is made from two identical rectangular waveguides of dimensions a,b joined by a **thin** common wall as shown. The coupling consists of three small holes in the common wall with nearest-neighbor spacing ℓ. The relative field strengths coupled by the three holes are in the ratios 1,2,1 and the radius of the largest hole is very small compared to the free-space wavelength. At the center frequency (1.5 times cutoff frequency of the TE_{10} mode), 1% of the incident power in guide A is coupled to guide B **through the center hole**.

a) If $a = 2b$, find ℓ/a so that at the center frequency **all** of the power coupled to guide B is traveling in the **same direction** as that in guide A.

b) Calculate (approximately) the coupling and directivity (in dB) as a function of frequency. (Directivity is the ratio of the coupled power traveling in the reverse direction to that in the forward direction.)

c) Consider an N-hole coupler (nearest neighbor spacing ℓ) with weak coupling $<$ -20 dB and relative coupling of the holes that follows a binomial distribution:

N					*COUPLING*				
2				1		1			
3			1		2		1		
4		1		3		3		1	
5	1		4		6		4		1
–	–	–	–	–	etc.	–	–	–	–

Estimate the directivity (dB) as a function of frequency and N.

d) What are the relative sizes of the holes?

HINT: The electromagnetic power scattered from a uniform plane wave of frequency ω that is normally incident upon the plane of a very thin perfectly conducting disk of radius R is proportional to $\omega^4 R^6$, provided that $kR \ll 1$ (Rayleigh scattering).

Problem 25.7-6

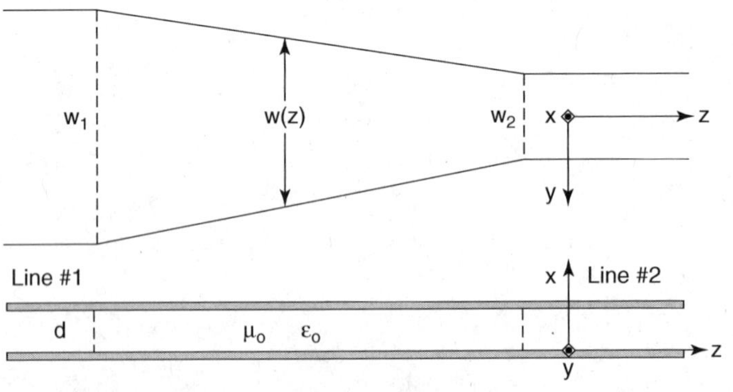

Two parallel-plate TEM transmission lines (made from perfect conductors) have the same plate spacing, d, but different plate widths, w_1 and w_2. They are connected by a tapered section of width $w(z)$ and length ℓ as shown above. Assume that line 2 is terminated with a matched load and $w_2 \gg d$.

a) Find the complex voltage and current within the tapered section and lines #1 and #2. (Assume that the WKB approximation is valid.)

b) For the sinusoidal steady-state frequency, ω_o, design an **exponential** taper [$w(z) \sim \exp(2\alpha z)$] of minimum length ℓ/λ_o so that for $w_1/w_2 = 5$, the VSWR in line 1 is ≤ 1.5.

Problem 25.7-7

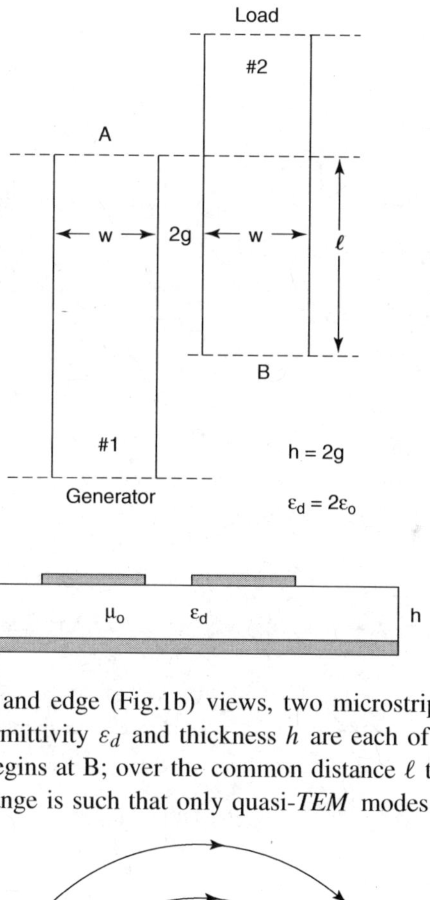

As shown in top (Fig. 1a) and edge (Fig.1b) views, two microstrip lines deposited on a dielectric substrate of permittivity ε_d and thickness h are each of width $w \gg h$. Line #1 ends at A and line #2 begins at B; over the common distance ℓ the line separation is $2g = h$. The frequency range is such that only quasi-*TEM* modes can propagate.

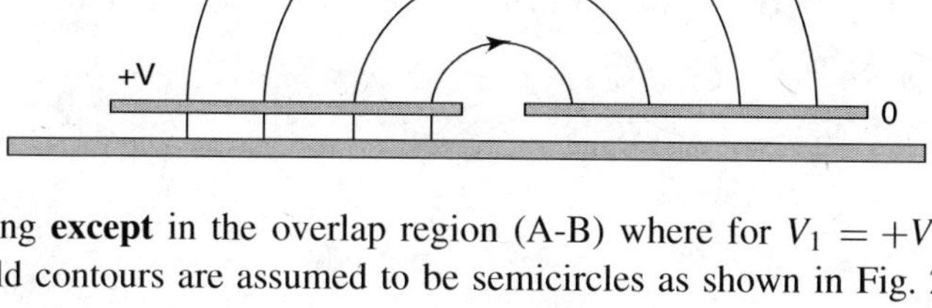

Neglect fringing **except** in the overlap region (A-B) where for $V_1 = +V$, $V_2 = 0$, the fringing **E**-field contours are assumed to be semicircles as shown in Fig. 2.

a) Develop separate transmission-line models valid in the overlap and nonoverlap regions. Evaluate (approximately) all line parameters.

b) Assume that line #1 is excited by an ideal voltage source with Thevenin resistance $R_s = Z_o = 30\ \Omega$ and that line #2 is terminated with $R_L = Z_o = 30\ \Omega$. Find (approximately) the coupling length ℓ such that at a center frequency of 10 *GHz* all of the power is transferred from line #1 to line #2 when A and B are both:

(i) short circuits

(ii) open circuits

(iii) matched loads

25.8 RESONATORS

Problem 25.8-1

An air-filled waveguide resonator is made in the shape of a cube with edges 1 *cm* long. It is filled with a linear isotropic dielectric having complex permittivity, $\varepsilon' - j\varepsilon''$.

a) Calculate the three lowest resonant frequencies (assuming no dissipation in the walls and $\varepsilon' = 4\varepsilon_0$, $\varepsilon'' = 0$). Characterize all of the modes that have the same lowest frequency.

b) The metallic walls of the waveguide are nonmagnetic and have an electrical conductivity, $\sigma = 2 \times 10^8\ Sm^{-1}$ (mhos/m); the dielectric loss tangent is $\varepsilon''/\varepsilon' = 10^{-3}$. Calculate the approximate internal Q of the lowest-order resonance.

Problem 25.8-2

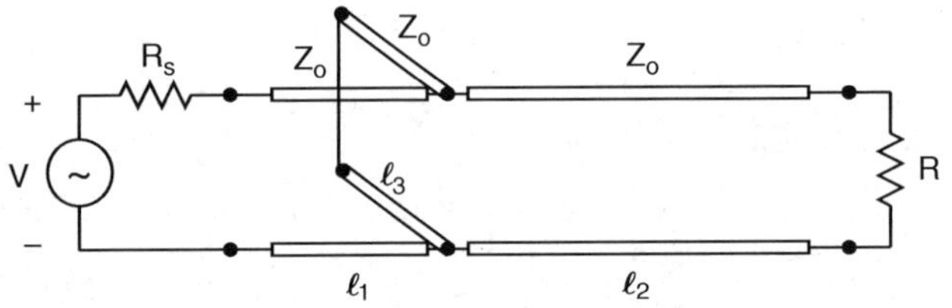

TEM transmission lines of characteristic impedance Z_o and lengths ℓ_1, ℓ_2, and ℓ_3 are connected as shown. The source frequency is such that $\ell_2 + \ell_3 \approx \lambda_o/2$.

a) If $R_s = Z_o$, $R = .01Z_o$, find the approximate ratio of $\ell_3/\ell_2 \ll 1$ that produces critical coupling. What effect does the value of ℓ_1 have on the resonance?

b) Calculate the **internal, external**, and **loaded** Q for the conditions of part (a).

c) Repeat part (b) when $\ell_1 = 0$ and ℓ_3 is reduced from its part (a) value by a factor of 2. What is the coupling factor?

Problem 25.8-3

A transmission cavity is made from air-loaded circular waveguide of radius $R = 2.4\ cm$ in which two identical thin irises are spaced a distance $\ell = 6\ cm$ apart; each iris has a small centered hole of radius $a = .8\ cm$. The cavity is excited with the TE^{o}_{11} mode. Neglect wall losses in the guide and irises. Use reasonable approximations to estimate:

a) the lowest frequency that will be transmitted through the cavity.

b) the bandwidth of the transmission resonance of part (a).

c) the coupling factor at resonance.

Problem 25.8-4

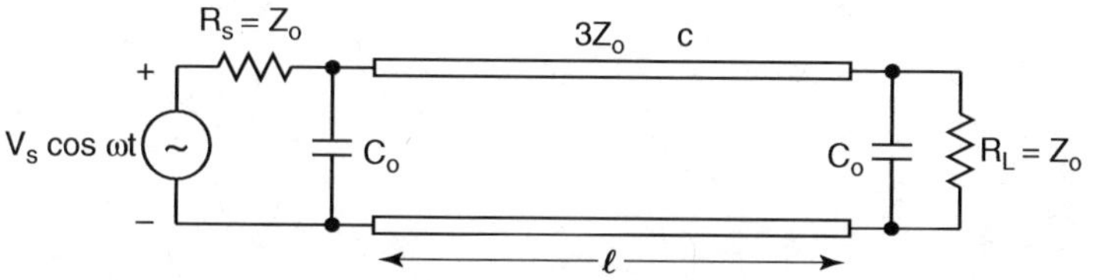

We wish to transfer the maximum possible power from the source to the load resistor at some specified frequency $\omega = \omega_0$.

a) Find the relationship among ω_0, ℓ, Z_0, and C_0 that will accomplish the objective.

b) Given that $Z_0 = 50\ \Omega$, $f_0 = 3 \times 10^9\ Hz$, and $C_0 = 1\ pF$ ($\omega_0 C_0 Z_0 = \frac{2}{3}\sqrt{2}$), find the value of ℓ nearest $\lambda_0/2$ that will satisfy the part (a) conditions.

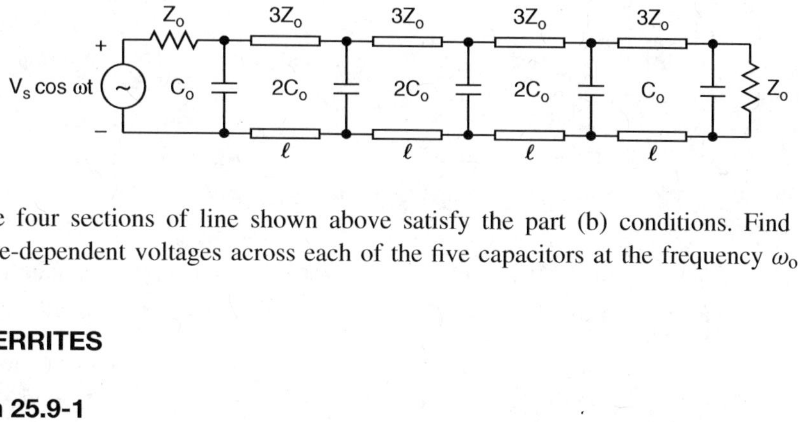

c) The four sections of line shown above satisfy the part (b) conditions. Find the time-dependent voltages across each of the five capacitors at the frequency ω_0.

25.9 FERRITES

Problem 25.9-1

Consider a ferrite medium of infinite extent that is magnetized to saturation along the z axis with a uniform dc magnetic field, H_z. The saturation magnetization is M_s; the gyromagnetic ratio is $-e/m_e$. The dielectric permittivity ε is uniform, linear and isotropic. The electric conductivity and dielectric and magnetic losses are all zero. A small-amplitude uniform plane wave of frequency ω propagates as $\exp(-j\mathbf{k} \cdot \mathbf{r})$ through the ferrite (which can be assumed a linear medium). Neglect magnetic anisotropy and quantum-mechanical exchange effects.

a) Derive the dispersion relation $\mathbf{k}(\omega)$ and find the polarization and relationships among $\mathbf{e}$, $\mathbf{h}$, $\mathbf{b}$, and $\mathbf{m}$ when the propagation vector $\mathbf{k}$ is oriented:

(i) parallel to the z axis.

(ii) perpendicular to the z axis.

b) For what portion(s) of these dispersion relations are the waves magnetostatic in character ($\nabla \times \mathbf{h} = 0$).

c) In the magnetostatic region, find the angle between the phase and group velocities. For what orientation of **k** with respect to $\mathbf{H}_{dc}$ is this "beam-steering" angle maximized?

Problem 25.9-2

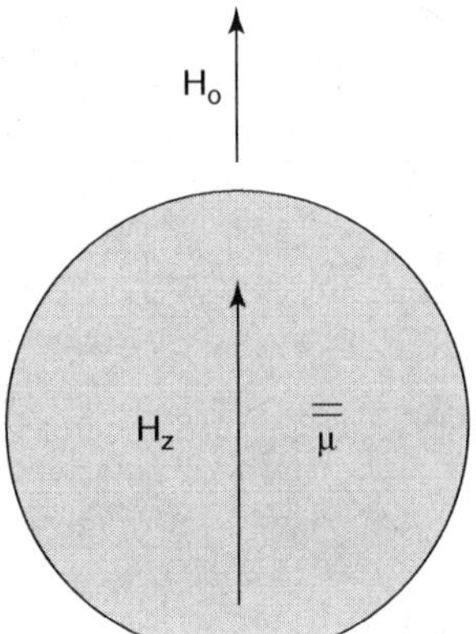

Consider a small ferrite sphere of radius R surrounded by free-space and magnetized to saturation along the z axis by an **externally applied** uniform dc field of strength H_o. Neglect anisotropy and loss.

For magnetostatic waves and modes in a lossless ferrite medium, the complex fields for the frequency, ω, are approximately $\nabla \times \underline{\mathbf{h}} = 0$, and $\underline{\mathbf{h}} = -\nabla \underline{\psi}$ with $\underline{\mathbf{b}} = \overline{\overline{\mu}} \cdot \underline{\mathbf{h}}$ and $\nabla \cdot \underline{\mathbf{b}} = 0$.

a) Find H_z inside the sphere. What is the minimum value of H_o required to saturate the ferrite? Find the complete dc magnetic field inside and outside the sphere.

b) Derive (in Cartesian coordinates) the equation (known as Walker's Equation) that governs the complex magnetostatic potential and is dependent upon the Polder permeability tensor. Show that outside the ferrite, it reduces to Laplace's Equation. What are the boundary conditions?

Consider the family of modes for which

$$\underline{\psi}_i = \underline{C}_i (x \pm j\, y)^{|n|} = \underline{C}_i\, r^{|n|} \sin^{|n|}\theta\, \exp(-jn\phi) \quad (r < R)$$

$$\underline{\psi}_o = \underline{C}_o\, r^{-\ell} \sin^{|n|}\theta\, \exp(-jn\phi) \quad (r > R)$$

c) Verify that these potentials satisfy Walker's Equation. Find the value of ℓ (in terms of the mode integer n) and the relationship between the constants $\underline{C}_i$ and $\underline{C}_o$. Show that inside the sphere, all of the small-signal fields are circularly polarized.

d) Apply the remaining boundary condition and find the frequency condition $\omega(n)$ that permits nontrivial solutions. Are both positive and negative integer values of n permitted?

HINT: Make use of scalar permeabilities in order to simplify the analysis.

e) Show that the solution for $n = 1$ represents the Kittel resonance—the so-called uniform precession (u.p.) mode. Calculate the numerical frequency for $\omega_M = \pi \cdot 10^{10}$ $rads^{-1}$, $H_o = 1000$ Oe (1 Oe $= 250/\pi$ Am^{-1}), and $\gamma = -e/m_e$ (2.8 MHz/Oe).

f) Show that as $n \to \infty$, the modes become more tightly bound to the surface and the frequencies approach a definite limit. Calculate the numerical value for the parameter values of part (e).

Problem 25.9-3

A transmission line is made of perfectly conducting parallel plates positioned at $y = 0$ and $y = b$ is completely filled with uniform ferrite material that is magnetized to saturation (M_s) by a uniform transverse dc magnetic field $\mathbf{H}_{dc} = \hat{\mathbf{y}}H_y$; the dielectric permittivity of the ferrite is ε. Consider *TE* modes of frequency ω that are without y variation and have small-signal complex fields that propagate in the z direction as $[\underline{c}_+ \exp(-jk_x x) + \underline{c}_- \exp(+jk_x x)]\exp(-jk_z z)$.

a) Show that these are *TEM* modes in the sense that $\underline{e}_z = 0$, $\underline{b}_z = 0$, but that, with respect to $\underline{\mathbf{h}}$, they are actually *TE* modes.

b) Define the transverse impedance $\underline{\eta}(x) = \underline{e}_y(x)/\underline{h}_z(x)$ and show that it is $Z_o\frac{1+\Gamma(x)}{1-\Gamma(x)}$ (Smith Chart form) in parallel with an inductance L located at x. Verify that $Z_o = \frac{\omega(\mu^2-\kappa^2)}{\mu\, k_x}$, $\Gamma(x) = \frac{\underline{c}_-}{\underline{c}_+}\exp(+j2k_x x)$, $L = \frac{\mu^2-\kappa^2}{\kappa\, k_z}$. Notice that $\underline{\eta}(x)$ is nonreciprocal with respect to the sign of k_z.

A rectangular waveguide is created by introducing perfectly conducting planes at $x = 0$ and $x = a$.

c) Use the results of part (b) to find the *TE* modes (without y variation) and the dispersion relation, $k_z(\omega, H_y)$. In what sense are the modes nonreciprocal?

d) Find the average power carried by the waveguide mode.

The loading of the rectangular waveguide is altered so that it is air-filled from $0 < x < a - d$ and ferrite-filled from $a - d \le x < a$; the dc magnetic field is unchanged.

e) Repeat part (c).

HINT: Use continuity of the transverse-impedance to match boundary conditions.

Problem 25.9-4

A thin film of ferrite serves as an analog to a dielectric waveguide with the Polder permeability tensor (instead of the dielectric permittivity) providing the wave confinement.

Small-signal linearized analysis in the ferrite film is assumed to be valid. For simplicity in modeling, a uniform thin film of ferrite is assumed to be of infinite extent along the y and z directions; the ferrite extends from $-d/2 < x < +d/2$ and is surrounded by free-space. A dc uniform magnetic field is applied so as to magnetically saturate the ferrite (M_s). In all of the cases to be considered, assume that nonuniform (evanescent) plane waves exist in the *free-space regions* that propagate without y variation in the z direction as $\exp(-\alpha_x|x|)\exp(-jk_z z)$. Use the magnetostatic-wave approximation to match boundary conditions at the surfaces of the film to find the small-signal mode pattern and dispersion relation, $k_z(\omega, H_{dc}, d)$, when the dc magnetic field is:

a) $\widehat{\mathbf{x}}H_{dc}$ (the modes are termed magnetostatic forward volume waves: *MSFVW*).

b) $\widehat{\mathbf{z}}H_{dc}$ (the modes are termed magnetostatic backward volume waves: *MSBVW*).

In cases (a) and (b) the modes belong to families that have symmetries equivalent to dielectric waveguide modes.

c) $\widehat{\mathbf{y}}H_{dc}$ (the single mode is termed a magnetostatic surface wave: *MSSW*).

d) Evaluate the group velocity of the lowest-order mode in cases (a)–(c) for $k_z \to 0$.

HINT: Evaluate the polarization of the free-space fringing fields and make use of the result to reduce the Polder tensor to an effective scalar.

Problem 25.9-5

Consider a small ferrite sphere magnetized to saturation along the $+z$ axis by a uniform field (far from the sphere) of strength H_{dc}. The saturation magnetization has a value $4\pi M_s = 1780\ G$ (in SI units $M_s = 1780 \cdot \frac{10^3}{4\pi} Am^{-1}$).

The Kittel uniform precession mode is excited and the precession cone angle equals one degree. If the total rf energy of the system is interpreted as being due to quasiparticle "magnons," each carrying energy $\hbar\omega$, find:

a) The magnon density (particles/cm^3).

b) The total energy in a 1-mm-diameter sphere if the precession frequency is 3 GHz.

Problem 25.9-6

Attempt to derive the "reciprocity theorem" when region(s) of uniformly magnetized ferrite (saturated) exist that are characterized by linear lossless Polder permeability tensor(s).

a) Show why the theorem <u>can</u> fail.

b) Does the theorem necessarily fail?

c) If all current sources and the ferrite(s) are located in a finite region surrounded by free-space, does the surface integral

$$\oint_{S_0} (\underline{\mathbf{E}}^a \times \underline{\mathbf{H}}^b - \underline{\mathbf{E}}^b \times \underline{\mathbf{H}}^a) \cdot \widehat{\mathbf{n}}\, da$$

still vanish when there is ferrite present within the volume? (The a fields are generated by $\mathbf{J}_S^a$, the b fields are generated by $\mathbf{J}_S^b$.)

Problem 25.9-7

When the wavelength of a magnetostatic wave is very small, the phase shift between neighboring magnetic moments becomes appreciable and quantum-mechanical exchange effects cannot be ignored. Assuming a uniform isotropic interaction, the exchange torques that are produced can be incorporated into the small-signal model by replacing the linearized rf magnetic field with

$$\mathbf{h} \rightarrow \mathbf{h} + \lambda_{ex} \nabla^2 \mathbf{m}$$

where λ_{ex} is a material-dependent constant [for yttrium iron garnet (YIG), $\lambda_{ex} \simeq 3 \cdot 10^{-12}$ cm^2, $\omega_M \simeq \pi \cdot 10^{10}$ $rads^{-1}$].

a) Derive the magnetostatic spin wave dispersion relation $\omega(\mathbf{k})$ that incorporates exchange.

b) Find the additional small-signal power-flux and energy density (due to exchange) that must be added to the small-signal Poynting theorem.

c) Show that for large wavenumber, k, the power flow is nonelectromagnetic, but still proportional to the group velocity times the energy density.

d) Find the spin wave group velocity in YIG, when $\mathbf{k}$ is parallel to $\mathbf{H}_{\text{dc}}$ and $k = 10^5 cm^{-1}$.

Problem 25.9-8

Consider a ferrite slab, of infinite extent along x and y and thickness d along z, that is magnetized to saturation along the $+z$ axis with a uniform dc magnetic field, H_z. The saturation magnetization is M_s; the gyromagnetic ratio is $-e/m_e$; the dielectric permittivity ε is uniform, linear, and isotropic. The electric conductivity and dielectric and magnetic losses are all zero. A small-amplitude uniform plane wave of frequency ω propagates in the $+z$ direction through the ferrite (which can be assumed a linear medium). At $z = 0$, the electric field of the wave is linearly polarized along the x direction with amplitude e_o. For parts (a)–(d), neglect any reflections at the $z = 0$ and $z = d$ boundaries. Also, neglect magnetic anisotropy and quantum-mechanical exchange effects.

a) Find expressions for **e**, **h**, **b**, and **m**.

b) Show that the plane of polarization at $z = d$ is rotated by an angle θ_F (known as the Faraday angle).

c) If $\varepsilon/\varepsilon_o = 10$, $\omega \gg e/m_e\mu_o H_z$, and $e/m_e\mu_o M_s = \pi \cdot 10^{10}$ $rads^{-1}$, find the value of d such that $|\theta_F| = \pi/2$.

d) What is the sign of θ_F for the conditions of part (c)?

e) If a perfectly conducting plane is placed at $z = d$ (causing the wave to be totally reflected), describe the fields at the plane $z = 0$.

Problem 25.9-9

a) Generalize the Maxwell–Poynting complex momentum theorem derived in Chapter 6, Section 6.4 to apply to a uniform ferrite of infinite extent that is characterized by a linearized Polder permeability tensor. The material is magnetized to saturation along the $+z$ axis with a uniform dc magnetic field, H_z; the saturation magnetization is M_s; the dielectric permittivity ε is uniform, linear, and isotropic. The electrical conductivity and dielectric and magnetic losses are all zero.

b) Apply the results of part (a) to find the time-averaged small-signal momentum density of a small-amplitude uniform plane wave of frequency ω that propagates as $\exp(-jkz)$. Compare the result to the value obtained from Eq. (23.20).

c) Find the time-averaged small-signal momentum density when a uniform plane wave of frequency ω is linearly polarized along the x direction (at $z=0$) with amplitude e_o, but propagates in the $+z$ direction, subject to Faraday rotation of the plane of polarization. Neglect losses and any reflections.

d) For what frequency is the average momentum density of part (c) maximized?

25.10 FOUR-DIMENSIONAL ELECTROMAGNETICS

Problem 25.10-1

Use Macsyma or Maxima to load (*4d-em.mac*) and use the 4d-emworkpad program to verify that (in free-space):

a) $\mathcal{V}\cdot\mathcal{V} = -c^2$

b) $\mathcal{E}\cdot\mathcal{H} = \mathbf{E}\cdot\mathbf{H}$

c) $\varepsilon_o\mathcal{E}\cdot\mathcal{E} - \mu_o\mathcal{H}\cdot\mathcal{H} = \varepsilon_o\mathbf{E}\cdot\mathbf{E} - \mu_o\mathbf{H}\cdot\mathbf{H}$

d) $\mu_o\mathcal{G}\cdot\mathcal{V} = \mathcal{E}$

e) $\varepsilon_o\mathcal{K}\cdot\mathcal{V} = \mathcal{H}$

f) $\Box\cdot\mathcal{G} = \mathcal{J}$

g) $\Box\cdot(\Box\times\mathcal{A}) = \mu_o\mathcal{J}$

h) $\Box^2\mathcal{G} = -\Box\times\mathcal{J}$

i) $\mu_o\mathcal{G}\cdot\mathcal{J} = \Box\cdot\frac{1}{2}(\mu_o\mathcal{G}\cdot\mathcal{G} + \varepsilon_o\mathcal{K}\cdot\mathcal{K}) = \mathcal{F}$

Problem 25.10-2

Assume that, with respect to the laboratory frame of reference, the velocity of the rest frame is $\mathbf{v} = [v_1, 0, 0]$ with v_1 a constant. Accordingly, the Lorentz transformation can be written in matrix form as

$$[x', y', z', ict'] = \mathbf{mat2vect}\,(LT\cdot[x,\ y,\ z,\ ict])$$

where the Macsyma/Maxima function **matrix** can be used to define

$$LT = \begin{bmatrix} \gamma & 0 & 0 & i\gamma \frac{v_1}{c} \\ 0 & 1 & 0 & 0 \\ 0 & 0 & 1 & 0 \\ -i\gamma \frac{v_1}{c} & 0 & 0 & \gamma \end{bmatrix}$$

Verify that the rest-frame values of the following rest-frame four-vectors agree with the transformations given in Appendix A, Section A.2. Use the Chu representation in part (b) and the 4d-em function, **mat2vect**.

a)

$$\mathcal{A}' = \textbf{mat2vect}(\ LT \cdot \mathcal{A}\ )$$
$$\mathcal{J} = \textbf{mat2vect}(\ LT \cdot \mathcal{J})$$

b)

$$\mathcal{E}' = \textbf{mat2vect}(\ LT \cdot \mathcal{E}\ )$$
$$\mathcal{H}' = \textbf{mat2vect}(\ LT \cdot \mathcal{H})$$

c) Generalize LT when $\mathbf{v} = [v_1, v_2, v_3]$ is a constant velocity and compare the result with the 4d-em.mac value **lorentz**.

Problem 25.10-3

a) Verify that the three-space vector fields in the Minkowski formulation are related to those in the Chu formulation by

$$\mathbf{E}^{\mathrm{o}} = \mathbf{E} - \mu_{\mathrm{o}}\mathbf{v} \times \mathbf{M}$$

$$\mathbf{H}^{\mathrm{o}} = \mathbf{H} + \mathbf{v} \times \mathbf{P}$$

$$\mathbf{P}^{\mathrm{o}} = \mathbf{P} + \frac{\mathbf{v} \times \mathbf{M}}{c^2}$$

$$\mathbf{M}^{\mathrm{o}} = \mathbf{M} - \mathbf{v} \times \mathbf{P}$$

b) Find the Chu fields in terms of the Minkowski fields

c) Verify that

$$\mathbf{D} = \varepsilon_{\mathrm{o}}\mathbf{E}^{\mathrm{o}} + \mathbf{P}^{\mathrm{o}} = \varepsilon_{\mathrm{o}}\mathbf{E} + \mathbf{P}$$

$$\mathbf{B} = \mu_{\mathrm{o}}(\mathbf{H}^{\mathrm{o}} + \mathbf{M}^{\mathrm{o}}) = \mu_{\mathrm{o}}(\mathbf{H} + \mathbf{M})$$

and that the four-vector fields, $\mathcal{E}$, $\mathcal{H}$, $\mathcal{P}$, $\mathcal{M}$, $\mathcal{D}$, and $\mathcal{B}$, are the same in either formulation.

Problem 25.10-4

Define the pair of six-vectors:

$$\mathcal{G}_o = \{\mathbf{H}^o;\ -ic\mathbf{D}\}$$

$$\mathcal{K}_o = \{-\mathbf{E}^o;\ -ic\mathbf{B}\}$$

and verify, in the Minkowski formulation, that:

a) Maxwell's Equations follow from

$$\Box \cdot \mathcal{G}_o = \mathcal{J}_u$$

$$\Box \cdot \mathcal{K}_o = 0$$

b) the Minkowski energy-momentum tensor and four-force density follow from

$$\mathrm{II}^{\text{mink}} = \frac{1}{2ic}(\mathcal{G}_o^{\dagger} \cdot \mathcal{K}_o - \mathcal{K}_o^{\dagger} \cdot \mathcal{G}_o)$$

$$\mathcal{F}^{\text{mink}} = \Box \cdot \mathrm{II}^{\text{mink}}$$

Problem 25.10-5

a) Express **D** and **B** in terms of $\mathbf{E}^o$ and $\mathbf{H}^o$ when $\mathcal{D} = \varepsilon_1 \mathcal{E}$, $\mathcal{B} = \mu_1 \mathcal{H}$ (ε_1 and μ_1 are constants), and $\mathbf{v} = [0,\ 0,\ v_3]$.

b) Show that the result is independent of v_3 when eps1 = epsilon and mu1 = mu.

c) Evaluate the Minkowski power-density term,

$$\frac{1}{2}\left(\mathbf{D} \cdot \frac{\partial \mathbf{E}}{\partial t} - \mathbf{E} \cdot \frac{\partial \mathbf{D}}{\partial t} + \mathbf{B} \cdot \frac{\partial \mathbf{H}}{\partial t} - \mathbf{H} \cdot \frac{\partial \mathbf{B}}{\partial t}\right)$$

(the fourth-component of $ic\mathcal{F}^{\text{mink}}$ when $\mathbf{J}_u = 0$) when $\frac{\partial}{\partial t} v_3 \neq 0$, but $v_3 = 0$.

HINT: Open either Macsyma or Maxima[1] and type the following entries:

```
(c1) load("c:\\empbook\\4d-em.mac")$
(c2) functions;
(c3) values;
(c4) eqs:append(eqlist(ddo=eps1*eeo),eqlist(bbo=mu1*hho));
(c5) eqdb:append(d,b);
(c6) linsolve(eqs,eqdb)$
(c7) eq:%,v[1]=0,v[2]=0;                    solution (a)
```

[1] With Maxima, load 4d-em-wxm.mac and type either %i, or c, whichever has been selected to be the inchar.

```
(c8) eq,eps1=epsilon,mu1=mu;
(c9) lratsubst(eqc2,%),factor;                solution (b)
(c10) %i*c*ffmink[4];
(c11) %,eqlist(j);
(c12) eqf4:%,diff;
(c13) lratsubst(eq,eqf4);
(c14) %,diff;
(c15) at(%,v[3]=0);
(c16) factor(%);                              solution (c)
```

Problem 25.10-6

Evaluate the four-dimensional divergences of $\mathcal{D}$ and $\mathcal{B}$ in a medium where $\mathcal{J}_u$ is nonzero and

a) $\mathbf{v} = 0$, but $\Box\mathcal{V} \neq 0$.

b) $\mathbf{v} \neq 0$, but $\Box\mathcal{V} = 0$.

Problem 25.10-7

Use Macsyma/Maxima to verify that in free-space that contains electric charge and current densities, ρ, $\mathbf{J}$:

a)

$$\text{II}^{\text{em}} = \varepsilon_o\mathcal{E}\,\mathcal{E} + \mu_o\mathcal{H}\,\mathcal{H} - \frac{1}{2}(\varepsilon_o\mathcal{E}\cdot\mathcal{E} + \mu_o\mathcal{H}\cdot\mathcal{H})\left(\mathcal{I} + 2\frac{\mathcal{V}\,\mathcal{V}}{c^2}\right)$$
$$+ \frac{i}{c^3}\left[\mathcal{V}\,(\mathcal{E}\times\mathcal{H})^{\dagger}\cdot\mathcal{V} + (\mathcal{E}\times\mathcal{H})^{\dagger}\cdot\mathcal{V}\,\mathcal{V}\right]$$

b)

$$\mathcal{F} = \mu_o\mathcal{G}\cdot\mathcal{J} = \frac{-\mathcal{J}\cdot\mathcal{V}}{c^2}\mathcal{E} + \frac{-i}{c}(\mathcal{J}\times\mu_o\mathcal{H})^{\dagger}\cdot\mathcal{V} + \frac{\mathcal{E}\cdot\mathcal{J}}{c^2}\mathcal{V}$$
$$= \rho\,\mathbf{E} + \mathbf{J}\times\mu_o\mathbf{H},\ \frac{i}{c}\mathbf{E}\cdot\mathbf{J}$$

c)

$$\Box\cdot\text{II}^{\text{o}} = \Box\cdot\text{II}^{\text{em}}$$

where II^{em} and II° (Lorenz gauge) are, respectively, the Maxwell–Poynting and Alternate energy-momentum tensors.

HINT: Use the predefined energy-momentum tensor values,

$$\textbf{nrgm} = \text{II}^{\text{em}}$$
$$\textbf{nrgmo} = \text{II}^{\text{o}}$$

d)

$$\frac{\partial}{\partial x_k}(\mathcal{R}_i \Pi^{em}_{jk} - \Pi^{em}_{ik}\mathcal{R}_j) = (\mathcal{R} \times \mathcal{F})_{ij}$$

$$\frac{\partial}{\partial x_k}(\mathcal{R}_i \Pi^{o}_{jk} - \Pi^{o}_{ik}\mathcal{R}_j) + (\mathcal{J} \times \mathcal{A})_{ij} = (\mathcal{R} \times \mathcal{F})_{ij}$$

where $\mathcal{R} = [x_1 = x,\, x_2 = y,\, x_3 = z,\, x_4 = ict]$, $\mathcal{A} = [\mathbf{A},\, \frac{i}{c}\Phi]$, $\mathcal{J} = [\mathbf{J},\, ic\rho]$, and summation over $k = 1, .., 4$ is assumed.

e) Evaluate the four-vector torque density, $\mathcal{T} = \frac{-i}{c}(\mathcal{R} \times \mathcal{F})^{\dagger} \cdot \mathcal{V}$, and show that, in the rest frame,

$$\mathcal{T}' = [\mathbf{r}' \times \mathbf{f}', 0]$$

HINT: Appendix B, Eqs. (B.16) are helpful identities.

Problem 25.10-8

According to the Theory of Special Relativity (as discussed in Appendix A), time dilation causes clocks in a moving inertial frame of reference to slow down when compared with their stationary counterparts. This result generated the famous "twin paradox" (mentioned in the Section A.1 footnote that is here repeated):

At some reference time, young identical twins are separated; one remains in the laboratory frame, the other is placed aboard a rocket ship that rapidly accelerates to—and then maintains—a velocity very close to c. After many (laboratory-frame) years, the trajectory is reversed and the ship is returned to the starting point. Because of the slowing of time aboard the rocket, the twin who traveled is much younger than the other. Yet by symmetry, the twin on the rocket considers that it is his/her sibling who has been in relativistic motion (with respect to the rocket) and thus should be younger!

The necessary accelerations have always clouded the issues and there are really three inertial frames that must be considered. Instead, we propose a two-frame scenario in which there is no acceleration:

Three surrogate expectant mothers are each carrying one of previously frozen triplet embryos. One woman is aboard a rocket ship traveling at velocity v, which is 80% of the speed of light. The other two remain in the stationary frame; one at the origin, the other a distance L from it, but on the path of the rocket ship. The timing of the pregnancies is such that at the instant of time the rocket ship passes close to the origin, all three women give birth to identical babies; then clocks (which keep perfect time) read zero at all three locations. The two stationary siblings A_1 (at the origin) and A_2 (at L) experience the same time t and grow at the same rate; sibling B experiences time t' in the moving frame and ages differently. At $t = L/v = 5$ years, the rocket ship passes very close to A_2 and both B and A_2 simultaneously take and exchange digitally transmitted photographs and compare their images. Notice that the contemporaneous image of A_1 (although not available) is identical to that of A_2.

a) Which of the triplets is the oldest? Do both A_2 and B agree?

Instead of the situation described above, assume that A_1 and A_2 are aboard separate rocket ships moving at identical velocities (.8c) in the same direction, but spaced in time by 5 years (as measured by B who remains at the origin of the stationary frame). Again, all three births occur when clocks (which keep perfect time) read zero at all three locations as the rocket carrying A_1 passes B. Five years later (as measured by B) A_2 passes B and they exchange images.

b) Which of the triplets is the oldest? Do both A_2 and B agree?

c) An observer in the stationary frame observed the birth of A_2 as the rocket passed by. Where was the observer located, and at what time was the observation made?

d) Can you reconcile the original "twin paradox"?

PART IV

BACKMATTER

SUMMARY

Maxwell's theory of electrodynamics was presented in differential form both when time as a parameter accompanies the three spatial coordinates (Part I) and in four-dimensional space-time where time becomes the fourth (imaginary) coordinate (Part II). Electromagnetic force, power, stress and energy were considered from within the conventional Maxwell–Poynting representation and quasistatic approximations that produce highly localized circuit representations. With appropriate modification, the approximations were altered so as to *exactly* satisfy both the stress-momentum and energy-power theorems. One alternate representation was given special attention because it connects directly with localized circuit-theory power and energy. For an electric charge q and vector line current $\mathbf{I}$, the power $\Phi \mathbf{I}$ and electromagnetic momentum $q\mathbf{A}$ are localized to the current and charge by the scalar and vector potentials. In general, components distributed in the space outside of the sources are also required, but for certain fields these are unnecessary.

Numerous electromagnetic examples are contained in Part III. Included are: quasistatic-field problems; electromagnetic-wave propagation in free-space and linear conducting and dielectric/magnetic materials; transmission lines and waveguides; radiation and diffraction from accelerating charges, antenna currents, and apertures. Equivalent circuits for transmission lines and waveguides are other important topics. The powers and energies in both the Maxwell–Poynting and Alternate representation are compared; in many (but not all) cases, analysis based upon the Alternate representation is simpler. Even when the complete vector and scalar potentials are too complicated to analyze, *any* particular solution that generates the correct electric and magnetic fields can be used to calculate

The Power and Beauty of Electromagnetic Fields, First Edition. F. R. Morgenthaler.

the total power and/or energy. Sometimes, a very simple particular solution is available; in such cases, the Alternate formulation remains advantageous.

Although our analysis and discussion has proceeded from the classical continuum Maxwell Equations, one can at least speculate on possible interpretations of the Alternate power-energy representation within the framework of quantum mechanics (qm) and pose obvious questions: Is the probability density of photons associated with *either* **S** and W^{em} *or* $\mathbf{S}^{\mathrm{o}}$and W^{o} or, in view of wave-particle duality, sometimes one and sometimes the other, *as influenced by the measurement of power*? Because both the $q\mathbf{A}$ conjugate momentum and the $\mathbf{A}\cdot\mathbf{J}$ Hamiltonian density have prominent roles in the qm treatment of charged particles and electromagnetic radiation, such speculations are natural. The Maxwell–Poynting representation seems to favor the wave interpretation; the Alternate-representation, the particle interpretation. In particular, azimuthally circulating modes of a circular waveguide, expressed in terms of the Alternate representation (for either the Lorenz or Coulomb gauge), reveal the particle-like energy, power, and linear momentum as well as $\mathbf{r}\times\rho\mathbf{A}$ and the spin and orbital angular momenta. From the latter, the macroscopic densities of spin ± 1 photons can be inferred. We have also showed that calculation of electromagnetic-torques, due to spin, often depends solely upon $\mathbf{J}\times\mathbf{A}$; this density joins the magnetic force density, $\mathbf{J}\times\mathbf{B}$, as a fundamental and useful concept. Also, we observe that the nonlocalized Alternate complex power flux:

$$-j\omega\frac{1}{4\mu_{\mathrm{o}}}\left(\underline{\underline{A}}\cdot\frac{\partial\underline{\underline{A}}^{*}}{\partial x_k}-\frac{\partial\underline{\underline{A}}}{\partial x_k}\cdot\underline{\underline{A}}^{*}\right)$$

is constructed of components, each of which is of exactly the same form as the probability current in quantum mechanics. The link would be even more compelling were there an equivalence between the vector potential and the qm wavefunction. But, in the case of superconductors, it is known that the vector potential is essentially the wave function elevated to the macroscopic level. Therefore, because of the coherence of the superconducting state, equivalence is made plausible for such materials.

Finally, because energy has a mass equivalent that is affected by gravitational fields, does a comprehensive theory force a choice between alternatives? (Possibly select the Lorenz gauge and allow regions of negative mass density, W^{o}/c^2 (although the total mass would have the same positive value as that calculated from W^{em}/c^2.) Because the electromagnetic power-flux, energy, and momentum densities all enter into the general theory of relativity, such speculation is natural, but certainly well beyond the scope of this text. Nevertheless, it is apparent that the consequences of the classical macroscopic Maxwell Equations run very deep.

The *difference* between the Alternate and Maxwell–Poynting energy-momentum tensor is itself a four-tensor, Π^{b}, that produces neither electromagnetic force nor $\mathbf{E}\cdot\mathbf{J}$ power density and is therefore an ephemeral quantity. In a bit of whimsy that honors the insights of Mason and Weaver, cited in the Preface, this author has dubbed Π^{b} the "electromagnetic-beauty" tensor; it satisfies the conservation law:

$$\Box\cdot\Pi^{\mathrm{b}}=0$$

which is equivalent to

$$\nabla\cdot(\mathbf{S}^{\mathrm{o}}-\mathbf{S}^{\mathrm{em}})+\frac{\partial}{\partial t}(W^{\mathrm{o}}-W^{\mathrm{em}})=0$$

$$\nabla\cdot(\overline{\overline{\mathbf{T}}}^{\mathrm{o}}-\overline{\overline{\mathbf{T}}}^{\mathrm{em}})-\frac{\partial}{\partial t}(\mathbf{G}^{\mathrm{o}}-\mathbf{G}^{\mathrm{em}})=0$$

These power-flux and energy-density differences can also be manipulated to produce (in any gauge) the divergence-free vector,

$$\mathbf{S}^{\text{beauty}} = \frac{1}{2\mu_o}\nabla \times \left(\mathbf{A} \times \frac{\partial \mathbf{A}}{\partial t}\right)$$

The "electromagnetic beauty" of a particular field pattern may be conserved, but what is its value? It is tempting to end this discussion by once again quoting Mason and Weaver [4, p. 326]:

"All statements are true if they are made about nothing."

The statement was made in connection with the properties of the "ether", but perhaps it is more instructive to try and calculate it. As a sample, we offer graphical representations of $\mathbf{S}^{\text{b}} = \mathbf{S}^{\text{o}} - \mathbf{E} \times \mathbf{H}$ for the radiation transient of a Hertzian dipole (at one instant of time and scaled by r^2) as well as the time-averaged values of S_z^{b} for the $TM_{01}{}^{\text{O}}$, $TM_{11}{}^{\text{O}}$, and $TE_{11}{}^{\text{O}}$ modes of a circular waveguide.

The final examples are plots of $\mathbf{S}^{\text{beauty}}$ calculated for a circularly polarized harmonic Gaussian beam (the central slice of this ϕ-directed vector field is the focal plane, $z = 0$) and "circularly-polarized" $TE_{nm}{}^{\text{O}}$ circular waveguide modes for $n, m = 1, 2, 3$. Since "beauty is in the eye of the beholder,"we leave it to the individual reader to judge for himself/herself whether or not it provides an apt description.

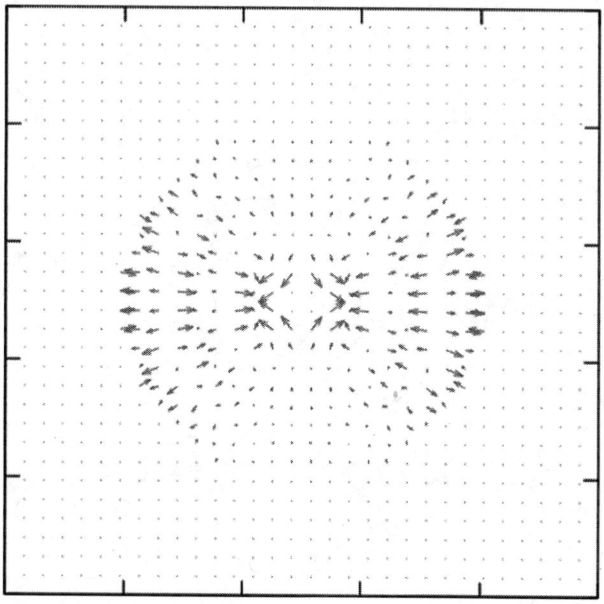

Hertzian dipole: Electromagnetic-beauty power flux

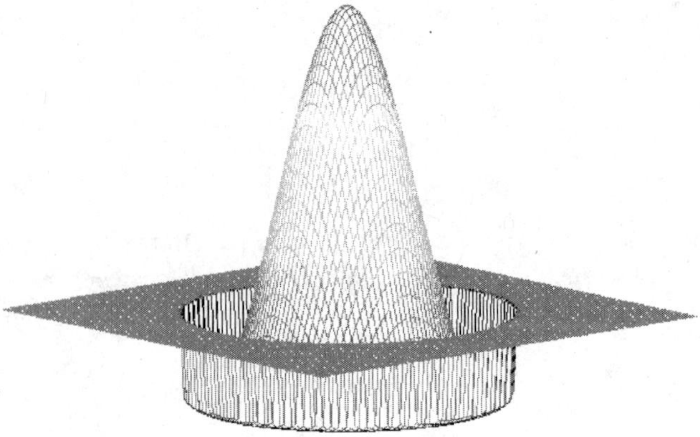

$TM_{01}{}^{\mathrm{O}}$mode: Electromagnetic-beauty power flux, $<S_z^b(x,\ y)>$

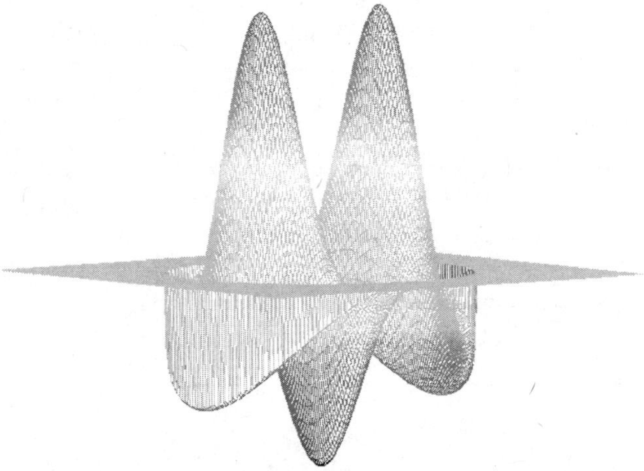

$TM_{11}{}^{\mathrm{O}}$mode: Electromagnetic-beauty power flux, $<S_z^b(x,\ y)>$

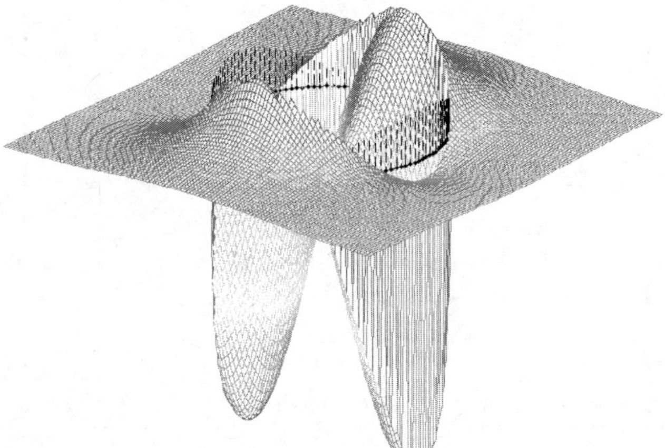

$TE_{11}{}^{\mathrm{O}}$mode: Electromagnetic beauty power flux, $<S_z^b(x,\ y)>$

(a) Vector field plot of $\mathbf{S}^{\text{beauty}} = \hat{\phi}\, S_{\phi}^{\text{beauty}}$

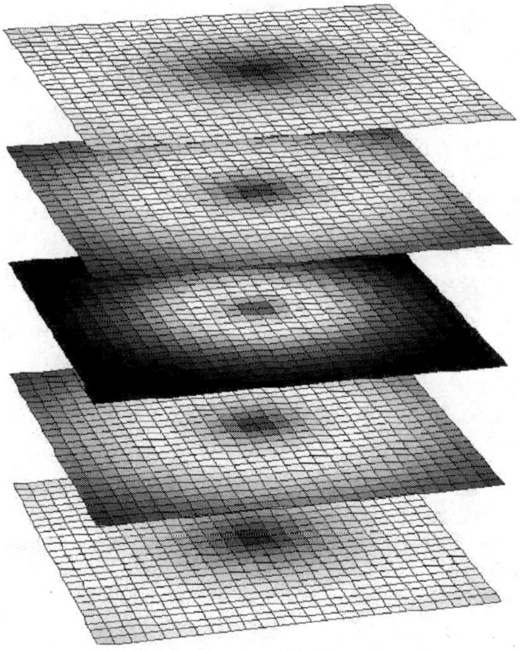

(b) Intensity contours of S_{ϕ}^{beauty}

Circularly polarized Gaussian beam: axial cross sections of $\mathbf{S}^{\text{beauty}}$

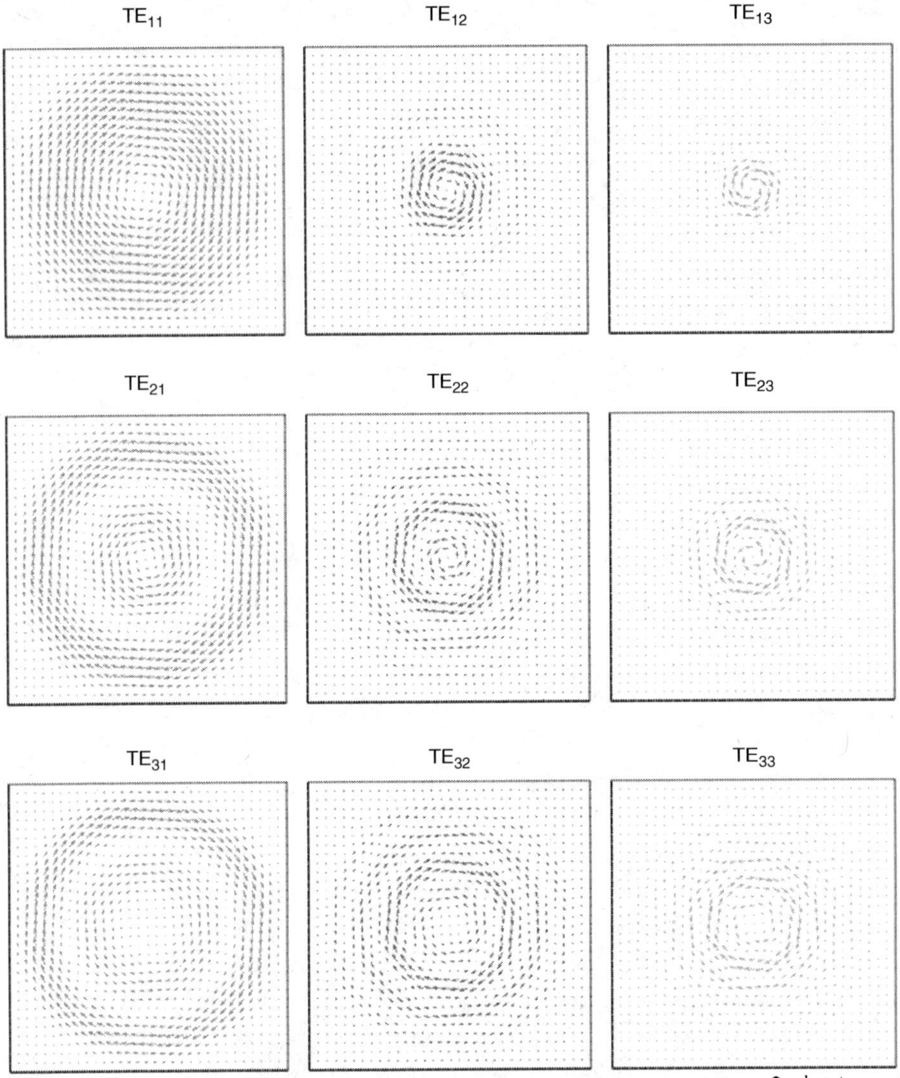

"Circularly polarized" TE_{nm}^{o} modes: Vector field plots of $S^{\text{beauty}} = \widehat{\phi}\, S_{\phi}^{\text{beauty}}$.

ELECTROMAGNETIC LUMINARIES

From a very long view of the history of mankind—seen from, say, ten thousand years from now—there can be little doubt that the most significant event of the 19th century will be judged as Maxwell's discovery of the laws of electrodynamics.

—Richard Feynman

The Power and Beauty of Electromagnetic Fields, First Edition. F. R. Morgenthaler.

Michael Faraday
1791–1867

Karl Friedrich Gauss
1777–1855

James Clerk Maxwell
1831–1879

John Henry
Poynting
1852–1914

Ludwig Valentine
Lorenz
1821–1891

Founders of Electromagnetic Theory

Charles-Augustin
Coulomb
1736–1806

Pierre-Simon Laplace
1749–1827

Simeon Dennis
Poisson
1781–1849

Hans Christian
Oersted
1777–1851

Andre-Marie Ampere
1775–1836

Joseph Henry
1799–1878

Heinrich Hertz
1847–1894

Wiiliam Thompson
(Lord Kelvin)
1824–1907

John William Strutt
(Lord Rayleigh)
1842–1919

Pioneers and Major Contributors

Albert Abraham
Michelson
1852–1931

Edward Williams
Morley
1838–1923

Albert Einstein
1879–1955

Hendrick Antoon
Lorentz
1854–1928

Hermann
Minkowski
1864–1904

Pioneers in Relativity

ABOUT THE AUTHOR

Frederic R. Morgenthaler, a native of Shaker Heights, Ohio, graduated from the Massachusetts Institute of Technology with S.B. and S.M. degrees in Electrical Engineering in June 1956. After serving for two years as a Lieutenant in the United States Air Force, he returned to M.I.T. for doctoral studies under Professor Lan Jen Chu and was granted the Ph.D. degree in June 1960.

Dr. Morgenthaler joined the M.I.T. faculty in 1960 and was promoted to the rank of Professor in 1968; he retired in 1996 and is currently Professor Emeritus of Electrical Engineering. He was Graduate Officer for the Department of Electrical Engineering and Computer Science during 1993–96 and held the Cecil H. Green Professorship for 1984–86. During his career, Professor Morgenthaler was a member of the Research Laboratory of Electronics, the Center for Materials Science and Engineering, and Director of the Microwave and Quantum Magnetics Group. He also held an appointment to the Harvard–M.I.T. Division of Health Sciences and Technology. His research, and graduate teaching, centered on the fields of microwave magnetics and the electrodynamics of waves and media. In addition, Professor Morgenthaler participated in an interdisciplinary course in medical and industrial ultrasonics. He also taught undergraduate electrical engineering core curriculum subjects in electromagnetic field theory, circuit theory, and semiconductor electronics.

The Power and Beauty of Electromagnetic Fields, First Edition. F. R. Morgenthaler.

Professor Morgenthaler served as a consultant to the U.S. Government as well as to private industry. He is a Fellow of the Institute of Electrical and Electronics Engineers and a member of the American Physical Society, the American Association for the Advancement of Science, and the Sigma Xi, Tau Beta Pi, and Eta Kappa Nu honorary professional societies. He is the author of more than 100 scientific publications and papers presented at technical conferences and has been granted approximately one dozen patents.

APPENDIX A

A.1 THEORY OF SPECIAL RELATIVITY

At the turn of the twentieth century, two contradictory statements were unresolved in physics:

1. According to classical mechanics, the velocity of any motion has different values for observers who are moving relative to one another.

2. Experimentally, the velocity of light, relative to ourselves, seems independent of our motion through free-space—a medium of propagation defined as the "ether."

In 1905 Albert Einstein resolved this dilemma by postulating (a) that physical laws are the same in all inertial reference systems (those moving at constant velocity with respect to one another) and (b) that the speed of light is invariant. Stupendous consequences flow from these two postulates.

The Theory of Special Relativity intermingles space and time. As a consequence, measuring rods contract and clocks slow down when in motion. In addition, the concept of absolute simultaneity is lost along with reality of the "ether." Finally, the equivalence between mass and energy, $E = mc^2$, emerges as perhaps the most famous formula in the history of physics.

The Power and Beauty of Electromagnetic Fields, First Edition. F. R. Morgenthaler.

Expanding spherical electromagnetic wave

Consider a Cartesian coordinate system (x', y', z') that is moving at constant velocity, v, with respect to the stationary system (x, y, z). Assume further that at $t = 0$ a spherical electromagnetic wave front is generated at the origin: $x = y = z = 0$, which expands isotropically at the speed of light, c.

With $\mathbf{r} = \widehat{\mathbf{x}}x + \widehat{\mathbf{y}}y + \widehat{\mathbf{z}}z$, it follows that

$$x^2 + y^2 + z^2 = \mathbf{r} \cdot \mathbf{r} = (ct)^2 \tag{A.1a}$$

or (following Minkowski) with $x_1 = x$, $x_2 = y$, $x_3 = z$, $x_4 = ict$,

$$\sum_{k=1}^{4} x_k^2 = \mathbf{r} \cdot \mathbf{r} + (ict)^2 = 0 \tag{A.1b}$$

According to Einstein's Theory of Special Relativity, the same wave front, if observed in the moving coordinate system for which $\mathbf{r}' = \widehat{\mathbf{x}}x' + \widehat{\mathbf{y}}y' + \widehat{\mathbf{z}}z'$, will also be represented by a sphere expanding at the *same* speed of light:

$$x'^2 + y'^2 + z'^2 = \mathbf{r}' \cdot \mathbf{r}' = (ct')^2 \tag{A.2a}$$

or, with $x_1' = x'$, $x_2' = y'$, $x_3' = z'$, $x_4' = ict'$,

$$\sum_{k=1}^{4} x_k'^2 = \mathbf{r}' \cdot \mathbf{r}' + (ict')^2 = 0 \tag{A.2b}$$

where t' is the time as measured in the moving frame. These results assume that the two spatial origins coincide at time $t' = t = 0$; this synchronization can be accomplished without any loss of generality.

Galilean transformation

Since by assumption the origin of the moving coordinates is located at vt and, *by tacit assumption*, $t' = t$, the "common-sense" relationship that expresses $\mathbf{r}'$ in terms of $\mathbf{r}$ is

$$\mathbf{r}' = \mathbf{r} - \mathbf{v}t \tag{A.3}$$

from which it follows that

$$\frac{d\mathbf{r}'}{dt'} = \frac{d\mathbf{r}}{dt} - \mathbf{v} \tag{A.4}$$

Expressed in terms of components $\mathbf{r}_{\parallel}$ and $\mathbf{r}_{\perp}$ that are, respectively, parallel and perpendicular to the velocity vector, $\mathbf{v}$, the so-called Galilean transformation becomes

$$\mathbf{r}_{\perp}' = \mathbf{r}_{\perp} \tag{A.5a}$$

$$\mathbf{r}_{\parallel}' = \mathbf{r}_{\parallel} - \mathbf{v}t \tag{A.5b}$$

$$t' = t \tag{A.5c}$$

Unfortunately, when substituted into Eq. (A.2a), Equations (A.5) do not satisfy the Theory of Special Relativity; fortunately the Lorentz transformation described in the next section does.

Lorentz transformation

Because the failure of the Galilean transformation to satisfy

$$\mathbf{r}_\perp \cdot \mathbf{r}_\perp + \mathbf{r}_\parallel \cdot \mathbf{r}_\parallel + (ict)^2 = \mathbf{r}'_\perp \cdot \mathbf{r}'_\perp + \mathbf{r}'_\parallel \cdot \mathbf{r}'_\parallel + (ict')^2 = 0 \tag{A.6}$$

is not caused by the perpendicular components, we tentatively retain Eq. (A.5a); then

$$r_\parallel^2 + (ict)^2 = r_\parallel'^2 + (ict')^2 \tag{A.7}$$

and we attempt to relate the coordinate pairs $(r'_\parallel, ict')$ and $(r_\parallel, ict)$ by a simple rotation through an angle θ that automatically satisfies the Pythagorean Theorem:

$$r'_\parallel = \cos\theta \; r_\parallel + \sin\theta \; ict \tag{A.8a}$$

$$ict' = -\sin\theta \; r_\parallel + \cos\theta \; ict \tag{A.8b}$$

But by assumption, $r_\parallel = vt$ when $r'_\parallel = 0$; therefore it follows that $\tan\theta = i\frac{\mathrm{v}}{c} = i \tanh \Theta$ is imaginary and

$$\cos\theta = \cosh\Theta = \frac{1}{\sqrt{1 - \frac{\mathrm{v}^2}{\mathrm{c}^2}}} = \gamma \tag{A.9a}$$

$$\sin\theta = i \sinh\Theta = i\gamma\frac{\mathrm{v}}{c} \tag{A.9b}$$

The complete Lorentz transformation is therefore

$$\mathbf{r}'_\perp = \mathbf{r}_\perp \tag{A.10a}$$

$$\mathbf{r}'_\parallel = \gamma(\mathbf{r}_\parallel - \mathbf{v}t) \tag{A.10b}$$

$$t' = \gamma\left(t - \frac{\mathbf{r}\cdot\mathbf{v}}{c^2}\right) \tag{A.10c}$$

or from

$$r_\parallel = \cos\theta \; r'_\parallel - \sin\theta \; ict' \tag{A.11a}$$

$$ict = \sin\theta \; r'_\parallel + \cos\theta \; ict' \tag{A.11b}$$

the inverse transformation is

$$\mathbf{r}_\perp = \mathbf{r}'_\perp \tag{A.12a}$$

$$\mathbf{r}_\parallel = \gamma(\mathbf{r}'_\parallel + \mathbf{v}t') \tag{A.12b}$$

$$t = \gamma\left(t' + \frac{\mathbf{r}'\cdot\mathbf{v}}{c^2}\right) \tag{A.12c}$$

where, in terms of the velocity vector $\mathbf{v}$, we have

$$\gamma = \frac{1}{\sqrt{1 - \frac{\mathbf{v} \cdot \mathbf{v}}{c^2}}} \tag{A.13}$$

Equations (A.10a)–(A.12c) can also be expressed in the equivalent vector forms:

$$\mathbf{r}' = \mathbf{r} + \gamma \mathbf{v} \left(\frac{\gamma}{\gamma + 1} \frac{\mathbf{r} \cdot \mathbf{v}}{c^2} - t \right) \tag{A.14a}$$

$$t' = \gamma \left(t - \frac{\mathbf{r} \cdot \mathbf{v}}{c^2} \right) \tag{A.14b}$$

and

$$\mathbf{r} = \mathbf{r}' + \gamma \mathbf{v} \left(\frac{\gamma}{\gamma + 1} \frac{\mathbf{r}' \cdot \mathbf{v}}{c^2} + t' \right) \tag{A.15a}$$

$$t = \gamma \left(t' + \frac{\mathbf{r}' \cdot \mathbf{v}}{c^2} \right) \tag{A.15b}$$

Notice that if $v/c \ll 1$, the Galilean transformation approximates the Lorentz transformation to a high degree; in the limit $c \to \infty$, $\gamma = 1$ and Eqs. (A.5) are exact. Evidently, our "common-sense" intuition[1] about space and time is correct only at the comparatively slow-speeds we term "nonrelativistic."

On the other hand, the range of velocity $|v|/c > 1$ makes no sense because an imaginary value of γ would lead to imaginary times and distances. This is prevented by Einstein's postulate that c is the maximum value of speed[2] for all ponderable matter. That postulate is consistent with the limit that Maxwell's Equations impose on all electrically charged particles.

Speed limit of a moving charge

Consider a bipolar pair of charges, $\pm q$ (surrounded by free-space), that are closely spaced for all negative time; their common location is taken as the origin of spherical coordinates. Evidently, for $t < 0$ and $r > 0$, the electric field is everywhere zero. Assume now that at $t = 0$ one of the charges remains stationary while the other accelerates very rapidly to some constant radial velocity, v_o. After time t, the charges are separated by the distance $v_o t$. Because of the finite speed of light, knowledge of the charge separation cannot have reached any observer located at a radius $r > ct$ and on or outside of that radius, $\mathbf{E} = 0$. However, Gauss' Law requires that

$$Q = \int \rho \, dV = \oint \varepsilon_o \mathbf{E} \cdot \mathbf{ds}$$

[1]The mixing of spatial and temporal coordinates is unsettling, and the loss of the concept of absolute time is especially so. Indeed, the "weird" nature of the result (the consequent shrinking of yardsticks and slowing down of clocks—when they are moving) causes many individuals to believe that the mathematics underlying the Theory of Special Relativity is too complicated for anyone—who is not an Einstein—to understand. This, despite the fact that understanding the Lorentz transformation requires only a knowledge of Pythagoras' Theorem and imaginary numbers (both concepts taught to nearly all high school students)!

[2]In this context, "speed" refers to the signal-velocity, which—except in highly dispersive (resonant) media—is the same as the group velocity. The phase velocity is often greater than c, but that is no violation of relativity. These matters were clarified by Sommerfeld and Brillouin and are discussed by Stratton [1, p. 339].

As required, the total charge, Q, within the sphere of radius ct, is zero provided both charges are still inside it; this is true only if $v_o < c$. This fundamental speed limit prohibits any charge from violating Maxwell's Equations and is, of course, consistent with the Theory of Special Relativity. Notice, however, that our argument does not apply to uncharged particles. Nevertheless, consider a single particle with an electric-dipole moment oriented parallel to the radial velocity vector. If $v_o > c$, one of the charges will reach and exceed $r = ct$ a moment before the other. But, if correct, Gauss' Law cannot be violated even for a brief time interval; therefore this or other macroscopic particles carrying arbitrary bipolar charges are not exempted from the speed limit—even when their net charge is zero.

Lorentz contraction

With respect to a moving coordinate system, a rod defined by $\mathbf{L}' = \mathbf{r}_2' - \mathbf{r}_1'$ is at rest. In the laboratory frame, the corresponding vectors are given by Eq. (A.15a) and therefore located at

$$\mathbf{r}_1 = \mathbf{r}_1' + \gamma\mathbf{v}\left(t' + \frac{\gamma}{\gamma+1}\frac{\mathbf{r}_1'\cdot\mathbf{v}}{c^2}\right)$$

$$\mathbf{r}_2 = \mathbf{r}_2' + \gamma\mathbf{v}\left(t' + \frac{\gamma}{\gamma+1}\frac{\mathbf{r}_2'\cdot\mathbf{v}}{c^2}\right)$$

The vector length of the rod is defined as $\mathbf{L} = \mathbf{r}_2 - \mathbf{r}_1$ provided the ends are measured at the *same* time, t'. Therefore,

$$\mathbf{L} = \mathbf{L}' + \gamma\left(\frac{\gamma}{\gamma+1}\frac{\mathbf{L}'\cdot\mathbf{v}}{c^2}\mathbf{v}\right) \tag{A.16}$$

or after resolving the vectors into components parallel and perpendicular to the velocity,

$$\mathbf{L}_{\parallel} = \gamma\mathbf{L}_{\parallel}'$$

$$\mathbf{L}_{\perp} = \mathbf{L}_{\perp}'$$

Although there is no change in the perpendicular component, the parallel component in the rest frame experiences a Lorentz contraction. Based upon measurements made when $\Delta t' = 0$, an observer concludes that the rest-frame value of the parallel length vector is reduced by

$$\mathbf{L}_{\parallel}' = \frac{1}{\gamma}\mathbf{L}_{\parallel} \tag{A.17}$$

As mentioned earlier, it follows that primed and unprimed quantities can be interchanged (provided the velocity is reversed); therefore, provided that the measurements are made at the *same* laboratory-frame time, t, we obtain

$$\mathbf{L}' = \mathbf{L} - \gamma\left(\frac{\gamma}{\gamma+1}\frac{-\mathbf{L}\cdot\mathbf{v}}{c^2}\mathbf{v}\right) \tag{A.18}$$

Now, based upon measurements made when $\Delta t = 0$, an observer concludes that it is the laboratory-frame value of the parallel length vector that has been contracted by

$$\mathbf{L}_{\parallel} = \frac{1}{\gamma}\mathbf{L}_{\parallel}' \tag{A.19}$$

Remarkably, *both* observers are correct!

Four-vector length

In the rest frame, the four-vector[3] length is

$$\mathcal{L}' = [\mathbf{L}', 0]$$

whereas in the laboratory frame, a Lorentz transformation produces

$$\mathcal{L} = \left\{ \begin{bmatrix} \gamma & 0 & -i\gamma\frac{v}{c} \\ 0 & 1 & 0 \\ i\gamma\frac{v}{c} & 0 & \gamma \end{bmatrix} \cdot \begin{bmatrix} \mathbf{L}'_{\shortparallel} \\ \mathbf{L}'_{\perp} \\ 0 \end{bmatrix} \right\}^{\text{transpose}} = \left[\gamma \mathbf{L}'_{\shortparallel} + \mathbf{L}'_{\perp},\ i\gamma \frac{\mathbf{L}' \cdot \mathbf{v}}{c} \right]$$

where $\mathcal{L}' \cdot \mathcal{L}' = \mathcal{L} \cdot \mathcal{L}$. We wish to express the length-vector in terms of the laboratory-frame values (where both ends of $\mathbf{L}'$ are measured at the same time, t) and accordingly use Eq. (A.19) and $\mathbf{L}'_{\perp} = \mathbf{L}_{\perp}$; the result is

$$\mathcal{L} = \left[\gamma^2 \mathbf{L}_{\shortparallel} + \mathbf{L}_{\perp},\ i\gamma^2 \frac{\mathbf{L} \cdot \mathbf{v}}{c} \right]$$

or equivalently

$$\mathcal{L} = \left[\mathbf{L} + \gamma^2 \frac{\mathbf{L} \cdot \mathbf{v}}{c^2} \mathbf{v},\ i\gamma^2 \frac{\mathbf{L} \cdot \mathbf{v}}{c} \right] \tag{A.20}$$

This form is valid in any inertial frame of reference.

Particle density

If particles of volume density n occupy a volume element $\Delta L_{\shortparallel} \Delta A_{\perp}$ that is aligned with $\Delta L_{\shortparallel}$ parallel to the velocity, the number of particles enclosed is $n \Delta L_{\shortparallel} \Delta A_{\perp}$. That number is the *same* when evaluated in the rest frame, therefore

$$n \Delta L_{\shortparallel} \Delta A_{\perp} = n' \Delta L'_{\shortparallel} \Delta A'_{\perp}$$

Based on measurements made at the same value of t, $\Delta L'_{\shortparallel} = \gamma \Delta L_{\shortparallel}$ and since all elements in the planes of area lie perpendicular to the velocity, $\Delta A_{\perp} = \Delta A'_{\perp}$. It therefore follows that

$$n = \gamma n' \tag{A.21}$$

Regardless of the shape chosen for the differential volume element, the Lorentz contraction of all parallel dimensions will cause $\Delta V' = \gamma \Delta V$.

Simultaneity and time dilation

Consider two observers moving at the same velocity, but located at different values of $\mathbf{r}'$; each carries a synchronized watch that keeps perfect time. In the laboratory frame, located along their path of motion, are a series of synchronized clocks which can be read by the observers. At the same time, t', each reads the time, t, displayed on the clock that is opposite him/her.

[3]Four-vector and six-vector notation (defined in Appendix B, Section B.3) is introduced in Part II, Chapter 7, in connection with four-dimensional electrodynamics.

According to Eq. (A.15b), repeated for convenience,

$$t = \gamma\left(t' + \frac{\mathbf{r}' \cdot \mathbf{v}}{c^2}\right)$$

each observer measures a *different* value of t. Events that are *simultaneous* in the moving frame are seen to occur at *different* times in the laboratory frame; because of Eq. (A.14b), the reverse statement is also true. The Theory of Special Relativity has thus dealt a mortal blow to the long cherished concept of simultaneity. Yet there are even stranger consequences. A clock located at some fixed position $\mathbf{r}'$ in the moving frame ($\frac{d\mathbf{r}'}{dt'} = 0$) will measure out intervals of time, $\Delta t'$, that are related to intervals in the laboratory frame, Δt, by

$$\Delta t' = \frac{1}{\gamma}\Delta t \tag{A.22}$$

This is the famous time-dilation effect by which moving clocks lose time compared with their stationary counterparts.

Again, symmetry of the Lorentz transformation requires a moving observer to conclude that it is the clock located at some fixed laboratory position, $\mathbf{r}$ ($\frac{d\mathbf{r}}{dt} = 0$) that will lose time according to

$$\Delta t = \frac{1}{\gamma}\Delta t' \tag{A.23}$$

Even more remarkably, *both* observers are again correct![4]

Addition of velocities

In general, $\frac{d\mathbf{r}}{dt}$ is not the same as the velocity of the moving frame; to avoid confusion we denote the latter by $\mathbf{v}_o$ and replace γ with

$$\gamma_o = \frac{1}{\sqrt{1 - \dfrac{\mathbf{v}_o \cdot \mathbf{v}_o}{c^2}}}$$

Then Eqs. (A.15a) and (A.15b) become

$$\mathbf{r} = \mathbf{r}' + \gamma_o\left(\frac{\gamma_o}{\gamma_o + 1}\frac{\mathbf{r}' \cdot \mathbf{v}_o}{c^2} + t'\right)\mathbf{v}_o$$

$$t = \gamma_o\left(t' + \frac{\mathbf{r}' \cdot \mathbf{v}_o}{c^2}\right)$$

[4]This result has spawned the famous "twin-paradox": At some reference time, young identical twins are separated; one remains in the laboratory frame, the other is placed aboard a rocket ship that rapidly accelerates to, and then maintains, a velocity very close to c. After many (laboratory frame) years, the trajectory is reversed and the ship is returned to the starting point. Because of the slowing of time aboard the rocket, the twin who traveled is much younger than the other. Yet by symmetry, the twin on the rocket considers that it is his/her sibling who has been in relativistic motion (with respect to the rocket) and thus should be younger!

Both of these equations may be differentiated with respect to t to generate $\frac{d\mathbf{r}}{dt}$ and $\frac{d\mathbf{r}'}{dt} = \frac{d\mathbf{r}'}{dt'}\frac{dt'}{dt}$. The results

$$\frac{d\mathbf{r}}{dt} = \left[\frac{d\mathbf{r}'}{dt'} + \gamma_o\left(\frac{\gamma_o}{\gamma_o+1}\frac{\frac{d\mathbf{r}'}{dt'}\cdot\mathbf{v}_o}{c^2} + 1\right)\mathbf{v}_o\right]\frac{dt'}{dt}$$

$$1 = \gamma_o\left(1 + \frac{\frac{d\mathbf{r}'}{dt'}\cdot\mathbf{v}_o}{c^2}\right)\frac{dt'}{dt}$$

may then be divided so as to eliminate $\frac{dt'}{dt}$ and produce the equation governing the combination of velocities.

$$\frac{d\mathbf{r}}{dt} = \frac{\frac{1}{\gamma_o}\frac{d\mathbf{r}'}{dt'} + \left(1 + \frac{\gamma_o}{\gamma_o+1}\frac{\frac{d\mathbf{r}'}{dt'}\cdot\mathbf{v}_o}{c^2}\right)\mathbf{v}_o}{1 + \frac{\frac{d\mathbf{r}'}{dt'}\cdot\mathbf{v}_o}{c^2}}$$

Finally, with $\frac{d\mathbf{r}}{dt} = \mathbf{v}$, $\frac{d\mathbf{r}'}{dt'} = \mathbf{v}'$ and decomposition of $\mathbf{v}$ and $\mathbf{v}'$ into components parallel and perpendicular to $\mathbf{v}_o$, there results (after simplification)

$$\mathbf{v} = \frac{\mathbf{v}'_{\parallel} + \frac{1}{\gamma_o}\mathbf{v}'_{\perp} + \mathbf{v}_o}{1 + \frac{\mathbf{v}'\cdot\mathbf{v}_o}{c^2}} \tag{A.24a}$$

As expected, the inverse formula is

$$\mathbf{v}' = \frac{\mathbf{v}_{\parallel} + \frac{1}{\gamma_o}\mathbf{v}_{\perp} - \mathbf{v}_o}{1 - \frac{\mathbf{v}\cdot\mathbf{v}_o}{c^2}} \tag{A.24b}$$

These formulas can also be derived by applying a Lorentz transformation (based on $\pm\mathbf{v}_o$ and γ_o) to the velocity four-vector $\mathcal{V}' = [\gamma'\mathbf{v}',\ i\gamma'c]$ or $\mathcal{V} = [\gamma\mathbf{v},\ i\gamma c]$.

Because the magnitudes of the individual velocities $\mathbf{v}'$ and $\mathbf{v}_o$ are bounded by the velocity of light, it is easy to verify that (regardless of orientation), the magnitude of their "sum" can never exceed the value c. For nonrelativistic velocities these formulas approximate the "common-sense" vector addition of everyday experience.

The reader should also verify that, in terms of the four-velocity $\mathcal{V}_o = [\gamma_o \mathbf{v}_o,\ i\gamma_o c]$,

$$Re\{\mathcal{V}\} = Re\left\{\mathcal{V}' + \frac{\gamma + \gamma'}{\gamma_o + 1}\mathcal{V}_o\right\} \tag{A.25a}$$

$$Re\{\mathcal{V}'\} = Re\left\{\mathcal{V} - \frac{\gamma + \gamma'}{\gamma_o + 1}\mathcal{V}_o\right\} \tag{A.25b}$$

$$\gamma = \frac{\mathcal{V}' \cdot \mathcal{V}_o^*}{c^2} = \frac{1}{\sqrt{1 - \frac{\mathbf{v}\cdot\mathbf{v}}{c^2}}} \tag{A.25c}$$

$$\gamma' = \frac{-\mathcal{V} \cdot \mathcal{V}_o}{c^2} = \frac{1}{\sqrt{1 - \frac{\mathbf{v}'\cdot\mathbf{v}'}{c^2}}} \tag{A.25d}$$

Velocity dependence of mass

Consider an observer A who is stationary in the laboratory frame of reference; he holds an elastic sphere. A second observer B moving at velocity $\mathbf{v}_o$ with respect to A holds an identical sphere. As B passes close by but slightly above A, *each* releases his sphere with a *very small velocity* (of equal magnitude) that is perpendicular to $\mathbf{v}_o$ and directed toward the other. From A's point of view, the velocity of his sphere (of mass m') is $\mathbf{v}'_\perp$ (up) while that of the other sphere (of mass m) is $\mathbf{v}_o - \mathbf{v}_\perp$. From B's point of view, the velocity of his sphere (of mass m') is $-\mathbf{v}'_\perp$ (down) while that of the other sphere (of mass m) is $-\mathbf{v}_o + \mathbf{v}_\perp$. The total perpendicular component of the linear momentum just before the elastic collision is $m'\mathbf{v}'_\perp - m\,\mathbf{v}_\perp$; just after, it is $-m'\mathbf{v}'_\perp + m\,\mathbf{v}_\perp$ because the velocities reverse in the energy conserving interaction. But linear momentum is also conserved, so that $m'\mathbf{v}'_\perp = m\,\mathbf{v}_\perp$. According to Eq. (A.24a), $\mathbf{v}' = \gamma_o \mathbf{v}_\perp$; therefore $m = \gamma_o m'$. Finally, because $|\mathbf{v}_\perp| \ll |\mathbf{v}_o|$ and $\mathbf{v}_{\shortparallel} = \mathbf{v}_o$, there is no distinction between γ_o and γ and

$$m = \gamma m' \tag{A.26}$$

Force, power, and energy

With a velocity-dependent mass, Newton's Law becomes

$$\frac{d}{dt}(m\mathbf{v}) = \frac{d}{dt}(\gamma m' \mathbf{v}) = \mathbf{F} \tag{A.27}$$

The power delivered by the force is $\mathbf{F} \cdot \mathbf{v} = \frac{dE}{dt}$, the time rate of change of the energy E that is associated with the mass. From Eq. (A.27) and $\mathbf{v} \cdot \mathbf{v} = c^2(1 - \gamma^{-2})$,

$$\mathbf{F} \cdot \mathbf{v} = \frac{d}{dt}(\gamma m' \mathbf{v} \cdot \mathbf{v}) - \gamma m' \frac{1}{2}\frac{d(\mathbf{v}\cdot\mathbf{v})}{d\gamma}\frac{d\gamma}{dt} \tag{A.28a}$$

$$\frac{dE}{dt} = \frac{d}{dt}(\gamma m' c^2) \tag{A.28b}$$

The last equation when integrated yields

$$E = mc^2 \tag{A.29}$$

although setting the integration constant to zero requires additional consideration.

These results can be conveniently represented in four-dimensional notation. The four-vector linear momentum $\mathcal{P}$ and force $\mathcal{F}^5$ are defined by

$$\mathcal{P} = m'\mathcal{V} = [\gamma m'\mathbf{v},\ i\gamma m'c] = \left[\mathbf{p}, \frac{i}{c}E\right] \tag{A.30}$$

and

$$\frac{\partial \mathcal{P}}{\partial \tau} = \gamma \frac{d\mathcal{P}}{dt} = \mathcal{F} = \left[\gamma \mathbf{F}, \frac{i}{c}\gamma \mathbf{F}\cdot\mathbf{v}\right] \tag{A.31}$$

where τ is the proper time. It follows that $\mathcal{F}\cdot\mathcal{V} = 0$, $\mathcal{P}\cdot\mathcal{P} = -(m'c)^2$ are Lorentz invariants; the latter can be expressed as

$$E^2 = c^2\mathbf{p}\cdot\mathbf{p} + (m')^2c^4 \tag{A.32}$$

When $m' > 0$, the energy can be expanded in powers of $|\mathbf{v}|/c$; if the velocity is nonrelativistic, the familiar kinetic energy term emerges from

$$E \simeq m'c^2 + \frac{1}{2}m'\mathbf{v}\cdot\mathbf{v} + \cdots \tag{A.33}$$

Since an object with finite rest mass would have infinite energy if it moved at the velocity of light, the fundamental speed limit—which is a property of space–time—is made understandable from yet another perspective.

For a free-space photon, $E = \hbar\omega$ and $\mathbf{p} = \hbar\mathbf{k}$. Because $\omega = ck$, it follows from Eq. (A.32) that m', the rest mass of the photon, is zero.

A.2 TRANSFORMATIONS BETWEEN FIXED AND MOVING COORDINATES

Electromagnetic fields and scalars

The straightforward method of transforming field vectors between stationary and moving inertial frames of reference is to apply a Lorentz transformation to the appropriate field tensor. This "rotation" in four-space produces a new six-vector with the transformed Cartesian field components. The transformation coefficients are based upon Eqs. (A.14a) and (A.14b), where, following common practice, all rest-frame quantities are denoted by primes. For example, if $\mathcal{G}^{\text{amp}}$ is transformed according to

$$\mathcal{G}'^{\text{amp}}_{ij} = \frac{\partial x'_i}{\partial x_k}\frac{\partial x'_j}{\partial x_l}\mathcal{G}^{\text{amp}}_{kl}$$

(summation over k and l is implied), so too are $\mathbf{E}$ and $\mathbf{B}$. Instead, we choose an alternate (easier) approach.

Because all of the four-vectors introduced in the four-dimensional formulation of electrodynamics are themselves Lorentz-invariant, the magnitude of each can be evaluated in terms of the three-space vectors defined in either the laboratory or the rest frame (which by definition is moving at the three-vector velocity, $\mathbf{v}$).

It is helpful to resolve each vector into components that are parallel and perpendicular to the velocity. As an example, we choose the four-vector electric field

$$\mathcal{E} = \left[\gamma(\mathbf{E}^{\text{o}} + \mathbf{v}\times\mathbf{B}),\ \gamma\frac{i}{c}\mathbf{E}^{\text{o}}\cdot\mathbf{v}\right] = \left[\gamma(\mathbf{E}^{\text{o}}_{\parallel} + \mathbf{E}^{\text{o}}_{\perp} + \mathbf{v}\times\mathbf{B}),\ \gamma\frac{i}{c}\mathbf{E}^{\text{o}}_{\parallel}\cdot\mathbf{v}\right]$$

[5]Not to be confused with the four-vector polarization and force density.

which in the (primed) rest-frame coordinates is

$$\mathcal{E}' = [\mathbf{E}^{o\prime},\ 0] = [\mathbf{E}^{o\prime}_{\parallel} + \mathbf{E}^{o\prime}_{\perp},\ 0] \tag{A.34}$$

It is important to realize that although $\mathcal{E} \neq \mathcal{E}'$ (unless $\mathbf{v} = 0$), their magnitudes are equal.

When evaluated in both frames and equated, the invariant can be written as $\mathcal{E}' \cdot \mathcal{E}' = \mathcal{E} \cdot \mathcal{E}$ or, equivalently,

$$E^{o\prime 2}_{\parallel} + E^{o\prime 2}_{\perp} = \gamma^2[E^{o2}_{\parallel} + (\mathbf{E}^o_{\perp} + \mathbf{v} \times \mathbf{B}) \cdot (\mathbf{E}^o_{\perp} + \mathbf{v} \times \mathbf{B})] - \gamma^2 E^{o2}_{\parallel} \frac{\mathbf{v} \cdot \mathbf{v}}{c^2} \tag{A.35a}$$

After substitution of $\gamma^2 \frac{\mathbf{v} \cdot \mathbf{v}}{c^2} = \gamma^2 - 1$, Eq. (A.35a) becomes

$$E^{o\prime 2}_{\parallel} + E^{o\prime 2}_{\perp} = E^{o2}_{\parallel} + \gamma^2(\mathbf{E}^o_{\perp} + \mathbf{v} \times \mathbf{B}) \cdot (\mathbf{E}^o_{\perp} + \mathbf{v} \times \mathbf{B}) \tag{A.35b}$$

and because the parallel and perpendicular components are orthogonal, we obtain

$$E^{o\prime 2}_{\parallel} = E^{o2}_{\parallel}$$

$$E^{o\prime 2}_{\perp} = \gamma^2(\mathbf{E}^o_{\perp} + \mathbf{v} \times \mathbf{B}) \cdot (\mathbf{E}^o_{\perp} + \mathbf{v} \times \mathbf{B})$$

Therefore,

$$\mathbf{E}^{o\prime} = \mathbf{E}^o_{\parallel} + \gamma(\mathbf{E}^o_{\perp} + \mathbf{v} \times \mathbf{B}) \tag{A.36a}$$

Because $\mathcal{B}$, $\mathcal{H}$, and $\mathcal{D}$ share similar forms, it follows that

$$\mathbf{B}' = \mathbf{B}_{\parallel} + \gamma\left(\mathbf{B}_{\perp} - \frac{\mathbf{v}}{c^2} \times \mathbf{E}^o\right) \tag{A.36b}$$

$$\mathbf{H}^{o\prime} = \mathbf{H}^o_{\parallel} + \gamma(\mathbf{H}^o_{\perp} - \mathbf{v} \times \mathbf{D}) \tag{A.36c}$$

$$\mathbf{D}' = \mathbf{D}_{\parallel} + \gamma\left(\mathbf{D}_{\perp} + \frac{\mathbf{v}}{c^2} \times \mathbf{H}^o\right) \tag{A.36d}$$

In all of the above, the Minkowski three-space fields were used to express the four-vector fields, but other choices can be made. For example, if $\mathcal{E}$ and $\mathcal{H}$ are written in terms of the Chu fields, **E** and **H**, the result is

$$\mathbf{E}' = \mathbf{E}_{\parallel} + \gamma(\mathbf{E}_{\perp} + \mathbf{v} \times \mu_o \mathbf{H}) \tag{A.37a}$$

$$\mathbf{H}' = \mathbf{H}_{\parallel} + \gamma(\mathbf{H}_{\perp} - \mathbf{v} \times \varepsilon_o \mathbf{E}) \tag{A.37b}$$

Invariants can be formed by taking the dot product of two different four-vectors. If $\mathcal{V}$ is chosen as one and the other has no fourth-component in the rest frame, orthogonality results. Thus, it follows that

$$\mathcal{V} \cdot \mathcal{E} = 0$$

$$\mathcal{V} \cdot \mathcal{H} = 0$$

$$\mathcal{V} \cdot \mathcal{D} = 0$$

$$\mathcal{V} \cdot \mathcal{B} = 0$$

$$\mathcal{V} \cdot \mathcal{P} = 0$$

$$\mathcal{V} \cdot \mathcal{M} = 0$$

On the other hand,

$$\mathcal{V}\cdot\mathcal{A}=\gamma(\mathbf{A}\cdot\mathbf{v}-\Phi)=-\Phi' \tag{A.38}$$

$$\mathcal{V}\cdot\mathcal{J}=\gamma(\mathbf{J}\cdot\mathbf{v}-c^2\rho)=-c^2\rho' \tag{A.39}$$

Other invariants which are the magnitudes of four-vectors are

$$\mathcal{V}\cdot\mathcal{V}=-c^2 \tag{A.40}$$

$$\mathcal{A}\cdot\mathcal{A}=\mathbf{A}\cdot\mathbf{A}-\frac{\Phi^2}{c^2}=\mathbf{A}'\cdot\mathbf{A}'-\frac{\Phi'^2}{c^2} \tag{A.41}$$

$$\mathcal{J}\cdot\mathcal{J}=\mathbf{J}\cdot\mathbf{J}-c^2\rho^2=\mathbf{J}'\cdot\mathbf{J}'-c^2\rho'^2 \tag{A.42}$$

If Φ' from Eq. (A.38) is substituted into Eq. (A.41) and ρ' from Eq. (A.39) into Eq. (A.42), the results after rearrangement are

$$\mathbf{A}_\perp\cdot\mathbf{A}_\perp+\gamma^2\left(\mathbf{A}_\parallel-\frac{\Phi}{c^2}\mathbf{v}\right)^2=\mathbf{A}'_\parallel\cdot\mathbf{A}'_\parallel+\mathbf{A}'_\perp\cdot\mathbf{A}'_\perp$$

$$\mathbf{J}_\perp\cdot\mathbf{J}_\perp+\gamma^2(\mathbf{J}_\parallel-\rho\mathbf{v})^2=\mathbf{J}'_\parallel\cdot\mathbf{J}'_\parallel+\mathbf{J}'_\perp\cdot\mathbf{J}'_\perp$$

Therefore,

$$\mathbf{A}'=\gamma\left(\mathbf{A}_\parallel-\Phi\frac{\mathbf{v}}{c^2}\right)+\mathbf{A}_\perp \tag{A.43a}$$

$$\Phi'=\gamma(\Phi-\mathbf{A}\cdot\mathbf{v})=-\mathcal{V}\cdot\mathcal{A} \tag{A.43b}$$

and

$$\mathbf{J}'=\gamma(\mathbf{J}_\parallel-\rho\mathbf{v})+\mathbf{J}_\perp \tag{A.44a}$$

$$\rho'=\gamma\left(\rho-\frac{\mathbf{J}\cdot\mathbf{v}}{c^2}\right)=-\frac{\mathcal{V}\cdot\mathcal{J}}{c^2} \tag{A.44b}$$

Another well-known invariant is

$$\varepsilon_0\mathcal{E}\cdot\mathcal{E}-\mu_0\mathcal{H}\cdot\mathcal{H}=\varepsilon_0\mathbf{E}\cdot\mathbf{E}-\mu_0\mathbf{H}\cdot\mathbf{H}=\varepsilon_0\mathbf{E}'\cdot\mathbf{E}'-\mu_0\mathbf{H}'\cdot\mathbf{H}'$$

Finally, it should be noted that because the Lorentz transformation is symmetric with respect to the primed and unprimed coordinates, all of the formulas that express rest-frame quantities in terms of their laboratory-frame counterparts can be inverted by simply interchanging the primed and unprimed values and reversing the sign of the velocity vector. For example, the inverses of Eqs. (A.37a) and (A.37b) are

$$\mathbf{E}=\mathbf{E}'_\parallel+\gamma(\mathbf{E}'_\perp-\mathbf{v}\times\mu_0\mathbf{H}')$$

$$\mathbf{H}=\mathbf{H}'_\parallel+\gamma(\mathbf{H}'_\perp+\mathbf{v}\times\varepsilon_0\mathbf{E}')$$

Sinusoidal steady-state plane waves

The previous results apply to all electric and magnetic fields including those representing sinusoidal steady-state plane waves of the form

$$\mathbf{E} = Re\{\underline{\mathbf{E}}_o \exp[j(\omega t - \mathbf{k}\cdot\mathbf{r})]\}$$

$$\mathbf{H} = Re\{\underline{\mathbf{H}}_o \exp[j(\omega t - \mathbf{k}\cdot\mathbf{r})]\}$$

which in the rest frame become

$$\mathbf{E}' = Re\{\underline{\mathbf{E}}'_o \exp[j(\omega' t' - \mathbf{k}'\cdot\mathbf{r}')]\}$$

$$\mathbf{H}' = Re\{\underline{\mathbf{H}}'_o \exp[j(\omega' t' - \mathbf{k}'\cdot\mathbf{r}')]\}$$

It follows that the phase is itself a Lorentz-invariant,

$$\omega' t' - \mathbf{k}'\cdot\mathbf{r}' = \omega t - \mathbf{k}\cdot\mathbf{r} \tag{A.45}$$

After substitution of Eqs. (A.14a) and (A.14b) and the resolution of vectors into parallel and perpendicular components, the rest-frame values of frequency and wavevector are found to be

$$\omega' = \gamma(\omega - \mathbf{v}\cdot\mathbf{k}) \tag{A.46a}$$

$$\mathbf{k}' = \gamma\left(\mathbf{k}_{\parallel} - \frac{\omega}{c^2}\mathbf{v}\right) + \mathbf{k}_{\perp} \tag{A.46b}$$

The inverse relationships are

$$\omega = \gamma(\omega' + \mathbf{v}\cdot\mathbf{k}') \tag{A.47a}$$

$$\mathbf{k} = \gamma\left(\mathbf{k}'_{\parallel} + \frac{\omega'}{c^2}\mathbf{v}\right) + \mathbf{k}'_{\perp} \tag{A.47b}$$

Energy-momentum tensors

A Lorentz transformation mixes electric and magnetic fields, scalar and vector potentials, and current and charge densities; when applied to any one of the energy-momentum tensors defined in Part II, Chapter 8, it mixes energy density with components of power flux, momentum density, and stress according to

$$\Pi'_{ij} = \frac{\partial x'_i}{\partial x_k}\frac{\partial x'_j}{\partial x_l}\Pi_{kl}$$

Although evaluation is straightforward, an alternate method is more instructive. When the medium is linear, the time-averaged power, energy, stress, and momentum components that are associated with a plane wave of frequency ω and wavenumber $\mathbf{k}$ can be represented by quasiparticles[6] of energy $\hbar\omega$, momentum $\hbar\mathbf{k}$, and power $\hbar\omega\mathbf{v}_{\text{group}}$, where

[6]In free-space, $|\mathbf{v}_g| = c$, and without dispersion the quasiparticles are photons with zero rest mass; in a dielectric/magnetic medium they become part of a collective excitation that may include mechanical attributes and are therefore dispersive (the group velocity is ω and/or $\mathbf{k}$-dependent).

$\mathbf{v}_{\text{group}} = \frac{\partial \omega}{\partial \mathbf{k}}$ is the group velocity. If n represents the volume density of the particles, the time-averaged energy-momentum tensor[7] can be expressed (in dyadic notation) as

$$\mathbf{II} = \begin{bmatrix} \overline{\mathbf{T}} & -ic\mathbf{G} \\ -\frac{i}{c}\mathbf{S} & W \end{bmatrix} = \begin{bmatrix} -n\hbar\mathbf{k}\,\mathbf{v}_{\text{group}} & -icn\hbar\mathbf{k} \\ -\frac{i}{c}n\hbar\omega\mathbf{v}_{\text{group}} & n\hbar\omega \end{bmatrix}$$

Previously, we learned how ω, $\mathbf{k}$, and $\mathbf{v}_{\text{group}}$ are altered by a Lorentz transformation of velocity $\mathbf{v}$. Repeated here for convenience are

$$\omega' = \gamma(\omega - \mathbf{v}\cdot\mathbf{k})$$

$$\mathbf{k}' = \gamma\mathbf{k}_{\parallel} + \mathbf{k}_{\perp} - \gamma\frac{\omega}{c^2}\mathbf{v}$$

$$\mathbf{v}'_{\text{group}} = \frac{\mathbf{v}_{\text{group}\parallel} + \frac{1}{\gamma}\mathbf{v}_{\text{group}\perp} - \mathbf{v}}{1 - \frac{\mathbf{v}_{\text{group}}\cdot\mathbf{v}}{c^2}}$$

The remaining task is to transform the particle density between inertial frames. With respect to the laboratory frame, the particles of density n are moving at velocity $\mathbf{v}_{\text{group}}$ and $n = \gamma_g n_g$; with respect to the primed frame, those of density n' are moving at velocity $\mathbf{v}'_{\text{group}}$ and $n' = \gamma'_g n_g$. From Eqs. (A.25c) and (A.25d), we obtain

$$\gamma_g = \frac{1}{\sqrt{1 - \frac{\mathbf{v}_{\text{group}}\cdot\mathbf{v}_{\text{group}}}{c^2}}} = \gamma\gamma'_g\left(1 + \frac{\mathbf{v}'_{\text{group}}\cdot\mathbf{v}}{c^2}\right)$$

$$\gamma'_g = \frac{1}{\sqrt{1 - \frac{\mathbf{v}'_{\text{group}}\cdot\mathbf{v}'_{\text{group}}}{c^2}}} = \gamma\gamma_g\left(1 - \frac{\mathbf{v}_{\text{group}}\cdot\mathbf{v}}{c^2}\right)$$

and n_g is the density in the frame in which the particles are motionless.[8] Using the second representation for γ'_g leads to

$$n' = \frac{\gamma'_g}{\gamma_g}n = \gamma n\left(1 - \frac{\mathbf{v}_{\text{group}}\cdot\mathbf{v}}{c^2}\right)$$

[7]For a nonlinear medium, linearization about an operating point is often possible; in that case the following analysis applies to the small-signal energy-momentum tensor. The transformation laws, once found, are generally applicable.

[8]If the particles (photons) move through free-space at the speed of light, c, there is no special "ether" frame, both γ_g and γ'_g are infinite, and $n_g = 0$. This presents no difficulties because, for any physical velocity, $\mathbf{v}$ (which defines the primed frame), $|\mathbf{v}| < c$, and n, n', and γ are all finite. On the other hand, if they propagate through dielectric/magnetic material at slower speeds, γ_g, γ'_g, and n_g are all finite.

In the primed (moving) frame it then follows that

$$\overline{\mathbf{T}}' = -n'\hbar\mathbf{k}'\,\mathbf{v}'_{\mathrm{group}} = -n\hbar\gamma\left(\gamma\mathbf{k}_{\shortparallel} + \mathbf{k}_{\perp} - \gamma\frac{\omega}{c^2}\mathbf{v}\right)\left(\mathbf{v}_{\mathrm{group}\shortparallel} + \frac{1}{\gamma}\mathbf{v}_{\mathrm{group}\perp} - \mathbf{v}\right)$$

$$\mathbf{G}' = n'\hbar\mathbf{k}' = n\hbar\gamma\left(\gamma\mathbf{k}_{\shortparallel} + \mathbf{k}_{\perp} - \gamma\frac{\omega}{c^2}\mathbf{v}\right)\left(1 - \frac{\mathbf{v}_{\mathrm{group}}\cdot\mathbf{v}}{c^2}\right)$$

$$\mathbf{S}' = n'\hbar\omega'\mathbf{v}'_{\mathrm{group}} = n\hbar\gamma^2(\omega - \mathbf{v}\cdot\mathbf{k})\left(\mathbf{v}_{\mathrm{group}\shortparallel} + \frac{1}{\gamma}\mathbf{v}_{\mathrm{group}\perp} - \mathbf{v}\right)$$

$$W' = n'\hbar\omega' = n\hbar\gamma^2(\omega - \mathbf{v}\cdot\mathbf{k})\left(1 - \frac{\mathbf{v}_{\mathrm{group}}\cdot\mathbf{v}}{c^2}\right)$$

and therefore

$$\overline{\mathbf{T}}' = \overline{\mathbf{T}}_{\perp\perp} + \gamma\left(\overline{\mathbf{T}}_{\shortparallel\perp} + \frac{\mathbf{v}\,\mathbf{S}_{\perp}}{c^2}\right) + \gamma\left(\overline{\mathbf{T}}_{\perp\shortparallel} + \mathbf{G}_{\perp}\mathbf{v}\right) + \gamma^2\left(\overline{\mathbf{T}}_{\shortparallel\shortparallel} + \mathbf{G}_{\shortparallel}\mathbf{v} + \frac{\mathbf{v}\,\mathbf{S}_{\shortparallel}}{c^2} - \frac{W\,\mathbf{v}\,\mathbf{v}}{c^2}\right) \tag{A.48a}$$

$$\mathbf{G}' = \gamma\left(\mathbf{G}_{\perp} + \frac{\overline{\mathbf{T}}_{\perp\shortparallel}\cdot\mathbf{v}}{c^2}\right) + \gamma^2\left(\mathbf{G}_{\shortparallel} + \frac{\overline{\mathbf{T}}_{\shortparallel\shortparallel}\cdot\mathbf{v}}{c^2} - \frac{W\,\mathbf{v}}{c^2} + \frac{\mathbf{S}\cdot\mathbf{v}}{c^4}\mathbf{v}\right) \tag{A.48b}$$

$$\mathbf{S}' = \gamma\left(\mathbf{S}_{\perp} + \mathbf{v}\cdot\overline{\mathbf{T}}_{\shortparallel\perp}\right) + \gamma^2[\mathbf{S}_{\shortparallel} + \mathbf{v}\cdot\overline{\mathbf{T}}_{\shortparallel\shortparallel} - W\,\mathbf{v} + (\mathbf{G}\cdot\mathbf{v})\,\mathbf{v}] \tag{A.48c}$$

$$W' = \gamma^2\left(W - \mathbf{G}\cdot\mathbf{v} - \frac{\mathbf{S}\cdot\mathbf{v}}{c^2} - \frac{\mathbf{v}\cdot\overline{\mathbf{T}}_{\shortparallel\shortparallel}\cdot\mathbf{v}}{c^2}\right) \tag{A.48d}$$

These results agree with those quoted by P-H [19, p. 115]. The subscripts $\shortparallel\shortparallel$, $\shortparallel\perp$, $\perp\shortparallel$, and $\perp\perp$ attached to the stress-tensor represent components, which in dyadic notation are parallel–parallel, parallel–perpendicular, perpendicular–parallel, or perpendicular–perpendicular with respect to the velocity, **v**. Here, too, the inverse formulas can be found by simply interchanging the primed and unprimed quantities and reversing the direction of the velocity.

Notice that the four-vectors

$$\mathcal{V}\cdot\Pi = [\gamma\,(\mathbf{S} + \mathbf{v}\cdot\overline{\mathbf{T}}),\; i\gamma c\,(W - \mathbf{G}\cdot\mathbf{v})]$$

$$\Pi\cdot\mathcal{V} = \left[\gamma\,(\mathrm{c}^2\mathbf{G} + \overline{\mathbf{T}}\cdot\mathbf{v}),\; i\gamma c\left(W - \frac{\mathbf{S}\cdot\mathbf{v}}{c^2}\right)\right]$$

when evaluated in the rest frame, are respectively $[\mathbf{S}', icW']$ and $[\mathrm{c}^2\mathbf{G}', icW']$. Insofar as their transform properties are concerned, the three-vectors take the role of **A**; the fourth components take that of Φ/c. Comparison with Eqs. (A.43a) and (A.43b) allows immediate validation of Eqs. (A.48b), (A.48c), and (A.48d). The latter can also be calculated directly from $W' = \frac{-\mathcal{V}\cdot\Pi\cdot\mathcal{V}}{c^2}$. Notice that these validations do not depend upon any assumption that the medium is linear (nor that a particle representation is valid).

APPENDIX B

B.1 THE UNIT STEP AND $U_K(T)$ FUNCTIONS

The unit-step function, $u_{-1}(t)$, is a member of the family of functions defined by

$$u_k(t) = \frac{du_{k-1}(t)}{dt}$$

where k is a positive or negative integer; if zero or positive, the functions are singular at the origin.

Although other representations are possible, we choose a Gaussian (with width controlled by the parameter α) to model the unit-impulse[1] ($k = 0$). In order to avoid mathematical difficulties, it is often advantageous to delay taking the limit $\alpha \to 0$ until the final stage of analysis. The unit ramp and its first three derivatives are defined below and plotted in Figure B.1 for the parameter $\alpha = .005$.

unit ramp: $$u_{-2}(t) = tu_{-1}(t) \tag{B.1a}$$

unit step: $$u_{-1}(t) = \lim_{\alpha \to 0} \frac{1}{2}\left[erf\left(\frac{t}{\sqrt{2\alpha}}\right) + 1\right] \tag{B.1b}$$

[1]When negative time is excluded, it is sometimes advantageous to shift the impulse slightly (or double its value) so that the integral over positive time maintains the value of unity. This is also done when u_0 is a function of a coordinate (such as the radial distance) that is defined to be positive.

The Power and Beauty of Electromagnetic Fields, First Edition. F. R. Morgenthaler.

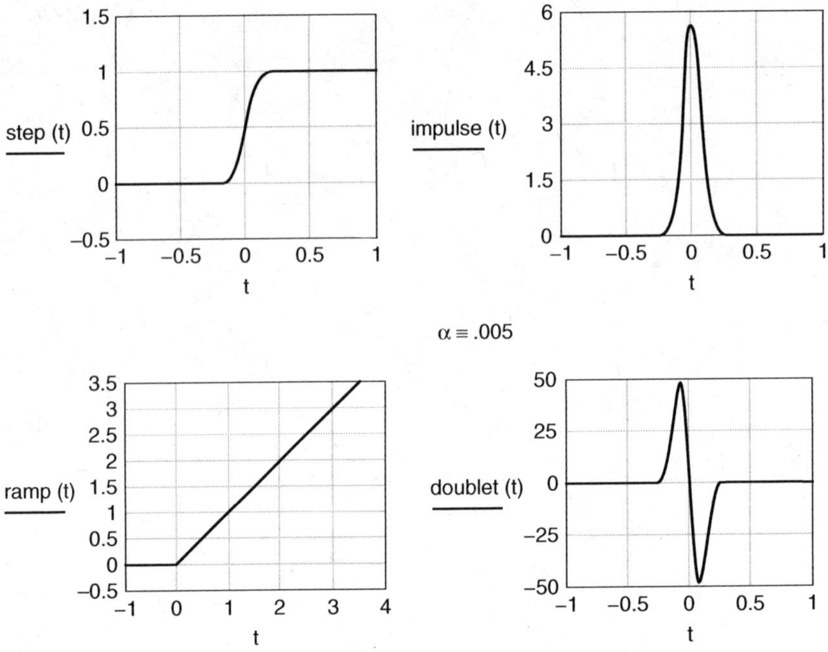

Figure B.1 Integral and derivatives of the unit-step function.

$$\textbf{unit impulse:} \qquad u_0(t) = \frac{du_{-1}(t)}{dt} = \lim_{\alpha\to 0}\sqrt{\frac{1}{2\pi\alpha}}\exp\left(\frac{-t^2}{2\alpha}\right) \qquad \text{(B.1c)}$$

$$\textbf{unit doublet:} \qquad u_1(t) = \lim_{\alpha\to 0}\sqrt{\frac{1}{2\pi\alpha}}\left(\frac{-t}{\alpha}\right)\exp\left(\frac{-t^2}{2\alpha}\right) \qquad \text{(B.1d)}$$

where the error function is defined by

$$erf(x) = \sqrt{\frac{1}{\pi}}\int_0^{2x}\exp\left[-(\tfrac{\xi}{2})^2\right]\,d\xi$$

B.2 THREE-DIMENSIONAL VECTOR IDENTITIES AND THEOREMS

Definitions

- The components of the vectors: **A**, **B**, and the scalar Ψ are assumed to be functions of space and time.
- The Kronecker delta is defined by

$$\delta_{ij} = \begin{cases} 1, & i = j \\ 0, & i \neq j \end{cases}$$

- The permutation symbol is defined by

$$\epsilon_{ijk} = \begin{cases} 1, & ijk = \text{even permutations } (123, 231, 312) \\ -1, & ijk = \text{odd permutations } (132, 213, 321) \\ 0, & \text{all other sets of } ijk \end{cases}$$

Basic operations

- The dot product of **A** and **B** is denoted by $\mathbf{A} \cdot \mathbf{B}$ and in Cartesian coordinates is

$$A_i B_i = A_1 B_1 + A_2 B_2 + A_3 B_3$$

- The cross product of **A** and **B** is denoted by $\mathbf{A} \times \mathbf{B}$, and in Cartesian coordinate index notation it is expressed as

$$(\mathbf{A} \times \mathbf{B})_k = \epsilon_{kij} A_i B_j$$

where, again, summation over repeated indices is implied.

- The del operator is defined, with respect to Cartesian coordinates, by

$$\nabla = \widehat{\mathbf{x}} \frac{\partial}{\partial x} + \widehat{\mathbf{y}} \frac{\partial}{\partial y} + \widehat{\mathbf{z}} \frac{\partial}{\partial z}$$

Curvilinear orthogonal coordinates

In terms of coordinates (u_1, u_2, u_3), the differential length, area, and volume are

$$ds = \sqrt{(h_1 du_1)^2 + (h_2 du_2)^2 + (h_3 du_3)^2}$$

$$d\mathbf{a} = \widehat{\mathbf{u}}_1 (h_2 du_2)(h_3 du_3) + \widehat{\mathbf{u}}_2 (h_3 du_3)(h_1 du_1) + \widehat{\mathbf{u}}_3 (h_1 du_1)(h_2 du_2)$$

$$dV = (h_1 du_1)(h_2 du_2)(h_3 du_3)$$

where (h_1, h_2, h_3) are the metric coefficients.

The divergence, curl, gradient, and Laplacian operations are defined,[2] respectively, by:

- In curvilinear coordinates (u_1, u_2, u_3), (h_1, h_2, h_3), we have

$$\mathbf{A} = \widehat{\mathbf{u}}_1 A_1 + \widehat{\mathbf{u}}_2 A_2 + \widehat{\mathbf{u}}_3 A_3$$

$$\text{div } \mathbf{A} = \frac{1}{h_1 h_2 h_3} \left[\frac{\partial}{\partial u_1} (h_2 h_3 A_1) + \frac{\partial}{\partial u_2} (h_3 h_1 A_2) + \frac{\partial}{\partial u_3} (h_1 h_2 A_3) \right] \quad \text{(B.2a)}$$

[2]For non-Cartesian-coordinate systems, it is a common but dangerous practice to use the notations $\nabla \cdot \mathbf{A}$ and $\nabla \times \mathbf{A}$ in place of div **A** and curl **A**. The danger is in using the generalized ∇ operator for evaluation; it is all too easy to ignore the spatial derivatives of the unit vectors themselves. There is no similar problem with grad $\Psi = \nabla\Psi$, but Laplacian Ψ should be evaluated as div grad Ψ.

$$\text{curl } \mathbf{A} = \frac{1}{h_1 h_2 h_3} \begin{vmatrix} h_1 \widehat{\mathbf{u}}_1 & h_2 \widehat{\mathbf{u}}_2 & h_3 \widehat{\mathbf{u}}_3 \\ \frac{\partial}{\partial u_1} & \frac{\partial}{\partial u_2} & \frac{\partial}{\partial u_3} \\ h_1 A_1 & h_2 A_2 & h_3 A_3 \end{vmatrix} \tag{B.2b}$$

$$\text{grad } \Psi = \widehat{\mathbf{u}}_1 \frac{1}{h_1} \frac{\partial \Psi}{\partial u_1} + \widehat{\mathbf{u}}_2 \frac{1}{h_2} \frac{\partial \Psi}{\partial u_2} + \widehat{\mathbf{u}}_3 \frac{1}{h_3} \frac{\partial \Psi}{\partial u_3} \tag{B.2c}$$

$$\text{Laplacian } \Psi = \frac{1}{h_1 h_2 h_3} \left[\frac{\partial}{\partial u_1} \left(\frac{h_2 h_3}{h_1} \frac{\partial \Psi}{\partial u_1} \right) + \frac{\partial}{\partial u_2} \left(\frac{h_3 h_1}{h_2} \frac{\partial \Psi}{\partial u_2} \right) + \frac{\partial}{\partial u_3} \left(\frac{h_1 h_2}{h_3} \frac{\partial \Psi}{\partial u_3} \right) \right] \tag{B.2d}$$

- In Cartesian coordinates (x, y, z) $(h_1 = h_2 = h_3 = 1)$, we have

$$\mathbf{A} = \widehat{\mathbf{x}} A_x + \widehat{\mathbf{y}} A_y + \widehat{\mathbf{z}} A_z$$

$$\text{div } \mathbf{A} = \nabla \cdot \mathbf{A} = \frac{\partial A_x}{\partial x} + \frac{\partial A_y}{\partial y} + \frac{\partial A_z}{\partial z} \tag{B.3a}$$

$$\text{curl } \mathbf{A} = \nabla \times \mathbf{A} = \widehat{\mathbf{x}} \left(\frac{\partial A_z}{\partial y} - \frac{\partial A_y}{\partial z} \right) + \widehat{\mathbf{y}} \left(\frac{\partial A_x}{\partial z} - \frac{\partial A_z}{\partial x} \right) + \widehat{\mathbf{z}} \left(\frac{\partial A_y}{\partial x} - \frac{\partial A_x}{\partial y} \right) \tag{B.3b}$$

$$\text{grad } \Psi = \nabla \Psi = \widehat{\mathbf{x}} \frac{\partial \Psi}{\partial x} + \widehat{\mathbf{y}} \frac{\partial \Psi}{\partial y} + \widehat{\mathbf{z}} \frac{\partial \Psi}{\partial z} \tag{B.3c}$$

$$\text{Laplacian } \Psi = \nabla \cdot \nabla \Psi = \nabla^2 \Psi = \frac{\partial^2 \Psi}{\partial x^2} + \frac{\partial^2 \Psi}{\partial y^2} + \frac{\partial^2 \Psi}{\partial z^2} \tag{B.3d}$$

- In cylindrical coordinates (r, ϕ, z), $(h_1 = 1,\ h_2 = r,\ h_3 = 1)$, we have

$$\mathbf{A} = \widehat{\mathbf{r}} A_r + \widehat{\phi} A_\phi + \widehat{\mathbf{z}} A_z$$

$$\text{div } \mathbf{A} = \frac{1}{r} \left(\frac{\partial\, r A_r}{\partial r} + \frac{\partial A_\phi}{\partial \phi} \right) + \frac{\partial A_z}{\partial z} \tag{B.4a}$$

$$\text{curl } \mathbf{A} = \widehat{\mathbf{r}} \left(\frac{1}{r} \frac{\partial A_z}{\partial \phi} - \frac{\partial A_\phi}{\partial z} \right) + \widehat{\phi} \left(\frac{\partial A_r}{\partial z} - \frac{\partial A_z}{\partial r} \right) + \widehat{\mathbf{z}} \frac{1}{r} \left(\frac{\partial\, r A_\phi}{\partial r} - \frac{\partial A_r}{\partial \phi} \right) \tag{B.4b}$$

$$\text{grad } \Psi = \widehat{\mathbf{r}} \frac{\partial \Psi}{\partial r} + \widehat{\phi} \frac{1}{r} \frac{\partial \Psi}{\partial \phi} + \widehat{\mathbf{z}} \frac{\partial \Psi}{\partial z} \tag{B.4c}$$

$$\text{Laplacian } \Psi = \frac{1}{r} \frac{\partial}{\partial r} \left(r \frac{\partial \Psi}{\partial r} \right) + \frac{1}{r^2} \left(\frac{\partial^2 \Psi}{\partial \phi^2} \right) + \frac{\partial^2 \Psi}{\partial z^2} \tag{B.4d}$$

- In spherical coordinates (r, θ, ϕ), $(h_1 = 1,\ h_2 = r,\ h_3 = r\sin\theta)$,

$$\mathbf{A} = \widehat{\mathbf{r}}\, A_r + \widehat{\theta}\, A_\theta + \widehat{\phi}\, A_\phi$$

$$\text{div } \mathbf{A} = \frac{1}{r^2}\frac{\partial\, r^2 A_r}{\partial r} + \frac{1}{r\sin\theta}\frac{\partial \sin\theta A_\theta}{\partial\theta} + \frac{1}{r\sin\theta}\frac{\partial A_\phi}{\partial\phi} \tag{B.5a}$$

$$\text{curl } \mathbf{A} = \widehat{\mathbf{r}}\,\frac{1}{r\sin\theta}\left(\frac{\partial \sin\theta A_\phi}{\partial\theta} - \frac{\partial A_\theta}{\partial\phi}\right) + \widehat{\theta}\,\frac{1}{r}\left(\frac{1}{\sin\theta}\frac{\partial A_r}{\partial\phi} - \frac{\partial\, rA_\phi}{\partial r}\right) + \widehat{\phi}\,\frac{1}{r}\left(\frac{\partial\, rA_\theta}{\partial r} - \frac{\partial A_r}{\partial\theta}\right) \tag{B.5b}$$

$$\text{grad } \Psi = \widehat{\mathbf{r}}\,\frac{\partial\Psi}{\partial r} + \widehat{\theta}\,\frac{1}{r}\frac{\partial\Psi}{\partial\theta} + \widehat{\phi}\,\frac{1}{r\sin\theta}\frac{\partial\Psi}{\partial\phi} \tag{B.5c}$$

$$\text{Laplacian } \Psi = \frac{1}{r^2}\left[\frac{\partial}{\partial r}(r^2\frac{\partial\Psi}{\partial r}) + \frac{1}{\sin\theta}\frac{\partial}{\partial\theta}\left(\sin\theta\frac{\partial\Psi}{\partial\theta}\right) + \frac{1}{\sin^2\theta}\frac{\partial^2\Psi}{\partial\phi^2}\right] \tag{B.5d}$$

Three-space identities

Important and useful identities involving three-space vectors, scalars, and the del operator:

$$\mathbf{A}\cdot(\mathbf{B}\times\mathbf{C}) = \mathbf{B}\cdot(\mathbf{C}\times\mathbf{A}) = \mathbf{C}\cdot(\mathbf{A}\times\mathbf{B}) \tag{B.6a}$$

$$\mathbf{A}\times(\mathbf{B}\times\mathbf{C}) = (\mathbf{A}\cdot\mathbf{C})\mathbf{B} - (\mathbf{A}\cdot\mathbf{B})\mathbf{C} \tag{B.6b}$$

$$(\mathbf{A}\times\mathbf{B})\cdot(\mathbf{C}\times\mathbf{D}) = (\mathbf{A}\cdot\mathbf{C})(\mathbf{B}\cdot\mathbf{D}) - (\mathbf{A}\cdot\mathbf{D})(\mathbf{B}\cdot\mathbf{C}) \tag{B.6c}$$

$$\nabla(\Psi_1\Psi_2) = \Psi_1\,\nabla\Psi_2 + \Psi_2\,\nabla\Psi_1 \tag{B.6d}$$

$$\nabla\cdot(\Psi\mathbf{A}) = \mathbf{A}\cdot(\nabla\Psi) + \Psi(\nabla\cdot\mathbf{A}) \tag{B.6e}$$

$$\nabla\times(\Psi\mathbf{A}) = (\nabla\Psi)\times\mathbf{A} + \Psi(\nabla\times\mathbf{A}) \tag{B.6f}$$

$$\nabla\times(\nabla\Psi) = 0 \tag{B.6g}$$

$$\nabla\cdot(\nabla\times\mathbf{A}) = 0 \tag{B.6h}$$

$$\nabla\times(\nabla\times\mathbf{A}) = \nabla\,(\nabla\cdot\mathbf{A}) - (\nabla\cdot\nabla)\ \mathbf{A} \tag{B.6i}$$

$$\nabla\cdot(\mathbf{A}\times\mathbf{B}) = \mathbf{B}\cdot(\nabla\times\mathbf{A}) - \mathbf{A}\cdot(\nabla\times\mathbf{B}) \tag{B.6j}$$

$$\nabla\times(\mathbf{A}\times\mathbf{B}) = \mathbf{A}\ (\nabla\cdot\mathbf{B}) - \mathbf{B}\ (\nabla\cdot\mathbf{A}) + (\mathbf{B}\cdot\nabla)\ \mathbf{A} - (\mathbf{A}\cdot\nabla)\ \mathbf{B} \tag{B.6k}$$

$$[(\nabla\times\mathbf{A})\times\mathbf{B}]_i = \left(\frac{\partial A_i}{\partial x_j} - \frac{\partial A_j}{\partial x_i}\right)B_j \tag{B.6l}$$

$$2[(\nabla \times \mathbf{A}) \times (\nabla \times \mathbf{B})]_i = \frac{\partial}{\partial x_j}[A_i(\nabla \times \mathbf{B})_j - B_i(\nabla \times \mathbf{A})_j + \nabla \cdot (\mathbf{A} \times \mathbf{B})\delta_{ij}]$$
$$+ \mathbf{A} \cdot \frac{\partial}{\partial x_i}(\nabla \times \mathbf{B}) - \mathbf{B} \cdot \frac{\partial}{\partial x_i}(\nabla \times \mathbf{A}) \qquad \text{(B.6m)}$$

In the last two identities, summation over the repeated index is understood for $j = 1, 2, 3$.

Vector theorems

As proved in Part I, Section 6.6, the divergence and curl of a vector, if specified completely, determine the vector itself. The basic definitions of these important quantities allow volume integrals of the divergence to be replaced by surface integrals and surface integrals of the curl replaced with contour integrals. The appropriate theorems that govern these substitutions are known, respectively, as the Divergence Theorem and Stokes' Theorem.

The Divergence Theorem

$$\int_V \nabla \cdot \mathbf{A}\, dV = \oint_{S_o} \mathbf{A} \cdot \widehat{\mathbf{n}}\, da \qquad \text{(B.7)}$$

The surface, S_o, encloses the volume V and $\widehat{\mathbf{n}}$ is the unit vector that is the outward normal to the surface. In the event that the surface is well-behaved and shrinks in all directions, toward a point that remains inside the enclosed volume, the value of the divergence at that point is

$$\nabla \cdot \mathbf{A} = \lim_{\Delta V \to 0} \frac{\oint_{S_o} \mathbf{A} \cdot \widehat{\mathbf{n}}\, da}{\Delta V}$$

where ΔV is the differential volume.

Stokes' Theorem

$$\int_S (\nabla \times \mathbf{A}) \cdot \widehat{\mathbf{n}}\, da = \oint_{C_o} \mathbf{A} \cdot \mathbf{ds} \qquad \text{(B.8)}$$

Here S is an arbitrary surface that spans the closed contour C_o. When S is a minimal surface (that of a soap bubble film anchored by the contour), its positive side is defined as that seen from above as we circle the contour in a counterclockwise direction. The unit vector, $\widehat{\mathbf{n}}$, normal to that surface (or any other that can be formed by simple deformation), points outward from its positive side. The vector differential area of S is defined by $d\mathbf{a} = \widehat{\mathbf{n}} da$; the vector differential distance along the contour, C_o, by $d\mathbf{s} = \widehat{\mathbf{t}} ds$, with $\widehat{\mathbf{t}}$ the unit vector tangential to the contour (and of positive polarity as we circle it).

In the event that the contour lies wholly in a plane that also serves as the surface S, and shrinks in all directions, toward a point that remains on the plane inside the contour, the value of the curl at that point satisfies

$$(\nabla \times \mathbf{A}) \cdot \widehat{\mathbf{n}} = \lim_{\Delta A \to 0} \frac{\oint_{C_o} \mathbf{A} \cdot \mathbf{ds}}{\Delta A}$$

where ΔA is the differential area of S. If the plane and its normal vector, $\widehat{\mathbf{n}}$, are chosen so that the latter points, in turn, along three mutually perpendicular directions, then the three components of the curl are defined at that point in Cartesian coordinates.

B.3 FOUR-DIMENSIONAL VECTOR AND TENSOR IDENTITIES

Definitions

- Any **antisymmetric** tensor of rank four with components T_{ij}

$$\begin{bmatrix} 0 & T_{12} & -T_{31} & T_{14} \\ -T_{12} & 0 & T_{23} & T_{24} \\ T_{31} & -T_{23} & 0 & T_{34} \\ -T_{14} & -T_{24} & -T_{34} & 0 \end{bmatrix}$$

is defined by six entries. These together form a **six-vector** denoted by $\{\mathbf{V}_1; \mathbf{V}_2\}$ and comprised of two three-vectors $\mathbf{V}_1 = (T_{23}, T_{31}, T_{12})$ and $\mathbf{V}_2 = (T_{14}, T_{24}, T_{34})$.

- The **dual six-vector** denoted by $\{\mathbf{V}_1; \mathbf{V}_2\}^\dagger$ is formed by simply interchanging these two vectors: $\{\mathbf{V}_1; \mathbf{V}_2\}^\dagger = \{\mathbf{V}_2; \mathbf{V}_1\}$.

  The result in matrix form is

$$\begin{bmatrix} 0 & T_{34} & -T_{24} & T_{23} \\ -T_{34} & 0 & T_{14} & T_{31} \\ T_{24} & -T_{14} & 0 & T_{12} \\ -T_{23} & -T_{31} & -T_{12} & 0 \end{bmatrix}$$

- Calligraphic-style letters such as $\mathcal{A}$ and $\mathcal{B}$ are used to denote arbitrary four-vectors

$$\mathcal{A} = [A_1,\ A_2,\ A_3,\ A_4] = [\mathbf{A},\ A_4]$$

and

$$\mathcal{B} = [B_1,\ B_2,\ B_3,\ B_4] = [\mathbf{B},\ B_4]$$

where $\mathbf{A}$, $\mathbf{B}$ are the three-space and A_4, B_4 the temporal components.

- In terms of the four-dimensional coordinates: $(x_1 = x, x_2 = y,\ x_3 = z, x_4 = ict)$, the generalization of the del operator is $\Box_j = \frac{\partial}{\partial x_j}$ or, in four-vector notation,

$$\Box = \left[\nabla,\ \frac{\partial}{\partial x_4}\right] = \left[\nabla,\ -\frac{i}{c}\frac{\partial}{\partial t}\right] \tag{B.9}$$

- The trajectory of a point in the continuum four-space has the form

$$\mathcal{R} = [\mathbf{r}(t) - \mathbf{r}_o,\ ic(t - t_o)]$$

where $\mathbf{r}_o$ is a constant vector defining the position of $\mathbf{r}$ at time t_o and $\mathcal{R} \cdot \mathcal{R} = 0$ is Lorentz-invariant.

- The four-velocity, $\mathcal{V}$, that describes the motion is defined as the derivative with respect to proper (rest frame) time, τ, of $\mathcal{R}$. Because $\mathcal{V}\cdot\mathcal{V} = -c^2$ is also Lorentz-invariant, we obtain

$$\mathcal{V} = \frac{\partial \mathcal{R}}{\partial t}\frac{\partial t}{\partial \tau} = [\gamma \mathbf{v},\ i\gamma c] \tag{B.10}$$

with $\mathbf{v} = \frac{d\mathbf{r}}{dt}$ the three-space velocity and $\gamma = \frac{\partial t}{\partial \tau} = 1/\sqrt{1 - \frac{\mathbf{v}\cdot\mathbf{v}}{c^2}}$.

- The derivative with respect to τ is an operator that can be expressed in terms of the four-velocity by

$$\mathcal{V}\cdot\Box = \gamma\left(\frac{\partial}{\partial t} + \mathbf{v}\cdot\nabla\right) = \gamma\frac{d}{dt} = \frac{\partial}{\partial \tau} \tag{B.11}$$

Notice that $\frac{\partial}{\partial t} + \mathbf{v}\cdot\nabla$ is often called the "substantial (material) derivative."

Basic operations

- Addition and multiplication

$$\mathcal{A} \pm \mathcal{B} = [\mathbf{A} \pm \mathbf{B},\ A_4 \pm B_4] \tag{B.12a}$$

$$\mathcal{A}\cdot\mathcal{B} = \mathbf{A}\cdot\mathbf{B} + A_4B_4 \tag{B.12b}$$

- The cross-product of $\mathcal{A}$ and $\mathcal{B}$ is the antisymmetric tensor:

$$\mathcal{A}\times\mathcal{B} = \mathcal{A}\,\mathcal{B} - \mathcal{B}\,\mathcal{A} = \{(\mathbf{A}\times\mathbf{B})\,;\ (B_4\mathbf{A} - A_4\mathbf{B})\} \tag{B.13}$$

where the four-dyad $\mathcal{A}\,\mathcal{B}$ is denoted in index notation by $(\mathcal{A}\,\mathcal{B})_{ij} = \mathcal{A}_i\mathcal{B}_j$.

- The dual of the cross-product is the dual six-vector:

$$(\mathcal{A}\times\mathcal{B})^\dagger = \{(B_4\mathbf{A} - A_4\mathbf{B});\ (\mathbf{A}\times\mathbf{B})\} \tag{B.14}$$

- The four-divergence and four-curl operations are defined, respectively, by

$$\Box\cdot\mathcal{A} = \frac{\partial A_j}{\partial x_j} \qquad \text{(summation over } j = 1, 2, 3, 4 \text{ implied)} \tag{B.15a}$$

$$(\Box\times\mathcal{A})_{ij} = \frac{\partial A_j}{\partial x_i} - \frac{\partial A_i}{\partial x_j} \tag{B.15b}$$

The dual of the four-curl is denoted by $(\Box\times\mathcal{A})^\dagger$.

B.4 FOUR-SPACE IDENTITIES

- Useful identities involving four-vectors and tensors:

$$(\mathcal{A}\times\mathcal{B})^\dagger\cdot\mathcal{C} = (\mathcal{B}\times\mathcal{C})^\dagger\cdot\mathcal{A} = \mathcal{B}\cdot(\mathcal{A}\times\mathcal{C})^\dagger$$
$$= \left[C_4\left(\mathbf{A} - \frac{A_4}{C_4}\mathbf{C}\right)\times\left(\mathbf{B} - \frac{B_4}{C_4}\mathbf{C}\right),\ (\mathbf{B}\times\mathbf{A})\cdot\mathbf{C}\right] \tag{B.16a}$$

$$(\mathcal{B} \times \mathcal{C})^\dagger \cdot (\mathcal{A} \times \mathcal{C})^\dagger = (\mathcal{C} \cdot \mathcal{C})\, \mathcal{A}\,\mathcal{B} + (\mathcal{A} \cdot \mathcal{B})\, \mathcal{C}\,\mathcal{C} - (\mathcal{B} \cdot \mathcal{C})\, \mathcal{A}\,\mathcal{C} \\ - (\mathcal{A} \cdot \mathcal{C})\, \mathcal{C}\,\mathcal{B} + [(\mathcal{A} \cdot \mathcal{C})(\mathcal{B} \cdot \mathcal{C}) \\ - (\mathcal{A} \cdot \mathcal{B})(\mathcal{C} \cdot \mathcal{C})]\mathcal{I} \tag{B.16b}$$

where $\mathcal{I}$ is the identity tensor or four-dimensional Kronecker delta.

If $\mathcal{A} \cdot \mathcal{V} = \mathcal{B} \cdot \mathcal{V} = 0$ $(A_4 = \frac{i}{c} A \cdot \mathbf{v},\ B_4 = \frac{i}{c} B \cdot \mathbf{v})$ and $\mathcal{C} = \mathcal{V}$,

$$-\frac{i}{c}(\mathcal{A} \times \mathcal{B})^\dagger \cdot \mathcal{V} = \left[\gamma\left(\mathbf{A} - \frac{\mathbf{A} \cdot \mathbf{v}}{c^2}\mathbf{v}\right) \times \left(\mathbf{B} - \frac{\mathbf{B} \cdot \mathbf{v}}{c^2}\mathbf{v}\right),\ \frac{i}{c}\gamma(\mathbf{A} \times \mathbf{B}) \cdot \mathbf{v}\right] \tag{B.17a}$$

$$\frac{i}{c}(\mathcal{B} \times \mathcal{V})^\dagger \cdot \frac{i}{c}(\mathcal{A} \times \mathcal{V})^\dagger = \mathcal{A}\,\mathcal{B} - \mathcal{A} \cdot \mathcal{B}\left(\frac{\mathcal{V}\,\mathcal{V}}{c^2} + \mathcal{I}\right) \tag{B.17b}$$

If $\mathcal{Q}$ is any antisymmetric four-tensor and $\mathcal{C}$ any four-vector,

$$(\mathcal{Q} \cdot \mathcal{C}) \times \mathcal{C} + [(\mathcal{Q}^\dagger \cdot \mathcal{C}) \times \mathcal{C}]^\dagger - \mathcal{C} \cdot \mathcal{C}\,\mathcal{Q} = 0 \tag{B.18a}$$

$$\mathcal{Q}_{ik}\frac{\partial \mathcal{Q}_{jk}}{\partial x_j} - \mathcal{Q}_{jk}\frac{\partial \mathcal{Q}_{ik}}{\partial x_j} = \mathcal{Q}^\dagger_{ik}\frac{\partial \mathcal{Q}^\dagger_{jk}}{\partial x_j} - \mathcal{Q}^\dagger_{jk}\frac{\partial \mathcal{Q}^\dagger_{ik}}{\partial x_j} \tag{B.18b}$$

(summation over $j, k = 1, 2, 3, 4$ implied)

- Useful identities involving four-space vectors, scalars, and the $\Box$ operator:

$$\Box \times (\Box\Psi) = 0 \tag{B.19a}$$

$$\Box(\Psi_1\Psi_2) = \Psi_1\,\Box\Psi_2 + \Psi_2\,\Box\Psi_1 \tag{B.19b}$$

$$\Box \cdot (\Psi\mathcal{A}) = \mathcal{A} \cdot (\Box\Psi) + \Psi(\Box \cdot \mathcal{A}) \tag{B.19c}$$

$$\Box \times (\Psi\mathcal{A}) = (\Box\Psi) \times \mathcal{A} + \Psi(\Box \times \mathcal{A}) \tag{B.19d}$$

$$\Box \cdot (\Box \times \mathcal{A}) = \Box(\Box \cdot \mathcal{A}) - (\Box \cdot \Box)\,\mathcal{A} \tag{B.19e}$$

$$\Box \cdot (\Box \times \mathcal{A})^\dagger = 0 \tag{B.19f}$$

$$\Box \cdot (\mathcal{A} \times \mathcal{B}) = \mathcal{A}\,(\Box \cdot \mathcal{B}) - \mathcal{B}\,(\Box \cdot \mathcal{A}) + (\mathcal{B} \cdot \Box)\,\mathcal{A} - (\mathcal{A} \cdot \Box)\,\mathcal{B} \tag{B.19g}$$

$$\Box \cdot (\mathcal{A} \times \mathcal{B})^\dagger = (\Box \times \mathcal{A})^\dagger \cdot \mathcal{B} - (\Box \times \mathcal{B})^\dagger \cdot \mathcal{A} \tag{B.19h}$$

APPENDIX C

C.1 STATIONARY SPATIALLY SYMMETRIC SOURCES

For stationary (time-invariant) sources, Eqs. (1.3a)–(1.5b) become

$$\oint_{C_o} \mathbf{H} \cdot d\mathbf{s} = I$$

$$\oint_{C_o} \mathbf{E} \cdot d\mathbf{s} = 0$$

$$\oint_{S_o} \mathbf{J} \cdot d\mathbf{a} = 0$$

$$\oint_{S_o} \varepsilon_o \mathbf{E} \cdot d\mathbf{a} = Q$$

$$\oint_{S_o} \mu_o \mathbf{H} \cdot d\mathbf{a} = 0$$

where Q is the total charge contained within the closed surface S_o and I is the net current passing through any surface that is bounded by the closed contour C_o.

Spherical symmetry

Spherical symmetry dictates that, when expressed in spherical coordinates (r, θ, ϕ), ρ and $\mathbf{J} = \hat{\mathbf{r}} J_r$ are functions only of the spherical radius; conservation of charge mandates

The Power and Beauty of Electromagnetic Fields, First Edition. F. R. Morgenthaler.

that without time-varying charge, $J_r = 0$. Consequently, both electric and magnetic fields can have only radial components.

Point charge ($\lim_{a\to 0} \int_0^a 4\pi r^2 \rho(r)\, dr = q$)

If the point charge, q, is located at the origin of the spherical coordinates and S_o is the spherical surface of radius r, we have

$$\varepsilon_o E_r \oint da = \varepsilon_o E_r 4\pi r^2 = q$$

$$\mu_o H_r \oint da = \mu_o H_r 4\pi r^2 = 0$$

$$\mathbf{E} = \widehat{\mathbf{r}} \frac{q}{4\pi \varepsilon_o r^2} \tag{C.1a}$$

$$\mathbf{H} = 0 \tag{C.1b}$$

Cylindrical symmetry with no axial variation

Because, in terms of cylindrical coordinates (r, ϕ, z), ρ and $\mathbf{J} = \widehat{\mathbf{r}} J_r + \widehat{\phi} J_\phi + \widehat{\mathbf{z}} J_z$ are functions only of the cylindrical radius, conservation of charge mandates that without time-varying charge, $J_r = 0$ but J_ϕ can be nonzero. Consequently, the magnetic field can have radial and axial components, but the electric field can have only a radial component.

Line charge density ($\lim_{a\to 0} \int_0^a 2\pi r \rho(r)\, dr = Q'$)

If the line charge of uniform density, Q', is located along the z axis and S_o is the surface of a cylinder of radius r and length ℓ, we obtain

$$\oint_{C_o} \varepsilon_o \mathbf{E} \cdot \ell \, d\mathbf{s} = \varepsilon_o E_r 2\pi r \ell = Q' \ell$$

$$\mathbf{E} = \widehat{\mathbf{r}} \frac{Q'}{2\pi \varepsilon_o r} \tag{C.2a}$$

$$\mathbf{H} = 0 \tag{C.2b}$$

Line current ($\lim_{a\to 0} \int_0^a 2\pi r J_z(r)\, dr = I = I_z$)

If the uniform line current, I, is located along the z axis and C_o is a circular contour of radius r, we have

$$\oint_{C_o} \mathbf{H} \cdot d\mathbf{s} = H_\phi 2\pi r = I_z$$

$$\mathbf{H} = \widehat{\phi} \frac{I_z}{2\pi r} \tag{C.3a}$$

$$\mathbf{E} = 0 \tag{C.3b}$$

Azimuthal current density $J_\phi(r)$

If C_0 is a rectangular contour in the x–z plane of width $\Delta x = r_2 - r_1$ and length Δz, we have

$$\oint_{C_0} \mathbf{H} \cdot d\mathbf{s} = \int_S \mathbf{J} \cdot d\mathbf{a}$$

$$H_r(r) = 0$$

$$[H_z(r_2) - H_z(r_1)]\Delta z = -\int_{r_1}^{r_2} J_\phi(r)\, dr\, \Delta z$$

Assuming that $J_\phi(r \to \infty) = 0$ and $H_z(\infty) = 0$,

$$\mathbf{H} = \widehat{\mathbf{z}} \int_r^\infty J_\phi(r)\, dr \tag{C.4a}$$

$$\mathbf{E} = 0 \tag{C.4b}$$

For a uniform surface current at $r = R$, we have $J_\phi(r) = K_s u_0(r - R)$ and

$$\mathbf{H} = \widehat{\mathbf{z}} K_s u_{-1}(R - r) \tag{C.5}$$

This field approximates that of a closely wound solenoid of radius R, length $\ell \gg R$, and n' turns per unit length, each carrying current I. The equivalent surface current is $K_s = n'I$.

Plane symmetry without planar variation

Assume that, when expressed in terms of Cartesian coordinates (x, y, z), all sources are uniform with respect to x and y, symmetric with respect to z, and any current is y-directed.

Apply Gauss' Law to the differential volume $2dxdy\,|z|$ centered on the plane $z = 0$. Apply Ampere's Law to the contours that are the perimeters of the areas $2dx\,|z|$ and $2dy\,|z|$.

Surface charge density $[\rho = \sigma^s u_0(z)]$

It follows that

$$\varepsilon_0[E_z(z) - E_z(-z)]\, dxdy = \sigma^s dxdy$$

$$E_z(-z) = -E_z(z)$$

$$E_x(-z) = E_x(z) = E_y(-z) = E_y(z) = 0$$

from which

$$\mathbf{E} = \widehat{\mathbf{z}}\, \text{sign}(z) \frac{\sigma^s}{2\varepsilon_0} \tag{C.6a}$$

$$\mathbf{H} = 0 \tag{C.6b}$$

Surface current density $[\mathbf{J} = \widehat{\mathbf{y}}\, K_y u_0(z);\ \mathbf{K} = \widehat{\mathbf{y}}\, K_y]$

It follows that

$$\mu_o[H_z(z) - H_z(-z)]\, dxdy = 0$$

$$[H_x(z) - H_x(-z)]\, dxdz = K_y dxdz$$

$$H_x(-z) = -H_x(z)$$

$$H_z(-z) = -H_z(z) = H_y(-z) = H_y(z) = 0$$

from which

$$\mathbf{H} = \widehat{\mathbf{x}}\,\mathrm{sign}(z)\frac{K_y}{2} \tag{C.7a}$$

$$\mathbf{E} = 0 \tag{C.7b}$$

Superposition of high-symmetry fields

Additional symmetric solutions can be generated by superposing two of these charge or current distributions after reversing one of the polarities and displacing each by $\pm\mathbf{d}_1/2$ from the common origin. This process can be repeated using any of the composite source distributions with a vector displacement, $\pm\mathbf{d}_2/2$ to form yet another composite source. Evidently, any number of iterations is possible. Often, each $\mathbf{d}_i$ is chosen parallel to any one of the Cartesian coordinate axes. When point charges are employed, the set of electric fields far from the origin ($r \gg |\mathbf{d}_i|$) are those of a monopole, dipole, quadrupole, octopole, etc. In the case of current sources, no magnetic-monopole field exists. Multipole expansions of arbitrary static fields are discussed in the next section.

C.2 MULTIPOLE EXPANSIONS OF STATIC FIELDS

When the charges and currents located at the source point q are confined to a finite volume and the field point p is located where $r_q \ll r_p$, the distance between the points approximates $r_{qp} \simeq \sqrt{r_p^2 - 2\mathbf{r}_q \cdot \mathbf{r}_p}$. The series expansion,

$$\frac{1}{r_{qp}} \simeq \frac{1}{r_p}\left[1 + \frac{\mathbf{r}_q \cdot \widehat{\mathbf{r}}_p}{r_p} + \frac{3}{2}\left(\frac{\mathbf{r}_q \cdot \widehat{\mathbf{r}}_p}{r_p}\right)^2 + \cdots\right]$$

can then be made where, for convenience, $\widehat{\mathbf{r}}_p = \mathbf{r}_p/r_p$.

Electrostatics

The electrostatic potential,

$$\Phi_p = \frac{1}{4\pi\varepsilon_o}\int \frac{\rho_q}{r_{qp}}\, dV_q$$

can then be written as a sum of monopole, dipole, quadrupole, ... terms defined by

$$\Phi_p = \frac{1}{4\pi\varepsilon_o}\frac{1}{r_p}\int \rho_q\, dV_q + \frac{1}{4\pi\varepsilon_o}\frac{1}{r_p^2}\widehat{\mathbf{r}}_p \cdot \int \mathbf{r}_q \rho_q\, dV_q + \cdots \tag{C.8}$$

The electric dipole-moment,

$$\mathbf{p} = \int \mathbf{r}_q \rho_q \, dV_q \tag{C.9}$$

is independent of the choice of origin whenever the total charge, $Q = \int \rho_q \, dV_q$, vanishes. In general,

$$\Phi_p = \frac{1}{4\pi\varepsilon_o}\frac{Q}{r_p} + \frac{1}{4\pi\varepsilon_o}\frac{\widehat{\mathbf{r}}_p \cdot \mathbf{p}}{r_p^2} + \cdots$$

In spherical coordinates, the potential and electric field of a z-directed dipole (comprised of charges $\pm q$, separated by $\mathbf{d} = \widehat{\mathbf{z}}d$) are

$$\Phi = \frac{p_z}{4\pi\varepsilon_o}\frac{\cos\theta}{r^2} \tag{C.10a}$$

$$\mathbf{E} = \frac{p_z}{4\pi\varepsilon_o r^3}(\widehat{\mathbf{r}}\, 2\cos\theta + \widehat{\theta}\, \sin\theta) \tag{C.10b}$$

where $\mathbf{p} = q\mathbf{d}$ and $r \gg d$.

In cylindrical coordinates, the potential and electric field of an x-directed line dipole (comprised of uniform line charges $\pm q'$, parallel to $\widehat{\mathbf{z}}$ and separated vertically by $\mathbf{d} = \widehat{\mathbf{x}}d$) are

$$\Phi = \frac{p'_x}{2\pi\varepsilon_o}\frac{\cos\phi}{r} \tag{C.11a}$$

$$\mathbf{E} = \frac{p'_x}{2\pi\varepsilon_o r^2}(\widehat{\mathbf{r}}\cos\phi + \widehat{\phi}\, \sin\phi) \tag{C.11b}$$

where $\mathbf{p}' = q'\mathbf{d}$ and $r \gg d$. In any plane of constant z, the $\pm$ equipotentials and electric-field line contours are orthogonal circles. This is shown in Figure C.1.

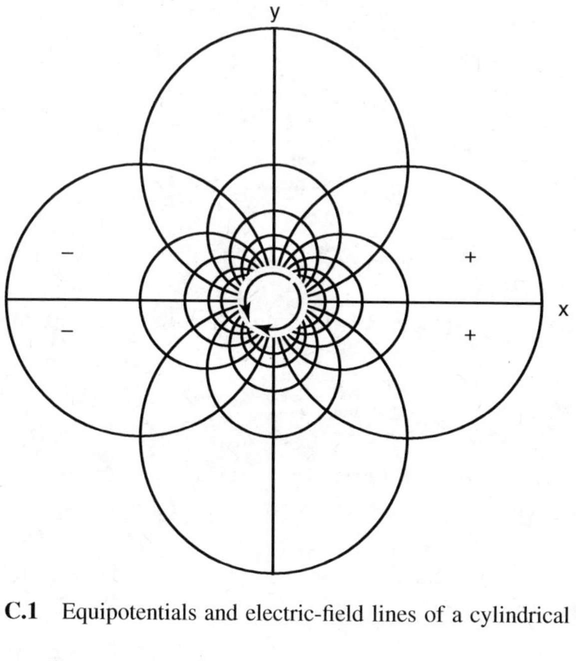

Figure C.1 Equipotentials and electric-field lines of a cylindrical dipole.

Magnetostatics

A similar expansion can be made for the magnetic vector potential,

$$\mathbf{A}_p = \frac{\mu_0}{4\pi} \int \frac{\mathbf{J}_q}{r_{qp}} \, dV_q$$

Instead, we choose to expand the magnetic field directly (as expressed in the Biot–Savart Law, derived in Section 1.5),

$$\mathbf{H}_p(\mathbf{r}_p) = \frac{1}{4\pi} \int \frac{\mathbf{J}_q \times (\mathbf{r}_p - \mathbf{r}_q)}{r_{qp}^3} \, dV_q$$

and because $\mathbf{r}_p$ can be taken outside of the integral, we find that

$$\mathbf{H}_p(\mathbf{r}_p) = \frac{-1}{4\pi r_p^2} \widehat{\mathbf{r}}_p \times \int \mathbf{J}_q (1 + 3\frac{\mathbf{r}_q \cdot \widehat{\mathbf{r}}_p}{r_p} + \cdots) \, dV_q + \frac{1}{4\pi r_p^3} \int \mathbf{r}_q \times \mathbf{J}_q \, dV_q + \cdots \tag{C.12}$$

Because $\nabla \cdot \mathbf{J} = 0$, it follows that $\int \mathbf{J}_q \, dV_q = 0$ (there is no magnetic monopole). Consequently, the leading terms are those of a uniquely defined magnetic dipole of moment, $\mathbf{m}$. If the coordinates are chosen so that the moment is z-directed, the vector potential and dipole field expressed in spherical coordinates are

$$\mathbf{A} = \frac{\mu_0 |\mathbf{m}|}{4\pi r^2} \widehat{\phi} \, \sin\theta \tag{C.13a}$$

$$\mathbf{H} = \frac{|\mathbf{m}|}{4\pi r^3} (\widehat{\mathbf{r}} \, 2\cos\theta + \widehat{\theta} \, \sin\theta) \tag{C.13b}$$

It therefore follows that

$$\mathbf{H}_p \cdot \widehat{\mathbf{r}}_p = \frac{2\mathbf{m}}{4\pi r_p^3} \cdot \widehat{\mathbf{r}}_p = \frac{1}{4\pi r_p^3} \widehat{\mathbf{r}}_p \cdot \int \mathbf{r}_q \times \mathbf{J}_q \, dV_q$$

and

$$\mathbf{m} = \frac{1}{2} \int \mathbf{r}_q \times \mathbf{J}_q \, dV_q \tag{C.14}$$

The higher-order terms in the electrostatic and magnetostatic multipole expansions can be developed in a similar way.

If a circular loop (of radius $R \ll r$) carries a ϕ-directed current I, Eq. (C.14) is expressed as

$$\mathbf{m} = \frac{1}{2} \mathrm{I} \oint_C \mathbf{r}_q \times \mathbf{ds}_q = \widehat{\mathbf{z}} \mathrm{I} \pi R^2$$

The equipotentials of A_ϕ are surfaces of revolution about the z axis; their intersection with the $x - z$ plane produces noncircular contours of the form shown in Figure C.2. Notice that, because the electric and magnetic dipoles are duals of one another, Eqs. (C.10b) and (C.13b) are of exactly the same form. Consequently, the $\mathbf{H}$ field can be produced by a magnetic-charge dipole, which that creates a scalar-potential similar to Eq. (C.10a).

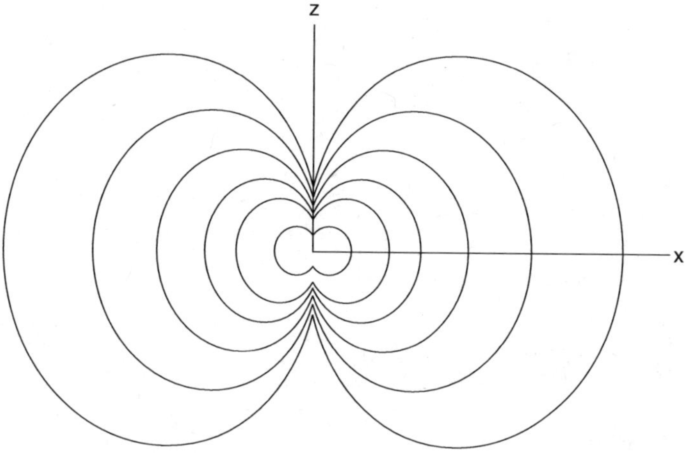

Figure C.2 Contours of $A_\phi(r,\theta)$ created by a tiny current loop.

C.3 AVERAGING PROPERTY OF LAPLACE'S EQUATION

We wish to prove that, provided that it is a solution of Laplace's Equation,

$$\nabla^2\Phi = 0$$

the potential at the center of a sphere of radius R is the average of the values of the potential on the surface of that sphere. We begin with Green's identity

$$\nabla \cdot (\Psi\nabla\Phi - \Phi\nabla\Psi) = \Psi\nabla^2\Phi - \Phi\nabla^2\Psi \tag{C.15}$$

where Φ is the Laplacian potential and Ψ any scalar, which we choose to be $\Psi = 1/r$ (with r the radial distance from the center of the sphere). This point-charge potential is itself Laplacian *except* at the origin where it is singular.

Applying the Divergence Theorem, Appendix B, Eq. (B.7), to $\nabla^2\Psi = \nabla \cdot (\nabla\Psi)$ produces

$$\oint_{S_0} \frac{\partial\Psi}{\partial r}\, da = \oint_{S_0} \frac{-1}{R^2}\, da = -4\pi = \int_V \nabla^2\Psi\, dV$$

Therefore, in terms of the symmetric three-dimensional unit impulse (delta function) $\delta(r)$, we have

$$\nabla^2\Psi(r) = \frac{1}{r^2}\frac{\partial}{\partial r}\left(r^2\frac{\partial\Psi}{\partial r}\right) = -4\pi\,\delta(r) = -\frac{1}{r^2}u_0(r)$$

where $u_0(r)$ is the unit-impulse function (defined by Appendix B, Eq. (B.1c)–but shifted to $r = 0_+$ so that $\int_0^\infty u_0(r)dr = 1$).

Finally, application of the Divergence Theorem to Eq. (C.15) results in

$$\frac{1}{R^2}\oint_{S_0}\left(R\frac{\partial\Phi}{\partial r} + \Phi\right) da = 4\pi\int_0^R u_0(r)\Phi(r,\theta,\phi)\, dr = 4\pi\,\Phi(0)$$

Because Φ is a Laplacian potential, it has no singularities within V and

$$\oint_{S_0}\frac{\partial\Phi}{\partial r}\, da = \int_V \nabla\cdot\nabla\Phi\, dV = 0$$

Therefore

$$\Phi(0) = \frac{1}{4\pi R^2} \oint_{S_o} \Phi(R,\theta,\phi)\, da$$

which proves that the potential at the center of a sphere, $\Phi(0)$, is the average of the values at its surface. Similar results can be derived for potentials that are essentially two-dimensional; in those cases the average is taken over a circle that is centered on the point of interest. If the averaging property is kept firmly in mind, a great deal of physical intuition can be developed that makes the mathematical solutions, which are presented next, more understandable.

C.4 SOLUTIONS OF LAPLACE'S EQUATION

Cartesian coordinates

In Cartesian coordinates (x,y,z), Laplace's Equation,

$$\frac{\partial^2\Phi}{\partial x^2} + \frac{\partial^2\Phi}{\partial y^2} + \frac{\partial^2\Phi}{\partial z^2} = 0 \tag{C.16}$$

is separable allowing product solutions $\Phi(x,y,z) = X(x)Y(y)Z(z)$ to be formed. These must satisfy

$$\frac{1}{X}\frac{d^2X}{dx^2} + \frac{1}{Y}\frac{d^2Y}{dy^2} + \frac{1}{Z}\frac{d^2Z}{dz^2} = 0$$

and this is possible only if each of the three terms is a constant. Then each single-coordinate function satisfies an ordinary differential equation

$$\frac{d^2X}{dx^2} + k_1^2 X = 0$$

$$\frac{d^2Y}{dy^2} + k_2^2 Y = 0$$

$$\frac{d^2Z}{dz^2} + k_3^2 Z = 0$$

where only two of the separation constants (written for convenience as k_i^2) are independent because $k_1^2 + k_2^2 + k_3^2 = 0$. At least one (but not all) of the separation constants must be imaginary. Assuming that $k_3 \rightarrow ik_3$, we have

$$X = \begin{cases} A_1\cos(k_1x) + A_2\sin(k_1x) \\ \text{or} \\ A_1 + A_2x \qquad (k_1 = 0) \end{cases} \tag{C.17a}$$

$$Y = \begin{cases} B_1\cos(k_2y) + B_2\sin(k_2y) \\ \text{or} \\ B_1 + B_2y \qquad (k_2 = 0) \end{cases} \tag{C.17b}$$

$$Z = \begin{cases} C_1' \exp(k_3 z) + C_2' \exp(-k_3 z) \\ \text{or} \\ C_1 \cosh(k_3 z) + C_2 \sinh(k_3 z) \\ \text{or} \\ C_1 + C_2 z \qquad\qquad (k_3 = 0) \end{cases} \tag{C.17c}$$

Of course, solutions with a second imaginary separation constant ($k_2 \rightarrow ik_2$) are also valid as well as those in which the coordinates x, y, z are permuted. The separation constants may be taken as complex, but that produces the same result as rotating the coordinate axes. The product functions arise from ordinary differential equations, but are restrictive. Nevertheless, linearity allows many terms (with different separation constants) to be superimposed to form a general solution that satisfies the relevant boundary conditions. One should never forget that $k_1 = k_2 = k_3 = 0$ produces the valid solution

$$\Phi = (A_0 + A_1 x)(B_0 + B_1 y)(C_0 + C_1 z) \tag{C.18}$$

Polar coordinates

In polar coordinates ($x = r\cos\phi,\ y = r\sin\phi$), Laplace's Equation,

$$\frac{\partial^2 \Phi}{\partial r^2} + \frac{1}{r}\frac{\partial \Phi}{\partial r} + \frac{1}{r^2}\frac{\partial^2 \Phi}{\partial \phi^2} = 0 \tag{C.19}$$

is separable allowing product solutions $\Phi(r, \phi) = R(r)\Psi(\phi)$ to be formed. These must satisfy

$$\frac{1}{Rr}\frac{d}{dr}\left(r\frac{dR}{dr}\right) + \frac{1}{\Psi}\frac{d^2\Psi}{d\phi^2} = 0$$

and this is possible only if each of the two terms is a constant. Then each single-coordinate function satisfies an ordinary differential equation

$$\frac{d^2\Psi}{d\phi^2} + n^2\Psi = 0$$

$$\frac{1}{r}\frac{d}{dr}\left(r\frac{dR}{dr}\right) = n^2 R$$

where there is only one separation constant, written for convenience as n^2.

$$R(r) = \begin{cases} A_1 r^n + A_2 r^{-n} & (n \neq 0) \\ A_1' + A_2' \ln r & (n = 0) \end{cases} \tag{C.20a}$$

$$\Psi(\phi) = \begin{cases} B_1 \cos(n\phi) + B_2 \sin(n\phi) & (n \neq 0) \\ B_1' + B_2'\phi & (n = 0) \end{cases} \tag{C.20b}$$

If all values of ϕ are included in the region of interest, and if the potential is single-valued, it follows that n is an integer.

Cylindrical coordinates

In cylindrical coordinates ($r = \sqrt{x^2 + y^2}, \phi = \tan^{-1}(y/x), z$), Laplace's Equation,

$$\frac{\partial^2 \Phi}{\partial r^2} + \frac{1}{r}\frac{\partial \Phi}{\partial r} + \frac{1}{r^2}\frac{\partial^2 \Phi}{\partial \phi^2} + \frac{\partial^2 \Phi}{\partial z^2} = 0 \tag{C.21}$$

is separable, allowing product solutions $\Phi(r, \phi, z) = R(r)\Psi(\phi)Z(z)$ to be formed. These must satisfy

$$\frac{1}{Rr}\frac{d}{dr}\left(r\frac{dR}{dr}\right) + \frac{1}{r^2\Psi}\frac{d^2\Psi}{d\phi^2} + \frac{1}{Z}\frac{d^2 Z}{dz^2} = 0$$

and this is possible only if the last term is a constant ($\pm k^2$). Then each single-coordinate function satisfies an ordinary differential equation

$$\frac{d^2 Z}{dz^2} \mp k^2 Z = 0$$

$$\frac{d^2\Psi}{d\phi^2} + n^2\Psi = 0$$

$$\frac{d^2 R}{dr^2} + \frac{1}{r}\frac{dR}{dr} - \left(\frac{n^2}{r^2} \pm k^2\right) R = 0$$

With k real and n an integer, general product solutions (that remain finite for all ϕ) are of the forms

$$[A_1 J_n(kr) + A_2 Y_n(kr)][B_1 \exp(kz) + B_2 \exp(-kz)] \sin(n\phi + \theta_o) \tag{C.22a}$$

$$[A_1 J_n(ikr) + A_2 Y_n(ikr)][B_1 \sin(kz) + B_2 \cos(kz)] \sin(n\phi + \theta_o) \tag{C.22b}$$

Here $J_n(kr)$, $Y_n(kr)$ are Bessel functions of the first and second kind and $J_n(ikr)$, $Y_n(ikr)$ are their imaginary argument counterparts (refer to Appendix F, Section F.1). The second linear combination can be replaced by $[A'_1 I_n(kr) + A'_2 K_n(kr)]$, comprised of modified Bessel functions of the first and second kind.

Other useful solutions (in general *not* product functions) are

$$P_n^*(r, z) = (z^2 + r^2)^{n/2} P_n^0(z/r) \tag{C.23a}$$

$$Q_n^*(r, z) = P_n^*(r, z) \ln(r) + N_n^*(r, z) \tag{C.23b}$$

where $P_n^0(z/r)$ is the associated Legendre function of degree n and order zero,[1] $N_n^*(r, z)$ is a polynomial of order n. With $P_0^* = 1$, $Q_0^* = \ln r$, these functions satisfy the recursion formulas:

$$P_n^* = \frac{2n-1}{n} z P_{n-1}^* - \frac{n-1}{n}(z^2 + r^2) P_{n-2}^* \tag{C.24a}$$

$$Q_n^* = \frac{2n-1}{n} z Q_{n-1}^* - \frac{n-1}{n}(z^2 + r^2) Q_{n-2}^* - \frac{2}{n}(P_n^* - z P_{n-1}^*) \tag{C.24b}$$

[1] In spherical coordinates, these polynomials are identical to the axial multipoles (those without a ϕ-dependence) and are products of the nth power of the spherical radius and $P_n^0(\cos\theta)$.

The first six of these are:

n	$P_n^*(r,z)$	$Q_n^*(r,z)$
0	1	$\ln r$
1	z	$z \ln r$
2	$z^2 - \frac{1}{2}r^2$	$(z^2 - \frac{1}{2}r^2)\ln r + z^2$
3	$z^3 - \frac{3}{2}zr^2$	$(z^3 - \frac{3}{2}zr^2)\ln r + z^3$
4	$z^4 - 3z^2r^2 + \frac{3}{8}r^4$	$(z^4 - 3z^2r^2 + \frac{3}{8}r^4)\ln r + z^4 - \frac{3}{16}r^4$
5	$z^5 - 5z^3r^2 + \frac{15}{8}zr^4$	$(z^5 - 5z^3r^2 + \frac{15}{8}zr^4)\ln r + z^5 - \frac{15}{16}zr^4$

Spherical coordinates

In spherical coordinates ($x = r\sin\theta\cos\phi,\ y = r\sin\theta\sin\phi,\ z = r\cos\theta$), Laplace's Equation,

$$\frac{1}{r^2}\frac{\partial}{\partial r}\left(r^2\frac{\partial\Phi}{\partial r}\right) + \frac{1}{r^2\sin\theta}\frac{\partial}{\partial\theta}\left(\sin\theta\frac{\partial\Phi}{\partial\theta}\right) + \frac{1}{r^2\sin^2\theta}\frac{\partial^2\Phi}{\partial\phi^2} = 0 \tag{C.25}$$

is separable allowing product solutions $\Phi(r,\phi,z) = R(r)\Theta(\theta)\Psi(\phi)$ to be formed. These must satisfy

$$\frac{1}{R}\frac{d}{dr}\left(r^2\frac{dR}{dr}\right) + \frac{1}{\Theta\sin\theta}\frac{d}{d\theta}\left(\sin\theta\frac{d\Theta}{d\theta}\right) + \frac{1}{\Psi\sin^2\theta}\frac{d^2\Psi}{d\phi^2} = 0$$

and this is possible only if the first term is a constant (conveniently taken as $\ell(\ell+1)$). The second separation constant associated with the ϕ -dependence is taken as $-m^2$. Then each single-coordinate function satisfies an ordinary differential equation

$$\frac{d}{dr}\left(r^2\frac{dR}{dr}\right) - \ell(\ell+1)R = 0$$

$$\frac{d^2\Psi}{d\phi^2} + m^2\Psi = 0$$

$$\sin\theta\frac{d}{d\theta}\left(\sin\theta\frac{d\Theta}{d\theta}\right) + [\ell(\ell+1)\sin^2\theta - m^2]\Theta = 0$$

General product solutions that remain finite when $\sin\theta = 0$, are of the form

$$\Phi(r,\theta,\phi) = (A_1 r^\ell + A_2 r^{-(\ell+1)})(B_1\cos m\phi + B_2\sin m\phi)(\sin\theta)^{|m|}P_\ell^m(\cos\theta) \tag{C.26}$$

where $P_\ell^m(\cos\theta)$ is an associated Legendre function of order ℓ that is related to the Legendre polynomial $P_\ell(x)$ by

$$P_\ell^m(\cos\theta) = \left.\frac{d^{|m|}P_\ell(x)}{dx^{|m|}}\right|_{x=\cos\theta} \tag{C.27a}$$

With $P_0 = 1$, the recursion formula for the polynomials is

$$(\ell+1)P_{\ell+1}(x) = (2\ell+1)xP_\ell(x) - \ell P_{\ell-1}(x) \tag{C.27b}$$

There are $2\ell + 1$ values of m associated with the separation integer, $\ell = 0, 1, 2, \cdots$; they are $m = 0, \pm 1, \cdots, \pm\ell$. Solutions for the first three values of ℓ are tabulated below; they include the point-charge, point-dipole, and point-quadrupole potentials $\ell = 0$

$$1 \quad \Bigg\| \quad \frac{1}{r}$$

$\ell = 1$

$r \sin\theta \cos\phi$	$r \cos\theta$	$r \sin\theta \sin\phi$
$\frac{1}{r^2} \sin\theta \cos\phi$	$\frac{1}{r^2} \cos\theta$	$\frac{1}{r^2} \sin\theta \sin\phi$

$\ell = 2$

$r^2 \sin^2\theta \cos 2\phi$	$r^2 \sin\theta \cos\theta \cos\phi$	$r^2(3\cos^2\theta - 1)$	$r^2 \sin\theta \cos\theta \sin\phi$	$r^2 \sin^2\theta \sin 2\phi$
$\frac{1}{r^3} \sin^2\theta \cos 2\phi$	$\frac{1}{r^3} \sin\theta \cos\theta \cos\phi$	$\frac{1}{r^3}(3\cos^2\theta - 1)$	$\frac{1}{r^3} \sin\theta \cos\theta \sin\phi$	$\frac{1}{r^3} \sin^2\theta \sin 2\phi$

C.5 LAPLACE'S EQUATION IN *N* DIMENSIONS

The generalization of Laplace's Equation to N (real) dimensions is

$$\sum_{i=1}^{N} \frac{\partial^2 \psi}{\partial x_i^2} = 0 \tag{C.28}$$

where we extend the definition of R^2 to

$$R^2 = \sum_{i=1}^{N} x_i^2$$

Consider symmetric solutions, $\psi(R)$, for which Eq. (C.28) simplifies to

$$\sum_{i=1}^{N} \frac{\partial}{\partial x_i}\left(\frac{\partial \psi}{\partial R}\frac{\partial R}{\partial x_i}\right) = \frac{N}{R}\frac{\partial \psi}{\partial R} + R\frac{\partial}{\partial R}\left(\frac{1}{R}\frac{\partial \psi}{\partial R}\right) = 0 \tag{C.29}$$

We choose $\psi = R^n$ as a trial function and find that it is a valid solution, provided that

$$n(n + N - 2) = 0$$

There are two solutions, $n = 0$ and $n = 2 - N$; the first of these is expected since a constant obviously satisfies Eq. (C.28) for *any* value of N. Unfortunately, since $n = 0$ is repeated for $N = 2$, we have lost a solution. It can be recovered by setting $R = \exp(\ln R)$ and evaluating

$$\lim_{n \to 0} \frac{(R^n - 1)}{n} = \ln R$$

Direct substitution into Eq. (C.29) verifies that $\ln R$ is indeed a solution. The linear potential of a uniform sheet charge ($N = 1$), the $\ln R$ potential of a uniform line charge ($N = 2$), and the $1/R$ potential of a point charge ($N = 3$) are all well known; the $N = 4$ solution (R^{-2}) is less familiar, but is of use in solving the inhomogeneous three-dimensional wave equation. This is because $\Psi(x_1, x_4)$, $\Psi(x_1, x_2, x_4)$, and $\Psi(x_1, x_2, x_3, x_4)$ satisfy one-, two-, and three-dimensional wave equations, provided that $x_4 = ict$. When one of the Cartesian coordinates is imaginary, Laplace's Equation becomes the wave equation.

C.6 ELLIPSOIDS IN UNIFORM FIELDS

If a finite volume of dielectric material of arbitrary shape is placed in a uniform static electric field, $\mathbf{E}^{\text{applied}}$, polarization charges will be generated within the material and on its surface. These, in turn, will generate fields that are nonuniform both inside and outside the material. Remarkably, if the material has the shape of a general ellipsoid defined by

$$\left(\frac{x}{a}\right)^2 + \left(\frac{y}{b}\right)^2 + \left(\frac{z}{c}\right)^2 = 1,$$

the interior field will be uniform (although the external field will be nonuniform in the vicinity of the ellipsoid). This result is of great practical importance because, for such shapes, measurements of material properties can be performed in uniform fields. A similar result holds when the material is magnetic and the uniform applied field is $\mathbf{H}^{\text{applied}}$. In all cases, the volume divergence of the polarization $\mathbf{P}$ (or magnetization $\mathbf{M}$) is zero; only surface electric (or "magnetic") charges create field perturbations.[2]

Here we give the result of the derivation (that can be found in Stratton [1, pp. 207–213]) that relates the inner field to the applied field; the coordinates used are the principal axes of the ellipsoid.

$$E_i^{\text{inner}} = E_i^{\text{applied}} - N_{(i)} \frac{P_i}{\varepsilon_0} \tag{C.30a}$$

$$H_i^{\text{inner}} = H_i^{\text{applied}} - N_{(i)} M_i \tag{C.30b}$$

where P_i (or M_i) are components of the polarization (magnetization) vector. The three dimensionless factors N_i are so-called depolarization (demagnetization) factors that depend upon the ratios a/b and a/c. The parentheses in $N_{(i)}$ are used to indicate that the repeated index is not to be summed. Although the inner fields are uniform, they are not, in general, parallel to the applied fields unless the latter are parallel to a principal axis of the ellipsoid.

In general, these factors are given in terms of elliptic integrals, and it can be proved that (with $N_1 = N_x$, $N_2 = N_y$, $N_3 = N_z$) they obey the sum rule[3]:

$$N_x + N_y + N_z = 1$$

When two of the semi-axes are equal, the ellipsoid becomes a solid of revolution termed a spheroid; for these cases, the elliptic integrals reduce to elementary functions. We

[2] When the sample is nonellipsoidal, it is sometimes reasonable to use the Sommerfeld approximation. This method *assumes* that $\mathbf{P}$ (or $\mathbf{M}$) is uniform, calculates the surface-charges from $\hat{\mathbf{n}} \cdot \mathbf{P}$ (or $\hat{\mathbf{n}} \cdot \mathbf{M}$), and integrates the appropriate Poisson Equation using these as the only source terms. The method works best when there is a permanent $\mathbf{P}$ (or $\mathbf{M}$) with constrained orientation.

[3] In the cgs-system of units, the magnetic-field equation is written as

$$H_i^{\text{inner}} = H_i^{\text{applied}} - N_{(i)} 4\pi M_i$$

The reader should be warned that some authors choose to absorb the factor of 4π into the definition of $N_{(i)}$. In this case, the factors sum to 4π. Also, because

$$E_i^{\text{inner}} = E_i^{\text{applied}} - N_{(i)} P_i / \varepsilon_0,$$

other authors (like Stratton) choose to absorb the factor of ε_0 into the definition of $N_{(i)}$. Because they are purely geometric factors, this author deems it best to avoid such practices.

limit our discussion to spheroids and choose z to be the symmetry axis; then $a = b$ and $N_x = N_y = N_t$, where N_t is the transverse factor; and since $2N_t + N_z = 1$, either factor provides a complete description. There are three cases to be considered: prolate spheroids defined by $c > a$ [Figure C.3(a)], oblate spheroids defined by $c < a$ [Figure C.3(b)], and spheres where $c = a$.

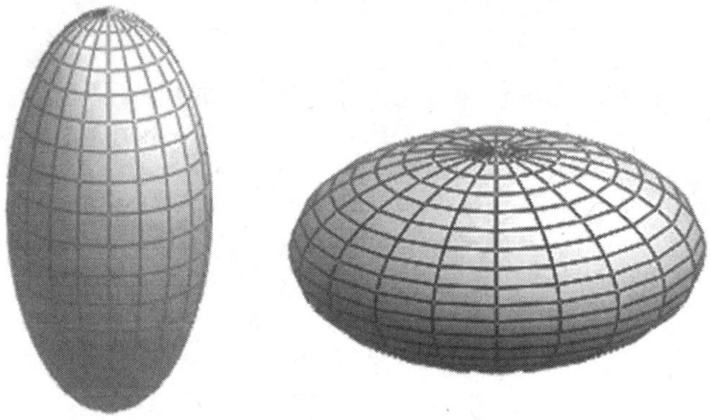

(a) Prolate (b) Oblate

Figure C.3 Spheroid geometries.

Prolate spheroid ($c > a$)

$$N_z = \frac{a^2}{c^2 - a^2}\left[\frac{c}{\sqrt{c^2 - a^2}}\tanh^{-1}\left(\frac{\sqrt{c^2 - a^2}}{c}\right) - 1\right] \tag{C.31}$$

When $c \gg a$, the prolate spheroid approaches an infinite cylinder and

$$N_z \to \left(\frac{a}{c}\right)^2\left[\ln\left(\frac{2c}{a}\right) - 1\right] \tag{C.32a}$$

$$N_t \to 1/2. \tag{C.32b}$$

If the applied field is parallel to the z axis, it is perturbed only slightly (there is little area to contain the surface charges and they are far apart, $N_z \simeq 0$); if perpendicular to the z axis, the perturbation outside the volume is that of a two-dimensional dipole.

Oblate spheroid ($c < a$)

$$N_z = \frac{a^2}{a^2 - c^2}\left[1 - \frac{c}{\sqrt{a^2 - c^2}}\tan^{-1}\left(\frac{\sqrt{a^2 - c^2}}{c}\right)\right] \tag{C.33}$$

When $c \ll a$, the oblate spheroid approaches a thin disk and

$$N_z \to 1 - \frac{\pi c}{2a} \tag{C.34a}$$

$$N_t \to \frac{\pi c}{4a} \tag{C.34b}$$

If the applied field is parallel to the z axis, the field inside the disk is reduced in magnitude (there is much area to contain the surface charges and they are close together, $N_z \simeq 1$), but only slightly perturbed outside the disk because of cancellation from the two planes of charge; if perpendicular to the z axis, the perturbation is very small everywhere because $N_t \simeq 0$. If the radius $a \to \infty$, the actual transverse shape is immaterial (except near the edges) and Eq. (C.30) with $N_z = 1$ remains valid for a very thin sheet of nearly arbitrary cross-sectional shape.

Sphere ($c = a$)

From symmetry considerations and the sum rule (or from taking the limit of either the prolate or oblate formula),

$$N_z = N_t = \tfrac{1}{3} \tag{C.35}$$

The perturbation outside the volume is that of a three-dimensional dipole.

APPENDIX D

D.1 ALTERNATE POWER, ENERGY, STRESS, AND MOMENTUM

In Part II, Chapter 9, Section 9.3, the Alternate Poynting theorem and the Alternate stress-momentum theorem (in the Amperian formulation) were found to be

$$\nabla \cdot \mathbf{S}^{\mathrm{o}} + \frac{\partial}{\partial t} W^{\mathrm{o}} = -\mathbf{E}^{\mathrm{o}} \cdot \mathbf{J}_{\mathrm{total}}$$

$$\nabla \cdot \overline{\mathbf{T}}^{\mathrm{o}} - \frac{\partial}{\partial t} \mathbf{G}^{\mathrm{o}} = \rho_{\mathrm{total}} + \mathbf{J}_{\mathrm{total}} \times \mathbf{B}$$

$$\mathbf{J}_{\mathrm{total}} = \mathbf{J}_{\mathrm{u}} + \frac{\partial \mathbf{P}^{\mathrm{o}}}{\partial t} + \nabla \times \mathbf{M}^{\mathrm{o}}$$

$$\rho_{\mathrm{total}} = \rho_{\mathrm{u}} - \nabla \cdot \mathbf{P}^{\mathrm{o}}$$

The quantities $\mathbf{S}^{\mathrm{o}}$, W^{o}, $\mathbf{G}^{\mathrm{o}}$, and $\overline{\mathbf{T}}^{\mathrm{o}}$ were derived for an arbitrary gauge using index notation to represent the Cartesian coordinates, x_i. The scalar

$$g = \Box \cdot \mathcal{A} = \nabla \cdot \mathbf{A} + \mu_{\mathrm{o}} \varepsilon_{\mathrm{o}} \frac{\partial \Phi}{\partial t}$$

The Power and Beauty of Electromagnetic Fields, First Edition. F. R. Morgenthaler.

defines the gauge; convenient values are

$$g = \begin{cases} 0 & \text{Lorenz} \\ \mu_0 \varepsilon_0 \dfrac{\partial \Phi}{\partial t} & \text{Coulomb} \\ -\mu_0 \sigma \Phi & \text{uniform conductor (stationary)} \\ & \mathbf{J}_u = \sigma \mathbf{E},\ \ \nabla \sigma = 0,\ \ \rho_u = 0 \end{cases} \tag{D.1}$$

Cartesian coordinates

$$S_x^o = c^2 G_x^o = \frac{1}{2\mu_0}\left(A_x \frac{\partial^2 A_x}{\partial x \partial t} - \frac{\partial A_x}{\partial x}\frac{\partial A_x}{\partial t} + A_y \frac{\partial^2 A_y}{\partial x \partial t} - \frac{\partial A_y}{\partial x}\frac{\partial A_y}{\partial t}\right) + \frac{1}{2\mu_0}\left(A_z \frac{\partial^2 A_z}{\partial x \partial t} - \frac{\partial A_z}{\partial x}\frac{\partial A_z}{\partial t}\right) - \frac{1}{2}\varepsilon_0\left(\Phi \frac{\partial^2 \Phi}{\partial x \partial t} - \frac{\partial \Phi}{\partial x}\frac{\partial \Phi}{\partial t}\right) \tag{D.2a}$$

$$S_y^o = c^2 G_y^o = \frac{1}{2\mu_0}\left(A_x \frac{\partial^2 A_x}{\partial y \partial t} - \frac{\partial A_x}{\partial y}\frac{\partial A_x}{\partial t} + A_y \frac{\partial^2 A_y}{\partial y \partial t} - \frac{\partial A_y}{\partial y}\frac{\partial A_y}{\partial t}\right) + \frac{1}{2\mu_0}\left(A_z \frac{\partial^2 A_z}{\partial y \partial t} - \frac{\partial A_z}{\partial y}\frac{\partial A_z}{\partial t}\right) - \frac{1}{2}\varepsilon_0\left(\Phi \frac{\partial^2 \Phi}{\partial y \partial t} - \frac{\partial \Phi}{\partial y}\frac{\partial \Phi}{\partial t}\right) \tag{D.2b}$$

$$S_z^o = c^2 G_z^o = \frac{1}{2\mu_0}\left(A_x \frac{\partial^2 A_x}{\partial z \partial t} - \frac{\partial A_x}{\partial z}\frac{\partial A_x}{\partial t} + A_y \frac{\partial^2 A_y}{\partial z \partial t} - \frac{\partial A_y}{\partial z}\frac{\partial A_y}{\partial t}\right) + \frac{1}{2\mu_0}\left(A_z \frac{\partial^2 A_z}{\partial z \partial t} - \frac{\partial A_z}{\partial z}\frac{\partial A_z}{\partial t}\right) - \frac{1}{2}\varepsilon_0\left(\Phi \frac{\partial^2 \Phi}{\partial y \partial t} - \frac{\partial \Phi}{\partial y}\frac{\partial \Phi}{\partial t}\right) \tag{D.2c}$$

$$W^o = -\frac{1}{2}\varepsilon_0\left[A_x \frac{\partial^2 A_x}{\partial t^2} - \left(\frac{\partial A_x}{\partial t}\right)^2 + A_y \frac{\partial^2 A_y}{\partial t^2} - \left(\frac{\partial A_y}{\partial t}\right)^2\right] - \frac{1}{2}\varepsilon_0\left[A_z \frac{\partial^2 A_z}{\partial t^2} - \left(\frac{\partial A_z}{\partial t}\right)^2\right] + \varepsilon_0 \frac{1}{2c^2}\left[\Phi \frac{\partial^2 \Phi}{\partial t^2} - \left(\frac{\partial \Phi}{\partial t}\right)^2\right] \tag{D.2d}$$

$$T_{ij}^o = \frac{1}{2\mu_0}\left(A_x \frac{\partial^2 A_x}{\partial x_i \partial x_j} - \frac{\partial A_x}{\partial x_i}\frac{\partial A_x}{\partial x_j} + A_y \frac{\partial^2 A_y}{\partial x_i \partial x_j} - \frac{\partial A_y}{\partial x_i}\frac{\partial A_y}{\partial x_j}\right) + \frac{1}{2\mu_0}\left(A_z \frac{\partial^2 A_z}{\partial x_i \partial x_j} - \frac{\partial A_z}{\partial x_i}\frac{\partial A_z}{\partial x_j}\right) - \frac{1}{2}\varepsilon_0\left(\Phi \frac{\partial^2 \Phi}{\partial x_i \partial x_j} - \frac{\partial \Phi}{\partial x_i}\frac{\partial \Phi}{\partial x_j}\right) \tag{D.2e}$$

The Alternate power flux, $\Phi \mathbf{J}_{\text{total}} + \dfrac{1}{2\mu_0}\left(g \dfrac{\partial \mathbf{A}}{\partial t} - \dfrac{\partial g}{\partial t}\mathbf{A}\right)$

energy density, $\dfrac{1}{2}(\rho_{\text{total}}\Phi + \mathbf{A} \cdot \mathbf{J}_{\text{total}}) + \dfrac{1}{2}\varepsilon_0\left(g \dfrac{\partial \Phi}{\partial t} - \Phi \dfrac{\partial g}{\partial t}\right)$

momentum density, $\rho_{\text{total}}\mathbf{A} + \dfrac{1}{2}\varepsilon_0(\Phi\ \nabla g - g\ \nabla \Phi)$

and stress tensor,[1]

$$-A_i (J_{\text{total}})_j + \frac{1}{2}(\mathbf{A} \cdot \mathbf{J}_{\text{total}} - \rho_{\text{total}}\Phi)\delta_{ij} + \frac{1}{2\mu_o}\left(g\frac{\partial A_j}{\partial x_i} - \frac{\partial g}{\partial x_i}A_j \right)$$

terms must be added to the preceding formulas. After this is done, the Alternate energy-momentum tensor is, in general, nonsymmetric ($\mathbf{S}^o \neq c^2\mathbf{G}^o$ and $T^o_{ij} \neq T^o_{ji}$).

When the vector and scalar potentials are expressed in other coordinates, transformations are necessary. The remainder of this section provides the appropriate formulas for the vector and scalar components when expressed in cylindrical, and spherical coordinates. Notice that the angular components of the fluxes have cross-terms that depend upon both the radial and angular components. As with the previous formulas, the terms due to electric charge and current and nonzero g must be added.

Cylindrical coordinates

The vector-potential components A_k are assumed to be $A_1 = A_r$, $A_2 = A_\phi$, $A_3 = A_z$. However, the formulas are also valid when Cartesian components are used, provided that all A_r and A_ϕ terms *outside* of the summations are set equal to zero.

$$S^o_r = c^2 G^o_r = \frac{1}{2\mu_o}\sum_{k=1}^{3}\left(A_k\frac{\partial^2 A_k}{\partial r\partial t} - \frac{\partial A_k}{\partial r}\frac{\partial A_k}{\partial t} \right) - \frac{1}{2}\varepsilon_o\left(\Phi\frac{\partial^2\Phi}{\partial r\partial t} - \frac{\partial \Phi}{\partial r}\frac{\partial \Phi}{\partial t} \right) \quad \text{(D.3a)}$$

$$S^o_\phi = c^2 G^o_\phi = \frac{1}{2\mu_o r}\sum_{k=1}^{3}\left(A_k\frac{\partial^2 A_k}{\partial \phi\partial t} - \frac{\partial A_k}{\partial \phi}\frac{\partial A_k}{\partial t} \right) + \frac{1}{\mu_o r}\left(A_\phi\frac{\partial A_r}{\partial t} - A_r\frac{\partial A_\phi}{\partial t} \right)$$
$$- \frac{1}{2r}\varepsilon_o\left(\Phi\frac{\partial^2\Phi}{\partial \phi\partial t} - \frac{\partial \Phi}{\partial \phi}\frac{\partial \Phi}{\partial t} \right) \quad \text{(D.3b)}$$

$$S^o_z = c^2 G^o_z = \frac{1}{2\mu_o}\sum_{k=1}^{3}\left(A_k\frac{\partial^2 A_k}{\partial z\partial t} - \frac{\partial A_k}{\partial z}\frac{\partial A_k}{\partial t} \right) - \frac{1}{2}\varepsilon_o\left(\Phi\frac{\partial^2\Phi}{\partial z\partial t} - \frac{\partial \Phi}{\partial z}\frac{\partial \Phi}{\partial t} \right)$$

$$W^o = -\frac{1}{2}\varepsilon_o\sum_{k=1}^{3}\left(A_k\frac{\partial^2 A_k}{\partial t^2} - \frac{\partial A_k}{\partial t}\frac{\partial A_k}{\partial t} \right) + \frac{1}{2}\frac{\varepsilon_o}{c^2}\left[\Phi\frac{\partial^2\Phi}{\partial t^2} - \left(\frac{\partial \Phi}{\partial t}\right)^2 \right] \quad \text{(D.3c)}$$

$$T^o_{rr} = \frac{1}{2\mu_o}\sum_{k=1}^{3}\left(A_k\frac{\partial^2 A_k}{\partial r^2} - \frac{\partial A_k}{\partial r}\frac{\partial A_k}{\partial r} \right) - \frac{1}{2}\varepsilon_o\left(\Phi\frac{\partial^2\Phi}{\partial r^2} - \frac{\partial \Phi}{\partial r}\frac{\partial \Phi}{\partial r} \right) \quad \text{(D.3d)}$$

[1] In dyadic notation, $g\,\nabla\mathbf{A} - \nabla g\,\mathbf{A}$ requires evaluation of the gradient of non-Cartesian-coordinate unit vectors. The appropriate gradients are listed at the ends of the next two subsections.

$$T^{\mathrm{o}}_{\phi\phi} = \frac{1}{2\mu_{\mathrm{o}} r^2}\sum_{k=1}^{3}\left(A_k\frac{\partial^2 A_k}{\partial\phi^2} - \frac{\partial A_k}{\partial\phi}\frac{\partial A_k}{\partial\phi} + rA_k\frac{\partial A_k}{\partial r}\right) + \frac{1}{\mu_{\mathrm{o}} r^2}\left[2\left(A_\phi\frac{\partial A_r}{\partial\phi} - A_r\frac{\partial A_\phi}{\partial\phi}\right) - (A_r^2 + A_\phi^2)\right] - \varepsilon_{\mathrm{o}}\frac{1}{2r^2}\left(\Phi\frac{\partial^2\Phi}{\partial\phi^2} - \frac{\partial\Phi}{\partial\phi}\frac{\partial\Phi}{\partial\phi}\right) + r\Phi\frac{\partial\Phi}{\partial r}) \quad \text{(D.3e)}$$

$$T^{\mathrm{o}}_{zz} = \frac{1}{2\mu_{\mathrm{o}}}\sum_{k=1}^{3}\left(A_k\frac{\partial^2 A_k}{\partial z^2} - \frac{\partial A_k}{\partial z}\frac{\partial A_k}{\partial z}\right) - \frac{1}{2}\varepsilon_{\mathrm{o}}\left(\Phi\frac{\partial^2\Phi}{\partial z^2} - \frac{\partial\Phi}{\partial z}\frac{\partial\Phi}{\partial z}\right) \quad \text{(D.3f)}$$

$$T^{\mathrm{o}}_{r\phi} = T^{\mathrm{o}}_{\phi r} = \frac{1}{2\mu_{\mathrm{o}} r^2}\sum_{k=1}^{3}\left[A_k\left(r\frac{\partial^2 A_k}{\partial r\partial\phi} - \frac{\partial A_k}{\partial\phi}\right) - r\frac{\partial A_k}{\partial r}\frac{\partial A_k}{\partial\phi}\right] + \frac{1}{\mu_{\mathrm{o}} r}\left(A_\phi\frac{\partial A_r}{\partial r} - A_r\frac{\partial A\phi}{\partial r}\right) - \varepsilon_{\mathrm{o}}\frac{1}{2r}\left[\Phi\left(\frac{\partial^2\Phi}{\partial r\partial\phi} - \frac{1}{r}\frac{\partial\Phi}{\partial\phi}\right) - \frac{\partial\Phi}{\partial r}\frac{\partial\Phi}{\partial\phi}\right] \quad \text{(D.3g)}$$

$$T^{\mathrm{o}}_{rz} = T^{\mathrm{o}}_{zr} = \frac{1}{2\mu_{\mathrm{o}}}\sum_{k=1}^{3}\left(A_k\frac{\partial^2 A_k}{\partial r\partial z} - \frac{\partial A_k}{\partial r}\frac{\partial A_k}{\partial z}\right) - \frac{1}{2}\varepsilon_{\mathrm{o}}\left(\Phi\frac{\partial^2\Phi}{\partial r\partial z} - \frac{\partial\Phi}{\partial r}\frac{\partial\Phi}{\partial z}\right) \quad \text{(D.3h)}$$

$$T^{\mathrm{o}}_{\phi z} = T^{\mathrm{o}}_{z\phi} = \frac{1}{2\mu_{\mathrm{o}} r}\left[\sum_{k=1}^{3}\left(A_k\frac{\partial^2 A_k}{\partial\phi\partial z} - \frac{\partial A_k}{\partial\phi}\frac{\partial A_k}{\partial z}\right) + 2\left(A_\phi\frac{\partial A_r}{\partial z} - A_r\frac{\partial A_\phi}{\partial z}\right)\right] - \varepsilon_{\mathrm{o}}\frac{1}{2r}\left(\Phi\frac{\partial^2\Phi}{\partial\phi\partial z} - \frac{\partial\Phi}{\partial\phi}\frac{\partial\Phi}{\partial z}\right) \quad \text{(D.3i)}$$

$$\nabla\hat{\mathbf{r}} = \frac{1}{r}\hat{\phi}\,\hat{\phi},\quad \nabla\hat{\phi} = \frac{-1}{r}\hat{\phi}\,\hat{\mathbf{r}},\quad \nabla\hat{\mathbf{z}} = 0$$

Spherical coordinates

The vector-potential components A_k are assumed to be $A_1 = A_r$, $A_2 = A_\theta$, $A_3 = A_\phi$. However, the formulas are also valid when Cartesian components are used, provided that all A_r, A_θ and A_ϕ terms *outside* of the summations are set equal to zero.

$$S^{\mathrm{o}}_r = \mathrm{c}^2 G^{\mathrm{o}}_r = \frac{1}{2\mu_{\mathrm{o}}}\sum_{k=1}^{3}\left(A_k\frac{\partial^2 A_k}{\partial r\partial t} - \frac{\partial A_k}{\partial r}\frac{\partial A_k}{\partial t}\right) - \frac{1}{2}\varepsilon_{\mathrm{o}}\left(\Phi\frac{\partial^2\Phi}{\partial r\partial t} - \frac{\partial\Phi}{\partial r}\frac{\partial\Phi}{\partial t}\right) \quad \text{(D.4a)}$$

$$S^{\mathrm{o}}_\theta = \mathrm{c}^2 G^{\mathrm{o}}_\theta = \frac{1}{2\mu_{\mathrm{o}} r}\left[\sum_{k=1}^{3}\left(A_k\frac{\partial^2 A_k}{\partial\theta\partial t} - \frac{\partial A_k}{\partial\theta}\frac{\partial A_k}{\partial t}\right) + 2\left(A_\theta\frac{\partial A_r}{\partial t} - A_r\frac{\partial A_\theta}{\partial t}\right)\right] - \varepsilon_{\mathrm{o}}\frac{1}{2r}\left(\Phi\frac{\partial^2\Phi}{\partial\theta\partial t} - \frac{\partial\Phi}{\partial\theta}\frac{\partial\Phi}{\partial t}\right) \quad \text{(D.4b)}$$

$$S_\phi^{\mathrm{o}} = \mathrm{c}^2 G_\phi^{\mathrm{o}} = \frac{1}{2\mu_{\mathrm{o}} r \sin\theta} \sum_{k=1}^{3} \left(A_k \frac{\partial^2 A_k}{\partial\phi \partial t} - \frac{\partial A_k}{\partial \phi} \frac{\partial A_k}{\partial t} \right) + \frac{1}{\mu_{\mathrm{o}} r} \left[A_\phi \frac{\partial A_r}{\partial t} - A_r \frac{\partial A_\phi}{\partial t} + \cot\theta \left(A_\phi \frac{\partial A_\theta}{\partial t} - A_\theta \frac{\partial A_\phi}{\partial t} \right) \right] - \varepsilon_{\mathrm{o}} \frac{1}{2r\sin\theta} \left(\Phi \frac{\partial^2 \Phi}{\partial \phi \partial t} - \frac{\partial \Phi}{\partial \phi} \frac{\partial \Phi}{\partial t} \right) \tag{D.4c}$$

$$W^{\mathrm{o}} = -\frac{1}{2}\varepsilon_{\mathrm{o}} \sum_{k=1}^{3} \left(A_k \frac{\partial^2 A_k}{\partial t^2} - \frac{\partial A_k}{\partial t} \frac{\partial A_k}{\partial t} \right) + \varepsilon_{\mathrm{o}} \frac{1}{2\mathrm{c}^2} \left[\Phi \frac{\partial^2 \Phi}{\partial t^2} - \left(\frac{\partial \Phi}{\partial t} \right)^2 \right]$$

$$T_{rr}^{\mathrm{o}} = \frac{1}{2\mu_{\mathrm{o}}} \sum_{k=1}^{3} \left(A_k \frac{\partial^2 A_k}{\partial r^2} - \frac{\partial A_k}{\partial r} \frac{\partial A_k}{\partial r} \right) - \frac{1}{2}\varepsilon_{\mathrm{o}} \left(\Phi \frac{\partial^2 \Phi}{\partial r^2} - \frac{\partial \Phi}{\partial r} \frac{\partial \Phi}{\partial r} \right) \tag{D.4d}$$

$$T_{\theta\theta}^{\mathrm{o}} = \frac{1}{2\mu_{\mathrm{o}} r^2} \sum_{k=1}^{3} \left(A_k \frac{\partial^2 A_k}{\partial \theta^2} - \frac{\partial A_k}{\partial \theta} \frac{\partial A_k}{\partial \theta} + r A_k \frac{\partial A_k}{\partial r} \right) + \frac{1}{\mu_{\mathrm{o}} r^2} \left[2 \left(A_\theta \frac{\partial A_r}{\partial \theta} - A_r \frac{\partial A_\theta}{\partial \theta} \right) - (A_r^2 + A_\theta^2) \right] - \varepsilon_{\mathrm{o}} \frac{1}{2r^2} \left(\Phi \frac{\partial^2 \Phi}{\partial \theta^2} - \frac{\partial \Phi}{\partial \theta} \frac{\partial \Phi}{\partial \theta} + r\Phi \frac{\partial \Phi}{\partial r} \right) \tag{D.4e}$$

$$T_{\phi\phi}^{\mathrm{o}} = \frac{1}{2\mu_{\mathrm{o}} r^2} \sum_{k=1}^{3} \left[\frac{1}{\sin^2\theta} \left(A_k \frac{\partial^2 A_k}{\partial \phi^2} - \frac{\partial A_k}{\partial \phi} \frac{\partial A_k}{\partial \phi} \right) + r A_k \frac{\partial A_k}{\partial r} + A_k \frac{\partial A_k}{\partial \theta} \cot\theta \right] + \frac{2}{\mu_{\mathrm{o}} r^2 \sin^2\theta} \left(A_\phi \frac{\partial A_\theta}{\partial \phi} - A_\theta \frac{\partial A_\phi}{\partial \phi} - A_r A_\theta \sin\theta \right) \cos\theta + \frac{1}{\mu_{\mathrm{o}} r^2} \left[(A_\theta^2 - A_r^2) - \frac{1}{\sin^2\theta} (A_\theta^2 + A_\phi^2) + \frac{2}{\sin\theta} \left(A_\phi \frac{\partial A_r}{\partial \phi} - A_r \frac{\partial A_\phi}{\partial \phi} \right) \right] - \varepsilon_{\mathrm{o}} \frac{1}{2r^2} \left[\frac{1}{\sin^2\theta} \left(\Phi \frac{\partial^2 \Phi}{\partial \phi^2} - \frac{\partial \Phi}{\partial \phi} \frac{\partial \Phi}{\partial \phi} \right) + r\Phi \frac{\partial \Phi}{\partial r} + \Phi \frac{\partial \Phi}{\partial \theta} \cot\theta \right] \tag{D.4f}$$

$$T_{r\theta}^{\mathrm{o}} = T_{\theta r}^{\mathrm{o}} = \frac{1}{2\mu_{\mathrm{o}} r} \sum_{k=1}^{3} \left(A_k \frac{\partial^2 A_k}{\partial r \partial \theta} - \frac{\partial A_k}{\partial r} \frac{\partial A_k}{\partial \theta} - \frac{A_k}{r} \frac{\partial A_k}{\partial \theta} \right) + \frac{1}{\mu_{\mathrm{o}} r} \left(A_\theta \frac{\partial A_r}{\partial r} - A_r \frac{\partial A_\theta}{\partial r} \right) - \varepsilon_{\mathrm{o}} \frac{1}{2r} \sum_{k=1}^{3} \left(\Phi \frac{\partial^2 \Phi}{\partial r \partial \theta} - \frac{\partial \Phi}{\partial r} \frac{\partial \Phi}{\partial \theta} - \frac{\Phi}{r} \frac{\partial \Phi}{\partial \theta} \right) \tag{D.4g}$$

$$T^{\text{o}}_{r\phi} = T^{\text{o}}_{\phi r} = \frac{1}{2\mu_{\text{o}} r \sin\theta} \sum_{k=1}^{3} \left[A_k \left(\frac{\partial^2 A_k}{\partial r \partial \phi} - \frac{1}{r} \frac{\partial A_k}{\partial \phi} \right) - \frac{\partial A_k}{\partial r} \frac{\partial A_k}{\partial \phi} \right]$$
$$+ \frac{1}{\mu_{\text{o}} r} \left[A_\phi \frac{\partial A_r}{\partial r} - A_r \frac{\partial A_\phi}{\partial r} + \left(A_\phi \frac{\partial A_\theta}{\partial r} - A_\theta \frac{\partial A_\phi}{\partial r} \right) \cot\theta \right]$$
$$- \varepsilon_{\text{o}} \frac{1}{2r \sin\theta} \left[\Phi \left(\frac{\partial^2 \Phi}{\partial r \partial \phi} - \frac{1}{r} \frac{\partial \Phi}{\partial \phi} \right) - \frac{\partial \Phi}{\partial r} \frac{\partial \Phi}{\partial \phi} \right] \quad \text{(D.4h)}$$

$$T^{\text{o}}_{\theta\phi} = T^{\text{o}}_{\phi\theta} = \frac{1}{2\mu_{\text{o}} r^2} \sum_{k=1}^{3} \frac{1}{\sin\theta} \left[A_k \left(\frac{\partial^2 A_k}{\partial \theta \partial \phi} - \frac{\partial A_k}{\partial \phi} \cot\theta \right) - \frac{\partial A_k}{\partial \theta} \frac{\partial A_k}{\partial \phi} \right]$$
$$+ \frac{1}{\mu_{\text{o}} r^2} \left[\left(A_\phi \frac{\partial A_r}{\partial \theta} - A_r \frac{\partial A_\phi}{\partial \theta} - A_\theta A_\phi \right) + \frac{1}{\sin\theta} \left(A_\theta \frac{\partial A_r}{\partial \phi} - A_r \frac{\partial A_\theta}{\partial \phi} \right) \right]$$
$$+ \frac{1}{\mu_{\text{o}} r^2} \left(A_\phi \frac{\partial A_\theta}{\partial \theta} - A_\theta \frac{\partial A_\phi}{\partial \theta} + A_r A_\phi \right) \cot\theta$$
$$- \varepsilon_{\text{o}} \frac{1}{2r^2 \sin\theta} \left[\Phi \left(\frac{\partial^2 \Phi}{\partial \theta \partial \phi} - \frac{\partial \Phi}{\partial \phi} \cot\theta \right) - \frac{\partial \Phi}{\partial \theta} \frac{\partial \Phi}{\partial \phi} \right] \quad \text{(D.4i)}$$

$$\nabla \widehat{\mathbf{r}} = \frac{1}{r} (\widehat{\theta}\,\widehat{\theta} + \widehat{\phi}\,\widehat{\phi}), \quad \nabla \widehat{\theta} = \frac{-1}{r} (\widehat{\theta}\,\widehat{\mathbf{r}} - \cot\theta\, \widehat{\phi}\widehat{\phi}), \quad \nabla \widehat{\phi} = \frac{-1}{r} (\widehat{\phi}\,\widehat{\mathbf{r}} + \cot\theta\, \widehat{\phi}\widehat{\theta})$$

In the event that the material is linear, isotropic, and *uniform* with scalar permittivity, ε, permeability, μ, and conductivity, σ, the previous formulas can be replaced by those in which the following substitutions have been made.

$$\varepsilon_{\text{o}} \to \varepsilon$$

$$\mu_{\text{o}} \to \mu$$

$$\mathbf{J}_{\text{total}} \to \mathbf{J}_{\text{u}} = \sigma \mathbf{E}$$

$$\rho_{\text{total}} \to \rho_{\text{u}} = 0$$

$$g \to \nabla \cdot \mathbf{A} + \mu\varepsilon \frac{\partial \Phi}{\partial t}$$

D.2 MINKOWSKI REPRESENTATIONS

When there is polarization and/or magnetization present, it is useful to start with the Minkowski form of Maxwell's Equations (Part I, Chapter 1, Section 1.7) and develop modified Maxwell–Poynting and Alternate representations (and where possible their duals) that are fully equivalent. That approach, carried out in Part II, Chapter 9, Section 9.3, is summarized below.

Maxwell–Poynting–Minkowski representation

With $\mathbf{P}^{\text{o}} = \mathbf{D} - \varepsilon_{\text{o}}\mathbf{E}^{\text{o}}$ and $\mathbf{M}^{\text{o}} = \mathbf{B}/\mu_{\text{o}} - \mathbf{H}^{\text{o}}$, the modified Maxwell–Poynting representation is

$$\nabla \cdot (\mathbf{E}^{\mathrm{o}} \times \mathbf{H}^{\mathrm{o}}) + \frac{\partial}{\partial t}\frac{1}{2}(\mathbf{E}^{\mathrm{o}} \cdot \mathbf{D} + \mathbf{B} \cdot \mathbf{H}^{\mathrm{o}}) =$$

$$-\mathbf{E}^{\mathrm{o}} \cdot \mathbf{J}_{\mathrm{u}} - \frac{1}{2}\left[\mathbf{E}^{\mathrm{o}} \cdot \frac{\partial \mathbf{P}^{\mathrm{o}}}{\partial t} - \frac{\partial \mathbf{E}^{\mathrm{o}}}{\partial t} \cdot \mathbf{P}^{\mathrm{o}} + \mathbf{B} \cdot \frac{\partial \mathbf{M}^{\mathrm{o}}}{\partial t} - \frac{\partial \mathbf{B}}{\partial t} \cdot \mathbf{M}^{\mathrm{o}}\right] \tag{D.5a}$$

$$\nabla \cdot \left[\mathbf{E}^{\mathrm{o}}\mathbf{D} + \mathbf{H}^{\mathrm{o}}\mathbf{B} - \frac{1}{2}(\mathbf{E}^{\mathrm{o}} \cdot \mathbf{D} + \mathbf{H}^{\mathrm{o}} \cdot \mathbf{B})\bar{\mathbf{I}}\right] - \frac{\partial}{\partial t}(\mathbf{D} \times \mathbf{B}) =$$

$$\rho_{\mathrm{u}}\mathbf{E}^{\mathrm{o}} + \mathbf{J}_{\mathrm{u}} \times \mathbf{B} + \frac{1}{2}(P_k^{\mathrm{o}}\nabla E_k^{\mathrm{o}} - E_k^{\mathrm{o}}\nabla P_k^{\mathrm{o}}) + \frac{1}{2}(M_k^{\mathrm{o}}\nabla B_k - B_k\nabla M_k^{\mathrm{o}}) \tag{D.5b}$$

In case $\rho_{\mathrm{u}} = 0$ and $\mathbf{J}_{\mathrm{u}} = 0$, it follows that the dual (refer to Part I, Chapter 6, Section 6.5) of this representation leads to identical equations.

Alternate–Minkowski representation

With $\mathbf{B} = \nabla \times \mathbf{A}$, $\mathbf{E}^{\mathrm{o}} = -\frac{\partial}{\partial t}\mathbf{A} - \nabla\Phi$, the equivalent modified Alternate representation can be expressed as

$$\nabla \cdot \mathbf{S}^{\mathrm{ou}} + \frac{\partial W^{\mathrm{ou}}}{\partial t} =$$

$$-\mathbf{E}^{\mathrm{o}} \cdot \mathbf{J}_{\mathrm{u}} - \frac{1}{2}\left[\mathbf{E}^{\mathrm{o}} \cdot \frac{\partial \mathbf{P}^{\mathrm{o}}}{\partial t} - \frac{\partial \mathbf{E}^{\mathrm{o}}}{\partial t} \cdot \mathbf{P}^{\mathrm{o}} + \mathbf{B} \cdot \frac{\partial \mathbf{M}^{\mathrm{o}}}{\partial t} - \frac{\partial \mathbf{B}}{\partial t} \cdot \mathbf{M}^{\mathrm{o}}\right] \tag{D.6a}$$

$$\nabla \cdot \overline{\mathbf{T}}^{\mathrm{ou}} - \frac{\partial}{\partial t}\mathbf{G}^{\mathrm{ou}} =$$

$$\rho_{\mathrm{u}}\mathbf{E}^{\mathrm{o}} + \mathbf{J}_{\mathrm{u}} \times \mathbf{B} + \frac{1}{2}(P_k^{\mathrm{o}}\nabla E_k^{\mathrm{o}} - E_k^{\mathrm{o}}\nabla P_k^{\mathrm{o}}) + \frac{1}{2}(M_k^{\mathrm{o}}\nabla B_k - B_k\nabla M_k^{\mathrm{o}}) \tag{D.6b}$$

where, with $g = \Box \cdot \mathcal{A} = \nabla \cdot \mathbf{A} + \mu_{\mathrm{o}}\varepsilon_{\mathrm{o}}\frac{\partial\Phi}{\partial t}$, we have

$$S_i^{\mathrm{ou}} = \left[\begin{array}{c} \Phi J_{\mathrm{u}i} + \dfrac{1}{2\mu_{\mathrm{o}}}\left(\mathbf{A} \cdot \dfrac{\partial^2 \mathbf{A}}{\partial x_i \partial t} - \dfrac{\partial \mathbf{A}}{\partial x_i} \cdot \dfrac{\partial \mathbf{A}}{\partial t}\right) \\ -\dfrac{1}{2}\varepsilon_{\mathrm{o}}\left(\Phi\dfrac{\partial^2\Phi}{\partial x_i \partial t} - \dfrac{\partial\Phi}{\partial x_i}\dfrac{\partial\Phi}{\partial t}\right) + \dfrac{1}{2\mu_{\mathrm{o}}}\left(\Phi\dfrac{\partial g}{\partial x_i} - \dfrac{\partial\Phi}{\partial x_i}g\right) \\ +\left[\left(\dfrac{1}{2}\dfrac{\partial\mathbf{A}}{\partial t} \times \mathbf{M}^{\mathrm{o}} - \mathbf{A} \times \dfrac{\partial \mathbf{M}^{\mathrm{o}}}{\partial t}\right) + \dfrac{1}{2}\left(\Phi\dfrac{\partial \mathbf{P}^{\mathrm{o}}}{\partial t} - \dfrac{\partial\Phi}{\partial t}\mathbf{P}^{\mathrm{o}}\right)\right]_i \end{array}\right] \tag{D.7a}$$

$$W^{\mathrm{ou}} = \left[\begin{array}{c} \dfrac{1}{2}(\mathbf{A} \cdot \mathbf{J}_{\mathrm{u}} + \rho_{\mathrm{u}}\Phi) - \dfrac{1}{2}\varepsilon_{\mathrm{o}}\left(\mathbf{A} \cdot \dfrac{\partial^2\mathbf{A}}{\partial t^2} - \dfrac{\partial\mathbf{A}}{\partial t} \cdot \dfrac{\partial\mathbf{A}}{\partial t}\right) \\ +\dfrac{1}{2}\dfrac{\varepsilon_{\mathrm{o}}}{c^2}\left(\Phi\dfrac{\partial^2\Phi}{\partial t^2} - \dfrac{\partial\Phi}{\partial t}\dfrac{\partial\Phi}{\partial t}\right) + \dfrac{1}{2}\varepsilon_{\mathrm{o}}(g\dfrac{\partial\Phi}{\partial t} - \Phi\dfrac{\partial g}{\partial t}) \\ +\dfrac{1}{2}\left(\mathbf{A} \cdot \dfrac{\partial\mathbf{P}^{\mathrm{o}}}{\partial t} - \dfrac{\partial\mathbf{A}}{\partial t} \cdot \mathbf{P}^{\mathrm{o}}\right) \end{array}\right] \tag{D.7b}$$

$$
T_{ij}^{\mathrm{ou}} = \begin{bmatrix}
-A_i J_{\mathrm{u}j} + \frac{1}{2}(\mathbf{A}\cdot\mathbf{J}_{\mathrm{u}} - \rho_{\mathrm{u}}\Phi)\delta_{ij} \\
+\frac{1}{2\mu_{\mathrm{o}}}\left(\mathbf{A}\cdot\frac{\partial^2\mathbf{A}}{\partial x_i\partial x_j} - \frac{\partial\mathbf{A}}{\partial x_i}\cdot\frac{\partial\mathbf{A}}{\partial x_j}\right) \\
-\frac{1}{2}\varepsilon_{\mathrm{o}}\left(\Phi\frac{\partial^2\Phi}{\partial x_i\partial x_j} - \frac{\partial\Phi}{\partial x_i}\frac{\partial\Phi}{\partial x_j}\right) + \frac{1}{2\mu_{\mathrm{o}}}\left(g\frac{\partial A_j}{\partial x_i} - A_j\frac{\partial g}{\partial x_i}\right) \\
-A_i(\nabla\times\mathbf{M}^{\mathrm{o}})_j - M_i^{\mathrm{o}}(\nabla\times\mathbf{A})_j - \frac{\partial\Phi}{\partial x_i}P_j^{\mathrm{o}} \\
+\frac{1}{2}[\nabla\cdot(\Phi\mathbf{P}^{\mathrm{o}}) + \mathbf{A}\cdot(\nabla\times\mathbf{M}^{\mathrm{o}}) + \mathbf{M}^{\mathrm{o}}\cdot(\nabla\times\mathbf{A})]\,\delta_{ij}
\end{bmatrix} \tag{D.7c}
$$

$$
G_i^{\mathrm{ou}} = \begin{bmatrix}
\rho_{\mathrm{u}}A_i + \frac{1}{2}\varepsilon_{\mathrm{o}}\left(\mathbf{A}\cdot\frac{\partial^2\mathbf{A}}{\partial x_i\partial t} - \frac{\partial\mathbf{A}}{\partial x_i}\cdot\frac{\partial\mathbf{A}}{\partial t}\right) \\
-\frac{1}{2}\frac{\varepsilon_{\mathrm{o}}}{\mathrm{c}^2}\left(\Phi\frac{\partial^2\Phi}{\partial x_i\partial t} - \frac{\partial\Phi}{\partial x_i}\frac{\partial\Phi}{\partial t}\right) \\
+\frac{1}{2}\left(\mathbf{P}^{\mathrm{o}}\cdot\frac{\partial\mathbf{A}}{\partial x_i} - \frac{\partial\mathbf{P}^{\mathrm{o}}}{\partial x_i}\cdot\mathbf{A}\right) + \frac{1}{2}\varepsilon_{\mathrm{o}}\left(\Phi\frac{\partial g}{\partial x_i} - g\frac{\partial\Phi}{\partial x_i}\right)
\end{bmatrix} \tag{D.7d}
$$

Dual Alternate–Minkowski representation

In case $\rho_{\mathrm{u}} = 0$ and $\mathbf{J}_{\mathrm{u}} = 0$, it follows that $\mathbf{D} = -\nabla\times\mathbf{A}_{\mathrm{m}}$, $\mathbf{H}^{\mathrm{o}} = -\frac{\partial}{\partial t}\mathbf{A}_{\mathrm{m}} - \nabla\Phi_{\mathrm{m}}$. Then the dual Alternate–Minkowski representation satisfies Eqs. (D.6a) and (D.6b), provided that the set $(\mathbf{S}^{\mathrm{ou}}, W^{\mathrm{ou}}, \overline{\overline{\mathbf{T}}}^{\mathrm{ou}}, \mathbf{G}^{\mathrm{ou}})$ is replaced by $(\mathbf{S}^{\mathrm{omu}}, W^{\mathrm{omu}}, \overline{\overline{\mathbf{T}}}^{\mathrm{omu}}, \mathbf{G}^{\mathrm{omu}})$ and

$$
S_i^{\mathrm{omu}} = \begin{bmatrix}
\frac{1}{2\varepsilon_{\mathrm{o}}}\left(\mathbf{A}_{\mathrm{m}}\cdot\frac{\partial^2\mathbf{A}_{\mathrm{m}}}{\partial x_i\partial t} - \frac{\partial\mathbf{A}_{\mathrm{m}}}{\partial x_i}\cdot\frac{\partial\mathbf{A}_{\mathrm{m}}}{\partial t}\right) \\
-\frac{1}{2}\mu_{\mathrm{o}}\left(\Phi_{\mathrm{m}}\frac{\partial^2\Phi_{\mathrm{m}}}{\partial x_i\partial t} - \frac{\partial\Phi_{\mathrm{m}}}{\partial x_i}\frac{\partial\Phi_{\mathrm{m}}}{\partial t}\right) + \frac{1}{2\varepsilon_{\mathrm{o}}}\left(\Phi_{\mathrm{m}}\frac{\partial g_{\mathrm{m}}}{\partial x_i} - \frac{\partial\Phi_{\mathrm{m}}}{\partial x_i}g_{\mathrm{m}}\right) \\
+\left[\frac{1}{2\varepsilon_{\mathrm{o}}}\left(\mathbf{A}_{\mathrm{m}}\times\frac{\partial\mathbf{P}^{\mathrm{o}}}{\partial t} - \frac{\partial\mathbf{A}_{\mathrm{m}}}{\partial t}\times\mathbf{P}^{\mathrm{o}}\right) + \frac{1}{2}\mu_{\mathrm{o}}\left(\Phi_{\mathrm{m}}\frac{\partial\mathbf{M}^{\mathrm{o}}}{\partial t} - \frac{\partial\Phi_{\mathrm{m}}}{\partial t}\mathbf{M}^{\mathrm{o}}\right)\right]_i
\end{bmatrix} \tag{D.8a}
$$

$$
W^{\mathrm{omu}} = \begin{bmatrix}
-\frac{1}{2}\mu_{\mathrm{o}}\left(\mathbf{A}_{\mathrm{m}}\cdot\frac{\partial^2\mathbf{A}_{\mathrm{m}}}{\partial t^2} - \frac{\partial\mathbf{A}_{\mathrm{m}}}{\partial t}\cdot\frac{\partial\mathbf{A}_{\mathrm{m}}}{\partial t}\right) \\
+\frac{1}{2}\varepsilon_{\mathrm{o}}\left(\Phi_{\mathrm{m}}\frac{\partial^2\Phi_{\mathrm{m}}}{\partial t^2} - \frac{\partial\Phi_{\mathrm{m}}}{\partial t}\frac{\partial\Phi_{\mathrm{m}}}{\partial t}\right) + \frac{1}{2}\mu_{\mathrm{o}}\left(g_{\mathrm{m}}\frac{\partial\Phi_{\mathrm{m}}}{\partial t} - \Phi_{\mathrm{m}}\frac{\partial g_{\mathrm{m}}}{\partial t}\right) \\
+\frac{1}{2}\mu_{\mathrm{o}}\left(\mathbf{A}_{\mathrm{m}}\cdot\frac{\partial\mathbf{M}^{\mathrm{o}}}{\partial t} - \frac{\partial\mathbf{A}_{\mathrm{m}}}{\partial t}\cdot\mathbf{M}^{\mathrm{o}}\right)
\end{bmatrix} \tag{D.8b}
$$

$$T_{ij}^{\text{omu}} = \begin{bmatrix} \frac{1}{2\varepsilon_o}\left(\mathbf{A}_m \cdot \frac{\partial^2 \mathbf{A}_m}{\partial x_i \partial x_j} - \frac{\partial \mathbf{A}_m}{\partial x_i} \cdot \frac{\partial \mathbf{A}_m}{\partial x_j}\right) \\ -\frac{1}{2}\mu_o\left(\Phi_m \frac{\partial^2 \Phi_m}{\partial x_i \partial x_j} - \frac{\partial \Phi_m}{\partial x_i}\frac{\partial \Phi_m}{\partial x_j}\right) + \frac{1}{2\varepsilon_o}\left(g_m \frac{\partial A_{mj}}{\partial x_i} - A_{mj}\frac{\partial g_m}{\partial x_i}\right) \\ +\frac{1}{\varepsilon_o}A_{mi}(\nabla \times \mathbf{P}^o)_j + \frac{1}{\varepsilon_o}P_i^o(\nabla \times \mathbf{A}_m)_j - \mu_o \frac{\partial \Phi_m}{\partial x_i}M_j^o \\ +\frac{1}{2}\left[\mu_o \nabla \cdot (\Phi_m \mathbf{M}^o) - \mathbf{A}_m \cdot \left(\nabla \times \frac{\mathbf{P}^o}{\varepsilon_o}\right) - \frac{\mathbf{P}^o}{\varepsilon_o} \cdot (\nabla \times \mathbf{A}_m)\right]\delta_{ij} \end{bmatrix} \tag{D.8c}$$

$$G_i^{\text{omu}} = \begin{bmatrix} \frac{1}{2}\mu_o\left(\mathbf{A}_m \cdot \frac{\partial^2 \mathbf{A}_m}{\partial x_i \partial t} - \frac{\partial \mathbf{A}_m}{\partial x_i} \cdot \frac{\partial \mathbf{A}_m}{\partial t}\right) \\ -\frac{1}{2}\frac{\mu_o}{c^2}\left(\Phi_m \frac{\partial^2 \Phi_m}{\partial x_i \partial t} - \frac{\partial \Phi_m}{\partial x_i}\frac{\partial \Phi_m}{\partial t}\right) + \frac{1}{2}\mu_o\left(\Phi_m \frac{\partial g_m}{\partial x_i} - g_m \frac{\partial \Phi_m}{\partial x_i}\right) \\ +\frac{1}{2}\mu_o\left(\mathbf{M}^o \cdot \frac{\partial \mathbf{A}_m}{\partial x_i} - \frac{\partial \mathbf{M}^o}{\partial x_i} \cdot \mathbf{A}_m\right) \end{bmatrix} \tag{D.8d}$$

Here, $g_m = \Box \cdot \mathcal{A}_m = \nabla \cdot \mathbf{A}_m + \mu_o \varepsilon_o \frac{\partial \Phi_m}{\partial t}$ is the dual of g.

D.3 STRESS-MOMENTUM REPRESENTATIONS OF TORQUE

Following the analysis of Chapter 3, Section 3.1, the torque density, $\mathbf{r} \times \mathbf{f}$, that arises from the Lorentz force density, $\mathbf{f} = \rho \mathbf{E} + \mathbf{J} \times \mu_o \mathbf{H}$, can be expressed as

$$\mathbf{r} \times \left(\nabla \cdot \overline{\mathbf{T}} - \frac{\partial}{\partial t}\mathbf{G}\right) = \mathbf{r} \times \mathbf{f}$$

or equivalently as

$$\left[\nabla \cdot (\mathbf{r} \times \overline{\mathbf{T}}) - \frac{\partial (\mathbf{r} \times \mathbf{G})}{\partial t} - \mathbf{r} \times \mathbf{f}\right]_k + \epsilon_{kji} T_{ji} = 0 \tag{D.9a}$$

where $k = 1, 2, 3$ is the Cartesian-coordinate index, ϵ_{kji} the permutation symbol (defined in Appendix B, Section B.2) and either the Maxwell–Poynting or an Alternate formulation of stress and momentum density is used. The four-space representation of this equation is derived in Chapter 8, Section 8.1.

The net torque, **Torque** $= \int_V \mathbf{r} \times \mathbf{f}\, dV$ acting on the volume V is *independent* of the choice of origin for the radial vector, $\mathbf{r}$ only when the net force, $\mathbf{F} = \int_V \mathbf{f}\, dV$, acting on that volume is zero. The integral form of Eq. (D.9a) is

$$\textbf{Torque} = \oint_S [\mathbf{r} \times (\overline{\mathbf{T}} \cdot \widehat{\mathbf{n}})]\, da + \int_V \epsilon_{kij} T_{ij}\, dV - \frac{d}{dt}\int_V \mathbf{r} \times \mathbf{G}\, dV \tag{D.9b}$$

where S is any surface in free-space that completely encloses the volume containing ρ and $\mathbf{J}$, with the positive normal $\widehat{\mathbf{n}}$ of the surface pointing outward.

When conduction currents, as well as polarization and/or magnetization vectors, are present, it is necessary to choose a specific electromagnetic model. To facilitate comparisons, we choose the Amperian model discussed fully in Chapters 9 and 10. Then, with

$$\mathbf{f}^{\mathrm{amp}} = \rho_{\mathrm{total}}\mathbf{E}^{\mathrm{o}} + \mathbf{J}_{\mathrm{total}} \times \mathbf{B}$$

and because the Amperian Maxwell stress tensor is symmetric,

$$\nabla \cdot (\mathbf{r} \times \overline{\overline{\mathbf{T}}}^{\mathrm{amp}}) - \frac{\partial\, \mathbf{r} \times (\varepsilon_{\mathrm{o}}\mathbf{E}^{\mathrm{o}} \times \mathbf{B})}{\partial t} = \mathbf{r} \times \mathbf{f}^{\mathrm{amp}} \tag{D.10}$$

is the final form of the Maxwell–Poynting representation. On the other hand, in the Alternate representation, the stress tensor may not be symmetric; therefore,

$$\nabla \cdot (\mathbf{r} \times \overline{\overline{\mathbf{T}}}^{\mathrm{o}}) - \frac{\partial(\mathbf{r} \times \mathbf{G}^{\mathrm{o}})}{\partial t} + \mathbf{J}_{\mathrm{total}} \times \mathbf{A} + \frac{1}{2\mu_{\mathrm{o}}}(g\nabla \times \mathbf{A} - \nabla g \times \mathbf{A}) = \mathbf{r} \times \mathbf{f}^{\mathrm{amp}} \tag{D.11a}$$

The integral form of Eq. (D.11a) is

$$\mathbf{Torque} = \begin{pmatrix} \oint_S [\mathbf{r} \times (\overline{\overline{\mathbf{T}}}^{\mathrm{o}} \cdot \widehat{\mathbf{n}})]\, da + \int_V \mathbf{J}_{\mathrm{total}} \times \mathbf{A}\, dV \\ + \frac{1}{2\mu_{\mathrm{o}}} \int_V (g\nabla \times \mathbf{A} - \nabla g \times \mathbf{A})\, dV \\ - \frac{d}{dt} \int_V \mathbf{r} \times \mathbf{G}^{\mathrm{o}}\, dV \end{pmatrix} \tag{D.11b}$$

In the Lorenz gauge these simplify because $g = 0$. In regions of a linear medium where there are only conduction currents, $\mathbf{J}_{\mathrm{total}} = \mathbf{J}_{\mathrm{u}} = \sigma\, \mathbf{E}^{\mathrm{o}}$ and $\mathbf{J}_{\mathrm{u}} \times \mathbf{A} = \frac{1}{\tau_d}\varepsilon_{\mathrm{o}}\mathbf{E}^{\mathrm{o}} \times \mathbf{A}$ where $\tau_d = \varepsilon_{\mathrm{o}}/\sigma$ is the dielectric relaxation time. The quantity $\varepsilon_{\mathrm{o}}\mathbf{E}^{\mathrm{o}} \times \mathbf{A}$ is the spin angular-momentum density that arises in the quantum theory of free-space photons (wherein $\Phi = 0$, $\mathbf{E}^{\mathrm{o}} = -\frac{\partial}{\partial t}\mathbf{A}$ and circularly polarized plane waves of frequency ω represent photons with spin of either ± 1).

Notice that $\mathbf{r} \times \mathbf{G}^{\mathrm{amp}} = \mathbf{L}^{\mathrm{amp}} = \varepsilon_{\mathrm{o}}\mathbf{r} \times [\mathbf{E}^{\mathrm{o}} \times (\nabla \times \mathbf{A})]$ can be written in Cartesian coordinates and index notation as $\varepsilon_{\mathrm{o}}\epsilon_{kij}x_i[\mathbf{E}^{\mathrm{o}} \times (\nabla \times \mathbf{A})]_j$. With the use of vector identity [Appendix B, Section B.2, Eq. (B.61)] $\partial x_i/\partial x_\ell = \delta_{i\ell}$ and $\varepsilon_{\mathrm{o}}\nabla \cdot \mathbf{E}^{\mathrm{o}} = \rho_{\mathrm{total}}$, we have

$$\begin{aligned} (\mathbf{r} \times \mathbf{G}^{\mathrm{amp}})_k &= -\varepsilon_{\mathrm{o}}\epsilon_{kij}x_i \left(\frac{\partial A_j}{\partial x_\ell} - \frac{\partial A_\ell}{\partial x_j}\right) E^{\mathrm{o}}_\ell \\ &= \varepsilon_{\mathrm{o}}\epsilon_{kij} \left[-\frac{\partial (x_i A_j E^{\mathrm{o}}_\ell)}{\partial x_\ell} + \frac{\partial x_i}{\partial x_\ell} A_j E^{\mathrm{o}}_\ell + x_i A_j \nabla \cdot \mathbf{E}^{\mathrm{o}} + x_i \mathbf{E}^{\mathrm{o}} \cdot \frac{\partial \mathbf{A}}{\partial x_j} \right] \\ &= -\nabla \cdot [\varepsilon_{\mathrm{o}}(\mathbf{r} \times \mathbf{A})_k \mathbf{E}^{\mathrm{o}}] + \epsilon_{kij}x_i \frac{\partial \mathbf{A}}{\partial x_j} \cdot \varepsilon_{\mathrm{o}}\mathbf{E}^{\mathrm{o}} + (\varepsilon_{\mathrm{o}}\mathbf{E}^{\mathrm{o}} \times \mathbf{A} + \mathbf{r} \times \rho_{\mathrm{total}}\mathbf{A})_k \end{aligned}$$

When integrated over a volume, V, bounded by the surface, S, and the Divergence theorem is used, the result is

$$\int_V \mathbf{L}^{\mathrm{amp}}\, dV = \int_V (\mathbf{L}^{\mathrm{spin}} + \mathbf{L}^{\mathrm{orbit}} + \mathbf{L}^{\mathrm{charge}})\, dV - \oint_S \varepsilon_{\mathrm{o}}(\mathbf{r} \times \mathbf{A})\mathbf{E}^{\mathrm{o}} \cdot \widehat{\mathbf{n}}\, da \tag{D.12}$$

where

$$\mathbf{L}^{\text{spin}} = \varepsilon_o \mathbf{E}^o \times \mathbf{A} \tag{D.13a}$$

$$L_k^{\text{orbit}} = \epsilon_{kij} x_i \frac{\partial \mathbf{A}}{\partial x_j} \cdot \varepsilon_o \mathbf{E}^o = \varepsilon_o \mathbf{E}^o \cdot \frac{\partial \mathbf{A}}{\partial \phi_k} \tag{D.13b}$$

$$\mathbf{L}^{\text{charge}} = \mathbf{r} \times \rho_{\text{total}} \mathbf{A} \tag{D.13c}$$

are defined, respectively, as the spin, orbital, and electric-charge angular-momentum densities. Equation (D.13b) is expressed in terms of the angles,

$$\phi_1 = \tan^{-1}(x_3/x_2), \qquad \phi_2 = \tan^{-1}(x_1/x_3), \qquad \phi_3 = \tan^{-1}(x_2/x_1) \tag{D.14}$$

The derivatives with respect to ϕ_k must be evaluated carefully whenever $\mathbf{A}$ is expressed in non-Cartesian components. The surface integral in Eq. (D.12) is expected to vanish as $S \to \infty$. Even when the surface integral vanishes, the Maxwell–Poynting (Amperian) angular-momentum density, $\mathbf{L}^{\text{amp}}$ is, in general, *not* identical to the sum, $\mathbf{L}^{\text{spin}} + \mathbf{L}^{\text{orbit}} + \mathbf{L}^{\text{charge}}$.

Although $\mathbf{L}^{\text{spin}}$ and $\mathbf{r} \times \rho_{\text{total}} \mathbf{A}$ are both hidden in the Maxwell–Poynting formulation, the Alternate formulation makes it easier to calculate the total torque arising from them. In this regard, we note that $\mathbf{r} \times \mathbf{f}^{\text{amp}}$ may be zero in regions where $\mathbf{J}_{\text{total}} \times \mathbf{A}$ is not. In many instances, the latter term accounts for nearly all of the average torque.

If the Minkowski formulation is used instead of the Amperian representation, the stress tensor is in general nonsymmetric. Terms $\mathbf{E} \times \mathbf{D} + \mathbf{H} \times \mathbf{B} = -\mathbf{P} \times \mathbf{E} - \mathbf{M} \times \mathbf{B}$ appear in Eq. (D.9a) (instead of $\mathbf{J} \times \mathbf{A}$); these vanish for linear isotropic (but *not* necessarily for anisotropic) material. Other results are similar except that $\mathbf{G}^{\text{mink}} = \mathbf{D} \times \mathbf{B}$ replaces $\mathbf{G}^{\text{amp}}$ in Eq. (D.12). When, $\mathbf{D} = \varepsilon \mathbf{E}^o$, both $\mathbf{L}^{\text{spin}}$ and $\mathbf{L}^{\text{orbit}}$ are increased by the factor $\varepsilon/\varepsilon_o$. Notice that in the Alternate formulation, $\mathbf{J}_u \times \mathbf{A}$ can still be written as $\frac{1}{\tau_d}\mathbf{L}^{\text{spin}}$, but now $\tau_d = \varepsilon/\sigma$ is the modified dielectric relaxation time. We recall that in a stationary medium, $\mathbf{E}^o = \mathbf{E}$ and the superscript o can be omitted.

Direct evaluation of $\mathbf{r} \times \mathbf{f}^{\text{amp}}$ is made difficult when uniform plane wave excitations are assumed and/or fringing fields are the source of the torque. In such cases, use of what appear to be reasonable approximations of $\mathbf{E}$ and $\mathbf{B}$ can lead to the erroneous conclusion that there is no net torque in configurations where there definitely is one (see Problem 25.4-16). When one integrates the left-hand side of Eq. (D.10) over any volume containing the currents and charges and employs the Divergence Theorem, the total torque can be found from the sum of the stress integral over the enclosing surface and the negative time rate of change of the total angular momentum. If, instead, Eq. (D.11a) (and $g = 0$) is used, there are similar integrations, but now the addition of the volume integral of $\mathbf{J}_{\text{total}} \times \mathbf{A}$. The latter is often the major component of the total torque and is straightforward to calculate using approximate fields [42]. For sinusoidal steady-state fields of a single frequency ω, there are no nonlocal reactive power or momentum densities and $\frac{\partial}{\partial t}\mathbf{G}^o = \frac{\partial}{\partial t}(\rho \mathbf{A})$; therefore in the Alternate formulation, we have

$$\textbf{Torque} = \oint_S [\mathbf{r} \times (\overline{\mathbf{T}}^o \cdot \widehat{\mathbf{n}})]\, da + \int_V \mathbf{J}_{\text{total}} \times \mathbf{A}\, dV - \frac{d}{dt}\int_V (\mathbf{r} \times \rho_{\text{total}} \mathbf{A}) dV \tag{D.15}$$

Evaluation is further simplified because, *unlike the corresponding term involving the Maxwell stress tensor*, this surface integral vanishes in cases where the $\mathbf{E}$ and $\mathbf{A}$ free-space fields are characterized by $\mathbf{L}^{\text{orbit}} = 0$. This assertion is proved in the last section of this Appendix. In such cases (which are common), the integrations are limited to the volumes containing current and charge. For harmonic fields, the time average of the torque only requires evaluation of the $\mathbf{J}_{\text{total}} \times \mathbf{A}$ integral which takes care of $\mathbf{L}^{\text{spin}}$. Because the Maxwell stress tensor is symmetric, the corresponding surface integral must account for torques produced by spin as well as orbital angular momentum.

Linear isotropic dielectric/magnetic conducting materials

In the Amperian formulation, the total electric current density is $\mathbf{J}_{\text{total}} = \mathbf{J}_{\text{u}} + \frac{\partial \mathbf{P}}{\partial t} + \nabla \times \mathbf{M}$. For a linear isotropic material that is stationary and conducting and has both polarization $\mathbf{P} = (\varepsilon - \varepsilon_{\text{o}})\mathbf{E}$ and magnetization $\mathbf{M} = (1/\mu_{\text{o}} - 1/\mu)\mathbf{B}$, we have

$$\mathbf{J}_{\text{total}} = \sigma \mathbf{E} + (\varepsilon - \varepsilon_{\text{o}})\frac{\partial \mathbf{E}}{\partial t} + \nabla \times [(1/\mu_{\text{o}} - 1/\mu)\mathbf{B}] \tag{D.16}$$

But $\nabla \times [(1/\mu_{\text{o}} - 1/\mu)\mathbf{B}] = (1/\mu_{\text{o}} - 1/\mu)\nabla \times \mathbf{B} - \nabla(1/\mu) \times \mathbf{B}$ and

$$\nabla \times \mathbf{B} = \frac{1}{c^2}\frac{\partial \mathbf{E}}{\partial t} + \mu_{\text{o}}\mathbf{J}_{\text{total}}$$

Therefore

$$\mathbf{J}_{\text{total}} = \frac{\mu}{\mu_{\text{o}}}\sigma \mathbf{E} + \left(\frac{\mu}{\mu_{\text{o}}}\frac{\varepsilon}{\varepsilon_{\text{o}}} - 1\right)\varepsilon_{\text{o}}\frac{\partial \mathbf{E}}{\partial t} - \frac{\mu}{\mu_{\text{o}}}\nabla(1/\mu) \times \mathbf{B}$$

and the torque arising from $\mathbf{J}_{\text{total}} \times \mathbf{A}$ can be evaluated as

$$\int_V \left[\frac{\mu}{\mu_{\text{o}}}\sigma \mathbf{E} + \left(\frac{\mu}{\mu_{\text{o}}}\frac{\varepsilon}{\varepsilon_{\text{o}}} - 1\right)\varepsilon_{\text{o}}\frac{\partial \mathbf{E}}{\partial t} + \nabla\left(\frac{\mu}{\mu_{\text{o}}}\right) \times \frac{\mathbf{B}}{\mu}\right] \times \mathbf{A}\, dV$$

For a spatially uniform material, $\nabla \mu = 0$ except at the boundaries where the normal derivative is infinite and Amperian surface currents, $\mathbf{K}_{\text{amp}} = \mathbf{M} \times \hat{\mathbf{n}}$, exist. In this case,

$$\textbf{Torque} = \left(\begin{array}{l} \oint_S [\mathbf{r} \times (\overline{\mathbf{T}}^{\text{o}} \cdot \hat{\mathbf{n}})]\, da + \int_V \left[\frac{\mu}{\mu_{\text{o}}}\sigma \mathbf{E} + \left(\frac{\mu}{\mu_{\text{o}}}\frac{\varepsilon}{\varepsilon_{\text{o}}} - 1\right)\varepsilon_{\text{o}}\frac{\partial \mathbf{E}}{\partial t}\right] \times \mathbf{A}\, dV \\ \qquad + \oint_{S_M} (\mathbf{M} \times \hat{\mathbf{n}}) \times \mathbf{A}\, da - \frac{d}{dt}\int_V (\mathbf{r} \times \mathbf{G}^{\text{o}})\, dV \end{array}\right) \tag{D.17}$$

Note that S_M is the surface defined by $|\nabla \mu| = \infty$.

If the conducting material is free to rotate, the linear constitutive laws apply to the fields in the moving frame (primed) of the rotating coordinates and $\mathbf{J}'_{\text{total}}$ and $\mathbf{A}'$ must be evaluated carefully. For nonrelativistic velocities, we have

$$\mathbf{J}_{\text{total}} \times \mathbf{A} \simeq \mathbf{J}'_{\text{total}} \times \left(\mathbf{A}' + \frac{\Phi'}{c^2}\mathbf{v}\right) - \rho'_{\text{total}}\mathbf{A}' \times \mathbf{v}$$

The inverse transformation follows if the primed and unprimed variables are interchanged and the sign of the velocity is reversed. Although the term involving the electric potential is first-order in the velocity, the division by c^2 renders it negligible in most cases. The dynamics of the motion will depend upon the torque, the moment of inertia, and both electrical and mechanical damping torques. Even without mechanical friction, electrical damping will limit the angular acceleration. If the excitation is a single–frequency harmonic field, the mechanical rotation of the current density may create field components (and currents) at frequencies that differ from the excitation; often these can be neglected.

The magnitude of the spin angular momentum of *any* single photon is $\hbar$. It follows that the generation of sizeable electromagnetic torques requires huge numbers of photons. Unless the frequency, and thus energy ($\hbar\omega$), of each is very small, the collective energy will be enormous. This is the reason that most motors and generators operate with low-frequency quasistatic (usually *MQS*) fields. However, for nanometer size devices, the relevant moment of inertia may be so small that only very tiny torques are required [43, 44]. In these cases, operation at microwave or even optical frequencies is feasible.

Torque contribution from the Alternate-stress integral

We are now in a position to prove that for single-frequency fields in the sinusoidal steady state, the contribution to the average torque from the surface integral

$$\oint_S [\mathbf{r} \times (\overline{\mathbf{T}}^{\mathrm{o}} \cdot \widehat{\mathbf{n}})]\, da$$

vanishes, provided that the closed surface S (surrounding the sources) is in free-space and there is no Alternate orbital angular momentum in the fields passing through it.

Expressed in Cartesian coordinates, $[\mathbf{r} \times (\overline{\mathbf{T}}^{\mathrm{o}} \cdot \widehat{\mathbf{n}})]_k = \epsilon_{kij} x_i T^{\mathrm{o}}_{j\ell} n_\ell$. From Eq. (D.2e), $\mathcal{A} = [A,\ \frac{i}{c}\Phi]$, and the rotational angles ϕ_k defined by Eqs. (D.14), we obtain

$$[\mathbf{r} \times (\overline{\mathbf{T}}^{\mathrm{o}} \cdot \widehat{\mathbf{n}})]_k = \frac{1}{2\mu_{\mathrm{o}}} \left(\mathcal{A} \cdot \frac{\partial^2 \mathcal{A}}{\partial \phi_k\, \partial x_\ell} - \frac{\partial \mathcal{A}}{\partial \phi_k} \cdot \frac{\partial \mathcal{A}}{\partial x_\ell} \right) n_\ell \tag{D.18}$$

which vanishes when $\frac{\partial \mathcal{A}}{\partial \phi_k} = 0$. This occurs in the Lorenz gauge wherever $L_k^{\mathrm{orbit}} = 0$ and the fields are time-dependent. Consequently, if there is no orbital angular momentum in the vicinity of the surface, the Alternate-stress integral must vanish. Of course, this result holds only for the Alternate stress tensor (the Maxwell stress tensor must account for spin as well as orbital angular momentum). Except that it is in free-space and encloses the volume of interest, there is no restriction on S. Nevertheless, when the surface is in the far field, an alternate analysis is instructive.

In the sinusoidal steady state, the four-vector potential can be expressed as $\mathcal{A} = Re\{\underline{\mathcal{A}}(x, y, z) \exp(j[\omega t - \Psi(x, y, z)]\}$. If the surface S is moved to the far field, the amplitude variations are negligible compared to those of the the phase. In terms of $\Theta(t, x, y, z) = \omega t - \Psi(x, y, z)$, the gradient operator can be expressed as $\frac{\partial}{\partial x_i} = -\frac{\partial \Psi}{\partial x_i}\frac{\partial}{\partial \Theta} = -\frac{1}{\omega}\frac{\partial \Psi}{\partial x_i}\frac{\partial}{\partial t}$. It follows that

$$\begin{aligned}
[\mathbf{r} \times (\overline{\mathbf{T}}^{\mathrm{o}} \cdot \widehat{\mathbf{n}})]_k &= -\frac{1}{2}\varepsilon_{\mathrm{o}} \left(\mathcal{A} \cdot \frac{\partial^2 \mathcal{A}}{\partial \phi_k\, \partial t} - \frac{\partial \mathcal{A}}{\partial \phi_k} \cdot \frac{\partial \mathcal{A}}{\partial t} \right) \frac{\mathrm{c}^2}{\omega} \nabla \Psi \cdot \widehat{\mathbf{n}} \\
&= -\frac{\mathrm{c}}{k}\frac{\partial \Psi}{\partial n} (\mathbf{r} \times \mathbf{G}^{\mathrm{o}})_k
\end{aligned} \tag{D.19}$$

where

$$(\mathbf{r} \times \mathbf{G}^{\mathrm{o}})_k = \varepsilon_{\mathrm{o}} \left[\frac{\partial^2}{\partial t\, \partial \phi_k} \left(\frac{1}{4} \mathcal{A} \cdot \mathcal{A} \right) - \frac{\partial \mathcal{A}}{\partial \phi_k} \cdot \frac{\partial \mathcal{A}}{\partial t} \right]$$

The three components $L_k^{\text{o orbit}} = (\mathbf{r} \times \mathbf{G}^{\text{o}})_k$ *define* the Alternate orbital angular-momentum density. The average values, $<L_k^{\text{o orbit}}> = <\varepsilon_{\text{o}} \frac{-\partial \mathcal{A}}{\partial t} \cdot \frac{\partial \mathcal{A}}{\partial \phi_k}>$, are identical to those of the average orbital angular-momentum density defined by Eq. (D.13b) when Φ is negligible. This will be the case if the surface S is far from all electric charge as was assumed. When $<\mathbf{L}^{\text{orbit}}> = 0$, the time-averaged surface integrand vanishes everywhere and the conjecture is proved for that S or any other that surrounds the volume containing ρ_{total} and $\mathbf{J}_{\text{total}}$. When $\mathbf{L}^{\text{orbit}} \neq 0$, the surface integral can be evaluated directly or, for the far field, after substituting Eq. (D.19). The quantity $\frac{c}{k}\frac{\partial \Psi}{\partial n} = \frac{c}{k}\nabla\Psi \cdot \widehat{\mathbf{n}}$ is the local group velocity at which $\mathbf{L}^{\text{o orbit}}$ is transported across the surface.

APPENDIX E

E.1 FIELDS OF SPECIFIED CHARGES AND CURRENTS

When the electric current density, **J**, and the associated charge density, ρ, are both specified, the four-dimensional formulation, developed in Part II, Chapter 7, Section 7.7, leads to the free-space vector and scalar potentials

$$\mathbf{A}_p(\mathbf{r}_p, t) = \int_{-\infty}^{\infty}\int_{-\infty}^{\infty}\int_{-\infty}^{\infty}\int_{-\infty}^{\infty} \frac{\mu_o \mathbf{J}_q \; dx_1^q dx_2^q dx_3^q d\xi}{4\pi^2(r_{qp}^2 + \xi^2)}$$

$$= \int_{-\infty}^{\infty}\int_{-\infty}^{\infty}\int_{-\infty}^{\infty} \frac{\mu_o \mathbf{J}_q(r_q, t^*) \; dx_1^q dx_2^q dx_3^q}{4\pi r_{qp}} \tag{E.1a}$$

$$\Phi_p(\mathbf{r}_p, t) = \int_{-\infty}^{\infty}\int_{-\infty}^{\infty}\int_{-\infty}^{\infty}\int_{-\infty}^{\infty} \frac{\rho_q \; dx_1^q dx_2^q dx_3^q d\xi}{4\pi^2 \varepsilon_o (r_{qp}^2 + \xi^2)}$$

$$= \int_{-\infty}^{\infty}\int_{-\infty}^{\infty}\int_{-\infty}^{\infty} \frac{\rho_q(r_q, t^*) \; dx_1^q dx_2^q dx_3^q}{4\pi \varepsilon_o r_{qp}} \tag{E.1b}$$

The Power and Beauty of Electromagnetic Fields, First Edition. F. R. Morgenthaler.

expressed either as four-dimensional integrals or (after integrating around the singularity at $\xi = -ir_{qp}$) as three-space integrals in terms of the retarded time, $t^* = t - \frac{r_{qp}}{c}$, where

$$r_{qp} = |\mathbf{r}_p - \mathbf{r}_q|$$

$$\xi = ic(t_q - t)$$

The electric and magnetic fields can be found either by differentiation of the potentials or directly from the four-dimensional integrals,

$$\mathbf{E}_p(\mathbf{r}_p, t) = \int_{-\infty}^{\infty}\int_{-\infty}^{\infty}\int_{-\infty}^{\infty}\int_{-\infty}^{\infty} \frac{\left[\rho_q(\mathbf{r}_p - \mathbf{r}_q) + \mathbf{J}_q(t_q - t)\right] dx_1^q dx_2^q dx_3^q d\xi}{2\pi^2\varepsilon_0\left(r_{qp}^2 + \xi^2\right)^2} \tag{E.2a}$$

$$\mathbf{H}_p(\mathbf{r}_p, t) = \int_{-\infty}^{\infty}\int_{-\infty}^{\infty}\int_{-\infty}^{\infty}\int_{-\infty}^{\infty} \frac{\mathbf{J}_q \times (\mathbf{r}_p - \mathbf{r}_q)\, dx_1^q dx_2^q dx_3^q d\xi}{2\pi^2\left(r_{qp}^2 + \xi^2\right)^2} \tag{E.2b}$$

E.2 FIELDS OF A MOVING POINT CHARGE

In the event that the charge distribution is that of a point charge, moving at velocity v, we have

$$\rho(\mathbf{r}', t') = qu_0[x' - x_0(t')]u_0[y' - y_0(t')]u_0[z' - z_0(t')]$$

$$\mathbf{J}(\mathbf{r}', t') = \rho(\mathbf{r}', t')\mathbf{v}(t')$$

$$\mathbf{v}(t') = \widehat{\mathbf{x}}\frac{dx_0(t')}{dt'} + \widehat{\mathbf{y}}\frac{dy_0(t')}{dt'} + \widehat{\mathbf{z}}\frac{dz_0(t')}{dt'}$$

Retarded potentials when the velocity is constant

For convenience, the coordinates are defined so that the charge travels along the z axis. Because the velocity is constant, we obtain $\mathbf{v}(t') = \widehat{\mathbf{z}}v$ and $x_0(t') = y_0(t') = 0$. Then

$$\Phi_p(\mathbf{r}, t) = \int_{-\infty}^{\infty}\int_{-\infty}^{\infty}\int_{-\infty}^{\infty} \frac{qu_0(x')u_0(y')u_0[z' - v(t - \frac{r_{qp}}{c})]\, dx'dy'dz'}{4\pi\varepsilon_0 r_{qp}}$$

$$= \int_{-\infty}^{\infty} \frac{qu_0[z' - vt + \frac{v}{c}r_{qp})]\, dz'}{4\pi\varepsilon_0 r_{qp}}$$

Because $r_{qp}^2 = x^2 + y^2 + (z - z')^2$, the charge impulse is located at the value of $z' = z^*$ where

$$z^* - vt + \frac{v}{c}\sqrt{x^2 + y^2 + (z - z^*)^2} = 0 \tag{E.3}$$

and for which $r_{qp} = r_{qp}^*$. The only values of z' that contribute to the integral are in the vicinity of z^*; there

$$z' - vt + \frac{v}{c}r_{qp} = k^*(z' - z^*) + \cdots$$

where

$$k = 1 + \frac{v}{c}\frac{\partial r_{qp}}{\partial z'} = 1 - \frac{v}{c}\frac{z - z'}{r_{qp}} = 1 - \frac{v}{c}\cos\theta$$

$$\cos\theta = \frac{z - z^*}{r_{qp}}$$

The factor k^* is important because although it does not alter the location of the impulse, the area under it is

$$\int_{-\infty}^{\infty} u_0[k^*(z' - z^*)]\, dz' = \frac{1}{k^*}$$

Therefore Eqs. (E.1a) and (E.1b) simplify to

$$\mathbf{A}_p(\mathbf{r}, t) = \frac{q\mu_0 \mathbf{v}}{4\pi r_{qp}^* \left(1 - \frac{v}{c}\cos\theta^*\right)} \tag{E.4a}$$

$$\Phi_p(\mathbf{r}, t) = \frac{q}{4\pi\varepsilon_0 r_{qp}^* \left(1 - \frac{v}{c}\cos\theta^*\right)} \tag{E.4b}$$

From Eq. (E.3), we obtain

$$z - z^* = \frac{z - vt + \frac{v}{c}\sqrt{(z - vt)^2 + \left(1 - \frac{v^2}{c^2}\right)(x^2 + y^2)}}{1 - \frac{v^2}{c^2}}$$

$$r_{qp}^* = \frac{\frac{v}{c}(z - vt) + \sqrt{(z - vt)^2 + \left(1 - \frac{v^2}{c^2}\right)(x^2 + y^2)}}{1 - \frac{v^2}{c^2}}$$

$$r_{qp}^*\left(1 - \frac{v}{c}\cos\theta^*\right) = \sqrt{(z - vt)^2 + \left(1 - \frac{v^2}{c^2}\right)(x^2 + y^2)}$$

which results in

$$A_z = \frac{v}{c^2}\Phi = \frac{\mu_0 q v}{4\pi}\frac{1}{\sqrt{(z - vt)^2 + \left(1 - \frac{v^2}{c^2}\right)(x^2 + y^2)}} \tag{E.5}$$

The electric and magnetic fields can easily be obtained by differentiation of the potentials; the result is

$$\mathbf{E} = \frac{q}{4\pi\varepsilon_0}\frac{\left(1 - \frac{v^2}{c^2}\right)[\widehat{\mathbf{x}}x + \widehat{\mathbf{y}}y + \widehat{\mathbf{z}}(z - vt)]}{[(z - vt)^2 + \left(1 - \frac{v^2}{c^2}\right)(x^2 + y^2)]^{3/2}} \tag{E.6a}$$

$$\mathbf{H} = \varepsilon_0 v\widehat{\mathbf{z}} \times \mathbf{E} \tag{E.6b}$$

Contour integration

We note that the evaluation of Φ and $\mathbf{A}$ can also be carried out by first integrating the four-dimensional forms of Eqs. (E.1b) and (E.1a) over the spatial coordinates. Then

$$A_z = \frac{v}{c^2}\Phi = \frac{\mu_0 q v}{4\pi^2}\int_{-\infty}^{\infty} \frac{d\xi}{[x^2 + y^2 + (z - vt + i\frac{v}{c}\xi)^2 + \xi^2]}$$

The electric and magnetic fields can also be obtained by direct integration of Eqs. (E.2a) and (E.2b), which leads to

$$\mathbf{E}(\mathbf{r}, t) = \frac{q(\mathbf{r} - \widehat{\mathbf{z}}vt)}{2\pi^2 \varepsilon_0} \int_{-\infty}^{\infty} \frac{d\xi}{[x^2 + y^2 + (z - vt + i\frac{v}{c}\xi)^2 + \xi^2]^2} \tag{E.7a}$$

$$\begin{aligned} \mathbf{H}(\mathbf{r}, t) &= \varepsilon_0 \widehat{\mathbf{z}} v \times \mathbf{E}(\mathbf{r}, t) \\ &= \frac{qv\widehat{\mathbf{z}} \times \mathbf{r}}{2\pi^2} \int_{-\infty}^{\infty} \frac{d\xi}{[x^2 + y^2 + (z - vt + i\frac{v}{c}\xi)^2 + \xi^2]^2} \end{aligned} \tag{E.7b}$$

The integral over ξ common to both $\mathbf{A}$ and Φ and that common to both $\mathbf{E}$ and $\mathbf{H}$ may be carried out by evaluating the residues of the singularities of the integrand and then choosing contours that encircle them. The result

$$\int_{-\infty}^{\infty} \frac{d\xi}{[x^2 + y^2 + (z - vt + i\frac{v}{c}\xi)^2 + \xi^2]} = \frac{\pi}{\sqrt{(1 - \frac{v^2}{c^2})(x^2 + y^2) + (z - vt)^2}}$$

validates Eq. (E.5) whereas

$$\int_{-\infty}^{\infty} \frac{d\xi}{[x^2 + y^2 + (z - vt + i\frac{v}{c}\xi)^2 + \xi^2]^2} = \frac{(1 - \frac{v^2}{c^2})\frac{\pi}{2}}{[(1 - \frac{v^2}{c^2})(x^2 + y^2) + (z - vt)^2]^{3/2}}$$

validates Eqs. (E.6).

Lorentz transformation

These formulas can be obtained very simply by using the Lorentz transformation to express the rest-frame electrostatic field of a stationary charge in terms of the laboratory-frame components. This leads to the four-vector potential,

$$\mathcal{A} = \frac{\mu_0 q \mathcal{V}}{4\pi r'}$$

where r' is the rest-frame value of the radial distance to the charge. From Appendix A, Section A.1, we have

$$r' = \sqrt{\mathbf{r}_\perp \cdot \mathbf{r}_\perp + \gamma^2(\mathbf{r}_\parallel - \mathbf{v}t) \cdot (\mathbf{r}_\parallel - \mathbf{v}t)}$$

$$\gamma^2 = \frac{1}{1 - \dfrac{v^2}{c^2}}$$

Alternatively, the electric and magnetic fields can be evaluated in the rest frame of the charge and then transformed to the laboratory frame by means of

$$\mathbf{E} = \mathbf{E}'_\parallel + \gamma\,(\mathbf{E}'_\perp - v \times \mu_0 \mathbf{H}')$$

$$\mathbf{H} = \mathbf{H}'_\parallel + \gamma\,(\mathbf{H}'_\perp + v \times \varepsilon_0 \mathbf{E}')$$

(derived in Appendix A, Section A.2) where, in this case, we have

$$\mathbf{E}'_\parallel = \frac{q}{4\pi\varepsilon_0}\frac{\widehat{\mathbf{z}}\,z'}{r'^3}, \qquad \mathbf{E}'_\perp = \frac{q}{4\pi\varepsilon_0}\frac{\widehat{\mathbf{x}}\,x' + \widehat{\mathbf{y}}\,y'}{r'^3}, \qquad \text{and} \qquad \mathbf{H}' = 0.$$

Unfortunately, this "trick" does not apply to the general problem of an accelerating charge, which we consider next. This is why we have considered methods that, while more complicated, do apply.

Liénard–Wiechert Potentials

Because the retarded potentials depend only upon r_{qp}^* and θ^* defined at the moment the point charge reaches z^*, we need to retain the Cartesian coordinates and the origin of t only for that instant. Should the direction of the velocity change, a new z axis and time origin ($ct = r_{qp}$) can be defined and v can be replaced by $v^* + \mathbf{a}^*(t - \frac{r_{qp}}{c})$. The retarded acceleration, $\mathbf{a}^*$, has no direct effect on Eqs. (E.4) because at $t^* = 0$, we have

$$\int_{-\infty}^{\infty} u_0 \left[x' - \frac{1}{2} a_x^* \left(t - \frac{r_{qp}}{c} \right)^2 \right] dx' = 1$$

$$\int_{-\infty}^{\infty} u_0 \left[y' - \frac{1}{2} a_y^* \left(t - \frac{r_{qp}}{c} \right)^2 \right] dy' = 1$$

$$\int_{-\infty}^{\infty} u_0 \left[z' - v^* \left(t - \frac{r_{qp}}{c} \right) - \frac{1}{2} a_z^* \left(t - \frac{r_{qp}}{c} \right)^2 \right] dz' = \frac{1}{1 - \dfrac{v^*}{c} \cos\theta^*}$$

The generalization of the constant velocity potentials is therefore

$$\mathbf{A}_p(\mathbf{r}, t) = \frac{q \mu_0 v^*}{4\pi r_{qp}^* \left(1 - \dfrac{v^*}{c} \cos\theta^* \right)} \tag{E.8a}$$

$$\Phi_p(\mathbf{r}, t) = \frac{q}{4\pi \varepsilon_0 r_{qp}^* \left(1 - \dfrac{v^*}{c} \cos\theta^* \right)} \tag{E.8b}$$

These are the famous Liénard–Wiechert potentials. Naturally, the acceleration affects the trajectory of the charge and therefore the values of v^*, θ^*, and r_{qp}^* as functions of position and time. In Figure E.1, which defines the geometry, the trajectory of the moving charge is indicated by the dashed line. At the retarded time, $t^* = 0$, it is located at the point q; the point of observation is p.

Fields of an accelerated charge

Equations (E.8) are deceptively simple; yet, when acceleration is present, evaluation by differentiation is complicated by the bookkeeping required to keep track of the retarded positions and times. However, when the velocity of the charge is very small compared to c, approximations are possible that greatly simplify the analysis. Consequently, we first consider the nonrelativistic case.

Approximate solution for nonrelativistic motion

We assume that at time $t = 0$ the charge is at or near the origin. At time t, the electromagnetic wave front has traveled a distance ct while the charge has traveled a

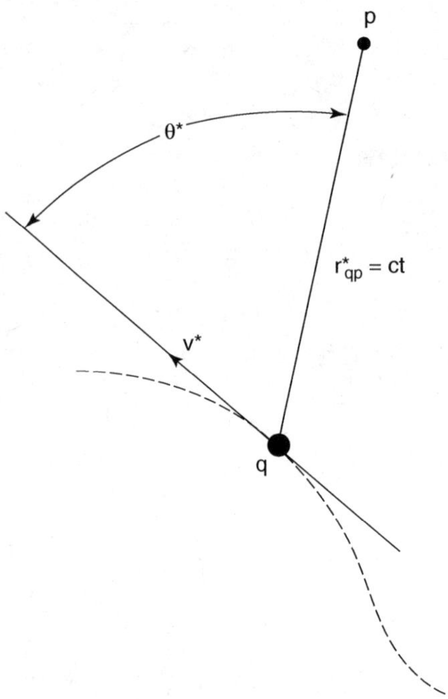

Figure E.1 Point-charge geometry.

very much smaller distance along a direction that we take to be the z axis. Because $r^*_{qp} = \sqrt{r^2 - 2z^*_q r\cos\theta + z^{*2}_q}$ and we assume $|z^*_q| \ll r$, it follows that $r^*_{qp} \simeq r - z^*_q \cos\theta$ and $\theta^* \simeq \theta$. Then Eq. (E.8b) becomes

$$\Phi_p \simeq \frac{q}{4\pi\varepsilon_0}\frac{1}{r - z^*_q\cos\theta}\frac{1}{1 - \frac{v^*_q}{c}\cos\theta} \simeq \frac{q}{4\pi\varepsilon_0}\left(\frac{1}{r} + \frac{z^*_q + v^*_q\frac{r}{c}}{r^2}\cos\theta\right) \tag{E.9a}$$

Except for the term proportional to v^*_q, this potential is the superposition of a point charge and a moving point dipole that are created by fixed charges $\pm q$ at the origin (which cancel each other) and the original charge at z^*_q. The latter in combination with $-q$ forms the dipole of moment qz^*_q. Although $v^*_q \ll c$, the $\frac{\cos\theta}{r}$ term that it generates is important. The leading term of Eq. (E.8a) is simply

$$A_p \simeq \frac{q\mu_0}{4\pi}\frac{v^*_q}{r} \tag{E.9b}$$

These solutions satisfy the Lorenz gauge and are valid as long as the charge is near the origin and moving along the z axis. Of course, new origins and axes can be assigned at later times if that is necessary. Both the retarded velocity and its time derivative generate electric and magnetic fields; the radiation field is due to the *acceleration* of the charge.

General solution

When the motion is relativistic or the general solution is required, accurate calculation of r^*_q is necessary. Rather than face the differentiations required to calculate the electric and magnetic fields, it is often easier to work directly with Eqs. (E.2a) and (E.2b) even though contour integration of functions of a complex variable is involved. The results obtained by Sommerfeld [16] using this method are given below without further proof.

It is convenient to resolve the fields into two parts: a velocity field (already solved) and an acceleration field. For the velocity field, one finds

$$\mathbf{E}_p = \frac{q}{4\pi\varepsilon_0}\frac{\gamma_0^{*3}}{r_{qp}^{*2}}\left(\widehat{\mathbf{r}}_0^* - \frac{v^*}{c}\right)\left(1 - \frac{v^{*2}}{c^2}\right) \tag{E.10a}$$

$$\mathbf{H}_p = \frac{q}{4\pi}\frac{\gamma_0^{*3}}{r_{qp}^{*2}}(v^* \times \widehat{\mathbf{r}}_0^*)\left(1 - \frac{v^{*2}}{c^2}\right) \tag{E.10b}$$

where

$$\gamma_0^* = \frac{1}{1 - \dfrac{v^* \cdot \widehat{\mathbf{r}}_0^*}{c}}$$

and $\widehat{\mathbf{r}}_0^*$ is the unit vector in the direction from charge to observer at the retarded instant of time.

The acceleration field is

$$\mathbf{E}_p = \frac{q}{4\pi\varepsilon_0}\frac{\gamma_0^{*2}}{r_{qp}^* c^2}\left[\gamma_0^*\left(\widehat{\mathbf{r}}_0^* - \frac{v^*}{c}\right)(\widehat{\mathbf{r}}_0^* \cdot \mathbf{a}^*) - \mathbf{a}^*\right] \tag{E.11a}$$

$$\mathbf{H}_p = \frac{q}{4\pi}\frac{\gamma_0^{*2}}{r_{qp}^* c}\left[\gamma_0^*\left(\frac{v^*}{c} \times \widehat{\mathbf{r}}_0^*\right)(\widehat{\mathbf{r}}_0^* \cdot \mathbf{a}^*) + \mathbf{a}^* \times \widehat{\mathbf{r}}_0^*\right] \tag{E.11b}$$

where $\mathbf{a}^*$ is the retarded value of $\mathbf{a} = \frac{dv}{dt}$. It should be noted that the trajectory of the charge will be affected by a radiation reaction caused by the acceleration field; usually, it is a negligible effect.

E.3 METHOD OF IMAGES

Infinite ground plane

When known electric charge and/or current distributions are placed in the half-space ($z > 0$) above a perfectly conducting ($\sigma = \infty$) ground plane, surface charges and/or surface currents are induced on the interface, $z = 0$. Together with the primary sources, they force both the electric and magnetic fields to be zero in and below the ground plane which can be of arbitrary thickness. The equivalence principle can be used to replace the surface components *and* the conductor with fictitious charges and currents in the lower half-space. Provided that

$$\rho(x, y, -z) = -\rho(x, y, z) \tag{E.12}$$

and

$$J_x(x, y, -z) = -J_x(x, y, z) \tag{E.13a}$$

$$J_y(x, y, -z) = -J_y(x, y, z) \tag{E.13b}$$

$$J_z(x, y, -z) = +J_z(x, y, z) \tag{E.13c}$$

the symmetry of the problem and the "image" sources insures that for $z > 0$, the resultant field will satisfy the necessary boundary conditions at $z = 0$: $E_x = E_y = 0$ and $H_z = 0$. Commonly, this method is applied to static or quasistatic point or line charges and to line currents.

Infinite-length conducting cylinder

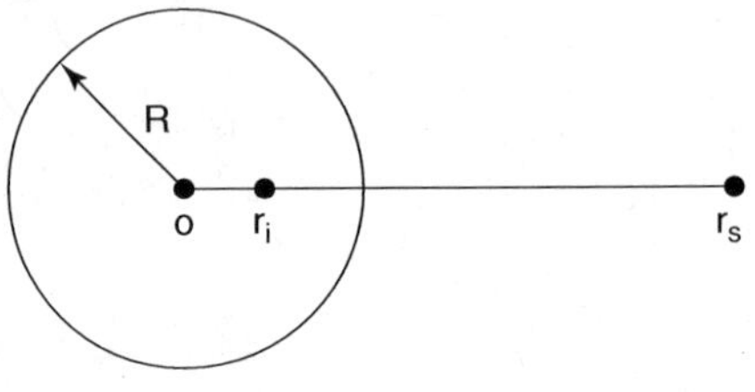

Figure E.2 Radial locations of source and image(s).

When a known static or quasistatic electric line charge or line current (oriented parallel to the z axis) is placed outside a perfectly conducting cylinder of radius R, the surface charges and/or surface currents induced on the boundary can be replaced by images positioned on the line between the source and the center of the cylinder. If the radius of the primary line charge or line current is $r_s \geq R$, that of the line image (of equal strength but opposite polarity) is (in cylindrical coordinates)

$$r_i = R^2/r_s \leq R$$

In addition, a line charge or current (located at the origin) may be required to maintain the correct value of the total charge or current on the cylinder. The cylinder, line source, and line image(s) are all perpendicular to the plane of Figure E.2. Of course, these images only provide equivalence at points *exterior* to the cylinder.

Conducting sphere

When a static electric point charge, q, is placed outside a perfectly conducting sphere[1] of radius R, the surface charge induced on the interface can be replaced by a point charge, q_i, positioned on the line between the source and the center of the sphere. Figure E.2 again applies, but now represents a plane containing the line between the center of the sphere and the charge. If the radius (in spherical coordinates) of the primary charge is $r_s \geq R$, that of the image (located along that radius) is

$$r_i = R^2/r_s \leq R$$

but now,

$$q_i = -q\frac{R}{r_s}$$

In addition, a point charge (located at the origin) may be required to maintain the correct value of the total charge on the sphere. Notice that for $r_s \to R$, the charge and its image

[1] As in the case of the cylinder, the conducting-sphere can be solid or have a hollowed-out center.

are both close to the surface which locally acts like a planar interface; consequently, $q_i \to -q$. On the other hand, for $r_s \gg R$, the image approaches $q_i \to 0$. Of course, these images only provide equivalence at points *exterior* to the sphere.

Image configurations for magnetic conductors

Use of the induction theorem (Part I, Chapter 6, Section 6.8) to analyze the scattering from large conducting surfaces led to consideration of the imaging of magnetic charges and/or magnetic currents by perfect electric conductors ($\sigma_e = \infty$), or, equivalently, the dual situation of the imaging of electric charges and/or electric-currents by perfect magnetic conductors ($\sigma_m = \infty$).

In the case of a magnetic ground plane of infinite extent, the altered boundary conditions require that when the primary sources are electric charges and electric currents, Equations (E.12)–(E.13c) be replaced by

$$\rho(x, y, -z) = \rho(x, y, z) \tag{E.14}$$

and

$$J_x(x, y, -z) = +J_x(x, y, z) \tag{E.15a}$$

$$J_y(x, y, -z) = +J_y(x, y, z) \tag{E.15b}$$

$$J_z(x, y, -z) = -J_z(x, y, z) \tag{E.15c}$$

For the infinite cylinder and sphere geometries, the image positions remain the same, but the polarities are reversed.

When the primary sources are magnetic charges and magnetic currents, consideration of the dual-field problems shows that the effects of electric and magnetic (perfect) conductors are reversed.

Figure E.3 summarizes the imaging of electric and magnetic charge/current in an electric perfect conductor.

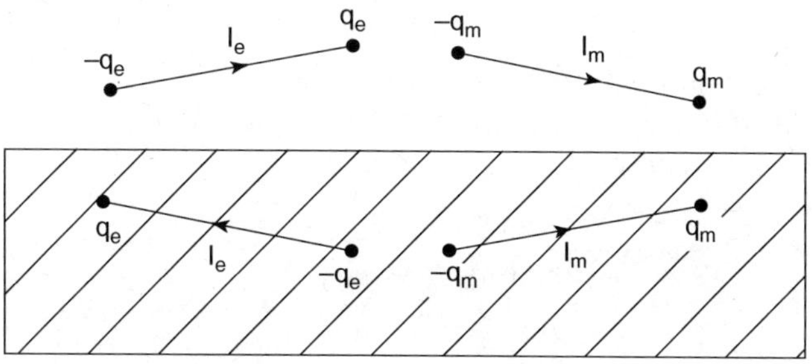

Figure E.3 Image charges/currents ($\sigma_e = \infty$).

Figure E.4 summarizes the imaging of electric and magnetic charge/current in a magnetic perfect conductor.

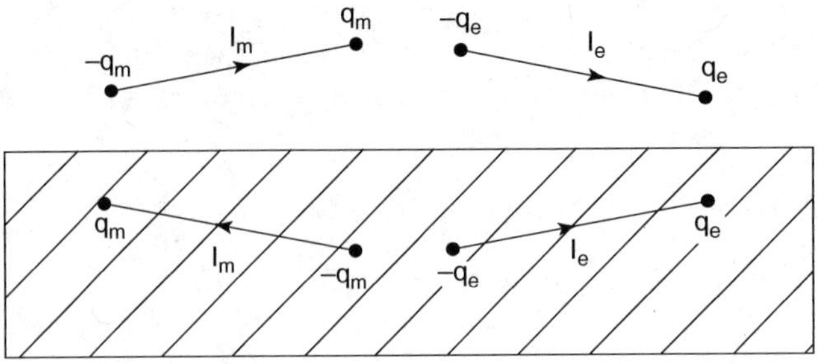

Figure E.4 Image charges/currents ($\sigma_m = \infty$).

E.4 CHARACTERISTIC IMPEDANCES OF TEM TRANSMISSION LINES

When a transmission line made from perfect conductors of uniform cross section is embedded in linear isotropic material that is uniform and lossless, the characteristic impedance of the *TEM* mode can be evaluated from

$$Z_o = \sqrt{\frac{L'}{C'}}$$

where C', the capacitance per unit length, can be evaluated from the **E** field of an open-circuited line when a static voltage is maintained across it and L', the inductance per unit length, can be evaluated from the **H** field of a short-circuited line carrying a static current. Because

$$\frac{Z_o}{\eta} = \frac{L'}{\mu} = \frac{\varepsilon}{C'} \tag{E.16a}$$

$$\eta = \sqrt{\frac{\mu}{\varepsilon}} \tag{E.16b}$$

only one of the static fields needs to be determined; we choose the electric field.

Coaxial line

The general coaxial transmission line is defined by Figure E.5.

If the inner conductor is maintained at the potential V and the outer conductor is grounded, the Laplacian potential between them is

$$\Phi(r) = V\,\frac{\ln(R_2/r)}{\ln(R_2/R_1)} \qquad (R_1 \le r \le R_2)$$

The value of Q' is $2\pi r\,\varepsilon E_r = C'V$; therefore from Eq. (E.16a) we obtain

$$Z_o = \eta\frac{1}{2\pi}\ln(1/\rho)$$

$$\rho = \frac{R_1}{R_2}$$

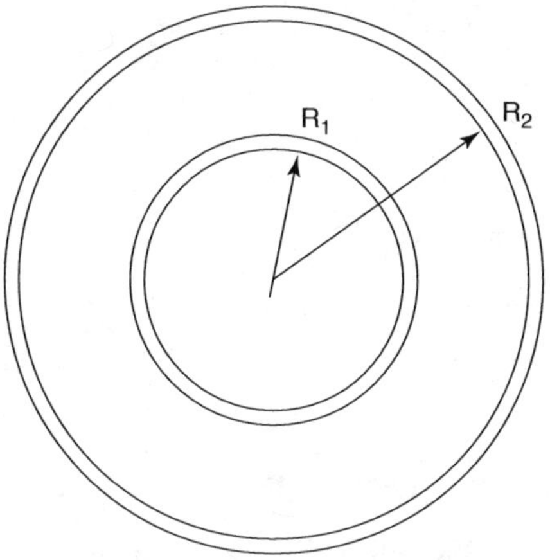

Figure E.5 Coaxial transmission line.

The normalized function,

$$Z_n(\rho) = \frac{1}{2\pi} \ln(1/\rho) \tag{E.17}$$

can be used with other two-conductor geometries, provided that an effective ρ can be evaluated in terms of the altered geometry. Often, this can be accomplished by either direct analysis or an appropriate conformal transformation, $W(Z)$, that maps the complex plane, $Z = x + iy$, into $W = u + iv$. Any *analytic function* of Z or W satisfies Laplace's Equation, but many important cases can be solved by the use of the elementary functions $\ln Z$, $1/Z$, $\sin Z$, and $\cos Z$ (singly or in combination). Others require the use of the Schwarz–Christoffel or other special transformations.

Lecher line

Parallel cylindrical conductors, which form the Lecher-line geometry defined by Figure E.6, can be analyzed using the method of images (Section E.3) for *both* cylinders.

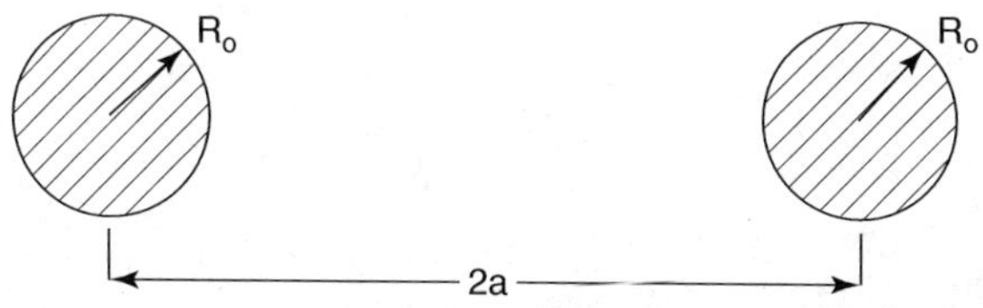

Figure E.6 Lecher transmission line.

Inside each cylinder, equal but opposite line charges are placed symmetrically on the line between the two centers. Outside of the cylinders, these image charges, if separated by the distance $2\sqrt{a^2 - R_o^2}$, are equivalent to the actual surface charges on the cylinders and allow calculation of $Q' = C'V$. Alternatively, the coaxial-line geometry can be conformally transformed by $W = K\ln(\frac{Z+1}{Z-1})$. Using either method, the final result is

$$\rho = \frac{a}{R_o} + \sqrt{\left(\frac{a}{R_o}\right)^2 - 1} \qquad (a \geq R_o)$$

$$Z_o = 2\eta Z_n(\rho) = \frac{\eta}{\pi}\cosh^{-1}\left(\frac{a}{R_o}\right)$$

If a ground plane is placed at the plane of symmetry of this (or any symmetric two-conductor line), the unbalanced half-circuit has a value of Z_o reduced by a factor of 2.

Elliptic-function-based transformations

The transmission-line geometries defined by Figures E.8–E.11 can be transformed from either the coaxial or Lecher line geometry by use of the formula,

$$L(\rho) = 2\rho \prod_{n=2}^{\infty} \left(\frac{1+\rho^{8n}}{1+\rho^{8n-4}}\right)^2 \tag{E.18}$$

This function is shown plotted in Figure E.7; it is derived from the properties of doubly periodic elliptic functions (see, for example, Nehari [45]). For $L < .5$, $\rho \simeq L/2$.

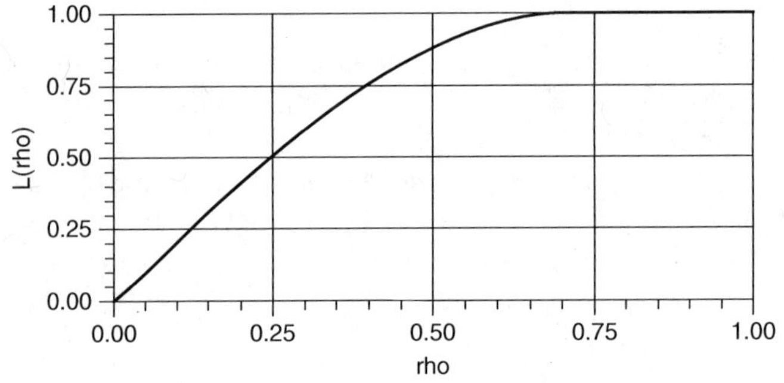

Figure E.7 Conformal mapping function $L(\rho)$.

Strip with cylindrical shield

$$L(\rho) = \frac{s}{R}$$

$$Z_o = \eta Z_n(\rho)$$

Coplanar strip line

$$b = \sqrt{\frac{g}{g+w}}$$

$$L(\rho) = \frac{1-b}{1+b}$$

$$Z_o = 2\eta Z_n(\rho)$$

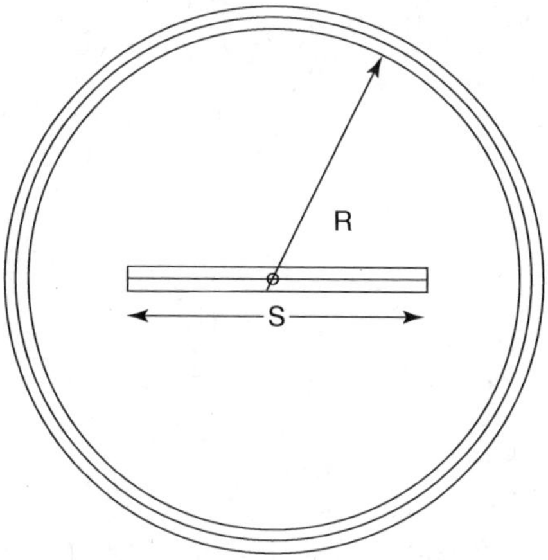

Figure E.8 Strip with cylindrical shield.

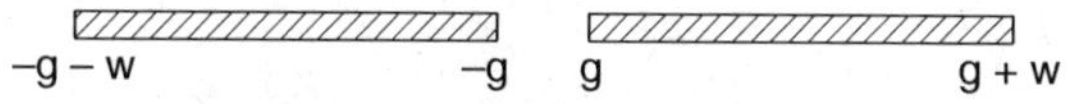

Figure E.9 Coplanar strips.

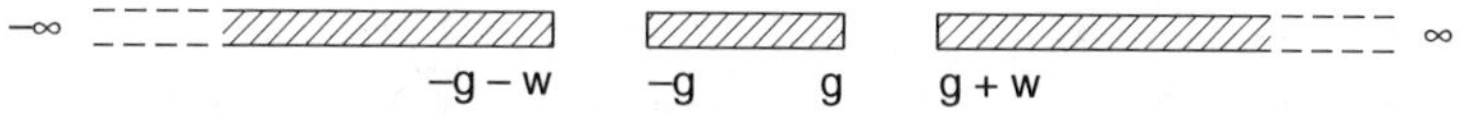

Figure E.10 Complement of coplanar strips.

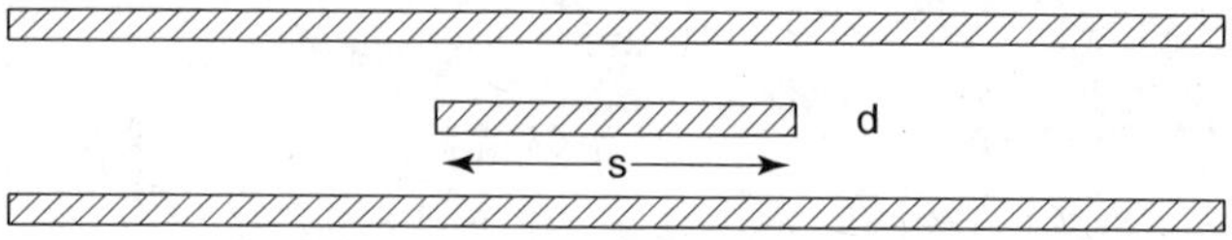

Figure E.11 Symmetric strip line.

The complement of the coplanar strip line (shown in Figure E.10) can most easily be evaluated in terms of the dual of the original structure, which employs perfect magnetic conductors. In accordance with the analysis presented in Chapter 22, Section 22.2, the characteristic impedance of the complement is

$$Z_o^{\text{comp}} = \frac{\eta^2}{4Z_o} = \frac{\pi\eta}{4\ln(1/\rho)}$$

The same result can be obtained through use of a conformal transformation.

Symmetric strip line

The symmetric strip line consists of a thin strip of width s placed midway between ground planes of infinite extent and separation d. Because the latter are maintained at the same potential, the line is a two-conductor system. Using $W = K \ln Z$ to transform the coplanar strip line into this geometry leads to

$$b = \exp\left(\frac{-\pi}{2}\frac{s}{d}\right)$$

$$L(\rho) = \frac{1-b}{1+b}$$

$$Z_o = \eta Z(\rho)$$

When $b \to 0$, the parameters ρ and L both approach unity. In this case, $s/d \gg 1$ and the electric field between the center strip and the ground planes is approximately uniform with values $\pm 2V/d$; in addition, the fringing field beyond the center strip region can be neglected.[2] Then $Q' \simeq \varepsilon(2V/d)2s$ and $Z_o/\eta \simeq d/(4s)$.

Because $L(\rho)$ saturates very rapidly for $\rho > .7$, Figure E.7 is of no use in evaluating $\rho(L)$ in that region. Instead, whenever $b = \frac{1-L}{1+L} \ll 1$, the formula

$$\rho(L) \simeq \exp\left[\frac{-(\pi/2)^2}{\ln\left(\dfrac{1+L}{1-L}\right)}\right] \tag{E.19}$$

(derived from the no-fringing limit of the symmetric strip-line impedance) provides a very good approximation. Based upon Eq. (E.19), the no-fringing limit of the coplanar strip-line impedance is

$$Z_o \simeq \eta \frac{\pi/2}{\ln\left(\dfrac{w+g}{g}\right)} \simeq \eta \frac{\pi/2}{\ln\left(\dfrac{w}{g}\right)}$$

which is consistent with an electric field that is ϕ-directed and proportional to $1/r$ for $g \leq r \leq g + w$ and zero elsewhere.

The tabulated values,

ρ	0	.05	.1	.15	.2	.25	.3	.35
$L(\rho)$	0	.1	.200	.300	.399	.496	.590	.680

ρ	.4	.45	.5	.55	.6	.65	.7	.75
$L(\rho)$	.762	.833	.892	.938	.969	.987	.996	.9993

are based upon Eq. (E.18) whereas

L	.9998	.99995	.999999	.9999995	.9999999	1
$\rho(L)$	.779	.792	.844	.850	.864	1

are based upon Eq. (E.19).

Parallel-plate line

If a single thin plate of width w is held at the voltage $+V/2$ (with respect to infinity), the two-dimensional Laplacian potential is

$$\Phi(x, y) = K \cosh^{-1} \xi + V/2$$

[2]The electric field is very large at the edges of a very thin strip. Nevertheless, the fringing energy makes a negligible contribution to C'.

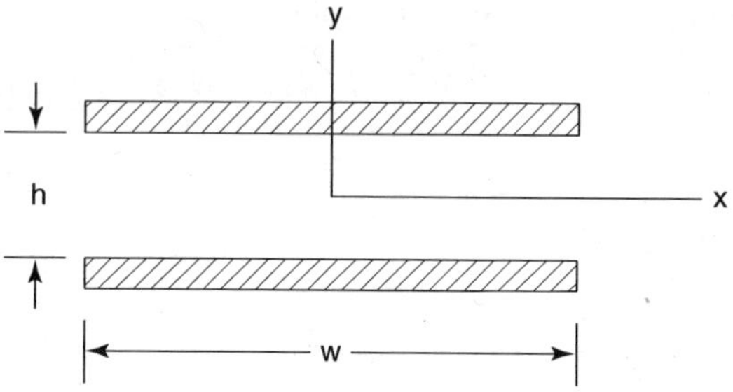

Figure E.12 Parallel-plate line.

where the equipotentials are ellipses[3] defined by

$$\frac{x^2}{\xi^2} + \frac{y^2}{\xi^2 - 1} = \left(\frac{w}{2}\right)^2$$

and the plate is located at $y = 0$, $|x| \le \frac{w}{2}$ ($\xi = 1$). Far from the plate, $\xi \gg 1$, $\frac{w}{2}\xi \to \sqrt{x^2 + y^2}$ and $\cosh^{-1}\xi \to \ln(2\xi)$. It follows that for $r \gg w$, the strip appears as a line charge with $K = -Q'/(2\pi\varepsilon)$. If the strip is shifted to $y = +h/2$ and a second strip of opposite charge $(-Q')$ and voltage $-V/2$ is placed at $y = -h/2$, the result is the parallel-plate line of Figure E.12

Provided that $h \gg w$, the resultant potential is approximately the sum of the two terms

$$\Phi(x, y) = K(\cosh^{-1}\xi_+ - \cosh^{-1}\xi_-)$$

where

$$\frac{x^2}{\xi_\pm^2} + \frac{(y \mp h/2)^2}{\xi_\pm^2 - 1} = \left(\frac{w}{2}\right)^2$$

The potential at $x = 0$, evaluates to

$$\Phi(0, y) = \frac{Q'}{2\pi\varepsilon}\left(\sinh^{-1}\left|\frac{2y + h}{w}\right| - \sinh^{-1}\left|\frac{2y - h}{w}\right|\right)$$

At $y = \pm h/2$, the values are $\pm V/2$ so because $Q' = C'V$, it follows that the characteristic impedance is approximately

$$Z_0 \simeq \frac{\eta}{\pi}\sinh^{-1}\left(\frac{2h}{w}\right) \tag{E.20}$$

Because each plate is not an exact equipotential, the formula is in error when the plates are closely spaced and $2h/w \ll 1$. However, the limit, $Z_0 \to \frac{2}{\pi}\eta\frac{h}{w}$, is correct except for the factor $2/\pi$. A more accurate approximation is

$$Z_0 \simeq \frac{\eta}{\pi}\sinh^{-1}\left(\frac{kh}{w}\right) \tag{E.21}$$

[3]The general solution can be expressed in terms of elliptical coordinates that employ the variables ξ and η defined by $x = c\xi\eta$ and $y = c\sqrt{\xi^2 - 1}\sqrt{1 - \eta^2}$ where, in this instance, $c = w/2$.

where the factor k interpolates between the limits of 2 and π. A reasonable choice is

$$k = (\pi - 2)\exp\left(-s\frac{h}{w}\right) + 2 \tag{E.22}$$

with the exponential factor governed by the constant s. Comparison with numerical (finite difference) solutions reveals that $s = 4$ provides an acceptable fit.

Microstrip transmission line

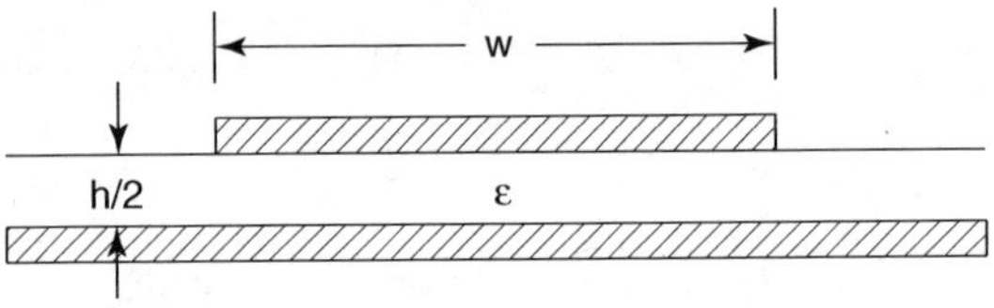

Figure E.13 Microstrip transmission line.

A ground plane (at $y = 0$) and dielectric substrate can be introduced to produce the microstrip transmission line shown in Figure E.13; such structures are of great technical importance—especially in the field of integrated circuits. If $\varepsilon = \varepsilon_0$, the value of Z_o is one-half that given by Eq. (E.21), but usually $\varepsilon > \varepsilon_0$. In that event, approximate formulas can be found in the literature [46]– [48]; in the last cited reference, the inductance and capacitance per unit length are approximated for $h/w \geq 1$ (using the method of images to account for the polarization surface charges) by

$$\frac{L'}{\mu_0} \simeq \frac{1}{2\pi}\sinh^{-1}\left(\frac{2h}{w}\right) \tag{E.23a}$$

$$\frac{\varepsilon_0}{C'} \simeq \frac{1}{(\varepsilon'+1)\pi}\left[\sinh^{-1}\left(\frac{2h}{w}\right) + f\left(\frac{\varepsilon'-1}{\varepsilon'+1}\right)\right] \tag{E.23b}$$

$$f(x) = \sum_{n=1}^{\infty}(-x)^n \ln\left(1+\frac{1}{n}\right) \simeq \frac{-3}{100}x(23 - 12x + 4x^2) \tag{E.23c}$$

and for $h/w \leq 1$, with $\alpha = \frac{2h}{\pi w}$ and $\varepsilon' = \varepsilon/\varepsilon_0$, by

$$\frac{L'}{\mu_0} \simeq \frac{\pi/4}{1/\alpha + (1/\sqrt{1+\alpha})\tanh^{-1}(1/\sqrt{1+\alpha}) + \alpha^{.15}} \tag{E.24a}$$

$$\frac{\varepsilon_0}{C'} \simeq \frac{\pi/4}{\varepsilon'/\alpha + (1/\sqrt{1+\alpha/\varepsilon'})\tanh^{-1}(1/\sqrt{1+\alpha/\varepsilon'}) + (\varepsilon')^{\frac{\varepsilon'-1}{\varepsilon'+1}}\alpha^{.15}} \tag{E.24b}$$

where the empirically generated $\alpha^{.15}$ terms permit h/w to be increased to unity. In both cases, the quasi-*TEM* approximation yields

$$Z_o \simeq \sqrt{\frac{L'}{C'}}$$

and the phase velocity,

$$v_p \simeq 1/\sqrt{L'C'}$$

It should be noted that when $\varepsilon' = 1$, the value $2Z_o$ based upon Eqs. (E.24) is a more accurate approximation than Eq. (E.21), although both agree in the limit $\alpha \ll 1$.

APPENDIX F

F.1 BESSEL FUNCTIONS

We restrict our discussion to Bessel functions of integer-order n. The linear combination, $Z_n(x) = C_1 J_n(x) + C_2 Y_n(x)$, is the general solution of Bessel's Equation

$$\frac{d^2 Z_n(x)}{dx^2} + \frac{1}{x}\frac{dZ_n(x)}{dx} + \left(1 - \frac{n^2}{x^2}\right) Z_n(x) = 0 \tag{F.1}$$

where $J_n(x)$ and $Y_n(x)$ are, respectively, Bessel functions of the first kind and second kind.

In Figures F.1 and F.2, $J_n(x)$ and $Y_n(x)$ are respectively plotted over the ranges $0 \leq x < 20$ and $2 \leq x < 20$ for the integers $n = 0, 1, 2, 3$.

$Z_n(ix)$ is a linear combination of the modified Bessel functions of the first and second kind, $I_n(x) = (-i)^n J_n(ix)$ and $K_n(x) = \frac{\pi}{2} i^{n+1}[J_n(ix) + iY_n(ix)]$. These real functions diverge respectively at $|x| \to \infty$ and $x \to 0$; neither is needed for the analysis of the examples considered in Part III of the text.

The Power and Beauty of Electromagnetic Fields, First Edition. F. R. Morgenthaler.

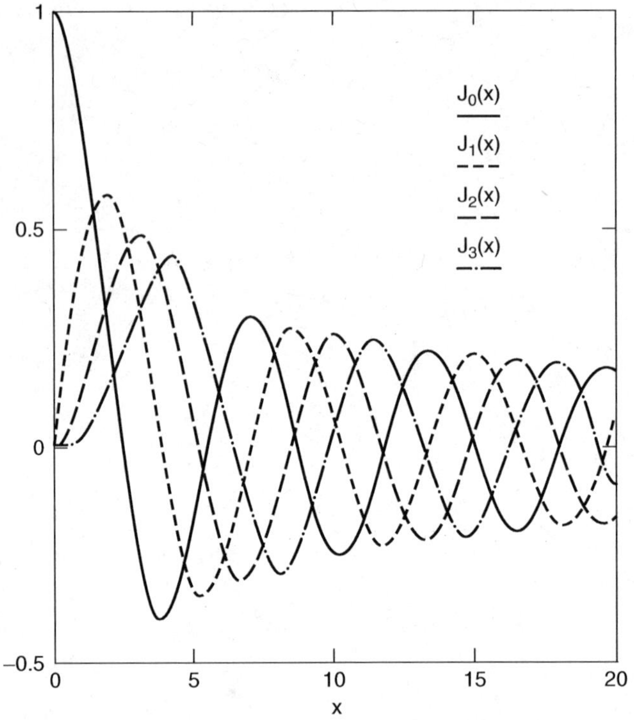

Figure F.1 Bessel functions of the first kind.

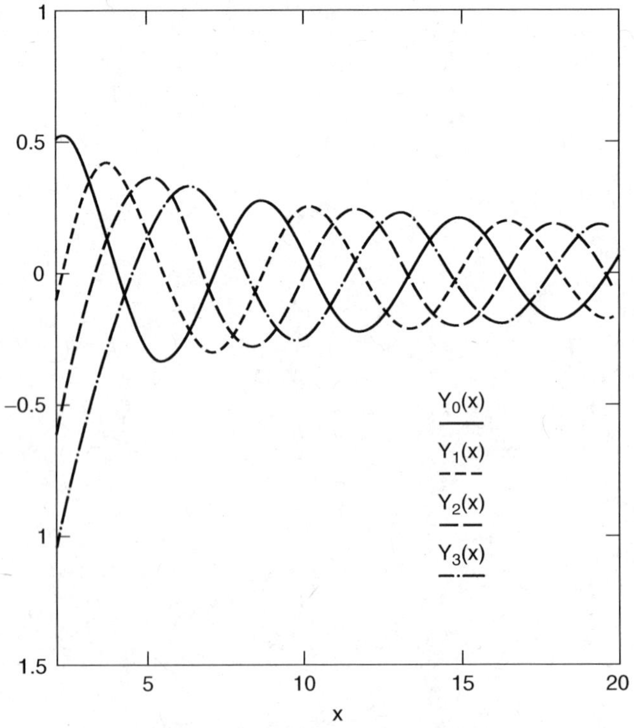

Figure F.2 Bessel functions of the second kind.

Integral definitions of zero-order functions

$$J_0(x) = \frac{1}{\pi}\int_0^{\pi} \cos(x \sin\phi)\, d\phi = \frac{1}{\pi}\int_0^{\pi} \cos(x\cos\phi)\, d\phi \tag{F.2a}$$
$$= \frac{2}{\pi}\int_0^{\infty} \sin(x\cosh t)\, dt \qquad (x > 0)$$

$$Y_0(x) = \frac{4}{\pi^2}\int_0^{\pi/2} \cos(x\cos\phi)[\gamma + \ln(2x\sin^2\phi)]\, d\phi \tag{F.2b}$$
$$= -\frac{2}{\pi}\int_0^{\infty} \cos(x\cosh t)\, dt \qquad (x > 0)$$
$$\gamma = .5772156649 \quad \text{Euler's constant}$$

Series solutions and asymptotic approximations

$$J_n(x) = \sum_{k=0}^{\infty} \frac{(-1)^k (\frac{x}{2})^{2k+n}}{k!(k+n)!} \tag{F.3}$$

$$Y_n(x) = \frac{2}{\pi}\left[\left(\ln\left(\frac{x}{2}\right) + \gamma\right) J_n(x) - \frac{1}{2}\sum_{k=0}^{n-1} \frac{(n-k-1)!(\frac{x}{2})^{2k-n}}{k!}\right.$$
$$\left. + \frac{1}{2}\sum_{k=0}^{\infty} (-1)^{k+1}[\varphi(k) + \varphi(k+n)] \frac{(\frac{x}{2})^{2k+n}}{k!(k+n)!}\right] \tag{F.4}$$

where

$$\varphi(k) = \sum_{m=1}^{k} \frac{1}{m}$$
$$\varphi(0) = 0$$

For small x,

$$J_0(x) = 1 - (\frac{x}{2})^2 + \cdots \tag{F.5a}$$

$$Y_0(x) = \frac{2}{\pi}\ln(x) + \cdots \tag{F.5b}$$

$$J_n(x) = \frac{1}{n!}\left(\frac{x}{2}\right)^n + \cdots \qquad (n \geq 0) \tag{F.5c}$$

$$Y_n(x) = \frac{-(n-1)!}{\pi}\left(\frac{2}{x}\right)^n + \cdots \quad (n > 0) \tag{F.5d}$$

For large x,

$$J_n(x) \sim \sqrt{\frac{2}{\pi x}} \cos\left(x - \frac{\pi}{4} - \frac{n\pi}{2}\right) \tag{F.6a}$$

$$Y_n(x) \sim \sqrt{\frac{2}{\pi x}} \sin\left(x - \frac{\pi}{4} - \frac{n\pi}{2}\right) \tag{F.6b}$$

Complex Hankel functions

Linear combinations of functions that (asymptotically) describe propagating cylindrical waves are

$$J_n(kx)\cos(\omega t) \pm Y_n(kx)\sin(\omega t) \sim \sqrt{\frac{2}{\pi x}} \cos\left(kx \mp \omega t - \frac{\pi}{4} - \frac{n\pi}{2}\right)$$

These results make evident the utility of defining complex Hankel functions of the first and second kind (Bessel functions of the third kind)[1]:

$$H_n^{(1)}(x) = J_n(x) + iY_n(x) \tag{F.7a}$$

$$H_n^{(2)}(x) = J_n(x) - iY_n(x) \tag{F.7b}$$

Recurrence relations

$$\frac{dJ_n(x)}{dx} = \frac{-n}{x} J_n(x) + J_{n-1}(x) \tag{F.8a}$$

$$\frac{dJ_n(x)}{dx} = \frac{n}{x} J_n(x) - J_{n+1}(x) \tag{F.8b}$$

$$\frac{dY_n(x)}{dx} = \frac{-n}{x} Y_n(x) + Y_{n-1}(x) \tag{F.8c}$$

$$\frac{dY_n(x)}{dx} = \frac{n}{x} Y_n(x) - Y_{n+1}(x) \tag{F.8d}$$

These relations may be combined to yield:

$$\frac{2n}{x} J_n(x) = J_{n+1}(x) + J_{n-1}(x) \tag{F.9a}$$

$$\frac{2n}{x} Y_n(x) = Y_{n+1}(x) + Y_{n-1}(x) \tag{F.9b}$$

Orthogonality and normalization integrals

$$\int_0^1 xJ_n(k_1x)J_n(k_2x)\,dx = \frac{1}{k_1^2 - k_2^2}[k_2J_n(k_1)J_{n-1}(k_2) - k_1J_{n-1}(k_1)J_n(k_2)] \tag{F.10}$$

[1]These functions are intended to be multiplied by $\exp(-i\omega t)$ before the real part is extracted. When, as in this text, $\exp(j\omega t)$ is used instead, one simply replaces i with $-j$. For either choice, the asymptotic $H_n^{(1)}$ is associated with the outward and $H_n^{(2)}$ the inward traveling wave.

When $k_1 = k_2 = k$, the limit of Eq. (F.10) is

$$\int_0^1 xJ_n^2(kx)\,dx = \frac{1}{2}\left\{\frac{1}{k^2}\left[\frac{dJ_n(kx)}{dx}\right]_{x=1}^2 + \left(1 - \frac{n^2}{k^2}\right)J_n^2(k)\right\} \tag{F.11}$$

The mth positive root of $J_n(x) = 0$ is defined by $x = u_{nm}$; that of $\frac{dJ_n(x)}{dx} = 0$ is defined by $x = u'_{nm}$. Tables of these roots are given below for $0 \le n \le 3$, $1 \le m \le 3$.

u_{nm}	$m = 1$	$m = 2$	$m = 3$
u_{0m}	2.405	5.520	8.654
u_{1m}	3.832	7.016	10.174
u_{2m}	5.136	8.417	11.620
u_{3m}	6.380	9.761	13.015

u'_{nm}	$m = 1$	$m = 2$	$m = 3$
u'_{0m}	3.832	7.016	10.174
u'_{1m}	1.841	5.331	8.536
u'_{2m}	3.054	6.706	9.969
u'_{3m}	4.201	8.015	11.346

If $k_1$and k_2 are both distinct values of u_{nm} (or of u'_{nm}), the value of Eq. (F.10) is zero.

From Eq. (F.11), it also follows that

$$\int_0^1 xJ_n^2(u_{nm}x)\,dx = \int_0^1 xJ_{n\pm1}^2(u_{nm}x)dx = \frac{1}{2}J_{n+1}^2(u_{nm}) \tag{F.12}$$

and

$$\int_0^1 xJ_n^2(u'_{nm}x)\,dx = \frac{1}{2}J_n^2(u'_{nm})\left[1 - \frac{n^2}{u'^2_{nm}}\right] \tag{F.13a}$$

$$\int_0^1 xJ_{n-1}^2(u'_{nm}x)\,dx = \frac{1}{2}J_n^2(u'_{nm})\left[1 - \frac{n(n-2)}{u'^2_{nm}}\right] \tag{F.13b}$$

$$\int_0^1 xJ_{n+1}^2(u'_{nm}x)\,dx = \frac{1}{2}J_n^2(u'_{nm})\left[1 - \frac{n(n+2)}{u'^2_{nm}}\right] \tag{F.13c}$$

Another useful result is

$$\int_1^\infty x\{\frac{1}{2}[J_{n+1}^2(u'_{nm}x) + Y_{n+1}^2(u'_{nm}x) + J_{n-1}^2(u'_{nm}x) + Y_{n-1}^2(u'_{nm}x)]$$

$$-[J_n^2(u'_{nm}x) + Y_n^2(u'_{nm}x)]\}\,dx = \left[\frac{-Y_n(x)}{x}\frac{dY_n(x)}{dx}\right]_{x=u'_{nm}} \tag{F.14}$$

Wronskian

$$Y_{n-1}(x)J_n(x) - J_{n-1}(x)Y_n(x) = \frac{2}{\pi x} \tag{F.15}$$

F.2 CHEBYSHEV POLYNOMIALS

Polynomials of the first kind

Chebyshev polynomials of the first kind, $T_n(x)$, are defined by

$$T_n(x) = \begin{cases} \cos(n\cos^{-1}x) = \cos(n\theta) & (|x| \le 1) \\ \cosh(n\cosh^{-1}x) & (|x| \ge 1) \end{cases} \tag{F.16}$$

where n is a positive integer.

Recurrence relation

$$T_{n+1}(x) = 2xT_n(x) - T_{n-1}(x) \qquad (n \ge 1) \tag{F.17}$$

The first nine polynomials are

$$\begin{aligned}
T_0(x) &= 1 \\
T_1(x) &= x \\
T_2(x) &= 2x^2 - 1 \\
T_3(x) &= 4x^3 - 3x \\
T_4(x) &= 8x^4 - 8x^2 + 1 \\
T_5(x) &= 16x^5 - 20x^3 + 5x \\
T_6(x) &= 32x^6 - 48x^4 + 18x^3 - 1 \\
T_7(x) &= 64x^7 - 112x^5 + 56x^3 - 7x \\
T_8(x) &= 128x^8 - 256x^6 + 160x^4 - 32x^3 + 1
\end{aligned}$$

Zeros

For $n \ge 1$, all the roots occur in the interval $-1 \le x \le 1$ where $T_n(x) = \cos n\theta$. Evidently, $n\theta = \frac{2m-1}{2}\pi$ and

$$x_m^{(n)} = \cos\left(\frac{2m-1}{n}\frac{\pi}{2}\right) \qquad (m = 1, 2, \ldots, n) \tag{F.18}$$

Orthogonality and normalization integrals

In the range $-1 \le x \le 1$, with $x = \cos\theta$, these polynomials are orthogonal with respect to the weighting factor $(\sin\theta)^{-1} = \frac{1}{\sqrt{1-x^2}}$, as is clear from

$$\begin{aligned}
\int_{-1}^{+1} \frac{1}{\sqrt{1-x^2}} T_n(x)T_m(x)\,dx &= \int_0^{\pi} \cos(n\theta)\cos(m\theta)d\theta \qquad \text{(F.19)} \\
&= \frac{\pi}{2}\delta_{nm}(1 + \delta_{n0})
\end{aligned}$$

The weighting emphasizes the endpoints of the interval.

Polynomials of the second kind

Recurrence relations

Chebyshev polynomials of the second kind, $U_n(x)$, share the recurrence relation Eq. (F.17), except that $U_0 = 1$, $U_1 = 2x$. They are also defined by

$$U_n(x) = 2T_n(x) + U_{n-2}(x) \qquad (n \geq 2) \tag{F.20}$$

The first nine polynomials are

$$\begin{aligned}
U_0(x) &= 1 \\
U_1(x) &= 2x \\
U_2(x) &= 4x^2 - 1 \\
U_3(x) &= 8x^3 - 4x \\
U_4(x) &= 16x^4 - 12x^2 + 1 \\
U_5(x) &= 32x^5 - 32x^3 + 6x \\
U_6(x) &= 64x^6 - 80x^4 + 24x^2 - 1 \\
U_7(x) &= 128x^7 - 192x^5 + 80x^3 - 8x \\
U_8(x) &= 256x^8 - 448x^6 + 240x^4 - 40x^2 + 1
\end{aligned}$$

Zeros

For $n \geq 1$, all the roots occur in the interval $-1 \leq x \leq 1$ where the polynomials, expressed in terms of $\theta = \cos^{-1} x$, are

$$U_n(x) = \begin{cases} 2\sum_{k=1}^{n} \cos[(2k-1)\theta] & (n \text{ odd}) \\ -1 + 2\sum_{k=0}^{n/2} \cos(2k\theta) & (n \text{ even}) \end{cases} \tag{F.21}$$

It follows that $U_n(x) = 0$ when $\theta = \frac{m}{n+1}\pi$ and

$$x_m^{(n)} = \cos(\tfrac{m}{n+1}\pi) \qquad (m = 1, 2, \cdots, n) \tag{F.22}$$

Orthogonality and normalization integrals

In the range $-1 \leq x \leq 1$, these polynomials are orthogonal with respect to the weighting factor $\sqrt{1-x^2}$, which emphasizes the midpoint rather than the endpoints of the interval.

$$\int_{-1}^{+1} \sqrt{1-x^2}\, U_n(x) U_m(x)\, dx = \frac{\pi}{2}\, \delta_{nm} \tag{F.23}$$

F.3 HERMITE POLYNOMIALS

The nth-order Hermite polynomial, $H_n(x)$ is a solution of

$$\frac{d^2H_n}{dx^2} - 2x\frac{dH_n}{dx} + 2nH_n = 0 \tag{F.24a}$$

If $u_n = \exp(-x^2/2)\, H_n(x)$, then

$$\frac{d^2u_n}{dx^2} + (2n + 1 - x^2)u_n = 0 \tag{F.24b}$$

Recurrence relations

$$H_n(x) = (-1)^n \exp(x^2)\frac{d^n}{dx^n}\exp(-x^2) \tag{F.25a}$$

$$H_{n+1}(x) = 2xH_n(x) - 2nH_{n-1}(x) \tag{F.25b}$$

$$\frac{dH_n(x)}{dx} = 2nH_{n-1}(x) \tag{F.25c}$$

The first nine polynomials are

$$H_0(x) = 1$$

$$H_1(x) = 2x$$

$$H_2(x) = 4x^2 - 2$$

$$H_3(x) = 8x^3 - 12x$$

$$H_4(x) = 16x^4 - 48x^2 + 12$$

$$H_5(x) = 32x^5 - 160x^3 + 120x$$

$$H_6(x) = 64x^6 - 480x^4 + 720x^2 - 120$$

$$H_7(x) = 128x^7 - 1344x^5 + 3360x^3 - 1680x$$

$$H_8(x) = 256x^8 - 3584x^6 + 13440x^4 - 13440x^2 + 1680$$

Hermite polynomials of the complex variable w find application in diffraction analysis that involves the complete set of Gaussian beam solutions.

Orthogonality and normalization integrals

In the range $-\infty \le x \le \infty$, these polynomials are orthogonal with respect to the weighting factor $\exp(-x^2)$, which emphasizes the midpoint of the interval.

$$\int_{-\infty}^{+\infty} \exp(-x^2)H_m(x)H_n(x)dx = 2^n n!\sqrt{\pi}\,\delta_{mn} \tag{F.26}$$

APPENDIX G

G.1 MACSYMA AND MAXIMA

This appendix describes the suite of three programs that are included on the DVD that contains the electronic version of this book. They are described below and were written to be used with **Macsyma 2.x**, a very powerful mathematics program capable of sophisticated symbolic, numerical, and graphical analysis. The program produces superior fancy font output and provides context-sensitive help.

At this writing, Macsyma, Inc., which sold and supported Macsyma, is no longer in existence,[1] nor is their web site macsyma.com. Fortunately, a similar (but not identical) program, **Maxima**, is freely available as licensed under the GNU Public License. Version **Maxima 5.23.2** is included on the CDROM as **maxima-5.23.2.exe** (both *xMaxima* and the *wxMaxima* interface are included). The file **GNUlicense.txt** contains a copy of the GNU Public License. An Introduction and the Maxima manual, both in pdf form, are also included. It is advisable to consult the **README** file before installing the program. Although either can be used, the new interface, *wxMaxima*, is a distinct improvement over *xMaxima*, especially with respect to the output.

[1]Note that as of the Spring of 2006, Macsyma 2.4 is again available for purchase from Symbolics—a reincarnation, of sorts, of the old Symbolics, Inc. that marketed Macsyma in the 1980s. Macsyma 2.4 is the last version of Macsyma Pro that was released by Macsyma, Inc., Arlington, MA. in 1998. The new offer includes PDEase2D, a partial differential equation solver. Interested readers should check the Internet at **http://www.symbolics.com/MacsymaSpecial.htm.**

The Power and Beauty of Electromagnetic Fields, First Edition. F. R. Morgenthaler.

Predefined functions, variables, and equations endow the Macsyma/Maxima program **4d-em.mac** with "expert knowledge" of four-dimensional vectors, tensors, and operators as applied to four-dimensional electrodynamics (in the Chu, Amperian, and Minkowski formulations). In addition, the full capabilities of Macsyma/Maxima are available. This *Four-dimensional Electrodynamics Workpad* is therefore an important aid to understanding the topics covered in Part II and facilitating the solution of the Part III, Chapter 25, Section 25.10 practice problems. The 4d-em program can be conveniently started from the Macsyma Notebook, **4d-emwpad.mfe**. Two Macsyma Notebooks, **4d-vectdemo.mfe** and **4d-emdemo.mfe,** demonstrate some of the capabilities of **4d-em.mac** to perform and apply symbolic four-dimensional calculus operations to four-dimensional electrodynamics.

Although **Maxima 5.x** will *not* run .mfe files, it can be used to run the interactive script: **4d-em.mac** (with either the **load** or **batchload** command).This is the most important program because the outputs of the two Macsyma 2.x demonstration programs are available as **4d-vectdemo.htm** and **4d-emdemo.htm**; both can be viewed with standard Internet browsers. To install Maxima,

1) run **maxima-5.23.2.exe** (select the *wxMaxima* option).

2) install into the directory **c:\Maxima-5.23.2**

3) from directory **c:\EMPBook**,

   copy **4d-em-wxm.mac** to

   c:\Maxima-5.23.2\share\maxima\5.23.2\share\contrib

and

4) copy **maxima-init.mac** to

   c:\Maxima-5.23.2\share\maxima\5.23.2\share.

G.2 MACSYMA PROGRAM DESCRIPTIONS

The suite of programs contained on the CDROM includes:

Four-Dimensional Vectors and Operators

- **4d-vectdemo.mfe**

  A non-interactive demonstration program with hyperlinked sections:

1. Four-vectors
2. Basic operations: aa + bb, aa - bb, aa · bb
3. Fourdyad and fourdelta
4. Fourcross
5. Sixvector, dualsixvector (dual)
6. rr(x, y, z, ict); vv(x, y, z, ict)
7. Fourgrad, fourdiv, fourcurl
8. vv · fourgrad(s)
9. Fourlaplacian
10. Four-vector identities

4d-emdemo
4d-emWorkpad

These present Appendix B, Section B.3 operators and formulas in the notation required by the other programs. For example, depending upon the context, single letters represent either scalars or normal three-vectors; double letters represent four-vectors; three or more letters represent four-tensors. Therefore, a = **a**, aa = $\mathcal{A}$, bb = $\mathcal{B}$, and $\mathcal{A} \times \mathcal{B} = \mathcal{AB} - \mathcal{BA}$ might be defined as ccc.

Four-Dimensional Electrodynamics (Free-Space)

- **4d-emdemo.mfe**

A non-interactive demonstration program with hyperlinked sections

1. Four-vector velocity: vv
2. Four-vector electric current density: jj
3. Conservation of electric charge
4. Four-vector magnetic potential: aa
5. Lorenz and Coulomb gauge conditions
6. Electromagnetic field tensors: ggg, kkk
7. Four-vector electric and magnetic fields: ee, hh
8. Maxwell's Equations
9. Electromagnetic energy-momentum tensor: nrgm
10. Four-vector electromagnetic force density: ff

4d-vectdemo
4d-emWorkpad

These constitute the key elements of the free-space electrodynamics that are presented in Part II of the text **4d-emwpad.mfe (4d-em.mac)**

Four-Dimensional Electrodynamics Workpad

Unlike the two preceding demonstration programs that require only the Macsyma front end (or the transcripts of Macsyma sessions provided by .htm files), the third requires a Macsyma 2.x engine (or Maxima 5.x) and provides the user with a fully interactive workspace in which to create, manipulate, and evaluate any expression arising from four-dimensional electromagnetics. In addition, all of the standard algebraic, numerical, and graphical capabilities of Macsyma are available. For convenience, the following four-vectors are predefined:

vv = $\mathcal{V}$, aa = $\mathcal{A}$, jj = $\mathcal{J}$, ee = $\mathcal{E}$, hh = $\mathcal{H}$, pp = $\mathcal{P}$, mm = $\mathcal{M}$, dd = $\mathcal{D} = \varepsilon_0\mathcal{E} + \mathcal{P}$, bb = $\mathcal{B} = \mu_0(\mathcal{H} + \mathcal{M})$

These four-vectors may be expressed in terms of either the Chu (**e**, **h**, **p**, and **m**) or the Minkowski ($\mathbf{e}^{\text{o}}$, $\mathbf{h}^{\text{o}}$, $\mathbf{p}^{\text{o}}$, and $\mathbf{m}^{\text{o}}$) three-vector fields. When the latter representation is chosen, an o is added to the four-vector (eeo, hho, ppo, mmo, ddo, bbo) and three-vector (eo, ho, po, mo, do, bo); these are exceptions to the single and double letter conventions. In addition, energy-momentum tensors (nrgm = nrgmchu, nrgmamp, nrgmmink, nrgmalt, nrgmo = ev(nrgmalt, g=0)) and force-density four-vectors (ff, ffc, ffchu, ffamp, ffmink, ffa) are predefined for the Maxwell-Poynting, Chu, Amperian, Minkowski, and the Alternate representations. Six-vectors (ggg, kkk, ...) and their duals (ggg1, kkk1, ...) are also predefined using a three or four letter convention.

The complete list of user values is available by typing: **values** at any input prompt; the result is the list of *value_names*:

[ag, bg, dg, Dg, eg, gg, Gg, hg, kg, lg, Lg, mg, ng, og, Og, pg, Pg, qg, Qg, rg, sg, Sg, tg, wg, Wg, xg, zg, greek, eqg, eq2g, eqc2, eq2c, eqec, eqmc, fourdelta, a, v, j, ja, je, jm, e, eo, h, ho, p, po, m, mo, d, b, r, rr, aa, vv, jj, jja, jje, jjm, ee, hh, pp, mm, dd, bb, eeo, hho, ddo, bbo, ppo, mmo, qqqa, jjamp, jjpchu, jjechu, jjmchu, eqp, eqpo, eqm, eqmo, eqe, eqeo, eqh, eqho, eqd, eqdo, eqb, eqbo, eqja, eqrhoa, eqje, eqrhoe, eqjm, eqrhom, ggg, kkk, ggg1, kkk1, ggga, kkka, gggo, kkko, gggo1, kkko1, nrgm, nrgmchu, nrgmamp, nrgmmink, nrgmo, nrgmalt, ff, ffchu, ffc, ffamp, ffa, ffmink, eqsfree, eqschu, eqsmink, eqsamp, eqeoa, eqba, eqap, lorentz]

To learn the definition of a value, type *value_name*; at any input prompt.

For example, to display the list of two-letter shortcuts that generate symbols in Greek script, type **greek.** Helpful functions and operations are also provided. The complete list of the program functions is available by typing: **functions** at any input prompt; the result contains the list of *function_names*:

[mat2vect(qq), fourdyad(qq, rr), fourcross(qq, rr), fourcurl(qq), fourdiv(qq), sixvector(q, r), dualsixvector(qqq), dual(qqq), eqlist(eq), vectfacsum(qq), matfacsum(qqq), make3vect(v), make4vect(v, v4), cross(q, r), div(q), curl(q), grad(s), fourgrad(s), laplacian(q), fourlaplacian(qq), tgo(i, j), expand_eqs(eqq)]

The operations involving four-vectors and tensors are defined in **4d-vectdemo**; the others have the following uses:

lratsubst(uuu,vvv)[2] substitutes the list of equations,

uuu = $[l_1 = r_1, l_2 = r_2, \cdots]$ into expression **vvv**.

[2]This function is always available whenever Macsyma 2.x is run but will *not* be listed with the user functions. The program "4d-em" loads "lrats.mac" that contains **lratsubst;** this and other related functions *will* be listed when Maxima is run.

w:make3vect(w) generates and labels the normal vector, w:[w_1,w_2,w_3]. Once the three-vector is defined,

ww:make4vect(w,w4) appends the scalar, w4, to generate and label the four-vector, ww:[w_1,w_2,w_3, w4].

- Three- and four-vector electromagnetic fields are predefined in free-space and the Chu, Amperian, and Minkowski representations.
- Lists of Maxwell's Equations can be expanded with **expand_eqs(eqq)** where **eqq** is one of **(eqsfree, eqschu, eqsamp, eqsmink).**
- Lists of **eo** and **b** in terms of **a** and Φ (or the reverse) are provided by **eqeoa, eqba, and eqap**.

Values:

eqc2, eq2c, eqmc, eqec, eqg, and eq2g provide useful identities that can be used with **lratsubst**.

mat2vect(aa) converts a column matrix to a row matrix

vectfacsum(qq) and **matfacsum(qqq)** cause **factorsum** to be applied to each component of the vector **qq** or matrix **qqq**

cross(q, r), div(q), curl(q), operate on three-vectors;

grad(s) operates on a scalar; these allow the program to be run with either Macsyma or Maxima.

When **eq** is a vector-equation of the form $[l_1, l_2, l_3] = [r_1, r_2, r_3]$ or $[l_1, l_2, l_3, l_4] = [r_1, r_2, r_3, r_4]$ then **eqlist(eq)** creates the list: $[l_1 = r_1, l_2 = r_2, l_3 = r_3]$ or $[l_1 = r_1, l_2 = r_2, l_3 = r_3, l_4 = r_4]$. If **eq** is a three-vector or a four-vector, then the function assumes that the missing right-hand side is zero and proceeds accordingly.

The definition of any of these functions can be found by typing **dispfun(***function_name***);**

G.3 MACSYMA NOTEBOOKS

Setup and execution

Assuming that Macsyma 2.x is installed, check that the files: **4d-vectdemo.mfe**, **4d-emdemo.mfe**, and **4d-emwpad.mfe** are located in the directory **c:\EMPBook\Macsyma** and **4d-em.mac** is located in **c:\EMPBook**.

1) Click below on the appropriate Macsyma icon to select **4d-vectdemo.mfe** (or **4d-emdemo.mfe)** and open the chosen Macsyma Notebook.
2) Click on any of the ten underlined Notebook hyperlinks that comprise the index to move to the desired section. Click on any hyperlink labelled index to return to the top of the notebook. There are hyperlinks labeled 4d-emdemo (4d-vectdemo) and the emWorkpad at the end of the index section and at the bottom of the notebook.

Click on the link to the other demo or the interactive program. If emWorkpad is selected, the directions

FOUR-DIMENSIONAL ELECTRODYNAMICS WORKPAD

1) Select CONNECT from the Macsyma Menu

2) Select ALL then REEXECUTE from the Edit Menu

3) Enter commands starting on Line (c4)

appear at the beginning of the new Macsyma Notebook and should be followed (after reviewing subsection *4d-em workpad*).

4d-vector and 4d-em demos

The icons below are hyperlinks that can be clicked to open the demo Macsyma Notebooks (.mfe). **Macsyma 2.x** is required.

[4d-vectdemo.mfe] [4d-emdemo.mfe]

Lines prefixed (**c**) are the inputs to the Macsyma program; those prefixed (**d**) are the corresponding outputs. Note however, that if an input line ends with a $, the output line is suppressed.

If Macsyma is *not* available, transcripts generated by converting the outputs of **4d-vectdemo.mfe** and **4d-emdemo.mfe** to html files are available and can be viewed from standard browsers (such as Internet Explorer or Netscape). Simply click on the .htm file icons:

[4d-vectdemo.htm] [4d-emdemo.htm].

The underlined hyperlinks between the index, sections, and the other Notebooks are, of course, *not* active in the .htm transcripts.

4d-em workpad

After the notebook **4d-emwpad.mfe** is opened (see program setup and execution above), and steps 1)–3) are executed, the first three (**c**) and (**d**) lines should be:

```
(c1) load("c:\\EMPBook\\4d-em.mac")$
(c2) functions

(d2) [mat2vect(qq), fourdyad(qq,rr), fourcross(qq,rr),
fourcurl(qq), fourdiv(qq), sixvector(q, r), dualsixvector(qqq),
dual(qqq), eqlist(eq), vectfactsum(qq), matfacsum(qqq),
make3vect(v), make4vect(v,v4), cross(q,r), div(q), curl(q),
grad(s), fourgrad(s), laplacian(q), fourlaplacian(qq), tgo(i, j),
expand_eqs(eqq)]

(c3) values

(d3) [ag, bg, dg, Dg, eg, gg, Gg, hg, kg, lg, Lg, mg, ng, og, Og,
pg, Pg, qg, Qg, rg, sg, Sg, tg, wg, Wg, xg, zg, greek, eqg, eq2g,
eqc2, eq2c, eqec, eqmc, fourdelta, a, v, j, ja, je, jm, e, eo, h, ho,
p, po, m, mo, d, b, r, rr, aa, vv, jj, jja, jje, jjm, ee, hh, pp, mm,
dd, bb, eeo, hho, ddo, bbo, ppo, mmo, qqqa, jjamp, jjpchu,
jjechu, jjmchu, eqp, eqpo, eqm, eqmo, eqe, eqeo, eqh, eqho, eqd,
eqdo, eqb, eqbo, eqja, eqrhoa, eqje, eqrhoe, eqjm, eqrhom, ggg,
kkk, ggg1, kkk1, ggga, kkka, gggo, kkko, gggo1, kkko1, nrgm,
nrgmchu, nrgmamp, nrgmmink, nrgmo, nrgmalt, ff, ffchu, ffc,
ffamp, ffa, ffmink, eqsfree, eqschu, eqsmink, eqsamp, eqeoa,
eqba, eqap, lorentz].
```

Special functions and values are listed in lines (d2) and (d3). Simply type a **value**[3] on any input line in order to learn how it is defined. Alternatively, one can review the program listing contained in the following section. The alternate energy-momentum tensor value **nrgmalt** is a function of the gauge, $g = \Box \cdot \mathcal{A}$; the value **nrgmo** assumes the Lorenz gauge, $g = 0$.

The icons below are hyperlinks that can be clicked to open the four-dimensional electromagnetics workpad from the Macsyma 2.x Notebook. The Maxima 5.x program must be opened manually.

[Start MACSYMA Notebook: **4d-emwpad.mfe**]

[Start wxMaxima *manually* from your computer],

At INPUT: type **load("4d-em-wxm")** and press **enter** *or* from the file menu, use **Load Package** and locate **"4d-em-wxm.mac"**.

[3]Uppercase letters must be entered in lowercase prefixed with a \.

G.4 TEXT OF MACSYMA/MAXIMA BATCH PROGRAM

4d-em.mac (4d-em-wxm.mac[4])

```
disp("FOUR-DIMENSIONAL ELECTRODYNAMICS IN CHU,
 MINKOWSKI, AMPERIAN, AND ALTERNATE FORMULATIONS");
block(ag:alpha, bg:beta, dg:delta, \dg:\delta, eg:epsilon, gg:gamma,
\gg:\gamma, hg:eta, kg:kappa, lg:lambda, \lg:\lambda, mg:mu,
ng:nu, og:theta, \og:\theta, pg:phi, \pg:\phi, qg:psi, \qg:\psi, rg:rho,
sg:sigma, \sg:\sigma, tg:tau, wg:omega, \wg:\omega, xg:chi, zg:zeta)$
greek:["%pi"=%pi, "%e"='%e, "%i"=%i, 'ag=ag, 'bg=bg, 'dg=dg,
    "\\dg"=\dg, 'eg=eg, 'gg=gg, "\\gg"=\gg, 'hg=hg, 'kg=kg, 'lg=lg,
    "\\lg"=\lg, 'mg=mg, 'ng=ng, 'og=og, "\\og"=\og, 'pg=pg,
    "\\pg"=\pg, "\\pi"=\pi, 'qg=qg, "\\qg"=\qg, 'rg=rg, 'sg=sg,
    "\\sg"=\sg, 'tg=tg, 'wg=wg, "\\wg"=\wg, 'xg=xg, 'zg=zg]$
load(lrats)$
declare([c, mu, epsilon], constant)$
assume(mu > 0, epsilon > 0, c > 0, gamma > 0)$
depends([a, b, d, e, eo, h, ho, j, ja, je, jm, m, mo, p, po, v,
    \phi, rho, rhoa, rhoe, rhom, g], [x, y, z, t])$
scalarmatrixp:all$
eqg:gamma=c/sqrt(c^2-v[1]^2-v[2]^2-v[3]^2)$
eq2g:reverse(1/eqg)^2*c^2$
eqc2:c^2=1/mu/epsilon$
eq2c:mu*epsilon=1/c^2$
eqec:epsilon*c=1/mu/c$
eqmc:mu*c=1/epsilon/c$
```

[4]The list, **Greek**, contains two-letter shortcuts that produce lowercase Greek letters and (when they differ from the English spellings) uppercase symbols – if the prefix \ is appended. For example, **dg** produces δ, whereas **\dg** produces Δ (because **\d** produces D).

The wxMaxima program, **4d-em-wxm.mac**, is nearly identical, except that capitalized letters do not require the \ prefix; therefore **Dg** produces Δ. In addition, **gamma**, **psi**, and **lambda** are each appended with a %.

```
mat2vect(qq):=block([rr], rr:qq, if rank(rr)#1 or length(rr)=1 then
    disp("ERROR: input not a column matrix") else
    makelist(if listp(rr[i]) then rr[i][1] else rr[i], i, 1, length(rr)))$
fourdelta:ident(4)$
fourdyad(qq, rr):=block([o6], (for i thru 4 do
    (for j thru 4 do o6[i, j]:qq[i]*rr[j]), genmatrix(o6, 4, 4)))$
fourcross(qq, rr):=block([o6], (for i thru 4 do (for j thru 4
    do o6[i, j]:qq[i]*rr[j]-qq[j]*rr[i]), genmatrix(o6, 4, 4)))$
fourcurl(qq):=block([x4, k4, p6], (x4[1]:x, x4[2]:y, x4[3]:z,
    x4[4]:t, k4[1]:k4[2]:k4[3]:1, k4[4]:-%i/c), for i thru 4 do
    (for j thru 4 do p6[i, j]:'diff(qq[j], x4[i])*k4[i]-
    'diff(qq[i], x4[j])*k4[j]), genmatrix(p6, 4, 4), ev(%%, diff))$
fourdiv(qq):=block([x4, k4, rr], (x4[1]:x, x4[2]:y, x4[3]:z, x4[4]:t,
    k4[1]:k4[2]:k4[3]:1, k4[4]:-%i/c, rr:transpose(qq)),
    ev(sum('diff(rr[i], x4[i])*k4[i], i, 1, 4), diff))$
sixvector(q, r):=block([q1, q2, m6], (q1:q, q2:r), if (length(q1)#3 or
    length(q2)#3) then disp("ERROR: input not a pair of three-
    vectors") else subst([m6[1, 1] = 0, m6[1, 2] = q1[3],
    m6[1, 3] = -q1[2], m6[1, 4] = q2[1], m6[2, 1] = -q1[3],
    m6[2, 2] = 0, m6[2, 3] = q1[1], m6[2, 4] = q2[2],
    m6[3, 1] = q1[2], m6[3, 2] = -q1[1], m6[3, 3] = 0,
    m6[3, 4] = q2[3], m6[4, 1] = -q2[1], m6[4, 2] = -q2[2],
    m6[4, 3] = -q2[3], m6[4, 4] = 0], genmatrix(m6, 4, 4)))$
dualsixvector(qqq):=block([q6, n6], q6:transpose(qqq)+qqq,
    if ratsimp(q6) # 0*fourdelta then disp("ERROR: input not
    a sixvector") else subst([n6[1, 1] = 0, n6[1, 2] = qqq[3, 4],
    n6[1, 3] = qqq[4, 2], n6[1, 4] = qqq[2, 3], n6[2, 1] = qqq[4, 3],
    n6[2, 2] = 0, n6[2, 3] = qqq[1, 4], n6[2, 4] = qqq[3, 1],
    n6[3, 1] = qqq[2, 4], n6[3, 2] = qqq[4, 1], n6[3, 3] = 0,
```

```
    n6[3, 4] = qqq[1, 2], n6[4, 1] = qqq[3, 2], n6[4, 2] = qqq[1, 3],
    n6[4, 3] = qqq[2, 1], n6[4, 4] = 0], genmatrix(n6, 4, 4)))$
dual(qqq):=dualsixvector(qqq)$
eqlist(eq):=block([qq, leq, req, neq], qq:eq, leq:lhs(ev(qq, diff)),
    if not listp(leq) then disp("ERROR: input not a list") else
    (neq:length(leq), if rhs(qq) = 0 then req:makelist(0, i, 1, neq)
    else req:rhs(ev(qq, diff)), makelist(leq[i] = req[i], i, 1, neq)))$
vectfacsum(qq):=block([rr], rr:qq, if not listp(rr) then
    disp("ERROR: input not a row vector") else (for i thru length(rr)
    do rr[i]:factorsum(rr[i]), rr))$
matfacsum(qqq):=block([rrr], rrr:qqq, if not matrixp(rrr) then
    disp("ERROR: input not a matrix") else (for i thru length(rrr)
    do (for j thru length(transpose(rrr))
    do rrr[i, j]:factorsum(rrr[i, j])), rrr))$
make3vect(v):=[v[1], v[2], v[3]]$
make4vect(v, v4):=append(v, [v4])$
cross(q, r):=[q[2]*r[3]-q[3]*r[2], q[3]*r[1]-q[1]*r[3], q[1]*r[2]-q[2]*r[1]]$
div(q):='diff(q[1], x)+'diff(q[2], y)+'diff(q[3], z)$
curl(q):=['diff(q[3], y) - 'diff(q[2], z), 'diff(q[1], z) - 'diff(q[3], x),
    'diff(q[2], x) - 'diff(q[1], y)]$
grad(s):=['diff(s, x), 'diff(s, y), 'diff(s, z)]$
fourgrad(s):=['diff(s, x), 'diff(s, y), 'diff(s, z), -%i/c*'diff(s, t)]$
laplacian(q):=block([x], x[1]:x, x[2]:y, x[3]:z,
    sum('diff(q, x[i], 2), i, 1, 3), ev(%%, diff))$
fourlaplacian(qq):=block([x, k], x[1]:x, x[2]:y, x[3]:z, x[4]:t, k[1]:k[2]:k[3]:1,
    k[4]:-1/c^2, sum(k[i]*'diff(qq, x[i], 2), i, 1, 4), ev(%%, diff))$
a:make3vect(a)$
v:make3vect(v)$
j:make3vect(j)$
```

```
ja:make3vect(ja)$

je:make3vect(je)$

jm:make3vect(jm)$

e:make3vect(e)$

eo:make3vect(eo)$

h:make3vect(h)$

ho:make3vect(ho)$

p:make3vect(p)$

po:make3vect(po)$

m:make3vect(m)$

mo:make3vect(mo)$

d:make3vect(d)$

b:make3vect(b)$

aa:make4vect(a, %i/c*\phi)$

vv:make4vect(gamma*v, %i*gamma*c)$

jj:make4vect(j, %i*c*rho)$

jja:make4vect(ja, %i*c*rhoa)$

jje:make4vect(je, %i*c*rhoe)$

jjm:make4vect(jm, %i*c*rhom)$

ee:make4vect(e + cross(v, mu*h), %i/c*e · v)*gamma$

hh:make4vect(h - cross(v, epsilon*e), %i/c*h · v)*gamma$

pp:make4vect(p/gamma^2 + (p · v)*v/c^2, %i/c*(p · v))*gamma$

mm:make4vect(m/gamma^2 + (m · v)*v/c^2, %i/c*(m · v))*gamma$

dd:epsilon*ee + pp$

bb:mu*(hh + mm)$

eeo:make4vect(eo + cross(v, b), %i/c*eo · v)*gamma$

hho:make4vect(ho - cross(v, d), %i/c*ho · v)*gamma$

ddo:make4vect(d + cross(v, ho)/c^2, %i/c*d · v)*gamma$

bbo:make4vect(b - cross(v, eo)/c^2, %i/c*b · v)*gamma$
```

```
ppo:make4vect(po - cross(v, mo)/c^2, %i/c*(po · v))*gamma$

mmo:make4vect(mo + cross(v, po), %i/c*(mo · v))*gamma$

qqqa:sixvector(mo, %i*c*po)$

jjamp:fourdiv(qqqa), factor$

eq1:fourdiv(fourcross(pp, vv))$

jjpchu:factorsum(eq1)$

jjechu:jj+jjpchu$

eq1:mu*fourdiv(fourcross(mm, vv))$

jjmchu:factorsum(eq1)$

eqp:'p = gamma^2*('po-'(po · v)*'v/c^2 + '(cross(mo, v/c^2)))$

eqpo:'po = 'p - '(cross(m, v/c^2))$

eqm:'m = gamma^2*('mo-'(mo · v)*'v/c^2 - '(cross(po, v)))$

eqmo:'mo = 'm + '(cross(p, v))$

eqe:'e = 'eo + gamma^2*mu*('(cross(v, mo)) +
      '(po · v)*'v - '(('v · 'v)*po))$

eqeo:'eo = 'e + mu*'(cross(m, v))$

eqh:'h = 'ho + gamma^2*('(cross(po, v)) +
      ('(mo · v)*'v - '(('v · 'v)*mo))/c^2)$

eqho:'ho = 'h - '(cross(p, v))$

eqd:'d = epsilon*'e + 'p$

eqdo:'d = epsilon*'eo + 'po$

eqb:'b = mu*('h + 'm)$

eqbo:'b = mu*('ho + 'mo)$

eqja:'ja = 'diff('po, t) + '(curl(mo))$

eqrhoa:'rhoa = -'(div(po))$

eqje:'je = 'j + 'diff('p, t) + '(curl(cross(p, v)))$

eqrhoe:'rhoe = 'rho - '(div(p))$

eqjm:'jm = mu*('diff('m, t) + '(curl(cross(m, v))))$

eqrhom:'rhom = -mu*'(div(m))$
```

```
ggg:sixvector(h, -%i*c*epsilon*e), factor$

kkk:sixvector(-e, -%i*c*mu*h), factor$

ggg1:dual(ggg)$

kkk1:dual(kkk)$

ggga:sixvector(b/mu, -%i*c*epsilon*eo), factor$

kkka:sixvector(-eo, -%i*c*b), factor$

gggo:sixvector(ho, -%i*c*d), factor$

kkko:sixvector(-eo, -%i*c*b), factor$

gggo1:dual(gggo)$

kkko1:dual(kkko)$

    eq1:1/2*(mu*ggg · ggg + epsilon*kkk · kkk)$

    eq2:lratsubst(eqc2, eq1)$

    eq3:lratsubst(eq2c, eq2)$

nrgm:matfacsum(eq3)$

nrgmchu:nrgm$

    eq:1/2*(mu*ggga · ggga + epsilon*kkka · kkka)$

    eq1:lratsubst(eqc2, eq)$

nrgmamp:matfacsum(eq1)$

    eq2:(gggo1 · kkko - kkko1 · gggo)/(%i*c*2)$

nrgmmink:matfacsum(eq2)$

tgo(i,j):=(-diff(g,r[i])*a[j]+g*diff(a[j],r[i]))/(2*mu)$

ggo:1/2*epsilon*(-g*grad(\phi)+\phi*grad(g))$

sgo:(-diff(g,t)*a+g*diff(a,t))/(2*mu)$

wgo:1/2*epsilon*(-\phi*diff(g,t)+g*diff(\phi,t))$

nrgmo:genmatrix(nrgmo, 4, 4)$

    eq1:aa · 'diff(aa, x, 2) - 'diff(aa, x) · 'diff(aa, x)$

    eq2:ev(eq1, diff)$

nrgmo[1,2]:eq2/2/mu-aa[1]*jj[2]+tgo(1,2)$

nrgmo[2,1]:eq2/2/mu-aa[2]*jj[1]+tgo(2,1)$
```

```
    eq1:aa.'diff(aa,x,1,z,1)-'diff(aa,x).'diff(aa,z)$
    eq2:ev(eq1,diff)$
nrgmo[1,3]:eq2/2/mu-aa[1]*jj[3]+tgo(1,3)$
nrgmo[3,1]:eq2/2/mu-aa[3]*jj[1]+tgo(3,1)$
    eq1:aa.'diff(aa,y,2)-'diff(aa,y).'diff(aa,y)$
    eq2:ev(eq1,diff)$
nrgmo[2,2]:eq2/2/mu+(aa.jj)/2-aa[2]*jj[2]+tgo(2,2)$
    eq1:aa.'diff(aa,y,1,z,1)-'diff(aa,y).'diff(aa,z)$
    eq2:ev(eq1,diff)$
nrgmo[2,3]:eq2/2/mu-aa[2]*jj[3]+tgo(2,3)$
nrgmo[3,2]:eq2/2/mu-aa[3]*jj[2]+tgo(3,2)$
    eq1:aa.'diff(aa,z,2)-'diff(aa,z).'diff(aa,z)$
    eq2:ev(eq1,diff)$
nrgmo[3,3]:eq2/2/mu+(aa.jj)/2-aa[3]*jj[3]+tgo(3,3)$
    eq1:-%i*(aa.'diff(aa,x,1,t,1)-'diff(aa,x).'diff(aa,t))/c$
    eq2:ev(eq1,diff)$
nrgmo[1,4]:eq2/2/mu-aa[1]*jj[4]-%i*c*ggo[1]$
nrgmo[4,1]:eq2/2/mu-aa[4]*jj[1]-%i/c*sgo[1]$
    eq1:-%i*(aa.'diff(aa,y,1,t,1)-'diff(aa,y).'diff(aa,t))/c$
    eq2:ev(eq1,diff)$
nrgmo[2,4]:eq2/2/mu-aa[2]*jj[4]-%i*c*ggo[2]$
nrgmo[4,2]:eq2/2/mu-aa[4]*jj[2]-%i/c*sgo[2]$
    eq1:-%i*(aa.'diff(aa,z,1,t,1)-'diff(aa,z).'diff(aa,t))/c$
    eq2:ev(eq1,diff)$
nrgmo[3,4]:eq2/2/mu-aa[3]*jj[4]-%i*c*ggo[3]$
nrgmo[4,3]:eq2/2/mu-aa[4]*jj[3]-%i/c*sgo[3]$
    eq1:diff(aa,t)$
    eq:ev(eq1,diff)$
    eq2:eq.eq$
```

```
    eq1:aa.'diff(aa,t,2)$
    eq3:ev(eq1,diff)$
    eq:(eq2-eq3)/c^2$
nrgmo[4,4]:1/2/mu*eq + 1/2*(aa.jj) - aa[4]*jj[4]+wgo$
    eq1:lratsubst(eqc2, nrgmo)$
    eq2:lratsubst(eqmc, eq1)$
nrgmo:matfacsum(eq2)$
nrgmalt:nrgmo$
nrgmo:ev(nrgmalt, g=0)$
    eq1:mu*(ggg . jj)$
    eq2a:lratsubst(eqg, eq1)$
    eq2b:lratsubst(eqmc, eq2a)$
    eq3:lratsubst(eqec, eq2b)$
ff:factorsum(mat2vect(eq3))$
    eq1:mu*(ggg . jje) + epsilon*(kkk . jjm)$
    eq2a:lratsubst(eqg, eq1)$
    eq2b:lratsubst(eqmc, eq2a)$
    eq3:lratsubst(eqec, eq2b)$
ffchu:vectfacsum(mat2vect(eq3))$
    eq1:mu*(ggg . jjechu) + epsilon*(kkk . jjmchu)$
    eq2a:lratsubst(eqg, eq1)$
    eq2b:lratsubst(eqmc, eq2a)$
    eq3:lratsubst(eqec, eq2b)$
ffc:vectfacsum(mat2vect(eq3))$
    eq1:mu*ggga . (jj+jja)$
    eq2:lratsubst(eq2c, eq1)$
ffamp:vectfacsum(mat2vect(eq2))$
    eq1:mu*ggga . (jj+jjamp)$
    eq2:lratsubst(eq2c, eq1)$
```

```
ffa:vectfacsum(mat2vect(eq2))$
    eq1:fourcross(jj, bbo)$
    eq2:-%i/c*dual(eq1) . vv + ((eeo . jj)*vv - (jj . vv)*eeo)/c^2$
    eq3a:lratsubst(eqg, eq2)$
    eq3b:lratsubst(eqmc, eq3a)$
    eq4:lratsubst(eqec, eq3b)$
ffm1:mat2vect(eq4)$
    eq1:('d . 'diff('eo, x) - 'eo . 'diff('d, x) + 'b . 'diff('ho, x) -
        'ho . 'diff('b, x) )/2$
    eq2:('d . 'diff('eo, y) - 'eo . 'diff('d, y) + 'b . 'diff('ho, y) -
        'ho . 'diff('b, y) )/2$
    eq3:('d . 'diff('eo, z) - 'eo . 'diff('d, z) + 'b . 'diff('ho, z) -
        'ho . 'diff('b, z) )/2$
    eq4:('d . 'diff('eo, t) - 'eo . 'diff('d, t) + 'b . 'diff('ho, t) -
        'ho . 'diff('b, t) )/2$
ffm2[1]: ffm1[1] + eq1$
ffm2[2]: ffm1[2] + eq2$
ffm2[3]: ffm1[3] + eq3$
ffm2[4]: factorsum(ffm1[4] - %i/c*eq4)$
ffmink:[ffm2[1], ffm2[2], ffm2[3], ffm2[4]]$
    eq1:'(curl(h)) - epsilon*'diff('e, t) = 'j$
    eq2:'(curl(e)) + mu*'diff('h, t) = 0$
    eq3:'(div(e)) = 'rho/epsilon$
    eq4:'(div(h)) = 0$
eqsfree:[eq1, eq2, eq3, eq4]$
    eq1:'(curl(h)) - epsilon*'diff('e, t) = 'je$
    eq2:'(curl(e)) + mu*'diff('h, t) = -'jm$
    eq3:'(div(e)) = 'rhoe/epsilon$
    eq4:'(div(h)) = 'rhom/mu$
```

```
eqschu:[eq1, eq2, eq3, eq4]$
    eq1:'(curl(ho)) - 'diff('d, t) = 'j$
    eq2:'(curl(eo)) + 'diff('b, t) = 0$
    eq3:'(div(d)) = 'rho$
    eq4:'(div(b)) = 0$
eqsmink:[eq1, eq2, eq3, eq4]$
    eq1:'(curl(b)) - 1/c^2*'diff('eo, t) = mu*('j + 'ja)$
    eq2:'(curl(eo)) + 'diff('b, t) = 0$
    eq3:'(div(eo)) = ('rho + 'rhoa)/epsilon$
    eq4:'(div(b)) = 0$
eqsamp:[eq1, eq2, eq3, eq4]$
eqeoa:'eo = -'diff('a, t) - '(grad(\phi))$
eqba:'b = '(curl(a))$
eqap:append(eqlist(reverse(eqeoa)), eqlist(reverse(eqba)))$
expand_eqs(eqq):=append(eqlist(eqq[1]), eqlist(eqq[2]),
    [ev(eqq[3])], [ev(eqq[4])])$
remvalue(eq, eq1, eq2, eq2a, eq2b, eq3, eq3a, eq3b, eq4, ffm1, ffm2)$
lt[i,j]:=gg^2/(1+gg)*v[i]*v[j]/c^2$
lt1:genmatrix(lt,3,3)$
lt2:%i/c*gg*v$
lt3:append(-lt2,[0])$
lt4:addcol(lt1,lt2)$
lt:addrow(lt4,lt3)$
ltd:matrix([1,0,0,0], [0,1,0,0], [0,0,1,0], [0,0,0,gg])$
lorentz:lt+ltd$
remvalue(lt,lt1,lt2,lt3,lt4,ltd)$
```

APPENDIX H

H.1 ANIMATED FIELDS OF SURFACE CURRENTS

Refer to Chapter 14

Planar surface current: step pulse with exponential decay

$S_z(z, t/T)$ and $S_z^{\mathrm{o}}(x, t/T)$ [Kstep-exp.avi]
Poynting and Alternate fluxes associated with

$$K_y(t) = K_{\mathrm{o}}[u_{-1}(t) + [\exp(-\alpha t) - 1]u_{-1}(t - T)]$$

The solid curves are for $\alpha T = 1$; the dotted curves for $\alpha T = 4$. A single frame ($t = 5T$) was used to create Figure 14.3.

Planar surface current: Gaussian pulse

$A_y(z, t), S_z(z, t), S_z^{\mathrm{o}}(z, t)$ [KyGaussian.avi]

The Power and Beauty of Electromagnetic Fields, First Edition. F. R. Morgenthaler.

Vector potential and Poynting and Alternate fluxes associated with

$$K_y(t) = K_o \exp(-\alpha c^2(t - T)^2)u_{-1}(t)$$

The parameters are $\alpha = 1, cT = 5$. A single frame $(t = 2T)$ was used to create Figure 14.6.

Alternate null power flux: interaction of Gaussian pulses

$f_1(z, t), f_2(z, t), S^{o}_{12}(z, t)$ [A1-A2-So12.avi]

As discussed in Chapter 16, Section 16.1, a null Alternate power flux (and null Alternate energy density) can be created by the *interaction* of nonharmonic vector-potential plane waves. This animation plots $f_{1,2}(z, t) = f_{\pm}(z \mp ct)$ and $S^{o}_{12}(z,\ ct)$ when both waves are Gaussian pulses, $f_{\pm} = \exp[-\frac{1}{2}\alpha(z \mp ct)^2]$ with $\alpha = 1$. Notice that the null flux has no average value and appears only when f_+ and f_-overlap. The Alternate fluxes $S^{o}_{11}(z,\ ct)$ and $S^{o}_{22}(z,\ ct)$ are not shown

Refer to Chapter 15, Section 15.3.

Strip surface current, $K_z(t) = K_o u_{-1}(t)$

$A_z(x, y, t)$ [Kz-stepAz.avi]

$-E_z(x, y, t)$ [Kz-stepEz.avi]

Selected frames are included in Figure 15.13.

$|H|(x, y, t)$ [Kz-stepHxy.avi]

Selected frames are included in Figure 15.14.

For similar animations that allow user interaction to alter the viewpoint, run the Macsyma file [AzEzHstep.mfe] in Section H.7.

H.2 ANIMATED FIELDS OF A CYLINDRICAL VOLUME CURRENT, $J_Z(T) = J_O U_{-1}(T)$

(refer to Chapter 15, Section 15.3)

$-E_z(x, y, t)\ (r \le R)$ [JzEz1-xy.avi]

$H_\phi(x, y, t)$ $(r \leq R)$ [JzHphi1-xy.avi]

$-E_z(x, y, t)$ $(r \leq 5R)$ [JzEz5-xy.avi]

$H_\phi(x, y, t)$ $(r \leq 5R)$ [JzHphi5-xy.avi]

$-E_z$ $(r \leq 5R, t)$ [JzEz-r.avi]

H_ϕ $(r \leq 5R, t)$ [JzHphi-r.avi]

$-E_z$, H_ϕ $(r \leq 5R, t)$ [JzEzHphi.avi]
Selected frames are included in Figure 15.10.

S_r $(r \leq 5R, t)$ [JzSr1-r.avi]

rS_r $(r \leq 5R, t)$ [JzrSr-r.avi]

H.3 ANIMATED FIELDS OF A CYLINDRICAL SURFACE CURRENT, $K_Z(T) = K_O U_{-1}(T)$

Refer to Chapter 15, Section 15.3

$-E_z(x, y, t)$ $(r \leq 5R)$ [KzEz-xy.avi]

$H_\phi(x, y, t)$ $(r \leq 5R)$ [KzHphi-xy.avi]

$-E_z$, H_ϕ $(r \leq 5R, t)$ [KzEzHphi-r.avi]
Selected frames are included in Figure 15.12

S_r $(r \leq 5R, t)$ [KzSr-r.avi]

rS_r $(r \leq 5R, t)$ [KzrSr.avi]

H.4 ANIMATED FIELDS OF LINE-CURRENT TRANSIENTS

For the following transients, the current waveform is sketched in Figure 15.15. The fields and Poynting flux generated by both single and double reversals are depicted. A similar, but triple, current reversal transient is analyzed in Chapter 15, Section 15.4; the fields and Poynting flux are plotted in Figure 15.16.

$-E_z(r,t)$, $H_\phi(r,t)$ for single current reversal [IzEH1reversals.avi]

$S_r(r,t)$ for single current reversal [IzS1rev.avi]

$-E_z(r,t)$, $H_\phi(r,t)$ for double current reversal [IzEH2reversals.avi]

$S_r(r,t)$ for double current reversal [IzS2rev.avi]

For the following transients, the current is the best least-squares trapezoidal approximation of the sine waveform sketched in Figure 15.17. The fields and Poynting flux generated by both single and double cycle pulses are depicted. A somewhat similar transient made up of five current reversals is analyzed in Chapter 15, Section 15.4 and plotted in Figure 15.18.

$-E_z(r,t)$, $H_\phi(r,t)$ for trapezoidal “sine” current pulses [IzEH2trapezoid.avi]

$S_r(r,t)$ for trapezoidal “sine” current pulses [IzS2trapezoid.avi]

$-E_z(r,t)$, $H_\phi(r,t)$ for trapezoidal “sine” [IzEH4trapezoid.avi]

$S_r(r,t)$ for trapezoidal “sine” current pulses [IzS4trapezoid.avi]

H.5 ANIMATED FIELD OF A RADIATING HERTZIAN DIPOLE

In the following animations the magnitude of the electric field of an electric dipole (or of the magnetic field of a magnetic dipole) is shown both as a contour plot and a vector plot. The Poynting and Alternate power fluxes follow; both are displayed as vector plots. The transients for both types of dipole are analyzed in Chapter 13, Sections 13.1 and 13.2. In each case, the dipole is oriented vertically. The horizontal and vertical extent of all plots is approximately 5λ and the region close to the origin is excluded.

Hertzian-dipole radiation (field contours) [Dipole-contours.avi]

Hertzian-dipole radiation (field vectors) [Dipole-vectors.avi]

Selected frames are the basis of Figures 13.1 and 13.2. Notice the reversal of direction of the field lines and how, in the radiation region, they close upon themselves rather than terminating on the dipole.

Hertzian-dipole radiation ($\mathbf{E} \times \mathbf{H}$ flux vectors) [ExH-vectors.avi]

Hertzian-dipole radiation ($\mathbf{S}^{\mathrm{o}}$ flux vectors) [So-vectors.avi]

Selected frames are the basis of Figures 13.3 and 13.4. The power fluxes are scaled by the factor r^2 in order to facilitate plotting over the entire range.

Hertzian-dipole ($\mathbf{S}^{\mathrm{o}} - \mathbf{E} \times \mathbf{H}$ flux vectors) [Dipole-EM-Beauty.avi]

Hertzian-dipole Alternate-energy density contours [Dipole-Wo.avi]

Successive frames "zoom" in on the polar plot of W^{o} calculated from Chapter 3, Eq. (3.55b). The z axis is vertical; the radial scale is in units of kr. Solid contours represent positive values of energy, color coded as $\mathrm{W}_{\mathrm{red,green,blue}}$; dotted contours represent negative values, $-\mathrm{W}_{\mathrm{red,green,blue}}$.

H.6 ANIMATED BEAUTY-POWER FLUXES OF CYLINDRICAL WAVEGUIDE MODES

In the following animations the power-flux of a cylindrical TM_{nm} mode alternates periodically in time between the Poynting and Alternate forms. The maximum is S_z^{o}; the minimum is $-S_z$. Midway between these limits, the superposition equals the beauty-power flux.

TM^o_{01} [TM01circBeauty.avi]

TM^o_{02} [TM02circBeauty.avi]

TM^o_{11} [TM11circBeauty.avi]

TM^o_{12} [TM12circBeauty.avi]

TM^o_{21} [TM21circBeauty.avi]

TM^o_{22} [TM22circBeauty.avi]

H.7 MACSYMA ANIMATIONS AND GRAPHICS

These **.mfe** files are Macsyma Notebooks that are viewable with the Macsyma Front End (without the Macsyma Engine being connected); they will **not** run under Maxima 5.x. Once a Notebook is open, select a figure to make the graph control toolbar appear. The rocketship tool icon launches the selected animation; the adjust graphic position tool icon controls the camera view. If Macsyma is *not* available, **non-interactive** transcripts generated by converting the outputs of **xyNullpower.mfe** and **CylinderNullpower.mfe** to html files are available and can be viewed from standard browsers (such as Internet Explorer or Netscape). Simply click on the .htm file icons.

Animations of A_z,-E_z,|H| [AzEzHstep.mfe]
Selected frames are included in Figures 15.13 and 15.14 (see also Section H.1)

Animations of fields generated by current transients [Iz-transients.mfe]
Selected frames are included in Figures 15.16 and 15.18 (see also Section H.4)

Cartesian null-power flux [xyNullpower.mfe]
Interactive version of Figure 19.6

[xyNullpower.htm]

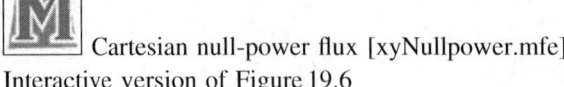

Cylindrical null-power flux [CylinderNullpower.mfe]
Interactive version of Figures 20.19 and 20.20

[CylinderNullpower.htm]

REFERENCES

1. J. A. Stratton, **Electromagnetic Theory**, McGraw-Hill Press, 1941.
2. Hector M. Macdonald, **Electric Waves**, Cambridge University Press, 1902.
3. George H. Livens, **The Theory of Electricity**, Cambridge University Press, 1926, p. 238.
4. Max Mason and Warren Weaver, **The Electromagnetic Field**, University of Chicago Press, 1929.
5. R. M. Fano, L. J. Chu, and R. B. Adler, **Electromagnetic Fields Energy and Forces**, John Wiley & Sons, 1960.
6. H. A. Haus and J. R. Melcher, **Electromagnetic Fields and Energy**, Prentice Hall, 1989; follows the Fano, Chu, Adler approach but includes the time-domain aspects of the skin-depth regime (magnetic diffusion).
7. C. Jeffries, *SIAM Rev.*, vol. **34** pp. 386–405, 1992; Proc. *PIERS* p. 154, 1993.
8. F. R. Morgenthaler, Notes for MIT Graduate Course 6.971, "Alternate Electromagnetic Power and Energy in the Context of Four-Dimensional Electrodynamics," March 1995 (unpublished), is the underlying basis for this text; they in turn evolved from a brief note distributed to students of MIT Graduate Course 6.642, Spring 1989.
9. D. H. Staelin, A. W. Morgenthaler, and J. A. Kong, **Electromagnetic Waves**, Prentice Hall, 1994.
10. R. B. Adler, L. J. Chu, and R. M. Fano, **Electromagnetic Energy Transmission and Radiation**, John Wiley & Sons, 1960.

The Power and Beauty of Electromagnetic Fields, First Edition. F. R. Morgenthaler.

11. Simon Ramo, John R. Whinnery, and Theodore Van Duzer, **Fields and Waves in Communication Electronics** 3rd ed., John Wiley & Sons, 1993 (early editions were titled **Fields and Waves in Modern Radio**)

12. H. Fröhlich, **Theory of Dielectrics**, Clarendon Press, Oxford, 1949, pp. 6 ff.

13. F. R. Morgenthaler, "Velocity Modulation of Electromagnetic Waves," *IRE Trans. on MTT*, pp. 167–172, April 1958.

14. E. A. Guillemin, **Introductory Circuit Theory**, John Wiley & Sons, 1953, pp. 340–352, 412–426; although nearly any modern text book on circuit theory will serve to introduce complex power and complex frequencies, few are as clear as those written by this legendary teacher.

15. E. A. Guillemin, **Theory of Linear Systems**, John Wiley & Sons, 1963.

16. A. Sommerfeld, in Riemann-Weber, "Partiellan Differentialgleichungen der mathematischen Physik," 8th ed., 1935, p. 786.

17. F. R. Morgenthaler, "On the Electrodynamics of a Deformable Ferromagnet Undergoing Magnetic Resonance," **The Mechanical Behavior of Electromagnetic Solid Continua IUTAM-IUPAP**, 1984, pp. 287–292.

18. D. M. Lipkin, *J. Math. Phys.* **5**, 696, 1964 deals with an electromagnetic field without sources. For this special case, termed the "zilch,"numerous conservation laws are possible because momentum-space density is constant in time. In the present formulation, the fields and potentials arise from real sources.

19. P. P. Penfield, Jr. and H. A. Haus, **Electrodynamics of Moving Media**, M.I.T. Press, 1967.

20. B. D. H. Tellegen, "Magnetic-Dipole Models," *Am. J. Phys.*, vol. **30**, pp. 650–652, September 1962.

21. F. R. Morgenthaler, "Exchange Energy, Stress and Momentum in a Rigid Ferromagnet," *J. Appl. Phys.*, vol. **38**, p. 1069, 1967.
————, "Predicted New Components of Magnetic Force on a Ferromagnet Undergoing Resonance," *AIP Conf. Proc.* No. **29**, 1975.

22. Jin Au Kong, **Electromagnetic Wave Theory**, John Wiley & Sons, 1986, pp. 220–225.

23. L. J. Chu, *J. Appl. Phys.*, **19**, pp. 1163–1175, December 1948.

24. S. A. Schelkunoff, *Bell Syst. Tech J.*, vol. **22**, p. 80, 1943.

25. C. L. Dolph, *Proc. of the IRE*, vol. **34**, p. 335, 1946.

26. E. A. Guillemin, **The Mathematics of Circuit Analysis**, John Wiley & Sons, p. 485, 1949.

27. An introduction to ferrimagnetism and ferromagnetic resonance may be found in many texts. Examples are: C. Kittel, **Introduction to Solid State Physics, 3rd ed.**, John Wiley 1967; B. Lax and K. Button, **Microwave Ferrites and Ferrimagnetics**, McGraw-Hill, 1962.

28. C. L. Hogan, "The Elements of Nonreciprocal Microwave Devices," *Proc. of the IRE*, 1956, pp. 1345–1368. The October 1956 issue of *Proc. IRE.* is devoted to ferrites as is the February 1988 issue of the *Proceedings of the IEEE.*

29. C. Kittel, "On the theory of ferromagnetic resonance absorption" *Phys. Rev.*, vol. **73**, p. 155, 1948.

30. L. R. Walker, *J. Appl. Phys.*, vol. **29**, p. 318, 1958

31. Frederic R. Morgenthaler, "Dynamic Magnetoelastic Coupling in Ferromagnets and Antiferromagnets," IEEE *Trans. on Magnetics*, vol. MAG-**8**, no. 1, pp. 130–151, March 1972.

32. H. Suhl, "The nonlinear. behavior of ferrites at high microwave signal levels," *Proc. of the IRE*, pp. 1270–1284, 1956.

33. Frederic R. Morgenthaler, "An Overview of Electromagnetic and Spin Angular Momentum Mechanical Waves in Ferrite Media," *Proc. of the IEEE*, vol. **76**, no. 2, pp. 138–150, February 1988.

34. Daniel D. Stancil, **Theory of Magnetostatic Waves**, Springer-Verlag, 1993.

35. R. W. Damon and J. Eschbach, *J. Phys. Chem. Solids*, vol. **19**, p. 308, 1961.

36. L. J. Chu, "A Kinetic Power Theorem," 1951 IRE Conference on Electron Devices, Durham, New Hampshire, June 1951.

37. P. H. Smith, "Transmission-Line Calculator," *Electronics*, **12**, 29–31, January 1939; "An Improved Transmission-Line Calculator," *Electronics*, **17**, 130, January 1944.

38. S. Schelkunoff, ***Applied Mathematics for Engineers and Scientists***, D. .Van Nostrand Co., 1948, p. 210.

39. **Principles of Microwave Circuits**, edited by C. G. Montgomery, R. H. Dicke, and E. M. Purcell, Dover edition, 1965, pp. 303–308, 447–448, an unabridged republication of McGraw-Hill, 1948, originally volume 8 of the Massachusetts Institute of Technology Radiation Laboratory.

40. **Waveguide Handbook**, edited by N. Marcuvitz, Dover edition, 1965, pp. 217–248, an unabridged republication of McGraw-Hill, 1951, originally volume 10 of the Massachusetts Institute of Technology Radiation Laboratory.

41. Robert E. Collin, **Field Theory of Guided Waves,** Chapter 8, McGraw-Hill, 1960.

42. F. R. Morgenthaler, "Simplified Electromagnetic-Torque Calculations," PIERS 2003, October 13–16, 2003.

43. Robert E. Tuzun, Donald W. Noid, and Bobby G. Sumpter, "Dynamics of a laser driven molecular motor," *Nanotechnology*, vol. **6**, 52–63, 1995.

44. M. E. J. Friese, H. Rubinsztein-Dunlop et al., *Appl. Phys. Lett.* **78**, 547, 2001.

45. Zeev Nehari, **Conformal Mapping**, McGraw-Hill, 1952, pp. 295–298.

46. E. O. Hammerstad and F. Beckkadal, "Microstrip Handbook" ELAB report STF 44A74169, University of Trondheim, Norway, 1975, pp. 98–110.

47. H. A. Wheeler, "Transmission-Line Properties of a Strip on a Dielectric Sheet on a Plane," *IEEE Trans.*, vol. **MTT-25**, pp. 631–647, 1977.

48. F. R. Morgenthaler, "Theoretical Studies of Microstrip Antennas volume I: General Design Techniques and Analyses of Single and Coupled Elements," Report FAA-EM-79-11, I, September 1979.

INDEX

accelerated point-charge, 581
addition of relativistic velocities, 527
Adler, R. B., 2
Alternate energy-momentum tensor,
  free-space, 122
Alternate power and energy,
  quasistatic limit, 53
Alternate power energy, momentum and stress
  Cartesian-coordinates, 564
  circuit-theory representation, 49
  cylindrical-coordinates, 565
  spherical-coordinates, 566
Alternate Poynting theorem,
  Amperian form, 563
  free-space, 46
Alternate representation,
  Amperian form, 563
  Minkowski form, 569
Alternate stress-momentum theorem,
  Amperian form, 563
  free-space form, 48
Alternate torque-density theorem, 51
Alternate-energy theorem, 82
Alternate-momentum theorem, 87
Alternate-stress theorem, 86
Ampere's Law, 7
Ampere, Andre-Marie, 1, 515
Amperian,
  four-force density, 145
Amperian current-density, 131
Amperian current-loops, 20
Amperian formulation, 43, 148
angular-momentum densities,
  spin, orbital, electric-charge , 572
animated electromagnetic fields, 619
antennas, 341
  array theory, 349
  dipole receiving circuit, 395
  directivity (gain), 350
  end-fire, 347
  gain-resistance product, 344
  half-wave dipoles, 342
  Hertzian electric-dipole,
    current-element, 16, 195
  Hertzian magnetic-dipole,
    current-loop, 17, 200
  Luneburg lens, 467
  self-complementary, 345
  slot in a ground-plane, 344
  super-gain, 347
  travelling-wave, 345
  uniform linear-arrays, 350
  Woodward synthesis, 470
antisymmetric field-tensor, 108

The Power and Beauty of Electromagnetic Fields, First Edition. F. R. Morgenthaler.

associated Legendre functions, 556
attenuation-factor,
  low-loss estimate, 288
averaging property of Laplace's Equation, 553
axial-current step transient,
  alternate representation of power and energy, 239
  differential law method, 237
  integral law & symmetry method, 223
  Maxwell-Poynting representation, 238
axial-current with multiple pulses, 246

Babinet's Principle, 98
beam-steering,
  magnetostatic waves, 391
Bessel functions, 593
Biot-Savart Law, 15, 112
  magnetostatics, 552
boundary conditions, 24
  gauge considerations, 125
  linear conductor, 117
Brewster Angle,
  TE waves, 266
  TM waves, 268

c, velocity of light, 8, 61, 232
Cartesian-components, 40
Čerenkov-radiation, 185
characteristic admittance,
  transmission-line, 400
characteristic impedance, 586
  free-space wave-impedance, 8, 233
  transmission-line, 400
  uniform plane wave,
    free-space, 255
    linear materials, 257
Chebyshev polynomials, 466, 598
Chu formulation, 43, 148
  criticism of, 97, 148
  Lorentz force-densities, 22
  magnetic-charge model, 22
Chu four-force density, 147
Chu, L. J., xxiii, 2, 22, 143, 341, 349, 389
circuit representations, 139
circuit-energy, 79
circuit-power, 78
circular waveguides, 305
  null Alternate power and energy, 323
  photon representation of modes, 323
  potentials and fields, 305
  TE_{nm} "circularly polarized", 324
  TM_{nm} "circularly polarized", 329
  TE mode power-fluxes, 314
  TE_{nm} modes, 310
  TM mode power-fluxes, 309
  TM_{nm} modes, 307
  waveguide-dispersion, 306
circular-aperture diffraction,
  large radius, $kR \gg 1$, 360
  small radius, $kR \ll 1$, 369

coaxial line, 586
Collin, R. E., 427
complementary structures, 98
complex Alternate-power theorem, 80, 137
complex Alternate-power vector, 81
complex Hankel-functions, 596
complex momentum theorems, 86
complex power & energy theorems, 78
complex Poynting theorem, 79
complex Poynting vector, 80
complex stress theorems, 84
conduction problems,
  EQS regime, 35
  MQS regime, 36
conductivity (electrical), 23, 67
conservation of charge, 25, 106
constitutive law, 67
continuum velocity, 130
convective current-density, 41
convolution-integral,
  superposition in linear systems, 252
coplanar strip-line, 588
cosine integral, Ci(x), 253
Coulomb's Law, 112
Coulomb, Charles Augustin de, 515
Coulomb-gauge, 11, 332
coupling-coefficient,
  resonator, 426
critical-angle, 265
cylindrical symmetry, 548
cylindrically-symmetric waves, 65

depolarization and demagnetization factors of a spheroid, 559
dielectric and magnetic materials, 23
dielectric constant,
  relative permittivity, 23
dielectric materials, 18
dielectric resonators, 430
dielectric waveguides, 335
  antisymmetric *TE* modes, 336
  dispersion relations, 337
  symmetric *TE* modes, 336
dielectric-relaxation time, 68
diffraction, 341
  complementary-screen,
    Babinet's Principle, 371
  large circular-aperture, 360
  rectangular-slit, 356
  small circular-aperture, 369
diffusion-length, 69
diffusion-time, 69
dipole-moment,
  electric-dipole, 551
  magnetic-dipole, 552
dipole-radiation,
  electric-dipole, 195
  magnetic-dipole, 200
directional coupler, 487
  transmission line, waveguide,

magic-T, 421
directivity (gain),
antenna arrays, 350
end-fire antennas, 347
half-wave dipoles, 343
uniform arrays, 352
dispersion relation,
harmonic plane waves, 63
dispersion relations,
relativistic, 137
Displacement current, Maxwell, 7
Divergence Theorem, 543
Dolph-Chebyshev, 466
antenna-array synthesis, 355
Doppler-shift,
uniform plane waves, 257
dual Alternate representation,
energy-momentum tensor,
free-space, 122
dual Alternate-power-flux,
plane waves,
oblique incidence, 268
dual six-vector, 113
duality, 88
boundary conditions, 91
dual Alternate energy-momentum tensors, 92
dual Alternate-power-flux, 268
dual electric and magnetic fields, 88
dual materials, 90
dual sources, 89
dual structures and their complements, 100
dual vector-potentials, 91

$E = mc^2$, 125, 521, 529
eikonal equation,
geometrical optics, 467
Einstein, Albert, viii, 105, 517, 521
electric energy-density, 43
electric line-dipole,
two-dimensional, static, 551
electric surface-charge, 24
electric surface-currents, 24
electric-charge angular-momentum density, 572
electric-charge density, 6
electric-current density, 6
electric-currents,
conduction, 21, 23
convective, 41
polarization and Amperian, 19
electric-dipole,
current-element, 16, 195
three-dimensional, static, 551
electric-dipole example,
Maxwell-Poynting and Alternate representations, 58
electrically-conducting materials, 23
electrodynamics, 41, 148, 149, 505
Amperian formulation, 21
Chu formulation, 22, 97
four-dimensional, 49, 141
four-space "statics", 110
Minkowski formulation, 21
electromagnetic angular-momentum density, 41
electromagnetic energy-density, 40
electromagnetic force on a moving charge,
Lorentz-force, 6, 51
electromagnetic force-density, 39
electromagnetic momentum-density, 40, 41
electromagnetic power-conversion, 39
electromagnetic power-flux, 40
electromagnetic radiation, 506
accelerating charge, 202
electric-dipole, 195
field contours, vectors, 200
magnetic-dipole, 200
power-flux vectors, 200
electromagnetic stress vectors, 40
electromagnetic torque-density, 40
$\mathbf{J} \times \mathbf{A}$, 506, 574
Amperian formulation linear μ, ϵ, σ, 574
electromagnetic-beauty power-fluxes,
graphical representations, 507
electromagnetic-beauty tensor,
energy-momentum tensor, 126, 507
electromagnetic-beauty theorem,
Poynting Theorem, 127
electroquasistatics (EQS), 44
electrostatics, 14
Coulomb's Law, 550
ellipsoids in uniform fields,
depolarization and demagnetization factors,
spheroids, 559
elliptically-polarized waves, 269
end-fire antennas,
Hansen-Woodyard condition, 347
energy theorems, 78
energy-momentum tensors, 120
Alternate representation,
free-space, 122
Amperian representation, 132, 145
Chu representation, 147
dual Alternate representation,
free-space, 123
Maxwell-Poynting representation,
free-space, 121
Minkowski representation, 146
transformations, 533
energy-velocity, 260
equivalence principle, 96, 148
equivalent circuits,
dipole receiving circuit, 395
distributed-circuit model, 398
TE waves incident upon a plane boundary,
oblique incidence, 411
TM waves incident upon a plane boundary,
oblique incidence, 412
waveguides, 413
ether, 105, 507, 521

Fano, R. M., 2
Faraday's Law, 7
Faraday, Michael, 1, 514
ferrimagnetic materials,
ferrites,
magnetic oxides, 378
ferrites, 491
dual Alternate power & energy, 390
ferrimagnetic materials, 378
yttrium iron garnet (YIG), 378
group velocity,
beam-steering, 391
large-signal equations, 380
large-signal Poynting Theorem, 389
magnetic oxides, 378
magnetic resonance, 379
magnetostatic-waves, 387
magnons, 393
plane waves, 384
small ellipsoid,
Kittel frequency, 384
Kittel uniform-precession mode, 384
small sphere,
Walker modes, 492
small-signal dual Alternate-power & energy, 390
small-signal equations, 381
small-signal power and energy, 388
small-signal Poynting Theorem, 389
small-signal stress & momentum, 391
torque-equation, 379
Walker-modes of a spheroid, 384
Zeeman energy-density, 389
fields of a moving point-charge,
accelerated motion, 581
constant velocity, 578
fields of specified charges and currents, 577
force-densities,
Amperian formulation, 145
Chu formulation, 147
discussion, 148
Lorentz, 6
Minkowski formulation, 146
four-dimensional "statics", 110
four-dimensional curl-operator, 109
four-dimensional current-vector, 106
four-dimensional electrodynamics, 49
4d-em.mac, 608
Alternate formulation, 122
Amperian formulation, 141
Chu formulation, 143
dual Alternate formulation, 123
Maxwell-Poynting (free-space), 143
Minkowski formulation, 142
four-dimensional electromagnetics, 496
four-dimensional force-density, 112
four-dimensional space-time, viii
four-dimensional vector & tensor identites, 544
four-dimensional vector-potential, 122
four-divergence, 108
four-Laplacian operator, 107
four-vector electric and magnetic fields, 113
four-vector identities, 106
four-vector length, 526
four-vector potential, 107
four-vector torque-density, 120
four-vectors, 46
fourth-coordinate, 48
free-charge,
unpaired-charge, 21, 129
free-space properties,
characteristic impedance, η_o, 8, 233
permeability, μ_o, 8
permittivity, ε_o, 8
velocity of light, c, 8, 232
frequency of Alternate-energy circulation,
TE^o_{nm} modes, 327
TM^o_{nm} modes, 331
frequency of Poynting-energy circulation,
TE^o_{nm} modes, 327
TM^o_{nm} modes, 331
Fresnel Integrals, 357
Fresnel-lens, 367
Fresnel-region, 364
Fresnel-ripples, 358, 364
Fresnel-zones, 365

Galilean-transformation, 105, 522
gauge transformations,
Coulomb-gauge, 11, 332, 471
linear conductor, 117
linear isotropic materials, 135
Lorenz-gauge, 11, 324, 329
radiation-gauge, 11
Gauss' Law, 8, 524
Gauss, Carl Friedrich, 514
Gaussian-beam solutions,
one and two dimensional confinement, 373
Gaussian-beams,
higher-order solutions, 374
geometrical optics,
eikonal equation, 467
Gibb's phenomenon, 180
group-velocity, 71, 84, 260, 295, 307
Guillemin, E. A., 78

half-wave dipole
directivity, 343
radiation resistance, 343
slot in a ground-plane, 344
wire antenna, 342
Hamilton's Equations, 73
hamiltonian, 73
Hansen-Woodyard condition,
end-fire antennas, 347
harmonic plane waves,
particle representation, 71
harmonic uniform plane waves,
dispersion relation, 63
frequency, wavelength, 63

Haus, H. A., 2, 148
Helmholtz Equation, 64, 372
Henry, Joseph, 515
Hermite polynomials, 374, 600
Hertz, Heinrich, 515
Hertzian dipoles,
  electric-dipole, 16
    current-element, 58, 195
  magnetic-dipole,
    current-loop, 17, 200
Hogan, C. L., 380
homogeneous waves, 61

impedance matching, 404
induction theorem, 97
inhomogeneous waves,
  scalar wave-equation, 66
isotropic media, 134

junctions and couplers, 485

Kittel frequency,
  small ferrite ellipsoid, 384
Kittel resonance,
  small ferrite sphere, 493
Kittel, Charles, 384
Kong, J. A., 3
Kramers-Krönig relations, 74
Kyhl, R. L., 421

laboratory coordinates, 105
Laplace's Equation, 558
  averaging property, 553
  solutions in N-dimensions, 558
  solutions in two and three dimensions, 554
Laplace, Pierre-Simon, 515
Laplacian approximations,
  quasistatics, 37
Law of Conservation of Charge,
  conservation of charge, 7
Law of reflection, 265
Lecher line, 587
Liénard-Wiechert Potentials, 581
line-dipole,
  two-dimensional plots, 551
linear conductor,
  boundary-conditions, 117
linear dielectric & magnetic materials,
  four-dimensional analysis, 134
linear isotropic materials,
  dielectric & magnetic, 23
linear-momentum,
  photon, 57
linearly-polarized
  plane wave, 71, 263
Liouville approximation,
  WKB method, 406
Livens, G. H., viii
London Equation, 455
London, F., 455
Lord Rayleigh, 377
Lorentz force-density, 6, 40, 112
  Amperian form, 21
  Chu form, 22
Lorentz four-force, 119
Lorentz torque-density, 40
Lorentz, Hendrick Antoon, 517
Lorentz-contraction, 525
Lorentz-force, 124
  on a magnetic-charge, 114
  on an electric-charge, 114
Lorentz-invariant, 106, 115, 530
Lorentz-transformation, 105, 497, 523, 526, 580
Lorenz, Ludwig Valentine, 514
Lorenz-gauge, 11, 107, 324, 329
Luneburg lens,
  geometrical optics, 467

Macdonald, H. M., viii
macroscopic, 506
Macsyma, 601
  animations and graphics, 624
  Notebooks, 605
  program descriptions, 602
Macsyma/Maxima,
  4d-em.mac, 608
magic-T, 486
  scattering-matrix,
    directional-coupler, 421
magnetic diffusion, 179
magnetic diffusion-length, 70
magnetic diffusion-time, 70, 179
magnetic energy-density, 43
magnetic induction, 130
magnetic materials, 18
magnetic resonance,
  ferrites, 379
magnetic torque-density, 379
magnetic-dipole,
  current-loop, 17, 200
  magnetic-field, 552
  vector-potential, 552
magnetite, 378
magnetization-vector, 21
magnetoquasistatics (MQS), 44
magnetostatic waves,
  ferrites, 387
magnetostatics, 14, 552
magnons,
  quasiparticle interpretation, 393
Marcuvitz, N., 427
Mason, M., viii, 506
mathematical notation, 5
Maxima, 601
Maxwell Displacement current, 7
Maxwell stress-tensor, 41–43
Maxwell's Equations, 130, 148
  Amperian form, 21
  Chu form, 22
  complex form, 26

Maxwell's Equations (*continued*),
  differential form, 10
  integral form, 6
  Minkowski form, 21
Maxwell, James Clerk, vii, 1, 514
Maxwell-Poynting energy theorem, 81
Maxwell-Poynting momentum theorem, 86
Maxwell-Poynting representation, 42
  Amperian form, 43
  Chu form, 43
  energy-momentum tensor,
    free-space, 121
  Minkowski form, 568
  power, energy, stress, and momentum, 43
Maxwell-Poynting stress theorem, 85
Melcher, J. R., 2
method of images, 583
  conducting-sphere, 584
  infinite ground-plane, 583
  infinite-length conducting cylinder, 584
  magnetic-conductors, 585
Michaelson, Albert Abraham, 105, 517
microstrip transmission-line, 592
Minkowski,
  energy-density, 132
  energy-momentum tensor, 133
  four-force density, 146
Minkowski formulation, 148
  criticism, 74
Minkowski fourth-coordinate,
  $x_4 = ict$, 49, 522
Minkowski representations,
  modified Alternate, 568
  modified Maxwell-Poynting, 568
Minkowski, Hermann, viii, 106, 517, 522
modified Alternate energy-momentum tensor,
  Minkowski form, 134
modified Alternate Poynting theorem,
  dielectric/magnetic materials, 132
modified Alternate stress-momentum,
  dielectric/magnetic materials, 133
modified energy-momentum tensor,
  dielectric/magnetic materials,
    linear, uniform, stationary, 136
  Minkowski form, 133
modified Poynting theorem,
  dielectric/magnetic materials,
    Minkowski form, 132
modified stress-momentum theorem,
  dielectric/magnetic materials,
    Minkowski form, 133
momentum theorems, 86
monopole,
  electric, 550
Morgenthaler, Ann W., 3
Morgenthaler, F. R., 2, 75, 149,
    388, 519
Morley, Edward Williams, 105, 517
moving point-charge, 51, 124, 578
  constant velocity,
    fields, power, and energy, 183
multipole expansion, 550

Newton's Second-Law, 124
nonuniform plane waves, 258
  TE−waves, 259
  TM−waves, 259
null Alternate power and energy,
  circular waveguide, 323, 331
  rectangular waveguide, 299

oblique incidence,
  complex Alternate-power-flux, 267
  reflected and transmitted waves, 263
    TE polarization, 264
    TM polarization, 267
Oersted, Hans Christian, 1, 515
one-dimensinal waves, 62
orbital angular-momentum density, 572

parallel-plate transmission-line, 274, 590
  linear dielectric/magnetic loading, 289
paraxial approximation, 372
paraxial wave-equation, 372
particle density, 526
particle representation, 71
Penfield, P., 148
perfect-conductor, 23
permeability, μ_o, 8
permittivity, ε_o, 8
permutation symbol, 539
phase-velocity, 71, 295, 307
photon, 506
  energy, momentum, 57, 72, 530
  rest-mass, 530
photon angular-momentum,
  spin and orbital densities,
    TE^o_{nm} modes, 326
    TM^o_{nm} modes, 330
photon energy,
  number density,
    TE^o_{nm} modes, 326
    TM^o_{nm} modes, 330
photon representation,
  of Alternate energy-momentum,
    circular waveguide modes, 323
Planck's Constant, 57
plane symmetry, 549
plane waves, 263, 458
  characteristic impedance,
    linear materials, 257
  Doppler-shifted,
    TEM, 257
  elliptically-polarized, 269
  nonuniform, 258
  oblique incidence, 263
  TE polarization, 264
  TM polarization, 267
  uniform, 73, 255
  uniform harmonic waves, 63

Poisson's Equation, 13
Poisson, Simeon-Denis, 515
polarization-vector, 19
Polder tensors, 382
power theorems, 78
Poynting Theorem, viii, 40, 42, 127
  large-signal form (ferrites), 381, 389
  small-signal form (ferrites), 388
Poynting vector, viii, 39, 43
Poynting, John Henry, 39, 514
principle of virtual-power, 150
probability current, 506
propagation constant,
  complex, 400

quality-factors,
  Q-factors,
    resonators, 422
quasiparticles,
  energy, momentum, density, 533
  magnons, 393
  wavepackets, 72
quasistatic approximations, 29, 138
quasistatic fields,
  EQS, 31
  MQS, 33
quasistatic power and energy, 44
quasistatics, 448

radar-range equation, 469
radiation and diffraction, 462
radiation-gauge, 11
Rayleigh scattering, 370, 488
reactive energy-density, 81
reciprocity theorem, 100
rectangular waveguides, 293
  cutoff wavelength, 295, 307
  null Alternate power and energy, 299
  periodic potentials and fields, 294
  TE mode power-fluxes, 297
  TE_{nm} modes, 296
  TM mode power-fluxes, 298
  TM_{nm} modes, 298
  waveguide-dispersion, 295
rectangular-slit diffraction, 356
reflection and transmission coefficients, 265
Reintjes, J. F., 296
relative permeability, 23
relative permittivity,
  dielectric constant, 23
relativistic energy, 529
relativistic mass, 529
resonators, 422, 490
  coupling-coefficient, 426
  dielectric, 430
  quality-factors,
    Q-factors, 422
  transmission-line, 423
  transmission-type, 426
  waveguide, 427
  YIG sphere filter,
    ferrite-resonator, 430
rest-mass,
  photon, 530

scattering-matrix, 418
  lossless junctions, 420
  N-port junction, 419
  single-port, 418
Schelkunoff, S. A., 355
Schell, A. C., 341
simultaneity and time-dilation, 526
sine integral, Si(x), 253
sinusoidal axial-current, 251
six-vector, 113, 530
skin-depth, 70, 181
  limited current, 261
small-signal theorems,
  dual Alternate-power (ferrite), 390
  stress & momentum (ferrite), 391
Smith, P. H., 403
Smith-Chart, 403
Snell's Law, 265
Sommerfeld approximation, 559
Sommerfeld, A., 112
speed limit of a moving charge, 524
spherical symmetry, 547
spherically-symmetric waves, 64
spin angular-momentum,
  photon, 57
spin angular-momentum density, 572
square-waveguide modes,
  TE_{nm} and TM_{nm}, 332
Staelin, D. H., 3
Stancil, D. D., 388
static and quasistatic fields, 157
statics, 435
Stoke's Theorem., 543
Stratton, J. A., viii, xxiii, 112
stress theorems, 84
stress-momentum representations of torque,
  Alternate formulation, 572
  Amperian formulation, 572
  Minkowski formulation, 573
stress-vectors, 40
strip with cylindrical shield, 588
Strutt, John William (Lord Rayleigh), 515
substantial derivative,
  material derivative, 125
Suhl, H., 381
super-gain antennas, 347
superposition,
  high-symmetry fields, 550
superposition of axial line-currents, 240
superposition of sources, 93
surface-current step,
  cylindrical-shell, 242
  planar-strip, 245

susceptibilities,
electric and magnetic, 23
symmetric strip-line, 589
symmetry of energy-momentum,
nonsymmetric tensor, II^{mink}, 146
nonsymmetric tensor, II°, 122
symmetric tensor, II^{amp}, 145
symmetric tensor, II^{chu}, 147
symmetric tensor, II^{em}, 121

tapered "horn" transformer, 280
TE,
transverse electric, 259, 264, 294, 305
TE modes,
waveguide equivalent circuits, 417
Telegrapher's Equations, 399
Tellegen, B. D. H., 148, 377
TEM,
transverse electric and magnetic, 255, 271
TEM mode power-energy, 273
TEM mode stress-momentum, 273
TEM parallel-plate line,
high conductivity, 282
TEM tapered-plate "horn", 280
TEM transmission-lines, 271
characteristic impedance, 586
coaxial line, 586
complex impedance & admittance, 400
complex power, 400
coplanar strip-line, 588
distortionless, 401
distributed-circuit model, 398
Lecher line, 587
line impedance,
characteristic-impedance, 402
low-loss, 401
microstrip, 592
parallel-plate line, 590
reflection-coefficient, 402
strip with cylindrical shield, 588
symmetric strip-line, 589
tapered lines, 406
transients,
time-domain and frequency-domain solutions, 408
TE^{o}_{nm} "circularly polarized" modes, 324
tensor identities,
four-dimensional, 544
Theory of Special Relativity,
Albert Einstein, viii, 521
Thompson, William (Lord Kelvin), 515
three-dimensional vector identities, 538
time-harmonic,
vectors & scalars, 26
TM,
transverse magnetic, 259, 267, 294, 305
TM modes,
waveguide equivalent circuits, 415
TM^{o}_{nm} "circularly polarized" modes, 329

torque contribution from the Alternate-stress integral, 575
torque-density,
electromagnetic
J x A, 506
four-vector form, 120
Lorentz torque-density,
Maxwell-Poynting and Alternate formulations, 571
six-vector, $\mathcal{R} \times \mathcal{F}$, 120
three-vector, $\mathbf{r} \times \mathbf{f}$, 40, 120
transformation-coefficients, 530
transformations,
electromagnetic fields and scalars, 530
energy-momentum tensors, 533
sinusoidal steady-state plane waves, 532
transmission-line resonator, 423
transmission-lines, 472
TEM, 271
transmission-resonator, 426
transverse wave-impedance, 284
travelling-wave antennas, 345

uniform linear-arrays, 350
uniform plane wave example,
Maxwell-Poynting and Alternate representations, 55
uniform plane waves, 71, 73
TEM, 255
characteristic wave impedance, (free-space), 255
Doppler-shifted,
TEM, 257
particle (photon) representation, 85
uniform surface-current,
pulse excitations, 207
uniqueness theorems, 94
Laplace's Equation, 95
sufficiency of the curl and divergence, 96
unit-step and $u_k(t)$ functions, 537

vector identities,
four-dimensional, 544
three-dimensional, 538
velocity dependence of mass, 529
velocity of light, c, 8, 61, 232
virtual-power, 150
volume-current step,
circular-cylinder, 240

Walker modes,
small ferrite sphere, 493
small ferrite spheroid, 384
Walker's Equation, 493
Walker, L. R., 384
wave equations,
inhomogeneous (**E**, **H**), 12
homogeneous (scalar), 61
wave-diffusion equation,
four-dimensional analysis, 135

wave-impedance,
  free-space, 233
waveguide resonators, 427
waveguides, 481
  circular, 305
  dielectric, 335
  equivalent circuits, 413
  rectangular, 293
wavepackets,
  quasiparticles, photons, 72
waves obliquely-incident, 411
Weaver, W., viii, 506

YIG sphere filter,
  resonators, 430
yttrium iron garnet (YIG), 378

Zeeman energy-density, 389, 390

IEEE PRESS SERIES ON ELECTROMAGNETIC WAVE THEORY

Andreas C. Cangellaris, *Series Editor*
University of Illinois, Urbana-Champaign, Illinois

---

1. *Field Theory of Guided Waves, Second Edition*
   Robert E. Collin
2. *Field Computation by Moment Methods*
   Roger F. Harrington
3. *Radiation and Scattering of Waves*
   Leopold B. Felsen, Nathan Marcuvitz
4. *Methods in Electromagnetic Wave Propagation, Second Edition*
   D. S. J. Jones
5. *Mathematical Foundations for Electromagnetic Theory*
   Donald G. Dudley
6. *The Transmission-Line Modeling Method: TLM*
   Christos Christopoulos
7. *The Plane Wave Spectrum Representation of Electromagnetic Fields, Revised Edition*
   P. C. Clemmow
8. *General Vector and Dyadic Analysis: Applied Mathematics in Field Theory*
   Chen-To Tai
9. *Computational Methods for Electromagnetics*
   Andrew F. Peterson
10. *Plane-Wave Theory of Time-Domain Fields: Near-Field Scanning Applications*
    Thorkild B. Hansen
11. *Foundations for Microwave Engineering, Second Edition*
    Robert Collin
12. *Time-Harmonic Electromagnetic Fields*
    Robert F. Harrington
13. *Antenna Theory & Design, Revised Edition*
    Robert S. Elliott
14. *Differential Forms in Electromagnetics*
    Ismo V. Lindell
15. *Conformal Array Antenna Theory and Design*
    Lars Josefsson, Patrik Persson
16. *Multigrid Finite Element Methods for Electromagnetic Field Modeling*
    Yu Zhu, Andreas C. Cangellaris
17. *Electromagnetic Theory*
    Julius Adams Stratton
18. *Electromagnetic Fields, Second Edition*
    Jean G. Van Bladel

19. *Electromagnetic Fields in Cavities: Deterministic and Statistical Theories*
David A. Hill
20. *Discontinuities in the Electromagnetic Field*
M. Mithat Idemen
21. *Understanding Geometric Algebra for Electromagnetic Theory*
John W. Arthur
22. *The Power and Beauty of Electromagnetic Theory*
Frederic R. Morgenthaler

Forthcoming

Modern Lens Antennas for Communications Engineering
John Thornton, Kao-Cheng Huang

README

1) All instances of **free space** should be hyphenated: **free-space**
All instances of **Appendix-L** (L= A,B,C,D,E,F,G,H) should be un-hyphenated: **Appendix L**
Most are correct but a global check should be made.

2) The edited versions pdf files on the Wiley ftp site that were downloaded from the directory MORGENTHALER have been uploaded to the subdirectory: **MORGENTHALER\Edited Files**

For example, the edited version of **c01.pdf** is **c01edit.pdf**; it contains appended Notes that detail the changes.

3) Supplementary pdf files with further details are included in Edited Files; these clarify edits and contain *highlights*, *footnotes*, or in one instance a *Chapter 25 problem*.

4) Because the Preface was missing from both the paper page proofs and the ftp pdf files, the author's non Post Script version **Preface-Acknowledgeedit.pdf** is included. It and the appended **Notes** provide all of the essential information.